FOREWORD

At its May, 1985 Council meeting, the Geochemical Society formed a committee on the future of the Geochemical Society (FOGS), composed of B. R. Doe, S. R. Hart, H. L. Barnes and E. Anders. One of this committee's first actions was to recommend that the Geochemical Society institute a book series, and this proposal was adopted by Council in November, 1985. At its May, 1986 meeting the Council, on recommendation from the Publications Committee, J. M. Hayes Chairman, approved A. A. Levinson as the inaugural editor of this series, to be called The Special Publication Series of the Geochemical Society, with the present volume as the first publication in this series.

The Society's constitution defines the object of the Society to be "to encourage the application of chemistry to the solution of geological and cosmological problems." One of the principal contributions of the Society to this goal over the years since its inception in 1955 has been sponsorship, in collaboration with the Meteoritical Society, of the journal *Geochimica et Cosmochimica Acta*; it is clearly appropriate that this effort now be joined by a special publication series, which will be devoted to the publication of reviews, conference and short course proceedings and the like. Our goal is to put forward these publications in a quality format, at accessible prices. The publications will not be on any regular basis, but as the opportunity and need arise.

I am pleased indeed to present this inaugural issue to the geochemical community.

Stanley R. Hart
President
The Geochemical Society
November, 1986

Book Series Editor: A. A. Levinson
Department of Geology and Geophysics
The University of Calgary
Calgary, Alberta T2N 1N4 Canada

Magmatic Processes: Physicochemical Principles

A volume in honor of Hatten S. Yoder, Jr.

B. O. Mysen

Editor

Geophysical Laboratory, Carnegie Institution of Washington, Washington, D.C. 20008 U.S.A.

Special Publication No. 1

THE GEOCHEMICAL SOCIETY

Library of Congress Catalogue Number 86-83155

ISBN 0-941809-005

Printed in The United States of America
Lancaster Press, Inc.

PREFACE

Magmatic Processes: Physicochemical Principles contains 30 invited papers presented as part of a week-long conference at the Kona Village Resort, Kona, Hawaii, June 16–22, 1986. The meeting was held to mark the retirement of Hatten S. Yoder, Jr. after 38 years as a staff-member, including 15 years as the Director of the Geophysical Laboratory. Characterization of the chemical and physical processes that govern formation, aggregation, ascent, emplacement and crystallization of magma in the Earth has been, and remains, his principal scientific interest. This theme was, therefore, the focus for both the conference and this Volume.

There is a tendency to subdivide research in the earth sciences into direct observation of natural phenomena, laboratory experiments and theoretical modelling. Research within individual subdisciplines has frequently evolved independently, despite the need to integrate and coordinate the efforts among them. In order to maintain and expand the scientific momentum, it is more important than ever to stimulate cooperation across traditional scientific boundaries. Thus, this Volume contains contributions from specialists with interests ranging from theoretical geophysics and fluid dynamics, observational geochemistry, petrology and geophysics to experimentalists with the common goal of understanding the physics and chemistry of igneous processes. The subject index also was organized with this integration in mind. Furthermore, the reader is alerted to the extensive cross-referencing to individual chapters in the Volume.

Largely as a result of technological and conceptual developments during the last 10–20 years, we are now poised to understand and describe quantitatively rock-forming processes in terms of the physico-chemical principles that govern them. The principal thermochemical and structural framework is in place, for example, to describe and predict crystal-liquid equilibria at least under near-surface pressure conditions. Moreover, an understanding of transport processes involving both aqueous solutions and silicate magmatic liquids is emerging. From geophysical studies and from detailed evaluation of phase relations and trace and isotope geochemistry of mantle-derived, ultramafic nodules, possible source regions for basaltic magmas can be characterized. The melting processes in the various tectonic regimes are being actively studied by means of both geochemistry of erupted basaltic products and phase equilibrium experiments. An understanding exists in regard to what liquid compositions may be derived from the upper mantle, and what parameters constrain their bulk compositions, yet considerable diversity in opinion remains regarding conditions of generation of basalt types. The problems associated with magma flow through the lithosphere and the crust is only beginning to be appreciated. The understanding of the fluid dynamic processes controlling emplacement and evolution of and fractionation in magma chambers in the crust of the earth is advancing. Theorical models to explain crystallization behavior in the chambers and the relationships between magma chamber evolution and subsequent magmatic eruption are being proposed. The kinetics of magmatic processes is becoming increasingly important. Nevertheless, the necessary experimental data required to constrain theoretical models are not yet available. It is clear that major additional efforts in experimental geophysics, geochemistry and petrology are necessary to realize the potential that now exists to characterize quantitatively magmatic processes.

This Volume presents the current state of progress, possible alternatives and solutions to these problems.

BJORN O. MYSEN
Washington, D.C., November, 1986

ACKNOWLEDGMENTS

The conference and, therefore, this book, was made possible thanks to generous support by the Carnegie Institution of Washington and the National Science Foundation (EAR-8519773) and by co-sponsorships by the Mineralogical Society of America and the Geochemical Society, all which is gratefully acknowledged. The Geochemical Society made available its resources to produce this Volume and make it the first issue of a series of *The Geochemical Society Special Publications*. Their commitment and support are deeply appreciated.

Both the conference and this Volume became a reality through extensive efforts by the staff at the Geophysical Laboratory and the Hawaiian Volcano Observatory, U.S. Geological Survey. Each of the chapters in this book were reviewed by two or more referees. The individuals involved in this process are listed below.

A. T. Anderson	University of Chicago	I. Kushiro	Geophysical Laboratory
D. L. Anderson	California Institute of Technology	A. C. Lasaga	Yale University
N. T. Arndt	Max-Planck Institut	P. Lipman	U.S. Geological Survey
R. K. Bailey	University of Reading	R. W. Luth	Geophysical Laboratory
A. L. Boettcher	University of California, Los Angeles	I. D. MacGregor	National Science Foundation
R. W. Carlson	Department of Terrestrial Magnetism	B. D. Marsh	The Johns Hopkins University
		G. Muncill	Geophysical Laboratory
R. G. Coleman	Stanford University	B. O. Mysen	Geophysical Laboratory
R. Criss	U.S. Geological Society	A. Navrotsky	Princeton University
H. J. B. Dick	Woods Hole Oceanographic Institution	H. Nekvasil	University of Arizona
		P. Olson	The Johns Hopkins University
M. Dickenson	Virginia Polytechnic Institute	M. Pichavant	CRPG, Nancy
D. B. Dingwell	University of Toronto	D. C. Presnall	University of Texas, Dallas
W. G. Ernst	University of California, Los Angeles	M. Ryan	U.S. Geological Survey
L. W. Finger	Geophysical Laboratory	C. M. Scarfe	University of Alberta
M. C. Gilbert	Texas A & M University	F. A. Seifert	Bayersiche Geoinstitut
M. S. Ghiorso	University of Washington	T. Simkin	Smithsonian Institution
D. H. Green	University of Tasmania	J. C. Stormer	Rice University
R. T. Helz	U.S. Geological Survey	F. J. Spera	University of California, Santa Barbara
C. T. Herzberg	Rutgers University		
J. R. Holloway	The Arizona State University	H. Taylor	California Institute of Technology
H. Huppert	Cambridge University	D. L. Turcotte	Cornell University
T. N. Irvine	Geophysical Laboratory	D. Virgo	Geophysical Laboratory
R. W. Kay	Cornell University	T. L. Wright	U.S. Geological Survey
P. Kelemen	University of Washington	H. S. Yoder, Jr.	Geophysical Laboratory

All of their time and effort are greatly appreciated. Finally, I wish to extend my special thanks to my wife, Susana, without whom neither the conference nor this book would have become a reality.

TABLE OF CONTENTS

Part D. Magma Ascent, Emplacement and Eruption

Part E. Crustal Felsic Magma Properties and Processes

Dr. Hatten S. Yoder, Jr. Staff member, Geophysical Laboratory, 1948–1986; Director, Geophysical Laboratory, 1971–1986.

Magmatic Processes: Physicochemical Principles
© The Geochemical Society, Special Publication No. 1, 1987
Editor B. O. Mysen

Hatten S. Yoder, Jr.

AN INTERNATIONAL CONFERENCE and field study was held on the Island of Hawaii in June, 1986, in honor of Hatten S. Yoder, Jr. on his retirement as Director of the Geophysical Laboratory of the Carnegie Institution of Washington. Very appropriately, the theme of the week's meeting was physicochemical principles of magmatic processes, to which Dr. Yoder has made major contributions during his 38 years of research as a member of the scientific staff of the Laboratory; and the setting of the conference on the 1801 Hualalai alkali basalt flow was especially fitting in view of his significant contributions bearing on the origin of this type of magma.

As Director of the Laboratory for the past 15 years, Dr. Yoder has been remarkably effective in seeing that the traditional high-quality, basic geochemical research was continued. He was very successful in his selection of able, young scientists as Fellows, who after a period of two or three years of productive research at the Laboratory, moved on to positions at universities throughout the world, carrying to these institutions the research spirit and motivation which has characterized the Geophysical Laboratory scientists. His recognition of the need for the most advanced research equipment has been one of Dr. Yoder's strong points. On arriving at the Laboratory in 1948, after serving for four years on active duty in the Navy, Dr. Yoder set out immediately to build an internally-heated, gas-media, high-pressure apparatus, an improved version of the 1933 model of P. W. Bridgman. With this facility, samples sealed in platinum or other metal capsules can be subjected to hydrostatic pressures for hours or even days at pressures up to 10,000 bar and temperatures to 1400°C. The massive steel pressure vessel is water-cooled, permitting a rapid quench of the sample when the power is turned off. One of his first published studies using this pressure equipment was the determination of the high-low quartz inversion up to 10,000 bar, the results of which dispelled contemporary speculation that discontinuities in the velocity of seismic waves with depth could be related to the existence of this inversion. His study of diopside-anorthite-water showed the existence of a large shift in the diopside-anorthite eutectic toward anorthite with increasing water pressure, of especial interest in later studies of fractional crystallization of water-containing olivine basaltic magmas, where depression of plagioclase precipitation because of this phenomenon can explain production of high-alumina basalts and andesites. This work was followed by a series of important studies of both synthesized mixtures and rock samples at elevated temperatures and pressures, both dry and with water as a component. His study of the system $MgO-Al_2O_3-SiO_2-H_2O$ was a notable contribution to the understanding of the conditions of origin of phase assemblages found in metamorphic rocks. Studies of phase relations in natural specimens, as for example the 1801 alkali olivine basalt flow of Hualalai and the 1921 olivine tholeiite flow of Kilauea were basic to an understanding of basalt genesis. Studies of these basalts were part of the research described in the classic 1962 paper by Yoder and Tilley on "Origin of Basalt Magmas: an Experimental Study of Natural and Synthetic Rock Systems". This extensive paper was followed, in 1976, by the very useful and timely book by Yoder on "Generation of Basaltic Magma". A further, major contribution was, "The Evolution of the Igneous Rocks—Fiftieth Anniversary Perspectives", published in 1979. He was editor as well as a contributor to this successor to Bowen's landmark book.

The door to the Director's office at the Geophysical Laboratory was always kept open so that colleagues, as well as other friends and visitors, could enter to see Dr. Yoder no matter how high the stack of papers on his desk. This friendly and helpful gesture has been characteristic of Hat Yoder. Rarely would one leave a discussion with him without learning something useful and absorbing some of his enthusiasm for research; or, if the office visit had to do with an equipment problem or demonstration of a new technique, he was off immediately to become involved. Recognition for his important contributions to science has included election to membership in the National Academy of Sciences, and his receiving the Mineralogical Society of America Award, the Day Medal of the Geological Society of America, the A. G. Werner Medal of the German Mineralogical Society, and the Wollaston Medal of the Geological Society of London. The new mineral Yoderite was named in his honor. It is to Hat Yoder that his colleagues and other friends respectfully and gratefully dedicate this volume.

Sept. 23, 1986   E. F. Osborn

Part A.
Structure and Properties
of Source Regions

Magmatic Processes: Physicochemical Principles
© The Geochemical Society, Special Publication No. 1, 1987
Editor B. O. Mysen

The depths of mantle reservoirs

DON L. ANDERSON

Seismological Laboratory, California Institute of Technology, Pasadena, California 91125, U.S.A.

Abstract—Many petrological studies are concerned with the temperature and pressure of final equilibrium of erupted magmas and residual crystals. The average composition of the source region and its original depth are also of interest but these cannot be determined unambiguously from petrology. Seismic techniques can be used to infer the mineralogy of various regions of the mantle and the probable depth extent of the low–velocity zones (LVZ) associated with high-temperature buoyant upwellings.

Oceanic ridges are characterized by broad deep LVZ's which extend locally to depths in excess of 400 km. Partial melting is implied to depths of at least 300 km. The same is true for some young continental regions such as northeast Africa and western North America, and some midplate regions such as the central Pacific. Hotspots occur on the edges of broad upper mantle low–velocity anomalies, often in regions of thick crust and/or old and thick lithosphere.

Continental shields have thick (150 km), cold, refractory lithospheres which are unlikely sources of voluminous plateau basalt outpourings. The rapid decrease in velocity between 150 and 200 km beneath shields implies a high thermal gradient and a change in mineralogy. From 200 to 400 km the seismic velocities beneath shields fall on the 1400°C adiabat. This suggests that the stable continental plate is 150 km thick and that it is underlain by a thermal boundary layer which grades downward into a convective gradient. Continental and oceanic basalts probably share a common source region which is deeper than 350 km. When hot, this source region becomes buoyant, because of thermal expansion and the reduction or elimination of dense phases, such as garnet, rises into the shallow mantle, adiabatically decompresses, becomes an LVZ and a potential source of magma.

INTRODUCTION

THE DEPTHS and compositions of the midocean basalt (MORB), ocean island basalt (OIB) and continental flood basalt (CFB) reservoirs cannot be inferred unambiguously from petrological and geochemical studies. Some of the minerals which were in equilibrium with observed basalts can be inferred from these studies and therefore minimum depths can be placed on the final equilibration between magmas and residual crystals. The proportions of the various minerals, however, cannot be determined. The source region itself may be much deeper or may itself have risen from greater depths before the melts were expelled.

Seismic studies can place some constraints on these petrogenesis processes. First, seismology can be used as a three–dimensional mapping tool. For example, it can be used to map shallow magma chambers and map the depth extent of the low–velocity region under ridges and other magmatic centers. It can also provide information about the depth extent of continental lithosphere and subduction zones.

Secondly, seismology provides inferential information about the mineralogy and physical state of the various regions of the mantle. In this paper we address both issues: the structure of the upper mantle: and its mineralogy and physical state.

THREE–DIMENSIONAL STRUCTURE OF THE UPPER MANTLE

The low–velocity zone (LVZ) has played a prominent role in most discussions of the location of the basalt source region. In most global seismic models the low–velocity zone occupies the depth interval between about 50 and 200 km. The most plausible explanation of the LVZ involves a partial melt content (ANDERSON and SAMMIS, 1970; ANDERSON and SPETZLER, 1970). In the first global surface wave inversions, it was found that the depth and nature of the LVZ varies from one tectonic province to another (ANDERSON, 1967; TOKSÖZ and ANDERSON, 1966). In shield areas the LVZ is deeper than 120 km and is less pronounced than in oceanic regions. Active tectonic regions also have shallow and pronounced low–velocity zones. These early results are shown in Figure 1.

More recently, high–resolution body wave studies have provided details about upper mantle velocity structures in several tectonic regions. Figure 2 shows some of these results for the Canadian shield (stable continent), western North America–East Pacific

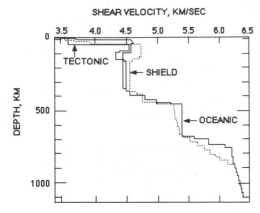

SHEAR VELOCITY, KM/SEC

FIG. 1. Shear velocities in upper mantle for different tectonic provinces derived from early surface wave studies (ANDERSON, 1967; TOKSÖZ and ANDERSON, 1966). Note the thick high–velocity shield LID and the shallow low–velocity zones under tectonic and oceanic regions.

Rise (tectonic–young ocean) and the western Atlantic (old ocean). Note that low–velocities extend to depths of about 390 km for the tectonic and oceanic structures. These regional studies confirm the general features of the earlier global studies (ANDERSON, 1967; TOKSÖZ and ANDERSON, 1966).

Shields have extremely high shear velocities extending to 150 km depth. It is natural to assume that this is the thickness of the stable continental plate and that the underlying mantle is free to deform and convect. The high–velocity layer, or LID, under tectonic and oceanic regions is much thinner, of the order of 30 to 50 km, and the shear velocities of the underlying mantle are much lower than under shields, implying higher temperatures and, possibly, the presence of a partial melt phase. The implication is that oceanic plates are much thinner and, possibly more mobile than continental plates.

JORDAN (1975) and SIPKIN and JORDAN (1976) made a radically different proposal. They suggested that the high seismic velocity associated with shields extended to depths in excess of 400 km and perhaps to 700 km and that the continental plates are equally thick. This hypothetical deep continental root was called the "tectosphere" (JORDAN, 1975). OKAL and ANDERSON (1975) and ANDERSON (1979) showed that the large differences in oceanic and continental ScS times (shear waves which reflect off the core), the data used in the development of the continental tectonosphere hypothesis, were mainly caused by differences shallower than 200 km. These waves have very little depth resolution and can only resolve differences below 400 km if the shallower mantle is independently constrained.

Although the largest variations (on the order of

10%) in seismic velocity occur in the upper 200 km of the mantle, the velocities from 200 to about 400 km under oceanic and tectonic regions are slightly less (on the order of 4% on average) than under shields. The question then arises, what is the cause of these deeper velocity variations? Is the continental plate 400 km thick or are the velocities between 150–200 and 400 km beneath shields appropriate for "normal" subsolidus convecting mantle?

ANDERSON and BASS (1984) attempted to answer this question by computing the seismic velocities for several plausible mineral assemblages along a variety of adiabats. The results are shown in Figure 3 along with several recent high–resolution seismic models for different tectonic regions. The heavy lines give the calculated velocities as a function of depth for adiabatic temperature gradients which start at surface temperatures ranging from 600° to 1800°C. The temperature gradient in convecting regions of the mantle is expected to be close to adiabatic. Near the surface and near chemical interfaces, i.e., in thermal boundary layers, heat is transferred by conduction and much higher temperature gradients can be maintained than elsewhere. In such regions the increase of velocity with depth is much

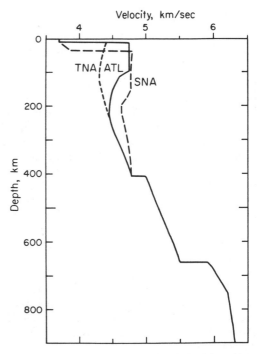

FIG. 2. Recent shear velocity profiles derived from high–resolution body wave studies for the Canadian Shield (SNA), western North America–East Pacific Rise (TNA) and the northeastern Atlantic (ATL). (from GRAND and HELMBERGER, 1984a,b).

less than along an adiabat. In Figure 3 the upper 150 km under shields and the upper 100 km under the North Atlantic (old ocean) have low velocity gradients, implying a rapid increase of temperature or a change in composition with depth.

The intersections of the adiabats with the dry solidus of peridotite are shown with dash–dot curves labelled "solidus" (Figure 3). Below these curves the calculated velocities represent upper bounds because the effect of the melt phase has not been taken into account. Measured seismic velocities which plot below the anhydrous solidus are presumably affected by the presence of melt and can be used only to infer upper bounds on the temperature.

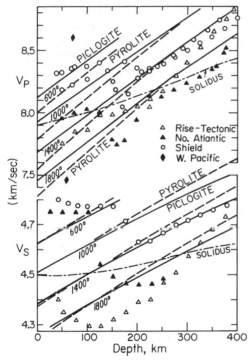

FIG. 3. Theoretical compressional velocity (V_p) and shear velocity (V_s) profiles along various adiabats for two mineral assemblages, pyrolite (mainly olivine and orthopyroxene) and piclogite (an olivine eclogite). A melt phase is probably present on the low–velocity side of the anhydrous solidus curves and this would lower the calculated velocities. The symbols are seismic results for various depths and tectonic provinces. The data are mostly from GRAND and HELMBERGER (1984b), WALCK (1984), L. ASTIZ and V. LEFEVRE (personal communications). Data which fall below the solidus curves are probably in the partial melt field. Note that the shield data fall near the 1400°C adiabat below 200 km depth. The low temperatures inferred for the shield lithosphere can be raised if the lithosphere is richer in olivine, particular forsterite–rich olivine (modified from ANDERSON and BASS, 1984).

In the upper 150 km of the shield mantle the derived temperature is low, 400–1000°C, and the inferred temperature gradient is higher than adiabatic, as appropriate for a conductive thermal boundary layer. Consistent temperatures can be obtained for both V_p and V_s if the shield lithosphere is more olivine–rich and more forsteritic than pyrolite (ANDERSON and BASS, 1984). The shield lithosphere therefore appears to be cold depleted peridotite or harzburgite. The rapid decrease of velocity between 150 and 200 km implies a high thermal gradient probably accompanied by a change in mineralogy toward a more fertile peridotite. By a change in mineralogy we mean a change in the relative proportions of the major mantle minerals; olivine, orthopyroxene, clinopyroxene and garnet. A garnet–poor depleted shield lithosphere is buoyant relative to fertile peridotite or pyrolite (BOYD and MCCALLISTER, 1976) unless offset by the colder temperatures. The presence of such a cold, thick, compositionally distinct layer extends the depth of the surface thermal boundary layer. Below 200 km the sub–shield velocities follow the 1400°C adiabat. This observation implies convection below this depth, and this is consistent with the flow behavior of olivine–rich rocks (e.g., KOHLSTEDT and GOETZE, 1974). Olivine flows easily at these temperatures and at stress levels thought to be appropriate for the upper mantle. Even though the seismic velocities under shields are higher than average mantle velocities to depths approaching 400 km, there is no reason to suppose that this material is not participating in mantle convection. As mentioned earlier, conductive layers are characterized by high thermal gradients. Stagnant regions of the mantle, regions where convection is not operational, will therefore have low or negative seismic gradients unless offset by a chemical or mineralogical gradient. The fact that seismic velocities in the sub–shield mantle fall along an adiabat either implies a convective gradient or a fortuitous combination of thermal and chemical gradients. In the latter case temperatures would rise more rapidly than in the former. The implied nature (temperature and composition) of the 150 km–thick shield LID make this an unlikely source for continental flood basalts or rift volcanics. Evidence from kimberlite inclusions suggests that the upper 150 km of the shield mantle has been stable for some time (RICHARDSON et al., 1984; BOYD et al., 1985). A refractory, chemically distinct shield lithosphere underlain by a thermal boundary layer is also consistent with the chemistry of mantle nodules and the "kinked geotherm" derived therefrom (e.g., BOYD, 1979). The deeper nodules are sheared fertile peridotites, consistent with flow below 150 km, and a change in chemistry near this depth. The shield sub-lithospheric temperature inferred from Figure 3 is consistent with the temperature between 150–200 km estimated from present–day heat flow (POLLACK and CHAMPMAN, 1977) and at 3 b.y. from thermobarometry (BOYD *et al.*, 1985).

Seismic velocities which are lower than those inferred for any plausible mineral assemblage provide evidence for anelastic phenomena such as partial melting, grain boundary relaxation or dislocation relaxation (ANDERSON and SAMMIS, 1970; MINSTER and ANDERSON, 1980, 1981). These are all high temperature phenomena but it is essentially impossible to distinguish between partial melting and the subsolidus mechanisms. On the other hand there is abundant evidence from other considerations that a large part of the upper mantle is near or above the melting point, particularly if volatiles are present, so we have cast our discussion in terms of the partial melt mechanism. Seismic velocities which fall well below the calculated curves are almost surely affected by anelastic processes, but even the higher velocities may also be affected to some extent, by processes such as partial melting. We have placed conservative estimates on the depth of inferred partial melting.

There are three obvious mechanisms for providing hot material to the base of the stable continental lithosphere, a process required to initiate volcanism. (1) The thermal boundary layer below 150 km can become unstable, allowing hot material to impinge the base of the lithosphere. (2) A diapir from greater depth can be brought into the shallow mantle. Such diapirs may initiate in a thermal boundary layer at 400 or 650 km, if these are chemical boundaries, or at a depth where the geotherm crosses the solidus. And finally, (3) the continent may drift over a hotter region of the mantle.

It is not clear whether continental rifting is a result of these deep processes or if rifting initiates the instability at depth. Continental convergence and collision may also trigger instability of the thermal boundary layer. Large and deep (>300 km) seismic anomalies are associated with the Red Sea–East African Rift and western North America as well as with midoceanic ridges and back–arc basins (See later sections.) This observation might suggest that none of these areas are entirely a result of passive rifting and pressure–release melting, although these effects can contribute to further melting and intrusion.

The geochemical nature of continental flood basalts, rift basalts, ocean island basalts and kimberlites suggests that they have either originated in, or at least have passed through or evolved in, a mantle that is enriched in the more incompatible elements. The shallow mantle may be the source of this enriched or hotspot signature (or the "metasomatic" fluids), even if the bulk of the magma comes from greater depth. Rising diapirs from a deep, depleted (low Rb, Sr, U, Th, LREE, Rb/Sr, Nd/Sm etc.) yet fertile (high CaO, Al_2O_3, Na_2O) reservoir will tend to expel their melts at shallow depths which, in turn, pond beneath the cold continental lithosphere where they cool, fractionate and ecome contaminated with shallow mantle and lower lithosphere melts prior to eruption. A similar mechanism may operate under old oceanic lithosphere, such as Hawaii, and thick crustal regions such as Iceland. Cooling, fractionation and contamination will also result if the parent magma must flow laterally for large distances before it can escape to the surface. On the other hand, basalts from the uplifted depleted reservoir which can flow directly to the surface through thin crust and lithosphere, as at midocean ridges, will be relatively uncontaminated and will have experienced less cooling and fractionation and the fractionation (*e.g.*, olivine, plagioclase) may be mainly shallow. It is necessary to distinguish between the trace element characteristics and the major element chemistry of basalts because MORBs are trace element depleted and fractionated, relative to chondrites, but obviously come from a fertile source whereas OIB and CFB are trace element enriched (sometimes called undepleted) and also come from a fertile source. The use of "depleted", "enriched", "fertile" etc. to refer to both the major element and minor element or isotopic chemistry can be confusing. A fertile, yet depleted, source could be for example, a garnet–clinopyroxene–rich cumulate or a garnet–clinopyroxene–rich layer that has had its incompatible elements stopped out in a prior stage of partial melting (ANDERSON, 1982, 1985).

THE OCEANIC MANTLE

Figure 3 shows that the seismic velocities in most of the upper mantle, down to at least 300 km, are so low that partial melting is implied for the tectonic and oceanic mantles. The high negative velocity gradients in the shallow mantle are probably the combined effects of a high temperature gradient and an increasing melt fraction with depth. At greater depth the high positive gradients are consistent with a decreasing melt content with depth. It appears that geographically large areas of the upper mantle are partially molten. Melting is not restricted to narrow, shallow zones associated with midoceanic and continental rifts although the pressure release associated with such near surface phenomena, and the resulting adiabatic ascent, may increase the amount of partial melting. We show later that large volumes of the shallow mantle down to 200 or 300 km in the vicinity of back–arc basins are also very slow. Thus, much of the upper mantle above 300 km is close to or above the melting point. Even the 1400°C adiabat characterizing the shield mantle crosses the dry solidus near 100 km. In the absence of a cold shield lithosphere, material advecting up the 1400°C adiabat would melt at ~100 km depth and even deeper if the wet solidus is more appropriate. The implied temperatures between 150 and 200 km are close to the wet solidus of peridotite.

The suboceanic geotherms may be on higher temperature adiabats. The velocities will not follow the calculated adiabats if the melt content varies with depth. The effect of melting on the seismic velocities is not included in the calculations of Figure 3.

In the next section we show that higher than average velocities between 200 and 400 km are not confined to stable shield areas. Old oceans, on average, many convergent regions and some tectonic regions also have fast velocities in this depth interval.

GLOBAL SURFACE WAVE TOMOGRAPHY

A global view of the lateral variation of seismic velocities in the mantle can now be obtained with surface wave tomography (NAKANISHI and ANDERSON, 1983, 1984a,b; TANIMOTO and ANDERSON 1984, 1985; NATAF et al. 1984, 1986; ANDERSON and DZIEWONSKI, 1984; DZIEWONSKI and ANDERSON, 1984; WOODHOUSE and DZIEWONSKI, 1984; TANIMOTO, 1984, 1985). There are two basic approaches. One is the regionalization approach (ANDERSON, 1967; TOKSÖZ and ANDERSON, 1966) which assumes that the velocities of surface waves are linearly dependent on the fracton of time spent in various tectonic provinces. The inverse problem then states that the velocity profile depends only on the tectonic classification. For example, all shields are assumed to be identical at any given depth. This assumption appears to be valid for the shallow structure of the mantle (NATAF et al., 1984, 1986; WOODHOUSE and DZIEWONSKI, 1984) but becomes increasingly tenuous for depths greater than 200 km. However, it probably provides a maximum estimate of the depth of tectonic features and it also provides a useful standard model with which other kinds of results can be compared. The second approach subdivides the Earth into cells or blocks or by some smooth function such as

spherical harmonics. NATAF et al. (1984, 1986) used spherical harmonics for the lateral variation and a series of smooth functions joined at mantle discontinuities for the radial variation. In this approach no a priori tectonic information is built in.

In both of these approaches the number of parameters that one would like to estimate far exceeds the information content (i.e., number of independent data points) of the data. It is therefore necessary to decide which parameters are best resolved by the data, what is the resolution, or averaging length, which parameters to hold constant and how the model should be parameterized (e.g., layers or smooth functions, isotropic or anisotropic). In addition, there is a variety of corrections that might be made (e.g., crustal thickness, water depth, elevation, ellipticity, attenuation). The resulting models are as dependent on these assumptions and corrections as they are on the quality and quantity of the data. This is not unusual in science. Data must always be interpreted in a framework of assumptions and the data are always, to some extent, incomplete and inaccurate. In the seismological problem the relationship between the solution, or the model, including uncertainties, and the data can be expressed formally. The effects of the assumptions and parameterizations, however, are more obscure but these also influence the solution. The hidden assumptions are the most dangerous. For example, most seismic modelling assumes perfect elasticity, isotropy, geometric optics and linearity. To some extent, all of these assumptions are wrong, and their likely effects must be kept in mind. Petrologists, of course, also design and interpret their experiments in terms of an array of prejudices ("the paradiam") about what the source region is like and what processes may or may not be important. The data itself may be consistent with quite a different scenario, e.g., the actual composition and evolution of the Earth.

NATAF et al. (1986) made an attempt to evaluate the resolving power of their global surface wave dataset and invoked physical a priori constraints in order to reduce the number of independent parameters which needed to be estimated from the data. For example, the density, compressional velocity and shear velocity are independent parameters but their variation with temperature, pressure and composition show a high degree of correlation, i.e., they are coupled parameters. Similarly, the fact that temperature variations in the mantle are not abrupt means that lateral and radial variations of physical properties will generally be smooth except in the vicinity of phase boundaries, including partial melting. Changes in the orientation of crystals in

the mantle will lead to changes in both the shear
wave and compressional velocity anisotropies.
These kinds of physical considerations can be used
in lieu of the standard seismological assumptions
which are generally made for mathematical convenience rather than physical plausibility.

The studies of WOODHOUSE and DZIEWONSKI
(1984) and NATAF et al., (1984, 1986) give upper
mantle models which are based on quite different
assumptions and data analysis techniques. WOOD-
HOUSE and DZIEWONSKI (1984) inverted for shear
velocity, keeping the density, compressional velocity
and anisotropy fixed. They also use a very smooth
radial perturbation function that ignores the presence of mantle discontinuities and tends to smear
out anomalies in the vertical direction. They corrected for near–surface effects by assuming a bimodal crustal thickness, continental and oceanic.

NATAF et al. (1986) corrected for elevation, water
depth, shallow mantle velocities and measured or
inferred crustal thickness. They inverted for shear
velocity and anisotropy but included physically
plausible accompanying changes in density, compressional velocity and anisotropy. Corrections were
also made for anelasticity. The radial perturbation
functions were allowed to change rapidly across
mantle discontinuities, if required by the data. In
spite of these differences, the resulting models of
NATAF et al. (1986) and WOODHOUSE and DZIE-
WONSKI (1984) are remarkably similar above about
300 km. The main differences occur below 400 km.
These differences seem to arise from differences in
the assumptions and parameterizations (crustal
corrections, radial smoothing functions) rather than
the data. The choice of an a priori radial perturbation function can degrade the vertical resolution
intrinsic to the dataset. The solution, in this case,
is overdamped or oversmoothed.

REGIONALIZED RESULTS

Figure 4 shows vertical shear velocity profiles,
expressed as differences from the average Earth, using the regionalization approach. Young oceans,
Region D, have slower than average velocities
throughout the upper mantle and are particularly
slow between 80 and 200 km, in agreement with
the higher resolution body wave studies. Old oceans,
Region A, are fast throughout the upper mantle.
Intermediate age oceans are intermediate in velocity
at all depths. Most of the oldest oceans are adjacent
to subduction zones and the subduction of cold
material may be partially responsible for the fast
velocities at depth. Notice that velocities converge
toward 400 km but differences still remain below

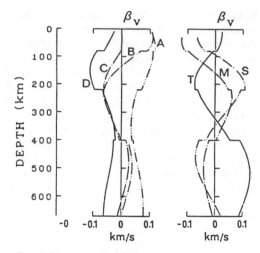

FIG. 4. Shear velocity (βv) vs. depth, expressed as differences from average Earth model PREM (DZIEWONSKI
and ANDERSON, 1981). The oceanic profiles are in order
of increasing age (A is oldest ocean, D is youngest). The
other profiles are shields (S), mountain–tectonic (M) and
trench–marginal seas (T) (after NATAF et al., 1986).

this depth (Figure 4). The continuity of the low velocities beneath young oceans, which include mid-
ocean ridges, suggests that the ultimate source region
for MORB is below 400 km. Shields are faster than
average and faster than all other tectonic provinces
except old ocean from 100 to 250 km. Below 220
km the velocities under shields decrease, relative to
average Earth, and below 400 km shields are among
the slowest regions. At all depths beneath shields
the velocities can be accounted for by reasonable
mineralogies and temperatures without any need
to invoke partial melting. Trench and marginal sea
regions, on the other hand, are relatively slow above
200 km, probably indicating the presence of a partial
melt, and fast below 400 km, probably indicating
the presence of cold subducted lithosphere. The
large size of the tectonic regions and the long wavelengths of surface waves require that the anomalous
regions at depth are much broader than the sizes
of slabs or the active volcanic regions at the surface.
This suggests very broad upwellings under young
oceans and abundant piling up of slabs under trench
and old ocean regions.

Maps of the regionalized results are shown at 250
and 350 km in Figure 5. Shields and young oceans
are still evident at 250 km. At 350 km the velocity
variations are much suppressed. Below 400 km,
most of the correlation with surface tectonics has
disappeared, in spite of the regionalization, because
shields and young oceans are both slow, and trench
and old ocean regions are both fast. Most of the

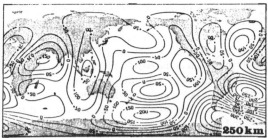
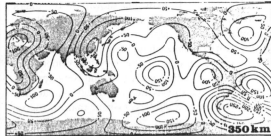

REGIONS

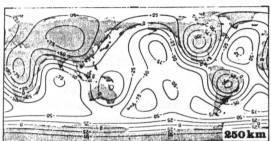
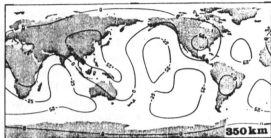

FIG. 5. Shear wave velocities (β_v, m/s) at depths of 250 and 350 km for spherical harmonic model (upper panels) and regionalized model (lower panels). The spherical harmonic model contains no *a priori* tectonic information but shields and platforms, in general, fall in high velocity regions and young oceans fall in low–velocity regions at 250 km and shallower depths. At 350 km the lowest velocity areas are under young oceans and in the central Pacific. Most hotspots lie on the periphery of these low–velocity regions. In the regionalized model all shield areas are assumed to be the same, all trench and marginal sea areas are assumed to be the same and oceanic profiles are assumed to be a function of age alone. The shields and stable platforms are still evident at 250 km in this parameterization but there is little lateral variation at 350 km (after NATAF *et al.*, 1986).

oceanic regions have similar velocities at depth. This is a severe test of the continental tectosphere hypothesis. Shields do not have higher velocities than some other tectonic regions below 250 km and definitely do not have "roots" extending throughout the upper mantle or even below 400 km. Results for other depths are given in NATAF *et al.* (1986). The maps shown (Figure 5) are spherical harmonic expansions of the regionalized results so that they can be compared with those in the following section. The minimum resolvable features with presently available data have a half–wavelength of about 2500 km. Only the long wavelength components of smaller features show up. In the high resolution studies discussed earlier we saw that subshield velocities dropped rapidly at 150 km depth although velocities remained relatively high to about 390 km. These high velocities could represent "roots" physically attached to the shield lithosphere, overridden cold oceanic lithosphere, or, simply, "normal" convecton mantle weakly coupled to the overlying shield lithosphere via a boundary layer at 150–200

km depth. We argued that the velocities below 200 km were on an adiabat and therefore probably represented normal convecting mantle. Therefore, it is the slow mantle under ridges and tectonic regions that is anomalous and, if anything, these are the regions with the roots. If the mantle under shields is convectively stagnant, as implied by the deep tectosphere hypothesis, a high thermal gradient would maintain over a large depth interval. This could lead to partial melting and a depression of the olivine–spinel phase boundary under shields. We therefore prefer the 150 km thick plate hypothesis, *i.e.*, a correspondence of the thickness of the plate with the seismic high–velocity layer.

SPHERICAL HARMONIC RESULTS

An alternate way to analyse the surface wave data is through a spherical harmonic expansion which ignores the surface tectonics. This provides a less biased way to access the depth extent of tectonic features. The results using this technique are shown in Figure 5. At 150 km (not shown) all the major shield areas fall near the centers of high–velocity anomalies and the ridges are in low–velocity regions. The main differences at this depth, compared to the regionalized models, are the very low velocities in eastern Asia, the Red Sea region, and New Zealand. At 250 km, the shields are less evident than at shallower depths and also less evident than in the regionalized model. On the other hand, the areas containing ridges are more pronounced low–velocity regions. The central Pacific and the Red Sea are also very slow. The highest velocity anomalies are in the far south Atlantic, northwest Africa to southern Europe and the eastern Indian Ocean to southeast Asia and are not confined to the older continental areas. At 350 km, most ridges are still evident although the slow region of the eastern Pacific has shifted off the surface expression of the East Pacific Rise. A central Pacific slow-velocity region persists throughout the upper mantle. Below 400 km there is little correlation with surface tectonics and in many areas the velocity anomalies are of opposite sign from those in the shallow mantle, in agreement with the regionalized results. The net result is that shields, on average, have very high velocities to 150–200 km and ridges, on average, have low velocities to 350–400 km. The Red Sea anomaly appears to extend to 400 km but the very slow velocities associated with western North America die out by 300 km.

CONCLUSIONS AND DISCUSSION

The high temperatures associated with midocean ridges appear to extend to depths of the order of 400 km. The low temperatures, or compositional differences, associated with stable continents attenuate rapidly below 200 km. The effects of temperature and composition on density must nearly compensate each other because shields are not evident in the geoid. Many shields, however, tend to fall in or near geoid lows but deglaciation effects may be partly responsible. Many hotspots occur in regions that overlie faster than average parts of the transition region. This throws doubt on the hypothesis that hotspots are deeply rooted. On the other hand, the most geochemically distinctive hotspots occur in regions of thick crust, thick lithosphere or where the upper 50 km are faster than average, and in regions where velocities are low between 150 and 400 km. Hotspots are generally on the edges of these low–velocity anomalies. Hawaii, for example, sits on old, cold Pacific lithosphere, which may be locally thinned and is on the northern boundary of the Pacific low–velocity anomaly. The local crust is also thick. The Cretaceous seamounts

and plateaus in the western Pacific were located over the central Pacific velocity anomaly when they were formed. The extensive volcanism may have been shut off by thick overriding Pacific lithosphere or by migration of the East Pacific Rise to its present location. Although separation of melt and residual crystals may occur at depths of the order of 100 km or shallower, the ultimate source region, and the instability that initiates adiabatic ascent into the shallow mantle, appear to be deeper than 350 km and, possibly, in the transition region. The effect of pressure on melt density and viscosity removes the necessity for postulating rapid removal of melt and small degrees of partial melting in deep reservoirs. At high pressure melt densities may approach the densities of olivine and pyroxene (RIGDEN et al., 1984). If so, the melt will not separate from the matrix, particularly if the melt viscosity is also high at high pressure or if the rock permeability is low. On the other hand, because the dense phase garnet is reduced or eliminated, upon partial melting, the partially molten rock will be less dense than it was before heating and melting. A deep garnet–rich source region can therefore become lighter than adjacent or overlying material when it is heated, even if the melt density is comparable to the density of olivine. Garnet exsolution at high temperature has a similar effect. These processes can be thought of as exaggerated forms of thermal expansion, which can initiate and substain buoyancy driven convection. A thermal boundary layer near 400 or 650 km or the intersection of the geotherm with the solidus near 400 km may cause the instability. If hotspots and ridges share the same deep fertile source, the contrasting geochemical characteristics of the basalts may reflect different degrees of contamination and crystal fractionation, including clinopyroxene and garnet, at shallow mantle depths (ANDERSON, 1985). Although the lateral extent of the high–velocity regions in the lower part of the upper mantle cannot be determined with the available resolution, they would not show up at all if they had the dimensions of slabs. The piling up of cold material over long periods of geological time seems to be implied. This in turn is consistent with stratified mantle convection.

The fact that subshield velocities lie on the 1400°C adiabat below 200 km indicate that the shield plate is of the order of thickness of the high-velocity LID (150 km) plus a 50–km thick thermal boundary layer. This is consistent with the thickness estimated from the geoid by TURCOTTE and MCADOO (1979).

There is no support from either high–resolution body wave studies or global tomography for a thick continental tectosphere. The relatively high velocities between 200 and about 400 km under shield regions indicate the absence of partial melt rather than a compositionally distinct buoyant region that is permanently attached to the shield. A small perturbation in temperature, or an instability in the thermal boundary layer, would raise the temperatures between 150–200 km above the wet solidus. This is perhaps the trigger for kimberlite intrusion. At depths below some 300 km, the fastest regions of the mantle are not all located beneath the shields. Below 400 km much of the Canadian Shield is slow (see also GRAND, 1986).

PETROLOGICAL IMPLICATIONS

There has always been cross–fertilization between seismology and petrology. For example, the standard petrological view that the source region of basalts is olivine–rich is partially based on the early observation that the compressional velocities in the upper mantle are similar to olivine. Hidden assumptions were that no other combination of minerals would satisfy the seismic data and that basalts sampled the same part of the mantle as the seismic data. More recent seismological studies and studies of mantle xenoliths have strengthened the conclusion that the shallow mantle at least, above 200 km, is olivine–rich. The seismic data include velocities and anistrophy. The continental lithosphere, the top of the oceanic lithosphere and the upper mantle below the lithosphere and above 400 km are consistent with an olivine–rich peridotite. The main unsampled region is the lower oceanic lithosphere. Unfortunately, the velocities of olivine and orthopyroxene, on the one hand, and clinopyroxene and garnet, on the other hand, both bracket the observed velocities in unmelted regions of the upper mantle. Therefore, one cannot discriminate between peridotitic and eclogitic assemblages on the basis of seismic velocities alone and the seismic constant is lost. The seismic anisotropy of the upper 200 km slightly favors the peridotitic hypothesis. In a large portion of the shallow mantle the velocities are too low to be explained by any combination of minerals. The presence of a small melt fraction can explain these results but it removes the possibility of inferring mineralogy.

Seismological evidence for a low–velocity zone and its interpretation as a partial melt zone seemed to provide petrologists with a source region for basalts. Most petrologists now assume that not only do basalts come from the shallow mantle but that the ultimate source is also shallow. The LVZ is equated with "the source region" generally for depleted MORB–type basalts. The alternative is that basalts or their source rocks are brought into the shallow mantle by deeper processes and that basalts only record the final stages of their evolution.

The presence results show that the LVZ is not a global layer of uniform thickness or velocity nor does it invariably terminate at some shallow depth such as 150 or 200 km. It can be traced to much greater depths under some ridges and tectonic regions suggesting that petrologists should start to be concerned about deeper processes. The broad areal extent of low–velocity regions, compared to the area of volcanic centers, suggests that sublithospheric cooling and fractionation, fractionation that may include garnet and clinopyroxene, may be important petrological processes. The possible great depth of the initiation of adiabatic ascent suggests that much larger degrees of partial melting might be involved than have previously been considered plausible. Large degrees of partial melting at depth, followed by large amounts of crystal fractionation at modest depth must be considered as a viable alternative to small degrees of partial melting, at shallow depth, of a primitive pyrolite–like reservoir.

In a way, we have taken a step backwards. The recognition that peridotites and eclogites can have similar seismic velocities, at least above 400 km, and that the lowest velocity regions of the mantle have velocities lower than any subsolidus assemblage mean that we cannot constrain the mineralogy of most of the upper mantle. The anisotropy of the upper oceanic lithosphere and the very high velocities and V_s/V_p ratio of the shield mantle are evidence that these regions are probably peridotitic but these regions are not pertinent to basalt petrogenesis. The anisotropy of the upper 200 km of the mantle is weak evidence that this region, on average, is olivine–rich. Olivine, being a refractory residual phase and being less dense than fertile peridotite, should concentrate in the shallow mantle as a result of basalt separation throughout Earth history. In contrast to garnet– and clinopyroxene–rich assemblages, olivine does not undergo any temperature or pressure induced phase changes at depths less than 400 km. It is therefore permanently trapped in the shallow mantle. Basalts rising from a deeper source must, of course, traverse this olivine–rich region and may, to some extent, be trapped in it, converting it to a fertile peridotite. Basalts, however, are less dense than olivine in the shallow mantle and they are more likely to be trapped below the lower density crust or colder impermeable, high–viscosity lithosphere, if they cannot proceed directly to the surface. If only a fraction of this melt escapes to the surface then there may be a shallow basalt–rich layer which will cool to eclogite at depths greater than about 50 km. Such a layer could contribute to the eventual instability of the oceanic lithosphere and, possibly, to midplate volcanism.

Acknowledgments—Much of the research reported here was done in collaboration with Ichiro Nakanishi, Henri–Claude Nataf, Jay Bass, Toshiro Tanimoto, and Adam Dziewonski. The author would like to acknowledge many enjoyable hours of cooperation and discussion with these colleagues. This research was supported by NSF Grants EAR-8509350 and EAR-8317623 and support from DARPA. Contribution 4354, Division of Geological and Planetary Sciences, California Institute of Technology, Pasadena, California 91125.

REFERENCES

ANDERSON DON L. (1967) Latest information from seismic observations. In *The Earth's Mantle*, (ed. T. F. GASKELL), Chap. 12, pp. 355–420. Academic Press.
ANDERSON DON L. (1979) The deep structure of continents. *J. Geophys. Res.* **84**, 7555–7560.
ANDERSON DON L. (1982) Isotopic evolution of the mantle: a model. *Earth Planet. Sci. Lett.* **57**, 13–24.
ANDERSON DON L. (1985) Hotspot magmas can form by fractionation and contamination of MORB. *Nature* **318**, 145–149.
ANDERSON DON L. and BASS J. D. (1984) Mineralogy and composition of the upper mantle. *Geophys. Res. Lett.* **11**, 637–640.
ANDERSON DON L. and DZIEWONSKI A. M. (1984) Seismic tomography. *Sci. Amer.* **4**, 60–68.
ANDERSON DON L. and SAMMIS C. (1970) Partial melting in the upper mantle. *Phys. Earth Planet. Int.* **3**, 41–50.
ANDERSON DON L. and SPETZLER H. (1970) Partial melting and the low-velocity zone. *Phys. Earth Planet. Int.* **4**, 62–64.
BOYD F. R. and MACALLISTER R. H. (1976) Densities of fertile and sterile garnet peridotites. *Geophys. Res. Lett.* **3**, 509–512.
BOYD F. R., GUERNEY J. J. and RICHARDSON S. H. (1985) Evidence for a 150–200 km thick Archaean Lithosphere from diamond inclusion thermobarometry. *Nature* **315**, 387–389.
BOYD F. R. (1979) Garnet lherzolite xenoliths from kimberlites of East Griqualand, South Africa. *Carnegie Inst. Wash. Yearb.* **78**, 488–492.
DZIEWONSKI A. M. and ANDERSON DON L. (1981) Preliminary reference Earth model. *Phys. Earth Planet. Int.* **25**, 297–356.
DZIEWONSKI A. M. and ANDERSON DON L. (1984) Seismic tomography of the Earth's interior. *Amer. Sci.* **72**, 483–494.
GRAND S. P. (1986) Shear velocity structure of the mentle beneath the North American plate. Ph.D. Thesis, Calif. Inst. Tech.
GRAND S. P. and HELMBERGER D. V. (1984a) Upper mantle shear structure of North America. *Geophys. J. Roy. Astron. Soc.* **76**, 399–438.
GRAND S. P. and HELMBERGER D. V. (1984b) Upper mantle shear structure beneath the northwest Atlantic ocean. *J. Geophys. Res.* **89**, 11,465–11,475.
JORDAN T. H. (1975) The continental tectosphere. *Rev. Geophys. Space Phys.* **13**, 1–12.
KOHLSTEDT D. and GOETZE C. (1974) Low-stress high-temperature creep in olivine simple crystals. *J. Geophys. Res.* **79**, 2045–2051.
MINSTER J. B. and ANDERSON DON L. (1980) Dislocations and nonelastic processes in the mantle. *J. Geophys. Res.* **85**, 6347–6352.
MINSTER J. B. and ANDERSON DON L. (1981) A model of dislocation-controlled rheology in the mantle. *Phil. Trans. Roy. Soc. London* **299**, 1449, 319–356.
NAKANASHI I. and ANDERSON DON L. (1983) Measurements of mantle wave velocities and inversion for lateral heterogeneity and anisotropy: Part I, Analysis of Great Circle phase velocities. *J. Geophys. Res.* **88**, 10,267–10,283.
NAKANISHI I. and ANDERSON DON L. (1984a) Aspherical heterogeneity of the mantle from phase velocities of mantle waves. *Nature* **307**, 117–121.
NAKANISHI I. and ANDERSON DON L. (1984b) Measurements of mantle wave velocities and inversion for lateral heterogeneity and anisotropy: Part II, Analysis by single-station method. *Geophys. J. Roy. Astron. Soc.* **78**, 573–617.
NATAF H.-C., NAKANISHI I. and ANDERSON DON L. (1984) Anisotropy and shear velocity heterogeneities in the upper mantle. *Geophys. Res. Lett.* **11**, 109–112.
NATAF H.-C., NAKANISHI I. and ANDERSON DON L. (1986) Measurements of mantle wave velocities and inversion for lateral heterogeneities and anisotropy: III, Inversion. *J. Geophys. Res.* **91**, 7261–7308.
OKAL E. A. and ANDERSON DON L. (1975) A study of lateral inhomogeneities in the upper mantle by multiple ScS travel-time residuals. *Geophys. Res. Lett.* **2**, 313–316.
POLLACK H. and CHAPMAN D. S. (1977) On the regional variation of heat flow, geotherms and lithosphere thickness. *Tectonophys.* **38**, 279–296.
RICHARDSON S., GURNEY J. J., ERLANK A. and HARRIS J. W. (1984) Origin of diamonds in old enriched mantle. *Nature* **310**, 198–202.
RIGDEN S., AHRENS T. and STOLPER E. (1984) Densities of liquid silicates at high pressures. *Science* **226**, 1071–1074.
SIPKIN S. A. and JORDAN T. H. (1976) Lateral heterogeneity of the upper mantle determined from the travel times of multiple ScS. *J. Geophys. Res.* **81**, 6307–6320.
TANIMOTO T. and ANDERSON DON L. (1984) Mapping convection in the mantle. *Geophys. Res. Lett.* **11**, 287–290.
TANIMOTO T. and ANDERSON DON L. (1985) Lateral heterogeneity and azimuthal anisotropy of the upper mantle: Love and Rayleigh waves 100–250 sec. *J. Geophys. Res.* **90**, 1842–1858.
TOKSÖZ M. N. and ANDERSON DON L. (1966) Phase velocities of long–period surface waves and structure of the upper mantle, 1. great circle Love and Rayleigh wave data. *J. Geophys. Res.* **71**, 1649–1658.
TURCOTTE DON L. and McADOO D. C. (1979) Geoid anomalies and the thickness of the lithosphere. *J. Geophys. Res.* **84**, 2381–2387.
WALCK M. C. (1984) The *P*-wave upper mantle structure beneath an active spreading center: the Gulf of California. *Geophys. J. Roy. Astron. Soc.* **76**, 697–723.
WOODHOUSE J. H. and DZIEWONSKI A. M. (1984) Mapping the upper mantle: three–dimensional modeling of earth structure by inversion of seismic waveforms. *J. Geophys. Res.* **89**, 5953–5986.

Magmatic Processes: Physicochemical Principles
© The Geochemical Society, Special Publication No. 1, 1987
Editor B. O. Mysen

Composition and structure of the Kaapvaal lithosphere, southern Africa

F. R. Boyd[1] and S. A. Mertzman[2]

[1] Geophysical Laboratory, Carnegie Institution of Washington,
2801 Upton St., N.W., Washington, D.C. 20008, U.S.A.

[2] Department of Geology, Franklin and Marshall College, P.O. Box 3003, Lancaster, PA 17604-3003, U.S.A.

Abstract—Bulk and mineral analyses have been used to calculate modes for 24 peridotite xenoliths from the Kaapvaal craton. Large samples for bulk analysis, weighing 0.5–1.0 kg, were prepared to avoid the effects of heterogeneities in grain distribution that are commonly present on a scale of several centimeters. These and comparable data obtained by Cox *et al.* (1987) are used as a basis for speculation about the igneous and tectonic processes through which the craton has formed.

Peridotites with low equilibration temperatures (<1100°C) that comprise rigid lithosphere within the craton are strongly depleted in Fe and moderately depleted in diopside and garnet but are relatively rich in modal enstatite. They have a range in modal olivine of 45–80 weight percent but have an average olivine content (61 weight percent) that is approximately the same as that of hypothetical fertile peridotite, such as pyrolite. Peridotites with high equilibration temperatures (>1100°C), believed to underlie the low–temperature suite, are characterized by higher modal olivine than is present in the low–temperature rocks, combined with lower Mg/(Mg + Fe) and more abundant diopside and garnet. The average composition of high–temperature peridotite from the Premier kimberlite in the central part of the craton is more fertile in elements concentrated in basalt than that of high–temperature peridotite from the southern margin of the craton. The more depleted high–temperature peridotites have compositions that are similar to residual oceanic peridotites.

The low–temperature peridotites may have originated as residues of partial fusion events that occurred at pressures above 50 kbar. Their wide range of modal olivine might be the product of cumulate processes. Alternatively, the compositions of these peridotites may reflect primordial compositions or events associated with accretion and core formation.

The high–temperature peridotites are speculated to have originated in an oceanic plate subducted beneath the southern margin of the Kaapvaal craton possibly at the time of formation of the Namaqua–Natal mobile belt, approximately 1 b.y. ago. Subsequent to subduction these oceanic(?) peridotites have been heated to temperatures in the range 1100–1500°C, and converted to rocks having the mechanical properties of asthenosphere.

INTRODUCTION

UNDERSTANDING of the igneous and tectonic processes by which cratons have formed has traditionally been sought through study of the greenstone belts and gneisses that comprise their crustal caps. It is now appreciated, however, that cratons have mantle roots extending to depths substantially greater than the base of oceanic lithosphere. The concept of craton roots had been developed by JORDAN (1978, 1979) who proposed that the density imbalance resulting from cooler temperatures in the mantle beneath continents was compensated by a greater degree of magmatic depletion of the sub-continental mantle. He has interpreted seismic evidence to suggest that craton root zones depleted in basaltic components may extend below the lithosphere to depths as great as 400 km.

Evidence in support of the existence of craton roots has also come from studies of kimberlite xenoliths and of diamond occurrences and inclusions. Regularities in temperatures and depths of equilibration estimated for kimberlite xenoliths erupted in southern Africa have been interpreted

to reflect an asthenosphere–lithosphere boundary that shelves from a depth of 180 km beneath the Kaapvaal craton to 140 km beneath the western and southern mobile belts (*e.g.,* BOYD and NIXON, 1978). Coarse peridotites with equilibration temperatures that are predominantly below 1100°C are interpreted to be characteristic of the lithosphere whereas peridotites with higher equilibration temperatures are commonly deformed and are believed to have originated in the asthenosphere (BOYD, 1987). Occurrences of diamonds and of associated subcalcic garnet xenocrysts in southern Africa also appear related to the presence of a thickened lithosphere (BOYD and GURNEY, 1986). The base of the Kaapvaal lithosphere contains peridotites as old as the crustal rocks that form the cap (KRAMERS, 1979; RICHARDSON *et al.* 1984), and by the late Archaean the thermal gradient within the craton was similar to that estimated from present–day heat flow (BOYD and GURNEY, 1986; POLLACK and CHAPMAN, 1977). How the Kaapvaal and other cratons formed, however, remains a matter for investigation and speculation.

13

Major element studies of xenoliths of mantle origin included in kimberlites have primarily been carried out by electron probe analysis of constituent minerals. Data obtained in these studies have become a basis for thermobarometry and the development of mantle stratigraphies for the Kaapvaal and other cratons. Additional insights relating to ultimate igneous origin of these rocks can be obtained, however, by investigations of modal and bulk chemical variations. Modes estimated by O'HARA et al. (1963) for a small suite of garnet peridotite xenoliths from kimberlites showed them to be distinctively enstatite–rich. MATHIAS et al. (1970) used modes to demonstrate the unique compositional ranges of peridotite, pyroxenite and eclogite xenoliths from Kaapvaal kimberlites and provided an average mode for garnet lherzolite (Table 1). MAALOE and AOKI (1977) used bulk analyses and modes to demonstrate differences in average composition between spinel peridotite and garnet lherzolite xenoliths, the former being primarily from basaltic volcanics and the latter from kimberlites. Their average mode for garnet lherzolite includes the data of MATHIAS et al. (1970) and is close to the values given by the latter authors (Table 1). A distinction between low– and high–temperature peridotites was not made in previous modal studies. The low–temperature peridotites are considerably more abundant, however, and the similarity between the earlier average modes for garnet peridotite and the average for low–temperature peridotite obtained in the present work (Table 1) is thus concordant. COX et al. (1987) have used modal data determined on large peridotite xenoliths to show that diopside and garnet in these rocks may have exsolved from Mg–rich pyroxene in ultra–coarse harzburgites that crystallized near the dry solidus.

Modal and chemical data for twenty–four large peridotite xenoliths from the Kaapvaal craton that

have been obtained in the present investigation are combined with comparable data for thirteen xenoliths analyzed by COX et al. (1987) as a basis for speculation about the igneous history of the craton. The desirability of having large analytical samples, explained hereafter, has constricted the choice of localities; most of the xenoliths studied in the present work are from the Jagersfontein and Letseng–la–Terai kimberlites with a lesser number from the Premier, Frank Smith, Mothae and Monastery kimberlites. Those studied by COX et al. (1987) are primarily from Bultfontein (Kimberley). Both high– and low–temperature garnet peridotites and one spinel peridotite are included. Detailed petrographic and electron probe data for these specimens are given in an Appendix that can be obtained on request from Boyd.

APPROACH

Modes for peridotite xenoliths have been determined by calculation from bulk and mineral analyses with a computer program (LSPX) written by Felix Chayes. With this approach estimates of modal composition can be obtained for substantially larger volumes of rock than is possible by point–counting with thin sections or slabs.

Most peridotite xenoliths erupted in kimberlites are free of mineral banding and other large–scale segregations, having textures that most visibly reflect variable deformation. Heterogeneities in mineral distribution are, nevertheless, present in most specimens and it is difficult to obtain bulk analyses and modes that represent true compositions on a xenolith scale (10–30 cm). COX et al. (1987) have demonstrated that the fabrics of peridotite xenoliths from the Kaapvaal craton consist of clusters of olivine grains interspersed with clusters of enstatite, diopside and garnet. Heterogeneities in distribution of a particular mineral produced by clustering are commonly on a scale that is approximately 2–3 cm. In some instances, however, heterogeneities exceed 3 cm, as can be seen in tracings of slabs cut from two of the specimens analyzed in the present study (Figure 1).

A sample of peridotite weighing 90 g has a volume equivalent to a cube 3 cm on an edge. Many bulk analyses of peridotite xenoliths have carried out on analytical samples that were smaller than 90 g and it now appears that variations between these analyses may be due in part to small scale heterogeneities of the kind described by COX et al. (1987). Samples weighing approximately 500 g were used by them for bulk analyses. Analytical samples with minimum weights near 500 g and ranging up to 950 g were prepared in the present investigation.

Xenolith samples were pulverized incrementally, beginning with a jaw crusher equipped with steel plates, then a pulverizer equipped with mullite ceramic plates followed by crushing in a shatterbox having an alumina grinding container so that all the resulting particles were less than 180 μm in diameter. Contamination via the crushing method is less than 0.1 weight percent Al_2O_3 and Fe_2O_3. The sample powders were split with a stainless steel sample splitter until a 10 to 15 g aliquot of powder remained. Initial samples were split into four aliquots of this volume with each being analyzed to establish homogeneity.

Table 1. Comparison of published average modes for garnet lherzolite with average modes obtained in the present study for low– and high–temperature garnet peridotite from the southern Kaapvaal, weight percent.

| | MATHIAS et al. (1970) | MAALOE and AOKI (1977) | Southern Kaapvaal | |
			Low–T	High–T
OLV	64	62	62	75
OPX	27	30	31	16
GAR	7	5	5	6
CPX	3	2	2	3

Abbreviations: OLV—olivine, OPX—orthopyroxene, GAR—garnet, CPX—clinopyroxene.

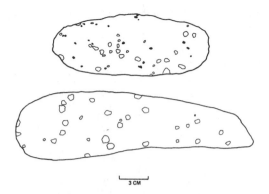

FIG. 1. Tracings of garnet and diopside grain distribution in slabs of two high–temperature lherzolite xenoliths from the Jagersfontein kimberlite pipe; top is FRB 997C and bottom is FRB 1031. Garnet grains are open and diopside grains are solid. Diopside is not shown in the tracing of 1031 because it has insufficient color contrast for consistent identification in the sawed face.

For XRF analysis, whole rock powder was mixed with $Li_2B_4O_7$, diluted with LiI solution to reduce the viscosity, and fused into a homogeneous glass disc. Working curves were determined by analyzing a suite of forty-two geochemical rock standards, data for each having been compiled by ABBEY (1983). Standards PCC-1, DTS-1, and NIM-P were prepared and analyzed as unknowns; the results essentially matched their published analyses.

Ferrous iron was titrated by a modified REICHEN and FAHEY (1962) method and loss on ignition (LOI) was determined by heating an exact aliquot (~ 1 g) of the sample at 900°C for one hour. Trace element briquettes were prepared by mixing 1.0000 g of whole rock powder with 0.50000 g of pure microcrystalline cellulose and pressing the mixture into a pellet. The mass absorption data correction method of HOWER (1959) was used for data reduction. The above standards together with BHVO-1 were prepared and analyzed as unknowns and the results agreed well with the published values.

Electron microprobe analyses were made on thin sections with supplemental grain mounts, employing techniques and standards long in use at the Geophysical Laboratory (FINGER and HADIDIACOS, 1972). Analyses of 5–8 grains were sufficient to establish average compositions for most minerals, but more detailed analysis was carried out on phases with marginal metasomatic alterations. Bulk analyses for the 24 xenoliths are given in Table 2; electron microprobe analyses, weights of individual analytical samples and petrographic descriptions are contained in an unpublished Appendix that will be supplied by Boyd on request.

METASOMATIC EFFECTS

Serpentine, phlogopite and fine–grained phases comprising kelyphite have crystallized subsequent to the primary igneous events in which the rocks studied were formed. Marginal zoning and minor amounts of secondary crystals of olivine, enstatite, diopside and spinel that are spatially associated with mica have also been found in several of the low–

temperature peridotites from Jagersfontein. The serpentine and kelyphite are present in varying concentrations in virtually all the xenoliths and they appear to have developed by reaction of primary phases in the presence of introduced H_2O. In calculating modes from chemical analyses on an H_2O–free basis the presence of serpentine and kelyphite have been ignored. The formation of phlogopite, however, has involved the introduction of elements in addition to water and has occurred with both the consumption and growth of anhydrous phases (Table 3). Evaluation of these effects is required for determination of primary modes.

Five low–temperature peridotites from Jagersfontein contain several percent of phlogopite that primarily forms interganular lenticles and clots. The grain size of the mica is predominantly of the order of 0.1 mm, but coarser flakes with a poikilitic habit are present in several of the peridotites. The phlogopite has commonly crystallized in contact with enstatite and envelopes all grains of garnet; it is not found along contacts between olivine grains.

Small amounts of secondary olivine and enstatite are present in the margins of some primary grains in association with interstitial mica. The occurrence of the zoned margins is erratic and they appear to have formed by alteration of primary grains rather than as overgrowths. Marginal zones in olivine are 50–100 μm thick. Enstatite is not zoned in some occurrences where zoning of olivine is present. Secondary diopside in mica forms subhedral granules ranging up to 0.3 mm. Secondary spinel occurs both as mantles on primary grains and as granules in mica.

Secondary phases are uniformly enriched in Ti and Fe. In addition, secondary enstatite is enriched in Ca, whereas secondary diopside is depleted in this element. These pyroxene relations may indicate that the zoning developed with increasing temperature associated with kimberlite magmatism, although compositional inhomogeneities are clear evidence that an approach to equilibrium was imperfect. Granules of secondary diopside and spinel in mica are relatively depleted in Al, perhaps because of concomitant crystallization of the Al–rich mica.

Comparison of bulk analyses calculated from the modes and primary mineral analyses with analyses of whole–rock samples shows that the principal elements introduced along with H_2O in the formation of phlogopite and other secondary phases are K, Na, Ti, and Fe (Table 3). Additional evidence for the introduction of Ti and Fe can be obtained by comparison of the compositions of secondary anhydrous phases with primary values (Table 4). The

Table 2. Bulk analyses for garnet peridotite xenoliths and one spinel peridotite from the Kaapvaal craton, southern Africa. Weight percent.

Spec	1 FRB-1007	2 FRB-1008	3 FRB-1012	4 FRB-1031	5 FRB-1033	6 FRB-76	7 FRB-999	8 FRB-1009	9 FRB-447	10 FRB-448	11 FRB-135	12 FRB-492
SiO_2	42.52	43.44	41.54	40.68	41.57	41.49	43.08	43.26	40.51	41.11	46.17	45.62
TiO_2	0.07	0.16	0.15	0.08	0.16	0.17	0.11	0.07	0.31	0.02_6	0.00	0.00
Al_2O_3	0.80	1.14	1.16	1.69	1.02	1.80	0.94	1.61	1.42	0.79	1.49	1.60
Fe_2O_3	2.08	2.02	2.94	3.39	1.70	2.34	2.18	2.42	2.77	1.57	1.24	0.91
FeO	3.78	4.11	5.01	3.53	6.08	5.62	4.60	4.70	6.04	5.97	4.66	5.15
MnO	0.10	0.11	0.13	0.12	1.14	0.13	0.12	0.13	0.12	0.12	0.11	0.12
MgO	44.38	42.74	42.53	40.47	43.99	41.92	43.36	41.01	40.82	44.41	43.00	42.88
CaO	0.53	0.71	0.85	1.03	0.94	1.30	0.68	0.96	1.20	0.40	0.56	0.77
Ha_2O	0.08	0.13	0.06	0.06	0.08	0.08	0.12	0.16	0.10	0.04	0.06	0.20
K_2O	0.22	0.44	0.03_2*	0.01_3	0.01	0.02_6	0.21	0.28	0.14	0.05_2	0.01	0.17
P_2O_5	0.02	0.01	0.00	0.00	0.00	0.01	0.02	0.02	0.04	0.00_3	0.00	0.05
L.O.I.	5.70	5.31	5.98	9.37	4.34	4.90	4.69	5.14	6.65	5.48	2.30	1.96
Total	100.28	100.32	100.38_2	100.43	100.02_8	99.78_6	100.11	99.76	100.12	99.97_1	99.60	99.43
$Fe_2O_3^T$	6.29	6.60	8.52	7.32	7.97	8.60	7.30	7.65	9.50	8.22	6.43	6.64
RB	13	26	6	2	5	2	6	13	4	2	1	4
Sr	55	92	17	15	15	25	48	76	128	20	8	55
Ba	142	80	42	30	16	51	55	67	936	63	11	52
Y	16	17	14	7	29	15	18	11	4	2	3	6
Zr	18	6	21	16	20	22	12	12	19	2	16	25
V	14	24	33	33	29	52	27	33	72	31	28	23
Ni	2180	1998	2168	2044	2262	2154	2067	1961	2257	2380	2177	2156
Cr	1970	2499	2601	2500	2750	2826	2491	2511	2272	2666	2696	2673

absolute magnitudes of these differences in bulk composition, however, are small and most of the components of phlogopite have been supplied by solution and recrystallization of primary phases rather than by subsolidus introduction.

Calculated modes that include analyzed mica differ systematically from those calculated on a mica-free basis. For all five mica–bearing peridotites from Jagersfontein that have been analyzed, the mica-present modes contain less enstatite and garnet and more diopside than the mica–absent modes (*e.g.,* Table 2). These differences are interpreted to indicate that the mica–forming reaction was:

garnet + enstatite + fluid →

phlogopite + diopside

This conclusion is supported by the textural evidence; mica in the Jagersfontein rocks is found most commonly in contact with enstatite and it appears to replace garnet. The development of secondary diopside is manifest in some specimens by crystallization of fine–grained crystals in mica. Secondary diopside is also present as a component of secondary enstatite. The suggested reaction is concordant with inferences based on studies of metasomites from the Kimberley pipes (ERLANK *et al.,* 1986).

The primary peridotite compositions are of particular interest in the present study. Accordingly, the modal values used are calculated on a mica-free basis. These calculated primary modes, however, contain substantially more garnet and enstatite and somewhat less diopside than the actual modes for specimens that are relatively rich in phlogopite. Both mica–free and mica–included modes are listed for such specimens in the unpublished Appendix, available on request from Boyd.

MODAL RESULTS

Values of modal olivine calculated for 500–1000 g analytical samples of coarse Mg–rich peridotites with equilibration temperatures below 1100°C have a range of 45–80 weight percent with a broad maximum near 60 (Figure 2). The dispersion is greater

Table 2. (Continued)

Spec	13 FRB- 983	14 FRB- 997B	15 FRB- 997C	16 FRB- 1003A	17 PHN- 4254	18 PHN- 4257	19 PHN- 4258	20 PHN- 4259	21 PHN- 4265	22 PHN- 4274	23 PHN- 5267	24 PHN- 5268
SiO_2	43.05	41.87	40.55	40.32	44.29	44.56	43.24	45.84	46.82	44.41	43.57	42.26
TiO_2	0.05	0.10	0.10	0.10	0.03_1	0.08_1	0.04	0.06_6	0.00_6	0.00_7	0.03	0.11
Al_2O_3	1.13	2.00	1.10	0.51	0.99	1.36	0.74	1.63	1.52	0.91	1.80	0.86
Fe_2O_3	1.66	1.72	2.64	2.61	2.06	1.87	2.03	1.30	1.27	1.61	1.78	2.63
FeO	4.80	6.90	4.88	5.34	3.41	4.26	3.96	4.62	4.30	4.14	5.55	4.79
MnO	0.11	0.13	0.12	0.13	0.10	0.11	0.10	0.11	0.11	0.10	0.13	0.10
MgO	44.60	39.77	42.75	44.80	43.77	42.55	44.87	42.57	41.96	43.28	41.44	44.46
CaO	0.44	1.45	0.49	0.31	0.32	0.52	0.14	0.76	0.67	0.40	1.94	0.42
Na_2O	0.10	0.11	0.07	0.08	0.03	0.08	0.03	0.08	0.08	0.06	0.13	0.09
K_2O	0.23	0.05	0.01	0.03	0.03_2	0.08_1	0.02_5	0.03_4	0.02_8	0.02	0.12	0.07_4
P_2O_5	0.01	0.00	0.00	0.00	0.00	0.00	0.00	0.00	0.00	0.00	0.00	0.02
L.O.I.	3.96	5.75	7.12	5.91	5.09	4.37	4.62	3.34	3.20	4.68	3.08	4.44
Total	100.14	99.87	99.84	100.05	100.15_3	99.83_2	99.79_5	100.29_0	99.96_4	99.61_7	99.57	100.25_4
$Fe_2O_3^T$	7.00	9.40	8.07	8.56	5.86	6.61	6.44	6.44	6.06	6.22	7.96	7.96
Rb	8	1	<1	2	<1	3	1	2	<1	1	4	5
Sr	39	21	7	14	21	16	11	12	2	16	27	29
Ba	84	43	9	43	28	38	21	9	15	18	120	184
Y	3	5	3	2	8	5	7	7	1	13	5	4
Zr	13	24	13	21	7	12	10	12	1	6	6	2
V	21	49	35	24	20	26	20	27	31	19	45	46
Ni	2327	2048	2297	2350	2123	2098	2330	2090	1924	2128	2136	2323
Cr	2190	2379	2295	1791	2978	2412	2413	2558	2704	2217	2853	2068

Garnet peridotites:
1–5, 7, 8, 13–16: Jagersfontein
6, 9, 10: Frank Smith
11: Mothae
17, 18,, 20–22: Letseng-la-Terai
23, 24: Premier
Spinel peridotite
19: Letseng-la-Terai
* Values in subscripts have lesser significance

for previous data based on analytical samples much smaller than 500 g and this relationship is believed due to the influence of small scale heterogeneities on the older data. These rocks are enstatite–rich with as much as 30–40 weight percent enstatite (Figure 3) but the amounts of garnet and diopside are commonly less than 5–10 weight percent. The average olivine content of the low–temperature peridotites is 61 weight percent, a value near that calculated for hypothetical, more Fe–rich, fertile lherzolite compositions such as pyrolite (57 weight percent; RINGWOOD, 1982).

The deformed, more Fe–rich peridotites with equilibration temperatures above 1100°C are more olivine–rich and enstatite–poor than the low–temperature rocks. Modal enstatite for the high–temperature peridotites from the southern Kaapvaal is predominantly less than 20 weight percent and barely overlaps the range for the low–temperature peridotites (Figure 3). There are differences in modal ranges for these minerals however, between the high–temperature suite from the Premier mine in the central part of the craton and suites from lo-

calities near the southern margin of the craton, including northern Lesotho, Kimberley and Jagersfontein (Figure 4). The Premier suite has a broader range of modal olivine with an average of 66 weight percent, whereas suites from the southern Kaapvaal have an average of 76 weight percent olivine. Most of the Premier analyses were carried out on samples of restricted size (DANCHIN, personal communication). Nevertheless, the pronounced difference in *average* modal olivine is unlikely to be related to sample size although small scale heterogeneities can affect the range of values found.

Plots of modal olivine against $Mg/(Mg + Fe)$ of olivine illustrate the pronounced contrast in ranges of $Mg/(Mg + Fe)$ between the Mg-rich low–temperature peridotites and the relatively more Fe–rich high–temperature rocks (Figure 5). A value of 0.915 for $Mg/(Mg + Fe)$ of olivine divides the two groups with only a few exceptions.

Garnet is commonly more abundant than diopside in both the high– and low–temperature peridotites (Figure 6). The high–temperature peridotites contain more diopside and garnet on the average

Table 3. Calculated modes of garnet, lherzolite xenolith FRB 1007, Jagersfontein, together with a comparison of the analysed bulk composition with bulk compositions calculated from the mineral analyses and modes, both with and without mica, weight percent.

Modes	No mica	With mica
Olivine	70.2	70.3
Enstatite	23.9	22.8
Garnet	2.4	0.8
Diopside	1.8	2.2
Spinel	0.3	0.3
Phlogopite	—	2.8
Totals	98.6	99.2

Bulk composition	Analyzed	Calculated	
		No mica	With mica
SiO_2	44.81	44.80	44.80
TiO_2	0.07	0.00	0.05
Al_2O_3	0.84	0.81	0.83
Cr_2O_3	0.30	0.28	0.29
FeO	5.95	6.09	6.01
MnO	0.10	0.09	0.08
MgO	46.76	46.73	46.74
CaO	0.56	0.57	0.56
Na_2O	0.08	0.04	0.07
K_2O	0.24	0.00	0.26
NiO	0.29	0.25	0.26
Totals	100.00	99.66	99.95

than do the low–temperature rocks, although there is much overlap in their ranges. The Premier high-temperature suite includes a larger number of more garnet– and diopside–rich lherzolites than do the suites for the southern Kaapvaal.

DISCUSSION

Metasomatism

The fine grain size and textural relations of the phlogopite and other secondary phases in the Jagersfontein peridotites studied in this investigation are evidence that the metasomatism occurred as a part of the kimberlite magmatic event. The association of the mica with marginal zoning of adjacent enstatite and olivine is further evidence of a late-stage metasomatism. It is unlikely that such zoning, involving Fe, persisted for long periods of time (SMITH and BOYD, 1987a,b). Secondary diopside in these rocks is less calcic than the primary phase, whereas secondary enstatite contains increased Ca. These secondary minerals are inhomogeneous and have not equilibrated; nevertheless, these relations may be evidence of mica formation in an environment of rising temperature, possibly during eruption. Other examples of metasomatism believed to

be associated with kimberlite magmatism include dunites from the Kampfersdam pipe at Kimberley that have secondary enrichments of Fe (BOYD et al., 1983), and mineralization of peridotites from the Matsoku pipe, Lesotho (HARTE and GURNEY, 1975).

Modal scatter

The compositions of residues of varying degrees of isobaric depletion of a given fertile parent should form a trend of increasing modal olivine with increasing Mg/(Mg + Fe). Plots of modal olivine against Mg/(Mg + Fe) for both the low– and high-temperature Kaapvaal peridotites show considerable scatter (Figure 5). Many low–temperature peridotites contain even less olivine than hypothetical fertile peridotite compositions, such as pyrolite or

Table 4. Compositions of minerals in a coarse, low–temperature garnet lherzolite from the Jagersfontein kimberlite pipe, FRB 1008, weight percent.

	Primary				
	OLV	OPX	CPX	SPN	GAR
SiO_2	41.6	47.6	55.6	0.07	41.9
TiO_2	<0.03	<0.03	<0.03	0.04	<0.03
Al_2O_3	<0.03	0.75	1.46	17.09	22.4
Cr_2O_3	<0.03	0.18	0.94	51.2	2.32
Fe_2O_3	—	—	—	2.81	—
FeO	6.87	4.36	1.21	12.9	7.99
MnO	0.05	0.11	0.08	0.08	0.50
MgO	52.7	37.3	17.2	13.7	20.4
CaO	<0.03	0.17	22.7	<0.03	5.15
Na_2O	n.d.*	<0.03	1.00	n.d.	<0.03
NiO	0.48	n.d.	0.05	0.04	n.d.
Totals	101.6	100.5	100.2	98.2	100.6

	Secondary				
	OLV	OPX	CPX	SPN	MICA
SiO_2	40.8	57.4	55.1	0.13	38.7
TiO_2	<0.03	0.30	0.59	3.82	2.27
Al_2O_3	<0.03	0.48	0.94	6.97	14.9
Cr_2O_3	0.03	0.20	1.07	52.7	1.65
Fe_2O_3	—	—	—	6.04	—
FeO	9.78	6.29	3.38	14.7	3.79
MnO	0.11	0.15	0.18	0.37	0.03
MgO	50.3	35.7	19.3	13.7	23.0
CaO	0.06	0.60	18.6	<0.03	<0.03
Na_2O	n.d.	0.05	1.05	n.d.	0.35
K_2O	n.d.	n.d.	n.d.	n.d.	9.53
NiO	0.39	0.11	0.04	0.16	0.19
Totals	101.5	101.3	100.3	98.6	94.4

* n.d. is not determined

Abbreviations: OLV—olivine, OPX—orthopyroxene, CPX—clinopyroxene, SPN—spinel, GAR—garnet

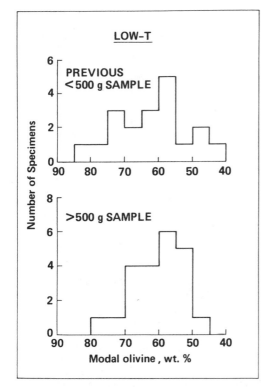

FIG. 2. Histograms of modal olivine contents of low–temperature garnet peridotite xenoliths from the Kaapvaal craton. New data calculated from analyses of samples larger than 500 grams are from this report and from Cox *et al.* (1987). Previous data for smaller analytical samples are calculated from analyses given by Nixon and Boyd (1973); Cox *et al.* (1973); Carswell *et al.* (1979) and Danchin (1979).

1611. The compositions of the Kaapvaal peridotites thus appear to have been modified by some process or combination of processes other than simple depletion. Pressure has a significant effect on the position of the Fo–En liquidus boundary, discussed hereafter, and large differences in the depth of partial melting could lead to variations in the proportions of olivine and enstatite in the residues. These variations, however, would not be expected to include modal olivine proportions less than that of fertile peridotite.

Any modal proportion of olivine and enstatite independent of Mg/(Mg + Fe) could form in a cumulate process, but would require relatively high degrees of partial melting. Layering is uncommon in peridotite xenoliths from kimberlites. Nevertheless, the size of the nodules is small and these rocks have undergone subsolidus recrystallization. The possibility that some of the wide variation in modal olivine is due to cumulate origin seems worth consideration.

Exsolution

Diopside and garnet are the first phases to melt in peridotite at moderate pressures within the garnet stability field (*e.g.*, Mysen and Kushiro, 1977). The reasons for the widespread occurrence of both these phases in Mg–rich, strongly depleted Kaapvaal peridotites are thus not immediately obvious. The pseudo-eutectic in the system Diopside–Forsterite–Pyrope has a composition near the mid–point on

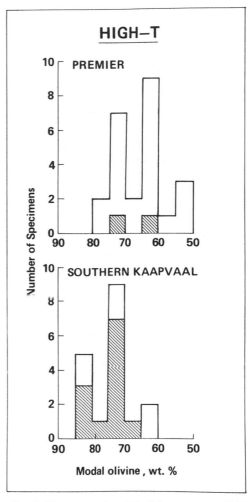

FIG. 3. Histograms of modal olivine contents of high–temperature garnet peridotite xenoliths from the Kaapvaal craton. Specimens from the Premier pipe in the central part of the craton are distinguished from those erupted in other pipes which occur along the southern margin of the craton; specimens from the southern Kaapvaal include xenoliths from the northern Lesotho suite, Jagersfontein, and Frank Smith. Data for analytical samples larger than 500 grams are shaded and are from this report and from Cox *et al.* (1987). Previous data for smaller analytical samples are calculated from analyses given by Nixon and Boyd (1973) and Danchin (1979).

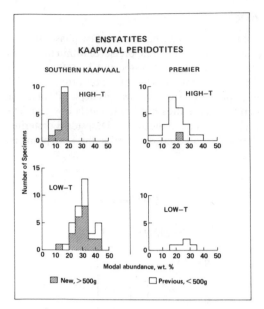

FIG. 4. Histograms of modal enstatite contents of Kaapvaal peridotites. Sources of the data are given in the legends for Figures 2 and 3.

the Diopside–Pyrope join at 40 kbar, (DAVIS and SCHAIRER, 1965) and residual liquids formed near the solidus would be expected to crystallize in approximately equal proportions of diopside and garnet. Garnet commonly is in excess (Figure 6), however, and the occurrence of substantial numbers of garnet harzburgites may be evidence that some garnet is a residual phase. Low–temperature peridotites in particular must have cooled from primary crystallization temperatures to ambient temperatures

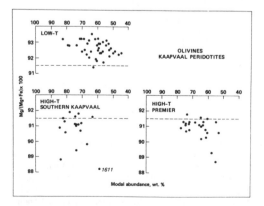

FIG. 5. Plots of Mg/(Mg + Fe) of olivine against modal olivine for garnet peridotite xenoliths from the Kaapvaal craton. Sources of the data are given in the legends for Figures 2 and 3. The point for fertile lherzolite PHN 1611 is distinguished. The dashed line is provided for reference and is drawn at an Mg value of 91.5 in all three plots.

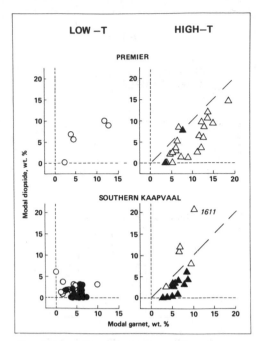

FIG. 6. Plots of modal diopside against modal garnet for Kaapvaal peridotite xenoliths. Solid points are for modal values calculated from analyses of samples larger than 500 grams (this report and COX *et al.*, 1987). Previous data from smaller samples are shown as open points. The point for fertile lherzolite PHN 1611 is identified. The dashed line at a 1:1 ratio is for reference.

and some of the diopside and garnet in them must have exsolved.

Exsolution may have been the dominant process in generating the small amounts of garnet and diopside that are commonly present in the low–temperature peridotites (DAWSON *et al.*, 1980; COX *et al.*, 1987). Orthopyroxene is eliminated in the melting interval of fertile peridotite (*e.g.*, PHN 1611) at pressures above 40 kbar (TAKAHASHI *et al.*, 1986). The clinopyroxene that crystallizes at higher pressures, however, is an Mg–rich pigeonitic phase with 6–7 weight percent CaO (GREEN *et al.*, 1986). This phase may have been the parent for the enstatite–rich clusters of pyroxene and garnet that are commonly present in low–temperature peridotites (COX *et al.*, 1987).

Exolved phases initially form lamellae and blebs, but with recrystallization they coalesce as interstitial grains. Exsolution textures are relatively abundant in eclogites and pyroxenites but only rare examples have been found in lherzolites (DAWSON and SMITH, 1973). COX *et al.* (1987) suggest that the rarity of exsolution features in peridotites is due to the lesser strength of olivine and the relatively

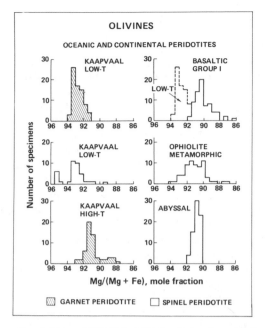

OLIVINES

OCEANIC AND CONTINENTAL PERIDOTITES

FIG. 7. Values of Mg/(Mg + Fe) for olivines in peridotite xenoliths from the Kaapvaal craton, Group I (Cr–diopside) basaltic xenoliths and peridotites from oceanic lithosphere. Spinel and garnet peridotites are distinguished. The outline of the low–temperature Kaapvaal histogram is reproduced in the basaltic Group I plot for easy comparison. This figure is reproduced from BOYD (1987) with the addition of data for Kaapvaal spinel peridotites.

greater ease with which olivine–bearing rocks are deformed and recrystallized.

Compositional comparisons

Both the high– and low–temperature peridotites are depleted in basaltic elements relative to hypothetical fertile peridotite (*e.g.,* pyrolite) and thus

both suites may be residues. Nevertheless, marked differences in their compositional ranges are evidence that the two are products of different magmatic processes. The abundance of olivine and the relative concentrations of Fe and of elements concentrated in diopside and garnet are commonly-used indices of depletion. These indices appear decoupled, however, in contrasting the compositional variations of the two Kaapvaal suites.

The low–temperature peridotites are enstatite-rich, containing an average of 31 weight percent modal enstatite in comparison to 16 weight percent for the high–temperature suite from the southern Kaapvaal and 20 weight percent for the high–temperature peridotites from the Premier mine in the central part of the craton. The high–temperature rocks are proportionately richer in olivine, but contrary to expected depletion relations they are also richer in Fe and are variably richer in diopside and garnet (Tables 1 and 5). Most high–temperature peridotites are markedly richer in Ti but there is a large dispersion of values for Ti in these rocks and it is possible that it has been irregularly introduced in the initial stages of kimberlite magmatism (BOYD, 1987; SMITH and BOYD, 1987a,b).

Peridotites with compositions characteristic of the low–temperature Kaapvaal suite may be confined to Archaean cratons. The high–temperature Kaapvaal peridotites, however, are strikingly similar in composition to peridotites from oceanic and orogenic regions. Comparison of Mg/(Mg + Fe) for olivines in low– and high–temperature Kaapvaal peridotites with values for abyssal peridotites, tectonites from ophiolites, and Cr diopside–bearing (Group 1) spinel lherzolite xenoliths from basaltic volcanics (Figure 7) shows that both spinel and garnet peridotites in the low–temperature suite forming the craton lithosphere are the most Mg–rich, with

Table 5. Modes and bulk chemical compositions of high–temperature peridotite xenoliths and depleted oceanic peridotites. See text for method of calculation of garnet facies modes of oceanic rocks, weight percent.

	High–T premier	High–T S. Kaapvaal	Abyssal average Dick and Fisher (1984)	MORB residue Green *et al.* (1979)
OL	66	75	79	75
OPX	20	16	13	17
GAR	9	6	5	6
CPX	5	3	4	3
Mg No*	909	911	908	917
TiO_2	0.24†	0.13	0.02	0.04
Cr_2O_3	0.30†	0.37	0.22	0.58
NiO	0.25†	0.30	0.22	0.32

* Mg no. is Mg/(Mg + Fe) × 100 olivine, mole percent
† Values from Danchin (1979)
Abbreviations: OL—olivine, OPX—orthopyroxene, CPX—clinopyroxene, GAR—garnet

Mg/(Mg + Fe) predominantly greater than 0.920. Olivines from the underlying high–temperature peridotites, oceanic peridotites and those from Group 1 (Cr–diopside) basalt xenoliths are more Fe–rich, having Mg/(Mg + Fe) primarily in the range 0.880–0.920. More detailed comparisons of the compositions of high–temperature peridotites with those originating in oceanic lithosphere can be made utilizing modes as well as bulk chemical analyses. To facilitate this comparison, modes for the average composition of abyssal peridotites (DICK and FISHER, 1984) and a hypothetical MORB residue (model D, GREEN et al., 1979) have been calculated as garnet peridotites (Table 5). Mineral compositions for these calculations were chosen by matching Mg/(Mg + Fe) of olivine in individual high–temperature peridotites with comparable values for the abyssal average and MORB model.

There is a striking similarity in mode between the high–temperature peridotite xenoliths from the southern Kaapvaal and the oceanic residues (Table 5). Comparably depleted rocks are included in the Premier high–temperature suite but the presence of more fertile peridotites at Premier causes the average to differ and the similarity is not as great.

Origin of low–temperature peridotites

The characteristics of the low–temperature peridotites that are particularly distinctive are Mg/(Mg + Fe) of olivine in the range above 0.920 and values for modal enstatite in the range 20–45 weight percent. Residues produced in the generation of basalt at relatively low pressures in oceanic environments have less modal enstatite and are commonly less magnesian (Table 5). There is, however, a possibility that the low–temperature peridotites are products of melting at high pressures. HERZBERG and O'HARA (1985) have proposed that peridotite melting may become eutectic-like at high pressures and they suggest that mantle peridotites may have originated as liquids rather than residues. Liquids and residues have similar compositions in eutectic-like melting, but the fact that the low–temperature peridotites have higher Mg/(Mg + Fe) than all other peridotite groups (e.g., Figure 7) makes it appear more likely that they are residues. Alternatively, their compositions might reflect primordial compositions or early Archean events.

Olivine and enstatite comprise over 90 weight percent of the mode of most low–temperature peridotites. If speculation that much of their diopside and garnet has exsolved is correct (see above), many of them were composed entirely of olivine and an Mg–rich pyroxene solid solution at the time of their

igneous crystallization. Experiments have shown that the composition of the eutectic in the system $MgSiO_3$–Mg_2SiO_4 shifts from a position near enstatite toward forsterite with increasing pressure in the range up to 30 kbar (KUSHIRO, 1968; CHEN and PRESNALL, 1975). The magnitude of the eutectic shift with increasing pressure decreases at pressures above 30 kbar and the extrapolated eutectic composition appears unlikely to become more forsterite–rich than 30 mole percent at pressures over 100 kbar (KATO and KUMAZAWA, 1985). The pseudobinary eutectic in the system $CaMgSi_2O_6$–Mg_2SiO_4 also shifts toward forsterite with increasing pressure, but experiments and calculations suggest that the shift may be much greater, extending almost to forsterite composition at pressures above 100 kbar (HERZBERG, 1983).

The phase relations determined for these binary systems are concordant with the discovery that the compositions of liquids formed by small degrees of partial fusion of natural peridotites become more forsterite–rich as pressures are increased to 140 kbar (TAKAHASHI and SCARFE, 1985). Nevertheless, enstatite is replaced by an Mg–rich clinopyroxene containing 6–7 weight percent CaO in the melting interval of fertile peridotite at pressures above 40 kbar (GREEN et al., 1986; TAKAHASHI et al., 1986). Variation in the position of the liquidus boundary between this phase and forsterite as a function of pressure is not yet known.

Removal of small amounts of more forsterite–rich melts at high pressures would leave residues that were more enstatite–rich than those formed at lower pressures. The phase boundary shifts are in the right direction to permit interpretation of the low–temperature peridotites as high–pressure residues; it is possible that the shifts are also of the right magnitude, but insufficient experimental data are available at present to test this hypothesis.

The crystallization ages of the low–temperature peridotites are not known but they are probably greater than 3.3 b.y., the oldest age determined for diamonds believed to have crystallized in the craton root (RICHARDSON et al., 1984). It is conceivable that the compositions of these ancient peridotites reflect primordial compositions or events associated with accretion and core–formation.

Tectonic considerations

Similarity in composition of the high–temperature Kaapvaal peridotites to residues of MORB formation may be evidence that they are the remains of subducted oceanic lithosphere originally formed at pressures near or below 10 kbar. Beneath the

rigid block composed of low–temperature peridotites, the oceanic rocks have been heated and transformed to peridotites having the mechanical properties associated with the asthenosphere. A subduction system is speculated to have developed along
the southern margin of the craton during the formation of the Namaqua–Natal mobile belt approximately 1 b.y. ago. This model is consistent
with the crustal geology in that rocks overthrust on
the Kaapvaal craton have been interpreted as ob-
ducted oceanic crust (MATTHEWS, 1972). Analogous subduction from the south in Mesozoic time
has been suggested as a precursor to kimberlite
eruption (SHARP, 1974; HELMSTAEDT and GUR-
NEY, 1982). A subducted origin for eclogite xenoliths in Kaapvaal kimberlites has been proposed
on the basis of $^{16}O/^{18}O$ relations (MACGREGOR and
MANTON, 1987) and appears consistent with the
model suggested here.

The configuration of Mg– and enstatite–rich low-
temperature peridotites overlying olivine–rich high-
temperature peridotites is gravitationally stable. The
density of average low–temperature peridotite is estimated to be 3.32 g/cm^3 in contrast to 3.38 g/cm^3
for average high–temperature peridotite and 3.39
g/cm^3 for fertile garnet lherzolite, PHN 1611 (BOYD
and MCCALLISTER, 1976).

The petrologic model for the Kaapvaal mantle
is consistent with more general, geophysical models
developed by ANDERSON (1987) and by JORDAN
(1978, 1979). In Anderson's model for the continental mantle, rigid lithosphere composed of depleted peridotite extends to a depth of 150 km. Between 150 and 200 km there is a decrease in seismic
velocity that is interpreted to reflect a high thermal
gradient accompanied by a change in composition
to more fertile peridotite. Below about 200 km the
velocities plot close to a 1400°C adiabat for olivine-
rich rock and heat transfer is believed to be convective. Jordan's model has a root zone of depleted
peridotite extending to a greater depth of 300–400
km, well beneath the lithosphere–asthenosphere
boundary.

The transition between low–temperature and
high–temperature peridotites is interpreted as the
lithosphere-asthenosphere boundary (BOYD and
GURNEY, 1986) and its depth beneath the Kaapvaal
craton is estimated to be 175–200 km (FINNERTY
and BOYD, 1987), in agreement with geophysical
estimates. The depth interval represented by the
high–temperature peridotites, however, is not more
than a few tens of kilometers. The depth of the
deeper transition to fertile peridotite is not represented in xenolith suites thus far discovered. Thus,
present petrologic results cannot be used to resolve

the difference between the models of Jordan and
Anderson.

Acknowledgments—Informal discussions with many colleagues, especially during the Fourth International Kimberlite Conference in Perth, have helped to clarify thoughts
expressed in this paper. Particular thanks are due to Keith
G. Cox, Bjorn Mysen, Douglas Smith and Hatten S. Yoder,
Jr., who have provided constructive suggestions for the
improvement of the manuscript. This research was supported by National Science Foundation Grant EAR-
8417437.

REFERENCES

ABBEY S. (1983) Studies in "Standard Samples" of silicate
rocks and minerals 1969–1982. *Geological Survey of
Canada Paper 83-15,* 114 pp.
ANDERSON D. L. (1987) The depth of mantle reservoirs.
In *Magmatic Processes: Physicochemical Principles,* (ed.
B. O. MYSEN), The Geochemical Society Spec. Publ.
No. 1, pp. 3–12
BOYD F. R. (1987) High– and low–temperature garnet
peridotite xenoliths and their possible relation to the
lithosphere–asthenosphere boundary beneath Southern
Africa. In *Mantle Xenoliths,* (ed. P. H. NIXON), John
Wiley, New York. (In press).
BOYD F. R. and GURNEY J. J. (1986) Diamonds and the
African lithosphere. *Science* 232, 472–477.
BOYD F. R. and MCCALLISTER R. H. (1976) Densities of
fertile and sterile garnet peridotites. *Geophys. Res. Lett.*
3, 509–512.
BOYD F. R. and NIXON P. H. (1978) Ultramafic nodules
from the Kimberley pipes, South Africa. *Geochim. Cos-
mochim. Acta* 42, 1367–1382.
BOYD F. R., JONES R. A. and NIXON P. H. (1983) Mantle
metasomatism: The Kimberley dunites. *Carnegie Inst.
Wash. Yearb.* 82, 330–336.
CARSWELL D. A., CLARKE D. B. and MITCHELL R. H.
(1979) The petrology and geochemistry of ultramafic
nodules from Pipe 200, northern Lesotho. In *The Mantle
Sample, Proc. 2d Int. Kimberlite Conf., Vol. 2,* (eds.
F. R. BOYD and H. O. A. MEYER), pp. 127–144, Amer-
ican Geophysical Union, Washington, D.C.
CHEN C. and PRESNALL D. C. (1975) The system
Mg$_2$SiO$_4$–SiO$_2$ at pressures up to 25 kilobars. *Amer.
Mineral.* 60, 398–406.
COX K. G., GURNEY J. J. and HARTE B. (1973) Xenoliths
from the Matsoku Pipe. In *Lesotho Kimberlites,* (ed.
P. H. NIXON), pp. 76–100, Lesotho National Devel-
opment Corporation, Maseru, Lesotho.
COX K. G., SMITH M. R. and BESWETHERICK S. (1987)
Textural studies of garnet lherzolites: Evidence of prob-
able exsolution origin from fertile harzburgite. In *Mantle
Xenoliths,* (ed. P. H. NIXON), John Wiley, New York.
DANCHIN R. V. (1979) Mineral and bulk chemistry of
garnet lherzolite and garnet harzburgite xenoliths from
the Premier Mine, South Africa. In *The Mantle Sample:
Inclusions in Kimberlites and Other Volcanics, Proc. 2d
Int. Kimberlite Conf., Vol. 2,* (eds. F. R. BOYD and
H. O. A. MEYER), pp. 104–126, American Geophysical
Union, Washington, D.C.
DAVIS B. T. C. and SCHAIRER J. F. (1965) Melting relations
in the join diopside–forsterite–pyrope at 40 kilobars and
at one atmosphere. *Carnegie Inst. Wash. Yearb.* 64,
123–126.

DAWSON J. B. and SMITH J. V. (1973) Garnet exsolution from stressed orthopyroxene in garnet lherzolite from the Monastery Mine. In *First Int. Conf. on Kimberlites, Extended Abstracts*, pp. 81–82, Rondebosch, South Africa.

DAWSON J. B., SMITH J. V. and HERVIG R. L. (1980) Heterogeneity in upper–mantle lherzolites and harzburgites. *Phil. Trans. Roy. Soc. London*, **A297**, 323–331.

DICK H. J. B. and FISHER R. L. (1984) Mineralogic studies of the residues of mantle melting: Abyssal and alpine-type peridotites. In *Kimberlites II: The Mantle and Crust–Mantle Relationships, Proc. 3rd Int. Kimberlite Conf.*, (ed. J. KORNPROBST), pp. 295–308, Elsevier.

ERLANK A. J., WATERS F. G., HAWKESWORTH C. J., HAGGERTY S. E., ALLSOPP H. L., RICKARD R. S. and MENZIES M. (1986) Evidence for mantle metasomatism in peridotite nodules from the Kimberley Pipes, South Africa. In *Mantle Metasomatism*, (eds. M. MENZIES and C. J. HAWKESWORTH), Academic Press, New York.

FINGER L. W. and HADIDIACOS C. G. (1972) Electron microprobe automation. *Carnegie Inst. Wash. Yearb.* **71**, 598–600.

FINNERTY A. A. and BOYD F. R. (1987) Thermobarometry for garnet peridotite xenoliths: A basis for upper mantle stratigraphy. In *Mantle Xenoliths*, (ed. P. H. NIXON), John Wiley, New York. (in press)

GREEN D. H., HIBBERSON W. O. and JAQUES A. L. (1979) Petrogenesis of mid-ocean ridge basalts. In *The Earth: Its Origin, Structure and Evolution*, (ed. M. W. McELHINNEY), pp. 265–299 Academic Press, London.

GREEN D. H., FALLOON T. J., BREY G. P. and NICKEL K. G. (1986) Peridotite melting to 6 GPa and genesis of primary mantle–derived magmas. *Fourth International Kimberlite Conference, Geol. Soc. Australia Abstract Series* **16**, 181–183.

HARTE B. and GURNEY J. J. (1975) Ore mineral and phlogopite mineralization within ultramafic nodules from the Matsoku kimberlite pipe, Lesotho. *Carnegie Inst. Wash. Yearb.* **74**, 528–536.

HELMSTAEDT H. and GURNEY J. J. (1982) Kimberlites of Southern Africa—are they related to subduction processes? *Terra Cognita* **2**, 272–273.

HERZBERG C. T. (1983) Solidus and liquidus temperatures and mineralogies for anhydrous garnet–lherzolite to 15 GPa. *Phys. Earth Planet. Int.* **32**, 193–202.

HERZBERG C. T. and O'HARA M. J. (1985) Origin of mantle peridotite and Komatiite by partial melting. *Geophys. Res. Lett.* **12**, 541–544.

HOWER J. (1959). Matrix corrections in the X–ray spectrographic trace element analysis of rocks and minerals. *Amer. Mineral.* **44**, 19–32.

JORDAN, T. H. (1978) Composition and development of the continental tectosphere. *Nature* **274**, 544–548.

JORDAN T. H. (1979) The deep structure of the continents. *Scientific Amer.* **240**, 92–107.

KATO T. and KUMAZAWA M. (1985) Effect of high pressure on the melting relation in the system Mg_2SiO_4–$MgSiO_3$. Part 1. Eutectic relation up to 7 GPa. *J. Phys. Earth* **33**, 513–524.

KRAMERS J. D. (1979) Lead, uranium, strontium, potassium, and rubidium in inclusion-bearing diamonds and mantle–derived xenoliths from Southern Africa. *Earth Planet. Sci. Lett.* **42**, 58–70.

KUSHIRO I. (1968) Compositions of magmas formed by partial zone melting of the Earth's upper mantle. *J. Geophys. Res.* **73**, 619–634.

MAALOE S. and AOKI K. (1977) The major element composition of the upper mantle estimated from the composition of lherzolites. *Contrib. Mineral. Petrol.* **63**, 161–173.

MACGREGOR I. D. and MANTON W. I. (1987) The Roberts Victor eclogites: Ancient oceanic crust. *J. Geophys. Res.* (in press).

MATHIAS M., SIEBERT J. C. and RICKWOOD P. C. (1970) Some aspects of the mineralogy and petrology of ultramafic xenoliths in kimberlite. *Contrib. Mineral. Petrol.* **26**, 73–123.

MATTHEWS P. E. (1972) Possible Precambrian obduction and plate tectonics in Southeastern Africa. *Nature* **240**, 37–39.

MYSEN B. O. and KUSHIRO I. (1977). Compositional variations of coexisting phases with degree of melting of peridotite in the upper mantle. *Amer. Mineral.* **62**, 843–865.

NIXON P. H. and BOYD F. R. (1973) Petrogenesis of the granular and sheared ultrabasic nodule suite in kimberlites. In *Lesotho Kimberlites*, (ed. P. H. NIXON), pp. 48–56, Lesotho National Development Corporation, Maseru, Lesotho.

O'HARA M. J. and E. L. P. MERCY (1963) Petrology and Petrogenesis of some Garnetiferous Peridotites. *Trans. Roy. Soc. Edin.* **65**, pp. 251–314.

POLLACK H. N. and CHAPMAN D. S. (1977) On the regional variations of heat flow, geotherms and lithosphere thickness. *Tectonophys.* **38**, 279–296.

RICHARDSON S. H., GURNEY J. J., ERLANK A. J. and HARRIS J. W. (1984) Origin of diamonds in old enriched mantle. *Nature* **310**, 198–202.

RINGWOOD A. E. (1982) Phase transformations and differentiation in subducted lithosphere: implications for mantle dynamics, basalt petrogenesis, and crustal evolution. *J. Geol.* **90**, 611–643.

REICHEN L. E. and FAHEY J. J. (1962) An improved method for the determination of FeO in rocks and minerals including garnet. *U.S. Geol. Surv. Bull. 1144B*, 1–5.

SHARP W. E. (1974) A plate–tectonic origin for diamond–bearing kimberlites. *Earth Planet. Sci. Lett.* **21**, 351–354.

SMITH D. and BOYD F. R. (1987a) Compositional heterogeneities in a high-temperature lherzolite nodule and implications for mantle processes. In *Mantle Xenoliths*, (ed. P. H. NIXON), John Wiley, New York. (in press).

SMITH D. and BOYD F. R. (1987b) Compositional heterogeneities in phases in sheared lherzolite inclusions from African kimberlites. *Proc. 4th Int. Kimberlite Conf., Perth, Australia*, (in press).

TAKAHASHI E. and SCARFE C. M. (1985) Melting of peridotite to 14 GPa and the genesis of komatiite. *Nature* **315**, 566–568.

TAKAHASHI E., ITO E. and SCARFE C. M. (1986) Melting and subsolidus phase relation of mantle peridotite up to 25 GPa. *Fourth International Kimberlite Conference, Geol. Soc. Australia, Abstracts Series* **16**, 208–210.

Magmatic Processes: Physicochemical Principles
© The Geochemical Society, Special Publication No. 1, 1987
Editor B. O. Mysen

Magma density at high pressure Part 1: The effect of composition on the elastic properties of silicate liquids

CLAUDE T. HERZBERG

Department of Geological Sciences, Rutgers University, New Brunswick, New Jersey, 08903 U.S.A.

Abstract—The effect of composition on the 1 atmosphere isothermal bulk modulus, K^0, of silicate liquids has been evaluated from a thermodynamic analysis of the fusion curves for β–quartz, forsterite, and fayalite in addition to measurements of ultrasonic velocities in silicate melts. Numerical simulations of the fusion curves in temperature–pressure space are very sensitive to the value of K^0 for the liquid phase. However, K^0 determined for liquid SiO_2, Fe_2SiO_4, and Mg_2SiO_4 depends also on the choice of parameters used in the equation of state and, most importantly, the value chosen for the thermal expansivity of the liquid phase. The wide range of reported values have been critically evaluated, and a set of model thermal expansivities is offered. Bounds have also been placed on the pressure derivative of the bulk modulus, and it is shown that it must be less than about 6, 4.7, and 4 for liquid SiO_2, Fe_2SiO_4, and Mg_2SiO_4, respectively. The K^0 for liquid SiO_2 determined from the fusion curve of β–quartz is in good agreement with that determined by light scattering methods. The K^0 for liquid Fe_2SiO_4 determined from the fusion curve of fayalite is exceptionally well constrained because it is based on a comprehensive set of thermodynamic and elastic parameters that are internally consistent with all experimental data, including the calorimetry work of STEBBINS and CARMICHAEL (1984); K^0 determined from this fusion curve analysis is much higher than that determined by RIVERS and CARMICHAEL (1986) from ultrasonic measurements.

Bulk moduli of the oxide components SiO_2, Al_2O_3, FeO, MgO, CaO, SrO, BaO, Li_2O, Na_2O, K_2O, Rb_2O, and Cs_2O in silicate liquids at 1400°C and 1 atmosphere have been estimated from complex melt compositions using the Voigt–Reuss–Hill mixing approximation. For liquid FeO and MgO there are two contradictory sets of data; bulk moduli for these deduced from the fusion curve analysis are over 200 percent higher than those deduced from the ultrasonic measurements of RIVERS and CARMICHAEL (1986) on liquid Fe_2SiO_4, and $MgSiO_3$.

It is shown that there exists an inverse relationship between bulk modulus and specific volume for all liquid oxide components in silicate melts when use is made of K^0 for liquid FeO and MgO derived from the fusion curve analysis. Similar observations have been made on crystalline binary compounds (ANDERSON and NAFE, 1965) for which compression can be described by a simple elastic law, the empirical law of corresponding states (ANDERSON, 1966). It is demonstrated that this elastic law holds also for metals in both the liquid and crystalline states. The implication is that the law of corresponding states can be generalized to substances in the liquid and crystalline states for all binary metal–anion chemical systems containing intermediate compounds (*e.g.*, oxides, tellurides, etc.). However, bulk moduli for liquid MgO and FeO deduced from ultrasonic studies on liquid $MgSiO_3$ and Fe_2SiO_4 violate this law.

Silicate liquids which compress in violation of the law of corresponding states support the hypothesis that there exists an olivine–liquid density crossover at upper mantle pressures. However, for a universal elastic law, the assumptions upon which this hypothesis rests are seriously in error and, indeed, are likely to be fatally flawed.

INTRODUCTION

ONE OF THE physical parameters which will determine the transport fate of magmas at depth is the density contrast between coexisting crystalline and liquid phases. Interest in this problem was stimulated by the hypothesis of STOLPER et al. (1981) that silicate melts may become neutrally buoyant or even denser than solid residua at high pressures (4 to 8 GPa). This unorthodox idea was based on an analysis of densities of basalts measured to 1.5 GPa (FUJII and KUSHIRO, 1977a,b), and the compressibilities of basaltic melts and various liquids in alkali oxide-SiO_2 systems deduced from ultrasonic measurements. These data demonstrate that

the compressibility of basaltic melts is considerably greater than that for typical crystalline silicates such as olivine. RIGDEN et al. (1984) corroborated these findings from shock wave measurements on liquid anorthite$_{36}$diopside$_{64}$ (mol percent), a basalt analogue. The possibility that olivine flotation may occur at high pressures has received attention in a number of theories dealing with partial melt processes in the mantle and the internal structure of the Earth (STOLPER et al., 1981; NISBET and WALKER, 1982; WALKER, 1983; HERZBERG, 1984; OHTANI, 1984; RIGDEN et al., 1984).

Another indirect method for determining the compressibility of silicate melts is by analysis of the fusion curves of minerals. Early attempts (*e.g.*,

CARMICHAEL *et al.*, 1977) were reexamined by HERZBERG (1983) and OHTANI (1984), and the results supported the suggestion that an olivine–magma density crossover may exist, although HERZBERG (1983) suggested it may be in the 10–15 GPa range. A more rigorous thermodynamic approach was provided by BOTTINGA (1985) who noted that the compressibility of silicate liquids depends on its bulk composition. Compressibility increases with SiO_2 content, a result supported by compressibility measurements in the system SiO_2–$NaAlO_2$–$CaAl_2O_4$ below the glass transition (SEIFERT *et al.*, 1982), and also by the ultrasonic data of SATO and MANGHNANI (1985).

The effect of bulk composition on the compressibility/bulk modulus of silicate liquids is crucial to the debate because anhydrous magmas at pressures in the mantle where the density crossover is suggested to occur are not basaltic in composition. Rather, initial melts have a considerable amount of olivine in the norm, and range from komatiite to lherzolite in composition (HERZBERG, 1983; TAKAHASHI and SCARFE, 1985; HERZBERG *et al.*, 1986; OHTANI *et al.*, 1986). Although the question of an olivine–basaltic magma density crossover is petrologically meaningless, there remains considerable merit to the question if it can be established that variations in bulk composition have no effect on the compressibility of silicate liquids. This was the position adopted by RIGDEN *et al.* (1984), based on the observation that the bulk modulus of an Fe_2SiO_4 liquid determined by RIVERS and CARMICHAEL (1986) from ultrasonic velocities is essentially the same as that for the basalt analogue used in their shock wave experiments. If correct, the existence of an olivine–magma density crossover at high pressures is assured. However, BOTTINGA (1985) calculated that the bulk modulus of liquid Mg_2SiO_4 is 300% higher than that determined from the ultrasonic data of RIVERS and CARMICHAEL (1986) on liquid Fe_2SiO_4. If this is correct, then the hypothesis of a density crossover is unconvincing (see HERZBERG, 1987).

In order to resolve the question of an olivine–magma density crossover, the compressibility of silicate liquids has been reevaluated with special attention focused on the effect of bulk composition. The examination involves both a thermodynamic analysis of the fusion curves of β–quartz, forsterite, and fayalite and compressibility information determined by light scattering and ultrasonic methods.

The melting curves have been simulated numerically and are compared to experimental data. From an error analysis of the parameters used in these calculations, it is demonstrated that previous attempts at a fusion curve analysis are seriously flawed

because they have assumed values of the thermal expansivity of silicate liquids which are in many cases demonstrably in error. It is shown here that their effect on the calculation of the bulk modulus of silicate liquids is huge. Owing to this difficulty, this work is divided into two parts. In Part 1 the effect of composition on the bulk modulus of silicate liquids is examined with considerable attention focused on an evaluation of the uncertainties arising from partial molar volumes and thermal expansivities of the oxide components in silicate melts. The hypothesis of an olivine-magma density crossover is considered in Part 2 (HERZBERG, 1987).

FUSION CURVE ANALYSIS

Thermodynamic considerations

The equations used to derive the isothermal bulk modulus of liquid SiO_2, Fe_2SiO_4, and Mg_2SiO_4 are almost identical to those used by BOTTINGA (1985), although small differences exist in the overall methodology. In this analysis, the fusion curves were calculated by means of a finite element solution to the Clausius-Clapeyron equation:

$$dT/dP = \Delta V/\Delta S \qquad (1)$$

$$= \frac{V_l(T,P) - V_s(T,P)}{\Delta S^0 + P(\alpha_s V_s(T,P) - \alpha_l V_l(T,P)) + \int_{T_f}^{T} \frac{\Delta C_p dT}{T}} . \qquad (2)$$

For the liquid(l) and solid(s) phases, the molar volume, $V(T, 1)$, at temperature T and 1 atmosphere is:

$$V(T,1) = V^0(1 + \alpha(T - T_r)) \qquad (3)$$

where α, the isobaric thermal expansivity, is expressed:

$$\alpha = 1/V^0(\partial V/\partial T)_P \qquad (4)$$

and T_r is the standard reference temperature for V^0. For crystalline materials T_r is 25°C; for liquid Mg_2SiO_4 and Fe_2SiO_4 T_r is the fusion temperature T_f, and for the oxides listed in Table 2 T_r is 1400°C. Molar volumes for liquid Mg_2SiO_4, Fe_2SiO_4, and SiO_2 listed in Table 1 were determined from the partial molar volumes of MgO, FeO, and SiO_2 listed in Table 2, assuming no excess volumes of mixing.

Accurate data are now available on the temperature derivative of the thermal expansivity for forsterite and fayalite (SUZUKI *et al.*, 1981, 1983). These data were fitted to the equation:

$$\alpha = \alpha^0 + (T - 25)(d\alpha^0/dT) \qquad (5)$$

where α^0 is valid at 25°C. The temperature derivative of the thermal expansivity of β–quartz is not known, but it is expected to be very small; consequently it was set at zero.

Table 1. Thermodynamic and elastic data

	Mg_2SiO_4	Fe_2SiO_4	SiO_2	Dimension
Crystalline Phases				
V^0	43.63[1,2,3]	46.26[1,6,7,8]	23.80[11]	cm^3/gfw
α^0	2.77[3,4]	2.57[9]	-0.65[11]	$\times 10^{-5}/°C$
$d\alpha^0/dT$	1.03[3,4]	0.97[9]	0.00	$\times 10^{-8}/°C^2$
K^0	127.77[3,4]	137.33[8]	37.10[12]	GPa
dK^0/dT	-0.0227[3,4]	-0.0278[8]	0.00	GPa/°C
dK^0/dP	5.39[5]	5[10]	6.2	—
Liquid Phases				
V^0	53.67	52.43	27.08	cm^3/gfw
α^0	10.84	12.42	1.00	$\times 10^{-5}/°C$
$d\alpha^0/dT$	0	0	0	$\times 10^{-8}/°C^2$
dK^0/dT	0	0	0	GPa/°C
Melting Parameters at 1 atmosphere				
T_f	1890[13]	1205[14]	1469[15]	°C
$(dT/dP)_i$	$55(5)$[16,17]	$68(2)$[18,19,20]	$400(25)$[15]	°C/GPa
ΔS^0 [21]	$113.1(13.5)$	$60.88(6.74)$	$8.7(.8)$	$J/(gfw \cdot °C)$

1 FISHER and MEDARIS (1969). 2 HAZEN (1976). 3 SUZUKI et al. (1983). 4 calculated from data in 3 by least squares regression. 5 KUMAZAWA and ANDERSON (1969). 6 HAZEN (1977). 7 SMYTH (1975). 8 SUMINO (1979). 9 SUZUKI et al. (1981). 10 assumed: CHUNG (1979) reports 5.97 which seems excessive (SAWAMOTO, personal communication; ANDERSON and ANDERSON, 1970; ANDERSON 1972). 11 metastable extension; SKINNER (1966). 12 LEVIEN et al. (1980). 13 BOWEN and ANDERSON (1914). 14 BOWEN and SCHAIRER (1932). 15 metastable melting point; JACKSON (1976). 16 DAVIS and ENGLAND (1964). 17 OHTANI and KUMAZAWA (1981). 18 LINDSLEY (1967). 19 AKIMOTO et al. (1967). 20 OHTANI (1979). 21 calculated; see text.

V^0, α^0, and K^0 for crystalline phases are at 25°C; V^0 and α^0 are model values for liquid phases at T_f.

Many contradictory values have been reported for the thermal expansivity of silicate liquids, and these will receive considerable attention in this analysis. These are listed in Table 2 and shown in Figure 1. New constraints on the thermal expansivities for MgO and FeO have been obtained, and these are discussed below; it is suggested that the thermal expansivities for the oxide components

Table 2. Molar volumes V^0 (cm^3/gfw) at 1400°C and thermal expansivities ($\times 10^5/°C$) for components in silicate liquids

	SiO_2	Al_2O_3	FeO	MgO	CaO	Na_2O	K_2O	
BW	26.80	37.96	12.80	11.57	16.50	28.90	46.00	V^0
	1.0	2.6	15.6	23.4	18.2	23.8	15.2	α
NC	27.20	36.26	12.64	12.08	16.16	28.57	45.49	V^0
	-1.5	15.1	21.4	15.8	39.9	24.6	26.3	α
MCRS	27.03	36.63	13.86	11.43	16.32	28.78	45.93	V^0
	-0.7	14.8	31.5	9.5	38.4	23.5	24.9	α
BWR	26.75	—	13.94	12.32	16.59	29.03	46.30	V^0
	0.1	—	33.4	12.2	17.0	25.9	37.3	α
SCM	26.75	37.69	13.67	11.95	16.89	29.05	46.52	V^0
	-1.5	2.7	30.4	18.4	23.5	25.6	25.8	α
SSC	27.08	36.83	—	—	—	28.58	—	V^0
	—	6.2	—	—	—	27.3	—	α
MODEL	27.08[1]	36.83[1]	13.31[2]	11.87[3]	16.49[3]	28.82[3]	46.05[3]	V^0
	.19	.47	.38	.37	.28	.22	.39	$\pm$
	1.0	6.2	23.6	23.4	24.3	24.7	26.1	α

1 from SSC; 2 this work; 3 average value & 1 standard deviation
BW: BOTTINGA and WEILL (1970)
NC: NELSON and CARMICHAEL (1970)
MCRS: MO, CARMICHAEL, RIVERS and STEBBINS (1982)
BWR: BOTTINGA, WEILL and RICHET (1982)
SCM: STEBBINS, CARMICHAEL and MORET (1984)
SSC: STEIN, STEBBINS and CARMICHAEL (1984)

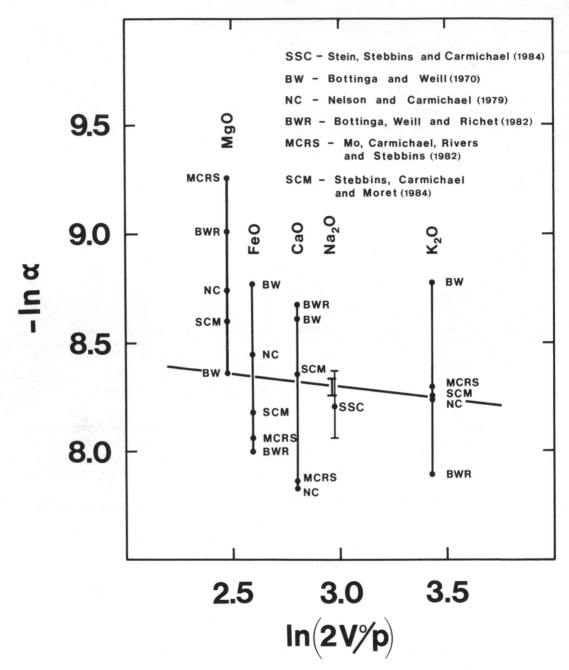

Fig. 1. Reported thermal expansivities of oxide components in silicate liquids (Table 2) and a recommended relation to specific volume.

shown are approximately the same. In light of these first order ambiguities, it is clear that the temperature derivative of the thermal expansivity for silicate liquids must be set at zero.

The pressure–volume relations were determined using a Birch-Murnaghan equation of state:

$$P = 3/2K \left\{ \left[\frac{V(T,1)}{V(T,P)} \right]^{7/3} - \left[\frac{V(T,1)}{V(T,P)} \right]^{5/3} \right\}$$

$$\times \left\{ 1 - 3/4(4-K') \left(\left[\frac{V(T,1)}{V(T,P)} \right]^{2/3} - 1 \right) \right\} \quad (6)$$

where K is the isothermal bulk modulus at T and 1 atmosphere. K is calculated from the 1 atmosphere isothermal bulk modulus, K^0, at reference temperature, T_r, and known values for dK^0/dT; $K' = dK^0/dP$. Again, the reference temperature for crystals is 25°C. It is important to know the temperature derivative of the bulk modulus for the crystalline phases because these calculations are done at melting temperatures which are well in excess of room temperature where K^0 is usually determined for most minerals. Because this information is only rigorously known for forsterite, fayalite, and MgO, this fusion curve analysis is restricted to the olivines and β–quartz, where reasonable bounds can be inferred for the latter.

There is a considerable amount of ultrasonic information that indicates that the temperature derivative of the bulk modulus for silicate liquids is very small (RIVERS and CARMICHAEL, 1986; SATO and MANGHNANI, 1985), particularly above 1400°C; therefore, dK^0/dT for liquids has been set at zero.

In all cases, ΔS^0 was determined from the volume change of melting at 1 atmosphere and the initial slope, $(dT/dP)_i$, from the relation $\Delta S^0 = \Delta V(T_f, 1)/(dT/dP)_i$. The ΔS^0 for fayalite melting is in excellent agreement with the calorimetrically determined value from STEBBINS and CARMICHAEL (1984; see Appendix I). Calorimetric data are not available for forsterite melting. Inspection of Table 1 shows ΔS^0 to be much larger than that for fayalite melting. Although the initial slopes for both are similar, the difference arises because the volume change of melting of forsterite is much larger than that for fayalite. The ΔS^0 for β–quartz was determined from $\Delta V(T_f, 1)$ and the initial slope of metastable melting at 1 atmosphere as given by JACKSON (1976).

For fayalite melting, the calorimetric work of STEBBINS and CARMICHAEL (1984) was used to evaluated the entropy change above the 1 atmosphere melting temperature by the expression:

$$\int_{T_f}^{T} \frac{\Delta C_p dT}{T} = [(119.91T - 3.1167$$

$$\times 10^{-2}T^2 - 21057)/T]_{T_f}^{T}. \quad (7)$$

The same expression was used to determine the entropy change of forsterite melting above its 1 atmosphere melting temperature. However, it should be noted that these terms are very small and, indeed, could be dropped all together with only a trivial contribution to the overall error. For example, removing this term increases K^0 determined for Fe_2SiO_4 liquid from 33.0 GPa to 34.9 GPa.

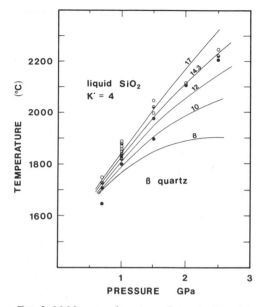

FIG. 2. Melting experiments on β–quartz (JACKSON, 1976) and simulated fusion curves for various values of K^0 for liquid SiO_2. Symbols: filled circles = β–quartz; open circles = glass. Thermal expansivity = $1.0 \times 10^{-5}/°C$.

Liquid SiO_2

Experimental data on the melting of β–quartz (JACKSON, 1976) are shown in Figure 2. Additionally, a number of simulated fusion curves are shown for SiO_2 liquid having a partial molar volume of 27.08 cm^3 (STEIN et al., 1984) and a thermal expansivity of $1.0 \times 10^{-5}/°C$. The simulated fusion curve for an SiO_2 liquid having $K^0 = 14.3$ GPa is in very good agreement with the experimental data. An error in the bulk modulus of about ± 2 GPa produces fusion curves which are resolvably at variance with the experimental data. In the hypothetical case of an SiO_2 liquid having $K^0 = 8$ GPa, the fusion curve is so strongly curved that dT/dP becomes zero at about 2.5 GPa, and SiO_2 liquid would have the same density as β–quartz.

A number of simulations were made in order to test the effect of uncertainties in the parameters on the derived bulk modulus. The results are:

a) Uncertainties in the partial molar volume of SiO_2 liquid at 1400°C result in an error which is small, and equivalent in size to those arising from the experimental data as shown by the range of values of Figure 3. This can be seen by comparing the bulk modulus of a STEBBINS et al. (1984) and a NELSON and CARMICHAEL (1979) SiO_2 liquid; the thermal expansivities of both are the same, but their partial molar volumes of 26.75 and 27.20 cm^3/gfw result in a bulk modulus of 21.8 and 20.1 GPa,

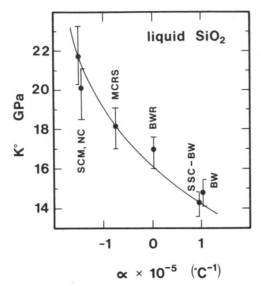

FIG. 3. The effect of thermal expansivity on the bulk modulus of liquid SiO_2 determined by a simulation of the observed melting experiments in the previous figure. See Figure 1 for key to acronyms.

respectively. A partial molar volume of 27.08 cm³/gfw from STEIN *et al.* (1984) yields a molar volume of liquid Fe_2SiO_4 which is most consistent with the calorimetric and high pressure melting data for fayalite (Appendix I). Consequently, the STEIN *et al.* (1984) partial molar volume for liquid SiO_2 is used as the model value throughout this paper.

b) The largest uncertainties in the estimated bulk modulus of SiO_2 liquid arise from variations in the partial molar volume of SiO_2 liquid along the solidus at temperatures higher than 1400°C. This occurs because of published values of the thermal expansivity of SiO_2 liquid which range from -1.5 to $+1.0 \times 10^{-5}/°C$.

c) The ΔS^0 determined from the Clapeyron relation is about 8.7 J/gfw · °C. This value compares with 5.5 J/gfw · °C determined from the enthalpy measurements of RICHET *et al.* (1982). This smaller value is incapable of producing a simulated fusion curve which is consistent with the data of JACKSON (1976) under any condition; the $(dT/dP)_i$ is too high and the slope at higher pressures cannot be sufficiently lowered to conform with the experimental data. Because of this difficulty, ΔC_p was set at zero rather than computed from the equations in RICHET *et al.* (1982). The bulk modulus will be overestimated from this approximation by an amount which, judging from the calorimetry results for Fe_2SiO_4 liquid discussed below, is likely to be small.

d) Because there exists no direct measurements of dK^0/dT for quartz, its value was set at zero. It can be shown, however, that values which are larger

in magnitude than -0.005 GPa/°C result in a simulated fusion curve for which d^2T/dP^2 becomes positive at pressures greater than about 3 GPa. This occurs largely because of the substantial reduction in the entropy change with increasing pressure. A bulk modulus of 14.3 GPa will be reduced to about 12.9 GPa for a hypothetical value of $dK^0/dT = -0.004$ GPa/°C.

e) The effect of uncertainties in K' for liquid SiO_2 is demonstrated in Figure 4. The experimental data do not permit a unique value for K' to be determined. However, it is shown that for K' greater than or equal to 6, d^2T/dP^2 becomes positive in the metastable region at pressures around 5 GPa. A value of K' equal to 6 is likely to be an uppermost bound. For $K' = 4$, d^2T/dP^2 assumes its negative value throughout the stable and metastable regions.

Although there is a number of important uncertainties in this fusion curve analysis, the values of K^0 obtained for positive α (*i.e.*, about 14.3 GPa) are close to those which have been reported by other methods. In particular, BUCARO and DARDY (1976) measured the bulk modulus of amorphous SiO_2 between the glass transformation and melting temperatures of cristobalite by a light-scattering technique, and reported a value of 11.8 GPa. This result compares favorably with their interpretation of the data of WEINBERG (1962; 12.8 GPa), RENNINGER and UHLMANN (1974; 14.3 GPa) and the unpublished data of SCHOEDER quoted by BUCARO and DARDY (1976; 13.0 GPa). It is most likely that this

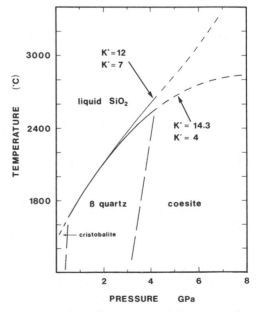

FIG. 4. Simulated fusion curves of β–quartz arising from $K' = 4$ to 7.

fusion curve analysis results in K^0 which is somewhat too high because of uncertainties (c) and (d) above. A thermal expansivity of $1 \times 10^{-5}/°C$ yields a fusion curve-derived bulk modulus in closest agreement with those determined from other methods. The mean for K^0 from all methods of determination is 13 ± 1.3 GPa, a model value that will be adopted throughout the remainder of this paper.

Liquid Fe₂SiO₄

All experimental data on the melting of fayalite to high pressures are shown in Figures 5 and 6. These data are from three sources (LINDSLEY, 1967; AKIMOTO *et al.*, 1967; OHTANI, 1979) and are remarkably consistent. Melting at 1 atmosphere is incongruent, and becomes congruent at pressures between 1 atmosphere and 1.7 GPa (AKIMOTO *et al.*, 1967). The heat of melting for both congruent and incongruent melting at 1 atmosphere are nearly the same (STEBBINS and CARMICHAEL, 1984). Consequently, the simulations shown are valid for congruent melting at all conditions. The bulk moduli for the simulated fusion curves shown in Figures 5 and 6 are valid for the parameters listed in Table 1. As discussed more fully below, these are model parameters which were derived in order to develop an internally consistent set of thermodynamic and elastic properties for this system. A bulk modulus of 33 ± 2 GPa is obtained. A number of other simulations show that small deviations from 33 GPa result in simulated fusion curves that are in major violation of the experimental data.

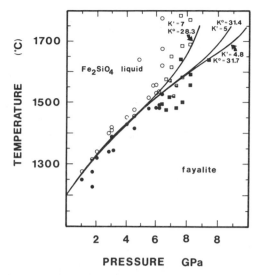

FIG. 6. Simulated fusion curves of fayalite arising from $K' = 4.8$ to 7. Symbols as for Figure 5.

Equally good simulations result from the raw data listed in Table 2 for liquid SiO_2 and FeO. However, the bulk modulus of liquid Fe_2SiO_4 ranges from about 44 GPa for the BOTTINGA and WEILL (1970; BW) parameters to about 25.6 GPa for a BOTTINGA *et al.* (1982; BWR) liquid (Figure 7), and depends largely on the value assumed for the thermal expansivity. The uncertainties in this analysis are small and arise from three sources, these being the initial slope (*i.e.*, $dT/dP_i = 68 \pm 2°C/GPa$), the experimental data that position the fayalite-spinel-liquid

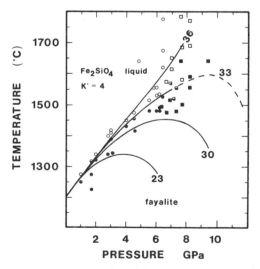

FIG. 5. Melting experiments on fayalite and simulated fusion curves for various values of K^0 for liquid Fe_2SiO_4. Symbols: filled circles = fayalite; open circles = quench fayalite; filled squares = spinel Fe_2SiO_4; open squares = quench spinel; other symbols are mixtures.

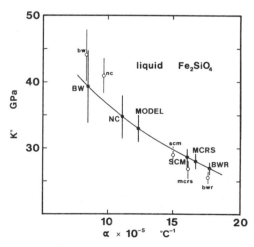

FIG. 7. K^0 vs thermal expansivity for liquid Fe_2SiO_4 determined from a fusion curve simulation of fayalite as a function of: A) volume and thermal expansivity of FeO and SiO_2 data reported in Table 2 (open circles, small lettering), and B) thermal expansivity of FeO, all other parameters kept constant (closed circles, large lettering). $K' = 4$. See Figure 1 for key to acronyms.

triple point at $1550° \pm 10°C$ and 7 GPa, and a $\pm 50°C$ error in temperature measurement at 7 GPa (OHTANI and KUMAZAWA, 1981).

It is clear that this or indeed any fusion curve analysis cannot yield a reliable value of the bulk modulus of the liquid phase in the absence of new constraints on the thermal expansivities of the liquid components. In the analysis which follows, improved constraints on the thermal expansivities for MgO and FeO indicate that there may be a regular and predictable relationship between thermal expansivity and specific volume, V_s. The specific volume is simply the volume per ion pair, defined as 2 (partial molar volume of i)/p, where p is the number of atoms per formula unit (ANDERSON; 1972). Its function is to permit an investigation of bulk-modulus-volume relations amongst materials having any number of atoms per formula unit (e.g., ANDERSON and NAFE, 1965; ANDERSON, 1972). For many crystalline compounds a linear inverse relationship exists between the log of the bulk modulus and the log of volume, a relationship in which compression occurs according to an empirical law of corresponding states (ANDERSON, 1966). By analogy, it is suggested here that variations in thermal expansivity may also be related to specific volume. A relationship is shown in Figure 1 and arises largely from the new constraints for MgO and FeO discussed below; these thermal expansivities are very similar to that for Na_2O. The relationship shown in Figure 1 is a model which requires further testing. However, based on liquid MgO, FeO, and Na_2O it is suggested that there is no significant change in the thermal expansivity of a network modifying component in a silicate liquid with variations in the specific volume. From a survey of thermal expansivities for crystalline compounds, and metals in the liquid and solid states (HERZBERG, in preparation), it can be demonstrated that thermal expansivity varies positively with specific volume for the entire population of data. However, for any particular class of substances (e.g., crystalline oxides, silicates, halides etc.) the variation is small or negligible. The model parameters of α for FeO, MgO, CaO, Na_2O, and K_2O listed in Table 2 were derived from Figure 1. Throughout the remainder of this analysis, a comparison will be made of these model expansivities and those reported for the raw experimental data.

Before we can proceed, there remains yet another problem concerning the choice of the correct value of the partial molar volume of FeO. Inspection of Table 2 shows that the range of published values is rather large (12.64 to 13.94 cm^3/gfw). Consequently, a model partial molar volume of FeO was determined (Appendix I), and has the value 13.31 ± 0.38 cm^3 at $1400°C$.

The effect of thermal expansivity on the value of K^0 derived for Fe_2SiO_4 liquid was evaluated by keeping constant V^0 and α for SiO_2 and V^0 for FeO (i.e., model values), and by using the range of reported thermal expansivities for FeO in Table 2. Simulation of the fusion curve of fayalite was repeated and the results are included in Figure 7. The uncertainties shown are from those listed above in addition to the partial molar volumes of liquid FeO and SiO_2 listed in Table 2. The results demonstrate unambiguously that the bulk modulus determined for liquid Fe_2SiO_4 depends mainly on the value assumed for the thermal expansivity of liquid FeO. The model molar volumes and thermal expansivities result in a bulk modulus of 33.0 ± 2.0 GPa.

The simulated fusion curves shown in Fig. 5 are valid for the model parameters of thermal expansivity and partial molar volumes discussed above with an additional restriction imposed; that is the pressure derivative of the bulk modulus is equal to 4. The effect of relaxing this restriction is shown in Figure 6. By increasing K' to 7, an upper bound for crystalline substances (ANDERSON, 1972), the simulated fusion curve is still consistent with the experimental data. However, the metastable extension inevitably develops the peculiar S–shape with d^2T/dP^2 becoming positive at high pressures. Because this configuration is contrary to all empirically determined fusion curves, it can be concluded that the high values of K' shown in Figure 6 must be in error. By trial and error it was determined that only those values of K' which are less than about 4.75 are capable of simulating a valid fusion curve, this being one in which d^2T/dP^2 is in all cases negative in both the stable and metastable regions.

Liquid Mg_2SiO_4

Experimental data on the melting of forsterite to high pressures are given in Figure 8, and originate from BOWEN and ANDERSEN (1914), DAVIS and ENGLAND (1964), and OHTANI and KUMAZAWA (1981). Again the interlaboratory agreement is excellent. Shown also are a number of simulated fusion curves having the model volume and thermal expansivity parameters listed in Tables 1 and 2, and the effect of varying the bulk modulus of Mg_2SiO_4 liquid. A value of $K^0 = 35.0 \pm 4.5$ GPa is most consistent with the experimental data. Again, small deviations from this value result in a fusion curve which is resolvably at variance with the experimental data.

Bulk moduli obtained from use of the raw data in Table 2 are shown in Figure 9. The errors shown

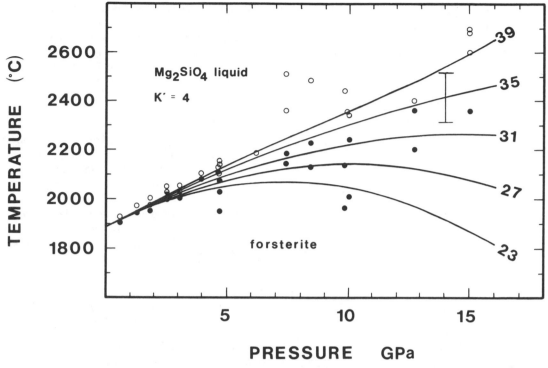

FIG. 8. Melting experiments on forsterite and simulated fusion curves for various values of K^0 for liquid Mg_2SiO_4. Symbols: filled circles = forsterite; open circles = quench forsterite.

are generated from uncertainties in the initial slope ($55° \pm 5°C/GPa$) and temperature measurement at 14 GPa ($\pm 100°C$; OHTANI and KUMAZAWA, 1981). As with liquid fayalite, the effects of uncertainties in the thermal expansivity of MgO was assessed by holding V^0 and α for SiO_2 constant, holding V^0 for MgO constant, and varying α for MgO; the results are included in Fig. 9. The model partial molar volume chosen for MgO is simply the mean of the reported range of values, and the errors arise from those described above in addition to the errors given for the partial molar volumes of MgO and SiO_2 in Table 2. A bulk modulus of 35.0 ± 4.5 GPa is calculated for the BOTTINGA and WEILL (1970) thermal expansivity and, as discussed below, is the model value chosen for liquid Mg_2SiO_4. Much higher values of the bulk modulus are derived for lower thermal expansivities, becoming as high as 55.1 ± 9.6 GPa. This is the main reason why the bulk modulus of the model Mg_2SiO_4 liquid derived in this fusion curve analysis is lower than those similarly derived by OHTANI (1984; 61.2 GPa) and BOTTINGA (1985; 58.8 GPa), although the choice of other parameters, particularly the temperature derivative of the thermal expansivity and bulk modulus for forsterite, also contribute to the dif-

ferences. It is clear that the range of reported thermal expansivities for MgO can result in a calculated bulk modulus for liquid Mg_2SiO_4 which varies by over 200%.

Finally, bounds have been established on the pressure derivative of the bulk modulus, and the results are shown in Figures 10 and 11. Again high values of K' result in S-shaped fusion curves. By trial and error it was found that K' and K must be less than 4 and 35.0 GPa respectively.

COMPONENT OXIDE COMPRESSIBILITY IN SILICATE MELTS

In order to examine more clearly the effect of composition on the compressibility of silicate liquids, it is useful to have an algorithm wherein the bulk modulus of a complex liquid can be calculated from the bulk moduli of its constituent oxides. Although no general theory has been developed for silicate liquids, an analogous problem has been considered for mixtures of isotropic solids (WATT et al., 1976) and for aqueous solutions in general (HAMANN, 1957; HAMMEL, 1985). Common to all these methods is that the bulk modulus of any complex substance K will be some function of the

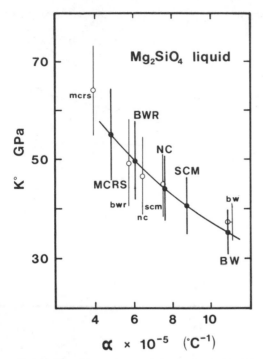

FIG. 9. K^0 vs thermal expansivity for liquid Mg_2SiO_4 determined from a fusion curve simulation of forsterite as a function of: A) volume and thermal expansivities of MgO and SiO_2 reported in Table 2 (open circles, small lettering), and B) thermal expansivity of MgO, all other parameters kept constant (closed circles, large lettering). See Figure 1 for key to acronyms.

bulk moduli of its components, K_i. These must be mixed in proportion to their volume fraction, V_i, not mol fractions, and thus are neither partial molar quantities nor extensive thermodynamic variables. For a Voigt average (i.e., K_v):

$$K_v = \Sigma V_i K_i \qquad (8)$$

and for a Reuss average (i.e., K_r):

$$K_r = (\Sigma(V_i/K_i))^{-1}. \qquad (9)$$

WATT et al. (1976) have demonstrated that the most satisfactory algorithm for reproducing experimental data on complex mixtures of isotropic crystalline substances from its endmember components is the average of the two, or the Voigt–Reuss–Hill (VRH) approximation:

$$K = (K_v + K_r)/2. \qquad (10)$$

The application of the Voigt–Reuss–Hill mixing relation to silicate liquids is fraught with theoretical weaknesses which are revealed in some of the experimental studies discussed below. However, it is

used in this analysis for want of a better algorithm, and because of its simplicity. Furthermore, it will be shown in Part 2 (HERZBERG, 1987) that it seems to be fairly successful in predicting the densities of complex melts which have been determined by independent methods.

Liquid FeO

With constraints now imposed on the bulk modulus of liquid Fe_2SiO_4 and SiO_2, K^0 for liquid FeO was calculated and the results are shown in Figure 12. The effect of uncertainties in thermal expansivity on the bulk modulus for liquid Fe_2SiO_4 (Figure 7) is propagated to liquid FeO, for which K^0 varies from about 54 to 100 GPa. The model thermal expansivity of liquid FeO yields a bulk modulus of 75.4 ± 9.0 GPa (Figure 12; Table 3).

Liquid MgO

The bulk modulus for liquid MgO was calculated from the range of possible values of the bulk modulus of liquid Mg_2SiO_4 and that for liquid SiO_2 (Figure 13). The calculations were done at 1400°C and 2825°C, the melting temperature of periclase, and the results are compared with the bulk modulus of crystalline MgO (SUMINO et al., 1983). It can be seen that K^0 for liquid MgO ranges considerably, from about 90 to 170 GPa at 1400°C, and depends on the value assumed for the thermal expansivity of liquid MgO. At both temperatures, K^0 for MgO liquid having the thermal expansivities of MO et al. (1982; MCRS) and BOTTINGA et al. (1982; BWR) are higher than those for periclase. This, of course, is rather curious, implying that these thermal expansivities are in error. Only the BOTTINGA and WEILL (1970; BW) thermal expansivity and possibly that reported by STEBBINS et al. (1984; SCM) can provide a bulk modulus of liquid MgO that is smaller than that of periclase at 2825°C. Accordingly, these comparisons provide a new way of constraining the thermal expansivity of liquid MgO. Clearly, the value from BOTTINGA and WEILL (1970) must be preferred because all others predict a higher compressibility for crystalline MgO than liquid MgO, particularly at the melting temperature of periclase; accordingly the BOTTINGA and WEILL (1970) value is the model value chosen for liquid MgO.

The bulk modulus of liquid MgO at 1400°C which is preferred in this analysis is therefore 88.8 ± 19.2 GPa (Figure 13; Table 3). At 2825°C this becomes 75.8 ± 16.2 GPa, implying an apparent temperature derivative of -0.009 GPa/°C. This is

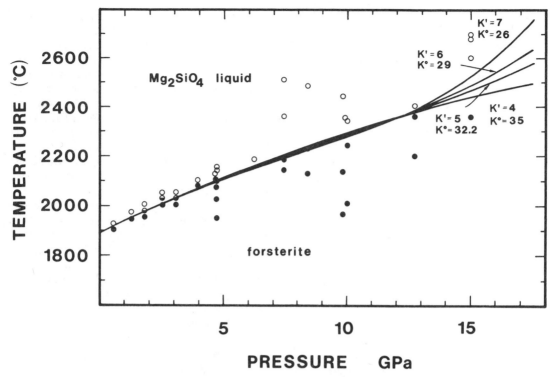

FIG. 10. Simulated fusion curves of forsterite arising from $K' = 4$ to 7. See Figure 8 for key to symbols.

not a true temperature derivative because it arises solely from the smaller volume fraction of the SiO_2 component in liquid Mg_2SiO_4 at 2825°C (*i.e.,* 0.465) compared to 1400°C (0.533).

Liquids of alkali oxides

A fairly thorough set of ultrasonic data is currently available on liquid compositions in the system $MO–SiO_2$ where MO is Li_2O, Na_2O, K_2O, Rb_2O, and Cs_2O (RIVERS and CARMICHAEL, 1986; BOCKRIS and KOJONEN, 1960; BAIDOV and KUNIN, 1968; and others), and the agreement amongst these

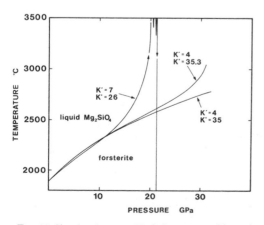

FIG. 11. Simulated metastable fusion curves of forsterite for various values of K'.

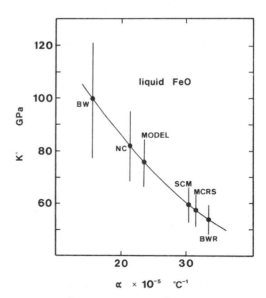

FIG. 12. K^0 for liquid FeO calculated from the bulk moduli of liquid Fe_2SiO_4 and SiO_2.

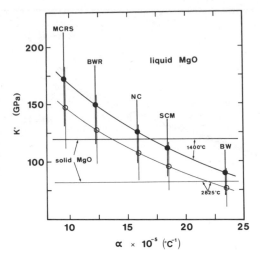

FIG. 13. K^0 for liquid MgO calculated from the bulk moduli of liquid Mg_2SiO_4 and SiO_2.

studies is generally very good (see RIVERS and CARMICHAEL, 1986, for detailed discussion). From the ultrasonic sound velocity, c, the adiabatic K_S and isothermal K_T bulk moduli are determined from the relations:

$$K_S = \rho c^2 \qquad (11)$$

$$K_T = 1/(K_S^{-1} + T\alpha^2/\rho Cp) \qquad (12)$$

where ρ is the density. For compositions on the join MO–SiO_2 α is determined from the thermal expansivities of the endmembers, α_i, (Table 2) by:

$$\alpha = \Sigma V_i \alpha_i. \qquad (13)$$

The bulk thermal expansion coefficient is the sum-

mation of the end-member thermal expansivities mixed in proportion to their volume fractions. Values of Cp used in Equation (12) were taken from the partial molar heat capacities given in STEBBINS et al. (1984). The results of these calculations are very similar to those reported in RIVERS and CARMICHAEL (1986), the small differences reflecting slightly different values calculated for α. The results at temperatures close to or at 1400°C are shown in Figure 14.

An attempt has been made to model the data on these binaries using the Voigt–Reuss–Hill mixing relation, and the results are also shown (Figure 14). The solid curves are mixing lines between the alkali oxide end-members and SiO_2 at about 1400°C. The data on the join Li_2O–SiO_2 can be modelled reasonably well, but problems begin to arise with the system K_2O–SiO_2. Indeed, for all joins, there appears to be a "fine structure" in that the bulk modulus of the MO end-member seems to be bulk composition-dependent. When K^0 for SiO_2 is fixed at 13.0 GPa, the mixing line must be moved up and down in order to satisfy all the data on each binary. Such deviations from a simple Voigt–Reuss–Hill averaging scheme can also be seen in some liquid alloys in the systems Sb–Te and Bi–Te (GLAZOV et al., 1983) and Mg–Bi, Sn–Bi and Sn–Pb (LYSOV and NOVIKOV, 1983). In these systems, the largest deviations of the data from the Voigt–Reuss–Hill equation occur for alloys having end-members which differ most in size and formal charge. Therefore, it is likely that this or indeed any other equation of the same form will only apply rigorously to mixtures having components which are structurally identical. The Voigt–Reuss–Hill approximation is, indeed, strictly an approximation. As discussed be-

Table 3. Isothermal bulk moduli (GPa) of oxide components in silicate liquids at 1400°C and 1 atmosphere

	K^0 *	Method of determination
SiO_2	13.0(1.3)	Light scattering & fusion curve
MgO	88.8(19.2)	Fusion curve—this work
	38.6(6.1)	Ultrasonic (K^0 for $MgSiO_3$ = 19.1; RIVERS and CARMICHAEL, 1986)
FeO	75.4(9.0)	Fusion curve—this work
	28.7(4.5)	Ultrasonic (K^0 for Fe_2SiO_4 = 19.3; RIVERS and CARMICHAEL, 1986)
CaO	49.4(5.4)	Ultrasonic (K^0 for $CaSiO_3$ = 22.9; RIVERS and CARMICHAEL, 1986)
SrO	43.1(7.9)	Ultrasonic (K^0 for $SrSi_2O_5$ = 18.7; RIVERS and CARMICHAEL, 1986)
BaO	34.6(6.1)	Ultrasonic (K^0 for $BaSi_2O_5$ = 18.1; RIVERS and CARMICHAEL, 1986)
Li_2O	26.8(12.2)	Ultrasonic (K^0 for $Li_2O.24$–$SiO_2.76$ = 14.8; BAIDOV and KUNIN, 1968)
Na_2O	13.5(3.0)	Ultrasonic (K^0 for $Na_2Si_2O_5$ = 13.2 GPa; RIVERS and CARMICHAEL, 1986)
K_2O	8.2(4.6)	Ultrasonic (K^0 for $K_2O.12$–$SiO_2.88$ = 11.9; BAIDOV and KUNIN, 1968)
Rb_2O	3.3(0.6)	Ultrasonic (K^0 for Rb_2SiO_5 = 7.2; RIVERS and CARMICHAEL, 1986)
Cs_2O	2.5(0.4)	Ultrasonic (K^0 for Cs_2SiO_5 = 6.0; RIVERS and CARMICHAEL, 1986)
Al_2O_3	20.9(4.5)	Ultrasonic (K^0 for $CaAl_2Si_2O_8$ = 20.0; RIVERS and CARMICHAEL, 1986)

* Except for SiO_2, oxide end-member values for K_0 were calculated from the mixture listed in Method of Determination and SiO_2 using Equation (10). Error is indicated in parentheses.

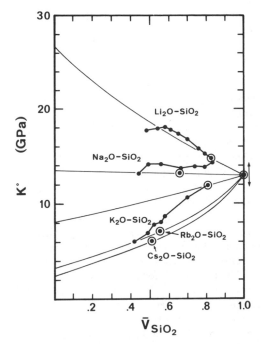

FIG. 14. Ultrasonic data on various liquid alkali oxide-SiO₂ binaries at $T = 1400°C$. Li₂O data from BAIDOV and KUNIN (1968; 1397°C); Na₂O data from BOCKRIS and KOJONEN (1960; 1300°C; circles), and RIVERS and CARMICHAEL (1986; 1420°C; dot–in–circle); K₂O data from BAIDOV and KUNIN (1968; 1397°C); Rb₂O and Cs₂O data from RIVERS and CARMICHAEL (1968; 1420°C). Solid curves are mixing lines between SiO₂ and the alkali oxide end-members listed in Table 3.

low, it will mask structural contributions to the elastic properties of silicate liquids.

With these mixing problems in mind, K^0 was calculated for each end-member using data from compositions closest to SiO₂. The philosophy behind this approach is that this should be most representative of the endmember bulk modulus of the alkali oxides in magmas occurring naturally; for all basic and ultrabasic compositions the concentrations are indeed small. Nevertheless, it should be noted that when the error in the bulk modulus of liquid SiO₂ is propagated in these calculations, K^0 for liquid Li₂O becomes valid for all data on the Li₂O–SiO₂ join. Similarly, K^0 for Na₂O should be valid for most of the Na₂O–SiO₂ binary if the spectrum of data at 1300°C (BOCKRIS and KOJONEN, 1960) can be shifted to the 1420° data point of RIVERS and CARMICHAEL (1986). However, an error propagation cannot satisfy the high potassium data on the join K₂O–SiO₂. Consequently, K^0 for liquid K₂O calculated for the silica–rich compositions will be higher than those for the K₂O–rich composition. The same problem is encountered for compositions

on the joins Rb₂O–SiO₂ and Cs₂O–SiO₂ (RIVERS and CARMICHAEL, 1986) where even larger differences exist in the SiO₂–molar volumes of these end-members. For these the bulk modulus will probably be considerably underestimated compared to compositions on those joins which are more SiO₂–rich, and this is revealed again in the analysis which follows.

For the Rb₂O and Cs₂O compositions studied by RIVERS and CARMICHAEL (1986), it should be noted that K^0 for these end-members can only have positive values when derived from the Voigt–Reuss–Hill or the Reuss approximations. Negative values are obtained when the endmembers are mixed either by volume using the simple Voigt scheme or by mol fractions. The results of these mixing calculations are summarized in Table 3.

Liquids of the alkali earth oxides

The bulk moduli of liquids CaO, SrO, and BaO were similarly calculated from the ultrasonic data of RIVERS and CARMICHAEL (1986) for the compositions CaSiO₃, SrSi₂O₅, and BaSi₂O₅ (Table 3). The value for CaO may be slightly lower than the desired value at 1400°C because the ultrasonic data for CaSiO₃ was obtained at 1563°C.

DISCUSSION OF RESULTS

The bulk moduli of the components in silicate liquids listed in Table 3 have been plotted in Figures 15 and 16. These are compared with elastic data for crystalline binary compounds, crystalline metals,

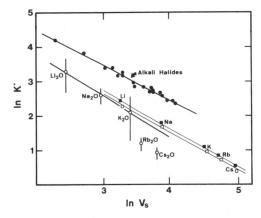

FIG. 15. The empirical law of corresponding states observed for: crystalline alkali halides (closed circles; Table AII–2); crystalline alkali metals (closed squares; Table AII–1) and liquid alkali metals (open squares; Table AII–1) in Group IA; and liquid alkali oxides of cations in Group IA (open circles) determined from Figure 14, with errors arising from uncertainties in K^0 for SiO₂.

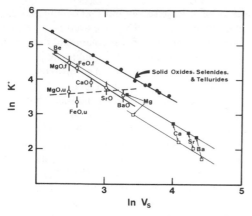

FIG. 16. The law of corresponding states observed (solid lines) for crystalline binary compounds having cations in Group IIA (closed circles; Table AII–2), crystalline metals (closed squares; Table AII–1) and liquid metals (open squares; Table AII–1) in Group IIA. For liquid oxides (open circles) the law of corresponding states is observed for MgO,f and FeO,f deduced from the fusion curve analysis, and violated (broken line) for MgO,u and FeO,u deduced from ultrasonic data.

and liquid metals in Groups IA and IIA of the periodic table listed in Appendix II. As noted above, there is a well-known regular and inverse relationship between the log of the bulk modulus and the log of the specific volume for a wide range of binary crystalline substances (ANDERSON and NAFE, 1965; ANDERSON and ANDERSON, 1970; ANDERSON, 1972), particularly those belonging to the same group in the periodic table. This relationship takes the form:

$$\ln K^0 = s \ln V_s + C \qquad (14)$$

where the slope, s, has a remarkably constant value of about -1 for most crystalline binary compounds. ANDERSON (1966) proposed that these substances compress in observation of the empirical law of corresponding states, in which the bulk modulus can be determined solely by the specific volume. The data in the original Anderson compilations have been updated, and are listed in Appendix II. Additionally, a large amount of ultrasonic data is now available on the elastic properties of liquid metals, and these have also been compiled in Appendix II.

The data in Figure 15 are restricted to substances of the monovalent cations in Group IA. Although the error is fairly large, there is a clear inverse relationship between bulk modulus and specific volume for liquid Li_2O, Na_2O, K_2O, Rb_2O, and Cs_2O. Using the linear least squares regression method of YORK (1966) for five data points, Equation (14) becomes:

$$\ln K^0 = -1.82(0.21) \ln Vs + 7.940(0.85). \qquad (15)$$

It was noted above, however, that K^0 for Rb_2O and Cs_2O are probably underestimated compared to those for more silica-rich compositions on the joins Rb_2O–SiO_2 and Cs_2O–SiO_2. By regressing only the data for Li_2O, Na_2O, and K_2O, the parameters in Equation (14) become: $s = -1.22(0.05)$ and $C = 6.24(0.14)$; the Rb_2O and Cs_2O data plot well below the line as anticipated. For the crystalline alkali halides in Appendix II, $s = -1.02(0.03)$ and $C = 6.55(0.26)$. These results suggest that there may be a relationship between compounds in the liquid and crystalline states; the constant C in Equation (14) may be similar but the slope s may always be larger for liquids compared to crystals. This, however, is not definitive at the present time because the errors in the regressions are fairly large.

The bulk moduli for liquid oxides of cations in Group IIA (Table 3) are plotted in Figure 16; the results for liquid FeO are included. Two possible linear regressions can be entertained. For MgO and FeO determined from the fusion curve analysis, together with the results for CaO, SrO, and BaO inferred from ultrasonic data, the regression is semi-parallel to the data for crystalline binary compounds and metals in the liquid and crystalline states; the regression yields $s = -1.23(0.12)$ and $C = 7.48(0.44)$. For crystalline binary compounds $s = -1.09(0.02)$ and $C = 7.78(0.14)$, again suggesting that C may be constant for all liquid and crystalline binary compounds having cations in Group IIA.

The other possible regression is very different. By using K^0 for MgO and FeO inferred from the ultrasonic data of RIVERS and CARMICHAEL (1986) for liquid $MgSiO_3$ and Fe_2SiO_4, respectively (see Table 3), the inverse relationship between K^0 and V_s seen for all other systems no longer holds. There appears to be no volume dependence to the bulk modulus and the law of corresponding states seems to be violated; a regression yields $s = +0.16(0.49)$ and $C = 3.20(1.83)$ with substantial errors.

Before the apparent violation of the law of corresponding states is discussed, it is useful to complete this survey by considering K^0 for liquid Al_2O_3. This was determined from the bulk modulus of liquid anorthite (RIVERS and CARMICHAEL, 1986), liquid SiO_2 (Table 3), and liquid CaO calculated from the regression for which the law of corresponding states applies. The bulk modulus of liquid Al_2O_3 is 20.9 ± 4.5 GPa (Table 3). This value is consistent with 25.0 GPa determined from an extrapolation of the ultrasonic velocity data of SLAGLE and NELSON (1970) to 1400°C. This value is plotted (Figure 17) together with other substances that have a mean atomic weight of about 20. ANDERSON

(1972) showed that this is a special case of the law of corresponding states, equivalent to Birch's Law, in which the slope of the regression for the crystalline oxides is now close to -4. A very similar relationship is apparent for the liquid oxides as well. It should be noted that a much better fit for liquid MgO, Al_2O_3 and SiO_2 could be obtained if the bulk modulus of anorthite liquid (RIVERS and CARMICHAEL, 1986) is slightly underestimated; that is, large underestimates of K^0 for liquid Al_2O_3 accrue from minor underestimates of K^0 for liquid $CaAl_2Si_2O_8$.

A single equation is all that is needed to describe fully bulk modulus–volume relations for crystalline alkali halides of cations in Group IA in addition to crystalline oxides, tellurides, and selenides of cations in Group IIA (see listing of these in Table AII-2); that is, it seems to be independent of the nature of the anion. Similarly, the law of corresponding states holds rigorously for crystalline metals in Group IIA. These relations are quite remarkable, and imply the existence of a simple elastic law which describes each crystalline binary metal–anion system and its intermediate compounds. With the law of corresponding states firmly established for liquid metals as well (see Figures 15, 16, and AII-1), a similar law is anticipated for liquids in any metal–anion binary system containing intermediate compounds. This is clearly indicated for all the liquid oxide

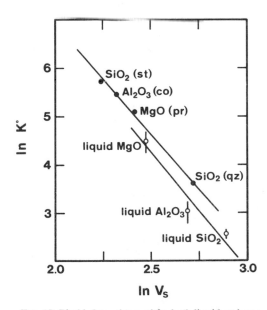

FIG. 17. Birch's Law observed for both liquid and crystalline oxides having similar mean atomic weights (≈ 20), with liquid MgO deduced from the fusion curve analysis. Data for stishovite, quartz, and periclase from WEIDNER *et al.* (1982), LEVIEN *et al.* (1980), and SUMINO *et al.* (1983) respectively.

components which include MgO and FeO determined from the fusion curve analysis. However, K^0 for these deduced from the ultrasonic measurements of RIVERS and CARMICHAEL (1986) on $MgSiO_3$ and Fe_2SiO_4 (see Table 3) are inconsistent with these laws. The K^0 for liquid Fe_2SiO_4 from the ultrasonic data is 19.3 GPa; inspection of Figures 5 and 6 demonstrates that the high pressure melting experiments on fayalite are inconsistent with this. Similarly, by using K^0 for liquid MgO determined from the ultrasonic data for $MgSiO_3$, K^0 for liquid Mg_2SiO_4 can be calculated to be 21.9 ± 2.1 GPa. Inspection of Figures 8 and 10 shows that the ultrasonic data are also inconsistent with the melting data on forsterite. Clearly, additional data would be useful in order to resolve the contradictory results for liquid compositions that are enriched in MgO and FeO.

For melts of geological interest that can have a variable MgO content (*e.g.*, basalts-komatiites) the effect of composition on compressibility is most clearly seen in Figure 17 where major differences in K^0 amongst MgO, Al_2O_3, and SiO_2 are demonstrated. Because MgO is a network modifier and the others are network formers, the simplest interpretation is that there is a structural control to the bulk modulus of silicate liquids, and this is reflected in Birch's Law. In crystals, glasses, and melts, the Si–O bond length is remarkably constant for a whole range of compositions (NAVROTSKY *et al.*, 1985), suggesting that the high bulk modulus of individual SiO_4 tetrahedra in crystals (*e.g.*, LEVIEN *et al.*, 1980; HAZEN and FINGER, 1979) is valid for liquids and glasses as well. However, in addition to reductions in the metal-oxygen bond length, it is the way in which SiO_4 and other polyhedra are structurally arranged that determines the aggregate compression of crystalline materials (HAZEN and FINGER, 1979), and this is best demonstrated by the low bulk modulus for quartz (LEVIEN *et al.*, 1980). For liquid compositions enriched in SiO_2, the high compressibility is most probably accomodated by the tilting or kinking of corner-sharing tetrahedral linkages, rather than any shortening in the Si–O bond length as is the case for quartz. However, crystalline MgO is relatively incompressible because it has a sodium chloride structure in which the MgO_6 octahedra share triangular faces. Opportunities for interpolyhedral bond angle distortion are minimized and compression is taken up largely by a reduction in the length of the Mg–O bond. Because liquid MgO is the most incompressible of all the liquid oxides considered, it can be inferred that it compresses in much the same way. Therefore, for complex liquids on the join $MgO-SiO_2$, it is suggested that the bulk modulus becomes reduced with increasing SiO_2

content because the mechanism of compression changes from intrapolyhedron bond length shortening to interpolyhedron bond angle distortions. In detail, however, this interpretation is bound to be a gross oversimplification because it ignores any possible effects arising from structural complexities in both short and intermediate range ordering; examples are the possibility of tetrahedrally coordinated Mg (YIN et al., 1983) and various possible anionic configurations (e.g., monomers, chains, rings etc.; MYSEN et al., 1982). If the effects of these structural changes on bulk modulus are measurable by experiment, they will most certainly be manifest as a fine structure superimposed on the Voigt–Reuss–Hill relation or, indeed, any other mixing approximation.

CONCLUSIONS

From an analysis of the fusion curves for β–quartz, forsterite, and fayalite in addition to light scattering and ultrasonic measurements on silicate liquids, the following conclusions have been drawn:

a) The bulk moduli for liquid SiO_2, Fe_2SiO_4, and Mg_2SiO_4 retrieved from any fusion curve analysis depend critically on the choice of suitable values of the thermal expansivities of these liquids. Accordingly, constraints have been provided on the wide range of values reported for the thermal expansivity of silicate liquids; the thermal expansivities for all components other than the network formers SiO_2 and Al_2O_3 appear to be largely independent of composition.

b) The pressure derivative of the bulk modulus, K', must be less than 6, 4.7, and 4 for liquid SiO_2, Fe_2SiO_4 and Mg_2SiO_4 respectively. It is suggested that K' for liquids and crystals of the same composition are similar, and that the very high values estimated by OHTANI (1984; 5.5 to 6.6 for liquids having the compositions of fayalite, pyroxenes, and garnets) and BOTTINGA (1985; 9.79 to 23.33 for liquids having compositions of pyroxene, garnets and albite) could be an artifact of the choice of parameters they used. In particular, the proper values of the thermal expansivity of the liquid phase, and temperature derivatives of the thermal expansivity and bulk modulus of the crystalline phases are essential; the latter have only been well characterized for MgO, Mg_2SiO_4, and Fe_2SiO_4.

c) The K^0 for liquid SiO_2 is in good agreement with that determined by light scattering methods. However, K^0 for liquid Fe_2SiO_4 and Mg_2SiO_4 are much higher than those obtained or deduced from the ultrasonic data of RIVERS and CARMICHAEL (1986).

The bulk modulus of the oxide components SiO_2, Al_2O_3, FeO, MgO, CaO, SrO, BaO, Li_2O, Na_2O, K_2O, Rb_2O, and Cs_2O in silicate liquids at 1400°C have been estimated. For liquid MgO and FeO which have been deduced from an analysis of the fusion curves of forsterite and fayalite, there is an inverse relationship between K^0 and the specific volume. ANDERSON (1966) noted that this relationship applies to crystalline binary compounds, and proposed that these substances compress according to the law of corresponding states. It is demonstrated that this elastic law applies also to liquid and crystalline metals. However, K^0 for liquid MgO and FeO which have been deduced from the ultrasonic data of RIVERS and CARMICHAEL (1986) on liquid $MgSiO_3$ and Fe_2SiO_4 contradict the law of corresponding states, demonstrating that either the law is not universal or the data are erroneous. Additional experimental data on silicate liquids having high MgO and FeO are needed to resolve these contradictions.

RIGDEN et al. (1984) concluded that silicate melts will be denser than olivine in the 6 to 10 GPa range. This was based on the assumption that the effect of composition on the bulk modulus of silicate liquids is small. At the time, this conclusion was certainly justified because the bulk modulus of their basalt analogue $An_{36}Di_{64}$ is so similar to that reported by RIVERS and CARMICHAEL (1986) for liquid Fe_2SiO_4. However, it has been shown that the RIVERS and CARMICHAEL (1986) determination is inconsistent with K^0 derived from an analysis of the melting curve of fayalite; it also contradicts the inverse relationship between K^0 and specific volume embodied in the law of corresponding states. The hypothesis of olivine flotation in ultrabasic magmas at high pressures is thus based upon an assumption of questionable merit. A universal elastic law has geological consequences on olivine-magma density relations in the upper mantle which are profoundly different from a law that is violated. These possibilities are explored in Part 2 (HERZBERG, 1987).

Acknowledgements—Thanks are extended to Bjorn Mysen for the invitation to participate in the Yoder conference, and to Bjorn, Larry Finger, and Yoshito Matsui for detailed comments and suggestions which helped to improve this paper. The completion of this paper was made possible in part by support from NSF under grant INT 84–18484.

REFERENCES

AKIMOTO S-I., KOMADA E. and KUSHIRO I. (1967) Effect of pressure on the melting of olivine and spinel polymorph of Fe_2SiO_4. *J. Geophys. Res.* **72**, 679–686.

ANDERSON D. L. and ANDERSON O. L. (1970) The bulk modulus–volume relationship for oxides. *J. Geophys. Res.* **75**, 3494–3500.

ANDERSON O. L. (1966) A proposed law of corresponding

states for oxide compounds. *J. Geophys. Res.* **71**, 4963–4871.

ANDERSON O. L. (1972) Patterns in elastic constants of minerals important to geophysics. In *Nature of the Solid Earth,* (ed. E. D. ROBERTSON), pp. 575–613, McGraw-Hill.

ANDERSON O. L. and NAFE J. E. (1965) The bulk-modulus volume relationship for oxide compounds and related geophysical problems. *J. Geophys. Res.* **70**, 3951–3963.

BAIDOV V. V. and KUNIN L. L. (1968) Speed of ultrasound and compressibility of molten silica. *Soviet Phys.—Dokl. (Engl. Transl.)* **13**, 64–65.

BOCKRIS J. O. and KOJONEN E. (1960) The compressibilities of certain alkali silicates and borates. *J. Amer. Chem. Soc.* **82**, 4493–4497.

BOEHLER R. and KENNEDY G. C. (1980) Equation of state of sodium chloride up to 32 kbar and 500°C. *J. Phys. Chem. Solids* **41**, 517–523.

BOTTINGA Y. (1985) On the isothermal compressibility of silicate liquids at high pressure. *Earth Planet. Sci. Lett.* **74**, 350–360.

BOTTINGA Y. and WEILL D. (1970) Densities of liquid silicate systems calculated from partial molar volumes of oxide components. *Amer. J. Sci.* **269**, 169–182.

BOTTINGA Y., WEILL D. and RICHET P. (1982) Density calculations for silicate liquids. I. Revised method for aluminosilicate compositions. *Geochim. Cosmochim. Acta* **46**, 909–919.

BOWEN N. L. and ANDERSEN O. (1914) The binary system MgO–SiO$_2$. *Amer. J. Sci.* (Fourth Series) **37**, 487–500.

BOWEN N. L. and SCHAIRER J. F. (1932) The system FeO–SiO$_2$. *Amer. J. Sci.* (Fifth Series) **24**, 177–213.

BUCARO J. A. and DARDY H. D. (1976) Equilibrium compressibility of glassy SiO$_2$ between the transformation and melting temperature. *J. Non-Cryst. Solids* **20**, 149–151.

CARMICHAEL I. S. E., NICHOLS J., SPERA F. J., WOOD B. J. and NELSON S. A. (1977) High temperature properties of silicate liquids; application to the equilibrium and ascent of basic magma. *Phil. Trans. Roy. Soc. London,* Ser. **A286**, 373–421.

CHANG Z. P. and GRAHAM E. K. (1977) Elastic properties of oxides in the NaCl-structure. *J. Phys. Chem. Solids* **38**, 1355–1362.

CHUNG D. H. (1971) Elasticity and equation of state of olivines in the Mg$_2$SiO$_4$–Fe$_2$SiO$_4$ system. *Geophys. J. Roy. Astron. Soc.* **25**, 511–538.

CRAWLEY A. F. (1974) Densities of liquid metals and alloys. *Int. Met. Rev.* **19**, 32–48.

DAVIS B. T. C. and ENGLAND J. L. (1964) The melting of forsterite up to 50 kbar. *J. Geophys. Res.* **69**, 1113–1116.

FISHER G. W. and MEDARIS L. G., JR. (1969) Cell dimensions and x-ray determinative curve for synthetic Mg–Fe olivines. *Amer. Mineral.* **54**, 741–753.

FUJII T. and KUSHIRO I. (1977a) Density, viscosity, and compressibility of basaltic liquid at high pressures. *Carnegie Inst. Wash. Yearb.* **76**, 419–424.

FUJII T. and KUSHIRO I. (1977b) Melting relations and viscosity of an abyssal tholeiite. *Carnegie Inst. Wash. Yearb.* **76**, 461–465.

GERWARD L. (1985) The bulk modulus and its pressure derivative for 18 metals. *J. Phys. Chem. Solids* **46**, 925–927.

GLAZOV V. M., PAVLOVA L. M., TIMOSHENKO V. I. and KIM S. G. (1983) Propagation velocity of ultrasound in AV–Te melts (AV = Sb, Bi). *High Temp. (Engl. Transl.)* **22**, 898–905.

GOL'TSOVA E. I. (1966) Densities of lithium, sodium, and potassium at temperature up to 1500–1600°C. *High Temp. (Engl. Transl.)* **4**, 348–351.

GONCHAROVA V. A., CHERNYSHEVA E. V. and VORONOV F. F. (1983) Elastic properties of silicon and germanium single crystals at pressures up to 8 GPa at room temperature. *Sov. Phys.—Solid State (Engl. Transl.)* **25**, 2118–2121.

GRIMSDITCH M. H. and RAMDAS A. K. (1975) Brillouin scattering in diamond. *Phys. Rev.* **B11**, 3139–3148.

GSCHNEIDNER K. A., JR. (1964) Physical properties and interrelationships of metallic and semimetallic elements. In *Solid State Physics,* (ed. F. SEITZ and D. TURNBULL), **16**, pp. 275–426 Academic Press.

HAMANN S. D. (1957) *Physico-Chemical Effects of Pressure.* 246 pp. Academic Press.

HAMMEL H. T. (1985) Exact relationship between compressibility of a solution and compressibilities of its pure constituents. *Phys. Chem. Liq.* **14**, 171–179.

HAZEN R. M. (1976) Effects of temperature and pressure on the crystal structure of forsterite. *Amer. Mineral.* **61**, 1280–1293.

HAZEN R. M. (1977) Effects of temperature and pressure on the crystal structure of ferromagnesian olivine. *Amer. Mineral.* **62**, 286–295.

HAZEN R. M. and FINGER L. W. (1979) Bulk modulus-volume relationship for cation-anion polyhedra. *J. Geophys. Res.* **84**, 6723–6728.

HELGESON H. C., DELANEY J. M., NESBITT H. W. and BIRD, D. K. (1978) Summary and critique of the thermodynamic properties of rock-forming minerals. *Amer. J. Sci.* **278-A**, 229 pp.

HERZBERG C. T. (1983) Solidus and liquidus temperatures and mineralogies for anhydrous garnet-lherzolite to 15 GPa. *Phys. Earth Planet. Inter.* **32**, 193–202.

HERZBERG C. T. (1984) Chemical stratification in the silicate Earth. *Earth Planet. Sci. Lett.* **67**, 249–260.

HERZBERG C. T., BAKER M. B. and WENDLANDT R. F. (1982) Olivine flotation and settling experiments on the join Mg$_2$SiO$_4$–Fe$_2$SiO$_4$. *Contrib. Mineral. Petrol.* **80**, 319–323.

HERZBERG C. T. (1987) Magma density at high pressure Part 2: A test of the olivine flotation hypothesis. In *Magmatic Processes: Physicochemical Principles* (ed. B. O. MYSEN) The Geochemical Society Spec. Publ. 1, pp. 47–58.

HERZBERG C. T., HASEBE K. and SAWAMOTO H. (1986) Origin of high Mg komatiites: constraints from melting experiments to 8 GPa. *Trans. Amer. Geophys. Union* **67**, 408.

JACKSON I. (1976) Melting of silica isotypes SiO$_2$, BeF$_2$ and GeO$_2$ at elevated pressures. *Phys. Earth Planet. Inter.* **13**, 218–231.

JEANLOZ R. and THOMPSON A. B. (1983) Phase transitions and mantle discontinuities. *Rev. Geophys. Space Phys.* **21**, 51–74.

KUMAZAWA M. and ANDERSON O. L. (1969) Elastic moduli, pressure derivatives, and temperature derivatives of single–crystal olivine and single crystal forsterite. *J. Geophys. Res.* **74**, 5961–5972.

LEVIEN L., PREWITT C. T. and WEIDNER D. J. (1980) Structure and elastic properties of quartz at pressure. *Amer. Mineral.* **65**, 920–930.

LINDSLEY D. H. (1967) Pressure–temperature relations in the system FeO–SiO$_2$. *Carnegie Inst. Wash. Yearb.* **65**, 226–230.

LYSOV V. I. and NOVIKOV V. N. (1983) Ultrasonic study

of the dependence of the compressibility of metal melts. *Sov. Phys.—Acoust. (Engl. Transl.)* **29**, 422–424.

MADHAVA M. R. and SAUNDERS G. A. (1979) Elastic constants of single crystals of the three phases of In–Pb alloys. *J. Phys. Chem. Solids* **40**, 923–931.

MCALISTER S. P., CROZIER E. D. and COCHRAN J. F. (1974) Sound velocity and compressibility in the liquid alkaline earth metals. *Can. J. Phys.* **52**, 1847–1851.

MCALISTER S. P., CROZIER E. D. and COCHRAN J. F. (1976) The effect of isotopic mass on the velocity of sound in liquid Li. *J. Phys. F.* **6**, 1415–1420.

MCSKIMIN H. J., ANDREATCH P. JR. and GLYNN P. (1972) The elastic stiffness moduli of diamond. *J. Appl. Phys.* **43**, 985–987.

MING L. C. and MANGHNANI M. H. (1984) Isothermal compression and phase transition in beryllium to 28.3 GPa. *J. Phys.* **F14**, L1–L8.

MO X., CARMICHAEL I. S. E., RIVERS M. and STEBBINS J. (1982) The partial molar volume of Fe_2O_3 in multicomponent silicate liquids and the pressure dependence of oyxgen fugacity in magmas. *Mineral. Mag.* **45**, 237–245.

MYSEN B. O., VIRGO D. and SEIFERT F. A. (1982) The structure of silicate melts: implications for chemical and physical properties of natural magmas. *Rev. Geophys. Space Phys.* **20**, 353–383.

NAVROSTKY A., GEISINGER K. L., MCMILLIAN P. and GIBBS G. V. (1985) The tetrahedral framework in glasses and melts—inferences from molecular orbital calculations and implications for structure, thermodynamics, and physical properties. *Phys. Chem. Minerals* **11**, 284–298.

NELSON S. A. and CARMICHAEL I. S. E. (1979) Partial molar volumes of oxide components in silicate liquids. *Contrib. Mineral. Petrol.* **71**, 117–124.

NISBET E. G. and WALKER D. (1982) Komatiites and the structure of the Archean mantle. *Earth Planet. Sci. Lett.* **60**, 105–113.

OHTANI E. (1979) Melting relation of Fe_2SiO_4 up to about 200 kbar. *J. Phys. Earth* **27**, 189–208.

OHTANI E. and KUMAZAWA M. (1981) Melting of forsterite (Mg_2SiO_4) up to 15 GPa. *Phys. Earth Planet. Inter.* **27**, 32–38.

OHTANI E. (1984) Generation of komatiite magma and gravitational differentiation in the deep upper mantle. *Earth Planet. Sci. Lett.* **67**, 261–272.

OHTANI E., KATO T. and HERZBERG C. T. (1986) Stability of lherzolite magmas at solidus temperatures and 20 GPa. *Trans. Amer. Geophys. Union* **67**, 408–409.

PASHUK E. G. and PASHAEV B. P. (1983) Temperature dependence of the velocity of the ultrasound and the volume dependence of the modulus of elasticity obtained from it for certain metals. *High Temp. (Engl. Transl.)* **21**, 362–366.

RENNINGER A. L. and UHLMANN D. R. (1974) Small angle x-ray scattering from glassy SiO_2. *J. Non-Cryst. Solids* **16**, 325–327.

RICHET P., BOTTINGA Y., DENIELOU L., PETITET J. P. and TEQUI C. (1982) Thermodynamic properties of quartz, cristobalite and amorphous SiO_2: drop calorimetry measurements between 1000 and 1800°K and a review from 0 to 200°K. *Geochim. Cosmochim. Acta* **46**, 2639–2658.

RIGDEN S. M., AHRENS T. J. and STOLPER E. M. (1984) Densities of liquid silicates at high pressures. *Science* **226**, 1071–1074.

RIVERS M. L. and CARMICHAEL I. S. E. (1986) Ultrasonic studies of silicate liquids. *J. Geophys. Res.* (In press).

ROBIE R. A., HEMINGWAY B. S. and FISCHER J. R. (1978) Thermodynamic properties of minerals and related substances at 298.15°K and 1 bar (10^5 pascals) pressure and at higher temperatures. *U.S. Geol. Surv. Bull. 1452*, 456 pp.

ROBIE R. A., BETHKE P. M., TOULMIN M. S. and EDWARDS J. L. (1966) X-ray crystallographic data, densities, and molar volumes of minerals. *Handbook of Physical Constants*, (ed. S. P. CLARK, JR.), pp. 27–74. The Geol. Soc. Amer. Mem. 97, 587 pp.

ROMAIN J. P., MIGAULT A. and JACQUESSON J. (1976) Relation between the Gruneisen ratio and the pressure dependence of Poisson's ratio for metals. *J. Phys. Chem. Solids* **37**, 1159–1165.

SATO H. and MANGHNANI M. H. (1985) Ultrasonic measurements of V_p and Q_p: relaxation spectrum of complex modulus on basalt melts. *Phys. Earth Planet. Inter.* **41**, 18–33.

SEIFERT F., MYSEN B. O. and VIRGO D. (1982) Threedimensional network structure of quenched melts (glass) in the systems SiO_2–$NaAlO_2$, SiO_2–$CaAl_2O_4$ and SiO_2–$MgAl_2O_4$. *Amer. Mineral.* **67**, 696–717.

SHIH W. H. and STROUD D. (1985) Thermodynamic properties of liquid Si and Ge. *Phys. Rev.* **B31**, 3751–3720.

SIMMONS G. and WANG H. (1971) *Single Crystal Elastic Constants.* Massachusetts Institute of Technology Press.

SKINNER B. J. (1966) Thermal expansion. In *Handbook of Physical Constants*, (ed. S. P. CLARK, JR.), 75–96 Geol. Soc. Amer. Mem. 97, 587 pp.

SLAGLE O. D. and NELSON R. P. (1970) Adiabatic compressibility of molten alumina. *J. Amer. Ceram. Soc.* **53**, 637–638.

SMITH C. S. and CAIN L. S. (1975) Born Model repulsive interactions in the alkali halides determined from ultrasonic data. *J. Phys. Chem. Solids* **36**, 205–209.

SMYTH J. R. (1975) High temperature crystal chemistry of fayalite. *Amer. Mineral.* **60**, 1092–1097.

STEBBINS J. F. and CARMICHAEL I. S. E. (1984) The heat of fusion of fayalite. *Amer. Mineral.* **69**, 292–297.

STEBBINS J. F., CARMICHAEL I. S. E. and MORET L. K. (1984) Heat capacities and entropies of silicate liquids and glasses. *Contrib. Mineral. Petrol.* **86**, 131–148.

STEIN D. J., STEBBINS J. F. and CARMICHAEL I. S. E. (1984) New measurements of liquid densities in the system Na_2O–Al_2O_3–SiO_2. *Trans. Amer. Geophys. Union* **65**, 1140.

STOLPER E., WALKER D., HAGAR B. and HAYS J. (1981) Melt segregation from partially molten source regions: the importance of melt density and source region size. *J. Geophys. Res.* **86**, 6261–6271.

SUMINO Y. (1979) The elastic constants of Mn_2SiO_4, Fe_2SiO_4 and Co_2SiO_4, and the elastic properties of olivine group minerals at high temperature. *J. Phys. Earth* **27**, 209–238.

SUMINO Y., ANDERSON O. and SUZUKI I. (1983) Temperature coefficients of elastic constants of single crystal MgO between 80 and 1300 K. *Phys. Chem. Minerals* **9**, 38–47.

SUMINO Y. and ANDERSON O. L. (1984) Elastic constants of minerals. In *Handbook of Physical Properties of Rocks*, Vol. III, (ed. R. S. CARMICHAEL), pp. 39–138, CRC Press, Boca Raton.

SUZUKI I., ANDERSON O. L. and SUMINO Y. (1983) Elastic properties of a single-crystal forsterite Mg_2SiO_4, up to 1200°K. *Phys. Chem. Minerals* **10**, 38–46.

SUZUKI I., SEYA K., TAKEI H. and SUMINO Y. (1981)

Thermal expansion of fayalite, Fe_2SiO_4. *Phys. Chem. Minerals* **7**, 60–63.

SWENSON C. A. (1985) Volume dependence of the Gruneisen parameter: alkali metals and NaCl. *Phys. Rev.* **B31**, 1150–1152.

TAKAHASHI E. and SCARFE C. M. (1985) Melting of peridotite to 14 GPa and the genesis of komatiite. *Nature* **315**, 566–568.

TALLON J. L. and WOLFENDEN A. (1979) Temperature dependence of the elastic constants of aluminum. *J. Phys. Chem. Solids* **40**, 831–837.

TSU Y., SUENAGA H., TAKANO K. and SHIRAISHI Y. (1982) The velocity of ultrasound in molten bismuth, aluminum, silver and copper. *Trans. Jpn. Inst. Met.* **23**, 1–7.

WALKER D. (1983) Lunar and terrestrial crust formation. *J. Geophys. Res.* **88**, B17–B25.

WATT J. P., DAVIES G. F. and O'CONNELL R. J. (1976) The elastic properties of composite materials. *Rev. Geophys. Space Phys.* **14**, 541–563.

WEBBER G. M. B. and STEPHENS R. W. B. (1968) Transmission of sound in molten metals. In *Physical Acoustics,* (ed. W. P. MASON), pp. 53–97 Academic Press.

WEIDNER D. J., BASS J. D., RINGWOOD A. E. and SINCLAIR W. (1982) The single-crystal elastic moduli of stishovite. *J. Geophys. Res.* **87**, 4740–4746.

WEINBERG D. L. (1962) Surface effects in small–angle x-ray scattering. *J. Applied Phys.* **33**, 1012–1013.

YIN C. D., OKUNO M., MORIKAWA H. and MARUMO F. (1983) Structural analysis of $MgSiO_3$ glass. *J. Non-Cryst. Solids* **55**, 131–141.

YORK D. (1966) Least squares fitting of a straight line. *Can. J. Phys.* **44**, 1079–1086.

APPENDIX I

The wide range of reported values for the partial molar volume of liquid FeO (Table 2) are propagated to important uncertainties in the molar volume of Fe_2SiO_4 used in the fusion curve analysis. Accordingly, an independent method for determining the partial molar volume of FeO

was developed and is discussed here. In Table AI–1 are listed the molar volumes at 25°C and 1 atmosphere for a wide range of different phases having compositions which are pure Mg and Fe end-members. These have been plotted in Figure AI–1. The data were obtained from Table 1, ROBIE *et al.* (1966, 1978), JEANLOZ and THOMPSON (1983), and HELGESON *et al.* (1978). It can be seen that an extremely good correlation exists, and this must be attributed to the different sizes of the Fe and Mg cations. A regression of the data yields:

$$V^0, \text{Fe phase} = 1.022\,(V^0, \text{Mg phase}) + 0.760 \quad \text{(AI–1)}$$

with a correlation coefficient of .99989.

The partial molar volume of liquid MgO was determined at 25°C from its model partial molar volume at 1400°C and model α in Table 2. By using Equation (AI–1) the partial molar volume of liquid FeO at 25°C was determined, then extended back to 1400°C using model α in Table 2 for liquid FeO, and yields 13.31 ± 0.38 cm³/gfw. This is close to the average of the values reported for liquid FeO in Table 2.

Constraints on the correct value for the partial molar volume of liquid FeO can also be evaluated using the experimental data of HERZBERG *et al.* (1982) on the join forsterite–fayalite. For a wide range of intermediate compositions on this join the liquids are more dense than coexisting olivines because of the extensive partitioning of the heavier element Fe into the liquid phase. It was demonstrated by experiment that for the bulk composition forsterite$_{30}$–fayalite$_{70}$ (mol percent) at 1425°C and 1 atmosphere, olivine (Fo$_{58.5}$) floats on its coexisting liquid (Fo$_{24.2}$). We now examine which of the reported values of V^0 for liquid FeO in Table 2 are consistent with these experimental observations.

From the raw data in Table 2 and the equation of state parameters for olivine derived from the forsterite and fayalite end-members in Table 1 (see Part 2, this volume) it can be demonstrated that only the data of BOTTINGA and WEILL (1970; BW), NELSON and CARMICHAEL (1979; NC), and STEBBINS *et al.* (1984; SCM) predict olivine floatation. The data of BOTTINGA *et al.* (1982; BWR) and MO

Table AI–1. Molar volumes of Mg and Fe phases at 25°C and 1 atmosphere

	V^0 Mg-phase (cm³/gfw)	V^0 Fe-phase (cm³/gfw)
(Mg, Fe)O	11.25	12.25
(Mg, Fe)$_2$SiO$_4$ α	43.63	46.26
(Mg, Fe)$_2$SiO$_4$ β	40.52	43.22
(Mg, Fe)$_2$SiO$_4$ γ	39.65	42.02
(Mg, Fe)SiO$_3$ opx	31.33	32.96
(Mg, Fe)SiO$_3$ pv	24.46	25.49
(Mg, Fe)SiO$_3$ il	26.35	26.85
Ca(Mg, Fe)Si$_2$O$_6$	66.09	68.27
K(Mg, Fe)(Si$_3$AlO$_{10}$)(OH)$_2$	149.66	154.32
(Mg, Fe)Al$_2$O$_4$	39.71	40.75
(Mg, Fe)$_3$Al$_2$Si$_3$O$_{12}$	113.29	115.28
(Mg, Fe)TiO$_3$	30.86	31.71
(Mg, Fe)CO$_3$	28.02	29.38
(Mg, Fe)Cr$_2$O$_3$	43.56	44.01
(Mg, Fe)$_5$Al(Si$_3$AlO$_{10}$)(OH)$_8$	207.11	213.42
7A(Mg, Fe)$_5$Al(Si$_3$AlO$_{10}$)(OH)$_8$	211.5	221.2
Na$_2$(Mg, Fe)$_3$Fe$_2$Si$_8$O$_{22}$(OH)$_2$	271.3	274.9
NaCa$_2$(Mg, Fe)$_4$Fe(Si$_6$Al$_2$O$_{22}$)(OH)$_2$	273.8	280.3
NaCa$_2$(Mg, Fe)$_4$Al(Si$_6$Al$_2$O$_{22}$)(OH)$_2$	273.5	279.9

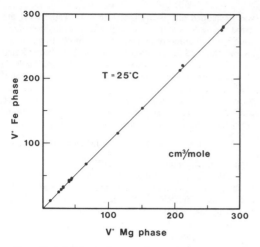

FIG. AI–1. Molar volumes of a wide range of Fe and Mg phases listed in Table AI–1.

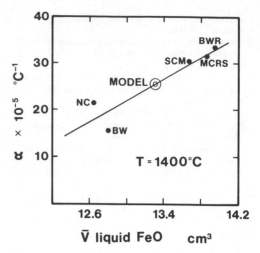

FIG. AI–3. Thermal expansivities and partial molar volumes of liquid FeO reported in Table 2.

et al. (1982; MCRS) yield the opposite result, predicting olivine denser than its coexisting liquid.

These calculations are now repeated keeping the volumes of the liquid components SiO_2 and MgO constant (*i.e.,* the model parameters in Table 2), and varying V^0

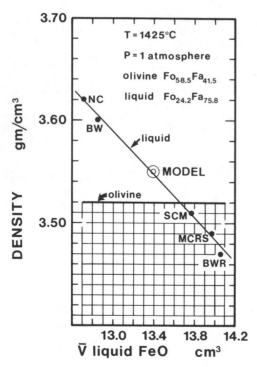

FIG. AI–2. Computed densities of olivine crystals and coexisting liquid for an experiment in the system forsterite–fayalite at 1 atmosphere where olivine flotation has been documented (HERZBERG *et al.,* 1982). Data of SCM, MCRS, and BWR in the hatched region are inconsistent with the experimental observation of olivine flotation.

and α for liquid FeO according to the raw and model parameters. The results are shown in Figure AI–2. It can be seen that only the model parameters for V^0 and α, in addition to those of NELSON and CARMICHAEL (1979; NC) and BOTTINGA and WEILL (1970; BW), are consistent with the experimental results. It is concluded that V^0 and α for liquid FeO reported by MO *et al.* (1982), BOTTINGA *et al.* (1982), and STEBBINS *et al.* (1984) are in error.

The range of reported values for α and V^0 for liquid FeO in Table 2 are shown in Figure AI–3. It appears that

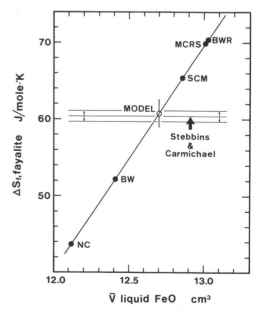

FIG. AI–4. Entropy of fusion of fayalite at 1 atmosphere determined calorimetrically (STEBBINS and CARMICHAEL, 1984) and computed using the volume data for FeO as shown. Only the Model V^0 and α for liquid FeO are in agreement with the calorimetric data.

a correlation exists, suggesting that experimental errors in the determination of the two are coupled. Regressing the raw data yields a value for α which is $25.6 \times 10^{-5}/°C$ for liquid FeO having $V^0 = 13.31$ cm^3/gfw, in excellent agreement with the model value in Table 2.

From the fusion curve data of fayalite (Figures 5 and 6) an initial slope of $68 \pm 2°C/GPa$ is tightly constrained; with slopes outside of these bounds it is difficult to simulate the high pressure melting experiments with any combination of adjusted values of K^0 and K' for liquid Fe$_2$SiO$_4$.

Table AII-1. Specific volumes and isothermal bulk moduli for liquid and crystalline metals and nonmetals

	$V_s \; 2x$ (cm^3/gfw)	K^0 (GPa)	Reference
Li, l	26.85	9.80	1, 2
Li, c	26.04	11.55	3
Na, l	49.61	5.38	4
Na, c	47.48	6.04	3
K, l	94.94	2.62	4
K, c	91.12	2.97	3
Rb, l	115.89	2.03	4
Rb, c	111.48	2.30	3
Cs, l	144.46	1.45	4
Cs, c	140.80	1.71	3
Be, c	9.77	119.0	5
Mg, l	30.59	19.8	6, 9
Mg, c	27.99	35.6	7
Ca, l	59.38	9.1	6, 9
Ca, c	54.37	15.5	8
Sr, l	73.79	7.6	6, 9
Sr, c	68.28	11.8	8
Ba, l	82.93	5.6	6, 9
Ba, c	76.41	10.5	8
Al, l	22.58	52.1	6, 10
Al, c	20.01	72.6	11
Ga, l	23.63	44.2	6, 12
Ga, c	23.59	58.0	8
In, l	32.74	33.5	6, 12
In, c	31.52	39.6	13
Tl, l	36.43	26.1	6, 4, 12
Tl, c	34.49	36.4	8, 14
C, diamond	6.84	442.0	15, 16
Si, l	22.20	78.7	17
Si, c	24.10	96.1	18
Ge, l	26.35	46.5	17
Ge, c	27.26	74.2	18
Sn, l	34.06	36.9	4
Sn, c	32.59	54.2	14

l = liquid; c = crystalline.

1 MCALISTER et al. (1976). 2 GOL'TSOVA (1966). 3 SWENSON (1985). 4 WEBBER and STEPHENS (1968). 5 MING and MANGHNANI (1984). 6 CRAWLEY (1974). 7 ROMAIN et al. (1976). 8 GSCHNEIDNER (1964). 9 MCALISTER et al. (1974). 10 TSU et al. (1982). 11 TALLON and WOLFENDEN (1979). 12 isothermal bulk modulus calculated from adiabatic data in PASHUK and PASHAEV (1983) with heat capacity and thermal expansivity parameters in 4 and 6. 13 MADHAVA and SAUNDERS (1979). 14 GERWARD (1985). 15 MCSKIMIN et al. (1972). 16 GRIMSDITCH and RAMDAS (1975). 17 theory from SHIH and STROUD (1985). 18 GONCHAROVA et al. (1983).

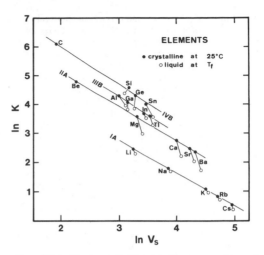

FIG. AII-1. The law of corresponding states observed for both liquid and crystalline metals and nonmetals occupying Groups IA, IIA, IIIB, and IVB of the periodic table.

Table AII-2. Specific volumes and isothermal bulk moduli of crystalline binary compounds at 25°C

	V_s (cm^3/gfw)	K^0 (GPa)	Reference
LiF	9.84	66.51	1
LiCl	20.45	29.68	1
LiBr	25.07	23.52	1
LiI	32.73	17.26	1
NaF	14.99	46.48	1
NaCl	27.02	23.68	1, 2
NaBr	32.17	19.47	1
NaI	40.83	14.87	1
KF	23.04	30.22	1
KCl	37.51	17.35	1
KBr	43.30	14.64	1
KI	53.13	11.51	1
RbF	27.19	26.68	1
RbCl	42.94	15.58	1
RbBr	49.26	13.24	1
RbI	59.60	10.49	1
CsCl	42.19	16.69	3
CsBr	47.75	14.57	3
CsI	57.40	11.56	3
BeO	9.30	220.1	4
MgO	11.25	162.8	5
CaO	16.76	11.2	6
SrO	20.69	89.28	6
BaO	25.59	74.06	6
CaSe	31.30	52.0	7
SrSe	36.68	46.7	7
BaSe	43.28	39.6	7
CaTe	38.66	47.4	7
SrTe	44.47	37.0	7
BaTe	51.65	34.1	7

1 SMITH and CAIN (1975). 2 BOEHLER and KENNEDY (1980). 3 SUMINO and ANDERSON (1984; averages at 25°C). 4 ANDERSON (1972). 5 SUMINO et al. (1983). 6 CHANG and GRAHAM (1977). 7 SIMMONS and WANG (1971).

The entropy of melting at 1 atmosphere has been calculated for the range of molar volumes of liquid Fe_2SiO_4, and the results are compared to the calorimetric value obtained by STEBBINS and CARMICHAEL (1984). (Incidently, it was discovered by trial and error that the model V^0 for FeO results in a molar volume for Fe_2SiO_4 which is most consistent with the calorimetric constraints if the 1400°C partial molar volume of liquid SiO_2 is 27.08 cm^3/gfw (Table 2; STEIN et al., 1984), although the other values for SiO_2 result in a good approximation as well.) Using the STEIN et al. (1984) value for SiO_2 together with the range of raw values of V^0 and α for liquid FeO in Table 2, a range values for ΔS^0 are computed. These are shown in Figure AI-4. It can be seen that the model values of α and V^0 result in $\Delta S^0 = 60.73 \pm 1.84$ J/gfw·K, in excellent agreement with the calorimetrically determined value of 60.42 $\pm$ 0.74 J/gfw·K. All other volumes give rise to entropies which are seriously in conflict with the calorimetric data.

It is concluded that the partial molar volume and the thermal expansivity of liquid FeO are now well constrained. All experimental data which bear on liquid fayalite, these being the melting experiments at high pressure, the data on the forsterite–fayalite join, the calorimetric data and the volume correlations between Mg and Fe phases are internally consistent with the model α and V^0 parameters of liquid FeO listed in Table 2.

APPENDIX II

ANDERSON (1972) demonstrated that the law of corresponding states applies to crystalline metals. Much of the bulk modulus data which were reported were from the original measurements, and have been reexamined by other methods (e.g., ultrasonic and Brillouin scattering). An updated data base has been compiled, together with bulk moduli for many metals in the liquid state at temperatures just above their melting points. These are listed in Table AII-1, portions of which have been shown above. The complete data base for metallic and nonmetallic elements is now plotted in Figure AII-1. There is no doubt that the bulk modulus of liquid metals varies systematically with specific volume, and this can be adequately described by the law of corresponding states.

Elastic data for crystalline binary compounds having cations in Groups IA and IIA are very well known. The compilations in ANDERSON (1972) have been updated and are listed in Table AII-2 and plotted in Figures 15, 16, and 17.

Magmatic Processes: Physicochemical Principles
© The Geochemical Society, Special Publication No. 1, 1987
Editor B. O. Mysen

Magma density at high pressure Part 2:
A test of the olivine flotation hypothesis

CLAUDE T. HERZBERG

Department of Geological Sciences, Rutgers University, New Brunswick, New Jersey, 08903 U.S.A.

Abstract—An algorithm is presented for computing the density of any complex magma composition to pressures in the 15–20 GPa range. It based on an estimation of the bulk modulus of the magma from the bulk moduli of its constituent oxide components, mixed in proportion to their volume fractions according to the Voigt-Reuss-Hill approximation. An examination is made of the effects of the contradictory elastic parameters reported in Part 1 (HERZBERG, this volume), and a test is made of the hypothesis that olivine may become less dense than nautrally occurring magmas at some finite depth in the upper mantle (STOLPER et al., 1981).

Two elastic models are considered. In the first, silicate liquids compress according to the law of corresponding states; the effect of composition on the compressibility of the melt is large, and the composition of the melt phase must be specified. In the second, silicate liquids compress in violation of the law of corresponding states, and all magma compositions compress by similar amounts over any given increment of pressure.

It is demonstrated that the olivine flotation hypothesis is unconvincing for silicate liquids which compress in observation of the law of corresponding states. Although an olivine-liquid density crossover is invariably computed, it is located at pressures where equilibrium between the two phases is an assumption which is either in doubt or demonstrably in error. For a liquid lherzolite composition, the density is calculated to be the same as the density of olivine at about 15 ± 2 GPa. However, this is likely to be an "apparent" olivine-liquid density crossover rather than a real one because it is within this pressure range where olivine and liquid lherzolite become unstable; olivine is replaced by majorite as the stable liquidus phase, and olivine is transformed to its high density modified spinel polymorph. For a high Mg komatiite composition, the density is also calculated to be the same as that for olivine at about 15 GPa. However, this density comparison is unambiguously erroneous because olivine is not in equilibrium with komatiite at 15 GPa; olivine is replaced by garnet as the liquidus phase in the 6 to 7 GPa range, and, again, the modified spinel polymorph of olivine is likely to be stable at about 15 GPa.

The olivine flotation hypothesis is supported for silicate liquids which compress in violation of the law of corresponding states. For komatiite and lherzolite liquids, this elastic law predicts that a real olivine-liquid density crossover will exist in the 5 to 6 GPa range.

The predicted pressures of the real and apparent density crossover points are so different for each elastic model, that one or the other should be easily falsifiable by experimentation. Presently available constraints from shock wave and Stokes' Law experiments are most consistent with liquids which compress in observation of the law of corresponding states.

INTRODUCTION

IN PART 1 of this paper (HERZBERG, 1987) it was demonstrated that the available information on the compressibility characteristics of silicate liquids is contradictory. Silicate liquids either do or do not compress in observation of the empirical law of corresponding states (HERZBERG, 1987). If the law is obeyed, the effect of composition on the compressibility of the melt is significant. If it is not obeyed, then all silicate melts will compress by the same amount over any specified increment of pressure.

The consequences of these two contradictory possibilities on the density of magmas at high pressures are profoundly different, and is the subject of this paper. With constraints now available on the partial molar volumes, thermal expansion coefficients, and bulk moduli of the oxide components,

it is possible, at least in principle, to calculate the density of any magma at any temperature and pressure of the geologist's choice. However, in practice there remains a number of ambiguities, the most serious of which was discussed in Part 1 (HERZBERG, 1987). That is, even if all the elastic parameters for the oxide components were unambiguously known, there is some uncertainty as to how they should be mixed to yield the desired result, this being the bulk modulus of a complex magma. The algorithm chosen is the Voigt-Reuss-Hill mixing scheme because it seems to work best for a number of alkali oxide-SiO_2 binary systems; it will be shown here that it also seems to work for complex magma compositions.

Two other important problems are encountered in any attempt to calculate the density of magmas at high pressures. These are: 1) the extent to which ideal mixing holds, and 2) the role of structural changes in the melt (*e.g.*, possible coordination changes of individual cations). In the calculations which follow it is assumed that the oxide components mix ideally with respect to volume, and that pressure-induced structural rearrangements will not significantly increase the density of natural ultrabasic magmas at high pressures. The former assumption seems relatively successful in the work of BOTTINGA and WEILL (1970) and NELSON and CARMICHAEL (1979), and was adopted in Part 1 of this paper. The latter assumption is certainly contentious for aluminous compositions such as albite and jadeite (*e.g.*, BOETTCHER *et al.*, 1982; OHTANI *et al.*, 1985; FLEET *et al.*, 1984). However, natural magmas generated by anhydrous melting in the mantle at high pressures are ultrabasic in composition, not basaltic (HERZBERG, 1983; TAKAHASHI and SCARFE, 1985), and their alumina contents are fairly low; consequently even if structural changes do occur in the 10 to 20 GPa pressure range the effect will probably be small.

In view of these difficulties, the density of magmas at high pressures estimated by calculation rather than by direct experimental determination are subject to important uncertainties. Accordingly, the approach taken here has been to adopt the Voigt-Reuss-Hill mixing algorithm in which the contradictory elastic parameters discussed in Part 1 (HERZBERG, 1987) are incorporated. This method is then tested by making predictions of the high pressure densities of several complex silicate melts for which independent constraints are available. Finally, an examination is made of the hypothesis that there exists a finite depth in the upper mantle below which natural magmas may be denser than coexisting olivine crystals (STOLPER *et al.*, 1981).

METHOD OF CALCULATION

For component i of interest its partial molar volume, $\bar{V}_i^0$, at 1400°C, thermal expansivity, α_i, and isothermal bulk modulus at 1 atmosphere, K_i^0, are listed in Table 1. The partial molar volumes and thermal expansivities are the model parameters developed in Part 1 (HERZBERG, 1987). The bulk moduli are valid for liquids which observe and violate the law of corresponding states.

At 1 atmosphere and temperature T the partial molar volume of i is $\bar{V}_i$ and is determined from:

$$\bar{V}_i = \bar{V}_i^0(+ \alpha_i(T - 1400)) \tag{1}$$

where T is in °C. The molar volume V is simply:

$$V = \Sigma X_i \bar{V}_i. \tag{2}$$

The density at temperature T and 1 atmosphere is ρ' and becomes:

$$\rho' = \Sigma X_i M_i / V \tag{2}$$

where M_i is the gram formula weight of i. At T and pressure of interest P, density ρ and pressure are related by a Birch-Murnaghan equation of state (BIRCH, 1952; STOLPER *et al.*, 1981; equation (6), HERZBERG, 1987).

HERZBERG (1987) suggested that $K' = dK^0/dP$ is close to 4 for liquid Mg_2SiO_4 and Fe_2SiO_4. This value is similar to that proposed by RIGDEN *et al.* (1984) for liquid $An_{36}Di_{64}$ (mol percent) and many crystalline compounds as well (ANDERSON, 1972). Accordingly, the value used throughout these calculations is 4.

The isothermal bulk modulus of a complex liquid composition at 1 atmosphere is evaluated from the bulk moduli of its end-members using the Voigt-Reuss-Hill relation:

$$K^0 = [\Sigma V_i K_i^0 + (\Sigma(V_i/K_i^0))^{-1}]/2 \tag{4}$$

Table 1. Volume and elastic parameters of oxide components in silicate liquids at 1400°C

	V_i^0 (cm³/gfw)	K_i^0 (GPa)	$(\alpha_i \times 10^5/°C)$
SiO₂	27.08 (0.19)	13.0 (1.3)	1.1
Al₂O₃	36.83 (0.47)	20.9 (4.5)	6.2
FeO	13.31 (0.38)	73.6 (61.6–85.9) O	23.6
		28.7 (4.5) V	
MgO	11.87 (0.37)	84.8 (70.9–99.1) O	23.4
		38.6 (6.1) V	
CaO	16.49 (0.28)	56.6 (47.4–65.7) O	24.3
		49.4 (5.4) V	
Na₂O	28.82 (0.22)	14.1 (8.0–19.9)	24.7
K₂O	46.05 (0.39)	8.0 (3.9–11.3)	26.1

O = law of corresponding states *o*bserved.
V = law of corresponding states *v*iolated.
Uncertainties are indicated in parentheses as ± from the mean, or a range about the mean.

where V_i is the volume fraction of i in the melt phase, and K_i are listed in Table 1. The temperature derivative of the bulk modulus dK^0/dT is small and was thus set at zero. Consequently, at 1 atmosphere the bulk modulus of component, i, at temperature, T, (i.e., K) is the same as the bulk modulus at the 1400°C reference temperature (i.e., K^0). The volume fraction of i can be evaluated from:

$$V_i = X_i \bar{V}_i / V. \qquad (5)$$

A computer program in BASIC which calculates the density of a magma according to the above algorithm is given in Appendix I.

TESTS OF THE MODELS

Shock wave studies on molten material having the composition $An_{36}Di_{64}$ have been obtained by RIGDEN et al. (1984), and these results are compared with the calculated densities (Figure 1). The experimental densities are those along an adiabat having an initial temperature of 1400°C at 1 atmosphere. The data point at 23.54 GPa is probably a minimum bound due to the effects of shock heating at pressures greater than 10 GPa (RIGDEN et al., 1984). From these Hugoniot data an isentropic bulk modulus of 22.6 ± .8 GPa and a pressure derivative $dK_s/dP = 4.15$ were obtained (RIGDEN et al., 1984). For liquids which observe the law of corresponding states, the isothermal bulk modulus is calculated to be 22.8 ± 3.3 GPa at 1400°C. The calculated pres-

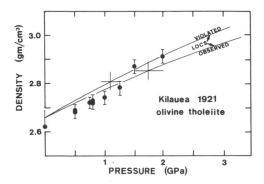

FIG. 2. Density of Kilauea 1921 olivine tholeiite calculated at 1400°C using elastic parameters with the "Law Of Corresponding States" (LOCS) observed and violated, and falling sphere data of FUJII and KUSHIRO (1977) and FUJII (1978).

sure-density relations shown are, therefore, in excellent agreement with the shock wave data. Differences in density stemming from the isothermal versus isentropic cases are small.

The amount of shock heating required to fit the 23.54 GPa point has been calculated by simply increasing the temperature of calculation until the density is sufficiently reduced so as to match the experimental data. The temperature is about 1700°C; given the adiabatic character of these data a temperature increase of no more than 300°C is indicated in these experiments.

The isothermal bulk modulus calculated for the liquid that violates the law of corresponding states is 20.0 ± 2.5 GPa, and the pressure-density relations are shown also in Figure 1. Although the fit to the experimental data is not as good, there is a slight overlap in the error bars between the calculated and experimentally observed densities.

The densities of liquid olivine tholeiite having the composition of Kilauea 1921 have been determined experimentally (FUJII and KUSHIRO, 1977; FUJII, 1978; Figure 2), and are compared with calculated values. If this liquid observes the law of corresponding states, its isothermal bulk modulus is calculated to be 21.8 ± 3.4 GPa. In all cases the error bars from the calculated and experimentally observed densities overlap. However, it can be seen that the experimental densities at pressures less than about 1.2 GPa are consistently lower than the calculated ones, and at higher pressures the reverse is true. Indeed, there appears to be a discontinuous change in the experimentally determined pressure-density relations at about 1.2 GPa, an observation which is consistent with a change in melt structure at that pressure. However, other falling sphere ex-

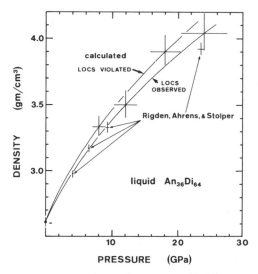

FIG. 1. Density of liquid $An_{36}Di_{64}$ (mol percent) calculated at 1400°C using elastic parameters with the "Law Of Corresponding States" (LOCS) observed and violated, and shock wave data from RIGDEN et al. (1984).

periments on a range of basaltic compositions similar to Kilauea 1921 do not show this feature (KUSHIRO, 1982).

For Kilauea 1921 which violates the law of corresponding states, K^0 is 18.1 ± 2.6 GPa. The calculated densities at 1.5 and 2.0 GPa are in better agreement with the experimental data; however, for the lower pressure data in the 0.5 to 1.2 GPa range, the agreement is poor.

Recent olivine flotation and settling experiments have been carried out with pure forsterite crystals and a basalt compositionally similar to Kilauea 1921 (T. FUJII, personal communication). The results are shown in Figure 3. Elastic parameters for forsterite and the temperature and pressure derivatives are taken from Part 1 (HERZBERG, 1987). Although reaction rims surrounding the forsterite crystals demonstrate unambiguously that this is not an equilibrium crystal-liquid pair, the well-bracketed settling and flotation experiments place important constraints on the contradictory elastic models. Densities calculated with the law of corresponding states observed are in excellent agreement with the experimental data which demonstrate forsterite settling at 6.5 GPa and flotation at 8.5 GPa; neutral buoyancy is calculated at 7.5 GPa. However, densities calculated in violation of the law of corresponding states seriously conflict with the experimental observations. Whereas forsterite flotation is predicted at 6.5 GPa, it has been observed to settle (T. FUJII, personal communication).

THE EFFECT OF TEMPERATURE, PRESSURE, AND COMPOSITION ON OLIVINE-MAGMA DENSITY RELATIONS

Olivine flotation has now been demonstrated in two separate sets of experiments. The first was the observation of flotation at atmospheric pressure for intermediate bulk compositions on the join Mg_2SiO_4-Fe_2SiO_4 (HERZBERG et al., 1982), and was explained by the extreme partitioning of the heavy element Fe into the melt phase. Secondly, as noted above, forsterite has been observed to float in a tholeiitic liquid composition at high pressures (T. FUJII, personal communication). Although useful information is obtained on the elastic properties of such melt compositions, the observation of forsterite flotation has geological limitations because forsterite and tholeiite are not in equilibrium under the experimental conditions. Indeed, there exist a wide range of temperatures, pressures, and compositions wherein olivine is calculated to float in magmatic liquids, and these are illustrated in this section. The geological restrictions to these conditions will be discussed in the following section.

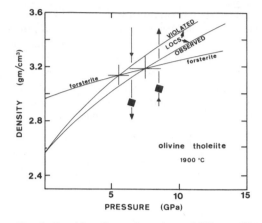

FIG. 3. Densities of pure forsterite and Kilauea 1921 olivine tholeiite calculated with the Law of Corresponding States (LOCS) observed and violated compared with forsterite settling and flotation experimental data (T. FUJII, personal communication). Arrows indicate olivine flotation (up) and settling (down).

The effects of temperature, pressure, and composition on olivine-magma density relations are explored systematically and in isolation from each other in Figures 4 and 5. The densities of crystalline olivines having the compositions Fo_{80}, Fo_{90}, and Fo_{100} have been determined by a linear interpolation of the elastic parameters of pure forsterite and fayalite reviewed in Part 1 (SUZUKI et al., 1981; 1983; SUMINO, 1979; KUMAZAWA and ANDERSON, 1969). This interpolation is justified because the elastic properties of the end-members are so similar. Additionally, volumes at 1 atmosphere and 25°C were obtained from the work of FISHER AND MEDARIS 1969). The algorithm used is listed in Appendix II as a computer program in BASIC. It should be noted that olivine densities so obtained are similar to those calculated from the equation offered by HAZEN (1977), but are about 0.05 gm/cm³ higher at 5 GPa, and 0.1 gm/cm³ lower at 17 GPa. The method chosen here is considered to be superior because d^2P/dP^2 is negative, whereas it is positive in the HAZEN algorithm.

Isothermal density relations at 1400°, 1800°, and 2200°C for two melt compositions which compress in conformity with the law of corresponding states are illustrated in Figures 4A–C. The case for liquids which compress in violation of this law is shown in Figures 5A–C. Two very different melt compositions were selected. These are average mid ocean ridge basalt (MORB; MELSON et al., 1976), and a melt having the composition of fertile garnet lherzolite PHN 1611 (NIXON and BOYD, 1973); both are listed in Table 2.

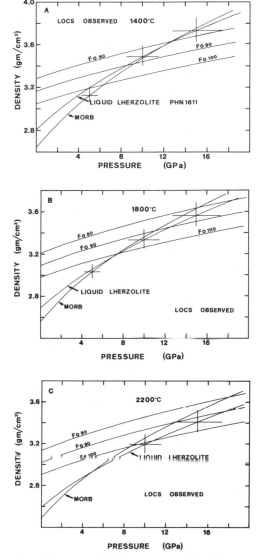

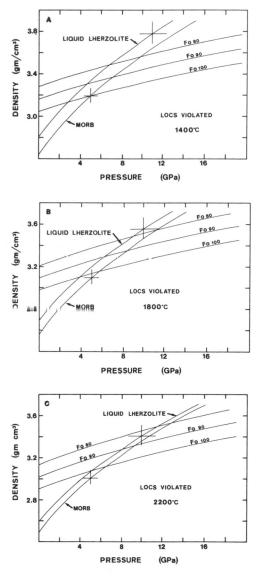

At 1400°C the olivine-melt density crossover is located at 4 to 6 GPa for both liquid compositions and pure Fo_{100} (Figure 4A). With increasing Fe content in the olivine crystals, the crossover shifts to progressively higher pressures, becoming 13 GPa for Fo_{80}. Because olivines crystallizing from lherzolite liquids are more forsterite-rich than those which crystallize from tholeiites, it is clear that the composition of olivine must be specified in any examination of olivine-liquid density comparisons.

FIG. 4. Densities of MORB tholeiite and liquid lherzolite PHN 1611 which observe the Law Of Corresponding States (LOCS) compared to various olivine compositions at 1400°C (4A), 1800°C (4b), and 2200°C (4C). Note that MORB liquids become denser than liquid lherzolite at high pressures.

At any temperature the density of MORB is less than that for the liquid lherzolite at low pressures. However, at high pressures MORB becomes denser than liquid lherzolite when the law of corresponding states is observed (Figure 4). This arises from the effect of composition on the bulk modulus. Where the law of corresponding states is violated, there is no compositional effect (Figures 5A–C), and the density-pressure curves for each composition become semiparallel.

FIG. 5. Densities of MORB tholeiite and liquid lherzolite PHN 1611 which violate the Law Of Corresponding States (LOCS) compared to various olivine compositions at 1400°C (5A), 1800°C (5B) and 2200°C (5C). Note that liquid lherzolite remains denser than MORB at all pressures.

Table 2. Selected magma compositions (weight percent) for density calculations

	Basalt	Komatiite	Liquid lherzolite
SiO_2	50.53	45.23	43.70
Al_2O_3	15.27	6.17	2.75
FeO	10.46	11.48	10.05
MgO	7.47	28.59	37.22
CaO	11.49	5.89	3.26
Na_2O	2.62	0.60	0.33
K_2O	0.16	0.04	0.14
K (GPa)	O 21.1 (3.2)	27.7 (4.2)	30.1 (4.5)
	V 18.0 (2.6)	19.9 (2.7)	20.5 (2.8)

All Fe_2O_3 where reported is converted to FeO.

Data sources: Basalt, MELSON et al. (1976); Komatiite, NESBITT and SUN (1976); Lherzolite is PHN 1611, NIXON and BOYD (1973).

O = law of corresponding states observed.

V = law of corresponding states violated.

Values of K are valid at 1400°C; these become reduced at higher temperatures due to changes in the volume fractions of the oxide components, taking the appearance of a temperature derivative; estimated uncertainties are given in parentheses.

Furthermore, the comparisons shown in Figure 4A are rather meaningless because it is well established that neither tholeiite nor liquid lherzolite can be generated at an anhydrous mantle at temperature as low as 1400°C in the pressure range where a density crossover is presumed to occur (see next section). At 1800°C and 2200°C all crossover pressures are located at higher pressures. This result arises from thermal expansivity differences between the olivine and melt phases. Indeed, whereas attention has been repeatedly focused on the consequences of the high compressibility of silicate liquids compared to olivine (e.g., STOLPER et al., 1981), the fact that these liquids have a thermal expansion coefficient which is about 3 times higher than crystalline silicates has been largely ignored. It is the tradeoff between liquid thermal expansivity and compressibility that will be important in determining the pressure at which any hypothetical olivine-melt density crossover may occur.

For liquids that compress in violation of the law of corresponding states, an olivine-liquid density crossover is assured for the range of olivine compositions considered (Figures 5A–C). At 1400°C the hypothetical density crossover is located at a pressure as low as 3 GPa for liquid lherzolite and 5 GPa for MORB (Figure 5A). Again, the density-pressure relations for both MORB and liquid lherzolite are parallel, owing to the similar bulk moduli for each. Furthermore, because the bulk moduli for MORB and liquid lherzolite are much lower than those used in Figs. 4A–C, liquid densities at high pressures are

much higher. For example, the density of liquid lherzolite at 10 GPa is 0.2 gm/cm^3 higher when the law of corresponding states is violated.

The conclusion to be drawn from the results in Figures 4 and 5 is that an olivine-liquid density crossover "appears" to exist. However, the pressure at which this apparently occurs is rather uncertain, and ranges from 3 to over 15 GPa, depending on temperature, melt composition, olivine composition, and whether the law of corresponding states is observed.

OLIVINE-MAGMA DENSITY RELATIONS WITH GEOLOGICAL RESTRICTIONS

The density of magmas having the compositions of liquid lherzolite, komatiite, and MORB (Table 2) and their coexisting olivine compositions will now be evaluated with a number of geological restrictions imposed. First, the equilibrium compositions of olivine crystals coexisting with the melts have been evaluated, and olivine densities adjusted according to changes in Fe content. Second, densities have not been calculated isothermally. Rather they have been calculated along the anhydrous mantle solidus, the lowermost temperature bound below which anhydrous magmas are not stable. The densities of MORB, komatiite, and liquid lherzolite have been calculated at all pressures along the solidus, although it is recognized that liquid lherzolite is not a stable initial melt composition at 1 atmosphere and MORB is not a stable initial melt composition at high pressures (e.g., OHTANI et al., 1986). Ideally the initial melt compositions at all pressures along the solidus would have to be known, and their densities calculated accordingly. This fine tuning, however, is not required in order to test the olivine flotation hypothesis.

Iron-Magnesium partitioning

For a melt of a known composition the Fe-Mg partition coefficient provides a means for calculating the equilibrium composition of coexisting olivine. The partition coefficient K_D takes the form:

$$K_D = \left[\frac{X_{MgO}}{X_{FeO}}\right]_{liquid} \left[\frac{X_{FeO}}{X_{MgO}}\right]_{olivine} \quad (6)$$

is about 0.3 ± 0.03 and relatively insensitive to changes in temperature and composition for basaltic and komatiitic liquids (ROEDER and EMSLIE, 1970; LONGHI et al., 1978; BICKLE, 1982). Pressure increases K_D (LONGHI et al., 1978; BICKLE, 1978; HERZBERG et al., 1982) by about 0.1/5 GPa (TAKAHASHI and KUSHIRO, 1983). For the species in the reaction:

Mg_2SiO_4,ol + Fe_2SiO_4,l

$$= Mg_2SiO_4,l + Fe_2SiO_4,ol \quad (7)$$

the pressure derivative of the distribution coefficient can be evaluated from the relation:

$$\frac{\partial \ln K_D}{\partial P} = \frac{\Delta V}{RT} \quad (8)$$

and the availability of the volume and elastic parameters given in Part 1 (HERZBERG, 1987). The ΔV term was calculated at 2000°C and 1 atmosphere to 20 GPa. The K_D was calculated iteratively with $K_D = 0.3$ at atmospheric pressure as the initial point. The results (Figure 6) are in overall agreement with the experimental observations of TAKAHASHI and KUSHIRO (1983) and BICKLE (1978) although the calculated K_D appears somewhat too high in the 2–3 GPa range. A number of important uncertainties will affect these calculations, including excess volume contributions to ΔV in the melt and olivine phases at high temperatures and pressures, and the composition of the melt (e.g., SiO_2). However, these uncertainties will be masked by the initial range of ±0.03 in K_D, and the effect propagated to olivine density calculations will be trivial. In the calculations that follow, the compositions of olivine in equilibrium with the coexisting liquids were determined from Figure 6.

Temperature

Both olivine and magma densities were calculated at temperatures along the anhydrous solidus. The solidus for anhydrous garnet lherzolite to 5 GPa, reviewed in HERZBERG (1983), has now been determined experimentally to 20 GPa (TAKAHASHI and SCARFE, 1985; TAKAHASHI, 1986; SCARFE and TAKAHASHI, 1986) for natural Fe-bearing compositions. Experimental data to 13 GPa for compo-

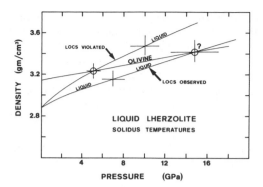

FIG. 7. Densities of liquid lherzolite PHN 1611 calculated with the Law Of Corresponding States (LOCS) observed and violated compared to densities of coexisting olivine. Temperatures are restricted to the anhydrous solidus. Where LOCS is observed the density crossover may be meaningless because olivine may not be a stable liquidus or solidus phase at these pressures.

sitions in the system $CaO-MgO-Al_2O_3-SiO_2$ place the solidus at significantly higher temperatures, indicating that FeO reacts with the graphite-diamond heater/container system and produces CO_2 vapor which reduces the anhydrous solidus in natural systems by over 200°C (HERZBERG, in preparation). The temperatures in °C and pressures in GPa of the anhydrous solidus used in these calculations are: 1400, 2; 1620, 4; 1800, 6; 1930, 8; 2020, 10; 2090, 12; 2150, 14; 2190, 16; 2220, 18 (note: the last three are extrapolated).

Numerical results

The densities of liquid lherzolite PHN 1611 and olivines which range from Fo_{95} at 1 atmosphere to Fo_{93} at 18 GPa are shown in Figure 7. Two completely different possibilities emerge, depending upon whether liquids compress according to or in violation of the law of corresponding states. For liquids which compress according to the law of corresponding states, a density crossover is computed at about 15 ± 2 GPa. However, this may be an apparent density crossover rather than a real one because it is within this pressure range where olivine and the liquid lherzolite become unstable. At these pressures, olivine is no longer the liquidus phase for PHN 1611 (HERZBERG, 1983; TAKAHASHI, personal communication). Rather, it is replaced by majorite, a pyroxene-garnet solid solution phase having a garnet crystal structure. Although the high temperature elastic properties for majorite are still poorly constrained, it can be shown that majorite is much denser than liquid lherzolite at these pressures (e.g., see HERZBERG, 1986). Additionally, at

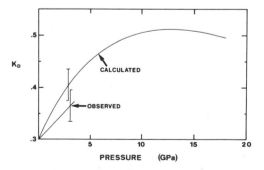

FIG. 6. Calculated and experimentally observed variations in K_D, the partition coefficient of Fe and Mg between olivine and melt, as a function of pressure.

about 16 GPa olivine is replaced by its higher pressure β-phase polymorph, which is denser than olivine by about 0.2 gm/cm³. It is concluded that the hypothesis of olivine flotation in lherzolite liquids which compress according to the law of corresponding states is unconvincing. However, owing to the size of the errors in these calculations, the hypothesis can neither be refuted nor supported at the present time. The possibility that olivine flotation may occur at some pressure slightly lower than that required to destabilize olivine cannot be precluded.

For the lherzolite liquid that compresses in violation of the law of corresponding states, the olivine-magma density crossover is located at about 5 ± 1.5 GPa. Clearly, this elastic model supports unambiguously the olivine flotation hypothesis.

Similar results are obtained for the komatiite composition (Figure 8). Olivine compositions are Fo_{94} at 1 atmosphere to Fo_{90} at 18 GPa. With the law of corresponding states maintained, an apparent density crossover is also located around 15 GPa. However, it has been established by experiment that for analogue compositions in the system CaO-MgO-Al_2O_3-SiO_2, olivine becomes replaced by garnet as the liquidus phase in the 6 to 7 GPa range (HERZBERG et al., 1986). Furthermore, the experimental results on lherzolite PHN 1611 indicate that olivines crystallizing from the komatiite will be transformed to the modified spinel polymorph in the 15 GPa range. Indeed, these density comparisons are rendered petrologically meaningless because olivine and komatiite are not in equilibrium at 15 GPa. At pressures where olivine and komatiite are stable (i.e., less than 6 to 7 GPa), olivine is predicted to remain considerably more dense than komatiite. For komatiite compression in violation of the law of corresponding states, the density crossover is located

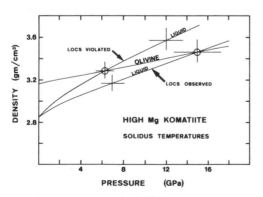

FIG. 8. Densities of high Mg komatiite calculated with the Law Of Corresponding States (LOCS) observed and violated compared to densities of coexisting olivine.

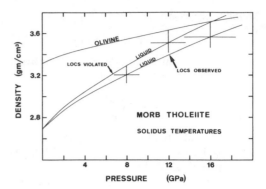

FIG. 9. Densities of MORB tholeiite calculated with the Law Of Corresponding States (LOCS) observed and violated compared to densities of coexisting olivine.

at 6 ± 1.5 GPa. At pressures less than those required to stabilize liquidus garnet, olivine flotation is a possibility for this elastic model.

The density relations for the MORB composition are shown in Figure 9. Olivine compositions range from Fo_{81} at 1 atmosphere to Fo_{71} at 18 GPa. Because these olivines are so iron-rich compared to those for the komatiite and lherzolite compositions, an apparent density crossover occurs only at extremely high pressures and for the condition where the law of corresponding states is violated. From the inference that olivine is neither a stable liquidus nor subliquidus phase at these pressures, it is predicted that olivine flotation will not be observed.

CONCLUSIONS

The elastic properties of silicate liquids reported in Part 1 (HERZBERG, 1987) have been applied to the calculation of magma densities to high pressures. The algorithm involves the calculation of the bulk modulus of the magma from the bulk moduli of its constituent oxide components mixed in proportion to their volume fractions according to the Voigt-Reuss-Hill approximation. Two elastic models have been entertained. In the first, silicate liquids compress according to the law of corresponding states; the effect of composition on the compressibility of silicate melts is large, and the composition of the melt must be specified. In the second, silicate liquids compress in violation of the law of corresponding states, and all magma compositions compress by similar amounts over any given increment of pressure.

The two contradictory elastic models have been tested. Both are consistent with the shock wave experiments of RIGDEN et al. (1984) on molten $An_{36}Di_{64}$ (mol percent), but the agreement is much better for the model that obeys the law of corresponding states. The same holds for the density experiments of FUJII and KUSHIRO (1977) and FUJII (1978) on an olivine tholeiite composition to 2.0 GPa. However, olivine flotation and settling experiments on a similar composition to 8.5 GPa is predicted by the model which obeys the law of corresponding states. The model which violates the law of corresponding states is also in major violation of these data.

The densities of a tholeiite, a komatiite, and a liquid lherzolite have been calculated along the anhydrous solidus to about 18 GPa, and have been compared to the densities of their equilibrium olivine compositions. It is demonstrated that the olivine flotation hypothesis is unconvincing for magmas which compress according to the law of corresponding states. Although an olivine-liquid density crossover is invariably computed, it is located at pressures where equilibrium between the two phases is an assumption which is either in doubt or demonstrably invalid. For liquid lherzolite, olivine is replaced by majorite as the liquidus phases at about the same pressure calculated for the apparent density crossover, this being 15 ± 2 GPa. For the komatiite the apparent density crossover is also about 15 GPa, well in excess of the 6 to 7 GPa pressure range required to stabilize garnet on the liquidus. Furthermore, at 15 GPa, olivine is likely to be transformed to its dense modified spinel polymorph for both the komatiite and liquid lherzolite.

The olivine flotation hypothesis is only plausible for silicate liquids which compress in violation of the law of corresponding states. For komatiite and lherzolite liquids, this elastic law predicts the existence of a real olivine-liquid density crossover at 5 to 6 GPa. The predicted pressures of the real and apparent density crossover points are so fundamentally different for each elastic model, that one or the other should be easily falsifiable by experimentation. Presently available constraints from shock wave and Stokes' Law experiments are most consistent with liquids which compress according to the law of corresponding states.

Acknowledgements—The author is indebted to T. Fujii at the Earthquake Research Institute in Tokyo for making available his experimental results prior to publication. Bjorn Mysen and Larry Finger are thanked for their critical reviews and detailed comments.

REFERENCES

ANDERSON O. L. (1972) Patterns in elastic constants of minerals important to geophysics. In *Nature of the Solid Earth,* (ed. E. D. ROBERTSON), pp. 575–613 McGraw-Hill.

BICKLE M. J. (1978) Melting experiments on peridotitic komatiites. In *Progress in Experimental Petrology,* Natural Environment Research Council, 4th Report, Series D, No. 11, 187–195.

BICKLE M. J. (1982) The magnesium contents of komatiitic liquids. In *Komatiites* (eds. N. T. ARNDT and E. G. NISBET), pp. 479–494, Allen and Unwin.

BIRCH F. (1952) Elasticity and constitution of the earth's interior. *J. Geophys. Res.* **57,** 227–286.

BOETTCHER A. L., BURNHAM C. W., WINDOM K. E. and BOHLEN, S. R. (1982) Liquids, glasses, and the melting of silicates to high pressures. *J. Geol.* **90,** 127–138.

BOTTINGA Y. and WEILL D. (1970) Densities of liquid silicate systems calculated from partial molar volumes of oxide components. *Amer. J. Sci.* **269,** 169–182.

FISHER G. W. and MEDARIS L. G., JR. (1969) Cell dimensions and x-ray determinative curve for synthetic Mg-Fe olivines. *Amer. Mineral.* **54,** 741–753.

FLEET M. E., HERZBERG C. T., HENDERSON G. S., CROZIER E. D., OSBORNE M. D. and SCARFE C. M. (1984) Coordination of Fe, Ga and Ge in high pressure glasses by Mössbauer, Raman and X-ray absorption spectroscopy, and geological implications. *Geochim. Cosmochim. Acta* **48,** 1455–1465.

FUJII T. (1978) Viscosity, density and ascending velocity of magma. *Bull. Volcanol. Soc. Japan,* (2nd Series) **23,** 117–130.

FUJII T. and KUSHIRO I. (1977) Density, viscosity, and compressibility of basaltic liquid at high pressures. *Carnegie Inst. Wash. Yearb.* **76,** 419–424.

HAZEN R. M. (1977) Effects of temperature and pressure on the crystal structure of ferromagnesian olivine. *Amer. Mineral.* **62,** 286–295.

HERZBERG C. T., BAKER M. B. and WENDLANDT R. F. (1982) Olivine flotation and settling experiments on the join Mg₂SiO₄-Fe₂SiO₄. *Contrib. Mineral. Petrol.* **80,** 319–323.

HERZBERG C. T. (1983) Solidus and liquidus temperatures and mineralogies for anhydrous garnet-lherzolite to 15 GPa. *Phys. Earth Planet. Inter.* **32,** 193–202.

HERZBERG C. T. (1986) Internal structures of the Earth and terrestrial planets: constraints from ultrahigh pressure magma density and phase equilibrium relations. In *Silicate Melts,* (ed. C. M. SCARFE), Ch. 10, pp. 279–304, Mineralogical Assoc. of Canada Short Course Handbook, Vol. 12.

HERZBERG C. T. (1987) Magma density at high pressure Part 1: The effect of composition on the elastic properties of silicate liquids. In *Magmatic Processes: Physicochemical Principles* (ed. B. O. MYSEN). The Geochemical Society Spec. Publ. 1, pp. 25–46.

HERZBERG C. T., HASABE K. and SAWAMOTO H. (1986) Origin of high Mg komatiites: constraints from melting experiments to 8 GPa. *Trans. Amer. Geophys. Union* **67,** 408.

KUMAZAWA M. and ANDERSON O. L. (1969) Elastic moduli, pressure derivatives, and temperature derivatives of single-crystal olivine and single-crystal forsterite. *J. Geophys. Res.* **74,** 5961–5972.

KUSHIRO I. (1982) Density of tholeiite and alkali basalt magmas at high pressures. *Carnegie Inst. Wash. Yearb.* **81,** 305–309.

LONGI J., WALKER D. and HAYS J. F. (1978) The distribution of Fe and Mg between olivine and lunar basaltic liquids. *Geochim. Cosmochim. Acta* **42,** 1545–1558.

MELSON W. G., VALIER T. L., WRIGHT T. L., BYERLY G. and NELEN J. (1976 Chemical diversity of abyssal volcanic glass erupted along Pacific, Atlantic, and Indian

Ocean sea-floor spreading centers. In *The Geophysics of the Pacific Ocean Basin and Its Margin,* pp. 351–367, American Geophysical Union.

NELSON S. A. and CARMICHAEL I. S. E. (1979) Partial molar volumes of oxide components in silicate liquids. *Contrib. Mineral. Petrol.* **71,** 117–124.

NESBITT R. W. and SUN S.-S. (1976) Geochemistry of Archaean spinifex-textured peridotites and magnesian an low-magnesian tholeiites. *Earth Planet. Sci. Lett.* **31,** 433–453.

NIXON P. H. and BOYD, F. R. (1973) Petrogenesis of the granular and sheared ultrabasic nodule suite in kimberlites. In *Lesotho Kimberlites* (ed. P. H. NIXON), pp. 48–56. Lesotho National Development Corporation, Maseru, Lesotho.

OHTANI E., TAULELLE F. and ANGELL C. A. (1985) Al^{3+} coordination changes in liquid aluminosilicates under pressure. *Nature* **314,** 78–81.

OHTANI E., KATO T. and HERZBERG C. T. (1986) Stability of lherzolite magmas at solidus temperatures and 20 GPa. *Trans. Amer. Geophys. Union* **67,** 408–409.

RIGDEN S. M., AHRENS T. J. and STOLPER E. M. (1984) Densities of liquid silicates at high pressures. *Science* **226,** 1071–1074.

ROEDER P. L. and EMSLIE R. F. (1970) Olivine-liquid equilibrium. *Contrib. Mineral. Petrol.* **29,** 275–289.

SCARFE C. M. and TAKAHASHI I. (1986) Melting of garnet peridotite to 13 GPa and the early history of the upper mantle. *Nature* **322,** 354–356.

STOLPER E., WALKER D., HAGAR B. and HAYS J. (1981) Melt segregation from partially molten source regions: the importance of melt density and source region size. *J. Geophys. Res.* **86,** 6261–6271.

SUMINO Y. (1979) The elastic constants of Mn_2SiO_4, Fe_2SiO_4 and Co_2SiO_4, and the elastic properties of olivine group minerals at high temperature. *J. Phys. Earth* **27,** 209–238.

SUZUKI I., ANDERSON O. L. and SUMINO Y. (1983) Elastic properties of a single-crystal forsterite Mg_2SiO_4, up to 1200°K. *Phys. Chem. Minerals* **10,** 38–46.

SUZUKI I., SEYA K., TAKEI H. and SUMINO Y. (1981) Thermal expansion of fayalite, Fe_2SiO_4. *Phys. Chem. Minerals* **7,** 60–63.

TAKAHASHI E. (1986) Melting of a dry peridotite KLB-1 up to 14 GPa: implications on the origin of peridotitic upper mantle. *J. Geophys. Res.* **91,** 9367–9382.

TAKAHASHI E. and KUSHIRO I. (1983) Melting of a dry peridotite at high pressures and basalt magma genesis. *Amer. Mineral.* **68,** 6261–6271.

TAKAHASHI E. and SCARFE C. M. (1985) Melting of peridotite to 14 GPa and the genesis of komatiite. *Nature* **315,** 566–568.

APPENDIX I

```
1  PRINT "THIS MAGMA DENSITY PROGRAM IS BASED ON AN ELASTIC MODEL WHEREIN"
2  PRINT "SILICATE LIQUIDS COMPRESS ACCORDING TO THE LAW OF CORRESPONDING"
3  PRINT "STATES. PARAMETERS FOR A VIOLATED LAW ARE RETRIEVABLE IN PART 1"
10 DIM N$(7),M9(7),WT(7),D1(7),D2(7),D3(7),CM1(7),CM2(7),CM3(7),ALFA(7),K1(7),K2(7),
   K3(7),KR(3),KV(3),VR(3),G(7)
20 FOR I=1 TO 7:READ N$(I):NEXT I
30 DATA SI02,AL203,FEO,MGO,CAO,NA20,K20
40 FOR I=1 TO 7:READ M9(I):NEXT I
50 DATA 60.085,101.962,71.846,40.311,56.079,61.979,94.203
60 PRINT:INPUT"SAMPLE NAME:"'$:PRINT
70 FOR I=1 TO 7
80 PRINT"WEIGHT % ";N$(I);" =";:INPUT WT(I)
90 WT(I)=WT(I)/M9(I)
100 SUMML=SUMML+WT(I)
110 NEXT I
120 FOR I=1 TO 7:WT(I)=WT(I)/SUMML:NEXT I
130 INPUT"TEMPERATURE IN CENTIGRADE";T
140 INPUT"PRESSURE IN KILOBARS";P
160 RESTORE 180
170 FOR I=1 TO 7:READ ALFA(I):NEXT I
180 DATA .000011,.000062,.000236,.000234,.000243,.000247,.000261
190 DELTAT=T-1400
200 FOR I=1 TO 7:READ D1(I),D2(I),D3(I):NEXT I
210 DATA 27.08,26.89,27.27,36.83,36.36,37.30,13.31,12.93,13.69,11.87,11.5,12.24
220 DATA 16.49,16.21,16.77,28.82,28.60,29.04,46.05,45.66,46.44
230 SUM1=0:SUM2=0:SUM3=O
240 FOR I=1 TO 7
250 CM1(I)=WT(I)*D1(I)*(1+ALFA(I)*DELTAT)
260 CM2(I)=WT(I)*D2(I)*(1+ALFA(I)*DELTAT)
270 CM3(I)=WT(I)*D3(I)*(1+ALFA(I)*DELTAT)
280 SUM1=SUM1+CM1(I):SUM2=SUM2+CM2(I):SUM3=SUM3+CM3(I)
290 NEXT I
300 FOR I=1 TO 7
310 V1(I)=CM1(I)/SUM1
320 V2(I)=CM2(I)/SUM2
330 V3(I)=CM3(I)/SUM3
340 NEXT I
350 FOR I=1 TO 7:READ K1(I),K2(I),K3(I):NEXT I
```

```
360 DATA 130,117,143,209,164,254,736,616,859,848,709,991,566,474,657,141,80,199,80,39,113
370 FOR I=1 TO 3:KR(I)=0:KV(I)=0:VR(I)=0:NEXT I
380 FOR I=1 TO 7
390 KR(1)=KR(1)+V1(I)/K1(I)
400 KR(2)=KR(2)+V2(I)/K2(I)
410 KR(3)=KR(3)+V3(I)/K3(I)
420 KV(1)=KV(1)+V1(I)*K1(I)
430 KV(2)=KV(2)+V2(I)*K2(I)
440 KV(3)=KV(3)+V3(I)*K3(I)
450 NEXT I
460 FOR I=1 TO 3
470 KR(I)=1/KR(I)
480 VR(I)=(KR(I)+KV(I))/2
490 NEXT I
500 SG=0
510 FOR I=1 TO 7
520 G(I)=WT(I)*M9(I)
530 SG=SG+G(I)
540 NEXT I
550 D1=SG/SUM1:D2=SG/SUM2:D3=SG/SUM3
560 DP1=D1:DD=.5
570 DP1=DP1+DD
580 Z1=Z
590 Z=1.5*VR(1)*((DP1/D1)^2.333-(DP1/D1)^1.667)
600 IF Z<P AND Z>Z1 THEN 570
610 IF Z<P AND Z<Z1 THEN DD=DD/2:GOTO 570
620 IF Z>P AND Z>Z1 THEN DD=DD/2:DP1=DP1-DD:GOTO 580
630 IF Z>P AND Z<Z1 THEN DP1=DP1-DD:GOTO 580
640 P2 = 3/2*VR(2)*((DP1/D1)^2.333 - (DP1/D1)^1.667)
650 P3 = 3/2*VR(3)*((DP1/D1)^2.333 - (DP1/D1)^1.667)
660 DP2 = DP1 - D1 + D3
670 DP3 = DP1 + D2 - D1
680 DP4 = D1:DD=.5
690 DP4 = DP4 + DD
695 Z1 = Z
700 Z = 1.5*VR(2)*((DP4/D1)^2.333-(DP4/D1)^1.667)
710 IF Z<P AND Z>Z1 THEN 690
720 IF Z<P AND Z<Z1 THEN DD=DD/2:GOTO 690
730 IF Z>P AND Z>Z1 THEN DD=DD/2:DP4=DP4-DD:GOTO 695
740 IF Z>P AND Z<Z1 THEN DP4=DP4-DD:GOTO 695
750 DP5 = D1:DD=.5
760 DP5 = DP5+DD
770 Z1 = Z
780 Z=1.5*VR(3)*((DP5/D1)^2.333-(DP5/D1)^1.667
790 IF Z<P AND Z>Z1 THEN 760
800 IF Z<P AND Z<Z1 THEN DD=DD/2:GOTO 760
810 IF Z>P AND Z>Z1 THEN DD=DD/2:DP5=DP5-DD:GOTO 770
820 IF Z>P AND Z<Z1 THEN DP5=DP5-DD:GOTO 770
830 DUP1 = DP3 - DP1
840 DUP2 = DP4 - DP1
850 DDWN1 = DP1 - DP2
860 DDWN2 = DP1 - DP5
870 DUP = (DUP1^2 + DUP2^2)^.5
880 DDWN = (DDWN1^2 + DDWN2^2)^.5
890 CLS
900 LPRINT"SAMPLE NAME:";NA$
910 LPRINT "TEMPERATURE ="; T;"C"
920 LPRINT USING "DENSITY AT 1 ATMOSPHERE = #.## GM/CC       RANGE = #.## TO #.##" · ,D3,D2
930 LPRINT "FOR PRESSURE ="; P;"KILOBARS"
940 LPRINT USING "DENSITY = #.## GM/CC       RANGE = #.## TO #.##"; DP1,DP1-DDWN,DP1+DUP
950 LPRINT USING "PRESSURE RANGE = ### TO ### KBAR",P3
960 LPRINT USING "BULK MODULUS = ### KILOBARS       RANGE = ### TO ###"(1),VR(2),VR(3)
970 LPRINT
980 PRINT "DO YOU WANT A CHANGE OF TEMPERATURE OR PRESSURE? (Y/N)"
990 Y$=INPUT$(1):IF Y$="" THEN 990
1000 IF Y$="Y" OR Y$="y" THEN 130
1010 END
```

APPENDIX II

```
10  INPUT "MOLE FRACTION FORSTERITE";XMG
20  INPUT "TEMPERATURE IN CENTIGRADE";T
30  INPUT "PRESSURE IN KILOBARS";P
40  MASS = 203.777 - 63.07*XMG
50  V = (307.23 - 15.49*XMG - 2.02*XMG^2)*.15055
60  ALFA = (9.7E-09 + XMG*6E-10)*T + .0000255 + XMG*.0000019
70  V0 = V*(1 + ALFA*(T-25))
80  K0L = (-.2775 + XMG*.0504)*T + 1380 - 96.7*XMG
90  KDL = 5.39
100 D1 = MASS/V0
110 DP1 = D1:DD = .5
120 DP1 = DP1 + DD
130 Z1 = Z
140 Z = 1.5*K0L*((DP1/D1)^2.333 - (DP1/D1)^1.667)*(1-.75*(4-KDL)*((DP1/D1)^.667 - 1))
150 IF Z < P AND Z > Z1 THEN 120
160 IF Z < P AND Z < Z1 THEN DD = DD/2:GOTO 120
170 IF Z > P AND Z > Z1 THEN DD = DD/2:DP1 = DP1 - DD:GOTO 130
180 IF Z > P AND Z < Z1 THEN DP1 = DP1 - DD:GOTO 130
190 LPRINT "MOLE FRACTION FORSTERITE IN OLIVINE =" XMG
200 LPRINT "TEMPERATURE =" T;"C"
210 LPRINT "PRESSURE = " P;"KILOBARS"
220 LPRINT USING "OLIVINE DENSITY AT 1 ATM = #.## GM/CC";MASS/V0
230 LPRINT USING "OLIVINE DENSITY AT P = #.## GM/CC";DP1
240 LPRINT
250 END
```

Magmatic Processes: Physicochemical Principles
© The Geochemical Society, Special Publication No. 1, 1987
Editor B. O. Mysen

Pressure dependence of the viscosity of silicate melts

CHRISTOPHER M. SCARFE*, BJORN O. MYSEN and DAVID VIRGO

Geophysical Laboratory, Carnegie Institution of Washington, 2801 Upton St.,
N.W., Washington, D.C., 20008 U.S.A.

Abstract—Most viscosity measurements on anhydrous melt compositions of geological interest have shown an isothermal decrease in viscosity with increasing pressure. These compositions (*e.g.*, $NaAlSi_3O_8$, $NaAlSi_2O_6$ and $Na_2Si_3O_7$), however, have highly polymerized melt structures at one atmosphere. The number of non–bridging oxygens to tetrahedrally coordinated cations (Si^{4+}, Al^{3+}, Fe^{3+}, Ti^{4+}, or P^{5+}) in these melts is less than 1 (NBO/T < 1).

In order to test possible relations between pressure dependence of viscosity and polymerization of the melt, three compositions ($CaMgSi_2O_6$, $Na_2Si_2O_5$ and Na_2SiO_3) all of which are relatively depolymerized at one atmosphere (NBO/T ≥ 1), have been investigated. An isothermal increase in viscosity of these melts to 20 kbar pressure indicates a fundamental difference in the behavior of highly polymerized versus depolymerized melts. Furthermore, as a function of pressure, density increases in highly polymerized melts are greater than increases observed for depolymerized melts.

The positive pressure dependence of the viscosity of melts with NBO/T ≥ 1 is in accord with the response of other inorganic liquids to the effects of pressure. Conversely, it is melts with NBO/T < 1 that appear to behave anomalously. Raman spectroscopic and X–ray data suggest that the decrease in viscosity is due to subtle changes in the anionic framework of melts. No spectroscopic evidence has been found for pressure–induced coordination changes in aluminum to 40 kbar.

Obsidian, andesite and most basalt melts have NBO/T < 1 and their viscosities are observed to decrease with increasing pressure under isothermal conditions. Basanites, picrites, komatiites and related silica–deficient, aluminum–poor melts with NBO/T > 1 will probably show an increase in their viscosities as a function of pressure. Several geological applications are briefly discussed in the light of this contrasting viscous behavior.

INTRODUCTION

THE VISCOSITY of silicate melts is an important parameter in problems related to the generation, evolution and emplacement of igneous rocks (*e.g.*, BARTLETT, 1969). Apart from the obvious effects of lava viscosity on land forms, it governs melt aggregation and segregation at source, the ascent of magmas to the surface, and xenolith transport. Viscosity also exerts important controls on transport processes, and through these processes affects nucleation and crystal growth, crystal settling and flotation of solids.

Although viscosities of many melt compositions are reasonably well known at 1 atmosphere (*e.g.*, BOTTINGA and WEILL, 1972; URBAIN *et al.*, 1982; SCARFE, 1986), knowledge of the effect of pressure on melt viscosity is more limited. KUSHIRO (1976, 1977, 1978, 1980) showed that the viscosities of several anhydrous silicate and aluminosilicate melts decrease when measured isothermally at pressures up to 20 kbar (Figure 1). KUSHIRO *et al.* (1976), FUJII and KUSHIRO (1977a, b) and SCARFE (1981)

have also shown that qualitatively similar behavior is exhibited by anhydrous rock melts of andesitic and basaltic composition (Figure 1). In the case of rock melts, sufficient data are available to indicate a viscosity decrease both isothermally and along the liquidus with pressure. Although the decrease in viscosity along the anhydrous liquidus may be due to structural changes in the melt, brought about by increasing temperature, the decrease in viscosity measured under isothermal conditions must be attributed to pressure–induced structural changes in the melt.

All the melts depicted in Figure 1 have highly polymerized structures at 1 bar (RIEBLING, 1966; TAYLOR and BROWN, 1979; MYSEN *et al.*, 1980a, 1982). If the number of non–bridging oxygens in these melts is used as a measure of the degree of polymerization (*e.g.*, BOTTINGA and WEILL, 1972; SCARFE, 1973; SCARFE *et al.*, 1979; MYSEN *et al.*, 1982; MYSEN, 1986), the number of non–bridging oxygens to tetrahedrally coordinated cations (NBO/T) is less than one. In order to test the compositional extent of the negative pressure dependence of viscosity, three compositions ($Na_2O \cdot 2SiO_2$, $Na_2O \cdot SiO_2$ and $CaMgSi_2O_6$) that are relatively depolymerized melts at 1 atmosphere (MYSEN *et al.*, 1980a), with NBO/T ≥ 1, were selected for study.

---

* *Permanent address:* Department of Geology and Institute of Earth and Planetary Physics, University of Alberta, Edmonton, Canada, T6G 2E3.

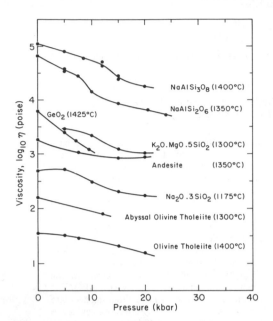

FIG. 1. Viscosity as a function of pressure for melts that have a highly polymerized structure at 1 bar. All measurements were made isothermally at the temperatures shown. Data sources are given in Table 3.

Sodium disilicate and metasilicate were selected because of their chemical simplicity, relatively low melting temperatures, and because data are available on the properties of their melts at 1 bar. Diopside was chosen because it is a major component of melts of geological interest.

This paper describes the determination of the viscosity of these melts at one bar and at high pressures and discusses the pressure dependence of viscosity in the light of previous work on synthetic and natural systems.

EXPERIMENTAL METHOD

Starting materials

Glasses and crystalline starting materials were prepared from mixtures of reagent grade Na_2CO_3, $CaCO_3$, MgO and SiO_2 at 1 bar. Fusions were performed for 1–2 hours in platinum crucibles at approximately 100°C above the respective melting points or liquidi of the compounds. Portions of the glasses were crystallized in 6–15 hours at approximately 100°C below the melting points.

For 1-bar viscosity measurements on diopside and sodium disilicate composition, the liquid was poured into a steel mold, the internal dimensions of which were identical to the internal dimensions of the outer cylinder of the viscometer (see below). Crystalline starting materials were used for all high

pressure experiments. Glasses were not used for high pressure experiments because they may soften and deform at temperatures below the melting point.

One–atmosphere viscometry

Viscosities were measured with a concentric–cylinder viscometer (DINGWALL and MOORE, 1954) with a cylinder design after DIETZEL and BRUCKNER (1955). Under run conditions the volume of the glass slug (approximately 8 cm^3) filled the cavity between the inner and outer platinum cylinders of the viscometer. The silicate liquid was sandwiched between the outer rotating cylinder and the inner stationary cylinder, which was suspended from a torsion wire. The torque transmitted to the inner cylinder by the silicate liquid was measured as an angular displacement and recorded by a light–spot deflection on a linear scale. Further details of the apparatus and method of operation can be found in SCARFE (1977). Viscosities were calculated from the equation for Newtonian liquids:

$$\eta = \frac{M}{4\pi\Omega(h+k)}\left(\frac{1}{r^2} - \frac{1}{R^2}\right), \qquad (1)$$

where M is the torque, Ω is the angular velocity of the outer cylinder, r and R are the respective radii of the inner and outer cylinders, and $(h + k)$ is the effective length of the inner cylinder (SCARFE, 1977).

The apparatus was calibrated under run conditions with a standard soda–lime–silica glass (U.S. National Bureau of Standards glass 710). Rotation speeds could be varied from 4 to 36 seconds per revolution. Viscosities were accurate to ±5% and temperatures to ±5°C. All experiments were conducted in air.

High-pressure viscometry

High pressure viscosities were measured by falling–sphere viscometry (SHAW, 1963; KUSHIRO, 1976) in a solid–media, high–pressure apparatus (BOYD and ENGLAND, 1960). The 0.75″–diameter furnace assembly incorporated a graphite heater with a 3–degree tapered inner wall, which reduces the temperature gradient along the sample to 10°C (KUSHIRO, 1976, 1978). Sealed platinum capsules, 5 mm in diameter and 10 mm long, were used to contain the samples in all experiments. The piston–out technique was used with a −4% correction to the pressure for friction. Pressures have an uncertainty of ±0.5 kbar, and temperatures, measured with Pt–Pt90Rh10 thermocouples, have an uncertainty of ±10°C. After the pressure was applied, the temperature was raised to approximately 100°C below the melting point for 5 min. to equilibrate the system thermally. The temperature was then rapidly raised to the desired experimental value within a few seconds with a programmable controller.

A variety of spheres of different density (chrome diopside, forsterite, ZrO_2, BN and Au–Pd alloy) with diameters generally <1.0 mm were used. Mineral and refractory spheres were made with the sphere–making device designed by BOND (1951). Alloy spheres were made by melting Au–Pd wire.

Under each set of temperature–pressure conditions a minimum of two experiments of different duration were made to determine the sinking or floating velocity of spheres of two different densities. After the charge was quenched (250°C/sec) and sectioned, the melt viscosity was calculated from the velocity and density of the spheres, with Stokes' equation combined with the Faxen correction for wall effects (SHAW, 1963):

$$\eta = \frac{2gr^2\Delta\rho}{9v(1+3.3r/h_c)}\left[1-2.104\left(\frac{r}{r_c}\right)\right.$$
$$\left.+2.09\left(\frac{r}{r_c}\right)^3-0.95\left(\frac{r}{r_c}\right)^5\right], \quad (2)$$

where η is the viscosity, g is the gravitational constant, r is the radius of the sphere, r_c is the radius of the container, h_c is the height of the container, $\Delta\rho$ is the density contrast between the sphere and the melt, and v is the sphere velocity. Melt densities were used exclusively in the calculation of melt viscosities. Viscosities have a maximum estimated uncertainty of ±15%.

Because two spheres of different density were used in each experiment (e.g., BN with $\rho = 2.29$ g/cm³, which floats, and chrome diopside with $\rho = 3.33$ g/cm³, which sinks), two Stokes' equations with two unknown parameters (η and ρ) provide the simultaneous determination of melt viscosity and density

Table 1. Viscosity of melts at one bar

Composition	Temperature (°C)	Viscosity (poise)
$Na_2O \cdot 2SiO_2$	1200	147
	1245	92
	1395	30
$CaMgSi_2O_6$	1401	9.3
	1432	7.4
	1461	6.5
	1480	5.6

Viscosities determined for three rotational speeds at each temperature. Measurements made under both cooling and heating paths. Uncertainties in viscosity and temperature are ±5% and ±5°C, respectively.

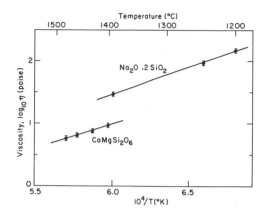

FIG. 2. Viscosity versus reciprocal temperature for melts of diopside and sodium disilicate composition at 1 bar.

(FUJII and KUSHIRO, 1977a). Sphere densities were corrected for compressibility and thermal expansion at the conditions of each experiment. The density of glass quenched from the melts at high pressures was measured on a torsion microbalance (BERMAN, 1939).

RESULTS

Viscosity measurements at 1 bar on melts of sodium disilicate and diopside composition are shown in Table 1 and Figure 2. At constant temperature there is a linear relationship between the rotation speed and the angular displacement of the inner cylinder. This relationship is equivalent to a plot of shear stress versus shear rate, which for Newtonian liquids is a straight line through the origin.

A linear relationship exists between the logarithm of the viscosity and reciprocal temperature. The temperature dependence of the viscosity can be described by an Arrhenius relationship of the form:

$$\eta = \eta_0 \exp(E_\eta/RT), \quad (3)$$

where η_0 is a constant, E_η is the activation energy for viscous flow, R is the gas constant, and T is the absolute temperature. Activation energies for $Na_2O \cdot 2SiO_2$ and $CaMgSi_2O_6$ were calculated as 41 ± 3 and 37 ± 3 kcal/mol, respectively. Results on sodium disilicate are in agreement with SHARTSIS et al. (1952) and BOCKRIS et al. (1955). Measurements on diopside are in accord with the data published by KIRKPATRICK (1974) and SCARFE et al. (1983) and are close to values calculated with the BOTTINGA and WEILL (1972) viscosity model.

Experiments were not performed at 1 bar on a melt of sodium metasilicate composition. This melt volatilizes readily at high temperatures and significant changes in bulk composition are observed in

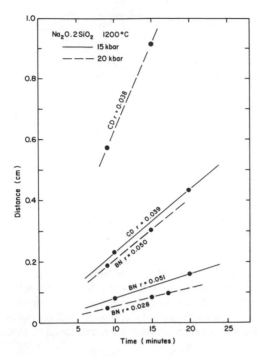

FIG. 3. Example of time–distance plots for spheres of boron nitride (ρ = 2.29 g/cm³) and chrome diopside (ρ = 3.33 g/cm³) in a melt of sodium disilicate composition.

Table 2. Viscosity and density of melts at 1 bar and at high pressure

Composition	Pressure (kbar)	Viscosity (poise)	Density (g/cm³)
$Na_2O \cdot 2SiO_2$	0.001	151	2.39
(1200°C)	10.0	257	2.44
	15.0	620	2.53
	20.0	274	2.54
	25.0	90	2.69
$Na_2O \cdot SiO_2$	0.001	2.0	2.25
(1300°C)	10.0	15	2.29
	20.0	15	2.40
$CaMgSi_2O_6$	0.001	3.0	2.77
(1640°C)	12.5	6.4	2.96
	15.0	11.2	3.00

Each value of viscosity and density represents 2–6 experiments using several types of sphere with different diameters. One atmosphere viscosities obtained by extrapolation on log η versus $1/T$ Arrhenius plots. One atmosphere densities from SHARTSIS et al. (1952), or by extrapolation. Densities of diopside composition glasses produced at 1725°C are recorded in Figure 5. Uncertainties in viscosity, temperature and pressure are ±15%, ±10°C and ±0.5 kbar, respectively.

short time durations. Viscosity and density data for this composition were taken from SHARTSIS et al. (1952) and BOCKRIS et al. (1955).

Measurements as a function of pressure were made isothermally at temperatures above the liquidus. In order to determine constant velocities of sinking or flotation of spheres, time–distance curves were constructed (Figure 3). All time–distance curves were straight lines, passing close to the origin or intersecting the time axis. In the latter case, a few seconds elapse before melting is complete and the spheres commence sinking or floating (KUSHIRO, 1976; KUSHIRO et al., 1976). Optically there was no evidence for reaction between spheres and melt in the short time duration of these experiments.

Viscosities initially increase with increasing pressure (Table 2 and Figure 4). Above 15 kbar, however, the viscosity of a melt of $Na_2O \cdot SiO_2$ composition decreased. Between 1 bar and 15 kbar, diopside and sodium disilicate melts show viscosity increases of about a factor of 4, whereas sodium metasilicate increases by a factor of 7.5. The decrease in the viscosity of sodium disilicate melt above 15 kbar cannot be attributed to water access to the charge. Care was taken to ensure complete dehydration of all starting materials, and all charges were heated to red heat with a torch prior to final welding of the capsules.

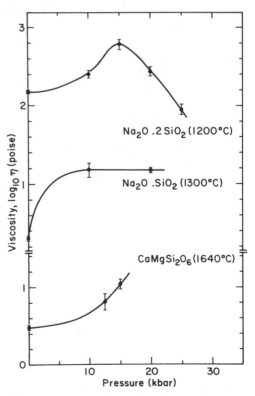

FIG. 4. Viscosity as a function of pressure for melts that have a relatively depolymerized structure at 1 bar. All measurements were made isothermally at the temperatures shown.

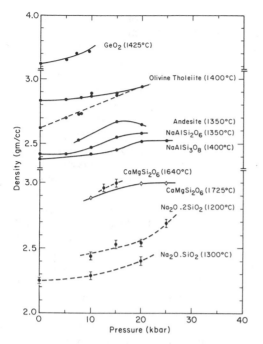

FIG. 5. Density as a function of pressure for melts and glasses quenched from the melts. All measurements were made isothermally at the temperatures shown. Data sources are given in Table 3. Glasses are shown as solid lines; melts as broken lines.

In the same pressure range, the density of melts and glasses increases (Table 2 and Figure 5). Up to 15 kbar, glasses formed from melts exhibit smaller

density changes than the melts themselves (Table 3). Melt densities between 1 bar and 20 kbar for sodium disilicate, sodium metasilicate and diopside increase by 6.3, 6.7 and 8.3%, respectively. Above 15 kbar, where the viscosity of sodium disilicate begins to decrease, the density of the melt shows a greater rate of increase than between 1 bar and 15 kbar. The compressibility of the melt, therefore, increases more rapidly at pressures greater than 15 kbar.

DISCUSSION

The positive pressure dependence of the viscosity of sodium disilicate, sodium metasilicate and diopside melts contrasts with the negative pressure dependence of the viscosity of albite, jadeite and sodium trisilicate melts (Figure 1). As discussed by MYSEN et al. (1980a), the structure of melts of sodium disilicate, sodium metasilicate and diopside composition are relatively depolymerized at 1 bar; whereas, melts of albite, jadeite and sodium trisilicate have highly polymerized, three–dimensional, anionic networks at 1 atmosphere.

KUSHIRO (1976, 1978) and VELDE and KUSHIRO (1978) explained the negative $(d\eta/dP)_T$ for albite and jadeite melts by pressure–induced coordination changes in aluminum. Earlier, WAFF (1975) had predicted on theoretical grounds that aluminum would change from four to six coordination at pressures below 35 kbar. From both infrared absorption spectroscopy and the wavelength shift of both Al

Table 3. Viscosity and density changes between 1 bar and 20 kbar

Composition	Ref.	NBO/T	Viscosity % change	Density % change		Temperature (°C)
				Glass	Melt	
CaMgSi$_2$O$_6$	(1)	2	+273 (15)	3.8 (10–20)	8.3 (15)	1640
Na$_2$O · SiO$_2$	(1)	2	+650		6.7	1300
Na$_2$O · 2SiO$_2$	(1)	1	+311 (15)		6.3	1200
K$_2$O · MgO · 5SiO$_2$	(2)	0.80	−64 (5–20)			1300
Na$_2$O · 3SiO$_2$	(3)	0.67	−65			1175
NaCaAlSi$_2$O$_7$	(4)	0.67	−29 (15)		12 (15)	1450
NaAlSi$_2$O$_6$	(3)	0	−90	6.8		1350
NaAlSi$_3$O$_8$	(5)	0	−84	5.7		1400
Ol. nephel.	(4)	1.17	−17		10	1450
Alk. ol. bas.	(4)	0.83	−57 (15)		10 (15)	1400
Ol. tholeiite	(6)	0.76	−57	3.9	12.2	1400
Abyssal thol.	(7)	0.66	−51 (12)		7.1 (12)	1300
Andesite	(8)	0.27	−51	4.7 (7.5–20)		1350
Obsidian	(9)	0.10	−80			1400

Data sources: (1) present study, (2) KUSHIRO (1977), (3) KUSHIRO (1976), (4) SCARFE (1981), (5) KUSHIRO (1978), (6) FUJII and KUSHIRO (1977a), (7) FUJII and KUSHIRO (1977b), (8) KUSHIRO et al. (1976), (9) SCARFE (unpubl.). For additional data not included in the Table see SHARMA et al. (1979b), KUSHIRO (1981, 1983, 1984 and 1986), SHARPE et al. (1983), BREARLEY et al. (1986) and DUNN and SCARFE (1986). Measurements made isothermally at the temperatures shown. Pressure in parentheses indicates that measurements were performed to that pressure or over that pressure range. Plus sign indicates increase; minus sign a decrease; all density changes are positive. NBO/T is the theoretical ratio of nonbridging oxygens to tetrahedrally coordinated cations in the melt (see text).

K_α and Al K_β, VELDE and KUSHIRO (1978) concluded that aluminum in the melt gradually changes from four–fold to six–fold coordination as the pressure increases from 1 atmosphere to 30 kbar.

SHARMA et al. (1979a), however, criticized several aspects of Velde and Kushiro's interpretation. They pointed out the lack of evidence for the presence in the infrared spectra of non–bridging oxygens that would be expected if aluminum transforms from four–fold to six–fold coordination, and the ambiguities introduced by overlap in Al K_β shifts for Al^{3+} in four and six coordination. Furthermore, studies of quenched melts of albite and jadeite composition by Raman spectroscopy (SHARMA et al., 1979a; MYSEN et al., 1980a,b, 1982, 1985; SEIFERT et al., 1982; FLEET et al., 1984) and by X–ray radial distribution function work (HOCHELLA and BROWN, 1985) show that there are only minor changes in the spectra between 1 bar and 40 kbar. These changes are a result of ordering of Al and Si in the three–dimensional network (SHARMA et al., 1979a) and a decrease of several degrees in the T–O–T angles (SEIFERT et al., 1982; MYSEN et al., 1983).

Three other important pieces of evidence can be cited against an explanation for viscosity decreases based upon pressure–induced aluminum coordination changes. The first is that other compositions not containing aluminum, show viscosity decreases similar to those of albite and jadeite with pressure. Sodium trisilicate (KUSHIRO, 1976) $K_2O \cdot MgO \cdot 5SiO_2$ (KUSHIRO, 1977) and GeO_2 (SHARMA et al., 1979b) all show negative $(d\eta/dP)_T$. The second is that the Raman spectra of glasses of GeO_2 composition between 1 bar and 18 kbar (SHARMA et al., 1979b) and $Na_2O \cdot 3SiO_2$ composition between 1 bar and 20 kbar (MYSEN et al., 1980a,b) show only minor changes. These changes are due to subtle modifications in the network structure of the melts and do not result from pressure–induced coordination changes in silicon or germanium. In fact, molecular dynamics calculations on the pressure effect on liquid SiO_2 (WOODCOCK et al., 1976) suggest that six–fold coordinated silicon is not present in the melt, at least to 200 kbar. Similar calculations for aluminosilicate melts suggest that >100 kbar are required to induce Al–coordination changes (ANGELL et al., 1982, 1983). Recent ^{27}Al solid–state NMR spectroscopy on albite glass quenched from approximately 2000°C indicates that ≥80 kbar may be required to induce an Al–coordination transformation (OHTANI et al., 1985). Finally, according to KUSHIRO (1981), for fully polymerized melts along the join SiO_2–$CaAl_2O_4$, the absolute value of $(d\eta/dP)_T$ decreases

with increasing Al/(Al + Si). Anorthite ($CaAl_2Si_2O_8$) and more aluminous compositions beyond Al/(Al + Si) > 0.5 show insignificant or slightly positive pressure dependence of melt viscosity. It is concluded, therefore, that pressure–induced coordination changes are not responsible for the decrease in viscosity of highly polymerized three–dimensional, network liquids as a function of pressure. Rather, it is suggested that the weakening of bridging T–O–T bonds resulting from the decrease of the T–O–T angles may explain the negative pressure dependence of highly polymerized silicate melts.

In contrast to the highly polymerized network liquids discussed above, most inorganic and polymer liquids exhibit increasing viscosities with pressure (BRIDGMAN, 1931; O'REILLY, 1964; SAKKA and MACKENZIE, 1969). Such behavior is predicted by the free–volume theory of silicate melts (COHEN and TURNBULL, 1959) as applied to the pressure–dependence of liquid viscosities (MATHESON, 1966). Under compression, free volume will be removed from the structure, giving rise to increases in viscosity and density. The pressure dependence of the viscosity of melts of sodium disilicate, sodium metasilicate and diopside composition presented in this paper follows the expected trend, all show positive $(d\eta/dP)_T$.

In the case of disilicate melts such as $Na_2O \cdot 2SiO_2$ (NBO/T = 1), the decrease in viscosity above 15 kbar may be explained by a collapse and weakening of the structure. Support for this interpretation comes from the small density increase between 1 bar and 15 kbar and the larger density change thereafter (see below). In this context, BREARLEY et al. (1986), in an investigation of the pressure dependence of melt viscosities on the diopside–albite join (NBO/T = 2–0), found that the intermediate composition $Di_{75}Ab_{25}$ (NBO/T = 1.2) passed through a minimum viscosity at approximately 12 kbar and 1600°C. Such behavior may be compared with the change of the viscosity of water with pressure at low temperatures (e.g., BRIDGMAN, 1931; ANGELL et al., 1983).

The densities of both types of melts increase with pressure, although the compressibility of highly polymerized melts is greater than the compressibility of relatively depolymerized melts (Tables 2 and 3). In both cases, compression induces some compaction and ordering of the melt structure. In the case of depolymerized melts, free volume is removed under pressure and both the density and viscosity increase; whereas, in highly polymerized melts, the free volume theory is violated, because density increases with pressure, but viscosity decreases.

Finally, it should be noted that for liquids where

the units involved in viscous flow and diffusion are the same, the diffusivity and viscosity are inversely related (COHEN and TURNBILL, 1954); hence, a decrease in viscosity would be accompanied by an increase in diffusion rate. Recently, SHIMIZU and KUSHIRO (1984) found an inverse relationship between the diffusion of oxygen (O^{2-}) anions and viscosity as a function of pressure in jadeite and diopside melt and DUNN and SCARFE (1986) found a similar relationship for andesite melt. These observations raise the question whether viscous flow may be rate limited by the diffusion of oxygen in silicate melts at high pressure.

GEOLOGICAL APPLICATIONS

The ratio of non–bridging oxygens to tetrahedrally coordinated cations (NBO/T) is a measure of the degree of polymerization of a silicate melt and it can be calculated from the chemical analysis. According to MYSEN et al. (1982), for multicomponent melts of geological interest, the major cations in tetrahedral coordination are Al^{3+} and Si^{4+}. However, provided that there is sufficient (Na^+ + K^+ + Ca^{2+} + Mg^{2+}) to charge balance both Al^{3+} and Fe^{3+}, ferric iron may be considered a network former (VIRGO and MYSEN, 1985) together with Ti^{4+} and P^{5+}. All other cations will be network modifiers in six–fold or higher coordination in the melt. The results of calculations of NBO/T for all compositions studied to date are given in Table 3.

It has already been noted that anhydrous rock melts of andesite and basalt composition exhibit decreases in viscosity with pressure similar to those described above for melts such as albite, jadeite and sodium trisilicate (Table 3). For example, olivine tholeiite and andesite melts show decreases in viscosity of approximately 50%, whereas decreases for albite and jadeite melts are about 90%. Both basalt and andesite melts have NBO/T < 1. The only rock composition with NBO/T > 1 that has been studied so far, is a melt of olivine nephelinite composition with an NBO/T = 1.17. This composition exhibits virtually no change in viscosity as a function of pressure under isothermal conditions (SCARFE, 1981). Thus, it seems that the correlation between the NBO/T of a melt and the pressure dependence of its viscosity, observed for melts of simple chemistry, may also hold for multicomponent rock melts (the only exception being two compositions measured by SHARPE et al., 1983). Obsidian, andesite and most basalt melts have NBO/T < 1, and their viscosities are observed to decrease with increasing pressure under isothermal conditions. On the other hand, picrites, komatiites and related silica–defi-

cient, aluminum–poor melts with NBO/T > 1, will probably show an increase in their viscosities as a function of pressure.

Because there are few aspects of igneous petrology that are not in some way related to the viscosity of silicate melts, the contrasting behavior of highly polymerized versus relatively depolymerized silicate melts, finds wide application. Apart from the effects of viscosity on the emplacement of volcanic and intrusive rocks (e.g., HARRIS et al., 1970), it plays an important part in many deep–seated magmatic processes. It is a factor in melt aggregation and segregation at source, the ascent of magmas toward the surface, and in xenolith transport. Viscosity also exerts a control on transport processes such as convection in magmas.

The decrease in viscosity with pressure of rhyolite, andesite and basalt melts, both along their liquidi and at constant temperature, indicates that most anhydrous magmas become more fluid with depth. Melts of picrite and komatiite composition, however, may become more viscous with depth, a factor which may place constraints on the ease with which these latter melts separate from the source, and their subsequent ascent to the surface. Crystal settling and crystal fractionation will be more effective at high pressures in melts with NBO/T < 1.

Finally, it is worth emphasizing that the preceding discussion has been limited to volatile–free melts and magmas that contain only a few percent crystals and are Newtonian in behavior. Magmas containing more than 10–20% crystals will show non–Newtonian viscous behavior (SHAW, 1969; SCARFE, 1973; MURASE and MCBIRNEY, 1973) and the considerations discussed in this paper will not be applicable. The way in which water reduces the viscosity of most melts of geological interest has been known for some time (e.g., SHAW, 1963; SCARFE, 1973; DINGWELL, 1987; DINGWELL and MYSEN, 1985); however, the effect of CO_2 on viscosity has yet to be determined.

Acknowledgments—Critical reviews by M. Brearley, D. B. Dingwell, I. Kushiro and H. S. Yoder, Jr. are appreciated. H. Rawson made available facilities for 1 bar viscometry at Sheffield University, U.K. Scarfe acknowledges support from the Carnegie Institution of Washington, the University of Alberta and Canadian NSERC grant A8384. This work was carried out while Scarfe was on leave from the University of Alberta as a guest investigator at the Geophysical Laboratory.

REFERENCES

ANGELL C. A., CHEESEMAN P. A. and TAMADDON S. (1982). Pressure enhancement of ion mobilities in liquid silicates from computer simulation studies to 800 kilobars. *Science* **218**, 885–888.

ANGELL C. A., CHEESEMAN P. A. and TAMADDON S. (1983) Water-like transport property anomalies in liquid silicates investigated at high T and P by computer simulation techniques. *Bull. Mineral.* **106**, 87–97.

BARTLETT R. W. (1969) Magma convection, temperature distribution, and differentiation. *Amer. J. Sci.* **267**, 1067–1082.

BERMAN H. (1939) A torsion microbalance for the determination of specific gravities of minerals. *Amer. Mineral.* **24**, 434–440.

BOCKRIS J. O'M., MACKENZIE J. D. and KITCHENER J. A. (1955) Viscous flow in silica and binary liquid silicates. *Trans. Faraday Soc.* **51**, 1734–1748.

BOND W. L. (1951) Making small spheres. *Rev. Sci. Instrum.* **22**, 344–345.

BOTTINGA Y. and WEILL D. F. (1972) The viscosity of magmatic silicate liquids: a model for calculation. *Amer. J. Sci.* **272**, 438–475.

BOYD F. R. and ENGLAND J. L. (1960) Apparatus for phase equilibrium measurements at pressures up to 50 kilobars and temperatures up to 1750°C. *J. Geophys. Res.* **65**, 741–748.

BREARLEY M., DICKINSON J. E. and SCARFE C. M. (1986) Pressure dependence of melt viscosities on the join diopside–albite. *Geochim. Cosmochim. Acta.* **50**, 2563–2570.

BRIDGMAN P. W. (1931) *The Physics of High Pressure.* 398 pp. Bell.

COHEN M. A. and TURNBULL D. (1959) Molecular transport in liquids and glasses. *J. Chem. Phys.* **31**, 1164–1169.

DIETZEL A. and BRUCKNER R. (1955) Aufbau eines Absolutviskosimetres fur hohe Temperaturen und Messung der Zahigkeit geschmolzener Borsaue fur Eichzwecke. *Glastech. Ber.* **28**, 455–467.

DINGWALL A. G. F. and MOORE H. (1954) The effects of various oxides on the viscosity of glasses of the soda–lime–silica type. *J. Soc. Glass Tech.* **37**, 316–372.

DINGWELL D. B. (1987) Melt viscosities in the system NaAlSi$_3$O$_8$–H$_2$O–F$_2$O$_{-1}$. In *Magmatic Processes: Physicochemical Principles,* (ed. B. O. MYSEN), The Geochemical Society Spec. Publ. No. 1, pp. 423–432.

DINGWELL D. B. and MYSEN B. O. (1985) Effects of water and fluorine on the viscosity of albite melt at high pressure: a preliminary investigation. *Earth Planet. Sci. Lett.* **74**, 266–274.

DUNN T. and SCARFE C. M. (1986) Variation of the chemical diffusivity of oxygen and viscosity of an andesite melt with pressure at constant temperature. *Chem. Geol.* **54**, 203–215.

FLEET M. E., HERZBERG C. T., HENDERSON G. S., CROZIER E. D., OSBORNE M. D. and SCARFE C. M. (1984) Coordination of Fe, Ga and Ge in high pressure glasses by Mossbauer, Raman and X-ray absorption spectroscopy, and geological implications. *Geochim. Cosmochim. Acta* **48**, 1455–1466.

FUJII T. and KUSHIRO I. (1977a) Density, viscosity and compressibility of basaltic liquid at high pressures. *Carnegie Inst. Wash. Yearb.* **76**, 419–424.

FUJII T. and KUSHIRO I. (1977b) Melting relations and viscosity of an abyssal tholeiite. *Carnegie Inst. Wash. Yearb.* **76**, 461–465.

HARRIS P. G., KENNEDY W. Q. and SCARFE C. M. (1970) Volcanism versus plutonism—the effect of chemical composition. In *Mechanism of Igneous Intrusion,* (eds. G. NEWALL and N. RAST), *Geol. J. Spec. Issue* **2**, 187–200.

HOCHELLA M. F. and BROWN G. E. (1985) The structures of albite and jadeite composition glasses quenched from high pressure. *Geochim. Cosmochim. Acta* **49**, 1137–1142.

KIRKPATRICK R. J. (1974) Kinetics of crystal growth in the system CaMgSi$_2$O$_6$–CaAl$_2$SiO$_6$. *Amer. J. Sci.* **274**, 215–242.

KUSHIRO I. (1976) Changes in viscosity and structure of melt of NaAlSi$_2$O$_6$ compositions at high pressures. *J. Geophys. Res.* **81**, 6347–6350.

KUSHIRO I. (1977) Phase transformations in silicate melts under upper mantle conditions. In *High Pressure Research: Applications in Geophysics,* (eds. M. H. MANGHNANI and S. AKIMOTO), pp. 25–37 Academic Press.

KUSHIRO I. (1978) Viscosity and structural changes of albite (NaAlSi$_3$O$_8$) melts at high pressures. *Earth Planet. Sci. Lett.* **41**, 87–90.

KUSHIRO I. (1980) Viscosity, density and structure of silicate melts at high pressures and their petrological applications. In *Physics of Magmatic Processes,* (ed. R. B. HARGRAVES), pp. 93–120 Princeton University Press.

KUSHIRO I. (1981) Change in viscosity with pressure of melts in the system CaO–Al$_2$O$_3$–SiO$_2$. *Carnegie Inst. Wash. Yearb.* **80**, 339–341.

KUSHIRO I. (1983) On the lateral variations in chemical composition and volume of Quaternary volcanic rocks across Japanese arcs. *J. Volcan. Geotherm. Res.* **18**, 435–447.

KUSHIRO I. (1984) Genesis of arc magmas: a case in Japanese arcs. *Proc 27th Int. Geol. Congr.* **9**, 259–281.

KUSHIRO I. (1986) Viscosity of partial melts in the upper mantle. *J. Geophys. Res.* **91**, 9343–9350.

KUSHIRO I., YODER H. S. JR. and MYSEN B. O. (1976) Viscosities of basalt and andesite melts at high pressures. *J. Geophys. Res.* **81**, 6351–6356.

MATHESON A. J. (1966) Role of free volume in the pressure dependence of the viscosity of liquids. *J. Chem. Phys.* **44**, 695–699.

MURASE T. and MCBIRNEY A. R. (1973) Properties of some common igneous rocks and their melts at high temperatures. *Bull. Geol. Soc. Amer.* **84**, 3563–3592.

MYSEN B. O. (1986) Structure and petrologically important properties of silicate melts relevant to natural magmatic liquids. In *Silicate Melts,* (ed. C. M. SCARFE) *Min. Assoc. Can. Short Course Handbook 12,* 180–209.

MYSEN B. O., VIRGO D., DANKWERTH P., SEIFERT F. A. and KUSHIRO I. (1983) Influence of pressure on the structure of melts on the joins NaAlO$_2$–SiO$_2$, CaAl$_2$O$_4$–SiO$_2$, and MgAl$_2$O$_4$–SiO$_2$. *Neues. Jahrb. Mineral. Abhand.* **147**, 281–303.

MYSEN B. O., VIRGO D. and SCARFE C. M. (1980a) Relations between the anionic structure and viscosities of silicate melts—a Raman spectroscopic study. *Amer. Mineral.* **65**, 690–710.

MYSEN B. O., VIRGO D., HARRISON W. J. and SCARFE C. M. (1980b) Solubility mechanisms of H$_2$O in silicate melts at high pressures and temperatures: a Raman spectroscopic study. *Amer. Mineral.* **65**, 900–914.

MYSEN B. O., VIRGO D. and SEIFERT F. A. (1982) The structure of silicate melts: implications for chemical and physical properties of natural magma. *Rev. Geophys. Space Phys.* **20**, 353–383.

MYSEN B. O., VIRGO D. and SEIFERT F. A. (1985) Relationships between properties and structure of aluminosilicate melts. *Amer. Mineral.* **70,** 88–105.

OHTANI E., TAULELLE F. and ANGELL C. A. (1985) Al^{3+} coordination changes in liquid aluminosilicates under pressure. *Nature* **314,** 78–80.

O'REILLY J. M. (1964) Effect of pressure on amorphous polymers. In *Modern Aspects of the Vitreous State,* (ed. J. D. MACKENZIE), Vol. 3, pp. 59–89 Butterworths.

RIEBLING E. F. (1966) Structure of sodium aluminosilicate melts containing at least 50 mole% SiO_2 at 1500°C. *J. Chem. Phys.* **44,** 2857–2865.

SAKKA S. and MACKENZIE J. D. (1969) High pressure effects on glass. *J. Non-Cryst. Solids.* **1,** 107–142.

SCARFE C. M. (1973) Viscosity of basic magmas at varying pressure. *Nature* **241,** 101–102.

SCARFE C. M. (1977) Viscosity of a pantellerite melt at one atmosphere. *Can. Mineral.* **15,** 185–189.

SCARFE C. M. (1981) The pressure dependence of the viscosity of some basic melts. *Carnegie Inst. Wash. Yearb.* **80,** 336–339.

SCARFE C. M., MYSEN B. O. and VIRGO D. (1979) Changes in viscosity and density of melts of sodium disilicate, sodium metasilicate and diopside composition with pressure. *Carnegie Inst. Wash. Yearb.* **78,** 547–551.

SCARFE C. M., CRONIN D. J., WENZEL J. T. and KAUFFMAN D. A. (1983) Viscosity–temperature relationships at 1 atm in the system diopside–anorthite. *Amer. Mineral.* **68,** 1083–1088.

SCARFE C. M. (1986) Viscosity and density of silicate melts. In *Silicate Melts,* (ed. C. M. SCARFE) *Min. Assoc. Can. Short Course Handbook 12,* 36–56.

SEIFERT F. A., MYSEN B. O. and VIRGO D. (1982) Three-dimensional network structure of quenched melts (glass) in the systems SiO_2–$NaAlO_2$, SiO_2–$CaAl_2O_4$ and SiO_2–$MgAl_2O_4$. *Amer. Mineral.* **67,** 696–717.

SHARMA S. K., VIRGO D. and MYSEN B. O. (1979a) Raman study of the coordination of aluminum in jadeite melts as a function of pressure. *Amer. Mineral.* **64,** 779–787.

SHARMA S. K., VIRGO D. and KUSHIRO I. (1979b) Relationship between density, viscosity and structure of GeO_2 melts at low and high pressures. *J. Non-Cryst. Solids* **33,** 235–248.

SHARPE M. R., IRVINE T. N., MYSEN B. O. and HAZEN R. M. (1983) Density and viscosity characteristics of melts of Bushveld chilled margin rocks. *Carnegie Inst. Wash. Yearb.* **82,** 300–305.

SHARTSIS L., SPINNER S. and CAPPS W. (1952) Density, expansivity and viscosity of molten alkali silicates. *J. Amer. Ceram. Soc.* **35,** 155–160.

SHAW H. R. (1963) Obsidian–H_2O viscosities at 1000 and 2000 bars and in the temperature range 700–900°C. *J. Geophys. Res.* **68,** 6337–6343.

SHAW H. R. (1969) Rheology of basalt in the melting range. *J. Petrol.* **10,** 510–535.

SHIMIZU N. and KUSHIRO I. (1984) Diffusivity of oxygen in jadeite and diopside melts at high pressure. *Geochim. Cosmochim. Acta* **48,** 1295–1303.

TAYLOR M. and BROWN G. E. (1979) Structure of mineral glasses—II. The SiO_2–$NaAlSiO_4$ join. *Geochim. Cosmochim. Acta* **43,** 1467–1473.

URBAIN G., BOTTINGA Y. and RICHET P. (1982) Viscosity of liquid silica, silicates and aluminosilicates. *Geochim. Cosmochim. Acta* **46,** 1061–1072.

VELDE B. and KUSHIRO I. (1978) Structure of sodium alumino–silicate melts quenched at high pressures; infrared and aluminum K–radiation data. *Earth Planet. Sci. Lett.* **40,** 137–140.

VIRGO D. and MYSEN B. O. (1985) The structural state of iron in oxidized vs. reduced glasses at 1 atm: a ^{57}Fe Mossbauer study. *Phys. Chem. Minerals* **12,** 65–76.

WAFF H. S. (1975) Pressure–induced coordination changes in magmatic liquids. *Geophys. Res. Lett.* **2,** 193–196.

WOODCOCK L. V., ANGELL C. A. and CHEESEMAN P. (1976) Molecular dynamics studies of the vitreous state: simple ionic systems and silica. *J. Chem. Phys.* **65,** 1565–1577.

Magmatic Processes: Physicochemical Principles
© The Geochemical Society, Special Publication No. 1, 1987
Editor B. O. Mysen

Physics of magma segregation processes

DONALD L. TURCOTTE

Department of Geological Sciences, Cornell University, Ithaca, New York 14853-1504, U.S.A.

Abstract—Partial melting occurs in the mantle due to pressure reduction or by volumetric heating. As a result magma is produced along grain boundaries at depths up to 100 km or more. Differential buoyancy drives this magma upwards by a porous flow mechanism. The small magma conduits on grain boundaries may coalesce to form rivulets of magma in the asthenosphere. The reduction in density due to the presence of magma may induce diapiric flows in the mantle. When the ascending magma reaches the base of the lithosphere it is likely to form pools of magma. Buoyancy driven magma fractures appear to be to the only mechanism by which this magma can penetrate through the lithosphere. How these fractures are initiated remains a subject of speculation.

INTRODUCTION

PLATE TECTONICS provides a framework for understanding magmatic activity on and in the Earth. However, many important aspects of magmatic processes are poorly understood. A large fraction of the Earth's volcanism is associated with mid-ocean ridges. Over a large fraction of the ridge system this is essentially a *passive process*. As the plates diverge mantle rock must ascend by solid state creep to fill the gap. Average upper mantle rock is sampled by this random, unsteady process. As the mantle rock ascends melting occurs due to the adiabatic decompression. A substantial fraction of the resulting magma ascends to near surface magma chambers due to the differential buoyancy. This magma solidifies to form the oceanic crust.

Volcanism is also associated with subduction zones. The origin of this volcanism is still a subject of controversy. Friction on the slip zone between the descending lithosphere and overlying plate provides some heat. However, as the rock solidus is approached, the viscosity drops to such low levels that direct melting is not expected (YUEN *et al.,* 1978). Offsets in the linear volcanic chains associated with subduction zones correspond with variations in the dip of the Benioff zones. This is strong evidence that the magma is produced on or near the boundary between the descending lithosphere and overlying plate. This magma then ascends vertically 150 ± 25 km to form the distinct volcanic edifices observed on the surface. How the magma collects, ascends and feeds the distinct centers is poorly understood.

Volcanism also occurs within plate interiors. A fraction of this volcanism may be associated with ascending plumes within the mantle. These plumes may be the result of instabilities of the thermal boundary layer at the base of the convecting upper mantle, either at a depth of 670 km for layered mantle convection or above the core–mantle boundary for whole mantle convection (ALLEGRE and TURCOTTE, 1985). Pressure release melting would produce the observed volcanism. However, in many cases, the magma must penetrate the full thickness of the cold lithosphere in order to produce surface volcanism.

We will consider three mechanisms for magma migration: (1) porous flow, (2) diapirism and (3) fracture. Aspects of the magma migration problem have been reviewed by SPERA (1980) and TURCOTTE (1982).

POROUS FLOW

Adiabatic decompression will lead to partial melting of the mantle on grain boundaries. Experimental studies (WAFF and DULAU, 1979, 1982; COOPER and KOHLSTEDT, 1984, 1986; FUJII *et al.,* 1986) and theoretical calculations (VON BARGEN and WAFF, 1986) have shown that the melt forms an interconnected network of channels along triple junctions between grains. Magma will migrate upwards along these channels due to differential buoyancy. This configuration suggests that a porous flow model may be applicable for magma migration in the region where partial melting is occurring.

Porous flow models for magma migration have been given by FRANK (1968), SLEEP (1974), TURCOTTE and AHERN (1978a,b), and AHERN and TURCOTTE (1979). Implicit in these models is the assumption that the matrix is free to collapse as the magma migrates upwards. The validity of this assumption has been questioned by MCKENZIE (1984). It has been shown by SCOTT and STEVENSON (1984) and by RICHTER and MCKENZIE (1984) that the coupled migration and compaction equations have unsteady solutions (solitons). However, it was shown approximately by AHERN and TURCOTTE (1979) and rigorously by RIBE (1985) that compaction is not important in mantle melting by decompression. The melting of the ascending mantle takes place sufficiently slowly that the crystalline matrix is free to collapse by solid state creep processes. It is appropriate to assume that the fluid pressure is equal to the lithostatic pressure. Any difference between the hydrostatic (fluid) pressure and the lithostatic pressure will drive the buoyant fluid upwards. The conclusion is that the unsteady solutions are not relevant to magma migration in the mantle.

In order to quantify magma migration in a porous solid matrix, Darcy's law gives (TURCOTTE and SCHUBERT, 1982, p. 414):

$$\Delta v = \frac{b^2 \phi \Delta \rho g}{24 \pi \eta_m} \tag{1}$$

where Δv is the magma velocity relative to the solid matrix, b is the grain size, ϕ is the volume fraction of magma (porosity), $\Delta \rho$ is the density difference between the magma and the solid matrix, g is the acceleration of gravity, and η_m is the magma viscosity. Taking $b = 2$ mm, $\Delta \rho = 600$ kg/m³ and $g = 10$ m/s², the magma migration velocity is given as a function of the magma viscosity in Figure 1 for several values of porosity.

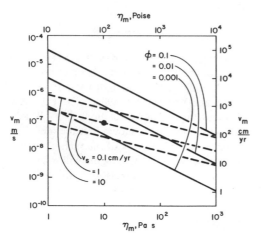

FIG. 1. The solid lines give the magma migration velocity v_m as a function of the magma viscosity η_m for several values of the volume fraction of magma (porosity) ϕ. This result from Equation (1) for the porous flow model assumes that the grain size $b = 2$ mm and density difference $\Delta \rho = 600$ kg/m³. The dashed lines give the magma migration velocity v_m as a function of the magma viscosity η_m for several values of the upward velocity of the mantle beneath the melt zone v_s. This result from Equations (1) and (2) assumes the melt fraction $f = 0.25$. The solid dot illustrates a particular example. Specifying $\eta_m = 10$ Pa s and $v_s = 1$ cm/year determines the position of the dot which gives $v_m = 10^{-7}$ m/s (300 cm/year). The corresponding value of the porosity is found from the intersecting solid line, for this example $\phi = 0.003$.

The upward flux of magma can be related to the upward flux of the component to be melted by

$$\frac{v_m \phi}{3} = v_s f \tag{2}$$

where v_m is the magma velocity, v_s is the upward velocity of the mantle prior to melting, and f is the fraction of the mantle that is melted. By using Equation (1) to relate v_m and ϕ and taking $f = 0.25$, the appropriate values of v_m and ϕ are given in Figure 1 for specified values of η_m and v_s. For example, with $v_s = 1$ cm/year and $\eta_m = 10$ Pa s, we find that $\phi = 0.003$ and $v_m = 10^{-7}$ m/s (300 cm/yr) as indicated by the solid dot in Figure 1.

Caution should be used in the application of the porous flow model. The theory assumes a uniform distribution of small channels and in fact the small channels may coalesce to form larger channels as the magma ascends in much the same way that streams coalesce to form rivers. There is some evidence that this occurs when ice melts.

DIAPIRISM

Diapirism has long been considered as a mechanism for magma migration (MARSH, 1978, 1982; MARSH and KANTHA, 1978). Two types of diapirism must be considered. First, consider magma that has been segregated from the mantle by the porous flow mechanism. The ascending magma is likely to pool at the base of the lithosphere. The question is whether the buoyancy can drive the pooled magma through the lithosphere as a diapir. In order for this to occur, the country rock must be displaced by solid state creep so that the viscosity of the medium through which the diapir is passing governs the velocity of ascent. A second question is whether the residual partial melt fraction in the asthenosphere can induce diapirism.

In order to consider these problems we study the idealized problem of a low viscosity, low density fluid sphere ascending through a viscous medium due to buoyancy. The velocity of ascent for this Stokes problem is given by (TURCOTTE and SCHUBERT, 1982, p. 267)

$$v_d = \frac{r^2 g \Delta \rho \phi}{3 \eta_s} \tag{3}$$

where, r, is the radius of the sphere and, η_s, is the viscosity of the medium. The ascent velocity is given as a function of the viscosity in Figure 2 for several values of the sphere radius and melt fraction and $\Delta \rho = 600$ kg/m³.

Because the lithosphere certainly has a viscosity of at least 10^{24} Pa s, the migration velocity spheres

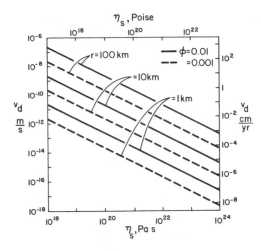

FIG. 2. The ascent velocity of a low viscosity, spherical diapir, v_d, as a function of the viscosity of the medium through which it is rising, η_s, for several values of the sphere radius, r, and porosity ϕ. This result from Equation (3) assumes $\Delta\rho = 600$ kg/m^3.

with a radius less than 100 km is extremely small. The diapir will solidify before it can migrate through a cold lithosphere. If the path through the lithosphere was heated so that the viscosity was lowered, then significant migration may occur. But what is the mechanism for heating if magma cannot penetrate the lithosphere? The conclusion is that diapirism is not a viable mechanism for the migration of magma through the lithosphere.

One possibility for this type of migration is that heat transported from the magma or heating by viscous dissipation, or both, softens the medium through which the diapir is passing. Studies of this problem (RIBE, 1983; EMERMAN and TURCOTTE, 1984; OCKENDON et al., 1985) show that these mechanisms do not significantly help in the penetration of cold lithosphere.

Results given in Figure 2 can also be used to determine whether the melt fraction in the asthenosphere can induce diapirism. Asthenosphere viscosities may be as low as 10^{18} Pa s although 10^{21} Pa s is probably a better estimate. With $\eta_s = 10^{21}$ Pa s, $\phi = 0.01$, and $r = 100$ km, we find $v_d = 1$ cm/yr. Under these conditions, diapirism due to partial melt is relatively unimportant. However, with $\eta_s = 10^{18}$ Pa s, the velocity may approach 10^3 cm/yr. Thus it is not possible to make a definitive conclusion on the relative role of diapirism within the asthenosphere. To treat this problem satisfactorily, the mantle convection problem must be solved simultaneously with the melting and porous flow migration problems and this has not been done so far.

MAGMA FRACTURE

A third mechanism for magma migration is fluid fracture. This is probably the only mechanism capable of transporting magma through the cold lithosphere. The common occurrence of dikes is direct observational evidence for the existence of magma fractures. The dynamics of pressure–driven fluid fractures has been studied by SPENCE and TURCOTTE (1985) and EMERMAN et al. (1986). These authors showed that dike propagation is restricted by the viscosity of the magma and by the fracture resistance of the media, but in most geological applications, the fracture resistance can be neglected.

The vertical transport of magma over large distances (10–100 km) is almost certainly driven by the differential buoyancy of the magma. One of the most spectacular examples of buoyancy driven magma fracture is a kimberlite eruption; the required velocities are estimated to be 0.5–5 m/s (PASTERIS, 1984). Buoyancy–driven fluid fractures as a mechanism for magma migration have been studied by WEERTMAN (1971), ANDERSON and GREW (1977), ANDERSON (1979) and SECOR and POLLARD (1975). The dynamics of a buoyancy driven fluid fracture have been given by SPENCE et al. (1986). These authors found that the stress intensity factor (fracture resistance) plays an essential role in the solution.

The propagation of a buoyancy-driven fluid fracture is governed by the equations for the fluid flow in the crack, the equation governing the deformation of the elastic medium, and the equation relating the empirically derived stress intensity factor to the tip curvature of the crack. Mathematical details for the two-dimensional problem have been given by SPENCE et al. (1986). These authors find a steady–state solution for a propagating crack. A universal shape is obtained in terms of non–dimensional variables. The non–dimensional crack width $2H$ is given as a function of the non–dimensional distance from the crack tip in Figure 3. The non–dimensional variables are related to actual variables by

$$H = \frac{h}{h_\infty}, \quad X = \left(\frac{[1-\nu]\Delta\rho g}{h_\infty \mu}\right)^{1/2} x \qquad (4)$$

where $2h_\infty$ is the width of the tail, μ the shear modulus, ν Poisson's ratio, $\Delta\rho$ the density difference and g the acceleration of gravity. One of the important results of the analysis is the requirement that the crack must have an infinitely long constant width tail. The flow in the tail is a balance between the buoyancy driving force and the viscous resisting

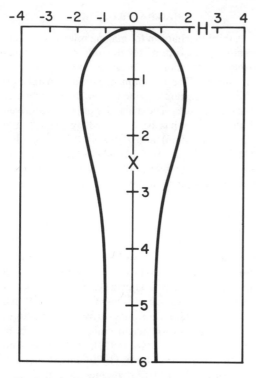

FIG. 3. Shape of an upward propagating magma fracture. The non–dimensional half–width H is given as a function of the non–dimensional distance from the crack tip X. The non–dimensional variables are defined in Equation (4).

force. Thus the velocity of propagation of the crack, c, is related to the half–width of the tail h_∞ by

$$c = \frac{gh_\infty^2 \Delta\rho}{3\eta_m} \quad (5)$$

for laminar flow in the crack and

$$c = \frac{7.71 h_\infty^{5/7}(\Delta\rho g)^{4/7}}{\rho_m^{3/7}\eta_m^{1/7}} \quad (6)$$

for turbulent flow.

For the tail of the crack, the non–dimensional half–width is $H_\infty = 1$. The maximum crack half–width is $H = 1.8975$ and this occurs at $X = 1.152$. A steady–state solution to this problem is found only for a single value of the non–dimensional stress intensity factor, λ, defined by

$$\lambda = \frac{1}{2^{1/2}}\left(\frac{(1-\nu)}{\mu h_\infty}\right)^{3/4}\frac{K_c}{(g\Delta\rho)^{1/4}} = 1.3078 \quad (7)$$

where K_c is the critical stress intensity factor. Solving for the half–width of the tail gives

$$h_\infty = 0.440\frac{K_c^{4/3}(1-\nu)}{\mu(g\Delta\rho)^{1/3}}. \quad (8)$$

When the critical stress intensity factor is specified, the half–width of the crack is obtained from Equation (8). The propagation velocity is then obtained from Equation (5) for laminar flow and Equation (6) for turbulent flow.

The critical stress intensity factor, K_c, is a measure of the fracture toughness of the material. Cracks can propagate at low velocities ($\sim 10^{-4}$ m/s) by the mechanism of stress corrosion; in this range $K < K_c$. However, when $K = K_c$, crack propagation becomes catastrophic and the propagation velocity may rapidly accelerate to a large fraction of the speed of sound. The values of K_c for a variety of rocks have been obtained in the laboratory. This work has been reviewed by ATKINSON (1984). Measured values for granite range from $K_c = 0.6$ to 2.2 MN/m$^{3/2}$ and from $K_c = 0.8$ to 3.3 MN/m$^{3/2}$ for basalts. A measured value for dunite is $K_c = 3.7$ MN/m$^{3/2}$. These values were obtained at atmospheric pressure. The influence of pressure on the stress intensity factor is difficult to predict. SCHMIDT and HUDDLE (1977) found a factor of four increase at a pressure of 62 MPa for Indiana limestone. Thus, it is difficult to specify values of the critical stress intensity factor for regions of the crust and mantle where buoyancy-driven magma fracture is occurring.

By taking $\mu = 2 \times 10^{10}$ Pa, $\nu = 0.25$, $\Delta\rho = 300$ kg/m^3 and $g = 10$ m/s^2 the half–width of the tail from Equation (8) is given as a function of the critical stress intensity factor in Figure 4. Propagation velocities from Equations (5) or (6) are given as a function of the critical stress intensity factor in Figure 5. Results are given for magma viscosities $\eta_m = 0.1, 3, 100$ Pa s. We use the laminar result equation (5) for Reynolds numbers Re $= \rho_m ch_\infty/\eta_m$ less than 10^3 and the turbulent result Equation (6) for Reynolds numbers greater than 10^3. The corre-

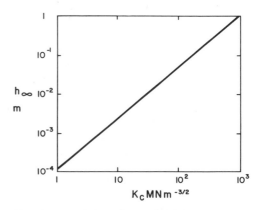

FIG. 4. The half width of the crack tail h_∞ as a function of the critical stress intensity factor K_c from Equation (8) by assuming $\mu = 2 \times 10^{10}$ Pa, $\nu = 0.25$, $\Delta\rho = 300$ kg/m^3.

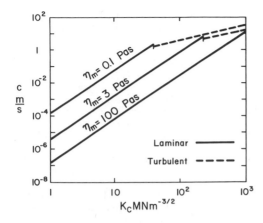

FIG. 5. Crack propagation velocity, c, as a function of the critical stress intensity factor, K_c, for several values of the magma viscosity, η_m. Laminar flow from Equation (5) and turbulent flow from Equation (6).

sponding values of the volumetric flow rate $V = 2ch_\infty$ per unit crack length are given in Figure 6.

CARMICHAEL et al. (1977) have shown that the entrainment of xenoliths in basaltic flows implies velocities, c, of at least $c = 0.5$ m/s. A variety of studies indicate that basaltic magma migrates upward at velocities in the range $c = 0.5$ to 5 m/s. A typical viscosity for a basaltic magma is $\eta_m = 0.1$ Pa s. From Figures 4 and 5 the corresponding range of the stress intensity factor is $K_c = 20$ to 100 MN/m$^{3/2}$ and the range of tail half-widths is $h_\infty = 5$ to 50 mm. From Figure 6 the corresponding range of flow rates per unit length is $V = 0.0025$ to 0.25 m^2/s.

Studies of the Kilauea Iki eruption on an island of Hawaii during 1959–1960 give flow rates of 50 to 150 m^3/s (WILLIAMS and MCBIRNEY, 1979, pp. 232–233). Taking a flow rate of 100 m^3/s, the range of flow rates per unit length given above, $V = 0.0025$ to 0.25 m^2/s, correspond to crack lengths between 400 m and 4 km. These appear to be reasonable lengths for the crack feeding Kilauea Iki. Although the above comparison appears reasonable, there is conflicting evidence whether the flow of magma at the surface represents the flow through the lithosphere. The role of shallow magma chambers in storing magma is poorly understood.

As a specific example of a magma fracture that transports magma through the lithosphere we take: $\mu = 2 \times 10^{10}$ Pa, $\nu = 0.25$, $\Delta\rho = 600$ kg/m^3, $g = 10$ m/s^2, $\eta_m = 0.1$ Pa s, and $K_c = 50$ MN/m$^{3/2}$. We find that $h_\infty = 0.0167$ m and $c = 2.86$ m/s. To convert the shape given in Figure 3 into actual distances, we require $h = 0.0167H$ m and $x = 272X$

m. If the length of the crack is 1 km the volume flux through the lithosphere is 95.5 m^3/s.

CONCLUSIONS

Pressure release melting produces magma on grain boundaries. Magma on triple junctions between grains produce an interconnected network of channels. Under mantle conditions the magma drains rapidly upwards due to differential buoyancy and the residual solid matrix collapses due to solid state creep. Magma that is produced over a depth range of approximately 50 km mixes during the vertical ascent to produce the magma reaching the base of the lithosphere.

At mid–ocean ridges there is no lithosphere and the ascending magma can form directly the oceanic crust. However, off the axis at ocean trenches and at intraplate volcanic centers, the magma must penetrate through the lithosphere. Magma fracture is a mechanism for the rapid transport of magma through the lithosphere. A theory for steady state magma fracture through the lithosphere is given. A typical fracture width is 2 cm and fracture velocity is 5 m/s. Under these conditions relatively little magma is lost by solidification during ascent. The existence of dikes is evidence that magma fracture is a pervasive mechanism. The extensive systems of dikes in deeply eroded terranes is evidence that dikes are the dominant mechanism for the ascent of magma through the crust. Kimberlite eruptions are direct evidences that magma fracture can transport magma through the lithosphere.

How magma fractures initiate remains a matter of speculation. It may be possible to form magma fractures as magmas perculating in small channels

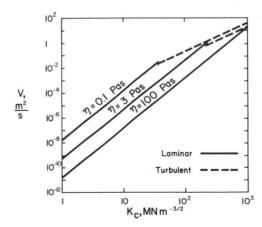

FIG. 6 Volumetric flow rate per unit crack length, V, as a function of the critical stress intensity factor, K_c, for several values of the magma viscosity, η_m.

coalesce to form larger channels as suggested by FOWLER (1985). Alternatively, magma may pool at the base of the lithosphere. When the pool reaches a critical size the buoyancy forces may initiate a magma fracture that subsequently drains the pool.

Acknowledgements—The author would like to acknowledge the many contributions of D. A. Spence to our understanding of the magma fracture problem. This research was supported by the Division of Earth Sciences, National Science Foundation under grant EAR-8518019. This is contribution 830 of the Department of Geological Sciences, Cornell University.

REFERENCES

AHERN J. L. and TURCOTTE D. L. (1979) Magma migration beneath an ocean ridge. *Earth Planet. Sci. Lett.* **45**, 115–122.

ALLEGRE C. J. and TURCOTTE D. L. (1985) Geodynamic mixing in the mesosphere boundary layer and the origin of oceanic islands. *Geophys. Res. Lett.* **12**, 207–210.

ANDERSON O. L. (1979) The Role of fracture dynamics in Kimberlite pipe formation. In *Kimberlites, Diatremes, and Diamonds: Their Geology, Petrology, and Geochemistry*, (eds. F. R. BOYD and H. O. A. MEYER), vol. 1, pp. 344–353 American Geophysical Union.

ANDERSON O. L. and GREW P. C. (1977) Stress corrosion theory of crack propagation with applications to geophysics. *Rev. Geophys. Space Phys.* **15**, 77–104.

ATKINSON B. K. (1984) Subcritical crack growth in geological materials. *J. Geophys. Res.* **89**, 4077–4114.

CARMICHAEL I. S. E., NICOLLS J., SPERA F. J., WOOD B. J. and NELSON S. A. (1977) High–temperature properties of silicate liquids: applications to the equilibrium and ascent of basic magma. *Phil. Trans. Roy. Soc. London* **286**, 373–431.

COOPER R. F. and KOHLSTEDT D. L. (1984) Solution-precipitation enhanced diffusional creep of partially molten olivine–basalt aggregates during hot–pressing. *Tectonophys.* **107**, 207–233.

COOPER R. F. and KOHLSTEDT D. L. (1986) Rheology and structure of olivine–basalt partial melts. *J. Geophys. Res.* **91**, 9315–9323.

EMERMAN S. H. and TURCOTTE D. L. (1984) Diapiric penetration with melting. *Phys. Earth Planet. Int.* **36**, 276–284.

EMERMAN S. H., TURCOTTE D. L. and SPENCE D. A. (1986) Transport of magma and hydrothermal solutions by laminar and turbulent fluid fracture. *Phys. Earth Planet. Int.* **41**, 249–259.

FOWLER A. C. (1985) A mathematical model of magma transport in the asthenosphere. *Geophys. Astrophys. Fluid Dyn.* **33**, 63–96.

FRANK F. C. (1968) Two-component flow model for convection in the Earth's upper mantle. *Nature* **220**, 350–352.

FUJII N., OSAMURA K. and TAKAHASHI E. (1986) Effect of water saturation on the distribution of partial melt in the olivine–pyroxene–plagioclase system. *J. Geophys. Res.* **91**, 9253–9259.

MARSH B. D. (1978) On the cooling of ascending andesitic magma. *Phil. Trans. Roy. Soc. London* A**288**, 611–625.

MARSH B. D. (1982) On the mechanics of igneous diapirism, stoping and zone melting. *Amer. J. Sci.* **282**, 808–855.

MARSH B. D. and KANTHA L. H. (1978) On the heat and mass transfer from an ascending magma. *Earth Planet. Sci. Lett.* **39**, 435–443.

OCKENDON J. R., TAYLOR A. B., EMERMAN S. H. and TURCOTTE D. L. (1985) Geodynamic thermal runaway with melting. *J. Fluid Mech.* **152**, 301–314.

PASTERIS J. D. (1984) Kimberlites: complex mantle melts. *Ann. Rev. Earth Planet. Sci.* **12**, 133–153.

RIBE N. M. (1983) Diapirism in the Earth's mantle: Experiments on the motion a hot sphere in a fluid with temperature–dependent viscosity. *J. Volcanol. Geotherm. Res.* **16**, 221–245.

RIBE N. M. (1985) The deformation and compaction of partial molten zones. *Geophys. J. Roy. Astron. Soc.* **83**, 487–501.

RICHTER F. M. and McKENZIE D. (1984) Dynamical models for melt segregation from a deformable matrix. *J. Geol.* **92**, 729–740.

SCHMIDT R. A. and HUDDLE C. W. (1977) Effect of confining pressure on fracture toughness of Indiana limestone. *Int. J. Rock Mech. Min. Sci. Geomech. Abstr.* **14**, 289–293.

SCOTT D. R. and STEVENSON D. J. (1984) Magma solitons. *Geophys. Res. Lett.* **11**, 1161–1164.

SECOR D. T. and POLLARD D. D. (1975) On the stability of open hydraulic fractures in the earth's crust. *Geophys. Res. Lett.* **2**, 510–513.

SLEEP N. H. (1974) Segregation of magma from a mostly crystalline mush. *Bull. Geol. Soc. Amer.* **85**, 1225–1232.

SPENCE D. A., SHARP P. W. and TURCOTTE D. L. (1986) Buoyancy driven crack propagation: a mechanism for magma migration. *J. Fluid Mech.*, (In press).

SPENCE D. A. and TURCOTTE D. L. (1985) Magma-driven propagation of cracks. *J. Geophys. Res.* **90**, 575–580.

SPERA F. J. (1980) Aspects of magma transport. In *Physics of Magmatic Processes*, (ed. R. B. HARGRAVES), pp. 263–323 Princeton University Press.

TURCOTTE D. L. (1982) Magma migration. *Ann. Rev. Earth Planet. Sci.* **10**, 397–408.

TURCOTTE D. L. and AHERN J. L. (1978a) A porous flow model for magma migration in the asthenosphere. *J. Geophys. Res.* **83**, 767–772.

TURCOTTE D. L. and AHERN J. L. (1978b) Magma production and migration within the moon. *Proc. Lunar Planet. Sci. Conf. 9th* 307–318.

TURCOTTE D. L. and SCHUBERT G. (1982) *Geodynamics*, John Wiley.

VON BARGEN N. and WAFF H. S. (1986) Permeabilities, interfacial areas and curvatures of partially molten systems: Results of numerical computations of equilibrium microstructures. *J. Geophys. Res.* **91**, 9261–9276.

WAFF H. S. and BULAU J. R. (1979) Equilibrium fluid distribution in an ultramafic partial melt under hydrostatic stress conditions. *J. Geophys. Res.* **84**, 6109–6114.

WAFF H. S. and BULAU J. R. (1982) Experimental determination of near–equilibrium textures in partially molten silicates at high pressures. In *High–Pressure Research in Geophysics*, (eds. S. AKIMOTO and M. H. MANGHNANI), pp. 229–236 Center for Academic Publications.

WEERTMAN J. (1971) Theory of water–filled crevasses in glaciers applied to vertical magma transport beneath oceanic ridges. *J. Geophy. Res.* **76**, 1171–1183.

WILLIAMS H. and McBIRNEY A. R. (1979) *Volcanology*, Freeman-Cooper.

YUEN D. A., FLEITOUT L., SCHUBERT G. and FROIDEVAUX C. (1978) Shear deformation zones along major transform faults and subducting slabs. *Geophys. J. Roy. Astron. Soc.* **54**, 93–119.

Part B.
Upper Mantle Partial Melting and Fractionation

Magmatic Processes: Physicochemical Principles
© The Geochemical Society, Special Publication No. 1, 1987
Editor B. O. Mysen

High pressure phase equilibrium constraints on the origin of mid–ocean ridge basalts*

D. C. PRESNALL and J. D. HOOVER[1]

Dept. of Geosciences, The University of Texas at Dallas, P.O. Box 830688, Richardson, Texas 75083, U.S.A.

Abstract—A search of about 1700 analyses of abyssal basalt glasses has produced a set of 40 very primitive compositions with *mg* numbers greater than 70. None are picritic. On an olivine–plagioclase–quartz normative triangle, the primitive compositions form an array elongated toward the quartz apex that mimics the array of abyssal basalt glasses in general. Fractional crystallization with or without magma mixing would produce residual liquids with lower *mg* numbers and higher concentrations of K_2O, TiO_2, and Na_2O than the parental magma. When these changes are used to test the suitability of various experimentally produced basaltic liquids as parental magmas to the primitive glasses, we find 9 suitable analyses synthesized at 10 kbar and one at 5 kbar. The compositions of these parental basalts are either close to those of the primitive glasses or fall within the field of the primitive MORB glass array. Picritic liquids produced to date at pressures above 15 kbar do not have sufficiently high *mg* numbers and sufficiently low TiO_2, Na_2O, and K_2O contents to be parental, but this may indicate only that the appropriate experimental conditions have not yet been found.

Melting relationships from 1 bar to 20 kbar for simplified lherzolite in the system CaO-MgO-Al_2O_3-SiO_2-Na_2O indicate that at pressures below 9.6 kbar where plagioclase lherzolite is stable, the composition of the liquid changes little with degree of melting up to 19 (at 0 kbar) to 32 (at 9.6 kbar) percent liquid. Above a transition inverval from 9.6 to 10.8 kbar, spinel lherzolite is stable, and the liquid composition changes more rapidly with percent melting from an alkalic composition below about 10 to 25% melting (depending on pressures) to a tholeiitic composition at higher melting percentages. For amounts of melting of simplified lherzolite less than about 5 to 15%, picritic primary magmas cannot be parental to any of the primitive mid–ocean ridge basalt glasses, and only if the maximum amount of melting is greatly in excess of 20 to 30% could picritic magmas be parental to the entire array of primitive MORB glasses. However, these very large amounts of melting exceed estimates based on trace element considerations. The phase relationships support a model involving generation of primary non–picritic magmas throughout the pressure interval 5–11 kbar coupled with polybaric fractional crystallization as the magmas ascend to the earth's surface. The majority of primary magmas appear to be generated at about 9 to 10 kbar where a low–temperature region occurs on the simplified lherzolite solidus.

INTRODUCTION

THE COMPOSITION and depth of origin of primary mid–ocean ridge basalts has been a matter of continuing debate. Some have advocated picritic primary magmas generated at pressures of about 15–30 kbar (O'HARA, 1968; GREEN *et al.,* 1979; STOLPER, 1980; JAQUES and GREEN, 1980; ELTHON and SCARFE, 1984; ELTHON, 1986), whereas others have advocated primary magmas generated at pressures of about 7–11 kbar that have non–picritic compositions similar or identical to the most primitive lavas observed at mid–ocean spreading centers (GREEN and RINGWOOD, 1967; KUSHIRO, 1973; PRESNALL *et al.,* 1979; FUJII and BOUGAULT, 1983; TAKAHASHI and KUSHIRO, 1983; FUJII and SCARFE, 1985; PRESNALL and HOOVER, 1984, 1986).

In this paper, we focus on the generation and crystallization of the most primitive mid–ocean ridge basalt (MORB) glasses because they represent the best candidates for primary magmas that have reached the earth's surface. The paper is divided into two main parts. We first compare the compositions of primitive MORB glasses to liquid compositions produced from existing melting experiments on natural starting materials in order to determine which of these experimentally produced liquids could represent magmas parental to the primitive glasses. Then we present new experimental data on melting relationships of simplified lherzolite in the system CaO-MgO-Al_2O_3-SiO_2-Na_2O (CMASN) from 1 bar to 20 kbar and discuss the bearing of these data on the generation and crystallization of primitive MORB glasses.

PRIMITIVE MORB COMPOSITIONS

As a criterion for a primitive MORB glass, we require that the *mg* number [$mg = 100$ Mg/(Mg + Fe^{2+})] be greater than 70, with the *mg* number

---

* Contribution no. 492, Department of Geosciences, The University of Texas at Dallas.

[1] Current address: Dept. Geological Sciences, The University of Texas at El Paso, El Paso, Texas, 79968.

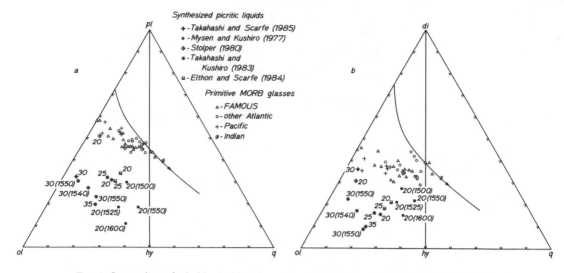

FIG. 1. Comparison of primitive MORB glass compositions and synthesized picritic liquids. Mineral proportions in mol percent are calculated according to the CIPW norm conventions. $pl = Na_2Al_2Si_6O_{16} + CaAl_2Si_2O_8$; $ol = Mg_2SiO_4 + Fe_2SiO_4$; $hy = MgSiO_3 + FeSiO_3$; $di = CaMgSi_2O_6 + CaFeSi_2O_6$; $q = SiO_2$. For all compositions in this and subsequent figures, $Fe^{2+}/(Fe^{2+} + Fe^{3+})$ is assigned the value 0.86 (PRESNALL et al., 1979). Numbers without parentheses indicate pressures in kbar. Numbers enclosed in parentheses are temperatures in °C. The curved line with an arrow indicating the direction of decreasing temperature indicates the trace of liquids in equilibrium with olivine, clinopyroxene, and plagioclase (WALKER et al., 1979).

being calculated on the assumption that $Fe^{2+}/(Fe^{2+} + Fe^{3+}) = 0.86$ (O'DONNELL and PRESNALL, 1980). This criterion may be so restrictive that we exclude some lavas with $mg < 70$ that are, in fact, just as close to primary magmas as the MORB glasses included in our sample but that are the result of a smaller degree of melting or are derived from a slightly less magnesian source. Although these exclusions may result in the loss of some information, we prefer to minimize the possibility of including samples with compositions that may be significantly changed by processes taking place subsequent to magma generation. Any model capable of explaining abyssal basalts in general must be capable of explaining our small subset of primitive MORB glasses.

Presnall and Hoover (1986) searched about 1700 analyses of glassy and aphyric abyssal basalts and

found 42 very primitive compositions (exclusive of glass inclusions in minerals) that have mg numbers greater than 70. We have used this same list of primitive compositions but have retained only the glass analyses. This criterion eliminates two analyses of aphyric basalts reported by FLOWER et al. (1983). For the remaining 40 analyses (see Appendix 1), CIPW norms have been calculated, and the resulting molecular proportions of olivine–plagioclase–quartz and olivine–diopside–quartz plotted in Figure 1.

These diagrams show several interesting features. (1) None of the primitive glass compositions are picritic,[2] and we are not aware of even a single analysis of a picritic MORB glass. Picritic abyssal basalts occur but they are always olivine-phyric. (2) The primitive glass array shows a pronounced elongation toward the quartz apex that mimics the trend of MORB compositions in general (not shown, but see PRESNALL et al., 1979, Figures 12 and 13). Analytical uncertainties enhance the elongation (PRESNALL and HOOVER, 1984) but account for only about one quarter of its length. (3) Because the temperature decreases to the right on the curve along which olivine, plagioclase, and diopside crystallize (Figure 1), less-fractionated basalts have commonly been assumed to lie near the quartz–poor[3] end of the MORB array. This relationship is supported in a general way by analyses of natural

---

[2] The term "picritic" will be used to indicate a basaltic glass composition that departs significantly from the main trend of MORB glasses (see, for example, PRESNALL et al., 1979, Figures 12 and 13) toward the olivine apex on an olivine–plagioclase–quartz or olivine–diopside–quartz normative triangle.

[3] "Quartz–poor" and "quartz–rich" compositions and "normative quartz" will refer to each triangle as a whole as if normative hypersthene were recalculated to equivalent olivine and quartz.

samples (for example, see PRESNALL and HOOVER, 1984; Figure 5). However, even though fractionation drives residual liquids toward the quartz apex, Figure 5 of PRESNALL and HOOVER (1984) and Figure 1 show that primitive MORB glasses are not confined to or even concentrated in the quartz–poor portion of the MORB array. Although the number of primitive glass samples is limited, there is a suggestion of a concentration of compositions only slightly to the quartz–poor side of the *pl-hy* (Figure 1a) and *di-hy* (Figure 1b) joins, a position close to the location of a similar centrally located concentration of MORB compositions in general (WYLLIE *et al.*, 1981, Figure 3.3.34). (4) The distribution of primitive glass compositions does not follow the 1-bar boundary line along which olivine, plagioclase, and diopside crystallize, but the quartz–rich end of the distribution appears to be limited by this boundary line. This relationship is similar to that observed for MORB compositions in general (O'DONNELL and PRESNALL, 1980; GROVE and BRYAN, 1983; ELTHON and SCARFE, 1984).

EXPERIMENTAL DATA ON NATURAL COMPOSITIONS

One way of helping to clarify the origin of the primitive MORB glasses plotted in Figure 1 is to test the suitability of various experimentally produced basalts as parental liquids. We test suitability in two ways. The first method, which has been used

extensively by others (for example, STOLPER, 1980; ELTHON and SCARFE, 1984), is to note the relative positions of presumed parental and derived liquids on the triangular normative diagrams *ol-pl-q* and *ol-di-q* and determine visually if liquids thought to be derived are capable of being produced by crystal fractionation from liquids thought to be parental. For example, in the case of low–pressure crystallization of olivine from picritic liquids generated at high pressures, the test is accomplished simply by determining if the experimentally-produced picritic compositions lie along joins between primitive basalt compositions and the olivine apex. The second method involves a comparison of *mg* numbers and K_2O, TiO_2, and Na_2O contents of the presumed parental and derived liquids. Fractional crystallization would decrease the *mg* number and increase the K_2O, TiO_2, and Na_2O percentages in a derived liquid relative to these values in a parental liquid. The same changes would be produced by magma mixing between a parental liquid and another magma fractionated from an earlier similar parental liquid.

Application of the first method suggests that crystallization of olivine from experimentally produced picritic parental liquids (Figure 1) is capable of producing derivative liquids in the primitive MORB glass field, as pointed out by O'HARA (1968), GREEN *et al.* (1979), JAQUES and GREEN (1980), STOLPER (1980), ELTHON and SCARFE

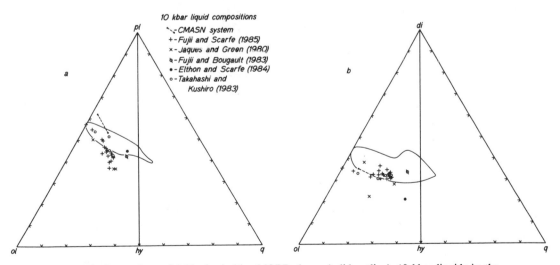

FIG. 2. Comparison of field of primitive MORB glasses (solid outline), 10 kbar liquids in the CMASN system (dashed line, taken from Figure 7), and 10 kbar liquids in equilibrium with olivine, orthopyroxene, and clinopyroxene in experiments on natural compositions. Positions of points are calculated in the same way as explained in the caption for Figure 1. In triangle *b*, the glass of JAQUES and GREEN (1980) that meets our criteria for a parental magma is the point at 28 mol percent *di* and 25 mol percent *q* (see text).

(1984), and ELTHON (1985). On the other hand, Figure 2 shows that experimentally produced melts in equilibrium with olivine, orthopyroxene, and clinopyroxene at about 10 kbar lie near the olivine–rich border of the primitive MORB glass array. These 10 kbar liquids would be capable of producing compositions in most of the primitive MORB glass field by at most only a small amount of olivine crystallization (GREEN and RINGWOOD, 1967; KUSHIRO, 1973; PRESNALL *et al.,* 1979; FUJII and BOUGAULT, 1983; TAKAHASHI and KUSHIRO, 1983; PRESNALL and HOOVER, 1984, 1986; FUJII and SCARFE, 1985). Thus, application of the first method does not lead to a resolution of the controversy over the composition and depth of origin of primary magmas at spreading ridges; both scenarios are equally viable.

Now consider the second method. Small variations in TiO_2, Na_2O, K_2O, and mg number are generally of little consequence when the results of melting experiments on natural compositions are used for the usual purpose of showing approximate phase relationships. Typically, the phase relationships show a rough consistency and are useful even when one or more of these components do not correspond exactly to values found in rocks. In spite of this, if experimental data could be found that not only show appropriate phase relationships but also show consistent mg numbers and minor element concentrations, a higher level of confidence could be placed in the results. With this goal in mind, we have examined the available experimental melting data on natural compositions to see if any of the data meet these higher standards.

PRESNALL and HOOVER (1986) pointed out that $mg = 67.7$ for the picritic liquid produced experimentally at 20 kbar by STOLPER (1980), is a value lower than that of any of the 40 primitive glasses. Also, at least one of the oxides, K_2O, TiO_2, and Na_2O, has a smaller percentage in 39 of these glasses than in STOLPER's (1980) 20 kbar picritic liquid. Thus, Stolper's picritic liquid is not a suitable parental magma for these glasses. PRESNALL and HOOVER (1986) also pointed out that for the 20 and 25 kbar picritic liquids of ELTHON and SCARFE (1984), the amount of at least one of the oxides, K_2O, TiO_2, and Na_2O, is higher than that of most of the primitive glasses. Thus, the picritic liquids of ELTHON and SCARFE (1984) are also unsuitable parental liquids for most of the primitive glasses.

We have carried out a more general review of experimental data on natural compositions to determine if picritic liquids produced experimentally by other investigators could be parental to the primitive MORB glasses even though these investigators may not have suggested that their picritic liquids are parental. Figure 3 shows that the picritic liquids of MYSEN and KUSHIRO (1977) are eliminated as parental magmas on the basis of their K_2O contents. Figure 4 shows that on the basis of their TiO_2 contents, only a small proportion of the primitive glasses could be derived from picritic liquids reported by other investigators. Thus, if both the major and minor element concentrations of the liquids are considered, not just the positions of the liquids on triangular normative diagrams, none of the picritic liquids produced to date in laboratory experiments is a suitable parent for most of the primitive glasses.

The same criteria can be used to test possible parental liquids observed in laboratory experiments on natural compositions at lower pressures. Those advocating the generation of primary magmas in the vicinity of 10 kbar (for example, PRESNALL *et al.,* 1979) have argued that these primary magmas occasionally appear at the earth's surface either unmodified or only slightly changed in composition by such processes as fractional crystallization and magma mixing. Thus, we consider an experimentally produced liquid to be a suitable model of a

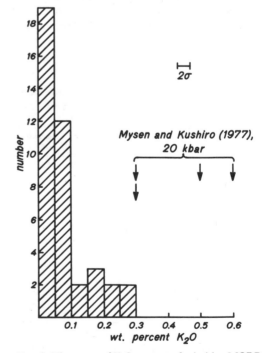

FIG. 3. Histogram of K_2O content of primitive MORB glasses (see Appendix 1, values normalized to a total analysis of 100 weight percent) and four picritic liquids produced experimentally by MYSEN and KUSHIRO (1977) at 20 kbar (arrows).

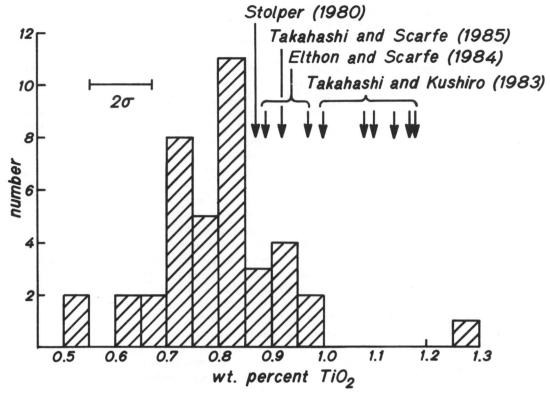

FIG. 4. Histogram of TiO$_2$ content of primitive MORB glasses (see Appendix 1, lower case values normalized to a total analysis of 100 weight percent) and picritic liquids produced experimentally at pressures from 20 to 35 kbar (arrows).

primary magma parental to the primitive glasses if its K$_2$O, TiO$_2$, and Na$_2$O contents are either within or below the most commonly observed range of values for these oxides in the primitive MORB glasses. On the basis of Figures 3, 4, and 5, we require the K$_2$O, TiO$_2$, and Na$_2$O contents to be less

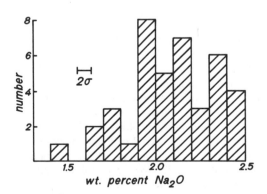

FIG. 5. Histogram of Na$_2$O content of primitive MORB glasses (see Appendix 1, values normalized to a total analysis of 100 weight percent).

than 0.15 weight percent, 0.9 weight percent, and 2.4 weight percent, respectively. Also, we require the *mg* number to be greater than 70. Any liquid is considered that is in equilibrium with olivine, orthopyroxene, and clinopyroxene at pressures between 5 and 15 kbar. FUJII and KUSHIRO (1977) listed such a liquid composition with an *mg* number of only 64.4. Therefore, this liquid could not be a parental magma, a conclusion reinforced by its high TiO$_2$ (1.64 weight percent) and Na$_2$O (2.70 weight percent) contents. TAKAHASHI and KUSHIRO (1983) listed 15 liquid compositions produced at pressures ranging from 5 to 15 kbar. All are eliminated as parental magmas by various combinations of high K$_2$O, TiO$_2$, and Na$_2$O contents, and all except one are eliminated by their low *mg* numbers (<68.8). ELTHON and SCARFE (1984) listed two liquid compositions, one at 10 kbar and another at 15 kbar. The 10–kbar liquid composition is eliminated as a parental magma by its high TiO$_2$ content (1.4 weight percent) and low *mg* number (64.1). The liquid produced at 15 kbar is eliminated by its high TiO$_2$ content (1.18 weight percent).

Three studies gave liquid compositions that meet all the criteria for a parental basalt. FUJII and BOUGAULT (1983) gave one such analysis of a glass at 10 kbar (their Table 1, column 1). JAQUES and GREEN (1980) listed three liquid compositions produced by melting pyrolite, one each at 5, 10, and 15 kbar. The TiO_2 (3.0–3.2 weight percent) and Na_2O (2.8–3.1 weight percent) contents of all three are too high for a parental magma. However, they also listed four liquids produced by melting Tinaquillo lherzolite, one each at 5 and 15 kbar and two at 10 kbar. The K_2O contents of the 15–kbar liquid and one of the 10–kbar liquids are too high (0.2 weight percent), but the remaining two liquids at 5 and 10 kbar could be parental (their Table 5, columns 7 and 11). FUJII and SCARFE (1985) listed 18 liquid compositions synthesized at 10 kbar. Five are eliminated as parental liquids by various combinations of high K_2O, TiO_2, and Na_2O percentages, and the remaining thirteen meet our criteria for a parental magma for abyssal basalts. However, seven of these thirteen analyses have extremely high mg numbers of 78 to 82.1, with calculated coexisting olivine compositions of $Fo_{92.2}$ to $Fo_{94.1}$ (ROEDER and EMSLIE, 1970). If DICK and FISHER (1984) are correct that olivines from abyssal peridotites, which lie in the composition range Fo_{89}–$Fo_{91.6}$ (HAMLYN and BONATTI, 1980; DICK and FISHER, 1984), are residual compositions remaining from partial fusion of the mantle at spreading centers, the very high mg numbers of the seven glasses of FUJII and SCARFE (1985) are indeed unrealistic. Six of the glass compositions remain as suitable parental magmas produced at 10 kbar. The analyses are listed in columns 4 and 7 of Table 5 and columns 2, 7, 8, and 9 of Table 6 in FUJII and SCARFE (1985).

On the basis of the above strict criteria for a parental magma, only a few experiments on natural compositions have yielded suitable parental magmas at any pressure. These few experiments indicate that magmas generated at pressures from about 5 to 10 kbar are suitable parents for observed primitive MORB glasses. All of the 10–kbar liquids are plotted (among others) on Figure 2 and lie either close to or within the field of primitive MORB glasses. The 5–kbar liquid (not shown) lies slightly toward the olivine apex from the silica–rich end of the primitive MORB field. The lack of similar experimental support for picritic parental magmas generated at higher pressures does not necessarily indicate that such magmas cannot be primary; it may simply indicate that the proper experimental conditions have not yet been found. For example, suitable picritic compositions might occur if experiments were carried out at very high melting

percentages or on lherzolite starting materials low in TiO_2, Na_2O, and K_2O. However, independent reasons would be needed for using starting compositions with reduced TiO_2, Na_2O, and K_2O to avoid the criticism that the conclusions were defined by the initial conditions of the experiment.

Our emphasis on the importance of the mg number and the minor components TiO_2, Na_2O, and K_2O might seem to be inconsistent with our use in the second part of the paper of phase relationships in the CMASN system, in which FeO, TiO_2, and K_2O are not even present. No inconsistency exists, however, because the type of information obtained from the CMASN system (partial melt compositions as a function of pressure and percent melting) would be affected little by the inclusion of small amounts of additional components. For experiments on natural compositions, we use the mg number and the concentrations of TiO_2, Na_2O, and K_2O to emphasize the fact that when the phase relationships (controlled mainly by the major elements), mg number, and certain minor elements are all considered, available experimental data on natural compositions are consistent only with the generation of non–picritic primary magmas at about 10 kbar. Scenarios involving picritic primary magmas must rely, at least at present, on supporting experimental data of a lower order of confidence in which the phase relationships are consistent but the mg number and certain minor element concentrations are not.

MELTING RELATIONSHIPS OF SIMPLIFIED LHERZOLITE IN THE SYSTEM CaO-MgO-Al_2O_3-SiO_2-Na_2O

The solidus curve and liquid compositions along the solidus for simplified lherzolite in the 4-component system CaO-MgO-Al_2O_3-SiO_2 (CMAS) have been determined by PRESNALL et al. (1979) over the pressure range of 1 bar to 20 kbar. We have extended this study by determining changes in the phase relationships for liquids containing up to 4 weight percent Na_2O. The data in the resulting 5-component system (CMASN) cannot be represented visually as a conventional phase diagram. However, the methods of PRESNALL (1986) can be used to calculate algebraically the solidus temperature, liquid path, crystal path, and phase proportions during partial fusion at each pressure for any arbitrarily chosen lherzolite composition within the 5-component system.

We present here preliminary results for equilibrium fusion of two model lherzolite compositions. The complete data set on which these calculated

Table 1. Mantle compositions

	Model lherzolite A	Model lherzolite B	Undepleted mantle (Carter, 1970)	Pyrolite 3 (Ringwood, 1975)
SiO_2	46.7	46.2	42.9	46.1
TiO_2	—	—	0.3	0.2
Al_2O_3	8.1	6.3	7.0	4.3
Fe_2O_3	—	—	0.4	—
FeO	—	—	9.0	8.2
MgO	39.3	43.1	35.1	37.6
CaO	5.3	4.1	4.4	3.1
Na_2O	0.54	0.36	0.45	0.4

results are based will be presented elsewhere. Model lherzolite composition A (Table 1) was calculated by mixing 70% of an assumed residue from partial fusion with 30% of our liquid composition in equilibrium with olivine, orthopyroxene, clinopyroxene, spinel, and plagioclase at 9.6 kbar in the CMASN system. Mineral proportions used for the residue are 74% olivine, 21% orthopyroxene, 4% clinopyroxene, and 0.5% spinel. These proportions are the same, within the stated uncertainties, as the average proportions of phases in residual periodotites from the ocean basins given by DICK and FISHER (1984). Mixing of 70% of this residue with 30% of the 9.6 kbar liquid composition yields a source mantle composition in which plagioclase, clinopyroxene, and spinel disappear nearly simultaneously from the crystalline assemblage during equilibrium partial melting at 9.6 kbar. This feature of the melting behavior is consistent with the observation of CARTER (1970) that clinopyroxene and spinel disappear from residual lherzolite xenoliths from Kilbourne's Hole, New Mexico, at approximately the same forsterite content of coexisting olivine. The oxide percentages of the model lherzolite A (Table 1) are similar to, but in all cases slightly higher than those of the corresponding oxides in the mantle compositions of CARTER (1970) and RINGWOOD (1975), a feature consistent with the absence of iron from the model composition (see Table 1).

Figures 6 and 7 show melting relationships for simplified lherzolite composition A. The subsolidus phase assemblage at pressures less than 9.6 kbar is ol + en + di + pl (plagioclase lherzolite) and at pressures above 10.8 kbar is ol + en + di + sp (spinel lherzolite). Between 9.6 and 10.8 kbar, both plagioclase and spinel coexist with olivine, enstatite, and diopside just below the solidus. At pressures less than 9.6 kbar, the first crystalline phase to disappear during equilibrium melting is diopside, and it does so at liquid percentages ranging from 19 at 1 bar to 32 at 9.6 kbar. Exactly at 9.6 kbar, all five crystalline phases (olivine, orthopyroxene, clino-

pyroxene, plagioclase, spinel) are retained until the amount of liquid reaches 32%, at which point plagioclase, diopside, and spinel disappear essentially simultaneously. Between 9.6 and 10.8 kbar, where the subsolidus assemblage ol + en + di + pl + sp exists, plagioclase is the first crystalline phase to be totally dissolved, and spinel follows at about 32% melting. From 10.8 to 20 kbar, spinel is the first crystalline phase to be totally dissolved, which occurs at melt percentages of about 32 at 10.8 kbar to about 30 at 20 kbar.

When the amount of liquid mixed with the assumed residue is changed from 30% to 20%, the composition of the model lherzolite composition changes to that given in column 2 of Table 1 (composition B). In this composition, the concentrations of SiO_2, Al_2O_3, CaO, and Na_2O are closer to the compositions of CARTER (1970) and RINGWOOD (1975). If it is considered appropriate to combine molar MgO and FeO in the real mantle composition and treat this sum as comparable to molar MgO in the model mantle, composition B might then be a preferred model mantle composition. The melting

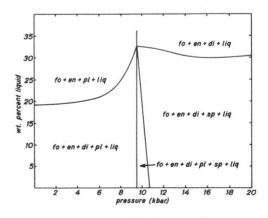

FIG. 6. Phase assemblages as a function of pressure and percent melting for simplified lherzolite in the CMASN system.

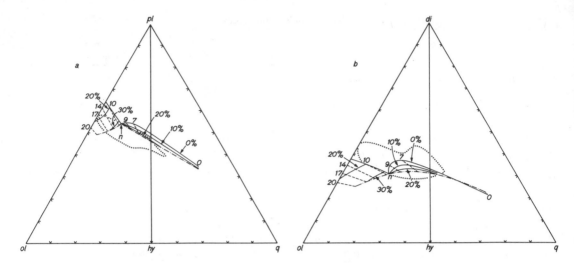

FIG. 7. Compositions of liquids as a function of pressure and weight percent melting for simplified lherzolite in the CMASN system. Solid lines are contours of constant liquid proportion. Dashed lines are isobars labeled according to their pressure in kbar. The dash–dot line shows the locus of points at which diopside disappears on melting at pressures below 9.6 kbar and the locus of points at which spinel disappears at pressures above 9.6 kbar. The dotted enclosure shows the field of primitive MORB glasses taken from Figure 1.

relationships for this composition would be similar to those shown in Figure 7 except that the 20% melting contour would be shifted approximately to the position of the 30% melting contour in Figure 7.

We have also calculated a model lherzolite on the assumption of 10% liquid mixed with the assumed residue. With this assumption, the Na_2O content of the model lherzolite is 0.19 weight percent, a value only about half that of the mantle compositions given by CARTER (1970) and RINGWOOD (1975). Thus, we consider this model composition to be unlikely.

EFFECT OF OTHER COMPONENTS ON MELTING RELATIONSHIPS IN THE SYSTEM CaO-MgO-Al₂O₃-SiO₂-Na₂O

Before applying phase relationships in the CMASN system to the melting behavior of the mantle, we consider briefly the probable effect of additional components on the liquid compositions. Inclusion of iron oxide, the most important component missing from the CMASN system, would reduce liquidus temperatures for the olivine and pyroxene fields but would leave the plagioclase field relatively unaffected. Thus, in the *ol-pl-q* triangle (Figure 7a), boundary lines along which liquid compositions move during crystallization would be expected to move slightly away from the plagioclase apex, but in the *ol-di-q* triangle (Figure 7b), little change would occur.

Figure 2 shows a comparison between liquid compositions produced at 10 kbar in the CMASN system and those observed in melting experiments on natural compositions. All liquids are in equilibrium with olivine, orthopyroxene, and clinopyroxene and the CMASN liquids are also in equilibrium with spinel. As expected, liquid compositions in the melting experiments on natural compositions are shifted away from the plagioclase apex in Figure 2a relative to the CMASN liquids, but the two sets of data are essentially coincident in Figure 2b. This comparison supports the contention that missing components other than iron oxide have only a small effect on liquid compositions and that the CMASN phase relationships can be meaningfully applied to melting processes in the mantle if allowance is made for the slight shift in liquid compositions that would be produced by the addition of iron oxide. That is, liquid compositions in Figure 7a would shift slightly away from the *pl* apex but those in Figure 7b would remain approximately as shown.

VARIABLE VERSUS CONSTANT PRIMARY MAGMA COMPOSITIONS FROM THE MANTLE

YODER and TILLEY (1962) suggested that the mantle melts in a manner analogous to the formation of liquid at an isobaric invariant point. They pointed out that this type of melting behavior provides a mechanism for producing a uniform magma composition at a given pressure from a heterogeneous peridotite mantle in which the proportions of phases vary. Subsequently, PRESNALL (1969, 1979) showed that if melting occurs at an invariant point, not only can compositional heterogeneity of the source rock be tolerated without changing the initial melt composition, but a large amount of melt of constant composition can be produced. Of course, the mantle cannot melt at a true invariant point because the number of phases is small and the number of components is large. However, situations commonly exist with higher variance in which the melting behavior approximates that at an invariant point. That is, a relatively constant liquid composition can be derived from a somewhat heterogeneous source even when the amount of melting varies widely. In other situations with higher variance, changes in liquid composition with percent melting and with variations in the bulk composition of the source can be very large (PRESNALL, 1979). The problem reduces to a determination of the specific situation relevant to the mantle at various pressures. Several studies have addressed this issue for both natural compositions and simplified model systems but have not reached agreement (MYSEN and KUSHIRO, 1977; PRESNALL *et al.*, 1978, 1979; JAQUES and GREEN, 1980; STOLPER, 1980; TAKAHASHI and KUSHIRO, 1983; ELTHON and SCARFE, 1984; FUJII and SCARFE, 1985; KLEIN and LANGMUIR, 1986).

Melting relationships of model lherzolite A in the CMA3N system (Figure 7) show that for pressures less than 9.6 kbar, the composition of the melt changes very little with percent melting up to the point of disappearance of the first crystalline phase, diopside. As pressure increases in the range 8 to 9.6 kbar, the magnitude of the change in composition for a given amount of fusion decreases and reaches zero at 9.6 kbar (point *n*). At this pressure, the melt composition remains fixed at *n* until the amount of liquid exceeds 32% (Figures 6 and 7). As pressure increases from 9.6 to 10.8 kbar (the pressure range in which plagioclase and spinel coexist in the simplified lherzolite just below the solidus), the melt composition remains fixed during isobaric melting until plagioclase disappears at melt percentages ranging from 32 at 9.6 kbar to zero at 10.8 kbar (Figure 6). Just as plagioclase completely dissolves, the liquid composition, shown in Figure 7, starts to move along the appropriate dashed isobaric line

(only the 10–kbar isobaric line is shown in this pressure interval), and it continues along this line until spinel is completely dissolved just as the liquid path reaches the dash–dot line at about 32% liquid. At pressures above 10.8 kbar, initial melt compositions are nepheline–normative (not shown in Figure 7), and as the percent liquid increases during isobaric equilibrium fusion, the liquid composition moves into the hypersthene–normative field (see GREEN and RINGWOOD, 1967, and TAKAHASHI and KUSHIRO, 1983, for similar results on natural compositions). A large amount of melting occurs prior to the disappearance of the first crystalline phase (spinel), but in this pressure range the change of liquid composition for a given amount of melting is much greater than it is at lower pressures.

These data indicate that generalizations applicable to all pressures cannot be made about the manner in which the mantle melts. At low pressures, especially near 9 to 10 kbar, the melting behavior approximates that at an invariant point; liquid is produced with a composition that changes only a small amount as a function of percent melting. At higher pressures, the liquid composition changes more rapidly as a function of percent melting. The transition between these two types of melting behavior occurs abruptly in a very narrow pressure interval between 9.6 and 10.8 kbar.

CONSTRAINTS ON PICRITIC PRIMARY MAGMAS

For model lherzolite A, the percent melting contours in Figure 7 show that if primary mid–ocean ridge basalts are produced by less than about 15% melting, picritic liquids with compositions lying toward olivine from the primitive MORB array do not exist at any pressure up to the 20–kbar limit of our data. Extrapolation of our data to higher pressures is straightforward up to about 22 kbar, the pressure of first appearance of garnet lherzolite along the solidus in the CMAS system (GASPARIK, 1984), and it is clear from Figure 7 that for this small additional extension of the pressure range, the constraint against picritic magmas at less than about 15% fusion of lherzolite A continues to hold. Compositions of liquids at higher pressures are more uncertain, but an indication of the changes to be expected is given by Figures 3.3.15 and 3.3.16 of WYLLIE *et al.* (1981). These diagrams show liquid compositions in equilibrium with simplified lherzolite in the CMAS system including a liquid at 40 kbar (DAVIS and SCHAIRER, 1965) that shows only a slight shift toward higher normative quartz[4] in comparison to the 22–kbar liquid.[5] This observa-

---

[4] See footnote 3.
[5] The estimated transition from spinel lherzolite to garnet lherzolite at the solidus was given in Figures 3.3.15 and 3.3.16 of WYLLIE *et al.* (1981) as 24 kbar rather than 22 kbar, as used here.

tion, coupled with the fact that the liquid at 15% fusion in the CMASN system lies deep within the alkalic (nepheline–normative) basalt field at 22 kbar, indicates that hypersthene–normative liquids at less than 15% fusion are unlikely to be produced from model lherzolite A at any pressure up to at least 40 kbar. If model lherzolite B is preferred, this conclusion would apply for amounts of fusion less than about 8%. Thus, although experimental confirmation of the phase relationships at high pressures is obviously important, present indications are that if primary magmas at spreading ridges are produced by less than about 8 to 15% fusion of the mantle source, these primary magmas are not picritic.

At 20 to 30% fusion of composition A (10 to 20% fusion for composition B), Figure 7 indicates that picritic liquids are possible if the pressure is higher than about 15 kbar. However, the composition range of these liquids is constrained to the quartz–poor portion of each triangle. Thus, low–pressure crystallization of olivine from picritic primary magmas produced by 10–30% fusion would be capable of yielding liquids in the quartz–poor end of the primitive MORB array but not in the quartz–rich end. This difficulty in producing primitive magma compositions in the quartz–rich end of the array would not be so constraining if an alternative mechanism could explain them. However, other mechanisms are not evident. As discussed below in the section on crystallization of primitive MORB glasses, primitive liquids near the quartz–rich end of the array could not be derived by fractional crystallization from primitive liquids near the quartz–poor end. Also, magma mixing with quartz–poor primitive liquids would be equally ineffective because the quartz–rich liquids would be one of the end members to be mixed.

The constraint against picritic primary magmas becomes less severe as the percent fusion increases, and olivine crystallization from picritic liquids might be capable of yielding the complete range of observed primitive liquids if the amount of fusion is allowed to range up to very high values considerably in excess of 30% for composition A or 20% for composition B. However, our data do not extend to these very high fusion percentages.

Because of the need for large amounts of fusion to produce picritic parental magma compositions, the amount of fusion responsible for the formation of mid–ocean ridge basalts is a critical parameter in determining the viability of picritic parental magmas. BENDER et al. (1978) estimated that basalts in the FAMOUS region were produced by about 20% melting of the mantle. For basalts in the Tamayo region, BENDER et al. (1984) concluded that

about 5% melting was required. In a more recent evaluation of ridge basalts worldwide, KLEIN and LANGMUIR (1986) concluded from both major– and trace–element considerations that the average amount of fusion varies between about 8 and 22%. If these calculations of the amount of partial melting are correct, magma compositions only in the quartz–poor end of the primitive MORB array could be explained as derivatives from picritic parental magmas, and the origin of primitive mid–ocean ridge basalts by fractionation from picritic parental magmas fails as a general process.

GENERATION OF PRIMITIVE MORB GLASSES

When allowance is made for the effect of iron oxide on liquid compositions (Figure 2), it can be seen in Figure 7 that liquids produced in the pressure range of about 5 to 11 kbar would have compositions that lie near the olivine–rich margin of the primitive MORB glass array. As pressure decreases, the olivine field enlarges (KUSHIRO, 1968; PRESNALL et al., 1978; O'DONNELL and PRESNALL, 1980, Figure 10), so primary magmas produced along the olivine–rich margin of the primitive MORB array would lie within the olivine field as they passed upward to the earth's surface. Thus, the phase relationships are consistent with a simple model for the origin of primitive MORB glasses that involves generation of non–picritic primary magmas over a range of pressures from about 5 to 11 kbar followed by a small amount of olivine crystallization at lower pressures. Alternatively, the primitive MORB liquids could be direct partial melts essentially unmodified by fractional crystallization. We feel that existing experimental data are not precise enough to distinguish between these two possibilities.

Previously, we argued that primary magmas at spreading ridges are generated at pressures from about 7 to 11 kbar (PRESNALL and HOOVER, 1984). We now believe the lower limit of this range must be decreased to about 5 kbar to account for the most silica–rich compositions in the primitive MORB array (Figure 7). The dashed isobars in Figure 7 show that variations in the amount of melting could contribute to the elongation of the primitive MORB array toward the quartz apex, but the elongation is much too large to be due mainly to this effect.

In the CMAS system, PRESNALL et al. (1979) emphasized the importance of a cusp on the mantle solidus at 9 kbar and argued that this cusp controlled the depth of generation and composition of primary mid–ocean ridge basalts. PRESNALL *et al.* (1979, p. 26) suggested that for mantle peridotite, the cusp would appear as a low–temperature region spread over a small pressure interval. For the model mantle composition used here in the CMASN system, the cusp appears as a low–temperature region on the solidus that extends over the pressure interval 9.6 to 10.8 kbar. Comparison of Figures 1 and 7 shows that the greatest concentration of primitive MORB glass compositions lies close to liquid compositions in the CMASN system at about 9 to 10 kbar. Also, pressures at which the largest amounts of liquid are produced with the smallest changes in liquid composition with percent melting cluster around 9.6 kbar. Thus, our data in the CMASN system substantiate the main features of the model proposed earlier based on the CMAS system (PRESNALL *et al.*, 1979) and provide a more refined estimate of the range of pressures over which primary MORB magmas are generated.

CRYSTALLIZATION OF PRIMITIVE MORB GLASSES

The strong divergence of the quartz–poor end of the primitive MORB array from the 1-bar boundary along which olivine, plagioclase, and clinopyroxene crystallize (Figure 1) shows that the primitive MORB glasses have not resided in a crustal magma chamber long enough for their compositions to be controlled by crystal fractionation in that chamber. The compositional trend of the primitive MORB array conforms instead to the trend expected for primary magmas generated over the pressure interval of about 5 to 11 kbar with at most only slight modification by subsequent crystallization of olivine (Figure 7).

O'DONNELL and PRESNALL (1980) pointed out that a similar trend observed for more fractionated MORB glasses is not consistent with fractional crystallization of olivine, plagioclase, and clinopyroxene at any constant pressure but could be explained by *polybaric* crystallization of these phases as the magmas rise to the earth's surface. Therefore, it is important to examine the possibility that the quartz–poor end of the primitive MORB array may represent magmas that are parental to compositions in the quartz–rich end. Because all the primitive MORB glasses have very high *mg* numbers, such an explanation would require the decrease of *mg* number with increasing fractionation toward the quartz–rich end of the primitive MORB array to be extremely small. The experimental data of

WALKER *et al.* (1979), shown in Figure 8, demonstrate that for isobaric crystallization of olivine, plagioclase, and clinopyroxene, the change of *mg* number with percent normative quartz is, in fact, very large. Note also that the decreasing *mg* number with increasing normative quartz[6] is consistent with the decreasing forsterite content of coexisting olivine in their experiments. If the change of *mg* number for polybaric crystallization of these same phases is assumed to be similar, a magma at the quartz–rich end of the primitive MORB array with an *mg* number of 71 would require a parental magma at the quartz–poor end with an *mg* number of about 91, an obviously absurd result. We conclude that crystallization processes cannot explain the extension of the *primitive* MORB array toward the quartz apex. Because the phase relationships suggest that the compositions of primary magmas lie near the olivine–rich margin of the primitive MORB array, the primitive MORB glasses may in some cases be slightly modified by olivine crystallization. However, crystallization of olivine alone would move residual liquid compositions away from the olivine apex at a high angle to the long direction of the primitive MORB glass array. Thus, the possible existence of a small amount of olivine crystallization is consistent with our conclusion that the elongation of the primitive MORB array toward quartz is not due to crystallization processes.

Many authors have presented evidence for mul-

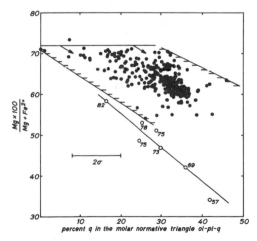

FIG. 8. Percent quartz in the molar normative triangle olivine–plagioclase–quartz as a function of *mg* numbers for FAMOUS and AMAR basalt glasses (filled circles, BRYAN and MOORE, 1977; BRYAN, 1983; STAKES *et al.*, 1984) and for basalt liquids in equilibrium with olivine, plagioclase, and diopside at 1 bar (open circles, WALKER *et al.*, 1979). Numbers beside the points of WALKER *et al.* are the forsterite contents of analyzed coexisting olivines.

---

[6] See footnote 3.

tiple parental magmas in the adjoining FAMOUS and AMAR areas of the Mid–Atlantic Ridge (BOUGAULT and HEKINIAN, 1974; BRYAN and THOMPSON, 1977; LANGMUIR *et al.*, 1977, BRYAN, 1979, 1983; LEROEX *et al.*, 1981; GROVE and BRYAN, 1983; STAKES *et al.*, 1984; PRESNALL and HOOVER, 1984). In Figure 8, glass compositions from the FAMOUS–AMAR area plot as a very wide band but with an overall slope close to that determined experimentally by WALKER *et al.* (1979). The correspondence of the two slopes indicates that crystallization processes were important in producing the less primitive FAMOUS–AMAR basalts. However, because of the great width of the FAMOUS–AMAR trend in Figure 8 and the existence of quartz–rich, very primitive glass compositions, crystallization from a single parental magma cannot explain the entire range of glass compositions. The broad trend is consistent, however, with the existence of multiple primary magmas of widely varying normative quartz content that crystallized olivine, clinopyroxene, and plagioclase under polybaric conditions as the magmas rose to the earth's surface. The arrows descending from the horizontal line in Figure 8 indicate expected crystallization trends from a range of primary magma compositions with *mg* numbers of about 70–73.

CONCLUSIONS

Future detailed data from a wider range of ridge segments may reveal differences that obviate a uniform petrogenetic model applicable to all ridges, but data available at present show sufficient similarities that we believe the following generalizations are justified.

(1) Most mid–ocean ridge basalts are not derived from picritic primary magmas and none require a picritic parent.

(2) Primitive mid–ocean ridge basalts are best explained as products of at most only a small amount of olivine crystallization from multiple primary magmas. These primary magmas are not picritic and have a wide range of normative quartz content.[7]

(3) Phase relationships support the generation of primary magmas beneath spreading ridges over a pressure range of about 5 to 11 kbar, with the majority being generated at a low–temperature region on the solidus at about 9 to 10 kbar.

(4) Generation of multiple primary magmas over the pressure range of 5 to 11 kbar and polybaric

fractional crystallization both cause the normative compositions of MORB magmas to deviate in a similar way from the 1–bar boundary line along which olivine, plagioclase, and clinopyroxene crystallize.

(5) In agreement with many previous studies, we concur that data from the FAMOUS–AMAR area show that multiple parental magmas are capable of being produced in the same geographic area.

(6) The compositions of mid–ocean ridge basalts are best explained as the combined result of generation over a range of pressures followed by polybaric fractional crystallization as the primary magmas pass upward to the earth's surface. The fact that virtually the entire polybaric magma generation and crystallization process is recorded in the compositions of the lavas implies the existence of an eruptive mechanism capable of interrupting and sampling magmas at all depths and stages of evolution and bringing them rapidly to the earth's surface in a relatively unmodified condition. In such a model, magma mixing (DUNGAN *et al.*, 1978; DUNGAN and RHODES, 1978; RHODES *et al.*, 1979; WALKER *et al.*, 1979; STAKES *et al.*, 1984) would also be expected to occur during the sampling process. However, magma mixing, acting alone, could not explain the elongation of the primitive MORB array toward the quartz apex. Our model is at variance with certain aspects of some models (for example, PALLISTER and HOPSON, 1981; STAKES *et al.*, 1984) that emphasize an extended residence time and consequent fractional crystallization in a crustal magma chamber. We do not argue against the existence of a crustal magma chamber, but we believe that fractional crystallization in such a magma chamber is a subordinate process in the chemical evolution of magmas erupted at the earth's surface.

Acknowledgements—We thank K. Krafft for a valuable editorial review. Also, several improvements in the manuscript were made as a result of reviews by B. Mysen and R. Helz. This work was supported by National Science Foundation Grants EAR-8018359, EAR-8212889, and EAR-8418685.

REFERENCES

BENDER J. F., HODGES F. N. and BENCE A. E. (1978) Petrogenesis of basalts from the project FAMOUS area: Experimental study from 0 to 15 kbars. *Earth Planet. Sci. Lett.* **41**, 277–302.

BENDER J. F., LANGMUIR C. H. and HANSON G. N. (1984) Petrogenesis of basalt glasses from the Tamayo region, East Pacific Rise. *J. Petrol.* **25**, 213–254.

BOUGAULT H. and HEKINIAN R. (1974) Rift valley in the Atlantic Ocean near 36°50′N, Petrology and geochemistry of basaltic rocks. *Earth Planet. Sci. Lett.* **24**, 249–261.

---

[7] See footnote 3.

BRYAN W. B. (1979) Regional variation and petrogenesis of basalt glasses from the FAMOUS area, Mid–Atlantic Ridge. *J. Petrol.* **20**, 293–325.

BRYAN W. B. (1983) Systematics of modal phenocryst assemblages in submarine basalts: Petrologic implications. *Contrib. Mineral. Petrol.* **83**, 62–74.

BRYAN W. B. and MOORE J. G. (1977) Compositional variations of young basalts in the Mid–Atlantic Ridge rift valley near lat. 36°49′N. *Bull. Geol. Soc. Amer.* **88**, 556–570.

BRYAN W. B. and THOMPSON G. (1977) Basalts from DSDP leg 37 and the FAMOUS area: Compositional and petrogenetic comparisons. *Can. J. Earth Sci.* **14**, 875–885.

CARTER J. L. (1970) Mineralogy and chemistry of the earth's upper mantle based on the partial fusion–partial crystallization model. *Bull. Geol. Soc. Amer.* **81**, 2021–2034.

DAVIS B. T. C. and SCHAIRER J. F. (1965) Melting relations in the join diopside–forsterite–pyrope at 40 kilobars and at one atmosphere. *Carnegie Inst. Wash. Yearb.* **64**, 123–126.

DICK H. J. B. and FISHER R. L. (1984) Mineralogic studies of the residues of mantle melting: Abyssal and alpine-type peridotites. In *Kimberlites II: The Mantle and Crust–Mantle Relationships,* (ed. J. KORNPROBST), pp. 295–308 Elsevier.

DUNGAN M. A., LONG P. E. and RHODES J. M. (1978) Magma mixing at mid–ocean ridges: Evidence from legs 45 and 46–DSDP. *Geophys. Res. Lett.* **5**, 423–425.

DUNGAN M. A. and RHODES J. M. (1978) Residual glasses and melt inclusions in basalts from DSDP legs 45 and 46: Evidence for magma mixing. *Contrib. Mineral. Petrol.* **67**, 417–431.

ELTHON D. (1986) Comments on "Composition and depth of origin of primary mid–ocean ridge basalts: by D. C. Presnall and J. D. Hoover." *Contrib. Mineral. Petrol.* (in press).

ELTHON D. and SCARFE C. M. (1984) High–pressure phase equilibria of a high–magnesia basalt and the genesis of primary oceanic basalts. *Amer. Mineral.* **69**, 1–15.

FLOWER M. F. J., PRITCHARD R. G., SCHMINCKE H. U. and ROBINSON P. T. (1983) Geochemistry of basalts: Deep Sea Drilling Project sites 482, 483, and 485 near the Tamayo Fracture Zone, Gulf of California. Init. Rept. DSDP 65, 559–578.

FREY F. A., BRYAN W. B. and THOMPSON G. (1974) Atlantic Ocean floor: geochemistry and petrology of basalts from legs 2 and 3 of the Deep Sea Drilling Project. *J. Geophys. Res.* **79**, 5507–5527.

FUJII T. and BOUGAULT H. (1983) Melting relations of a magnesian abyssal tholeiite and the origin of MORBs. *Earth Planet. Sci. Lett.* **62**, 283–295.

FUJII T. and KUSHIRO I. (1977) Melting relations and viscosity of an abyssal tholeiite. *Carnegie Inst. Wash. Yearb.* **76**, 461–465.

FUJII T. and SCARFE C. M. (1985) Composition of liquids coexisting with spinel lherzolite at 10 kbar and the genesis of MORBs. *Contrib. Mineral. Petrol.* **90**, 18–28.

GASPARIK T. (1984) Two–pyroxene thermobarometry with new experimental data in the system CaO-MgO-Al_2O_3-SiO_2. *Contrib. Mineral. Petrol.* **87**, 87–97.

GREEN D. H., HIBBERSON W. O. and JAQUES A. L. (1979) Petrogenesis of mid–ocean ridge basalts. In *The Earth: Its Origin, Structure and Evolution,* (ed. M. W. McELHINNY), pp. 265–299 Academic Press.

GREEN D. H. and RINGWOOD A. E. (1967) The genesis of basaltic magmas. *Contrib. Mineral. Petrol.* **15**, 103– 190.

GROVE T. L. and BRYAN W. B. (1983) Fractionation of pyroxene-phyric MORB at low pressure: An experimental study. *Contrib. Mineral. Petrol.* **84**, 293–309.

HAMLYN P. R. and BONATTI E. (1980) Petrology of mantle-derived ultramafics from the Owen Fracture Zone, northwest Indian Ocean: implications for the nature of the oceanic upper mantle. *Earth Planet. Sci. Lett.* **48**, 65–79.

JAQUES A. L. and GREEN D. H. (1980) Anhydrous melting of peridotite at 0–15 kb pressure and the genesis of tholeiitic basalts. *Contrib. Mineral. Petrol.* **73**, 287–310.

KLEIN E. M. and LANGMUIR C. H. (1986) Ocean ridge basalt chemistry, axial depth, crustal thickness and temperature variations in the mantle. *J. Geophys. Res.* (In press).

KUSHIRO I. (1968) Compositions of magmas formed by partial zone melting of the earth's upper mantle. *J. Geophys. Res.* **73**, 619–634.

KUSHIRO I. (1973) Origin of some magmas in oceanic and circumoceanic regions. *Tectonophys.* **17**, 211–222.

LANGMUIR C. H., BENDER J. F., BENCE A. E., HANSON G. N. and TAYLOR S. R. (1977) Petrogenesis of basalts from the FAMOUS area, Mid–Atlantic Ridge. *Earth Planet. Sci. Lett.* **36**, 133–156.

LEROEX A. P., ERLANK A. J. and NEEDHAM H. D. (1981) Geochemical and mineralogical evidence for the occurrence of at least three distinct magma types in the FAMOUS region. *Contrib. Mineral. Petrol.* **77**, 24–37.

MELSON W. G., VALLIER T. L., WRIGHT T. L., BYERLY G. and NELEN J. (1976a) Chemical diversity of abyssal volcanic glass erupted along Pacific, Atlantic, and Indian Ocean sea–floor spreading centers. In *The Geophysics of the Pacific Ocean Basin and its Margin,* (eds. G. H. SUTTON, M. H. MANGHNANI and R. MOBERLY), Mon. 19, pp. 351–367 American Geophysical Union.

MELSON W. G., BYERLY G. R., NELEN J. A., O'HEARN T., WRIGHT T. L. and VALLIER T. (1976b) A catalog of the major element chemistry of abyssal volcanic glasses. In *Mineral Sciences Investigations, 1974–1975,* (ed. B. MASON), Smithsonian Contrib. Earth Sciences 19, 31–60.

MYSEN B. O. and KUSHIRO I. (1977) Compositional variations of coexisting phases with degree of melting of peridotite in the upper mantle. *Amer. Mineral.* **62**, 843–865.

NATLAND J. H. and MELSON W. G. (1980) Compositions of basaltic glasses from the East Pacific Rise and Siqueiros Fracture Zone, near 9°N. Init. Rept. DSDP 54, 705–723.

O'DONNELL T. H. and PRESNALL D. C. (1980) Chemical variations of the glass and mineral phases in basalts dredged from 25°–30°N along the Mid–Atlantic Ridge. *Amer. J. Sci.* **280-A**, 845–868.

O'HARA M. J. (1968) Are ocean floor basalts primary magma? *Nature* **220**, 683–686.

PALLISTER J. S. and HOPSON C. A. (1981) Samail ophiolite plutonic suite: field relations, phase variation, cryptic variation and layering, and a model of a spreading ridge magma chamber. *J. Geophys. Res.* **86**, 2593–2644.

PERFIT M. R., BATIZA R., ALLEN J., SMITH T. and FORNARI D. J. (1985) Geochemistry of basalts from a seamount group at 09 53′N near the East Pacific Rise: Possible implications for off-axis magma chambers. *EOS* **66**, 1079.

PRESNALL D. C. (1969) The geometrical analysis of partial fusion. *Amer. J. Sci.* **267**, 1178–1194.

PRESNALL D. C. (1979) Fractional crystallization and partial fusion. In *The Evolution of the Igneous Rocks— Fiftieth Anniversary Perspectives,* (ed. H. S. YODER, JR.), pp. 59–75 Princeton Univ. Press.

PRESNALL D. C. (1986) An algebraic method for determining equilibrium crystallization and fusion paths in multicomponent systems. *Amer. Mineral.* **71**, 1061–1070.

PRESNALL D. C. and HOOVER J. D. (1984) Composition and depth of origin of primary mid–ocean ridge basalts. *Contrib. Mineral. Petrol.* **87**, 170–178.

PRESNALL D. C. and HOOVER J. D. (1986) Composition and depth of origin of primary mid–ocean ridge basalts—reply to D. Elthon. *Contrib. Mineral. Petrol.,* (in press).

PRESNALL D. C., DIXON J. R., O'DONNELL T. H. and DIXON S. A. (1979) Generation of mid–ocean ridge tholeiites. *J. Petrol.* **20**, 3–35.

PRESNALL D. C., DIXON S. A., DIXON J. R., O'DONNELL T. H., BRENNER N. L., SCHROCK R. L. and DYCUS D. W. (1978) Liquidus phase relations on the join diopside–forsterite–anorthite from 1 atm to 20 kbar: Their bearing on the generation and crystallization of basaltic magma. *Contrib. Mineral. Petrol.* **66**, 203–220.

RHODES J. M., DUNGAN M. A., BLANCHARD D. P. and LONG P. E. (1979) Magma mixing at mid–ocean ridges: Evidence from basalts drilled near 22°N on the Mid–Atlantic Ridge. *Tectonophys.* **55**, 35–61.

RINGWOOD A. E. (1975) *Composition and Petrology of the Earth's Mantle.* 618 pp. McGraw–Hill.

ROEDER P. L. and EMSLIE R. F. (1970) Olivine–liquid equilibrium. *Contrib. Mineral. Petrol.* **29**, 275–289.

SIGURDSSON H. (1981) First–order major element variation in basalt glasses from the Mid–Atlantic Ridge: 29°N to 73°N. *J. Geophys. Res.* **86**, 9483–9502.

STAKES D. S., SHERVAIS J. W. and HOPSON C. A. (1984) The volcanic–tectonic cycle of the FAMOUS and AMAR Valleys, Mid–Atlantic Ridge (36°47'N): Evidence from basalt glass and phenocryst compositional variations for a steady state magma chamber beneath the valley midsections, AMAR 3. *J. Geophys. Res.* **89**, 6995–7028.

STOLPER E. (1980) A phase diagram for mid–ocean ridge basalts: preliminary results and implications for petrogenesis. *Contrib. Mineral. Petrol.* **74**, 13–27.

TAKAHASHI E. and KUSHIRO I. (1983) Melting of a dry peridotite at high pressures and basalt magma genesis. *Amer. Mineral.* **68**, 859–879.

TAKAHASHI E. and SCARFE C. M. (1985) Melting of peridotite to 14 GPa and the genesis of komatiite. *Nature* **315**, 566–568.

WALKER D., SHIBATA T. and DELONG S. E. (1979) Abyssal tholeiites from the Oceanographer Fracture Zone II, Phase equilibria and mixing. *Contrib. Mineral Petrol.* **70**, 111–125.

WYLLIE P. J., DONALDSON C. H., IRVING A. H., KESSON S. E., MERRILL R. B., PRESNALL D. C., STOLPER E. M., USSELMAN T. M. and WALKER D. (1981) Experimental petrology of basalts and their source rocks. In *Basaltic Volcanism Study Project,* Chap. 3, pp. 493–630 National Aeronautics and Space Adm., Houston.

YODER H. S. JR. and TILLEY C. E. (1962) Origin of basalt magmas: An experimental study of natural and synthetic rock systems. *J. Petrol.* **3**, 342–532.

Appendix 1. Primitive MORB glass compositions[1]

Sample No.	SiO₂	TiO₂	Al₂O₃	Cr₂O₃	FeO[2]	MnO	MgO	CaO	Na₂O	K₂O	P₂O₅	Total	mg[3]	Reference
Atlantic Ocean														
519-4-1	49.07	0.74	16.44	0.03	8.86	0.16	10.15	11.65	2.13	0.07	n.d.[4]	99.30	70.4	BRYAN & MOORE (1977)
519-4-2	48.81	0.73	16.13	0.04	8.89	0.16	10.32	11.56	2.15	0.07	n.d.	98.86	70.6	BRYAN & MOORE (1977)
522-2-1	49.65	0.87	16.05	0.06	8.72	0.15	9.82	11.76	2.25	0.09	n.d.	99.42	70.0	BRYAN & MOORE (1977)
525-5-1	49.19	0.85	16.12	0.08	8.74	0.14	10.41	11.91	2.35	0.09	n.d.	99.88	71.2	BRYAN & MOORE (1977)
525-5-2	48.93	0.84	16.14	0.09	8.75	0.15	10.49	11.84	2.41	0.09	n.d.	99.73	71.3	BRYAN & MOORE (1977)
525-5-3	48.96	0.82	15.75	0.08	8.52	0.13	9.79	11.92	1.97	0.11	n.d.	98.05	70.4	BRYAN & MOORE (1977)
526-1-1B	51.36	0.78	15.03	0.08	7.86	0.10	9.01	12.38	2.04	0.10	n.d.	98.77	70.0	BRYAN & MOORE (1977)
527-1-2	49.50	0.66	16.51	0.11	8.76	0.17	9.86	12.68	2.08	0.04	n.d.	100.37	70.0	BRYAN & MOORE (1977)
528-4-1	49.93	0.72	15.92	0.08	8.37	0.15	9.78	12.67	1.96	0.05	n.d.	99.53	70.8	BRYAN & MOORE (1977)
529-3-2	50.82	0.92	15.92	0.06	7.91	0.13	9.38	12.07	2.30	0.16	n.d.	99.67	71.1	BRYAN & MOORE (1977)
530-3-1	49.98	0.77	15.24	0.00	8.60	0.13	10.11	12.45	2.23	0.07	n.d.	99.58	70.9	BRYAN & MOORE (1977)
534-2-1	50.30	0.84	15.99	0.08	8.70	0.13	9.91	12.33	2.16	0.09	n.d.	100.53	70.2	BRYAN & MOORE (1977)
534-2-2A	50.37	0.84	15.86	0.09	8.57	0.16	9.82	12.33	2.18	0.09	n.d.	100.31	70.4	BRYAN & MOORE (1977)
ARP-73-10-3	50.6	1.05	15.0	0.00	8.97	0.18	8.70	12.5	2.43	0.11	n.d.	99.54	73.4	BRYAN (1979)
AII-77-76-61	51.2	0.84	15.0	0.08	7.98	0.16	9.02	13.3	1.89	0.08	n.d.	99.55	70.1	BRYAN (1979)
AII-77-76-71	51.4	0.83	15.0	0.07	8.05	0.16	9.15	13.2	1.97	0.09	n.d.	99.92	70.2	BRYAN (1979)

ARP-74-14-31	48.2	0.51	17.0	n.d.	8.51	0.16	10.1	12.7	2.34	0.04	n.d.	99.56	71.1	BRYAN (1979)
ARP-74-14-33	48.2	0.50	16.9	0.01	8.50	0.15	9.91	12.8	2.30	0.04	n.d.	99.31	70.7	BRYAN (1979)
3-14-10-1[5]	50.0	0.79	17.4	n.d.	8.22	0.13	9.81	12.9	2.28	0.02	n.d.	101.55	71.2	FREY et al. (1974)
3-18-7-1,1[5]	50.3	0.73	16.6	n.d.	7.99	0.12	10.2	13.2	2.00	0.01	n.d.	101.15	72.6	FREY et al. (1974)
3-18-7-1,4	50.4	0.79	17.1	n.d.	7.97	n.d.	9.88	12.9	1.62	0.02	n.d.	100.68	72.0	FREY et al. (1974)
A27.99S1	49.86	0.71	16.93	n.d.	8.12	n.d.	9.49	13.16	1.71	0.02	0.04	100.04	70.8	MELSON et al. (1976a)
A42.96N1	50.57	0.92	15.62	n.d.	7.96	n.d.	9.21	13.32	1.99	0.26	0.10	99.95	70.6	MELSON et al. (1976a)
72-17-44	49.68	0.86	16.75	n.d.	7.67	0.13	9.19	12.23	2.31	0.05	0.10	98.97	71.3	O'DONNELL & PRESNELL (1980)
TR-154-14-D1	51.36	1.29	16.48	n.d.	7.21	0.06	8.39	12.02	2.49	0.25	0.09	99.64	70.7	SIGURDSSON (1981)
ARP-74-10-16[6]	49.47	0.82	15.23	0.09	8.15	0.14	10.66	12.21	1.94	0.16	0.10	98.97	73.1	FUJII & BOUGAULT (1983)
CYP-31-35[7]	49.57	0.85	15.26	0.09	8.23	0.14	10.63	12.11	2.11	0.19	0.11	99.34	72.7	FUJII & BOUGAULT (1983)
VG-202	50.73	0.92	15.61	n.d.	7.97	n.d.	9.34	13.29	1.94	0.24	0.10	100.14	70.8	MELSON et al. (1976b)
VG-203	50.41	0.92	15.64	n.d.	7.95	n.d.	9.09	13.35	2.05	0.28	0.11	99.80	70.3	MELSON et al. (1976b)
VG-297	50.14	0.74	16.81	n.d.	8.25	n.d.	9.59	13.01	1.71	0.04	0.04	100.33	70.7	MELSON et al. (1976b)
VG-416	50.73	0.73	16.88	n.d.	8.22	n.d.	9.84	13.33	1.68	0.04	0.05	101.50	71.3	MELSON et al. (1976b)
VG-607	49.97	0.73	17.26	n.d.	8.39	n.d.	9.51	12.92	1.99	0.01	0.03	100.81	70.1	MELSON et al. (1976b)
VG-611	49.95	0.69	17.29	n.d.	7.91	n.d.	9.14	13.23	1.76	0.00	0.04	100.01	70.5	MELSON et al. (1976b)
Indian Ocean														
VG-649	50.41	0.92	15.64	n.d.	7.95	n.d.	9.09	13.35	2.05	0.28	0.11	99.80	70.3	MELSON et al. (1976b)
Pacific Ocean														
F	49.15	0.96	17.58	n.d.	8.01	n.d.	9.58	12.14	2.38	0.04	0.10	99.94	71.2	NATLAND & MELSON (1980)
H	49.21	0.97	17.54	n.d.	7.97	n.d.	9.12	12.35	2.43	0.04	0.09	99.72	70.3	NATLAND & MELSON (1980)
1561-1622	48.47	0.88	17.39	0.04	8.03	0.16	9.83	12.26	2.47	0.03	n.d.	99.56	71.7	PERFIT et al. (1985)[8]
F2-1	49.85	0.86	17.51	0.05	8.12	0.21	9.80	12.92	2.15	0.02	n.d.	101.49	71.4	PERFIT et al. (1985)[8]
F3-2	49.46	0.83	17.42	0.04	8.11	0.14	9.85	12.72	2.14	0.04	n.d.	100.75	71.6	PERFIT et al. (1985)[8]
F1[9]	47.2	0.77	16.9	0.07	8.80	0.14	10.90	12.2	2.05	0.02	n.d.	99.05	72.0	PERFIT et al. (1985)[8]

[1] All analyses by electron microprobe except as indicated.
[2] Total Fe as FeO.
[3] $mg = 100 \, Mg/(Mg + Fe^{2+})$ when $Fe^{2+}/(Fe^{2+} + Fe^{3+})$ is assigned the value 0.86 (PRESNALL et al., 1979).
[4] n.d. = not determined.
[5] FREY et al. (1974) also list trace element concentrations.
[6] X-ray fluorescence analysis. Sample also contains 247 ppm Ni.
[7] X-ray fluorescence analysis. Sample also contains 255 ppm Ni, 0.52 ppm Ta, 1.32 ppm Hf, and 0.42 ppm Tb.
[8] Also M. R. PERFIT (personal communication, 1985).
[9] Direct current plasma analysis by C. H. LANGMUIR (M. R. PERFIT, personal communication, 1985).

Magmatic Processes: Physicochemical Principles
© The Geochemical Society, Special Publication No. 1, 1987
Editor B. O. Mysen

Dry peralkaline felsic liquids and carbon dioxide flux through the Kenya rift zone

D. K. BAILEY

Department of Geology, The University, Whiteknights,
Reading RG6 2AB, Berkshire, U. K.

and

R. MACDONALD

Department of Environmental Sciences, University of Lancaster, Lancaster LA1 4YQ, U. K.

Preamble—When Hat Yoder was in London to receive the Wollaston Medal of the Geological Society the chance arose for him to see the Michaelangelo Tondo in the Royal Academy. His reaction, naturally, was to marvel that such beauty could be wrought from stone. Along with other participants at the meeting we pay tribute to Hat, who in his own way has continually given petrologists new perceptions of rocks.

Abstract—Around Lake Naivasha, at the topographic culmination of the Kenya rift, a wide range of peralkaline felsic magmas erupted in the Holocene. Glassy samples are abundant. Volcanological and petrographic evidence indicates that these glasses reflect characteristically low concentrations of H_2O in the melts, and this relationship is confirmed by experimental and chemical information.

Distinct magma types, trachytes, comendites and pantellerites, erupted almost exclusively from different centres, even though these are sufficiently close for intercalation of the erupted products from adjacent volcanoes. This individuality is maintained in chemical variation diagrams where the products from different centres form separate groups, with no gradational compositions. In some diagrams smaller clusters appear within the main groups. These distinctive patterns indicate that there is no continuous evolution of the melts below the rift, and that each centre represents the outpouring of a different melt. Batches within the same centre could also represent different episodes of melt generation.

Generation of these melts in a *regional* context can best be explained by melting of different sources during the influx of a CO_2–rich fluid. Trachytes, pantellerites and comendites could then be the products of a flux/melting cycle as it climbs progressively through mantle, mafic deep crust, and sialic crust. The high concentrations of sodium and iron in the melts call for enhancement of these elements in the source rocks: this could be achieved by regional metasomatism resulting from prolonged activity through the rift segment over the past 23 Ma. Incompatible trace element variations cannot be attributed to solid-melt interactions but could be imparted by the fluid during melting.

INTRODUCTION

FOR A VARIETY of reasons alkaline rocks have always held a fascination for igneous petrologists, but in recent years have assumed extra significance as it has emerged that their richness in incompatible elements and radio-isotopes are signalling special conditions of source, or melt–forming processes, or both. In most alkaline provinces peralkalinity has developed in varying degrees. Paradoxically, peralkalinity is strongest when magma penetrates continental lithosphere, where the crustal rocks are typically peraluminous. Such magmatism is especially evident in stable plate interiors where igneous activity of any kind must call for special explanation. Even where eruption is localised by an evident split in the lithosphere, there is still no ready explanation of why, for instance, continental fissuring may give vent to tholeiite floods in one region or epoch, and alkaline magmatism in another.

Aphyric volcanic glasses provide the closest samples to the composition of peralkaline melts, where it is clear that crystallisation, and indeed devitrification or hydration, leads notably to alkali and halogen losses (NOBLE *et al.,* 1967; MACDONALD and BAILEY, 1973). Glasses may also provide extra insight into the gases associated with strongly alkaline volcanism. This information is of value because the only direct, high temperature volcanic gas collection available is from the melilite nephelinite volcano of Mount Nyiragongo (CHAIGNEAU *et al.,* 1960): no direct, high temperature collections have been made of an oversaturated peralkaline volcano. The aim here is to examine and explore the potential information content of a range of oversaturated volcanic glasses from the central part of Kenya (Gregory) rift valley. All the samples are fresh, unaltered glasses; most are obsidians in which phenocrysts are absent or less than 2 percent in the mode; vesicles, if present, are widely separated. Pumice samples are excluded because of their susceptibility to post-eruptive hydration and alteration.

INTRINSIC H₂O CONTENTS OF PERALKALINE GLASSES

With one exception all peralkaline glass analyses in the compilation by MACDONALD and BAILEY (1973) have H_2O^+ contents of 0.67 weight percent or less. Many more analyses (published and unpublished) have since been added (BAILEY, 1978; 1980) and it is clear that the vast bulk of the H_2O^+ values are around the limit of detection ($\sim$0.1 weight percent): hence, most glasses are virtually anhydrous. These data are in complete accord with the case put forward earlier by NICHOLLS and CARMICHAEL (1969) that oversaturated peralkaline liquids are essentially dry. Other studies (see EWART, 1979) seem to indicate consistently low f_{O_2} and high temperatures in calc-alkaline rhyolites and dacites, which would be in keeping with the generally low H_2O^+ values in the glasses. Recently, however, TAYLOR et al. (1983) have reported H_2O contents up to 3.1 weight percent in fresh calc-alkaline rhyolite tephra, leading to the suggestion that these are the general levels in such melts, and that glasses with low H_2O are degassed. They propose that the main degassing of the melt takes place during foaming within the vent, so that later flows and domes with low H_2O represent extrusion of the collapsed foam. They report that there appear to be no differences in the compositions of the calc-alkaline glasses except for H_2O^+ and deuterium contents. In view of the manifest compositional changes in peralkaline glasses during devitrification and hydration, and during high temperature fusion experiments (HAMPTON and BAILEY, 1985), any massive loss of several percent H_2O from a high temperature peralkaline melt would render the consequent anhydrous glass suspect as a sample of the natural melt composition. It is essential, therefore, to examine the evidence with this in mind before considering the chemical variations among the glasses from the Kenya rift province.

General relationships and field evidence

The case presented here is part of a continuing study of alkaline glasses, encompassing the widest possible range of bulk compositions and samples from all regions and all different modes of eruption. The following list outlines the range and scope.

(1) Bulk compositions range from nephelinite through phonolite to comendite, i.e., the maximum range of silica values.

(2) Samples are from all geographic and tectonic settings.

(3) Wide variations in vesicularity, and in phenocryst types and contents, are represented.

(4) Eruptive mode varies widely. Samples come from flow margins (top and bottom); flow interiors; angular blocks in agglomerate; angular fragments in pumice deposits; fiamme from welded ash flows; splatter; and agglutinated spatter.

Significantly, tephra fragments ranging from small chips in pumice deposits to large blocks in agglomerates are all low in H_2O^+; there is no evidence of high speed ejectamenta that could represent an H_2O–rich melt.

Experimental crystallisation and comparative petrography

In marked contrast with calc-alkaline rhyolites, which fail to crystallise in the laboratory when dry, pantellerite glasses start to crystallise in as little as thirty minutes at appropriate temperatures without addition of water. Experimental study of the melting and crystallisation of peralkaline glasses under a range of conditions is, therefore, possible (COOPER, 1975; BAILEY and COOPER, 1978) with results as shown in Figure 1. Details of the experimental conditions and the phases developed were given in BAILEY and COOPER (1978), which also contains a discussion of the consequences of the contrasting behaviour of wet and dry pantellerite melts. Relevant points from this discussion are summarised below.

(1) The exsolution of vapour under near–liquidus conditions in the dry experiments (Figure 1) indicates that some natural volatiles were quenched in at partial pressures at least as high as 0.3 kbar. A synthetic rhyolite melt would contain 2.5 weight percent of H_2O at an equivalent pressure (LUTH, 1976). Addition of H_2O to the natural pantellerite profoundly changes the melting behaviour indicating that it is capable of dissolving a similar quantity of H_2O at 0.3 kbar. As H_2O is one of the most soluble gases in silicate melts it hardly seems credible that a pantellerite melt could lose 2–3 weight percent H_2O whilst effectively retaining other gases at 0.3 kbar solubility levels. Hence, although there must have been some degassing from some samples (in decompression) this gas was not simply H_2O.

(2) Crystallisation of an H_2O–rich pantellerite results in acmitic pyroxenes and potassic feldspars quite different from those observed as natural phenocrysts, and in the wrong sequence.

(3) Kenyan obsidians with modal phenocryst

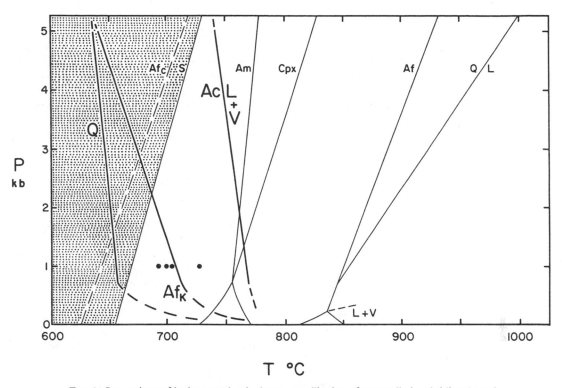

FIG. 1. Comparison of hydrous and anhydrous crystallisation of a pantellerite obsidian (sample KE12: Eburru, Kenya). Hydrous boundaries, heavy lines; anhydrous boundaries, light. The dry subsolidus region is shaded. V = fluid, L = liquid; Q = quartz; Af = alkali feldspar; Af_K = K-rich alkali feldspar; Af_c = alkali feldspar critical line (LUTH, 1976); Cpx = sodic hedenbergite; Am = arfvedsonitic amphibole; Ac = acmitic pyroxene; S = dry solidus. The points shown at 1 kbar are the range of liquidus minima (quartz + feldspar + liquid) in related synthetic systems at P_{H_2O} = 1 kbar (CARMICHAEL and MACKENZIE, 1963; THOMPSON and MACKENZIE, 1967). Taken from BAILEY and COOPER (1978).

contents greater than 5% are rare: most have less than 2% and, hence, have a near–liquidus aspect. This characteristic, combined with low H_2O^+ contents, effectively rules out an origin from a magma at, or below, the H_2O-saturated liquidus in Figure 1. Firstly, the large differences in temperature between the wet and dry systems, require that subvolcanic degassing of any wet pantellerite magma should have been accompanied by extensive crystallisation. Secondly, the negative slopes of the phase field boundaries in the wet system would imply that any wet pantellerite magma at depth could get to the surface only as a pyroclastic eruption composed of H_2O–saturated glass, or as degassed crystalline lava.

If pantelleritic melts ever were wet at depth then some special conditions of eruption would be called for to account for the dry, near–liquidus aspect of the glasses. Suppose the original melt had previously contained as much as 3 weight percent H_2O (e.g.,

approximately saturated at 2 km depth) and had somehow lost H_2O at a lower pressure. To achieve a dry near-liquidus condition its required starting temperature would be at least 50°C above the wet liquidus. Alternatively, it might have risen towards the surface from a point on a 3 weight percent H_2O liquidus, requiring a starting pressure of greater than 5 kbar. To make the transition from wet to dry melt therefore would seem to require an unusual eruption or thermal history. Non-systematic (vagarious) variation within the sample population would be expected in H_2O contents, oxidation states, alkalies and halogens. Furthermore, when crystals are present, at least some should carry a record of this eruption history. In fact, they are remarkably consistent in composition (NICHOLLS and CARMICHAEL, 1969; SUTHERLAND, 1974) even to the extent of showing scarcely any sign of zoning in the feldspars, the dominant phenocryst phase.

Peralkaline glasses are still puzzling in the light of their laboratory behaviour. Dry pantellerite held just above its solidus temperature for about an hour starts to crystallise; if taken to the liquidus temperature it bubbles. But both these states seem to have been largely missed out in the cooling of natural glass samples. In order to preserve such a glass it must somehow pass through the vesiculation and crystallisation temperature range (below about 700°C at low pressure) in a short time. Whereas this cooling history can be envisaged for particular instances such as flow margins, it is less easy in other cases. Glass clasts in pumice deposits are typically non-vesicular, non-porphyritic and *angular.* Hence, they were "through" this temperature range before incorporation in the pumice eruption. Our experiments with gas extraction indicate that foaming is a near–liquidus phenomenon (HAMPTON and BAILEY, 1984). It is difficult, therefore, to see how the obsidian clasts could have come from the part of the melt that was foaming. If they are chilled melt from the border zone of the foaming mass they should presumably record the highest H_2O contents available for peralkaline liquids.

Glass fiamme in welded ash flows also pose problems if, as suggested by TAYLOR *et al.* (1983), they "remain at magmatic temperatures for weeks or months". According to our experiments pantellerite fiamme could not survive as glass under such conditions. Almost wholly glassy flows and domes are also extruded from peralkaline silicic volcanoes and must pose similar problems. There is a need for some means of super-cooling peralkaline melts prior to eruption such that the viscosity is sufficiently high to inhibit crystallisation and bubbling, but flow is permitted.

Chemical variations and H_2O^+ levels

If peralkaline magmas have previously contained several percent H_2O, they must clearly have suffered massive vapour loss prior to glass formation. This must leave its mark on the melt composition, as other volatile species should be partitioned into the vapour. In high–temperature, gas extractions from glasses in the laboratory, H_2O is always accompanied by other species, either in the form of gases, or precipitated as sublimates. Most typically, the high–temperature sublimates include alkali halides (HAMPTON and BAILEY, 1985). These products mirror those observed at high temperature fumaroles on volcanoes. Similarly UNNI and SCHILLING (1978) have recorded that concentrations of Cl and H_2O in submarine basalts correlate with ocean depth, and are dependent on degree of vesiculation and loss of vapour on eruption at different pressures. In contrast to these observations, peralkaline silicic

obsidians show no correlation between H_2O and Cl (Figure 2). In fact, no correlation between H_2O and any other variable has been found in these rocks, certainly not with alkalies and halogens where it might be most expected. Present evidence offers no case for massive sub–volcanic loss of water-rich vapour from peralkaline melts. The low and variable H_2O levels in the glasses may be essentially "accidental", *i.e.,* unrelated to the melt generation process. On this basis it is appropriate to treat the glasses as melt samples, in an attempt to unravel their chemical relationships and try to discern their petrogenesis.

PERALKALINE VOLCANOES IN THE CENTRAL KENYA RIFT

Volcanism has been active along the Kenya rift for the past 23 Ma, ranging from carbonatites and nephelinites to rhyolites. Felsic rocks constitute about half the total erupted volume, about 150,000 km³ (WILLIAMS, 1982). The bulk of the Tertiary-Recent felsic rocks are peralkaline making it the most voluminous and variegated peralkaline province in the world. Holocene activity in the central part of the rift, around its topographic culmination in the vicinity of Lake Naivasha, has been largely expressed by felsic central volcanoes on the rift floor. Figure 3 shows the largely monotonic magma compositions emitted from the main centres. Eruptives from adjacent centres may interfinger around the margins, but are always petrographically distinct. The general geology around Lake Naivasha has been described by THOMPSON and DODSON (1963), and there have been individual descriptions of Suswa (JOHNSON, 1969); Longonot (SCOTT, 1980); and

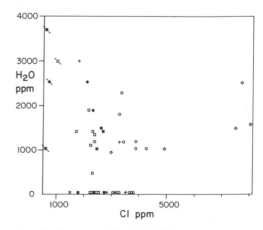

FIG. 2. Variation of Cl with H_2O in oversaturated peralkaline glasses. Diamonds indicate samples from Pantelleria, all other samples from the Kenya rift (for details see Figure 4).

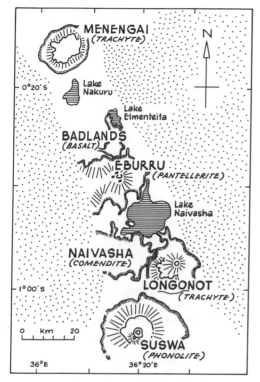

FIG. 3. Sketch map showing the chief Holocene volcanic centres in the Lake Naivasha section of the Kenya rift. Rift shoulders indicated by stipple.

south west Naivasha (MACDONALD *et al.,* 1986). In addition to the felsic volcanoes, there is a small basalt field forming the Badlands, north east of Eburru (MCCALL, 1957). Small basalt cones pepper the fringes of the comendite field of south west Naivasha, and mixed basalt/felsic lavas have been erupted from Longonot (SCOTT and BAILEY, 1984) and Eburru. North of the Badlands the next major central volcano is Menengai, erupting only trachyte (LEAT, 1984).

Lake Naivasha is located on the crest of the classic continental rift valley first described by GREGORY (1894) and its volcanism has a special place in igneous petrogenesis. Many of the lavas used by BOWEN (1937) to illustrate his concept of Petrogeny's Residua System (nepheline–kalsilite–silica) were from this region. The Holocene volcanism around Lake Naivasha is an almost perfect case for a petrographic province, fulfilling all the criteria laid down by WILCOX (1979).

The Lake Naivasha region has the added advantage that glassy samples, covering most of the range of oversaturated peralkaline compositions are abundant. The first pantelleritic trachyte glasses were recognised at Menengai and Longonot

(MCCALL, 1967; MACDONALD *et al.,* 1970), and the East African rift valley has so far provided the only samples of this melt. Trachyte glasses having clear chemical affinities with comendites are absent.

The extent and persistence of peralkaline activity through the East African lithosphere make it imperative to look for the causes, and the wide range of geology and sample compositions offer hope for eventual solutions to the questions of petrogenesis. It is appropriate first to look at chemical variations in the glasses with changing levels of peralkalinity.

Major element variations with peralkalinity

Some distinctions are seen immediately. For instance, the comendites from south west Naivasha separate from the pantellerites and the pantelleritic trachytes in terms of SiO_2, TiO_2, FeO, MnO and total iron (Figures 4 and 6). They also appear as a distinct group on most other variation diagrams. Like all other continental comendites, they appear to fall close to the quartz-feldspar cotectic, as extended into the peralkaline region (BAILEY and MACDONALD, 1970, Figure 5; MACDONALD *et al.,* 1986, Figure 7), with the implication that the source contained free quartz. This characteristic, together with many other features of their major element composition would be most obviously satisfied by a crustal or sialic source (BAILEY and MACDONALD, 1970) and more detailed studies on trace elements and isotopes are consistent with this interpretation (MACDONALD *et al.,* 1980, DAVIES and MACDONALD, 1986).

Figure 4 also reveals separate clustering of trachytes and pantellerites, which persists in Al_2O_3 (Figure 5). As might be anticipated, there is a general negative correlation between alumina and peralkalinity, but the comendites, pantellerites and trachytes form separate groups. This separation is emphasized when molecular Al_2O_3 is plotted; the comendites and pantellerites appear to define one trend, while the trachytes form parallel trends at higher levels. This feature might be taken to reflect different series of liquids with various levels of silica, but it is noteworthy that trachyte liquids show no direct connection with rhyolites.

Just as it might be anticipated that Al_2O_3 would correlate negatively with peralkalinity, so a positive correlation might be expected for the alkalies (Figure 5). In comendites and pantellerites, K_2O is almost constant around 4.4 weight percent, but there are signs of slight negative distributions, which are emphasised when the trachytes are taken into account. The general fall in K_2O levels with increasing peralkalinity may be linked in some way to the "or-

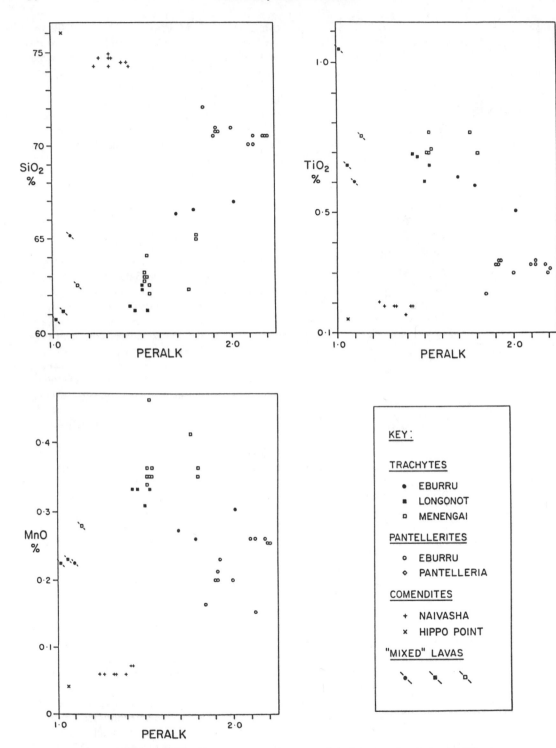

FIG. 4. Variations of SiO_2, TiO_2, and MnO with peralkalinity [Mol ($Na_2O + K_2O/Al_2O_3$)] in glassy samples from the Naivasha region. Plus signs, comendites from the main Naivasha field; cross, indicates tight group of comendites from the Hippo Point area on Lake Naivasha; open circles, pantellerites, closed circles, trachytic pantellerites from Eburru; filled squares, trachytes from Longonot; open squares, trachytes from Menengai; symbols with diagonal line indicate mixed lavas (possibly accumulitic in the case of Menengai).

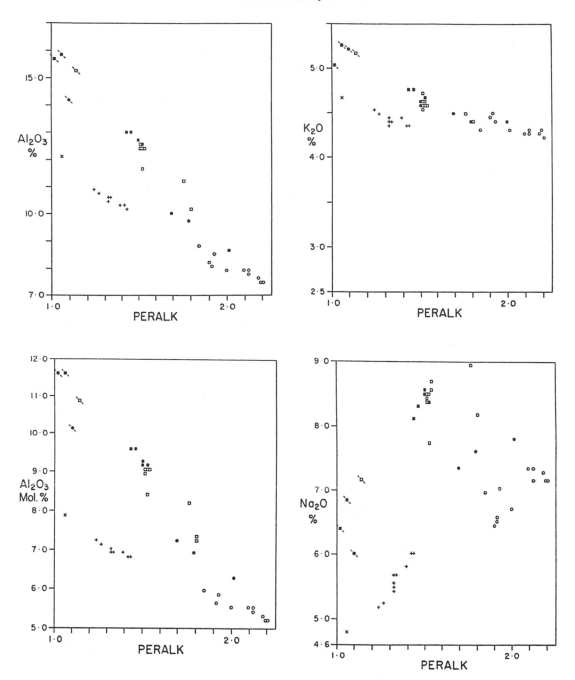

FIG. 5. Variations of Al_2O_3 (weight percent and mol percent), Na_2O and K_2O with peralkalinity. Mol percent Al_2O_3 can be used as an indication of variation in potential feldspar in the samples. Symbols as in Figure 4.

thoclase effect" (BAILEY and SCHAIRER, 1964) whereby alkali feldspars tend to be richer in K_2O than their coexisting melts. It is not easy, however, to reconcile the relatively constant K_2O in the comendites and the pantellerites with this effect if indeed

they represent series of liquids controlled by closed-system fractionation.

An even more striking pattern emerges for Na_2O (Figure 5). Trachytes from Longonot and Eburru fall on quite separate positive trends, which them-

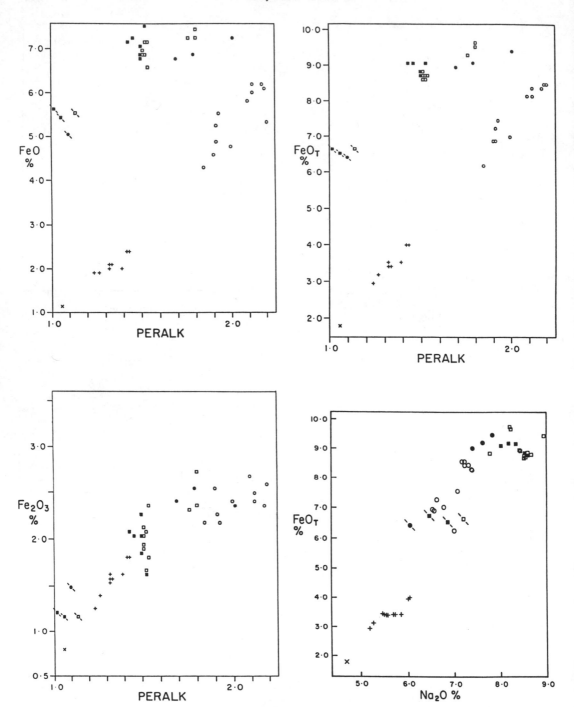

FIG. 6. Variations of soda and iron for Kenyan glasses. Symbols as in Figure 4.

selves are distinct from the distributions of pantellerites and comendites. This emphasizes the distinctions seen in Al_2O_3 and is mirrored by variation in other major elements, particularly FeO and total iron (Figure 6). Discrete populations of liquids are identified, some of which have been erupted even from the same volcano (Eburru). Within any series there is a general positive correlation between Na_2O and total iron, which may reflect some internal consistency in melt generation within that series, but the separation of the series, especially trachytes from rhyolites, would seem to require generation at least in distinct batches. Increasing soda correlates with iron content, and *within* each group with peralkalinity. Between trachytes and pantellerites, however, there is a broad negative distribution which seems to defy explanation in terms of interactions between melts and their observed phenocryst phases.

Where these rocks are not aphyric, the ubiquitous phenocryst is alkali feldspar, and the bulk compositions of the melts require that this phase must dominate in any crystallisation sequence. Separation of potassic feldspar from the trachytic liquids must lead to increasing sodicity beyond that actually observed in the pantellerites (MACDONALD *et al.,* 1970). Therefore, *if* these liquids were linked in a crystallisation sequence, a sodic phase was removed from the magma system as crystallisation progressed: no appropriate composition appears as a phenocryst in the trachytes.

The distribution of soda and iron (Figures 5 and 6), confirm the earlier suspicion that there is no chemical continuum between trachytes and pantellerites, and that they developed under different conditions. Pantellerite may represent a quartz-feldspar cotectic composition for a specific range of source rock compositions and pressures. At higher pressures (and in different source compositions?) the cotectic melt is closer to trachyte, as with increasing pressure in the "granite system" (TUTTLE and BOWEN, 1958) The special requirement in the "pantellerite system" would be that there is a phase boundary (marking the appearance of a phase rich in soda and iron) imposing a limit on the peralkalinity of the liquid that can be reached at any given pressure (see BAILEY, 1974).

A plot of Fe_2O_3 vs peralkalinity (Figure 6) reveals a steady progression through the comendites that could reflect an increasing contribution of acmite to the melts. There is no such regular variation of ferric iron and peralkalinity in the trachytes or pantellerites, and the latter appear to mark a "ceiling" in the possible levels of Fe_2O_3 in peralkaline melts. As the trachytes also have more iron than comendites or pantellerites, it is not possible to appeal simply to an acmitic component to account for the Fe_2O_3 variations in this part of the array. Melting of a source containing a combination of phases similar to the observed phenocrysts (feldspar, fayalite, sodic hedenbergite, aenigmatite, and oxide) could produce the trachytes. As discussed later, pantellerite melt generation may need to involve an arfvedsonitic amphibole in addition.

Formation of these peralkaline trachyte and pantellerite liquids seems to require melting of appropriate source compositions. This is indicated by the marked composition gaps, which have persisted despite intensive sampling, and must depict a real distribution of natural melts. The gaps must militate also against generation of trachytes and pantellerites by progressive melting of a single source. The separation of the trachytes from Longonot, Menengai and Eburru in various diagrams also renders unlikely the possibility of fractional melting at separate trachyte and rhyolite invariant points, for which there is no experimental evidence in any case. The separation of trachytic and pantelleritic liquids is further emphasized by the fact that in many diagrams the variations within the groups run *across* any expected trend from trachyte to pantellerite (Figure 4, 5 and 6). Further discussion of the questions of melt generation is best deferred until trace element variations have been examined.

Trace element versus major element functions

When variation in incompatible trace element concentrations is compared with peralkalinity the patterns are different (Figure 7). The main Naivasha comendites show a positive variation, but no other simple relationship emerges between trace elements and peralkalinity. Patterns in the trachytes are not well defined but some signs of positive trends may be detected within groups. The pantellerites, however, are scattered, with the highest levels of incompatible elements near the lowest levels of peralkalinity. Essentially, the trachytes, pantellerites and comendites form separate populations (c.f., the major element data), but distributions within groups are hinting at further differences between them. One major difference becomes clear when traces are compared with soda and iron. For the comendites there is a positive correlation of trace elements with Na_2O, but in the pantellerites the broad distribution is negative (Figure 8). Because FeO and FeO_T correlate positively with Na_2O, their relationships with trace elements are similar. What also emerges in Figure 8 is distinct clustering of different parts of the pantellerite population.

Variations between trace elements

Trace element correlations are much stronger. The strong correlation of Zr with Rb and F that was first noted for Eburru pantellerites (BAILEY and MACDONALD, 1975) appears in all groups of liquids even though they may separate from each other in the diagrams (Figure 9). This relationship has been found also in sample suites from the Azores and from Pantelleria (BAILEY, unpubl.).

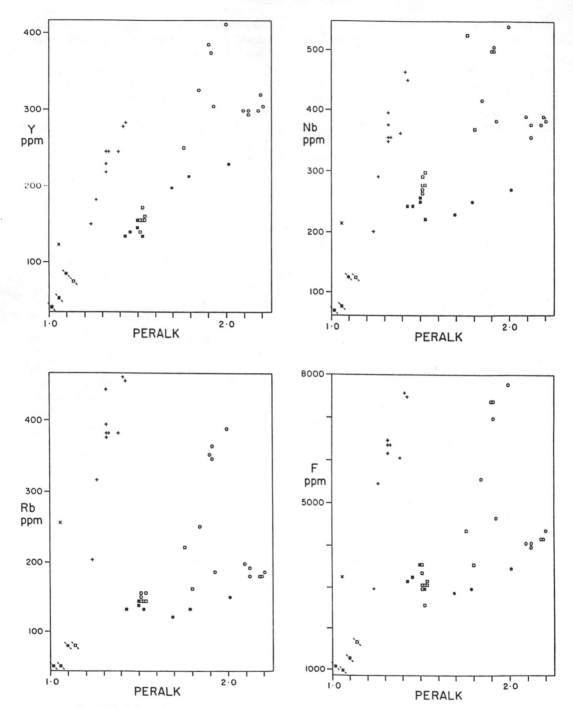

FIG. 7. Variations with peralkalinity for a selection of trace elements, Y, Nb, Rb and F, in Kenyan glasses. Symbols as in Figure 4.

Coherence also exists for Nb and Y through all the groups, and it had been noted previously (BAILEY and MACDONALD, 1975) that in the Eburru obsidians these two elements were more strongly correlated with Cl, as distinct from the strong co-variance in F, Zr and Rb. When all groups of glasses are compared, the pattern of Cl distribution is less clear cut, and there is a strong suggestion in Figure 9 of decoupling of the Cl distribution from varia-tions in the other trace elements in the higher con-

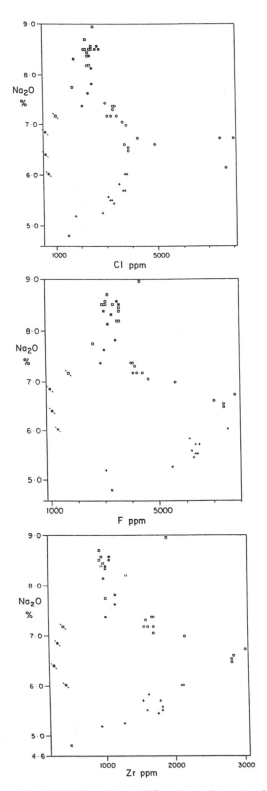

FIG. 8. Variations between different trace elements and Na₂O for peralkaline glasses from Kenya. Diamonds indicate samples from Pantelleria. Other symbols as in Figure 4.

centration ranges. There appears to be a "ceiling" for Cl in Kenyan glasses, which also shows up dramatically in the plot of Na_2O vs. Cl (Figure 8), where the negative distribution of the pantellerites and the positive distribution in the comendites appear to curve towards a common maximum. This contrasts with the Na_2O vs F diagram in which the more linear distributions might be pointing to a maximum outside the observed range. This limitation in the chlorine appears to be a regional effect because liquids from Pantelleria achieve much higher levels of Cl, describing a linear array with no obvious "ceiling" (Figures 8 and 9). The Kenyan rift liquids point to a limit in available Cl, but not F, in the source. The simple conclusion reached previously for the Eburru pantellerites that F, Zr and Rb formed one chemically coherent group, whilst Nb, Y and Cl formed another, therefore needs modification. A more reasonable conclusion would be that Zr and Rb cohere very strongly with F, whereas Nb and Y partly cohere with F, with part of their distribution connected with availability of Cl.

Summary of the chemical distributions

Comendites, pantellerites and trachytes form distinct populations, and for some elements, subgroups of these populations emerge. Three recurring sub-groups in the trachytes relate directly to the source volcanoes, Menengai, Longonot and Eburru. Eburru trachytes are distinct from Eburru pantellerites in most plots. *Within* individual groups there is a general trend of increasing soda and iron with peralkalinity, with the pantellerite groups lying on the extension of the comendite trend (Figure 6). Superimposed on this, however, there is a broad negative distribution from trachyte to pantellerite. Furthermore, although incompatible elements show positive covariance *within* the groups, they correlate negatively with soda and iron in the pantellerites, in contrast with trachytes and comendites. Viewed as a whole, it is clear that more than one process has been at work in generating these liquids. Soda (and iron) and peralkalinity seemingly can vary independently of incompatible trace elements.

THE SIGNIFICANCE OF SEPARATE POPULATIONS OF FELSIC LIQUIDS

The chemical variation diagrams emphasise the individuality of different volcanic centres by showing that each has erupted distinct and separate felsic magmas. Furthermore, there are contemporaneous basalts and hawaiites, but no extruded rocks with silica percentages between 54 and 61 weight percent that are not mixed lavas.

Different volcanoes have, therefore, erupted products that are not only chemically distinct, but

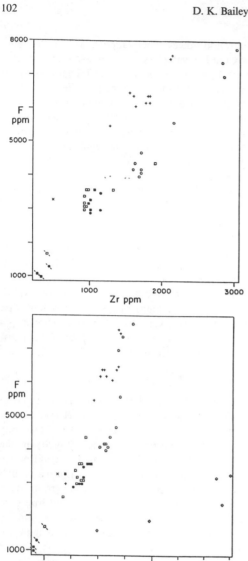

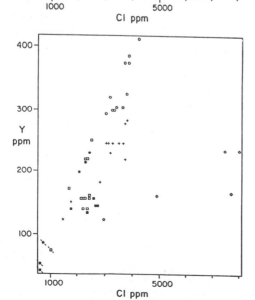

between which there are no bridging compositions. These distinct felsic magmas are all products of a major igneous cycle working through this segment of lithosphere. They are coeval and share some distinctive chemical features, especially peralkalinity. What is their relationship? If a basaltic parent were to be invoked, then each felsic batch has to evolve quite separately to produce the batches under the different volcanoes. In all cases the parent and the process must lead to strong peralkalinity and great enrichment in incompatibles. But if each magma type represents merely a stage in a common evolutionary process why should any one stage be restricted to one volcano? The most ready alternative is that the different volcanoes do not represent various degrees of melt differentiation but *are different melts*.

Closed-system magma evolution can be discounted even for single complexes, such as Eburru, where Rb varies as if it were an incompatible element (BAILEY and MACDONALD, 1975). This relationship cannot be reconciled with generation of the liquids solely by crystal–liquid interactions because the distribution coefficient for Rb in alkali feldspar is greater than 0.3. The behaviour of Rb in a series of liquids in equilibrium with alkali feldspar, based on measured distribution coefficients for Rb and Zr and the Rayleigh fractionation law is shown for the Eburru sequence in Figure 10. Also shown are the actual and calculated distributions for the Naivasha comendites and pantellerite glasses from Pantelleria. In all three cases the predicted curves depart from the actual distributions, and it must be assumed that in each series some factor other than evolution by feldspar–liquid interaction is necessary to give the observed distribution pattern for rubidium. As halogen contents typically rise above one weight percent in the more siliceous liquids, no mineral seen in the rocks could provide the source for incompatible elements. One possibility is that a halogen-rich fluid was contributing to the source region from the onset of melting.

Alternatively the high levels of incompatible elements might perhaps be attributed to late-stage processes after the melt had filled a shallow magma chamber. Some limited variation is possible by crystal fractionation, as inferred for Longonot (SCOTT, 1980) but the Rb factor rules this process out for Eburru (BAILEY and MACDONALD, 1975) and for Naivasha comendites (MACDONALD *et al.,* 1986). Diffusion, or "liquid state" processes, may account for part of the variation, as suggested from

FIG. 9. Variations between trace elements for peralkaline glasses. Diamonds indicate samples from Pantelleria. Other symbols as in Figure 4.

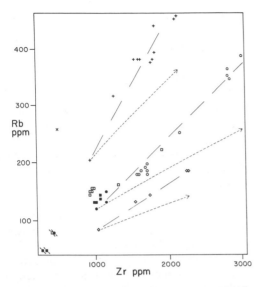

FIG. 10. Variation between Rb and Zr in peralkaline glasses. For three groups (comendites, and pantellerites from Eburru and from Pantelleria) the calculated alkali feldspar fractionation curves are shown by short dashed lines with arrows. The distribution coefficient for Rb (D = 0.3) is a mean from values given in BAILEY and MACDONALD (1975) and additional determinations, in which the distribution of Zr (D = 0.046) was also measured (using separated feldspars and glass matrices: Eburru samples; BAILEY, unpubl.). Symbols as in Figure 4.

the evidence of zoned ash flows on Menengai (LEAT et al., 1984), but if such a process has given rise to zoned magma chambers, the products lie within the range for the magma type, which itself still remains tightly constrained in any one complex. Concentration of incompatible elements solely by diffusion also runs into difficulty with low and non-systematic variations of H_2O, and the apparent ceiling on Cl levels in the Kenya province. Difficulties are further compounded by the fact that the distribution pattern between Na, Fe, and incompatible elements is reversed in pantellerites compared with trachytes and comendites.

High level, magma chamber processes offer no solutions to the fundamental *geological* questions of how these distinct felsic magma types have been produced, how they have preserved their individuality, and what *regional* conditions have led to *all* the individual centres developing such striking richness in volatile and incompatible elements. An explanation that aims to encompass these factors must consider melt sources and the possible influence of fluids in magma generation.

CARBON DIOXIDE FLUX THROUGH THE RIFT

In previous papers attention has been drawn to the evidence for massive emission of CO_2 through

the Kenya rift zone (BAILEY, 1978, 1980). The main points, with some additions, may be briefly summarised as follows.

(1) Igneous activity was initiated by nephelinite/carbonatite volcanism, which has continued intermittently for the past 23 Ma. An example of the early activity is the nephelinite volcano of Tinderet, with associated Miocene carbonatite ashes (DEANS and ROBERTS, 1984) 50 km to the west of the trachyte volcano of Menengai. The active carbonatite volcano of Oldonyo Lengai at the southern end of the rift is also part of regional igneous activity that includes felsic volcanism. In nephelinite volcanic gases, CO_2 dominates over H_2O; in nephelinite glasses from Nyiragongo, CO_2 is about ten times as abundant as H_2O; and carbonatite lavas from Oldonyo Lengai are anhydrous (but rich in F and Cl) (BAILEY, 1980).

(2) At the present time, all along the rift there is escape of CO_2, which is sufficiently abundant to be exploited in commercial wells at several places. This CO_2 flux has led to the suggestion by Kenyan geologists that present day geothermal fields along the rift are the result of heating of the ground water by hot juvenile CO_2 (WALSH, 1969).

(3) Due to the low solubility of CO_2 in salic magmas, evidence of the former presence of CO_2 in Kenyan peralkaline rocks is bound to be scarce, nevertheless, high-temperature carbonates are recorded in small pockets in some of the trachyte lavas on Longonot (SCOTT, 1982). A mixed carbonate/glass ash flow and carbonate lapillae tuffs have been discovered on Suswa (SKILLING and MACDONALD, personal communication, 1986). The main commercial CO_2 well also happens to be on the edge of the rift just to the east of Suswa.

The Kenya rift is fizzing with CO_2 and has been for at least the past 23 million years. This narrow gash in the lithosphere acts as a release channel for volatiles escaping from a large mantle volume below, analogous to the function of a pie funnel (BAILEY, 1980). The general thermal, chemical, and geodynamic consequences of such fluid focussing have been explored elsewhere (BAILEY, 1983).

SYNTHESIS

In the absence of any evidence for continuous melt evolution, the objective procedure would be to treat each magma batch as an entity, or the product of a separate event. Any links within or between the major groups, trachytes, pantellerites, and comendites, must then be attributable to common features in the melt generating regime. From this starting point, the following deductions are possible for the melting system:

(1) Comendites show regular increase in Fe_2O_3 with peralkalinity, consistent with control by melting of an acmitic pyroxene in the source. A sodic pyroxene could exert some influence on the trachyte variations. Any such pyroxenes could be formed by metasomatism prior to melting.

(2) By contrast, pantellerites show no regular variation in Fe_2O_3, even though Na_2O and FeO_T rise steadily through the range. This relationship could describe an increasing contribution from a stable amphibole (fluor–arfvedsonite?) at increased melt levels. Such a melting pattern would give rise to higher sodium and iron with greater melt volumes, consistent with the negative correlation between Na_2O and incompatible elements in the pantellerites. It would imply also that any fluorine contribution from the amphibole was insufficient to produce a noticeable effect on the distribution pattern of F in the pantellerites as a whole.

(3) Halogens and other incompatible elements are mainly contributed from a fluid entering the melt zone. The anticipated general result would be higher concentrations in earlier, smaller melt volumes, but some variations in the melt samples could be due to varying amounts of fluid flux through the melt zone (see 4.d).

(4) Part of the fluid has very low solubility in siliceous liquids, with the effect that it is not wholly consumed in the initial melt. Consequently:

(a) Elements are partitioned between melt and fluid.

(b) Partitioning is dependent on melt composition, *e.g.,* trachyte or rhyolite.

(c) Partitioning is dependent on pressure-temperature conditions, and source mineralogy.

(d) Because part of the fluid is insoluble in the melt there can be fluid flux *through the melt region.*

Thus each melt may be buffered for incompatible elements by the fluid.

(5) The fluid is essentially anhydrous and CO_2-rich. It also carries other volatile elements such as alkalies (which can be contributed to the melt) but its most striking effect is on the incompatible element budget of the melt. Partitioning works both ways and the fluid scavenges alkaline earths (notably Ba and Sr) from the more siliceous melts. This would account for the extremely low Ba and Sr concentrations in these melts, which defy explanation by closed-system crystal fractionation or partial melting hypotheses.

(6) The three main magma types are determined by three types of metasomatised source rocks:

(a) comendite from metasomatised sialic rocks;

(b) pantellerite from metasomatised basic rocks;

(c) trachyte from metasomatised mantle containing felsic minerals (BAILEY, 1986).

(7) Different magma types show major differences in their chemical patterns. These might simply reflect fundamental differences in the fluid source, but there is a strong possibility that some of the differences may be a consequence of interaction between the fluid and rocks in its path before entering the melt zone. For instance, the higher Rb in comendites may be related to earlier scavenging of this element as the fluid moved through crustal rocks before entering the melt zone. On the other hand, the generally lower levels of incompatible elements in the trachytic melts may be more a function of variations in the partitioning between the CO_2-rich fluid and a less siliceous melt at higher pressures.

(8) Individual magma types erupt from distinct centres depending on the source level in the lithosphere. Trachyte centres tap a melt source in metasomatised mantle. As the trachyte source is exhausted, and the melting cycle climbs higher in lithosphere, pantellerite can become the dominant eruptive. Comendite represents the ultimate stage of the rifting/melting process, when the melting cycle finally impinges on metasomatised sialic crust. Where an igneous cycle is well established, a range of sources, from mantle to crust, may be providing melts to a series of contemporaneous volcanoes. This scenario is consistent with the late development of the Lake Naivasha felsic magma centres, almost symmetrically disposed around the topographic culmination of the Gregory Rift.

Acknowledgments—We thank our friends and colleagues in the Department of Geology, University of Reading, whose skills and cooperation have been vital in this study. Fieldwork in Kenya was made possible by grants from the Natural Environment Research Council.

Tables of analysis are available from the first author.

REFERENCES

BAILEY D. K. (1974) Experimental petrology relating to oversaturated peralkaline volcanics: a review. *Bull. Volcanol. Special Vol.* **38**, 635–652.

BAILEY D. K. (1978) Continental rifting and mantle degassing. In *Petrology and Geochemistry of Continental Rifts,* (eds. E. –R. NEUMANN and I. B. RAMBERG), pp. 1–13 Reidel.

BAILEY D. K. (1980) Volcanism, earth degassing and replenished lithosphere mantle. *Philos. Trans. Roy. Soc. London* **A297**, 309–322.

BAILEY D. K. (1983) The chemical and thermal evolution of rifts. *Tectonophys.* **94**, 585–597.

BAILEY D. K. (1986) Fluids, melts, flowage and styles of eruption in alkaline ultramafic magmatism. *Trans. Geol. Soc. S.A.* (in press).

BAILEY D. K. and SCHAIRER J. F. (1964) Feldspar–liquid equilibria in peralkaline liquids—the orthoclase effect. *Amer. J. Sci.* **262**, 1198–1206.

BAILEY D. K. and MACDONALD R. (1970) Petrochemical variations among mildly peralkaline (comendite) obsidians from the oceans and continents. *Contrib. Mineral. Petrol.* **28**, 340–351.

BAILEY D. K. and MACDONALD R. (1975) Fluorine and chlorine in peralkaline liquids and the need for magma generation in an open system. *Mineral. Mag.* **40**, 405–14.

BAILEY D. K. and COOPER J. P. (1978) Comparison of the crystallisation of pantelleritic obsidian under hydrous and anhydrous conditions. In *Progress in Experimental Petrology, NERC Publications Series D* **11**, 230–233 Eaton Press.

BOWEN N. L. (1937) Recent high–temperature research on silicates and its significance in igneous geology. *Amer. J. Sci.* **33**, 1–21.

CARMICHAEL I. S. E. and MACKENZIE W. S. (1963) Feldspar–liquid equilibria in pantellerites: an experimental study. *Amer. J. Sci.* **261**, 382–396.

CHAIGNEAU M., TAZIEFF H. and FABRE R. (1960) Composition des gaz volcaniques du lac de lave permanent du Nyiragongo. *Acad. Sci., Paris, C. R.* **250**, 2482–2485.

COOPER J. C. (1975) Experimental melting and crystallization of oversaturated peralkaline glassy rocks. Ph.D. Thesis, University of Reading.

DAVIES G. R. and MACDONALD R. (1986) Crustal influences in the petrogenesis of the Naivasha basalt-rhyolite complex: combined trace elements and Sr–Nd–Pb isotope constraints. (Submitted to *J. Petrol.*).

DEANS T. and ROBERTS B. (1984) Carbonatite tuffs and lava clasts of the Tinderet foothills, western Kenya: a study of calcified natrocarbonatites. *J. Geol. Soc. London* **141**, 3, 563–580.

EWART A. (1979) A review of the mineralogy and chemistry of Tertiary-Recent dacitic, latitic, rhyolitic, and related salic volcanic rocks. In *Trondhjemites, Dacites, and Related Rocks* (ed. F. BARKER), pp. 13–121 Elsevier.

GREGORY J. W. (1894) Contributions to the physical geography of British East Africa. *Geogr. J.* **4**, 290.

HAMPTON C. M. and BAILEY D. K. (1984) Gas extraction experiments on volcanic glasses. *J. Non-Cryst. Solids* **67**, 147–168.

HAMPTON C. M. and BAILEY D. K. (1985) Sublimates obtained during fusion of volcanic glasses. *J. Volcanol. Geotherm. Res.* **25**, 145–155.

JOHNSON R. W. (1969) Volcanic geology of Mt. Suswa, Kenya. *Philos. Trans. Roy. Soc. London* **265A**, 383–412.

LEAT P. T. (1984) Geological evolution of the trachytic caldera volcano Menengai, Kenya Rift Valley. *J. Geol. Soc. London* **141**, 6, 1057–1069.

LEAT, P. T., MACDONALD R. and SMITH R. L. (1984) Geochemical evolution of the Menengai Caldera Volcano, Kenya. *J. Geophys. Res.* **89**, 8571–8592.

LUTH W. C. (1976) Granitic rocks. In *The Evolution of the Crystalline Rocks,* (eds. D. K. BAILEY and R. MACDONALD), pp. 335–417 Academic Press.

MCCALL G. J. H. (1957) Geology and groundwater conditions in the Nakuru area. *Tech. Rept. 3,* Hydraulic Branch, M.O.W. Nairobi.

MCCALL G. J. H. (1967) Geology of the Nakuru-Thomson's Falls—Lake Hannington area. *Geol. Surv. Kenya, Report No. 78.*

MACDONALD R. and BAILEY D. K. (1973) The chemistry of the peralkaline oversaturated obsidians. *U.S. Geol. Surv. Prof. Paper 440N-1.*

MACDONALD R., BAILEY D. K. and SUTHERLAND D. S. (1970) Oversaturated peralkaline glassy trachytes from Kenya. *J. Petrol.* **11**, 507–17.

MACDONALD R., BLISS C. M., LEAT P. T., BAILEY D. K. and SMITH R. L. (1986) Geochemical evolution of the Naivasha peralkaline rhyolitic complex, Kenya: I. Mineralogy and Geochemistry. (submitted to *J. Geol.*).

NICHOLLS J. and CARMICHAEL J. S. E. (1969) Peralkaline acid liquids: a petrological study. *Contrib. Mineral. Petrol.* **20**, 268–294.

NOBLE D. C., SMITH V. C. and PECK L. C. (1967) Loss of halogens from crystallized and glassy silicic volcanic rocks. *Geochim. Cosmochim. Acta* **31**, 215–223.

SCOTT S. C. (1980) The geology of Longonot volcano, Central Kenya: a question of volumes. *Philos. Trans. Roy. Soc. London* **296**, 437–465.

SCOTT S. C. (1982) Evidence from Longonot volcano, Central Kenya, lending further support to the argument for a coexisting CO_2 rich vapour in peralkaline magma. *Geol. Mag.* **119**, 215–217.

SCOTT S. C. and BAILEY D. K. (1984) Coeruption of contrasting magmas and temporal variations in magma chemistry at Longonot Volcano, Central Kenya. *Bull. Volcanol.* **47**, 849–873.

SUTHERLAND D. (1974) Petrography and mineralogy of the peralkaline silicic rocks. *Bull. Volcanol. Special Vol.* **38**, 517–547.

TAYLOR D. E., EICHELBERGER J. C. and WESTRICH H. R. (1983) Hydrogen isotopic evidence of rhyolitic magma degassing during shallow intrusion and eruption. *Nature* **306**, 541–545.

THOMPSON A. O. and DODSON R. G. (1963) Geology of the Naivasha area. *Geol. Surv. Kenya Report No. 55.*

THOMPSON R. N. and MACKENZIE W. S. (1967) Feldspar-liquid equilibria in peralkaline acid liquids: an experimental study. *Amer. J. Sci.* **265**, 714–34.

TUTTLE O. F. and BOWEN N. L. (1958) Origin of granite in the light of experimental studies in the system $NaAlSi_3O_8-KAlSi_3O_8-H_2O$. *Geol. Soc. Amer. Mem.* **74**, 153 pp.

UNNI C. K. and SCHILLING J.-G. (1978) Cl and Br degassing by volcanism along the Reykjanes Ridge and Iceland. *Nature* **272**, 19–23.

WALSH J. (1969) Mineral and thermal waters of Kenya. *XXIII International Geol. Cong.* **19**, 105–110.

WILCOX R. E. (1979) The liquid line of descent and variation diagrams. In *The Evolution of the Igneous Rocks: Fiftieth Anniversary Perspectives,* (ed. H. S. YODER, JR.), Chap. 7, pp. 205–232, Princeton University Press.

WILLIAMS L. A. J. (1982) Physical aspects of magmatism in continental rifts. *Continental and Oceanic Rifts Geodynamics,* Series 8, 193–222.

Magmatic Processes: Physicochemical Principles
© The Geochemical Society, Special Publication No. 1, 1987
Editor B. O. Mysen

Transfer of subcratonic carbon into kimberlites and rare earth carbonatites

PETER J. WYLLIE

Division of Geological and Planetary Sciences, California Institute of Technology, Pasadena, CA 91125, U.S.A.

Abstract—Carbon and volatile components involved in the genesis of kimberlites and carbonatites rise from a mantle reservoir below the asthenosphere. Vapors include the components C–H–O–S–K, in molecular form dependent on the oxygen fugacity, a parameter that varies as a function of depth in ways not yet fully understood. Kimberlites are generated where upward percolating reduced volatile components cross the solidus for peridotite–C–H–O. The depleted, refractory base of the lithosphere, 200–150 km deep, is a collecting site for kimberlite magma at temperatures above its solidus; this layer has been intermittently invaded by small bodies of carbonated kimberlite, through billions of years; most of these aborted and gave off vapors enriching the lower lithosphere by metasomatism, but some reached the surface, through vapor–enhanced crack propagation. Nephelinites and associated carbonatites require upward movement of solid mantle as a plume. Thinning of the lithosphere above a mantle plume, beneath a rift, results in magma trapped in the asthenosphere–lithosphere boundary layer rising with the isotherms, without crossing the solidus until the magma reaches the depth interval 90–65 km, where the solidus for peridotite–CO_2–H_2O becomes subhorizontal, with low dP/dT, pressure independent, and forming a ledge or phase equilibrium barrier. At this level, magma chambers form, and crystallization is accompanied by evolution of vapors, enhancing crack propagation and the eruption of nephelinitic magmas that differentiate to carbonatites. The release of vapors at this level generates another metasomatic layer, at depths known to contain metasomatic RE–titanates. These metasomes may be the source of the REE in carbonatites. Liquidus studies in the system $CaCO_3$–$Ca(OH)_2$–$La(OH)_3$ at 1 kbar demonstrate that residual carbonatite magmas may contain more than 20 weight percent $La(OH)_3$, as long as the REE were not removed at earlier stages of differentiation by other minerals.

INTRODUCTION

CARBON IS A TRACE element in most igneous rocks, but it becomes a major component in carbonatites, and it plays a significant role in kimberlites. Possible sources of carbon in igneous rocks include shallow-level material of the biosphere, sedimentary or metamorphic limestones, subducted limestones or carbonated ocean–floor basalt, or primordial carbon from the deep mantle.

A general scheme illustrating the inter-relationships of various types of igneous activity in different tectonic environments is presented in Figure 1. The parts of the scheme relevant for this paper are outlined here, and discussed in more detail in following pages. Oceanic basalts contain small concentrations of H_2O and CO_2 certainly derived from the mantle. The submarine basalts react with sea water, with the formation of calcite and hydrated minerals. The oceanic crust when subducted may carry with it in addition pelagic sediments, including slabs of limestone (during the relatively brief period of geological history that pelagic limestones have been formed). During metamorphism and melting of the subducted oceanic crust, it appears that most of the water escapes from the crust to participate in convergent boundary processes. Some of the carbonate may escape these processes and be carried deeper into the mantle for long–term storage of carbon.

Kimberlites intrusions rise from deep within the mantle, near the lithosphere–asthenosphere boundary layer. The diamonds that they transport to the surface have been resident in this region through thousands of millions of years, providing irrefutable evidence that carbon does exist in the mantle. Although diamonds occur only as trace minerals in kimberlites, there is now little doubt that some of the other carbon in kimberlites, consolidated in the form of carbonate, is derived from at least this depth. Carbonatites are formed by differentiation of nephelinitic magmas, commonly in rifted cratonic environments. Limestone assimilation has been invoked in the petrogenesis of alkaline magmas, but most investigators now agree that the carbon in carbonatites was derived from the deep source of the alkalic magmas. Mantle carbon may be primordial, or derived from subducted oceanic crust.

The geochemistry of the magmas represented in Figure 1 is currently interpreted in terms of melts derived from distinct mantle and crustal source reservoirs, contaminated to greater or lesser extent by metasomatic fluids. Some magmas have trace element and isotope fingerprints indicating that they have components derived from more than one source. The geochemistry of carbon with respect to the location and evolution of these reservoirs is an important problem still poorly understood.

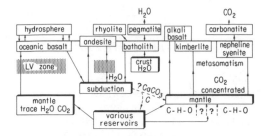

FIG. 1. Generalized cycle illustrating the distribution of H_2O and CO_2 among igneous rocks and their crust and mantle sources. From left to right, the tectonic environments represented are: divergent oceanic plate boundary, convergent plate boundary, and continental platform, with rifting.

SUBDUCTION OF OCEANIC CRUST

The main attention to subducted igneous crust in the context of igneous petrology has been in connection with the origin of calc–alkaline igneous rock series. Water is generally considered to be the dominant volatile component, although the large masses of carbon dioxide given off from andesitic cones have been attributed to sources in the subducted crust. The phase relationships required to trace the fate of calcite or dolomite in subducted oceanic crust have been determined (HUANG *et al.*, 1980; HUANG and WYLLIE, 1984, Figure 7). As a first step in study of melting relationships between eclogite and carbonate, OTTO (1984) and OTTO and WYLLIE (1983) determined the melting relationships of calcite in contact with albite–jadeite at 25 kbar. The near-solidus liquid contains more than 50% dissolved $CaCO_3$. HUANG *et al.* (1980) concluded that there is a prospect that some carbonates in subducted oceanic crust would escape dissociation and melting, be converted to aragonite, and be carried to considerable depths for long–term storage in the mantle. Calcite should react with adjacent peridotite to yield dolomite or magnesite (WYLLIE *et al.*, 1983), but according to current views on the redox state of the upper mantle, carbonate carried into the mantle should eventually become reduced to graphite or diamond (RYABCHIKOV *et al.*, 1981; HAGGERTY, 1986).

THE GENERATION AND ERUPTION OF KIMBERLITES

In order to locate the positions within the upper mantle for the generation of kimberlite magmas, we need to know the compositions of the rocks at various depths, their phase relationships as a function of depth and volatile contents, with particular attention to the solidus curves and the compositions

of near–solidus liquids, and the position of the geotherm. A compilation of some of these data is given in Figure 2. In order to decipher the process of eruption of kimberlites, we need to know the thermal history of the region at depth, which is a consequence of physical processes and the physics of the solid–melt–vapor systems.

Geotherms beneath cratons

From the geothermometry and geobarometry of peridotite nodules transported to the surface by kimberlites and alkali basalts, the geotherm for cratons is consistent with geotherms calculated from heat loss. The fossil geotherm for many kimberlites

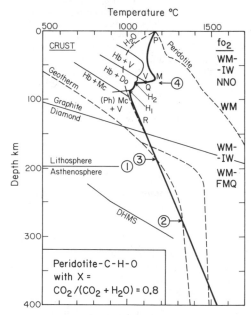

FIG. 2. Solidus curves for peridotite–H_2O and for peridotite–CO_2–H_2O with defined ratio of CO_2/H_2O (see Figure 10), compared with cratonic geotherm, with and without inflection (BOYD and GURNEY, 1986). For phase equilibrium background and sources, see WYLLIE (1977, 1978, 1979) and EGGLER (1978b). The positions of selected phase boundaries relevant to mantle processes and the origin of kimberlites and nephelinites are given, and discussed in the text. Level (1) corresponds to the depth of the lithosphere–asthenosphere boundary, points (2) and (3) are depth levels limiting the interval within which magma can be generated (in the presence of fluids) for the particular geotherm, and level (4) corresponds to the change in slope of the solidus curve near point Q. The oxygen fugacities at various levels are indicated by the buffers listed (HAGGERTY, 1987). Peridotite solidus from TAKAHASHI and SCARFE (1985). For DHMS (dense hydrated magnesian silicates) see RINGWOOD (1975, p. 295). Abbreviations: Hb = amphibole, Ph = phlogopite, Do = dolomite, Mc = magnesite, V = vapor. See also Figures 7 and 8 for discussion of R, H_1, H_2, V, Q, M and P.

(corresponding to the time of eruption) is inflected to higher temperatures at a depth of about 175 km, somewhat deeper than the graphite to diamond transition. Two geotherms are depicted in Figure 2, one with and one without an inflection (BOYD and GURNEY, 1986). The inflection has been interpreted in terms of uprise of mantle diapirs associated with the generation of kimberlites, or in terms of local magma chambers, or as due to materials having different thermal conductivities.

Phase relationships

The phase relationships for the system peridotite–CO_2–H_2O, involving the minerals amphibole, phlogopite, dolomite and magnesite (see Figure 2 for an example), provide the framework for upper mantle petrology. There is abundant evidence in fluid inclusions for the passage of H_2O and CO_2 through the lithosphere (*e.g.,* ANDERSEN *et al.,* 1984). The redox state of the deeper mantle, however, appears to be such that the components C–H–O exist as carbon and H_2O with CH_4, rather than as CO_2 and H_2O (DEINES, 1980; RYABCHIKOV *et al.,* 1981). HAGGERTY (1986) considered fluids in the asthenosphere to be relatively oxidized (between FMQ and WM buffers), within the field for CO_2 (and therefore, presumably, of carbonate), and concluded that these fluids would be reduced to microdiamonds when released into the more reduced lithosphere, between the WM and IW buffers. HAGGERTY (1987) concluded that a layer of the lithosphere between 60 and 100 km depth has been metasomatized and oxidized up to the NNO buffer. Figure 2 summarizes the possible variations of oxygen fugacity in terms of standard buffers down to 250 km. The phase relationships in the reduced system peridotite–C–H–O have been analyzed in detail by WOERMANN and ROSENHAUER (1985) (see also GREEN *et al.,* 1987). MEYER (1985) reviewed the distribution and redox state of C–H–O in the mantle in connection with the origin of diamonds. HOLLOWAY and JAKOBSSON (1986) presented data on the distribution of C–H–O components between liquid and vapors, with applications to oxygen fugacities in mantle and volcanic gases.

With low oxygen fugacities at high pressures, fluids in the system C–H–O exists as H_2O with CH_4 or graphite/diamond. Under these conditions, it appears that the peridotite–C–H–O solidus may remain close to that for peridotite–H_2O, with carbonate ions being generated in the melt when CH_4 or graphite/diamond dissolves (EGGLER and BAKER, 1982; RYABCHIKOV *et al.,* 1981; WOERMANN and ROSENHAUER, 1985). In the following

discussion, therefore, the extrapolated solidus for peridotite–H_2O, which I believe to be very close to that for peridotite–CO_2–H_2O at pressures greater than about 35 kbar (WYLLIE, 1978; ELLIS and WYLLIE, 1980), is adopted as the solidus at depth for peridotite–C–H–O. The extrapolated solidus curves given in Figure 2 correspond closely to those adopted previously in consideration of the origin of kimberlite (WYLLIE, 1980), and the qualifications stated then will not be repeated. The solidus for mixed volatile components is for $CO_2/(CO_2 + H_2O)$ = 0.8. With variation in this ratio the solidus temperature changes at pressures shallower than point Q at about 75 km, but remains unchanged at higher pressures (see WYLLIE, 1978, 1979; EGGLER, 1978b, for details). More detailed evaluation of the phase relationships at pressures between 50 and 100 km depth will be presented below, in connection with the origin of nephelinites and carbonatites.

Critical depth levels

There are four levels in the upper mantle, identified in Figure 2, where critical changes occur in the physical processes that control the composition and mode of migration of the low–SiO_2, volatile–rich magmas erupted in cratonic environments. The first critical level, (1), is the depth of the lithosphere–asthenosphere boundary layer, through which the mantle flow regime changes from convective (ductile) to static (brittle). A depth of 200 km is commonly adopted for this level in subcratonic mantle, corresponding approximately to the 1200°C isotherm. The two depths where the solidus is intersected by the local geotherm, (2) and (3), limit the depth interval within which magmas can be generated. The fourth level, (4), is the narrow depth interval near Q in Figure 2 where the solidus changes slope, and becomes sub-horizontal, with low dP/dT. The depth of level (4) differs according to different investigators (see review below). Levels (2), (3) and (4) are different for lherzolites and harzburgites (WYLLIE *et al.,* 1983); compared with lherzolite the solidus for harzburgite is higher, and therefore the levels (2) and (3) are deeper, and level (4) is also deeper. The depths of levels (1), (2) and (3) vary from place to place and from time to time, as a function of the geotherm and local history.

Other important levels in the mantle are those where volatile components would be released by activation of a dissociation reaction. One such reaction is represented in Figure 2 by the line DHMS giving the boundary for the reaction of olivine to form dense hydrous magnesian silicates in the presence of H_2O. The estimated position of the reaction

for the conversion of forsterite to brucite and enstatite in $MgO-SiO_2-H_2O$ is close to the DHMS
curve at 300 km (ELLIS and WYLLIE, 1979, 1980).

Processes

An interpretation of the processes involved in
the origin of kimberlites is given in Figure 3 (WYL-
LIE, 1980). The solidus corresponds to that in Figure
2, with somewhat higher estimated temperature for
M. Reduced volatile components rising up the geotherm cross point *a* (level 2), with the formation of
a trace of interstitial melt. Increased concentration
of melt results in density instability, followed by
uprise of diapirs. In this current modification of the
1980 treatment, motion of the diapirs is slowed or
stopped as they enter the base of the lithosphere at
level (1). If the melt cools sufficiently to release vapor, a crack may propagate from level *c*, releasing
kimberlite magma for intrusion into the crust or
eruption. Under these circumstances, the magma
passes right through level (4) without being affected
by the peridotite phase relationships. A more common fate for such small magmatic excursions, according to SPERA (1984), is solidification at depth
through thermal death.

If the melts are not stopped by the rheological
change at level (1), the asthenosphere–lithosphere
boundary, they may continue to rise along adiabats,
as represented by path *a–f* in Figure 4 (in which
case, only minor melting occurs), or they may follow
a path only slightly elevated above the geotherm,
represented by *a–e* (SPERA, 1984). As the magmas
approach the solidus from below, remaining in
equilibrium with host lherzolite, crystallization

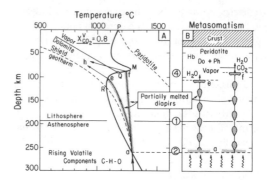

FIG. 4. Solidus controls on the upward migration of
partly melted diapirs through lithosphere, following adiabatic or shallower paths of uprise. Compare Figure 3.
Crystallization and evolution of vapors occurs at the solidus, M–Q–R, and the mantle above can become strongly
metasomatized (modified after WYLLIE, 1980).

proceeds at *f* or *e* and vapor is evolved. The overlying mantle is subjected to extensive metasomatism. HAGGERTY (1987) concluded independently
from study of mantle nodules that there is a metasomatized horizon between 100 km and 60 km
depth. Magma chambers may develop at positions
e and *f*. The point Q is significant because at shallower levels the composition of vapor released is
the same as the ratio of CO_2/H_2O in the system,
whereas at greater depths the vapor is progressively
enriched in H_2O (WYLLIE, 1978, 1979, 1980). The
compositions of melt and vapors associated with
the solidus in the region of level (4), MQR in Figure
4A, vary widely for quite small changes in conditions. Many alkalic magmas are probably derived
from parental magmas generated in this region.
Note that adiabatic paths for diapirs of higher temperature, rising from greater depths, would miss the
solidus at M; these might reach the solidus for peridotite, with generation of basaltic magmas. Similarly, if the ratio CO_2/H_2O was lower, the temperature of point M would be lower and again, the
diapirs could rise past this level without encoun-
tering the solidus.

Consider the craton in Figure 5 with normal, undisturbed geotherm, composed largely of lherzolite
with a concentration of harzburgite in its lower part,
between about 170 and 200 km depth, along with
pods of eclogite (SOBOLEV, 1977; BOYD and GUR-
NEY, 1986; HAGGERTY, 1986). No magma is generated unless volatile components are present or
are introduced into the depth level between 270 km
and 185 km (levels 2 and 3 in Figure 2). For the
geotherm and solidus given in Figures 2, no vapors
or carbonate can exist between (2), 275 km, and
(3), 180 km, because they would be dissolved in

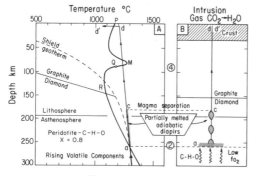

FIG. 3. Solidus and geotherm controls on the generation
of magma in mantle peridotite with a flux of volatile components inducing adiabatic uprise of partially melted diapirs. The diapirs fail to penetrate far into the lithosphere.
Under suitable tectonic conditions, with release of vapor
enhancing crack propagation, kimberlite magma is intruded into the crust from depth *c* (modified after WYLLIE,
1980).

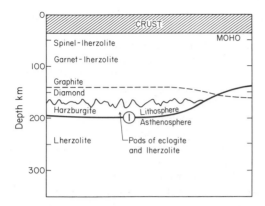

FIG. 5. Simple petrological cross-section of a craton, with depleted keel of harzburgite containing macrodiamonds (SOBOLEV, 1977; BOYD and GURNEY, 1982, 1986).

melt (WYLLIE, 1980, Figure 4). Therefore, the generation of new melt is dependent on the uprise of volatile components from depths greater than (2). Note that the reactions of olivine with water to form DHMS or brucite probably intersect the geotherm somewhere between 300–400 km, suggesting that perhaps only highly reduced gas, CO, CH_4 and H_2, can exist at deeper levels.

Assume that a part of the lithosphere is in the early stages of rifting, initiated by an increase in heat flow supplied by a mantle plume from the deep mantle. Sparse volatile components entrained in the rising plume, enhanced by release of H_2O from DHMS and brucite (WOERMANN and ROSEN-HAUER, 1985), if present, will generate interstitial melt at level (2), where the lherzolite is transported across the solidus curve (compare Figures 3 and 4). Figure 6 illustrates the plume diverging laterally below level (1), the asthenosphere–lithosphere boundary. As it diverges, the melt becomes concentrated in layers or chambers in the boundary layer above the plume. This aggregation is associated with local uprise of the geotherms, and thinning of the lithosphere. Lateral divergence of the asthenosphere transports some of the entrained plume melt, which later penetrates the lithosphere, forming small dikes or magma chambers.

The magmas entering the depleted lithosphere, both above the plume and laterally beneath the undisturbed craton, remain sealed within the more rigid lithosphere, maintained at temperatures above the solidus for lherzolite–C–H–O. The magmas have no tendency to crystallize nor to evolve vapors unless they reach level (3), 10–15 km above the asthenosphere–lithosphere boundary. Contact of the lherzolite–derived magma with harzburgite, however, should result in reaction and the precip-

itation of minerals through magma contamination. This slow process could lead to the more or less isothermal growth of large minerals resembling the discrete nodules in kimberlites.

Those magmas managing to insinuate their way near to level (3), the solidus, will evolve H_2O–rich vapors. CO_2–rich vapors cannot exist in this part of the mantle (WYLLIE, 1978, 1979; EGGLER, 1978b). They will either be reduced by thermal cracking (HAGGERTY, 1986) yielding microdiamonds that join the old macrodiamonds resident in the depleted keel of the craton through thousands of millions of years, or cause partial melting in lithosphere lherzolite, or they will react with harzburgite to produce magnesite and consequent enrichment of the vapor in H_2O. The vapors may also promote crack propagation, resulting in rapid uprise of the kimberlite magma. Many intrusions from this level will solidify before rising far (SPERA, 1984), but others will enter the crust as kimberlite intrusions (ARTYUSHKOV and SOBOLEV, 1984). Kimberlites may be erupted either from the early magma accumulating above the plume, or from the lateral magma chambers in the lithosphere base. Magnesite–harzburgite nodules would be disrupted by explosive dissociation of the carbonate during uprise,

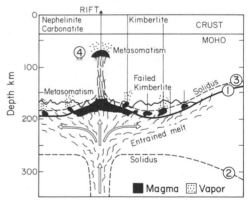

FIG. 6. Magmatic consequences of a deep mantle plume carrying volatile components across the solidus at level (2), with concentration of magma chambers above the plume at the asthenosphere–lithosphere boundary, and entrainment of melt in flowing lherzolite of the asthenosphere. The latter melt enters the boundary layer of the lithosphere in small dikes or magma chambers. If the magma approaches the solidus, level (3), evolution of vapor enhances crack propagation, permitting some kimberlite magmas to reach the crust (see Figure 3). Thinning of the lithosphere above the plume results in upward migration of the magma to level (4), where magma chambers develop, and vapor is evolved with strong metasomatic effects (compare Figure 4). This level is the source of parent magmas for nephelinitic volcanism, and the associated carbonatites.

which provides a satisfactory explanation for the correlation between low-calcium garnets and diamonds in the kimberlites of the Kaapvaal craton (BOYD and GURNEY, 1982; WYLLIE et al., 1983), and in USSR (SOBOLEV, 1977).

THE SOURCE OF NEPHELINITIC VOLCANISM AND CARBONATITES

Alkalic magmatism is commonly associated with doming and rifting on cratons. There is evidence that beneath major rifts the lithosphere is thinned (GLIKO et al., 1985). In Figure 6, the continued heat flux from the rising plume, and the concentration of hotter magma at the asthenosphere-lithosphere boundary, will promote further thinning of the lithosphere. According to GLIKO et al. (1985), it takes only several million years for lithosphere thickness to be halved when additional heat flow of reasonable magnitude is applied to the base of the lithosphere. The igneous sequence associated with this process was discussed by WENDLANDT and MORGAN (1982). The magma near the boundary layer at levels (1)–(3) will rise with the boundary layer, either percolating through the newly deformable matrix, or as a series of diapirs, with the amount of liquid increasing in amount as the boundary layer rises, extending further above the solidus for peridotite–C–H–O. This magma intersects the shallower solidus for peridotite–CO_2–H_2O at level (4) (compare Figures 2, 3, and 4), in the depth range 90–70 km. Magma chambers may be formed as the magma solidifies, and vapors will be evolved causing metasomatism in the overlying mantle, and causing intermittent crack propagation, which releases magmas through the lithosphere. The metasomatic vapors will be aqueous if released at depths greater than Q (Figures 2 and 4), and CO_2-rich if released at depths shallower than Q (HAGGERTY, 1987). A variety of alkalic magmas may be generated at level (4), depending sensitively upon conditions. Magmas rising from this level may include the parents of olivine nephelinites, melilite–bearing lavas, and other igneous associations differentiating at shallower depths to carbonatites.

Phase relationships

There are large changes in the compositions of melts and vapors in the region of level (4) for small changes in pressure and temperature (WYLLIE, 1978; WENDLANDT and EGGLER, 1980; WENDLANDT, 1984). Detailed knowledge of the phase relationships in this region are, therefore, essential for understanding the petrogenesis of alkaline rocks. In previous interpretations of the geometrical ar-

rangements of divariant surfaces in this region, based on available experimental data, it appeared that the amphibole surface would not overlap the dolomite–peridotite–vapor solidus curve, M–Q–R in Figures 3 and 4 (WYLLIE, 1978, 1979). The geometrical relationships for the system were illustrated as a guide for rock systems in which these two–phase elements might overlap. The relationships are shown in Figures 7A and B from WYLLIE (1978). Figure 7C represents the phase relationships for a rock with less H_2O (0.4 weight percent) and CO_2 (5 weight percent) than that required to make the maximum amount of amphibole and dolomite, respectively, in the peridotite. The vapor–absent field for amphibole–dolomite–peridotite extends across the solidus curve defined by the amphibole buffer, m_2–H_2, and the dolomite buffer, H_1–N (part of QR in Figures 3 and 4). For a rock with a higher ratio of CO_2/H_2O, a small field for peridotite + vapor would appear in the region of m_2–H_2 where the amphibole field does not reach the solidus (see Figure 7 in WYLLIE, 1979, and Figures 10C and D). EGGLER (1978a) constructed the dolomite peridotite-vapor solidus curve I_6–N (Figure 7B) with a pressure minimum, which carried it across the amphibole stability surface at a lower pressure than depicted in Figure 7.

Two experimental studies on amphibole–dolomite–peridotite indicate that the amphibole stability volume does overlap the solidus with dolomite, with results shown in Figures 8A and B. The points H_1 and H_2 correspond to the same points in Figure 7. There is a difference of several kbar between the results of BREY et al. (1983) in Figure 8A and those from the detailed investigation of OLAFSSON and EGGLER (1983) in Figure 8B. I have interpreted these two sets of experimental results in terms of the topology given in Figure 7 (WYLLIE, 1987), and conclude that the simplest interpretation indicates that there is a discrepancy related to the position of the invariant point I_6 in Figure 8C for the system peridotite–CO_2. The point E76 is EGGLER's (1978b) determination of the point in the model system CaO–MgO–SiO_2–CO_2, where the low-pressure solidus curve for peridotite–CO_2 is succeeded by the high-pressure solidus curve for dolomite–peridotite–CO_2. It is this fundamental change that is responsible for the solidus ledge on the peridotite–CO_2–H_2O solidus, described above as level (4) (Figures 2, 3, and 4). According to the present interpretation of the results in Figure 8A, the point is near 25 kbar (G in Figure 8C), and for the results in Figure 8B it is near 17 kbar (OE in Figure 8C). This difference in pressure amounts to almost 30 km within the mantle.

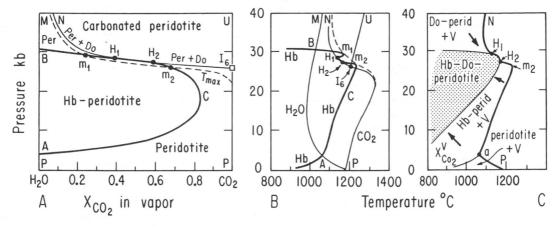

FIG. 7. An example of the topology of the phase relationships for peridotite–CO_2–H_2O if the amphibole stability range overlaps the solidus curve I_6–N (Q–N) for dolomite–peridotite–vapor. Figures 7A and 7B are after WYLLIE (1978). For abbreviations, see Figure 8. A. PX projection of the solidus surface and vapor buffer lines. Between H_1 and H_2 amphibole–dolomite–peridotite begins to melt in the presence of the specified buffered vapor phase composition (mol fraction). B. PT projection showing the amphibole buffer line ACB passing over the temperature maximum on the solidus surface (m_1–m_2), and down to the dolomite buffer line I_6–N between H_1 and H_2. C. Isopleth for peridotite with CO_2 and H_2O below amounts required to generate maximum possible dolomite and amphibole, for the condition of overlapping amphibole stability field in Figures 7A and 7B. The solidus is the heavy line: compare with the reported experimental solidus curves in Figures 8A and 8B. The heavy arrows show direction of increasing CO_2/H_2O in subsolidus areas. The shaded area is vapor–absent.

In an effort to resolve this discrepancy, the position of I_6 has been determined using an assemblage of natural minerals corresponding to a lherzolite

(WYLLIE and RUTTER, 1986). The preliminary experimental value at 21 kbar is between the two values deduced from the experiments in Figures 8A

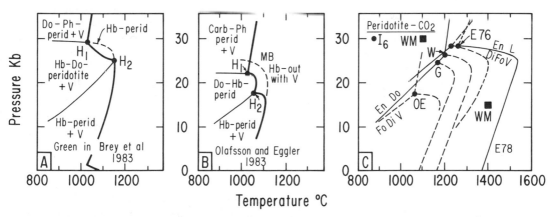

FIG. 8. A and B. Experimentally determined solidus curves for defined compositions in peridotite–H_2O–CO_2, with some interpretation of subsolidus phase fields for comparison with Figure 3. Note the subsolidus assemblage between H_1 and H_2 with vapor, (A), and without vapor, (B). The dashed curve, MB, is the amphibole–out curve in the same rock with H_2O. C. Locations for the invariant point I_6 on the solidus for peridotite–CO_2 deduced or estimated from experimental data in other systems. E78–E76 (solid lines) was experimentally determined in the model system CaO–MgO–SiO_2–CO_2 by EGGLER (1978b), and he estimated the adjacent dashed curves for peridotite from these results (EGGLER, 1976). Point W was estimated by WYLLIE (1978). G and OE are deduced from the experimental results given in Figures 8A and 8B, respectively. WM are two experimental points on the solidus for peridotite–CO_2 (WENDLANDT and MYSEN, 1980). Abbreviations: Hb = amphibole, Do = dolomite, Ph = phlogopite, Di = diopside, En = enstatite, Fo = forsterite, V = vapor, L = liquid, perid = peridotite. MB = MYSEN and BOETTCHER (1975).

and 8B. This result, in conjunction with those in Figures 8A and 8B, contributes to the reconstruction of the phase relationships in the system amphibole–dolomite–peridotite for a variety of volatile contents and ratios of CO_2/H_2O. All previous investigators have assumed that the point I_6 for peridotite is in the range 27–29 kbar, close to that in the synthetic system. Determination of its position at lower pressures results in significant changes in the shapes of the overlapping surfaces for dolomite, amphibole and the solidus, although the topology remains the same. In particular, the highest temperature on the peridotite–CO_2 solidus is reduced, and consequently the temperature of the maximum on the solidus for peridotite–CO_2–H_2O (M at level 4 in Figures 3 and 4) is also reduced.

Consider first the shapes of the divariant solidus surfaces for peridotite–vapor and for dolomite–peridotite–vapor, without consideration for amphibole. These are depicted by contours on the surfaces for constant vapor-phase compositions (WYLLIE, 1978). Figure 9 shows the solidus curves for peridotite–CO_2, peridotite–H_2O, and for peridotite in the presence of the successive vapor phase compositions indicated. The line QR is the dolomite–peridotite–vapor buffer line shown in Figures 3 and 4. Note that the extent of the sub-horizontal level (4) is reduced with increasing H_2O/CO_2, as Q migrates along the curve I_6–R. EGGLER's (1978a) interpretation has Q passing through a pressure minimum between I_6 and R.

Figure 10 is a similar diagram showing the intersections of the amphibole stability surface with the divariant surfaces for dolomite formation, and

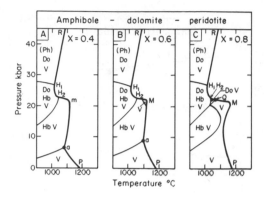

FIG. 10. Buffered solidus curves for amphibole–dolomite–peridotite–vapor with amphibole stability volume crossing the dolomite–peridotite–vapor buffer line I_6–N (Figure 7), I_6–R (Figure 9). Revised from WYLLIE (1979, Figures 6B and E, 7) with mol ratios of $CO_2/(CO_2 + H_2O)$ = X. Note the section of solidus H_1–H_2 with vapor, contrasted with H_1–H_2 in Figures 7C and 8B without vapor. For abbreviations, see Figure 8.

for the solidus with excess vapor. The amphibole stability field overlaps the solidus surface, generating a buffered solidus curve for amphibole–peridotite–vapor when CO_2/H_2O is somewhat reduced (Figures 10B and 10A). The field for amphibole–dolomite–peridotite is situated between the amphibole and dolomite fields (compare Figures 7 and 8). Figure 10C is the diagram used in Figure 2. Comparison of Figure 10C with the curves PMQR in Figures 3 and 4 indicates that changing the position of I_6 from about 28 kbar to 21 kbar reduces the maximum temperature of the subhorizontal solidus ledge at level (4), and reduces the sharpness of the temperature maximum on the solidus at M (for high CO_2/H_2O; compare Figure 9). According to Figure 10, however, there is a substantial ledge on the solidus at this level for every peridotite containing H_2O and CO_2, regardless of the ratio, which is borne out by the two sets of experimental data in Figures 8A and 8B. If H_2O and CO_2 contents are less than those required to make the maximum amphibole and dolomite, respectively, then the vapor-absent solidus for amphibole–dolomite–peridotite in Figure 10 would form another ledge between H_1 and H_2 (see Figures 7C and 8B).

ORIGIN OF CARBONATITES

As outlined above, the most likely source for the parent magmas of the alkaline complexes with carbonatite intrusions is in the metasomatized upper mantle at depth level (4), associated with the abrupt change in slope of the peridotite–H_2O–CO_2 solidus (Figures 2, 4, and 6). Although the parent melts are

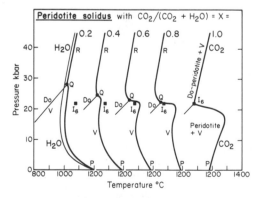

FIG. 9. The system peridotite–CO_2–H_2O, with excess CO_2 and H_2O. Contours for two parts of the solidus surface, the peridotite-vapor surface from low pressures to the line I_6–Q–R, and the dolomite–peridotite–vapor surface extending to higher pressures from the line I_6–Q–R. Contours are plotted in terms of mol fraction of CO_2 in the vapor phase. Revised from WYLLIE (1978).

derived from reduced materials at deeper levels (*e.g.,* Figures 4 and 6), they are relatively oxidized by the time they reach the 85–65 km depth level (4) (HAGGERTY, 1987).

Carbonatites are magmatic in origin, commonly derived by differentiation from a parent nephelinitic magma. Fractional crystallization plays a role in differentiation, but liquid immiscibility, *i.e.,* splitting a nephelinitic magma into an ijolite and a carbonatite magma, is another important process. LE BAS (1977) reviewed these topics in detail. The discovery that near–solidus melts in dolomite–peridotite have compositions corresponding to those of carbonatites (WYLLIE and HUANG, 1975; WENDLANDT and MYSEN, 1980) has led to renewed speculation that some carbonatites might be primary, but in most alkaline complexes the volume relations and the time sequence of intrusion appear to favor the derivative origin.

Rare earth carbonatites and phase relationships

Light rare earth elements are concentrated in carbonatites. PECORA (1956) distinguished two varieties of carbonatites, the apatite–magnetite and the rare–earth mineral varieties. Many carbonatites of the apatite–magnetite type, however, do contain REE in monazite. HEINRICH (1966, p. 157) noted that if the term "rare–earth carbonate type" is used, the schism is more pronounced and paragenetically more significant. Bastnaesite is the most abundant RE carbonate in carbonatites, containing up to 64 weight percent REE. There are two problems associated with rare–earth carbonate carbonatites. One is the original source of the rare earth elements, and the other is the process of their concentration.

The source is probably the layer of metasomatized subcontinental lithosphere represented at level (4) in Figures 4 and 6, and characterized in terms of chemistry and mineralogy by HAGGERTY (1987). Haggerty concluded that the existence of upper–mantle metasomes is virtually a prerequisite to the genesis of alkali melts with elevated (silicate–) incompatible element signatures. Among the new minerals discovered in association with spinels in depleted mantle peridotites less than 100 km are titanates enriched in rare earth elements (HAGGERTY *et al.,* 1986; HAGGERTY, 1987). Involvement of these minerals in magmatic processes at level (4) may provide the initially elevated rare earth element concentrations that can lead through differentiation to the formation of rare–earth carbonate carbonatites.

The RE–carbonate type of carbonatite usually represents the last stage in a series of carbonatitic

differentiates, and is normally subordinate to earlier carbonatite. Under some conditions, as in the Mountain Pass carbonatite, with 15 volume percent of the ore body composed of bastnaesite, very high concentrations of RE carbonates are produced.

JONES AND WYLLIE (1983, 1986) have approached the petrogenetic problem in two ways. The first is to build from simple to more complex phase diagrams in order to establish precisely the behavior of rare earth elements in carbonate–rich melts, as a guide for the interpretation of the more complex systems and of the rocks themselves. The second approach is to melt complex mixtures approximating the composition of the ore body at Mountain Pass, to follow paths of crystallization, and to determine the conditions for precipitation of bastnaesite.

The phase relationships in the join $CaCO_3$–$Ca(OH)_2$–$La(OH)_3$ at 1 kbar pressure are given in Figure 11. The synthetic carbonatite magma represented by eutectic E_1 dissolves about 20 weight percent $La(OH)_2$ at the ternary eutectic E at 610°C. With increasing CO_2/H_2O in the liquid, represented by the field boundary along E–a, the solubility of $La(OH)_2$ rises to 40% at the 700°C piercing point.

In the second set of experiments, JONES and WYLLIE (1983) and WYLLIE and JONES (1985) made synthetic mixtures approximating the composition of the Mountain Pass ore body as illustrated in Figure 12. The shaded area represents the chemistry of the ore excluding the rare earth elements. $Ca(OH)_2$ was added to generate the low-temperature synthetic magma (base of the triangle in Figure 12), and the mixture E is estimated to be close in composition to the quaternary eutectic. The phase fields

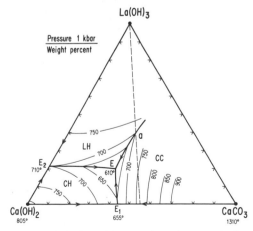

FIG. 11. Preliminary interpretation of ternary liquidus relations in the join $CaCO_3$–$Ca(OH)_2$–$La(OH)_3$ at 1 kbar pressure (JONES and WYLLIE, 1986). Abbreviations: CH = $Ca(OH)_2$, LH = $La(OH)_3$, CC = $CaCO_3$.

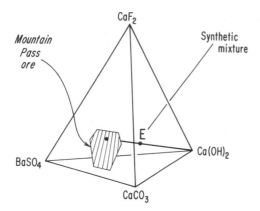

FIG. 12. The mixture E contains 39 weight percent CaCO$_3$, 17 weight percent CaF$_2$, and 13 weight percent BaSO$_4$, representing the Mountain Pass carbonatite ore without rare earth elements, together with 31 weight percent Ca(OH)$_2$ (JONES and WYLLIE, 1983; WYLLIE and JONES, 1985).

intersected by the composition join E–La(OH)$_2$ at 1 kbar were determined. The liquidus has a minimum piercing point at 625°C, with 18 weight percent dissolved La(OH)$_3$. Phase fields including La-bearing minerals extend from this piercing point. The solidus is at 543°C. Bastnaesite is stable to temperatures of 580°C at this pressure, well above the solidus temperature in Figure 12. Although bastnaesite was not encountered in this composition join, the results suggest that bastnaesite could crystallize along with calcite and barite from a melt of similar composition with suitable proportions of CO$_2$, H$_2$O and F.

Given these experimental results, one might expect that rare earth carbonatites would be more abundant. Probably the biggest factor in precluding the concentration of REE in residual carbonatites is their efficient extraction from the magma by higher–temperature REE–bearing minerals such as perovskite, monazite, and apatite. Therefore, if a parent alkali magma is low in phosphorus and titanium, the potential for differentiation to rare-earth-enriched carbonatite might be enhanced. The occurrence of apatite–rich upper–mantle xenoliths in some locations (WASS *et al.*, 1980) suggests that the heterogeneity of the metasomatized upper-mantle sources might influence the degree of phosphorus–depletion and REE–enrichment in parental magmas derived from HAGGERTY's (1987) mantle metasome layers at level (4) in Figure 6.

SUMMARY OF MAGMATIC AND METASOMATIC EVENTS

(1) The source of volatile components for the genesis of kimberlites and carbonatites is a mantle reservoir, deeper than the lithosphere and asthenosphere.

(2) The reservoir may include carbon derived from calcite in subducted oceanic basalt, or from limestones (recent subduction only).

(3) Kimberlites may be generated by upward percolation of volatile components, but nephelinites and associated carbonatites require upward movement of solid mantle materials, as a plume or in some other convective mode, followed by lithosphere thinning, and surface rifting.

(4) The asthenosphere–lithosphere boundary layer (near 200 km depth beneath cratons), where the rheology changes from deformable to rigid, is a collecting site for kimberlite magma that is trapped in small dikes or chambers. Temperatures remain too high for solidification of the magma by normal crystallization. Slow reaction of the lherzolite–derived magma with host harzburgite under these conditions, however, is conducive to the growth of large phenocysts, perhaps corresponding to some of the discrete nodules of kimberlites.

(5) Magmas migrating 5–15 km through the boundary layer approach the solidus, with evolution of a vapor phase. Under suitable stress conditions, the vapor promotes crack propagation, and the kimberlite enters the lithosphere as a dike. Abortion by thermal death and solidification is the fate of most dikes, but sufficiently large reservoirs can yield successive magma pulses through dikes reaching or puncturing the upper crust.

(6) A mantle plume transporting volatile components results in thinning of the lithosphere above it, and accumulation of magma in the asthenosphere–lithosphere boundary layer. Smaller amounts of magma transported by lateral flow in the asthenosphere percolate upward until they lodge in the boundary layer.

(7) Thinning of the lithosphere above the plume, equivalent to upward extension of the deformable asthenosphere, results in the accumulated melt to rise with the isotherms, without crossing the solidus boundary. Intermittent, local excursions of magma to the overlying solidus may be accompanied by vapor release, crack propagation, and magma escape into the developing rift system.

(8) For magma-mantle bodies following a range of depth-temperature trajectories, the ledge on the solidus for peridotite–CO$_2$–H$_2$O is a phase equilibrium barrier to further uprise located at a depth near 75 km. Magmas reaching this barrier evolve vapor and crystallize. The compositions of the vapors envolved, and the crystallization paths of the magmas, vary sensitively according to the position on the solidus boundary reached by the depth-tem-

perature trajectory. Release of vapor promotes crack propagation and eruption of magmas into the rift zone.

(9) Magma-mantle bodies following a trajectory at somewhat higher temperatures miss the phase equilibrium barrier, and these may generate basaltic magmas as they approach the volatile-free peridotite solidus.

(10) Kimberlites and other magmas freed from their mantle host at greater depth are not affected by the phase equilibrium barrier (which is relevant only for melts in equilibrium with peridotite).

(11) Metasomatism may result from reaction of mantle with melts or with vapors, entering or passing through a region. Metasomatic vapors (or solutions) in the subcratonic lithosphere are given off by magmas.

(12) There are two levels in the lithosphere where metasomatic effects should be most prominent. The depleted, refractory base of the lithosphere, 150–200 km deep, has been intermittently invaded by small bodies and dikes of kimberlite, through billions of years, and most of these aborted and gave off vapors. Whenever the craton was rifted, lithospheric thinning was accompanied by the release of abundant vapors at depth near 75 km. Those released at greater depths were enriched in H_2O, and those at shallow depths were enriched in CO_2.

(13) Parent magmas for melilitites and nephelinites are derived from depths near the phase equilibrium ledge, where extensive metasomatism has occurred.

(14) Nephelinites differentiate to carbonatites, and carbonatites themselves differentiate. Rarely, carbonatites differentiate to rocks with extraordinary enrichment in rare earth elements.

(15) The occurrence of rare earth titanates in mantle metasomatites from 70–100 km depth could be an important factor in the availability of rare earth elements in nephelinites and carbonatites derived from this level.

(16) The extreme concentration of rare earth elements in some differentiated carbonatites is probably due to the paucity of phosphorus in parental melts, and this also could be related to metasomatic events at the 75 km level, events involving apatite.

(17) Kimberlites, and rift valley magmas, are melts with histories involving more than one mantle source material. Both start as melts formed in fertile asthenosphere lherzolite, enriched by migration of vapors from deeper levels. Both spend time residing in contact with the depleted keel of the lithosphere, comprising harzburgite and lherzolite. Kimberlites are erupted from this level. Nephelinites and related magmas are developed by progressive evolution of

kimberlite–like magma as it rises through the lithosphere, increasing in melt fraction until it reaches the 75 km level. The magma may here be enriched by solution of metasomatites formed during a previous occurrence of rifting and magmatic processes.

CONCLUDING REMARKS

The details of what happens where, in the process of transfer of carbon from the mantle to the surface, obviously depend critically on the oxygen fugacity as a function of depth (Figure 2). The changing estimates of mantle oxygen fugacity through the past decade suggest that a complete answer has not yet been achieved. It may become established that the solidus for peridotite–C–H–O depicted in Figure 2 in fact lies at higher temperature, and is not intersected by the normal geotherm. Even if it proves, however, that the deep gases are too reduced to initiate melting at deep level (2), there is petrological evidence that kimberlite melts are erupted from depths of 200 km or so, the lithosphere–asthenosphere boundary level (1) in Figure 2. Therefore, the geotherm and solidus must intersect at 200 km or deeper from time to time and from place to place, either by local oxidation (lowering solidus temperatures) or by local heating (raising the geotherm).

The composition of magmas is controlled not only by the composition of the source material and the degree of melting and fractionation, but by the physical properties of rock-fluid systems. At every stage in the processes outlined in Figure 6, one needs to know the conditions for melt or vapor to flow through, to accumulate in, or to escape from a rock matrix, for rocks that are themselves deforming, as in a mantle plume, or for rocks that are rigid and cooler, as in the lithosphere. These topics are addressed by other papers in the Conference Proceedings, but it was evident from discussions during the Conference that there is as yet no consensus about the rheology of mantle–fluid systems.

Acknowledgments—This research was supported by the Earth Sciences section of the U.S. National Science Foundation, grant EAR84-16583. I thank R. J. Floran of Union Oil Research for his encouragement and the Union Oil Company of California Foundation for a tangible contribution to the program.

REFERENCES

ANDERSEN T., O'REILLY S. Y. and GRIFFIN W. L. (1984) The trapped fluid phase in upper–mantle xenoliths from Victoria, Australia: Implications for mantle metasomatism. *Contrib. Mineral. Petrol.* **88,** 72–85.
ARTYUSHKOV E. V. and SOBOLEV S. V. (1984) Physics of the kimberlite magmatism. In *Kimberlites. I: Kim-*

berlites and Related Rocks, (ed. J. KORNPROBST), pp. 309–322. Elsevier.

BOYD F. R. and GURNEY J. J. (1982) Low-calcium garnets: keys to craton structure and diamond crystallization. *Carnegie Inst. Wash. Yearb.* **81**, 261–267.

BOYD F. R. and GURNEY J. J. (1986) Diamonds and the African lithosphere. *Science* **232**, 472–477.

BREY G., BRICE W. R., ELLIS D. J., GREEN D. H., HARRIS K. L. and RYABCHIKOV I. D. (1983) Pyroxene–carbonate reactions in the upper mantle. *Earth Planet. Sci. Lett.* **62**, 63–74.

DEINES P. (1980) The carbon isotopic composition of diamonds: relationship to diamond shape, color, occurrence and vapor compositions. *Geochim. Cosmochim. Acta* **44**, 943–961.

EGGLER D. H. (1976) Does CO_2 cause partial melting in the low-velocity layer of the mantle? *Geology* **4**, 69–72.

EGGLER D. H. (1978a) Stability of dolomite in a hydrous mantle, with implications for the mantle solidus. *Geology* **6**, 397–400.

EGGLER D. H. (1978b) The effect of CO_2 upon partial melting of peridotite in the system $Na_2O–CaO–Al_2O_3–MgO–SiO_2–CO_2$ to 35 kb, with an analysis of melting in a peridotite–H_2O–CO_2 system. *Amer. J. Sci.* **278**, 305–343.

EGGLER D. H. and BAKER D. R. (1982) Reduced volatiles in the system C–O–H: implications to mantle melting, fluid formation, and diamond genesis. In *High Pressure Research in Geophysics, Advances in Earth and Planetary Sciences 12,* (eds. S. AKIMOTO and M. H. MANGHNANI), pp. 237–250. Reidel Publishing Co.

ELLIS D. and WYLLIE P. J. (1979) Carbonation, hydration, and melting relations in the system $MgO–H_2O–CO_2$ at pressures up to 100 kilobars. *Amer. Mineral.* **64**, 32–40.

ELLIS D. and WYLLIE P. J. (1980) Phase relations and their petrological implications in the system $MgO–SiO_2–CO_2–H_2O$ at pressures up to 100 kbar. *Amer. Mineral.* **65**, 540–556.

GLIKO A. O., GRACHEV A. F. and MAGNITSKY V. A. (1985) Thermal model for lithospheric thinning and associated uplift in the neotectonic phase of intraplate orogenic activity and continental rifts. *J. Geodynam. Res.* **3**, 137–153.

HAGGERTY S. E. (1986) Diamond genesis in a multiply-constrained model. *Nature* **320**, 34–38.

HAGGERTY S. E. (1987) Source regions for oxides, sulfides and metals in the upper mantle: clues to the stability of diamonds, and the genesis of kimberlites, lamproites and carbonatites. *Proc. Int. Kimb. Conf. 4th, Perth, Australia, 1986,* (In press).

HAGGERTY S. E., ERLANK A. J. and GREY I. E. (1986) Metasomatic mineral titanate complexing in the upper mantle. *Nature* **319**, 761–763.

HEINRICH E. W. (1966) *The Geology of Carbonatites,* 555 pp. Rand McNally.

HOLLOWAY J. R. and JAKOBSSON S. (1986) Volatile solubilities in magmas: transport of volatiles from mantles to planet surfaces. *J. Geophys. Res.* **91**, #B4, D505–508.

HUANG W. –L., WYLLIE P. J. and NEHRU C. E. (1980) Subsolidus and liquidus phase relationships in the system $CaO–SiO_2–CO_2$ to 30 kbar with geological applications. *Amer. Mineral.* **65**, 285–301.

HUANG W. –L. and WYLLIE P. J. (1984) Carbonation reactions for mantle lherzolite and harzburgite. *Proc.*

27th Int. Geol. Cong., Moscow 9, 455–473, VNU Science Press.

JONES A. P. and WYLLIE P. J. (1983) Low-temperature glass quenched from a synthetic rare–earth carbonatite: implications for the origin of the Mountain Pass deposit, California. *Econ. Geol.* **78**, 1721–1723.

JONES A. P. and WYLLIE P. J. (1986) Solubility of rare earth elements in carbonatite magmas, indicated by the liquidus surface in $CaCO_3–Ca(OH)_2–La(OH)_3$ at 1 kbar pressure. *Appl. Geochem.* **1**, 95–102.

LE BAS M. J. (1977) *Carbonatite–Nephelinite Volcanism.* Wiley and Sons. 347 pp.

MEYER H. O. A. (1985) Genesis of diamond: a mantle saga. *Amer. Mineral.* **70**, 344–355.

MYSEN B. O. and BOETTCHER A. L. (1975) Melting of a hydrous mantle. *J. Petrol.* **16**, 520–593.

OLAFSSON M. and EGGLER D. H. (1983) Phase relations of amphibole, amphibole–carbonate, and phlogopite–carbonate peridotite: petrologic constraints on the asthenosphere. *Earth Planet. Sci. Lett.* **64**, 305–315.

OTTO J. (1984) Melting relations in some carbonate–silicate systems: sources and products of CO_2–rich liquids. Ph.D. Dissertation, Univ. of Chicago.

OTTO J. W. and WYLLIE P. J. (1983) Phase relations on the join calcite–nepheline–albite at 25 kb: Subducted basalt limestone as magmatic source (abstr.). *EOS* **64**, 897.

PECORA W. T. (1956) Carbonatites: a review. *Bull. Geol. Soc. Amer.* **67**, 1537–1556.

RINGWOOD A. E. (1975) *Composition and Petrology of the Earth's Mantle,* McGraw-Hill.

RYABCHIKOV I. D., GREEN D. H., WALL W. J. and BREY G. P. (1981) The oxidation state of carbon in the reduced-velocity zone. *Geochem. Int. 1981,* 148–158.

SOBOLEV N. V. (1977) *Deep–Seated Inclusions in Kimberlite and the Problem of the Composition of the Upper Mantle.* Amer. Geophys. Union. 279 pp.

SPERA F. J. (1984) Carbon dioxide in petrogenesis III: role of volatiles in the ascent of alkaline magma with special reference to xenolith–bearing mafic lavas. *Contrib. Mineral. Petrol.* **88**, 217–232.

TAKAHASHI E. and SCARFE C. M. (1985) Melting of peridotite to 14 GPa and the genesis of komatiite. *Nature* **315**, 566–568.

WASS S. Y., HENDERSON P. and ELLIOTT C. (1980) Chemical heterogeneity and metasomatism in the upper mantle—evidence from rare earth and other elements in apatite–rich xenoliths in basaltic rocks from eastern Australia. *Philos. Trans. Roy. Soc. London* **A297**, 333–346.

WENDLANDT R. F. (1984) An experimental and theoretical analysis of partial melting in the system $KAlSiO_4–CaO–MgO–SiO_2–CO_2$ and applications to the genesis of potassic magmas, carbonatites and kimberlites. In *Kimberlites. I: Kimberlites and Related Rocks,* (ed. J. KORNPROBST), pp. 359–369. Elsevier.

WENDLANDT R. F. and EGGLER D. H. (1980) The origins of potassic magmas: II. Stability of phlogopite in natural spinel lherzolite and in the system $KAlSiO_4–MgO–SiO_2–H_2O–CO_2$ at high pressures and high temperatures. *Amer. J. Sci.* **280**, 421–458.

WENDLANDT R. F. and MORGAN P. (1982) Lithospheric thinning associated with rifting in East Africa. *Nature* **298**, 734–736.

WENDLANDT R. F. and MYSEN B. O. (1980) Melting phase relations of natural peridotite + CO_2 as a function of degree of partial melting at 15 and 30 kbar. *Amer. Mineral.* **65**, 37–44.

WOERMANN E. and ROSENHAUER M. (1985) Fluid phases and the redox state of the Earth's mantle: extrapolations based on experimental, phase–theoretical and petrological data. *Fortschr. Mineral.* **63**, 263–349.

WYLLIE P. J. (1977) Mantle fluid compositions buffered by carbonates in peridotite–CO_2–H_2O. *J. Geol.* **85**, 187–207.

WYLLIE P. J. (1978) Mantle fluid compositions buffered in peridotite–CO_2–H_2O by carbonates, amphibole, and phlogopite. *J. Geol.* **86**, 687–713.

WYLLIE P. J. (1979) Magmas and volatile components. *Amer. Mineral.* **64**, 469–500.

WYLLIE P. J. (1980) The origin of kimberlites. *J. Geophys. Res.* **85**, 6902–6910.

WYLLIE P. J. (1987) Metasomatism and fluid generation in mantle xenoliths: experimental. In *Mantle Xenoliths*, (ed. P. H. NIXON), Wiley, (In press).

WYLLIE P. J. and HUANG W. –L. (1975) Peridotite, kimberlite, and carbonatite explained in the system CaO–MgO–SiO_2–CO_2. *Geology* **3**, 621–624.

WYLLIE P. J. and JONES A. P. (1985) Experimental data bearing on the origin of carbonatites, with particular reference to the Mountain Pass rare earth deposit. In *Applied Mineralogy*, (eds. W. C. PARK, D. M. HAUSEN and R. D. HAGNI), 935–949. Amer. Inst. Mining, Metallurgical, and Petroleum Engineers, Inc., New York.

WYLLIE P. J. and RUTTER M. (1986) Experimental data on the solidus for peridotite–CO_2, with applications to alkaline magmatism and mantle metasomatism (abstr.). *EOS* **67**, 390.

WYLLIE P. J., HUANG W. –L., OTTO J. and BYRNES A. P. (1983) Carbonation of peridotites and decarbonation of siliceous dolomites represented in the system CaO–MgO–SiO_2–CO_2 to 30 kbar. *Tectonophys.* **100**, 359–388.

Magmatic Processes: Physicochemical Principles
© The Geochemical Society, Special Publication No. 1, 1987
Editor B. O. Mysen

The petrogenetic role of methane: Effect on liquidus phase relations and the solubility mechanism of reduced C–H volatiles

W. R. TAYLOR and D. H. GREEN

Geology Department, University of Tasmania, GPO Box 252C, Hobart, Australia 7001

Abstract—Methane and other reduced volatiles may be an important species in the Earth's mantle. The nature of mantle partial melting under reduced conditions in the presence of such volatiles is, however, largely unknown. To evaluate the petrogenetic role of C–H volatiles experimental liquidus studies were undertaken in the system nepheline(Ne)–forsterite(Fo)–silica(Q)–C–O–H at 28 kbar. Methane-dominated fluids are conveniently generated in high pressure experiments from a mixture of Al_4C_3 and $Al(OH)_3$. Compared to the volatile-absent system, the effect of CH_4-rich fluids is to expand the Fo phase field relative to En_{ss} (melt depolymerization) and to bring phases of high octahedral aluminum content onto the liquidus (leading to early appearance of garnet on the liquidus of other compositions). This contrasts with the effect of CO_2 which gives rise to the expansion of the En_{ss} phase field and the effect of H_2O which results in a greater expansion of the Fo field than CH_4. The magnitude of liquidus temperature depressions for C–H fluid-saturated Fo–Ne–Q melts are comparable to that of pure CO_2 ($\sim 90°C$ at 28 kbar).

Infrared (IR) spectroscopic investigations of C–H fluid saturated jadeite and sodamelilite glasses quenched from 30 kbar, establish the presence of both dissolved, oxidized, and reduced components as also required by charge-balance constraints. The former occurs as O–H groups and the latter is consistent with a reduced network component of O:Si stoichiometry <2 ("silicon monoxide" units). There is no spectroscopic evidence for the presence of dissolved molecular CH_4, other C–H groups or carbonate. Pyrolysis/gas chromatographic analyses give dissolved H contents equivalent to ~ 3 weight percent H_2O and reduced C contents of 1000–2000 ppm (wt) in a form yet to be characterized. Aluminosilicate melts have a strong preference for dissolved H over C under reduced conditions.

INTRODUCTION

MAJOR ADVANCES in understanding the role of volatile components in igneous petrogenesis have taken place over the last ten years. Much of this effort has been directed at determining the petrogenetic role of the oxidized volatiles CO_2 and H_2O which are known to have a significant effect on super-solidus phase relations at upper mantle pressures (see review by HOLLOWAY, 1981).

Because H_2O and CO_2 are the most abundant species in volcanic gases (*e.g.*, ANDERSON, 1975), it has been commonly assumed that they will also be the most abundant species at depth in the Earth's upper mantle. If this basic assumption is incorrect, and this will depend to a large extent on the mantle's oxidation state, then the role of volatiles other than H_2O and CO_2 must be considered (RYABCHIKOV *et al.*, 1981).

A variety of evidence exists to support the idea that at least part of the upper mantle is reduced enough to stabilize CH_4 at depth ($f_{O_2} \leq IW + 2$ log units). This evidence includes intrinsic oxygen fugacity measurements on mantle-derived minerals indicating the prevalence of low f_{O_2} conditions amongst "type A" upper mantle (ARCULUS and DELANO, 1981) and the finding of primordial (^{3}He correlated) CH_4 as a significant component of fluids outgassing at mid-ocean ridge hydrothermal centers

(WELHAN and CRAIG, 1981, 1983). Current views on the redox state of the Earth's mantle have recently been summarized by WOERMANN and ROSENHAUER (1985, in particular see pp. 317–322). Although this area is still one of active debate and inquiry, it seems likely that a range of oxidation states from relatively oxidized ($f_{O_2} \sim FMQ$) to reduced ($f_{O_2} \sim IW$) are applicable to the upper mantle. Volatiles in the reduced part of the system C–O–H may therefore be of considerable importance in igneous petrogenesis.

If magma generation involving volatile components takes place in a reduced environment, for example at f_{O_2}'s near the iron-wustite (IW) oxygen buffer, then in the model system "peridotite"–C–O–H, volatiles will be dominantly $CH_4 > H_2O > H_2 > C_2H_6$ mixtures and crystalline carbonates will not be stable relative to diamond or graphite (RYABCHIKOV *et al.*, 1982; EGGLER and BAKER, 1982). To explore the nature of mantle melting under reduced conditions an adequate understanding of the thermodynamic properties of C–O–H fluids at elevated pressures and temperatures is required. Reduced volatile interactions with silicate melts can then be investigated by experimental and spectroscopic means. The first aspect has been considered by TAYLOR (1985, 1987) and the second is the purpose of this paper.

The contrasting effects of oxidized versus reduced volatiles on liquidus phase relations are investigated in the model peridotite system nepheline(Ne)–forsterite(Fo)–silica(Q) under conditions of CH_4, CO_2 and H_2O volatile saturation and in the absence of volatiles. To place constraints on the mechanism of reduced C–H volatile dissolution in aluminosilicate melts, CH_4-saturated, graphite-free glasses of sodamelilite ($NaCaAlSi_2O_7$, Sm) and jadeite ($NaAlSi_2O_6$, Jd) composition have been analysed for carbon and hydrogen and investigated by Fourier Transform infrared (FTIR) spectroscopic methods.

Iron–bearing compositions have not been considered in this study primarily because f_{O_2} conditions in the presence of a C–H fluid will lie well within the Fe-metal stability field. The effect of reduced volatiles on these compositions must instead be investigated in the presence of mixed H_2O–CH_4–H_2 fluids at higher f_{O_2}'s (near the IW buffer for example). In this study we concentrate on identifying the mechanism of reduced C–H volatile dissolution free from interference by other species and thus provide the necessary basis for extension into natural systems.

PREVIOUS WORK ON REDUCED C–H VOLATILE INTERACTIONS WITH SILICATE MELTS

Little is known of how reduced volatiles interact with silicate melts. EGGLER and BAKER (1982) conducted a number of reconnaissance experiments determining the effect of C–H volatiles on the melting and liquidus phase relations of diopside and the composition diopside $(Di)_{35}$ pyrope $(Py)_{65}$ at $P > 20$ kbar. Experimental f_{O_2} conditions were believed to be near the Si–SiO_2 buffer but could not be determined directly. EGGLER and BAKER (1982) found liquidus depressions of $\sim 100°C$ in diopside at 21 kbar and a large liquidus field of olivine plus garnet extending to at least 40 kbar in $Di_{35}Py_{65}$ coexisting with C–H fluid. In the latter case this differs from the effect of H_2O which does not bring garnet onto the liquidus and the effect of CO_2 which stabilizes orthopyroxene to high pressures. The presence of "depolymerized" phases such as olivine and garnet on the C–H volatile saturated liquidus led to the suggestion that reduced volatiles have a depolymerizing effect on silicate melts. Quench problems did not allow spectroscopic investigation of the glasses produced in their study and a detailed solubility mechanism could not be ascertained. In the only other study involving CH_4–bearing fluids, JAKOBSSON (1984), using a pyrolysis/mass spectrometry technique, determined the dissolved volatile content of quenched albite glasses equilibrated with H_2O–CH_4 fluids at 10–25 kbar ($f_{O_2} \sim$ IW). Gases released at 1200°C consisted mainly of H_2O (>80 mol percent) with smaller amounts of reduced volatiles (CO, CH_4 and H_2). The extent to which dispersed graphite in the samples affected results was not assessed; low total C/H ratios (<0.06), however, suggest reduced carbon solubilities are not large under these conditions. Raman and infrared (IR) spectroscopic measurements confirmed the dominance of dissolved water; carbonate and possibly molecular methane were detected in some samples. Recently, LUTH and BOETTCHER (1986) have demonstrated that H_2, a significant component of C–H fluids at high pressures, interacts strongly with silicate liquids and may be quite soluble in silicate melts.

In the light of the above studies it is of importance to establish the nature and mechanism of reduced C–H volatile interaction with silicate systems. Problems of graphite contamination, quench effects and uncertainty in the nature of the C–H fluids actually generated have limited previous studies. Resolution of these problems is essential for obtaining unambiguous spectroscopic and analytical results. The f_{O_2} conditions chosen for the present study are such that CH_4 is the dominant fluid species. At higher f_{O_2}'s, where H_2O becomes important, the effects of reduced C–H volatiles may not be clearly resolved from those of H_2O.

EXPERIMENTAL TECHNIQUES

High pressure experiments

All high pressure experiments were performed with 0.5″–diameter (1.27 cm) solid-media, high-pressure apparatus using techniques similar to those of GREEN and RINGWOOD (1967). Temperatures were recorded with a Pt/$Pt_{90}Rh_{10}$ thermocouple automatically controlled to within ±5–7°C of the set value. All experiments were carried out by using the "piston-in" technique and applying a -10% pressure correction to nominal load pressures; quoted pressures are accurate to ±1 kbar.

Silicate compositions in the system nepheline–forsterite–silica (Ne–Fo–Q) were prepared from analytical reagent grade Al_2O_3, SiO_2, Na_2CO_3 and MgO fired at 900°C after thorough mixing and repeated grinding under acetone. Compositions were checked by microprobe analysis of glass beads prepared from the sintered oxide mixes on an Ir-strip heater. Sodamelilite and jadeite starting materials were powdered glasses prepared from sintered oxides. Samples were melted at 1250°C in Pt–crucibles and quenched in air to colorless glasses; their composition and homogeneity were checked by microprobe analysis.

Methane-saturated experiments

Calculations by TAYLOR (1985, 1987) reveal that under upper mantle pressure-temperature conditions C–H fluids in equilibrium with graphite contain CH_4 as a major (>80 mol percent) component and H_2 and C_2H_6 as minor (<10 mol percent) components. Such fluids are stable only at f_{O_2}'s below $\sim$IW–4 log f_{O_2} units. Under these conditions f_{H_2} is large so that maintenance of fluid excess conditions during an experiment requires minimization of diffusive H_2-loss by employing run durations as short as possible while still achieving fluid–melt equilibrium. In preliminary studies, we found that the use of complex organic compounds as a methane-source (such as those used by HOLLOWAY and REESE, 1974; EGGLER and BAKER, 1982) results in the formation of disordered graphite which persists in experiments for run durations of less than a few hours. To overcome this problem in short duration experiments that do not incorporate a solid f_{O_2} buffer (as those undertaken here), methane should ideally be produced rapidly and by direct reaction.

Methane, together with a small amount of H_2, can be readily generated at 1 bar by the action of H_2O on Al_4C_3 (WADE and BANISTER, 1973). Because this reaction proceeds rapidly to completion and produces a C–H fluid with a $CH_4:H_2$ ratio of ~9:1 (similar to that calculated for high pressure-temperature equilibrium), it is an ideally suited generation reaction. By including $Al(OH)_3$ as the source of H_2O a convenient solid reactant can be prepared. Overall the methane generation reaction may be written:

$$4Al(OH)_3 + Al_4C_3 \rightarrow 4Al_2O_3 + 3CH_4 (+H_2 + C_2H_6). \quad (1)$$

To give the fluid–generating carbide/hydroxide mixture sufficient bulk, alumina was used as a diluant giving a mixture capable of generating 0.2 mg CH_4/10 mg mix. To eliminate the possibility that adsorbed water or excess air in the capsule might lead to oxidation of methane and production of H_2O, a 1:1 molar ratio of $Al(OH)_3$ to Al_4C_3 (three times carbide excess over reaction stoichiometry) was employed. The presence of excess carbide was confirmed optically and by x–ray diffraction at the end of each experiment. All mixtures containing Al_4C_3 (98 weight percent purity, Goodfellows Metals #AL516010) were stored under vacuum desiccation to prevent reaction with atmospheric moisture.

Experiments with C–H volatiles were carried out in large capacity "buffer" assemblies using Pt outer capsules (3.5 mm O.D.) and inner unsealed Pt capsules containing ~15 mg of silicate sample. Sample–containing capsules were heated briefly to red-heat prior to loading the methane-source to ensure removal of all traces of adsorbed water. The inner capsule was surrounded with sufficient fluid-generating mix to produce ~1 mg of methane. For jadeite and sodamelilite compositions, experiments were of 30 minutes duration at 30 kbar, 1320°C and 1350°C, respectively; run details for system Ne–Fo–Q experiments are listed in the Appendix. Talc was chosen as the pressure transmitting medium to maximize the external f_{H_2} of the system and thus limit H_2 loss from the capsule.

Prior to experiments with silicates a "blank" run was performed at 30 kbar, 1300°C for 15 minutes with the aim of determining whether fluid phase equilibrium is achievable in short run times by this technique. Analysis of the fluid phase following the run was accomplished by mass spectrometry using a capsule piercing technique. The presence of a vapor phase is readily recognized by the distended nature of the capsule. The device used for capsule piercing consists of a modified regulating valve (Whitey #SS–1VS6) with a redesigned stem tip fashioned into a hardened needle point. A removable cradle serves to position and hold the capsule in place during piercing. Gases were released under vacuum (~10^{-6} torr) and directed into the ion-source of a VG–micromass 7070 double focusing mass spectrometer via a modified probe insertion technique. To achieve low background levels, particularly for H_2O caused chiefly by absorbed molecules on metal surfaces, it was found necessary to evacuate the whole system (probe plus piercer) for ~12 hours prior to taking measurements. After piercing, mass spectra were acquired by multiple scans of ~2 sec duration over the mass range 10–70 m/z. The total ion current was monitored during the piercing experiment and both background and sample spectra were recorded at various sensitivities. With the data acquisition system used by the VG instrument, H_2 was below the lower mass–range limit of recorded spectra. The presence of both H^+ and H_2^+ ions was, however, confirmed qualitatively by oscillographic traces down to low mass numbers. Mass spectra are normalized to zero background (mainly residual air and water vapor in the instrument) by reference to the m/z 32 (O_2^+) or 40 (Ar^+) peak. X–ray diffraction (XRD) and optical examination of the solid product showed the presence of Al_2O_3, ordered graphite and excess Al_4C_3. No oxycarbides were identified either optically or by XRD. The presence of abundant ordered graphite indicates that an f_{H_2} buffer reaction of the type: C(graphite) + $2H_2$ = CH_4, has been operative during the experiment. The form of the mass spectra for the blank experiment and for those experiments with silicate present are identical except a finite amount of H_2O (<0.2 mol percent) is found in the latter. Figure 1 shows a typical methane (m/z 12–16) and ethane (m/z 24–30, m/z 28 overlaps with background N_2^+) spectrum for an experiment with silicate present. Trace amounts of C_{3-4} hydrocarbons are also present. Although the C_2H_6/CH_4 ratio of the fluid is slightly higher than predicted by the MRK–equation calculations of TAYLOR (1986): ~25 vs. 18 expected, overall the agreement with theory is good. Derived fluids are thus believed to represent those at graphite-fluid equilibrium. The absence of oxygen–containing volatiles (i.e., H_2O) places an upper limit on the log f_{O_2} of the system at near IW-5 log f_{O_2} units under experimental conditions for the "blank" experiments and near IW-4.5 log f_{O_2} units for those containing silicate. The use of a carbide/hydroxide mixture is thus a rapid and convenient method for preparing fluids dominated by CH_4 at high pressures.

Water, carbon dioxide and volatile–absent experiments

These experiments were performed in talc/pyrex or talc–only (for water–saturated runs) sleeved assemblies using 2.3 mm O.D. Pt or $Ag_{50}Pd_{50}$ capsules. H_2O (~30 weight percent) was added via microsyringe and CO_2 (~15 weight percent) was generated from $Ag_2C_2O_4$. Run times for CO_2 and volatile–absent experiments were kept to <20 min. (see Appendix) to avoid the risk of H_2O formation by hydrogen diffusion into the capsule. All mixes and assembly parts were dried at 120°C, 24 hours prior to use.

Spectroscopic methods

Spectra of C–H fluid—saturated glasses were investigated by Fourier Transform infrared (FTIR) spectroscopy using a Digilab model FTS–20E spectrometer. FTIR spectroscopy offers significant advantages over conventional instrumentation, including increased signal–to–noise ratio, increased sensitivity due to high energy throughput and ready access to computer-based data manipulation procedures such as band fitting and spectrum subtraction. Spectra of both powdered glass and crystalline samples were obtained by the conventional KBr disc method. Aproximately 2 mg (1 mg for crystalline compounds) of sample was ground together with 200 mg of IR–grade KBr in an agate mortar for 10 minutes. After drying the powder at 120°C, pellets were pressed between 1 cm diameter polished stainless steel dies; any cloudy discs were remade. Discs were then dried under P_2O_5 desiccant overnight to remove traces of adsorbed water. Spectra were acquired by signal averaging 200 scans at 4 cm^{-1} resolution referenced against a blank KBr disc.

Difficulties with KBr powder spectroscopy rest largely with its reproducibility. Differences in sample preparation, reflected mainly in particle size distribution and orientation effects may lead to changes in bandwidths and relative

A.

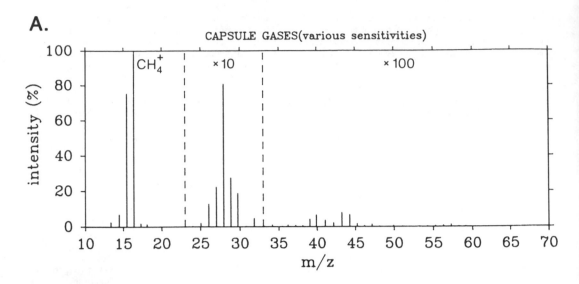

B.

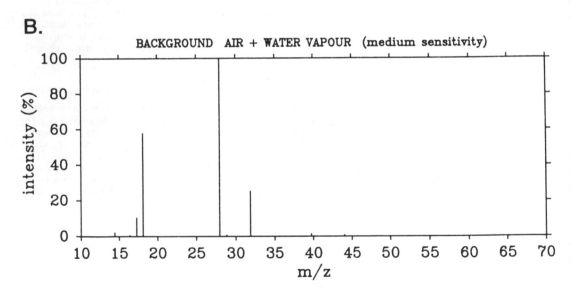

FIG. 1. (A) Mass spectrum of capsule gases released from run T–1341 at low (m/z 10–23), medium (×10, m/z 23–33) and high (×100, m/z >33) sensitivities. Methane: m/z 13–17; ethane: m/z 25–30; C_{3-4} hydrocarbons m/z >36; background water vapour and air: m/z 17–18, 28, 32. (B) Instrumental background at medium sensitivity. Note suppression of background ion intensity due to input of capsule gases (ratios of individual gases, however, remain unchanged).

intensities (MCMILLAN, 1985). Such changes are of critical importance in quantitative studies but are of lesser importance in the interpretative work undertaken here. Nevertheless, reproducibility checks were performed by the preparation of duplicate discs, in all cases using identical preparative methods. Duplicates showed close similarity in spectral features (as illustrated in Figures 7A and 7B). It is believed that the reproducibility of the KBr disc method, provided sample preparation techniques remain consistent, is not a problem in the interpretation of the silicate glass and crystal spectra considered here.

Analytical Methods

Pyrolysis/gas chromatography

Quantitative carbon and hydrogen analyses of the C–H fluid–saturated glasses were performed with a Hewlett–Packard 185B CHN analyser at the Analytical Services Section, Research School of Chemistry, Australian National University). The method has been previously applied by BREY (1976) in the analysis of ∼mg quantities of silicate glasses containing dissolved carbonate and water. For

C–H fluid–saturated glasses that may contain dissolved reduced components, ~0.5 to 1.5 mg of sample was intimately mixed with 85 mg of $MnO_2/WO_3/Cr_2O_3$ oxidant. Samples were then combusted at 1080°C for 90 seconds following an initial heating at 70°C to remove atmospheric gases.

In the pyrolysis of reduced C–bearing glasses, the possibility exists that crystallites of silicon carbide might form during the combustion process and remain unreactive to oxidation (SiC being known for its chemical inertness and high thermal stability: ROCHOW, 1973). To determine whether the pyrolysis/gas chromatography technique gives accurate results for the analysis of substances containing (or with the potential to form) Si–C bonds, three standard compounds consisting of mixtures of finely ground SiC and silica powder were analysed. Results for unknowns and standards are listed in Table 1. Standards were combusted for 90 seconds and 120 seconds (in the latter case there was a considerable loss in sensitivity).

Electron microprobe

Electron microprobe analyses for carbon were performed at the Electron Optical Centre (E.O.C.), University of Adelaide (JEOL 733 microprobe, B. J. Griffin analyst) and Central Science Laboratory, University of Tasmania (JEOL JXA50A microprobe). Measurements were obtained with a light element (STE crystal) wavelength dispersive spectrometer using 15 kV accelerating voltage in the former case and 10 kV in the latter. The beam in both cases was defocussed to a 10–20 μm diameter spot with 50 nA beam current to minimize specimen damage. Polished epoxy mounts containing the samples, blanks and an array of carbon standards (graphite, SiC and various carbonates) were coated with aluminum following the method of MATHEZ and DELANEY (1981). Analyses at E. O. C., Adelaide, were performed by 60 second counts on the peak and 20 seconds on the backgrounds with the spectrometer positioned 10 mm either side of the peak

position. Full ZAF corrections were applied. Surface carbon contamination, derived from cracking of vacuum pump oils under the beam, resulted in a detection limit for carbon of ~0.2 weight percent. Analyses of standards, blanks and unknowns gave results of high consistency mainly due to the good precision obtainable with long counting times at reasonably high count rates. Absolute accuracy is difficult to judge due to the large matrix corrections used and the effects of surface contamination. These uncertainties could amount to errors of as much as ±30%.

CONTRASTING LIQUIDUS PHASE RELATIONS IN THE SYSTEM Ne–Fo–Q–C–O–H

At pressures >25 kbar the system nepheline(Ne)–forsterite(Fo)–silica(Q) contains liquidus phase fields of forsterite (Fo), enstatite$_{ss}$ (En$_{ss}$) and jadeite$_{ss}$ (Jd$_{ss}$) analogous to the major minerals of upper mantle peridotite. This system forms the base of the simplified basalt tetrahedron (YODER and TILLEY, 1962) and as such offers a useful model for investigating small degrees of partial melting of mantle peridotite. Liquidus phase relationships in the system Ne–Fo–Q have therefore been widely studied particularly as a function of pressure and activity of volatile species (KUSHIRO, 1968, 1972; EGGLER, 1978; WINDOM and BOETTCHER, 1981). The position of the Fo–En$_{ss}$ two–phase boundary can be used as an indicator of melt–phase silica activity reflecting the relative degree of silicate melt polymerization/depolymerization (as discussed most recently by RYERSON, 1985). The aim of this study is to locate the Fo–En$_{ss}$ two–phase boundary in the system Ne–Fo–Q under conditions dominated by a CH_4–rich vapor phase. Comparisons can then be made with the boundary position under volatile–absent, H_2O–saturated and CO_2–saturated conditions. The resultant phase diagrams can be used to predict the nature of mantle melting under conditions of saturation with different volatile species and hence over a range of f_{O_2}'s.

A pressure of 28 kbar was chosen to allow incorporation of the volatile–absent experimental data of WINDOM and BOETTCHER (1981) along the joins jadeite–enstatite (Jd–En) and jadeite–forsterite (Jd–Fo) as well as duplicating pressure conditions near the top of the oceanic LVZ where ≳2% partial melt is believed to exist (GREEN and LIEBERMANN, 1976). Experiments were conducted along the join Ne$_{55}$Q$_{45}$–Ne$_{55}$Fo$_{45}$ to provide approximately 90° intersection with the two–phase boundary at liquidus temperatures that are not prohibitively high (*i.e.*, <1500°C). Vapor saturation was confirmed in each experiment by piercing the sample capsule and noting the weight loss at 25°C and 110°C. All

Table 1. Pyrolysis/gas chromatography analytical results*

Sample #	Comp.	C (Weight percent)	H_2O (Weight percent)	Combustion time (sec)
T–1296	Jd	0.13 ± 0.03	2.9 ± 0.4	90
T–1341	Sm	0.09 ± 0.07	3.0 ± 0.5	90
T–1318	Sm	0.12 ± 0.04	1.7 ± 0.1	90
T–1250	Sm	not analysed	5.5 ± 0.2	90

Standards (SiC + SiO_2)	Actual weight percent C	Analysed weight percent C	Combustion time (sec)
A	0.5	0.24 ± 0.04	90
		0.18 ± 0.02	120
B	1.0	0.32 ± 0.03	90
		0.21 ± 0.06	120
C	1.5	0.51 ± 0.07	90

* All samples analysed in duplicate, averages and std deviations quoted.

Detection limits: C 0.05 weight percent; H as H_2O 0.18 weight percent.

water–saturated runs produced a fine white precipitate surrounding the fluid exit hole; this arises from solid-phase solubility in high pressure–temperature aqueous fluids as discussed by RYABCHIKOV et al. (1982).

Experimental charges were examined optically and by electron microprobe analysis; all showed quench effects of variable extent, mainly as $MgAl_2$-SiO_6(MgTs)–rich pyroxene overgrowths sometimes extending to Jd–rich compositions and as individual crystallites on and about En_{ss}. Skeletal olivines are present in some runs. In selected experiments, electron microprobe analyses established liquid and crystal compositions as recorded in the Appendix and Table 2. In practice, only for the volatile–absent runs, showing the least quench effects and only small Na/Al ratio deviations from 1, were we able to analyse liquid compositions directly. This allowed the volatile–absent Fo–En_{ss} boundary to be well constrained with three experiments at $Ne_{55}Fo_{25}Q_{20}$. A liquid composition was estimated for CO_2–saturated run T–1227 from analysis of a large glass–rich area. The composition was projected back into the Na/Al = 1 plane from average quench En. In both H_2O–saturated and C–H fluid-saturated OH–containing glasses the combined effects of quench crystal growth and Na–volatilization under the electron microprobe beam precluded any estimate of liquid composition.

The position (in terms of weight percent Fo) of experimentally determined Fo–En_{ss} two–phase boundaries for the Ne_{55} compositions at $P = 28$ kbar are listed below:

Volatile species	Fo Weight percent	Estimated liquidus temperature (°C)
absent	24	1495 ± 5
H_2O–saturated	18 ± 1	1120 ± 15
CH_4–H_2–saturated	21 ± 1	1410 ± 15
CO_2–saturated	34 ± 1	1410 ± 15

These data allow delineation of Fo and En_{ss} liquidus phase fields on the ternary Ne–Fo–Q diagram (Figure 2); the volatile–absent fields for Jd_{ss}, Sp ($MgAl_2O_4$ spinel) and Ne_{ss} (nepheline$_{ss}$) are from GUPTA et al. (1987).

Volatile–absent boundary

The position of the volatile–absent two–phase boundary is consistent with the experimental results of KUSHIRO (1968) on the composition NFA–1 ($Ne_{62}Fo_{18}Q_{20}$) which has Fo on the liquidus up to 30 kbar. The intersection of the boundary with the Fo–Q binary system is estimated from the data of CHEN and PRESNALL (1975) to occur near Fo_{24}. These constraints place the Fo–En_{ss} boundary at 28 kbar close to the composition $Jd_{32}Fo_{68}$ on the Jd–Fo join. WINDOM and BOETTCHER (1981, Figure 2) inferred that this point should lie near $Jd_{50}Fo_{50}$ but this position could not be adequately constrained by their experimental data. Recent experimental work in the dry system by GUPTA et al. (1987) has refined the position of the three phase Fo–En_{ss}–Jd_{ss} point. Although differing from WINDOM and BOETTCHER's (1981) interpretation, it is in agreement with their experimental data along

Table 2. Sm and Jd composition glasses: experimental results at 30 kbar

Run #	Comp.	Volatile	T°C	Product
T–1158	Sm	absent	1420	Glass
T–1159	Sm	absent	1370	Sm crystals
T–1160	Sm	absent	1395	Glass
T–1178	Sm	CO_2 ~8 weight percent	1300	Clear glass (MH buffer)
T–1250	Sm	H_2O ~6 weight percent	1300	Clear glass (graphite capsule)
T–1318	Sm	C–H 9 weight percent	1320	Glass + grossular crystals
T–1341	Sm	C–H 9 weight percent	1350	Clear glass containing small fluid inclusions <1 μm diam.
T–1442	Sm/SiC	—	1500	Glass + disseminated graphite
T–1296	Jd	C–H 10 weight percent	1320	Clear glass, inclusion-free
Microprobe analyses:	SiO_2	Al_2O_3	CaO	Na_2O	Total
T–1341 (Av. glass)	44.2 (.4)*	18.6 (.4)	20.9 (.3)	11.8 (.2)	95.5
T–1296 (Av. glass)	56.0 (.6)	23.6 (.5)	—	14.7 (.2)	94.3
T–1318 (grossular)	39.48	22.56	37.56	0.00	99.6

* Figures in brackets are 1σ standard deviations.

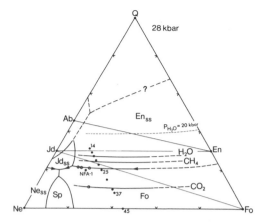

FIG. 2. Ternary liquidus phase diagram (weight percent) for the system Ne–Fo–Q at 28 kbar showing the position of the Fo–En$_{ss}$ two-phase boundary under volatile–absent and CO$_2$, H$_2$O and CH$_4$–H$_2$ volatile-saturated conditions. Volatile–absent phase boundaries are from GUPTA *et al.* (1986) and this work. Dashed phase boundaries are inferred. Compositions studied are indicated by the filled circles on the join Ne$_{55}$Fo$_{45}$–Ne$_{55}$Q$_{45}$, adjacent numbers indicate weight percent Fo component. Composition NFA-1 is from KUSHIRO (1968) and the 20 kbar water-saturated Fo–En$_{ss}$ boundary is from KUSHIRO (1972). Double circles are analysed liquid compositions.

the Jd–Fo and Jd–En joins as well as the earlier data of KUSHIRO (1968). Figure 3 presents the amended phase diagram for the Jd–Fo pseudobinary join. The main change from WINDOM and

BOETTCHER (1981) is the expansion of the Fo + L phase field (note also that two points are mislabelled on their original figure).

CO$_2$-saturated boundary

The shift in position of the Fo–En$_{ss}$ boundary relative to the volatile–absent system is greatest for CO$_2$-saturation (expansion of En$_{ss}$ field from Fo$_{24}$ to Fo$_{34}$ at Ne$_{55}$) reflecting CO$_2$'s strong melt polymerizing role. Equilibrium and quench En$_{ss}$ compositions on this boundary are rich in MgTs component (equilibrium crystals contain 15–17 weight percent Al$_2$O$_3$) and poor in Na$_2$O (<3.5 weight percent). This behaviour may arise from a decrease in Na$_2$O activity in the melt due to sodium–carbonate complex formation and accords with the CO$_2$ solubility mechanisms discussed by MYSEN and VIRGO (1980b,c).

H$_2$O-saturated boundary

The depolymerizing role of H$_2$O dominated fluids is clearly illustrated in Figure 2. Expansion of the Fo phase field at Ne$_{55}$ relative to the volatile–absent system is from Fo$_{24}$ → Fo$_{18}$. This shift is not as large as that for CO$_2$ in the opposite direction; however, there appears to be a substantial pressure effect associated with the H$_2$O-saturated boundary. At P_{H_2O} = 20 kbar this boundary is found near Ne$_{55}$Fo$_7$Q$_{38}$ (KUSHIRO, 1972) a difference of Fo$_{11}$

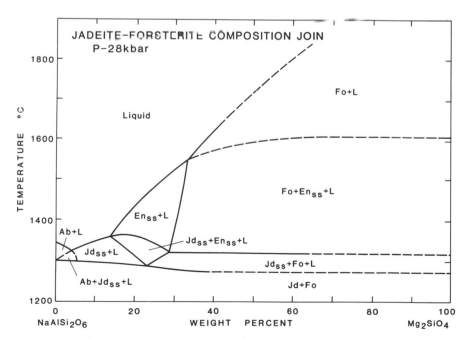

FIG. 3. Amended pseudobinary phase diagram for the join Jd–Fo at 28 kbar.

from the volatile–absent boundary at the same pressure (*cf.* the Fo_6 difference at 28 kbar). This effect may be ascribed to a diminishing ability of H_2O to depolymerize melts at higher pressures and could be due to a number of factors:

(a) changes in the solubility of solid components in co-existing aqueous fluid with pressure;

(b) dissolution of a higher proportion of H_2O in molecular form with increasing P_{H_2O} as shown by STOLPER (1982); because molecular water dissolves without (Si, Al)–O–Si bond cleavage there will be no accompanying network depolymerization;

(c) change in behaviour of H_2O with pressure toward that of a network polymerizer by formation of a greater proportion of alkali cation–hydroxy complexes relative to Si–OH bonds in an analogous fashion to CO_2 dissolution forming carbonate complexes.

If H_2O tends to be more polymerizing with increasing pressure via any of the above mechanisms, then there are important consequences for magma genesis. Thus, CO_2 may not be required in the genesis of highly silica undersaturated magmas (as proposed by EGGLER, 1978, and other workers) if H_2O can perform a melt polymerizing role. Further investigation of the role of H_2O at high pressures is clearly warranted.

C–H fluid—saturated boundary

The $Fo-En_{ss}$ two-phase boundary for saturation with reduced C–H fluids falls between the H_2O and the volatile–absent boundaries implying a depolymerizing role for reduced C–H volatiles. The compositions of equilibrium, quench rim and quench crystal pyroxenes for volatile–absent, C–H fluid, H_2O and CO_2-saturated runs in which Fo and En_{ss} equilibrium crystals coexist are compared in Figures 4A and 4B. The proportion of jadeite component in equilibrium pyroxenes varies with volatile species in the order: $CH_4–H_2 >$ volatile–absent $> CO_2 > H_2O$ (circled in Figure 4B). This order is generally retained for quench rims and quench crystals. In the volatile–absent, CO_2 and C–H fluid—saturated cases, temperature and hence silica activity (buffered by coexisting Fo and En_{ss}), are of similar magnitude. On this basis, we would interpret differences in pyroxene chemistry as due largely to changes in the activity of network modifying oxides in the liquid. The observed enrichment in pyroxene of the jadeite component under conditions of C–H fluid saturation may, therefore, reflect an increase in the activity of network modifying Na_2O and Al_2O_3 relative to the volatile–absent system.

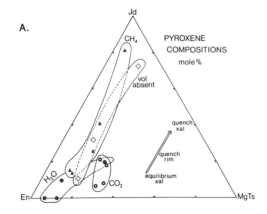

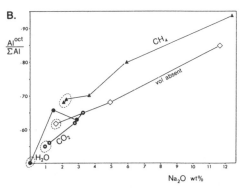

FIG. 4. A. Plot of pyroxene compositions (mol percent) coexisting with forsterite for selected runs in the system Ne–Fo–Q. Quench pyroxenes extend to very Jd–rich compositions in the CH_4 fluid—saturated and volatile–absent systems. ▲ C–H fluid—(*i.e.*, $CH_4–H_2$) saturated (T–1291); ◇ volatile–absent (T–1210); ◉ CO_2-saturated (T–1227); ● H_2O-saturated (T–1175). Arrow shows the direction of pyroxene quench trends. B. Plot of Al occupying octahedral sites/ΣAl versus Na_2O (weight percent) for pyroxenes of Figure 4A. Dotted areas indicate equilibrium pyroxenes.

The results presented here indicate that C–H volatile dissolution leads to network depolymerization accompanied by changes in the activity of network modifying oxides. In order to write a mechanism to describe the dissolution process, however, more detailed information on an atomic or molecular basis is needed than can be supplied by liquidus studies. This requires spectroscopic and analytical data on samples free from contamination as considered below.

SOLUBILITY MECHANISM OF METHANE: ANALYTICAL AND SPECTROSCOPIC CONSTRAINTS

Sodamelilite and jadeite compositions have been chosen for this investigation because their volatile–

free, H_2O and CO_2–containing glasses have been characterized structurally by x-ray diffraction and vibrational spectroscopic methods (*e.g.*, TAYLOR and BROWN, 1979; MYSEN and VIRGO, 1980a; MYSEN *et al.*, 1980; SHARMA and YODER, 1979). In addition, both compositions show good quenching behaviour in the presence of volatiles (MYSEN and VIRGO, 1980a).

Volatile–absent melting of sodamelilite and jadeite at 30 kbar

To determine the magnitude of liquidus depressions in the presence of C–H volatiles the volatile-free melting points of sodamelilite and jadeite are required at 30 kbar. Crystalline sodamelilite is stable above 4–5 kbar and at 10 kbar has a melting interval of ~50°C (YODER, 1964). At 20 kbar, the solidus and liquidus may be regarded as coincident within experimental precision (KUSHIRO, 1964). Experiments at 30 kbar (Table 2) show that crystalline sodamelilite melts congruently at 1380 ± 10°C. Jadeite melts congruently to liquid at 1370 ± 10°C, 30 kbar (BELL, 1964).

C–H saturated melting at 30 kbar

Results of the melting experiments at 30 kbar are listed in Table 2. Product glasses are clear and graphite-free. Interaction of C–H fluids and silicate melts leads to liquidus depressions of ~40°C in sodamelilite and at least 50°C in jadeite. This observation together with the consistently low electron microprobe totals reported in Table 2 implies a significant solubility of a reduced volatile component or components. All experimental charges retain excess carbide and mass spectra of quenched vapor indicate the presence of only trace quantities of bulk oxygen as H_2O. Thus the observed effects cannot be ascribed to absorbed H_2O or other external sources of oxidation.

In some experiments, small blebs of Pt–Si alloy were occasionally observed at the Pt capsule/silicate interface. This could be a potential source of oxygen (and hence H_2O) via the reaction:

$$x SiO_2(\text{melt}) + y Pt(\text{capsule}) \rightarrow$$

$$Si_x Pt_y(\text{alloy}) + x O_2, \quad \text{where} \quad 0 < x, y < 1. \quad (2)$$

Electron microprobe analyses for silica (see Table 2), however, show no detectable Si loss to a Pt–Si alloy within analytical error: compare 46.5 weight percent SiO_2 in the sodamelilite starting material with 46.3 weight percent (std deviation 0.4) found in the C–H saturated glass (total normalized to 100%). The amount of oxygen (expressed in the

form of dissolved H_2O) that could enter a sodamelilite melt by this process and remain undetected by silica analysis is <0.4 weight percent H_2O.

Under C–H fluid excess conditions the liquidus phase for the sodamelilite composition is grossular and not crystalline sodamelilite as observed under volatile–absent conditions. This observation is consistent with the early appearance of garnet on the liquidus of the C–H volatile—saturated $Di_{35}Py_{65}$ composition (EGGLER and BAKER, 1982). Combined with the results in the system Ne–Fo–Q these observations suggest that, in general, liquidus phases of higher octahedral aluminum content are favored by C–H fluid dissolution.

Spectroscopic results

Fourier Transform infrared (FTIR) spectra of sodamelilite and jadeite glasses are presented over

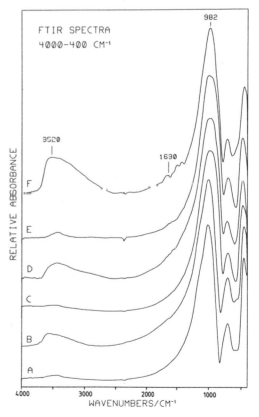

FIG. 5. FTIR spectra (4000–400 cm^{-1}): A. Volatile–free jadeite glass (quenched at 1 bar); B. C–H fluid—saturated jadeite glass (T–1296); C. Volatile–free sodamelilite glass (quenched at 30 kbar); D. C–H fluid—saturated sodamelilite glass (T–1341); E. Sodamelilite glass reduced by interaction with silicon carbide at 1500°C, 30 kbar (see text); F. Hydrous sodamelilite glass (T–1250). Weak positive or negative bands near 2350 cm^{-1} are due to atmospheric CO_2 vapour.

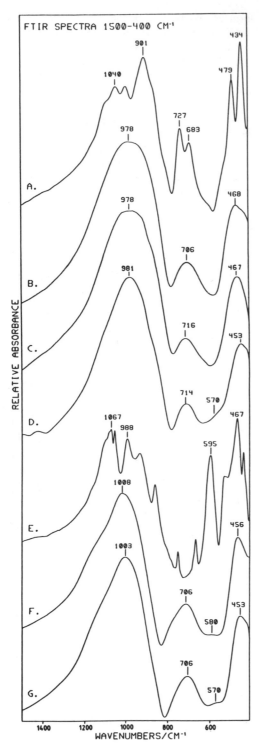

the ranges 4000–400 cm^{-1} and 1500–400 cm^{-1} in Figures 5 and 6. For jadeite, comparisons have been made between C–H fluid–saturated glass quenched from 30 kbar, 1320°C and the 1 bar volatile–absent jadeite glass. Such comparisons are valid because the Raman study of MYSEN *et al.* (1980) has shown that the spectroscopically resolvable structure of quenched jadeite melt remains essentially unaffected by pressure up to at least 38 kbar. For sodamelilite, the volatile–free glass used for comparison was quenched from 30 kbar, 1395°C. The following features distinguish spectra of the C–H fluid saturated glasses (Figs 5B, D and 6C, G) from the volatile–free glasses (Figures 5A, C and 6B, F):

High–frequency region 4000–1400 cm^{-1}

A broad, asymmetric O–H stretch band centered at ∼3580 cm^{-1} is the most prominent feature in the high–frequency region. Comparison of O–H peak areas with hydrous sodamelilite glass (Figure 5F) containing 5.5 ± 0.2 weight percent H_2O gives an estimated dissolved hydrogen content equivalent to 2.7 ± 0.3 weight percent H_2O (if all H is derived from CH_4 then this would correspond to a methane solubility of ∼1 weight percent). The area under the O–H envelope in C–H fluid—saturated jadeite suggests a similar dissolved OH content. There are no absorption bands at ∼2900 cm^{-1} that could be ascribed to C–H bond stretching in dissolved molecular methane or other hydrocarbon groups such as –CH_3 or –CH_2–. No absorptions appear in the frequency range 2600–1700 cm^{-1}. A weak band appears at ∼1630 cm^{-1} (ν_2 H–O–H bending vibration) due to the presence of dissolved molecular H_2O (STOLPER, 1982). There is no evidence for dissolved carbonate which has a characteristic ν_3 absorption band or bands at ∼1600–1380 cm^{-1}.

Aluminosilicate envelopes 1200–400 cm^{-1}

Changes occur in both the high–frequency and mid-range envelopes (centered at ∼1000 cm^{-1} and ∼700 cm^{-1} respectively) and in the spectral region near 570 cm^{-1}. These changes reflect structural rearrangements in the aluminosilicate network that result from volatile dissolution. They are more clearly illustrated in difference spectra (volatile-saturated minus volatile–absent glasses) presented in Figure 7. Strong positive features appear at

FIG. 6. FTIR spectra (1500–400 cm^{-1}): A. Crystalline sodamelilite (prepared at 30 kbar); B. Volatile–free sodamelilite glass (quenched at 30 kbar); C. C–H fluid—saturated sodamelilite glass (T–1341); D. Sodamelilite glass containing 5.5 ± 0.2 weight percent dissolved H_2O (T–

1250). The doublet at 1500–1400 cm^{-1} is due to trace dissolved carbonate; E. Crystalline jadeite (prepared at 25 kbar); F. Volatile–free jadeite glass (quenched at 1 bar); G. C–H fluid—saturated jadeite glass (T–1296).

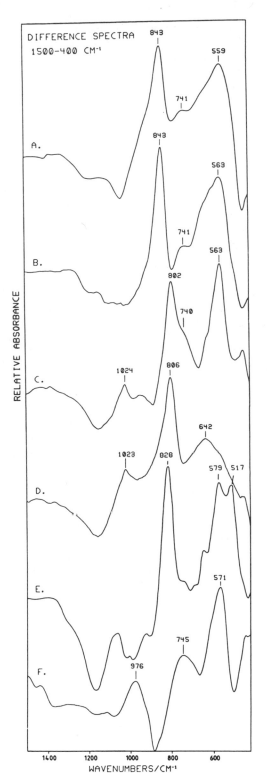

FIG. 7. Difference spectra (1500–400 cm^{-1}) (component-bearing) minus (component–absent): A. (C–H fluid—saturated jadeite glass T–1296) minus (1 bar volatile–free

~850–800 and ~570 cm^{-1} in C–H fluid—saturated sodamelilite and jadeite glasses. Weaker features are present as shoulders at ~950, ~740 and ~640 cm^{-1} and at ~1020 cm^{-1} in sodamelilite only. Both strong positive difference features appear in the spectrum of Ne–Fo–Q glass (sample from run T–1289) which also has a strong band at 517 cm^{-1} due the presence of crystalline enstatite. Negative difference features are much less pronounced and occur in spectra of both C–H fluid–saturated sodamelilite and jadeite at ~1160 cm^{-1} and in jadeite near 1050 cm^{-1}. These may be due to a decrease of Si–O–(Si, Al) bridging bonds but it is difficult to make specific assignments because of the complex overlapping nature of Si–O symmetric and asymmetric stretching modes of both bridging and non-bridging oxygen in this region of the infrared spectrum. There is little change in the low–frequency envelopes (centred near 450 cm^{-1}) except at <425 cm^{-1} where KBr background effects become noticeable.

Positive difference features related to OH dissolution

The prominent 570 cm^{-1} band is present in the FTIR difference spectrum of hydrous sodamelilite glass (Figure 7F) and therefore may be ascribed to changes in the aluminosilicate network resulting from OH dissolution. Bands in this region in the Raman spectra of aluminosilicate glasses have usually been assigned to in–plane Si–O–Si bridge bending motions or to the presence of 3– or 4–membered ring structures (MCMILLAN, 1984). However, in the 500–600 cm^{-1} region in the infrared TARTE (1965, 1967), FARMER et al. (1979) and SERNA et al. (1979) have assigned strong absorption bands in aluminosilicate glasses, crystals and gels to vibrations of AlO$_6$ polyhedra. While such an assignment will require confirmation (perhaps using more definitive techniques such as solid–state NMR), TAYLOR (1985) noted that strong IR bands in the 620–520 cm^{-1} region are found in all crystalline aluminosilicates containing AlO$_6$ polyhedra

jadeite glass); B. Same as A but duplicate KBr discs were used to record both component-bearing and component–absent spectra (reproducibility check); C. (C–H fluid—saturated sodamelilite glass T–1341) minus (30 kbar volatile–free sodamelilite glass); D. (SiC reduced sodamelilite glass) minus (30 kbar volatile-free sodamelilite glass) [the "reduced component"]; E. (C–H fluid—saturated Ne$_{55}$Fo$_{25}$Q$_{20}$ glass containing ~10% enstatite crystals: run T–1289) minus (Ne$_{55}$Fo$_{25}$Q$_{20}$ 1 bar glass); F. (Hydrous sodamelilite glass T–1250) minus (30 kbar volatile–free sodamelilite glass) [the "oxidized component"].

but are absent in structurally related minerals lacking Al or in those containing only AlO_4 polyhedra (exceptions are minerals based on small ring framework structures such as the feldspars or scapolites). This is illustrated (Figures 6A & 6E) in the spectra of crystalline jadeite ($NaAl^{VI}Si_2O_6$, strong band at 595 cm^{-1}) and sodamelilite ($NaCaAl^{IV}Si_2O_7$, no strong absorptions 620–520 cm^{-1}). The weaker 950 and 740 cm^{-1} features identified in the C–H fluid —saturated glasses are also present in the hydrous glass and could be due, respectively, to Si–OH stretching (MYSEN and VIRGO, 1980a) and vibrations of Al–O–Al linkages in aluminate condensates or "clusters" (TARTE, 1967; SERNA et al., 1977).

850–800 cm^{-1} positive difference feature

The ~850–800 cm^{-1} feature is located on the low–frequency limb of the high–frequency envelope in jadeite, sodamelilite and Ne–Fe–Q glasses. This band is not present in hydrous sodamelilite glass and therefore is unlikely to be associated with OH dissolution or network depolymerization. Instead, this band is characteristic of reduced C–H volatile dissolution and may be due to a dissolved reduced component. Spectroscopic identification of such a component in aluminosilicate glasses is verifiable experimentally as discussed later. Possible assignments for the 850–800 cm^{-1} feature are considered below after applying necessary theoretical constraints.

Theoretical constraints

Mechanisms for the dissolution of reduced volatiles in silicate melts have been suggested by very few authors. In a study that investigated the solubility behaviour of CO as a CO_2–CO volatile mixture, EGGLER et al. (1979) suggested that CO dissolves by a carbonation reaction of the type:

$$CO + 2O^- = CO_3^{2-} \quad \cdots \cdots A, \quad (3)$$

where "O^-" = non-bridging oxygen. EGGLER and BAKER (1982) proposed two mechanisms for CH_4 dissolution based on analogous reactions for H_2O, i.e.

$$4Si–O–Si + CH_4 = 4Si–OH + C \quad \cdots \cdots B \quad (4)$$

$$Si–O–Si + CH_4 = Si–CH_3 + Si–OH \quad \cdots \cdots C. \quad (5)$$

The validity of mechanisms A and B is, however, questionable because neither reaction can be correctly charge–balanced (at least in a chemically realistic sense in Equation A, "O^-" cannot be equated with NBO for usual oxidation states of oxygen in silicate melts). Because electroneutrality should always be obeyed when balancing any chemical reaction, it is evident that if a reduced volatile dissolves in a silicate melt to give an "oxidized bond" such as O–H or O–C, then this must be balanced at equilibrium by concurrent production of a "reduced bond." Such a bond is one involving an element in a lower oxidation state or one excluding oxygen or both. Candidates in the case of a C–H fluid could include: Si–H, Si–C, C–H, Si–Si and analogous bonds involving Al and other metal cations.

Reference to the FTIR spectra of sodamelilite and jadeite C–H fluid–saturated glasses immediately eliminates metal–hydrogen or C–H bonds as candidates for the "reduced bond." This is because bonds of this type have characteristic IR stretching frequencies in the range ~3000–1700 cm^{-1} where no absorption was noted. C–H bonds are expected at 3050–2850 cm^{-1} and Si–H bonds at 2250–2100 cm^{-1} (POUCHERT, 1981). Thus reaction C, suggested by EGGLER and BAKER (1982), while being properly charge balanced and involving Si–C and C–H as reduced bonds, is not consistent with observed spectroscopic results.

Possible mechanisms of C–H volatile dissolution

The theoretical and spectroscopic constraints discussed above, greatly limit the choice of a reduced component. We suggest that the most likely candidates are either (I) a network unit containing Si–C bonds or (II) a non-stoichiometric network component containing units having an O/Si ratio less than 2, such as found in amorphous silicon monoxide. The former alternative implies that only hydrocarbon species (i.e., dominantly CH_4) are involved in the dissolution process, whereas the latter alternative is a general reduction of the silicate network and could involve H_2 as well as CH_4. Support for candidate (I) is given by the known range of IR active Si–C bond stretching frequencies for molecular compounds (WELTNER and MCLEOD, 1964; POUCHERT, 1981) e.g., organosilicon compounds (Si–CH$_3$ bonds): 680–740 cm^{-1}; matrix–isolated SiC_2 and SiC molecules: 835 and 1226 cm^{-1} respectively. The range ~700 cm^{-1} to 1226 cm^{-1} encompasses the region that includes the strongest difference spectrum features at ~850–800 cm^{-1}. However, these features would also be consistent with the presence of silicon monoxide or related units. Compared with pure silica glass, the high–frequency envelope in amorphous silicon monoxide is shifted down frequency by some 100 cm^{-1} (PLISKIN and LEHMAN, 1965; KHANNA et al., 1981). A similar shift in band components in aluminosilicate glasses would be sufficient to extend the high–frequency envelope to lower wavenumbers and hence result in the observed $\sim$850–800 cm^{-1} positive difference features.

Mechanisms (I) and (II) may be described by the following balanced equations (for $[SiO_3]^{2-}$ and sodamelilite melt-units):

Mechanism I: (Silicon–carbon bond formation)

$$2[SiO_3]^{2-} + CH_4$$
$$= \{Si-C\} + 4[OH]^- + [SiO_2]^0 \quad (6)$$
$$2[Al^{IV}Si_2O_7]^{3-} + CH_4 = 2[Al^{NN}O_2]^-$$
$$+ 4[OH]^- + \{Si-C\} + 3[SiO_2]^0 \quad (7)$$

$\left.\begin{array}{c} \\ \\ \\ \end{array}\right\}$ ....D.

Mechanism II: (Network reduction by CH_4 or H_2, or both)

$$2[SiO_3]^{2-} + CH_4$$
$$= 2\{Si-O\} + 4[OH]^- + \{C^0\} \quad (8)$$
$$2[Al^{IV}Si_2O_7]^{3-} + CH_4$$
$$= 2[Al^{NN}O_2]^- + 4[OH]^-$$
$$+ 2\{Si-O\} + 2[SiO_2]^0 + \{C^0\} \quad (9)$$

$\left.\begin{array}{c} \\ \\ \\ \end{array}\right\}$ ....E

$$[SiO_3]^{2-} + H_2 = \{Si-O\} + 2[OH]^- \quad (10)$$
$$[Al^{IV}Si_2O_7]^{3-} + H_2 = [Al^{NN}O_2]^-$$
$$+ 2[OH]^- + \{Si-O\} + [SiO_2]^0 \quad (11)$$

$\left.\begin{array}{c} \\ \\ \\ \end{array}\right\}$ ....F

where

$\{\quad\}$ = unidentified location in silicate network
$[\quad]$ = melt-unit or complex
$\{C^0\}$ = graphite, diamond or carbon dissolved in the melt in unspecified form
NN = Al in non–network sites (*e.g.*, AlO_6 polyhedra).

For both mechanisms additional equilibria can be written to relate spectroscopically observed species such as dissolved molecular water:

$$[SiO_2] + 2[OH]^- = [H_2O]_{molecular} + [SiO_3]^{2-}. \quad (12)$$

Mechanisms (I) and (II) are similarly constructed; both reactions produce 4 moles of O–H bonds (the "oxidized component") per mole of dissolved methane. For the sodamelilite melt-unit, the solution process is written to accommodate a redistribution of Al between network and non–network sites. The major difference between the mechanisms is in the nature of the reduced bond formed (*i.e.*, the "reduced component"). In mechanism (I), a significant melt–phase solubility of reduced carbon as Si–C bonds is implied. This mechanism is directly

analogous to the structural role of nitrogen in Na_2O–CaO–silicon oxynitride glasses recently investigated by BROW and PANTANO (1984). Based on FTIR and x–ray photoelectron spectroscopic results those authors concluded that nitrogen is present in the silicate network of oxynitride glasses in the form of Si–N bonds with N in three–fold and possibly two–fold co–ordination sites. With increasing N content the principle changes in the FTIR spectra are seen in the high–frequency envelope which broadens and shifts to lower wavenumbers (similar to that observed in the C–H fluid—saturated glasses). The maximum amount of nitrogen dissolved in the Na_2O–CaO–silicon oxynitride glasses at 1 bar was $\sim$2.2 weight percent N but the effects of N substitution on the network are clearly seen in the FTIR spectrum at much lower levels.

Mechanism (II) requires the presence of a silicate network unit with an O:Si ratio <2; this is represented in Equations E and F by a $\{SiO\}$ or "silicon monoxide" group. It is conceivable that such units might resemble those found in amorphous silicon monoxide. The RDF study of YASAITIS and KAPLOW (1972) favors a structure for this compound based on puckered $(SiO)_n$ rings where the average co–ordination number about each Si atom does not deviate significantly from two. Such a structure gives each Si atom a formal valency of II.

Characterization of the "reduced component"

For the sodamelilite composition the "reduced component" was characterized spectroscopically by reducing the silicate network at high pressure under anhydrous conditions. The starting material consisted of sodamelilite glass in which a portion of the SiO_2 was substituted by –SiC [total C = 1.5 weight percent, equivalent to a methane solubility of 2 weight percent via mechanism (I)]. An inner graphite capsule was used to separate the silicate/carbide mix from the outer Pt capsule to prevent Pt–Si alloy formation. Over a run time of 90 min at 30 kbar and 1500°C, SiC was fully decomposed producing a product consisting of clear glass and disseminated graphite. This experiment does not distinguish between mechanisms (I) and (II) because SiC may dissolve to form reduced Si–C bonds or may reduce the silicate network directly via the reaction: $SiC + SiO_2 \rightarrow 2\{SiO\} + C$ [analogous to mechanism (II)]. FTIR spectra are shown in Figures 5E and 7D. Graphite is essentially IR inactive over this spectral range and does not contribute to the observed bands. The difference spectrum Figure 7D has a major positive feature at 806 cm^{-1} and weaker features at 1023 and 642 cm^{-1} corresponding closely to those present in C–H fluid—saturated sodamelilite. In fact a combination of the hydrous sodamelilite glass difference spectrum (the "oxidized component") and Figure 7D would be almost indistinguishable from the C–H fluid—saturated spectrum. Thus separate "reduced" and "oxidized" components can be characterized spectroscopically; the latter is associated with O–H bond formation and the former is best interpreted as a reduction of the silicate network. Whether or not this reduction involves formation of reduced bonds to carbon must be decided by analytical means.

Analytical constraints

Results of pyrolysis/gas chromatography analyses for C and H for both samples and standards are presented in Table 1. The analysed carbon contents of the standards (SiO_2–SiC mixtures) are between half and one third of the known amounts present. This suggests incomplete oxidation of SiC has occurred with greatest discrepancy occurring at C contents >0.5 weight percent. Carbon values for the glass unknowns must therefore be regarded as minimum quantities only; if α–SiC forms metastably in these samples then analyses may be low by as much as a factor of two for total C contents ≤0.5 weight percent. Hydrogen analyses (converted to weight percent H_2O) are in good agreement with those deduced spectroscopically for sodamelilite. C–H fluid equilibrated glasses therefore contain a minimum of ~1000 ppm C and a maximum of ~2000 ppm C. These results are confirmed by quantitative and semi–quantitative electron microprobe analyses for carbon (Table 3). Carbon contents in the C–H fluid—saturated glasses are near or below the limit of detection (<2000 ppm) consistent with the gas chromatographic analyses.

For mechanism (I) to operate a reduced carbon solubility of 8000–12000 ppm is required. Thus mechanism (II) is supported as the dominant process for reduced volatile interaction with aluminosilicate melts. The small amount of carbon detected could dissolve either by mechanism (I) or perhaps in the form of atomic carbon occupying interstitial sites or cation vacancies as suggested FREUND *et al.* (1980) for oxide and silicate lattices. Carbon dissolved in this manner would be undetectable by IR and Raman spectroscopic methods.

DISCUSSION

The recognition of mechanism (II) as the dominant process of C–H volatile solubility does not discriminate between the particular reduced volatile species involved. The mechanism is a general reduction of the silicate network that may take place in the presence of any of the reduced volatiles H_2, CH_4 or C_2H_6. In the presence of CH_4–rich fluids the H/C ratio of the melt phase greatly exceeds H/C of the coexisting fluid suggesting that reduced C–H volatile solubility will largely be a function of f_{H_2} and governed by equilibria similar to F above [Equations (10) and (11)]. In the system aluminosilicate–C–O, carbon monoxide is an important reduced volatile at $P < 20$ kbar and is believed to dissolve in melts by a carbonation reaction (EGGLER *et al.,* 1979). Charge balance constraints dictate a solubility mechanism that must involve a reduced component and we can propose reactions analogous to Equations E and F to describe CO dissolution, *e.g.,*

$$CO + [SiO_3]^{2-} = CO_3^{2-} + \{Si-O\}. \quad (12)$$

If carbonate ions and a reduced melt component such as {SiO} cannot coexist stably in silicate melts, there are real difficulties in proposing CO as a melt-soluble species other than in molecular form. Choice of mechanism (II) also helps rationalize the experimental results of LUTH and BOETTCHER (1986) which show that H_2 gives rise to significant depressions of the albite and diopside solidii implying a strong interaction between H_2 and aluminosilicate liquids as predicted by model F above.

Liquidus phase relations in the system Ne–Fo–Q suggest that dissolution of a C–H fluid, compared to the volatile–absent case, raises melt activities of network modifying Al_2O_3. This is in accord with the observed expansion of the garnet phase volume in other systems. The idea that carbon–rich eclogites could be the products of fractional crystallization of mantle melts under conditions of CH_4–H_2O–H_2 volatile saturation as proposed by EGGLER and BAKER (1982) is supported by the data presented here. KUSHIRO and YODER (1974) stated that ". . . in the presence of water . . . it should be possible for eclogite to form from garnet lherzolite" at depths

Table 3. Microprobe analyses for carbon*

Sample	Expected weight percent	Weight percent analysed
Graphite (treated as unknown)	100.00	101.22
T–1178 (Sm + CO_2)	2.1 ± 0.2	2.15
T–1296 (Jd + C–H fl.)		
T–1341 (Sm + C–H fl.)	~0.1–0.2	≤0.2†
T–1318 (Sm + C–H fl.)		

* B.J. Griffin, analyst.
† Limit of detection ≃0.2 weight percent.

greater than the 26 kbar limit in the volatile–absent case, *i.e.,* to within the stability field of diamond. Because both H_2O and CH_4–H_2 fluids have similar melt depolymerizing behaviour, the findings of KUSHIRO and YODER (1974) will apply equally well to CH_4–H_2O–H_2 volatile mixtures as to pure H_2O. The added advantage in the case of carbon–rich eclogites is that the preference of the melt for H compared with C will drive a coexisting CH_4–bearing fluid phase toward carbon saturation leading to precipitation of diamond or graphite via equations similar to E. This is in accord with the origins of graphite-diamond eclogite from the Roberts Victor kimberlite as discussed by HATTON and GURNEY (1979). Those authors propose an origin based on rapid crystallization of melts produced by volatile-induced partial melting of garnet lherzolite where melt volumes are such that gravitational separation of diamond or graphite is ineffective. Whereas the authors propose that carbon is a result of reduction of a CO_2-bearing vapor during cooling, an alternative mechanism in which CH_4–H_2O–H_2 fluids give rise to melting accompanied by carbon precipitation from the fluid phase can equally well explain the origin of these rocks.

The finding that the network portion of silicate melts is quite susceptible to reduction via formation of groups with O/Si < 2, has not been demonstrated previously. For the observed H_2O content of C–H fluid–saturated sodamelilite glass at 30 kbar, 1350°C, we calculate that ∼20% of the Si should be present as Si(II) (equivalent to 6–7 weight percent SiO) by mechanism (II). Reduced systems such as the enstatite chondrite group of meteorites have intrinsic oxygen fugacities that lie below IW. Measured values for the equilibrated EL enstatite chondrite group give f_{O_2}'s in the range ∼IW–3 to IW–4 log f_{O_2} units (BRETT and SATO, 1984; WALTER and DOAN, 1969), near the redox conditions of the present C–H fluid experiments. In these rocks Si is distributed amongst three phases: metal (kamacite with ∼1–4 weight percent Si), silicate (mainly enstatite), and silicon oxynitride (sinoite, Si_2N_2O) with graphite and not silicon carbide as an important accessory (SEARS *et al.,* 1982). This solid phase distribution of Si redox environments (at least at relatively low pressures) is consistent with the interpretation of C–H fluid—equilibrated aluminosilicate glasses where a reduced silicate network plus elemental carbon is evidently more stable than the equivalent melt structure containing Si–C bonds. It is possible to express the relationship between solid phases and a corresponding melt at f_{O_2} ∼ IW–4 × log f_{O_2} units via a disproportionation equilibrium of the type:

$$2[Si^{2+}O_x]_{\substack{\text{silicate} \\ \text{melt}}}$$
$$= (2 - x)[Si^0]_{\substack{\text{metallic} \\ \text{phase}}} + x[Si^{4+}O_2]_{\substack{\text{solid} \\ \text{silicate}}} \quad (13)$$

where $1 < x < 2$.

In view of recent hypotheses supporting the early incorporation of large amounts of reduced enstatite chondritic components into the Earth's mantle (SMITH, 1982; JAVOY and PINEAU, 1983; ITO *et al.,* 1984) such equilibria are expected to have an important bearing on the mantle–core segregation of Si.

SUMMARY

In this investigation, a mechanism for C–H volatile solubility in aluminosilicate melts has been proposed from the interpretation of FTIR spectra within theoretical and analytical constraints. The mechanism is supported by phase relations determined in the system Ne–Fo–Q–C–O–H which indicates a melt depolymerizing role for C–H fluids.

On spectroscopic and theoretical grounds reduced C–H volatile dissolution can be resolved into soluble oxidized and reduced components. The former is represented by O–H bonds (as hydroxyl groups and molecular water) and affects melt structure much in the manner of H_2O dissolution. There is no spectroscopic evidence for the presence of dissolved molecular CH_4 or carbonate. The reduced component is somewhat enigmatic. Analytical data establish that the reduced component is consistent with a network silicon-oxygen unit where the formal valency on silicon is reduced from IV to II (O:Si stoichiometry of the system <2). At the f_{O_2} conditions of these experiments, ∼IW–4.5 log f_{O_2} units, a general reduction of the silicate network is evidently favored over formation of Si–C bonds. This is not the case, however, in analogous reduced systems containing nitrogen where Si–N bonding in silicon oxynitride glasses is well characterized. Nevertheless, the reduced carbon solubility in the aluminosilicate melts studied is ∼1000 ppm (minimum) in a form yet to be characterized.

These results have been used to (1) propose solubility mechanisms for other reduced volatiles such as CO; (2) help rationalize the observed strong interaction between H_2 and silicate melts; (3) suggest an alternative interpretation for the origin of carbonaceous eclogites as the product of CH_4–H_2O fluid-induced partial melts of garnet lherzolite; (4) suggest a mechanism for mantle–core partitioning of Si and (5) provide a basis for investigating the nature of melting in a reduced mantle as discussed in the accompanying paper by GREEN *et al.* (1987).

Acknowledgements—We are grateful to Dr. B. J. Griffin (University of Adelaide) for much time spent at the microprobe analysing our samples for carbon. We thank J. Bignell, N. Davis, W. Jablonski and K. L. Harris for invaluable technical assistance. This study was supported financially by a Commonwealth Postgraduate Scholarship and Australian National Research Fellowship awarded to WRT and an ARGS grant to DHG.

REFERENCES

ANDERSON A. T. (1975) Some basaltic and andesitic gases. *Rev. Geophys. Space Phys.* **13**, 37–55.

ARCULUS R. J. and DELANO J. W. (1981) Intrinsic oxygen fugacity measurements: techniques and results for spinels from upper mantle peridotites and megacryst assemblages. *Geochim. Cosmochim. Acta* **45**, 899–913.

BELL P. M. (1964) High-pressure melting relations for jadeite composition. *Carnegie Inst. Wash. Yearb.* **63**, 171–174.

BRETT R. and SATO M. (1984) Intrinsic oxygen fugacity measurements on seven chondrites, a pallasite, and a tektite and the redox state of meteorite parent bodies. *Geochim. Cosmochim. Acta* **48**, 111–120.

BREY G. (1976) CO_2 solubility and solubility mechanisms in silicate melts at high pressures. *Contrib. Mineral. Petrol.* **57**, 215–221.

BROW R. K. and PANTANO C. G. (1984) Nitrogen coordination in oxynitride glasses. *J. Amer. Ceram. Soc. Comm.* **67**, C72–C74.

CHEN C-H. and PRESNALL D. C. (1975) The system Mg_2SiO_4–SiO_2 at pressures up to 25 kilobars. *Amer. Mineral.* **60**, 398–406.

EGGLER D. H. (1978) The effect of CO_2 upon partial melting of peridotite in the system Na_2O–CaO–Al_2O_3–MgO–SiO_2–CO_2 to 35 kb, with an analysis of melting in a peridotite–H_2O–CO_2 system. *Amer. J. Sci.* **278**, 305–343.

EGGLER D. H. and BAKER D. R. (1982) Reduced volatiles in the system C–O–H: Implications to mantle melting, fluid formation and diamond genesis. In *High Pressure Research in Geophysics,* (eds. S. AKIMOTO and M. H. MANGHANI), pp. 237–250 Center for Academic Publications.

EGGLER D. H., MYSEN B. O., HOERING T. C. and HOLLOWAY J. R. (1979) The solubility of carbon monoxide in silicate melts at high pressures and its effect on silicate phase relations. *Earth Planet. Sci. Lett.* **43**, 321–330.

FARMER V. C., FRASER A. R. and TAIT J. M. (1979) Characterisation of the chemical structures of natural and synthetic aluminosilicate gels and sols by infrared spectroscopy. *Geochim. Cosmochim. Acta* **43**, 1417–1420.

FREUND F., KATHREIN H., WENGELER H., KNOBEL R. and HEINEN H. J. (1980) Carbon in solid solution in forsterite—a key to the intractable nature of reduced carbon in terrestrial and cosmogenic rocks. *Geochim. Cosmochim. Acta* **44**, 1319–1333.

GREEN D. H. and LIEBERMANN R. C. (1976) Phase equilibria and elastic properties of the pyrolite model for the oceanic upper mantle. *Tectonophys.* **32**, 61–92.

GREEN D. H. and RINGWOOD A. E. (1967) The genesis of basaltic magmas. *Contrib. Mineral. Petrol.* **15**, 103–190.

GREEN D. H., FALLOON T. J. and TAYLOR W. R. (1987) Mantle derived magmas—roles of variable source peridotite and variable C–H–O fluid compositions. In *Magmatic Processes: Physicochemical Principles,* (ed. B. O. MYSEN), The Geochemical Society Spec. Publ. No. 1, pp. 139–154.

GUPTA A. K., TAYLOR W. R. and GREEN D. H. (1987) Experimental study of the system forsterite–nepheline–jadeite at variable temperatures under 28 kbar pressure. *Amer. J. Sci.,* (Submitted).

HATTON C. J. and GURNEY J. J. (1979) A diamond-graphite eclogite from the Roberts Victor Mine. *Proc. Int. Kimberlite Conf., 2nd,* Vol. 2, 29–36.

HOLLOWAY J. R. (1981) Volatile interactions in magmas. *Adv. Phys. Geochem.* **1** (ed. S. K. SAXENA), pp. 273–293. Springer-Verlag.

HOLLOWAY J. R. and REESE R. L. (1974) The generation of N_2–CO_2–H_2O fluids for use in hydrothermal experimentation. I. Experimental method and equilibrium calculations in the C–O–H–N system. *Amer. Mineral.* **59**, 587–597.

ITO E., TAKAHASHI E. and MATSUI Y. (1984) The mineralogy and chemistry of the lower mantle: an implication of the ultrahigh-pressure phase relations in the system MgO–FeO–SiO_2. *Earth Planet. Sci. Lett.* **67**, 238–248.

JAKOBSSON S. (1984) Melting experiments on basalts in equilibrium with a graphite-iron-wustite buffered C–O–H fluid. Ph.D. Dissertation, Arizona State Univ.

JAVOY M. and PINEAU F. (1983) Stable isotope constraints on a model Earth from a study of mantle nitrogen. *Meteoritics* **18**, 320.

KHANNA R. K., STRANZ D. D. and DONN B. (1981) A spectroscopic study of intermediates in the condensation of refractory smokes: Matrix isolation experiments of SiO. *J. Chem. Phys.* **74**, 2108–2115.

KUSHIRO I. (1964) The join akermanite-soda melilite at 20 kilobars. *Carnegie Inst. Wash. Yearb.* **63**, 90–92.

KUSHIRO I. (1968) Compositions of magmas formed by partial zone melting of the Earth's upper mantle. *J. Geophys. Res.* **73**, 619–634.

KUSHIRO I. (1972) Effect of water on the composition of magmas formed at high pressures. *J. Petrol.* **13**, 311–334.

KUSHIRO I. and YODER H. S. JR. (1974) Formation of eclogite from garnet lherzolite: liquidus relations in a portion of the system $MgSiO_3$–$CaSiO_3$–Al_2O_3 at high pressures. *Carnegie Inst. Wash. Yearb.* **73**, 266–269.

LUTH R. W. and BOETTCHER A. L. (1986) Hydrogen and the melting of silicates. *Amer. Mineral.* **71**, 264–276.

MATHEZ E. A. and DELANEY J. R. (1981) The nature and distribution of carbon in submarine basalts and peridotite nodules. *Earth Planet. Sci. Lett.* **56**, 217–232.

MCMILLAN P. (1984) Structural studies of silicate glasses and melts—applications and limitations of Raman spectroscopy. *Amer. Mineral.* **69**, 622–644.

MCMILLAN P. (1985) Vibrational spectroscopy in the mineral sciences. In *Reviews in Mineralogy, 14* (eds. A. NAVROTSKY and S. KIEFFER), pp. 9–63 Mineralogical Society of America.

MYSEN B. O. and VIRGO D. (1980a) Solubility mechanism of water in basalt melt at high pressures and temperatures: $NaCaAlSi_2O_7$–H_2O as a model. *Amer. Mineral.* **65**, 1176–1184.

MYSEN B. O. and VIRGO D. (1980b) The solubility behaviour of CO_2 in melts on the join $NaAlSi_3O_8$–$CaAl_2Si_2O_8$–CO_2 at high pressures and temperatures: A Raman spectroscopic study. *Amer. Mineral.* **65**, 1166–1175.

MYSEN B. O. and VIRGO D. (1980c) Solubility mechanisms of carbon dioxide in silicate melts: A Raman spectroscopic study. *Amer. Mineral.* **65,** 885–899.

MYSEN B. O., VIRGO D. and SCARFE C. M. (1980) Relations between anionic structure and viscosity of silicate melts—a Raman spectroscopic study. *Amer. Mineral.* **65,** 690–710.

PLISKIN W. A. and LEHMAN H. S. (1965) Structural evaluation of silicon oxide films. *J. Electrochem. Soc.* **112,** 1013–1019.

POUCHERT C. J. (1981) *The Aldrich Library of Infrared Spectra.* 3rd ed., 1203 pp. Aldrich Chemical Co.

ROCHOW E. G. (1973) The chemistry of silicon. In *Comprehensive Inorganic Chemistry,* Vol. 9, pp. 1323–1467 Pergamon Press.

RYABCHIKOV I. D., GREEN D. H., WALL V. J. and BREY G. P. (1981) The oxidation state of carbon in the reduced–velocity zone. *Geochem. Int.* **18,** 148–158.

RYABCHIKOV I. D., SCHREYER W. and ABRAHAM K. (1982) Compositions of aqueous fluids in equilibrium with pyroxenes and olivines at mantle pressures and temperatures. *Contrib. Mineral. Petrol.* **79,** 80–84.

RYERSON F. J. (1985) Oxide solution mechanisms in silicate melts: Systematic variations in the activity coefficient of SiO_2. *Geochim. Cosmochim. Acta* **49,** 637–650.

SEARS D. W., KALLEMEYN G. W. and WASSON J. T. (1982) The compositional classification of chondrites: II. The enstatite chondrite group. *Geochim. Cosmochim. Acta* **46,** 597–608.

SERNA C. J., VELDE B. D. and WHITE J. L. (1977) Infrared evidence of order–disorder in amesites. *Amer. Mineral.* **62,** 296–303.

SERNA C. J., WHITE J. L. and VELDE B. D. (1979) The effect of aluminium on the infra–red spectra of 7 Å trioctahedral minerals. *Mineral. Mag.* **43,** 141–148.

SHARMA S. K. and YODER H. S. JR. (1979) Structural study of glasses of akermanite, diopside and sodium melilite compositions by Raman spectroscopy. *Carnegie Inst. Wash. Yearb.* **78,** 526–532.

SMITH J. V. (1982) Heterogeneous growth of meteorites and planets, especially the Earth and moon. *J. Geol.* **90,** 1–125.

STOLPER E. (1982) Water in silicate glasses: an infrared spectroscopic study. *Contrib. Mineral. Petrol.* **81,** 1–17.

TARTE P. (1965) The determination of cation co-ordination in glasses by infra–red spectroscopy. In *Physics of Non-Crystalline Solids,* (ed. J. A. PRINZ), pp. 549–565. John Wiley.

TARTE P. (1967) Infra–red spectra of inorganic aluminates and characteristic vibrational frequencies of AlO_4 tetrahedra and AlO_6 octahedra. *Spectrochim. Acta* **23A,** 2127–2143.

TAYLOR M. and BROWN G. E. (1979) Structure of mineral glasses—II. The SiO_2–$NaAlSiO_4$ join. *Geochim. Cosmochim. Acta* **43,** 1467–1473.

TAYLOR W. R. (1985) The role of C–O–H fluids in upper mantle processes: a theoretical, experimental and spectroscopic study. Ph.D. Thesis, Univ. of Tasmania.

TAYLOR W. R. (1987) A 5-parameter modified Redlich-Kwong equation of state for C–O–H fluids at upper mantle pressures and temperatures, (In prep.).

WADE K. and BANISTER A. J. (1973) The chemistry of aluminium, gallium, indium and thallium. In *Comprehensive Inorganic Chemistry,* Vol. 12, pp. 993–1064 Pergamon Press.

WALTER L. S. and DOAN A. S. (1969) A determination of oxygen fugacities of chondritic meteorites (abstr.). *Geol. Soc. Amer. Abstr. Prog.* **1,** 232–233.

WELHAN J. A. and CRAIG H. (1979) Methane and hydrogen in East Pacific Rise hydrothermal fluids. *Geophys. Res. Lett.* **6,** 829–831.

WELHAN J. A. and CRAIG H. (1983) Methane, hydrogen and helium in hydrothermal fluids at 21°N on the East Pacific Rise. In *Hydrothermal Processes at Seafloor Spreading Centers,* (eds. P. A. RONA, K. BOSTROM, L. LAUBIER and K. L. SMITH), pp. 391–409 Plenum Press.

WELTNER W. and MCLEOD D. (1964) Spectroscopy of silicon carbide and silicon vapor trapped in neon and argon matrices at 4 K and 20 K. *J. Chem. Phys.* **41,** 235–245.

WINDOM K. E. and BOETTCHER A. L. (1981) Phase relations for the joins jadeite-enstatite and jadeite-forsterite at 28 kb and their bearing on basalt genesis. *Amer. J. Sci.* **281,** 335–351.

WOERMANN E. and ROSENHAUER M. (1985) Fluid phases and the redox state of the Earth's mantle; extrapolations based on experimental, phase theoretical and petrological data. *Fortschr. Mineral.* **63,** 263–349.

YASAITIS J. A. and KAPLOW R. (1972) Structure of amorphous silicon monoxide. *J. Appl. Phys.* **43,** 995–1000.

YODER H. S. JR. (1964) Soda melilite. *Carnegie Inst. Wash. Yearb.* **63,** 86–89.

YODER H. S. JR. and TILLEY C. E. (1962) Origin of basalt magma: an experimental study of natural and synthetic rock systems. *J. Petrol.* **3,** 342–532.

APPENDIX

Experimental results along the join $Ne_{55}Fo_{45}$–Ne_{55}–O_{45} P = 28 kbar

VOLATILE–ABSENT

Run number	Fo Weight percent	Duration (min.)	T (°C)	Approx. % Cryst.	Products*
T–1210**	25	15	1440	40	Fo, Opx, Liq, Qx
T–1213**	25	10	1500	—	Liq, Qx
T–1215**	25	10	1475	30	Fo, Opx, Liq, Qx
T–1218**	25	12	1490	10	Fo, Liq, Qx

H_2O–SATURATED (~30 Weight percent H_2O added as liquid)

T–1175	18	60	1100	5–10	Fo, Opx, Liq, Qx
T–1237	14	60	1040	20	Opx, Liq, Qx
T–1304	20	60	1100	5	Fo, Liq, Qx
T–1305	16	60	1060	10	Opx, Liq, Qx

CO_2–SATURATED (~15 Weight percent CO_2 generated from $Ag_2C_2O_4$)

T–1226	30	12	1390	10	Opx, Liq, Qx
T–1227†	35	12	1390	—	Fo, Opx, Liq, Qx
T–1334	37	12	1400	<2	Fo, Liq, Qx, Qcarb

C–H FLUID—SATURATED (~7 Weight percent CH_4 generated from $Al(OH)_3/Al_4C_3$ mix)

T–1174	18	25	1280	25	Opx, Liq, Qx, Fo (tr)
T–1208	25	20	1350	30	Fo, Opx, Liq, Qx
T–1284	22	20	1380	20	Fo, Opx, Liq, Qx
T–1289††	20	25	1380	10	Opx, Liq, Qx
T–1291	20	25	1360	20	Opx, Fo, Liq, Qx
T–1315	22	30	1395	5	Fo, Liq, Qx

* Fo = forsterite; Opx = enstatite s.s.; Liq = glass; Qx = quench crystals; Qcarb = quench carbonate; tr = trace.
† Liquid composition obtained (see below).
†† FTIR spectrum and difference spectrum obtained (see Figure 7)

Liquid Compositions (Weight percent)

Oxide	T–1213	T–1218	T–1215	T–1210	T–1227*
SiO_2	54.01	54.39	54.55	54.52	48.81
Al_2O_3	19.85	21.51	22.79	23.19	22.93
MgO	14.28	11.03	8.81	8.19	14.32
Na_2O	11.86	13.07	13.85	14.10	13.94
Fe	24.9 (25)†	19.3	15.4	14.3	25.0
Ne	55.3 (55)	59.9	63.5	68.6	63.9
Q	19.8 (20)	20.8	21.1	21.1	11.1

* Liquid projected back into Na/Al = 1 plane from average quench pyroxene: $En_{55}MgTs_{30}Jd_{15}$ (mol percent).
† Expected composition in brackets.

Magmatic Processes: Physicochemical Principles
© The Geochemical Society, Special Publication No. 1, 1987
Editor B. O. Mysen

Mantle-derived magmas—roles of variable source peridotite and variable C–H–O fluid compositions

D. H. Green, T. J. Falloon and W. R. Taylor

Geology Department, University of Tasmania, GPO Box 252C, Hobart, Australia 7001

Abstract—The system forsterite–nepheline–quartz is a useful simple system analogue of melting relations in upper mantle peridotite. The liquidus phase fields at 28 kbar differ from those at low pressure by expansion of the enstatite field at the expense of forsterite. The system illustrates a large field of liquid compositions, from model basanites to model quartz tholeiites, which can be derived from one peridotite source. More refractory source compositions permit a greater compositional range of derivative liquid compositions than more fertile compositions and in particular are required as source or parent compositions for enstatite-rich liquids.

The effects of C–H–O volatiles on melting relationships have been explored with H_2O, CH_4 and CO_2–vapour saturated experiments. The effect of water is to expand the olivine field and depress liquidus temperatures by 350–400°C, but liquids at low degrees of melting of a model peridotite remain nepheline–normative. The effect of CO_2 is most marked with liquids moving to increasingly undersaturated compositions. Methane saturation produces a similar liquidus depression but results in OH^- solution, low carbon solubility and a reduced melt structure, *i.e.*, Si:O < 1:2.

The studies in the simple Fo–Ne–Qz system are matched by melting studies of several peridotite compositions and by liquidus studies on a variety of magnesian primary magmas from different tectonic settings. Mid–ocean ridge basalts are most commonly derivative from picritic parents at 15–20 kbar although some low–olivine to quartz tholeiite liquids are probably primary from approximately 8 kbar pressure. However, other primary magmas such as high–magnesium quartz tholeiites and olivine–poor tholeiites and the very siliceous, low–calcium boninite liquids, are derived from much more refractory source rocks than MORB and require two–stage or multistage melting processes. The role of C–H–O fluids in fluxing such multi–stage melting on convergent margins is very important. An additional source of water, accompanied by low f_{O_2} conditions (i.e., $H_2O > CH_4$ fluids) is identified in the redox–interaction of oxidized lithosphere with $CH_4 > H_2$ fluids degassing from the deep earth.

INTRODUCTION

In seeking to unravel the complexities of magma genesis in the earth's upper mantle, experimental petrologists have successfully demonstrated the diversity of basaltic magmas which can arise from the same source composition by variation of pressure (depth of magma segregation) and temperature (degree of partial melting) (see Basaltic Volcanism Study Project, 1981, and references therein). In addressing the same problem, isotope geochemists have demonstrated that mantle–derived magmas have formed from isotopically different sources or reservoirs which have remained isolated and with considerably different ratios of radiogenic elements for long periods of time (Basaltic Volcanism Study Project, 1981, and references therein). Some studies of trace element geochemistry have helped to link these two approaches (*e.g.*, Frey *et al.*, 1978), particularly by emphasizing the role played by large differences in partition coefficients between residual crystals and liquids for different elements and have introduced concepts of distinctive behaviour for 'compatible' vs 'incompatible' elements, 'LIL-element', 'HFS' elements, 'light' vs 'heavy' rare earth elements etc.

It is possible that the mantle source regions for primary magmas may be relatively homogeneous in major and compatible elements but quite widely variable in incompatible elements, *i.e.*, those which are perceived as mobile because of high solubility in mantle fluid phase(s) or in small, volatile–rich melt phases. This view finds support from studies of mantle lherzolite samples which provide evidence for multievent histories including late stage enrichment in incompatible elements (Basaltic Volcanism Study Project, 1981) and from the recognition of mantle–derived peridotites which are extremely depleted in incompatible elements yet retain major element chemistry with large 'basaltic component' (CaO, Al_2O_3 > 3 weight percent; Frey and Green, 1974; Menzies, 1983; Kurat *et al.*, 1980; Frey *et al.*, 1985).

In this paper our major concern is to demonstrate that the mantle source regions for basaltic magmas are inhomogeneous in terms of major elements and that the inhomogeneity arises principally from multistage melting of more primitive mantle lherzolite. Our further purpose is to elaborate the role of C–H–O fluids in controlling the presence or absence of melting and to present arguments for redox–interactions between oxidized ($H_2O + CO_2$)

lithosphere and reduced ($CH_4 + H_2$) deeper mantle, within the tectonic framework of large-scale subduction and transform faulting along convergent margins of lithospheric plates.

We will firstly present data on a simple system forsterite (Fo)–nepheline (Ne)–silica (Qz), as a convenient means of illustrating the principles involved in the study of the multicomponent natural system.

A SIMPLE SYSTEM ANALOGUE FOR BASALT GENESIS (Fo–Ne–Qz)

YODER and TILLEY (1962) introduced the concept that the broad family of basalts, from olivine nephelinites, through basanites, alkali olivine basalts, picrites, olivine tholeiites to quartz tholeiites and basaltic andesites, could be pictured as a continuum of compositions within the 4–component tetrahedron with apices represented by olivine, quartz, cinopyroxene and feldspathoids. Since that time experimental petrologists have devoted much effort to defining, as functions of pressure and volatile content, the liquidus fields and cotectics, thermal divides and liquid evolution paths within the tetrahedron. Parallel approaches to the problem have used natural basaltic compositions and simple system analogues.

The system forsterite–nepheline–quartz (Fo–Ne–Qz) serves as a simplified analogue of basalt/peridotite in that it contains low melting liquids enriched in sodium aluminosilicate and large liquidus fields for olivine and enstatite. The system has been studied at 28 kbar under dry conditions by WINDOM and BOETTCHER (1981) and GUPTA *et al.* (1986). A pressure of 28 kbar approximates that near the top of the Earth's Low Velocity Zone beneath oceanic lithosphere, *i.e.,* depths of 80–90 km. In terms of mantle magma genesis, the most important boundary in Fo–Ne–Qz is the cotectic between olivine and enstatite which defines the range of liquids formed by increasing degrees of melting of mantle peridotite in the presence of residual olivine and enstatite.

Initial melts (Figure 1) of a model mantle, *i.e.,* olivine + enstatite + jadeite, are 'basanitic' with similar normative nepheline and albite contents. The partial melting involves the reaction Forsterite (Fo) + Jadeite (Jd) → Enstatite (En) + Liquid as the enstatite liquidus surface crosses the Fo–Jd join. With increasing temperature, liquids traverse the base of the simplified basalt tetrahedron passing from nepheline–normative compositions into hypersthene–normative compositions.

We wish to consider melting of two model peridotite compositions, a relatively 'fertile' peridotite

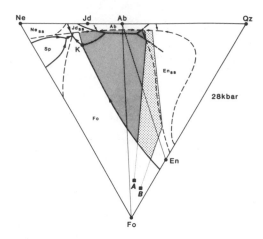

FIG. 1. Liquidus surface of the system Ne–Fo–Qz [weight percent] at 1 bar (dashed lines) and 28 kbar (solid lines) (see text for source references). Liquids formed by melting of a jadeite–enstatite–forsterite assemblage vary from composition K along the olivine–enstatite cotectic until enstatite or olivine is eliminated from the residue. The heavily shaded area includes all liquid compositions which can be derived from model peridotite A by single stage melting at 28 kbar to 1 bar. The more refractory peridotite composition B can yield liquids within the same area but *also* within the lightly shaded area.

(A) (enriched in the low–melting components), and a relatively 'refractory' peridotite (B). It is clear from Figure 1, that although the initial melt composition in each peridotite is the same (at the Fo–En–Jd invariant point) the more refractory peridotite will traverse further along the olivine–enstatite cotectic before enstatite is eliminated as a residual phase. The liquid then moves to 100% melting along the olivine control line passing through the bulk compositions A or B. The positions of the liquidus phase boundaries at 1 bar pressure (SCHAIRER and YODER, 1961) are also illustrated (boundaries to shaded fields of Figures 1 and 2). Liquids formed at 28 kbar along the olivine–enstatite cotectic will crystallize olivine at lower pressure and move towards the 1 bar olivine–enstatite reaction boundary or olivine–albite cotectic as appropriate. The shaded area in Figure 1 illustrates the range of derivative liquids which could be formed by anhydrous melting of both compositions A and B at 28 kbar, followed by crystal fractionation at lower pressure. The area marked by lighter shading includes derivative liquid compositions which could *not* be formed from peridotite A by anhydrous melting followed by crystal fractionation at lower pressure but *could* be formed from the more refractory composition B. In a later section we use this simple system analogue approach to demonstrate that very different

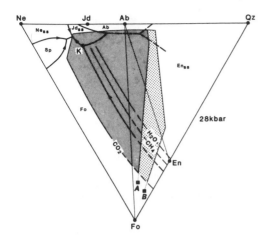

FIG. 2. As for Figure 1 but with the positions of the vapour–saturated cotectic between olivine and enstatite shown for H_2O, CH_4 and CO_2 saturated conditions at 28 kbar. The effect of C–H–O volatiles is to expand the field of possible liquids from peridotites A and B but the restriction of liquids lying in the lightly shaded areas to source compositions such as B, remains valid.

source peridotite compositions are required for MORB and boninite parent magmas.

The effect of C–H–O fluids on melting in the Fo–Ne–Qz system

In Figure 2, the position of the forsterite-enstatite cotectic at 28 kbar in the vapour–absent Fo–Ne–Qz system is compared with the positions of the same cotectic under vapour–saturated conditions, the three specific cases illustrated being H_2O–saturated, CO_2–saturated and CH_4–saturated (EGGLER, 1978; TAYLOR, 1985; GREEN *et al.*, 1986). The effect of water is to expand the olivine field at the expense of enstatite (KUSHIRO, 1972) but liquids at low degrees of partial melting remain strongly silica–undersaturated and nepheline–normative (EGGLER, 1975). Water enters the melt as OH^-, causing depolymerization by hydrolysis of Al–O–Si and Si–O–Si bonds and depression of the liquidus by 350–400°C. In contrast, CO_2 enters the melt at CO_3^{2-} in the form of metallocarbonate complexes resulting in melt polymerization and marked expansion of the liquidus field of enstatite at the expense of forsterite (see earlier discussions by EGGLER, 1978; BREY and GREEN, 1975, 1976; MYSEN and BOETTCHER, 1975; MYSEN *et al.*, 1980, and others on related systems). Liquids at low degrees of melting are nepheline–rich (Ne > Ab) and approach simple system analogues of olivine nephelinite.

Under CH_4–saturated conditions, liquidus temperatures are depressed by 80–90°C at 28 kbar (see also EGGLER and BAKER, 1982). Electron microprobe, gas chromatographic analyses and infrared spectroscopic studies show carbon solubilities up to 0.2% carbon (*i.e.*, ~0.25% CH_4 or ~0.7% CO_2 equivalent) and OH solution equivalent to 2–3 weight percent H_2O. The effect of methane–saturation is to expand the field of olivine at the expense of enstatite (Figure 2) but to a lesser extent than H_2O–saturation (see also EGGLER and BAKER, 1982). The infrared spectroscopic study of quenched glasses shows changes in the absorption bands attributed to the aluminosilicate network which suggest an increase in the Si:O ratio (*i.e.*, >1:2) of the network (TAYLOR, 1985; TAYLOR and GREEN, 1987). Thus, under methane–saturated conditions, melts may become *reduced* and contain small dissolved carbon and larger OH^- contents. On decompression, without redox–interaction with wall–rocks, such melts will exsolve $CH_4 + H_2$. It is noteworthy that initial melts near the low temperature minimum melting for jadeite–enstatite–forsterite, remain nepheline normative in all three cases illustrated in Figure 2. However, the degree of undersaturation is sensitively dependent on the fluid phase present.

In considering melting in the upper mantle, the evidence from mantle–derived liquids and xenoliths, from the presence of graphite and diamond and fluid inclusions within mantle samples, argue that mantle fluids are dominated by the C–H–O system. In considering melting in the upper mantle, it is necessary to consider possible variations on fluid phase compositions within the C–H–O system and whether particular source regions may be characterized by particular fluid phase characteristics. The simple system analogue illustrated in Figure 2 shows that variation in fluid phase composition will not invalidate one of the major conclusions from Figure 1, that is, that some specific liquids and potential source peridotite compositions cannot be related to each other because of limitations imposed by their major element compositions and phase relations.

FLUID-CONTROLLED MELTING IN THE UPPER MANTLE

The importance of water in determining the pressure–temperature conditions and form of the mantle solidus is well determined from studies on peridotite compositions (KUSHIRO, 1968; GREEN, 1973; MYSEN and BOETTCHER, 1975). The model 'pyrolite' composition studied (GREEN, 1973) may

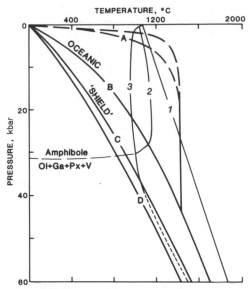

FIG. 3. Solidi for mantle peridotite ('Hawaiian' pyrolite) (GREEN, 1973): (1) Anhydrous (2) $0 < H_2O < 0.4$ weight percent (Amphibole dehydration solidus) (3) $H_2O > 0.5$ weight percent = water–saturated solidus. Also shown are geothermal gradients for mature oceanic crust (B), 'shield' regions (C, D) and near–mid ocean spreading centres (A) (BASALTIC VOLCANISM PROJECT, 1981; Figure 9.5.7). Ol, olivine; Px, pyroxene; Ga, garnet; V, vapour.

be referred to as 'Hawaiian pyrolite' composition as the 'melt' component in the 'melt + residue' model used to calculate the model 'pyrolite' composition (RINGWOOD, 1966), was based on Hawaiian olivine tholeiite liquids, *i.e.,* enriched in TiO_2, K_2O and incompatible elements relative to MORB. In Figure 3, the solidi for three distinctive water contents are illustrated:

(1) Anhydrous solidus
(2) Dehydration solidus; pargasitic hornblende is present on the solidus below 28–29 kbar but the water content is less than approximately 0.4 weight percent.
(3) Water–saturated solidus; pargasitic hornblende is present on the solidus but water content exceeds approximately 0.4 weight percent so that water is present in excess and sub-solidus assemblages coexist with a water–rich fluid.

In Figure 3 we illustrate geothermal gradients appropriate to thermal upwelling beneath spreading centres (A), beneath 'old' oceanic crust and lithosphere (B) and beneath ancient Archean crust and lithosphere (C, D) (BASALTIC VOLCANISM PROJECT,

* $X_C = x_{CO} + x_{CO_2} + \frac{1}{3}x_{CH_4} + \frac{2}{5}x_{C_2H_6}$.

1981; Figure 9.5.7, p. 1180). The presence or absence of a region of partial melting along geotherms B, C and D and the depth to such a region is clearly sensitively controlled by the role of water, specifically by the activity of water.

Recognising that carbon, as well as water, is a component of mantle fluids adds further complexity to the determination of the mantle solidus. It is necessary to determine the activity of water in the C–H–O system under mantle conditions. TAYLOR (1985, 1986) has presented a thermodynamic model of this system appropriate to mantle pressures and temperatures. Results may be presented, following FROST (1979) as isobaric, isothermal diagrams plotting $\log f_{O_2}$ vs X_C^* defined as the mole fraction of carbon relative to hydrogen in the bulk fluid.

In Figure 4 and Table 1 we illustrate results at 30 kbar, 1327°C showing the carbon–saturation surface bounding the metastable carbon–oversaturated fluid field (TAYLOR 1985, 1986). The strong dependence of fluid composition on oxygen fugacity is very clear in Figure 4 and Table 1 and it is noteworthy that graphite coexists with fluid approaching pure water in composition at $f_{O_2} = IW + 1$ to 2 log units.

The composition of C–H–O fluids present along geothermal gradients in Archean shield and in oceanic crust regions (taken from Figure 9.5.7, p. 1180, BASALTIC VOLCANISM STUDY PROJECT, 1981) can be calculated from this thermodynamic

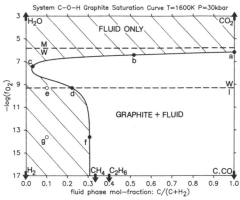

FIG. 4. $\log f_{O_2}$ vs $X_C \left[= \text{mol fraction } \frac{C}{(C + H_2)} \right]$ diagram at 1600 K (1327°C), 30 kbar showing the graphite saturation surface, stable fluid–only pseudo–divariant region (blank) and X_C–independent traces of IW and MW oxygen buffers. Arrows indicate the off–scale positions of the six C–O–H volatile species, filled circles ('a' to 'f') on the graphite saturation surface correspond to fluid compositions given in Table 1. The maximum mol fraction of H_2O on the graphite saturation curve is reached at point C.

Table 1. Fluid compositions at selected points in Figure 6; $T = 1600$ K, $P = 30$ kbar

EQUILIBRIUM FLUIDS

Mol percent species*

	$-\log f_{O_2}$	X_C	X_O	H_2O	CO_2	CO	H_2	CH_4	C_2H_6	$V^{mix}cm^3/mol$
a**	6.15	1.000	1.000	0	97.1	2.9	0	0	0	30.27
b	6.40	0.510	0.671	48.5	49.3	1.6	0.4	0.1	0	24.21
c†	7.40	0.032	0.339	94.1	2.3	0.2	1.5	1.9	0	17.90
d‡	9.30	0.220	0.091	26.9	0.1	0.1	7.8	63.3	1.8	25.60
e‡	9.30	0.100	0.192	57.4	0	0	12.7	29.7	0.2	21.05
f	13.60	0.306	0.001	0.2	0	0	8.7	87.7	3.4	28.69
g	13.60	0.100	0.004	0.2	0	0	68.8	29.9	0	19.83

** f_{O_2} = GCO, † f_{O_2} = GW(max.x_{H_2O}), ‡ f_{O_2} = IW.
* round-off error = ±0.1 mol percent.

model provided that we assume that graphite or diamond is a stable accessory mantle mineral. The C–H–O fluid composition at any pressure and temperature is only fixed if f_{O_2} is also known—in Figure 5 (GREEN et al., 1986) we calculate fluid compositions for the two geothermal gradients assuming f_{O_2} = FMQ, IW + 2 log units and IW, respectively. The results in Figures 4 and 5 demonstrate that if mantle f_{O_2} conditions are near IW to IW + 2 or 3 log units, then mantle fluids are dominated by water or water + methane. In the presence of a mixed fluid, (H_2O + CH_4), pargasitic hornblende will reach a maximum thermal stability near $a_{H_2O} \sim 0.5$ (see study by HOLLOWAY (1973) on pargasite + CO_2 + H_2O)—an interpretation supported by studies in progress in which subsolidus amphibole + garnet lherzolite crystallized at 25 kbar, 1100°C in the presence of graphite and an f_{O_2}-buffered fluid phase of CH_4:$H_2O \sim 1$:1.

There is evidence from measurements of intrinsic oxygen fugacity of mantle minerals, of deep–seated igneous intrusions and from identification of reduced gases (CH_4) evolving from MORB glasses and from submarine volcanic centres (ARCULUS et al. 1984; SATO and VALENZA, 1980; WELHAN and CRAIG, 1979, 1983) that the earth's mantle in the source regions for basaltic magmas is reduced with $f_{O_2} \sim$ IW to IW + 1 to 2 log units. These interpretations are not unchallenged and others, for example, EGGLER (1983) and MATTIOLI and WOOD (1986) infer higher mantle f_{O_2} conditions. At near-surface and shallow crustal conditions, mineral assemblages of igneous rocks indicate $f_{O_2} \geq$ FMQ (HAGGERTY, 1978; SATO, 1978). Thus, the earth's lithosphere may be seen broadly as a transition zone or boundary layer between deeper mantle with $f_{O_2} \sim$ IW and mantle fluids dominated by CH_4, H_2O + H_2 and near–surface conditions with $f_{O_2} \sim$ FMQ and fluids dominated by H_2O, CO_2. The conse-

quences of this redox interaction, coupled with the sensitivity of the mantle solidus to a_{H_2O}, are explored more fully elsewhere (GREEN et al., 1986) but are presented in summary form in the following section and Figures 6 and 7 which also build upon ideas canvassed by others including WYLLIE (1978, 1979, 1980), WOERMANN and ROSENHAUER (1985), EGGLER (1978), EGGLER and BAKER (1982), RYABCHIKOV et al. (1981) and GREEN (1973).

Mantle degassing and magma genesis: Oceanic lithosphere

We adopt the view that the earth's upper mantle *beneath the asthenosphere* is at $f_{O_2} \sim$ IW ± 1 log unit and that this region contains C–H–O fluids, possibly remnant from primordial entrapped volatiles, and minor carbon (diamond) contents. (In adopting this view we recognize that alternative views of higher oxygen fugacity (QFM to MW) are also current and our purpose is to explore the 'reduced mantle' model.) The fluids will be dominated by $CH_4 > H_2O > H_2$, C_2H_6 (cf. GOLD and SOTER, 1980, 1983). The transition from these reduced conditions towards more oxidized lithosphere will result in fluid change with increase in H_2O:CH_4 (Figures 4 and 5) but the oceanic geothermal gradient will lie within a field of partial melting at $P > 28$–29 kbar (Figure 3) due to instability of pargasitic hornblende and the role of water in depressing the mantle solidus. Provided that fluid concentrations are low, this region of partial melting will be a fluid–absent region in which a minor melt phase can accommodate continuously variable f_{O_2} via melt structure redox changes (SiO variation) with solubility of OH^- and minor carbon, and at higher f_{O_2}, by increasing solubility of CO_3^{2-}. At depths < 90 km, the stability of pargasitic hornblende at $f_{O_2} \sim$ IW + 1 to 3 log units creates a lid

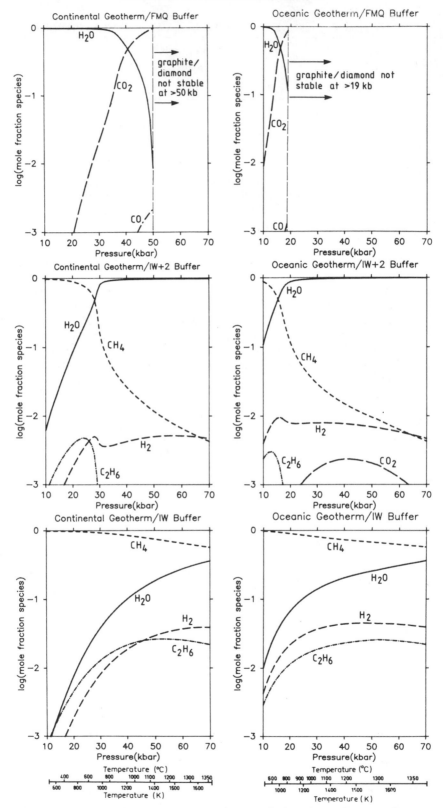

FIG. 5. Fluid compositions along model 'oceanic' and 'shield' geotherms (BASALTIC VOLCANISM STUDY PROJECT, 1981; Figure 9.5.7, p. 1180) assuming fluid coexists with graphite/diamond and with specific f_{O_2}'s corresponding to IW, IW + 2 log units and FMQ buffers.

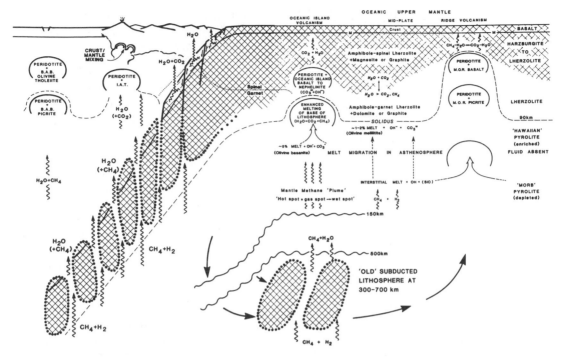

FIG. 6. Schematic model of the oceanic lithosphere and asthenosphere. The lithosphere is variable in composition due to melt extraction and fluid enrichment events and is envisaged as variably oxidized from FMQ (shaded areas) to ~IW + 1 to 2 log units (blank areas of lithosphere). The asthenosphere is a fluid–absent region in which C–H–O volatiles are dissolved in a small melt fraction (1–2% melt) which migrates towards the top of the asthenosphere or base of the lithosphere. Source regions for Hawaiian–type island chains (hot–spots) are in the upper part of the asthenosphere or base of the lithosphere. Source regions for MORB are in the lower part of the asthenosphere. The mantle is degassing reduced fluids (CH_4 + H_2) and the asthenosphere is a fluid–absent region with intrinsic oxygen content and fugacity varying from ~IW to IW + 2 to 3 log units—this variation is accommodated in change of melt structure from reduced (Si:O > 1:2) to contain OH^- and CO_3^{2-} (f_{O_2} ~ IW + 2 to 3 log units). The patterned symbol on the oceanic lithosphere signifies f_{O_2} ~ FMQ, i.e., 'oxidized' lithosphere. In the convergent margin or subduction plate boundary scenario, the oxidized lithosphere is portrayed as penetrated along fracture zones by CH_4 + H_2 fluids from subjacent mantle. The redox–interactions produce H_2O–rich fluids which strongly influence melting relations above or within the subducted lithosphere. Thus, even at depths below those at which hydrated minerals of the oceanic crust and lithosphere have been lost, the 'slab' may be a source of H_2O–rich fluids. Abbreviations for primary magmas developed within the schematic diapirs are: M.O.R. = mid–ocean ridge; I.A.T. – island arc tholeiite; B.A.B. = back–arc basin. See text for further discussion, including the genesis of boninite parent magmas.

to the partially molten layer because any interstitial melt migrating upwards from the asthenosphere will crystallize to pargasite–bearing lherzolite containing graphite or dolomite ± graphite (GREEN, 1973; EGGLER, 1978; WYLLIE, 1978; GREEN et al., 1986).

In Figure 6, the asthenosphere beneath oceanic crust is a fluid–absent region separating methane–rich fluids in the deep mantle from H_2O–rich and H_2O + CO_2 fluids in the lithosphere. It is a zone of decoupling between lithospheric plates and underlying mantle, a zone of increased seismic attenuation and decreased seismic velocity (GREEN and LIEBERMANN, 1976; GREEN, 1973; EGGLER, 1978;

WYLLIE, 1978) and most importantly, a zone of chemical differentiation rather than homogenization due to the migration of an interstitial melt phase towards the top of the asthenosphere. The character of the interstitial melt is extremely undersaturated olivine melilitite to olivine nephelinite (GREEN, 1971; GREEN and LIEBERMANN, 1976; BREY and GREEN, 1976; WYLLIE, 1979; EGGLER, 1978; WENDLANDT and MYSEN, 1980) and apart from enriching the upper part and depleting the lower part of the asthenosphere in incompatible elements, the high Ca/Al ratio of olivine melilitite will decrease the Ca/Al ratio of the lower part of the astheno-

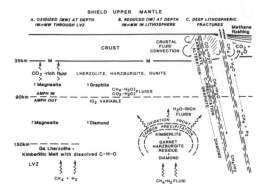

FIG. 7. Schematic model for the continental lithosphere beneath Archean/Proterozoic shield regions. See text for detailed discussion. The deep mantle is degassing (CH_4 + H_2) fluids that interact with oxidized lithosphere. (A) suggests a deep, thin asthenospheric layer (partial melting) in which the f_{O_2} change from mantle or lithosphere is accommodated within the fluid–absent melt. (B) suggests a mechanism of 'redox' melting (see text) in which diamond-bearing garnet harzburgite is left as a residue from oxidation of CH_4 + H_2 to H_2O + C with extraction of water–rich kimberlitic melt. (C) suggests a role for deep lithosphere fractures in localizing mantle fluid release and interaction of these fluids with oxidized crustal fluids at shallow depths.

sphere and increase the Ca/Al ratio of the uppermost part. The extent of chemical differentiation will be related to: (a) the age of the lithosphere, (b) the extent to which volumes of old lithosphere/asthenosphere are preserved near transform fault/ spreading centre intersections or at spreading axes themselves, and (c) depth within the asthenosphere. The base of the lithosphere will also become enriched in incompatible elements, hosted particularly in pargasitic hornblende, phlogopite and apatite as minor and accessory phases. In Figure 6, the lower part of the asthenosphere is depicted as a fluid–absent region at low f_{O_2} (IW to IW + 1 log unit) which has the incompatible element contents appropriate to MORB source (abbreviated as 'MORB' pyrolite) whereas the upper part of the asthenosphere has f_{O_2} near IW + 2 to 3 log units and incompatible element contents appropriate to source regions for tholeiites of oceanic island chains such as Hawaii ('hot spot' source regions, abbreviated as 'Hawaiian pyrolite').

The relationships between magmatism at spreading centres, at 'hot spots' or migrating tension fractures (see Figure 6) and the chemically zoned asthenosphere has been fully explored elsewhere (GREEN, 1971, 1972; GREEN and LIEBERMANN, 1976; GREEN et al., 1986) and in this paper we address the possibilities created by interaction of an

oxidized subducted lithospheric slab with reduced mantle below the asthenosphere (Figure 6).

It is generally accepted that the peridotite of the lithosphere is variable in composition from lherzolite to harzburgite, probably as a result of prior extraction of basaltic to picritic melts. Local, more extreme chemical variation results from trapped liquids, high-pressure accumulates or metasomatic enrichment from migratory fluid phases. Two examples of chemical variation are illustrated in Figure 8, showing chemical trends in spinel lherzolite inclusions from W. Victoria (NICKEL and GREEN, 1984) and from garnet and spinel lherzolites within the Ronda high–temperature, high–pressure peridotite intrusion (FREY et al., 1985). Even more refractory harzburgite compositions are seen in some ophiolite complexes. These are commonly partially or completely serpentinized and during serpentinization, became highly oxidized with abundant magnetite development.

The evidence from deep focus earthquakes suggests that subducted lithospheric slabs retain their identity to depths of at least 600–700 km. Dehydration reactions in subducted basaltic crust and serpentinized peridotite may lead to migration of H_2O-rich or H_2O + CO_2 fluids from the subducted slab into the overlying peridotite wedge, instigating melting on the water–saturated peridotite solidus (solidus 3 of Figure 3, KUSHIRO 1968; GREEN, 1973). Melts formed in this way will have geochemical characteristics reflecting the partially residual character of the remobilized lithosphere together with the imprint of components partitioned into

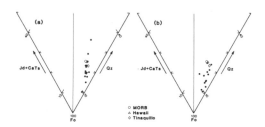

FIG. 8. Two examples of peridotite suites of variable composition from lherzolite to harzburgite, reflecting prior extraction of basaltic to picritic melts, plotted in the normative system Fo–Jd + CaTs–Qz–Di, projected from Di onto the plane Fo–Jd + CaTs–Qz. (a) Ronda, high temperature, high pressure peridotite intrusion FREY et al., 1985; (b) Spinel lherzolite inclusions from Lake Bullenmerri, W. Victoria, NICKEL and GREEN, 1984. Also plotted are 'MORB' pyrolite, 'Hawaiian' pyrolite and Tinaquillo lherzolite compositions, which have been studied experimentally under high pressures and temperatures (JAQUES and GREEN, 1980; FALLOON and GREEN, unpublished data).

and transported by the fluid phase migrating from the subducted slab. A related model suggests that temperatures on the upper part of the subducted slab (the subducted basaltic/eclogitic crust particularly) may be high enough for fluid–saturated or dehydration melting and for melt to migrate from the subducted slab to modify and melt the overlying peridotite mantle wedge. The models have been explored and evaluated in many previous publications (see RINGWOOD (1974) and GILL (1981) for review and references). However the concept of redox–interaction between deep mantle degassing CH_4 + H_2–rich fluids and lithospheric slab with mineral assemblages reflecting f_{O_2} near FMQ to MW conditions adds a further dimension to subduction. It is possible for dehydrated, very refractory peridotite or eclogite of the subducted slab to generate H_2O–rich fluid by reaction of anhydrous oxidized mineral assemblages (Fe^{3+} in silicates, spinels, carbonates) with primary $CH_4 + H_2$ fluids. The increased a_{H_2O} will cause melting, even of refractory harzburgite (KUSHIRO, 1972; MYSEN and BOETTCHER, 1975; GREEN, 1973). A second, important effect will be that any redox front bounding a residual slab of subducted lithosphere in contact with normal MORB pyrolite will be a locus for high a_{H_2O}—the mechanical effect of this will be to concentrate stress release (earthquake activity) at such redox fronts. Lower temperatures within the subducted slab will generally inhibit melting of the slab itself but this may be counteracted by high a_{H_2O} contrasting with high a_{CH_4} and thus high solidus temperatures in surrounding mantle. A particularly favourable location for redox interaction coupled with higher temperatures at the edges of a subducted slab will be at trench/transform intersections, e.g., the south end of the Marianas Trench and the northern end of the Tonga Trench, where the subducted slab abuts, across a transform fault zone, against normal oceanic lithosphere/asthenosphere to deep levels. The movement zone will probably act as a channel-way for enhanced mantle degassing and by providing a temperature contrast across the transform fault zone, will act as a heat input to the subducted slab.

In Figure 6, we suggest that the immediate source peridotite for Hawaiian or 'hot spot' volcanism lies in the uppermost part of the asthenosphere or alternatively within the base of the lithosphere, where this has been enriched by migration, reaction and crystallization of an interstitial melt. The success of the 'fixed hot spot' frame of reference in analyzing relative plate movement over the past 100 million years shows that the cause or trigger for island chain or 'hot spot trace' volcanism does not lie within the moving lithospheric plate but lies beneath this plate and reflects a 'fixed' mantle or a deep mantle in which the convective flow is much slower than and decoupled from the lithospheric plate motions (MORGAN, 1972; DUNCAN, 1981). If large slabs of subducted lithosphere are carried to depths > 600 km and embedded within the return flow or deeper mantle convection pattern (HOFFMAN and WHITE, 1982; RINGWOOD, 1986), then they may act to oxidize CH_4 + H_2 fluids passing through them to CH_4 + H_2O. This effect could feed CH_4 + H_2O plumes into overlying asthenosphere + lithosphere and trigger 'hot spot volcanism' as depicted in Figure 6. For example we suggest that the great frequency of sea mounts and island chain volcanoes in the West Pacific may be due to the Pacific plate over-riding relict subducted slabs embedded in the 150–650 km depth interval of the upper mantle and reflecting material originally subducted on Palaeozoic-Mesozoic subduction zones along the East Asian margin.

Mantle degassing and magma genesis: Continental lithosphere

In Figure 7 (cf. GREEN et al., 1986) we present a summary diagram illustrating mantle degassing of CH_4+ H_2 beneath a very old, thick lithosphere characterized by very low geothermal gradient (Shield geotherms C and D of Figure 3). The lithosphere has a zone of amphibole stability in lherzolitic compositions at $P < 28$-29 kbar but at deeper levels subsolidus assemblages contain phlogopite and may contain other hydrous phases such as titanoclinohumite (MCGETCHIN et al., 1970). Carbon will be present as diamond or graphite, or at depths > 150 km, as magnesite if $f_{O_2} >$ FMQ (Figure 5; EGGLER 1975, 1978; GREEN et al., 1986; BREY et al., 1983; WYLLIE, 1978, 1979). Samples of deep continental lithosphere occur as xenoliths in kimberlite pipes and illustrate heterogeneity and variations from possible trapped liquids or relict subducted oceanic crust (eclogite, grosspydite, lherzolite to highly refractory harzburgite, and extremely K– and Ti–rich phlogopite rocks).

If mantle temperatures approach the solidus 3 (Figure 3), then redox melting of lithosphere may occur by interaction of reduced CH_4–rich fluids from deeper levels with oxidized lherzolite, or harzburgite. GREEN et al. (1986) have suggested that the distinctive mineral inclusion suite within South African diamonds which is characterized by extremely magnesian olivine, refractory low–Al enstatite and extremely low calcium, Cr–rich pyrope garnet, is a consequence of redox melting (see Figure 7, scenario B).

Although Archaean and Proterozoic cratons preserve a remarkably long history of stability as recorded by cover of undeformed Proterozoic sediments, they are frequently cut by mafic dyke swarms. In the post–Permian breakup of Gondwanaland, stable cratons with thick underlying lithosphere, have rifted and drifted apart so that regions of old, thick lithosphere are now juxtaposed against young lithosphere of oceanic type. An initial situation of rapid temperature change at the base of the lithosphere of a newly rifted craton (cf. BOYD 1973; BOYD and NIXON, 1975; NICKEL and GREEN, 1985) will be followed by heating of the sub–craton lithosphere so that the solidus of phlogopite–bearing lherzolite and harzburgite will be exceeded. Small melt fractions of olivine lamproite to leucite lamproite composition (FOLEY, 1986; FOLEY et al., 1986) reflect phlogopite harzburgite source rocks and depths of segregation from 150 km to ~70 km respectively. Source rocks of phlogopite–bearing garnet lherzolite will produce potassic olivine nephelinites to basanites (WEDEPOHL, 1985) with the particular liquid composition being determined by $CO_2:H_2O$ as well as by source rocks and depth of magma segregation (FREY et al., 1978; EGGLER, 1978; GREEN, 1976; BREY and GREEN, 1976; WYLLIE, 1979). If upwelling and diapirism occurs from within the heated former sub–cratonic lithosphere, then magmas produced at lower pressures will retain trace element and isotopic compositions reflecting their source compositions but the magmas will be picritic to tholeiitic or alkali olivine basaltic in character, reflecting higher degrees of melting and shallower depths of magma segregation.

MAGMA SOURCE COMPOSITIONS

Mantle–derived peridotite suites

Compositions of two suites ranging from lherzolite to harzburgite and with high pressure mineral assemblages requiring crystallization at pressures in excess of 10 kbar are projected in Figure 8. The Ronda peridotite suite (FREY et al., 1985) defines a linear trend consistent with extraction of magma lying within the olivine–hypersthene–plagioclase–diopside volume, i.e., an olivine tholeiite or tholeiitic picrite liquid (FREY et al., 1985). The Bullenmerii suite shows more scatter but again a trend implying an olivine tholeiite or tholeiitic picrite extraction (cf. NICKEL and GREEN, 1984). Model Hawaiian pyrolite, MORB pyrolite and Tinaquillo lherzolite compositions are also plotted in Figure 8. All three have CaO and Al_2O_3 contents > 3–4 weight percent but the suites differ in Na_2O, TiO_2, K_2O, and incompatible element contents with Ti-

naquillo being the most refractory composition (JAQUES and GREEN, 1980; GREEN et al., 1979).

The trends in Figure 8 defined by the mantle samples (which could be matched by other xenolith suites or high–temperature, high–pressure intrusive peridotite suites) are considered to result from partial melting and incomplete melt extraction (FREY et al., 1985; NICKEL and GREEN, 1984) and thus we infer that mantle source compositions, prior to melt extraction are at least as 'fertile' as the most plagioclase–rich and diopside–rich limits of the trends shown in Figure 8. If the processes outlined in previous sections, which lead to one or more remelting events, particularly under the fluxing effects of C–H–O fluids, are operative in the earth, then the range of potential source peridotite compositions covers at least the range of the samples plotted, i.e., trending from lherzolite towards harzburgite with 80 weight percent Ol, 20 weight percent En.

Source for mid-ocean ridge basalts

Compositions of (>70) glasses from dredged ocean floor basalts are plotted in Figures 9 and 10— all have 100 Mg/(Mg + Fe^{2+}) > 68, i.e., they are primitive in the sense of suitability as melts in equilibrium with olivine > Mg_{87}. It may be noted that

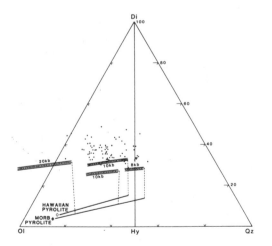

FIG. 9. Equilibrium partial melts of MORB pyrolite at 8, 10 and 20 kbar (FALLOON and GREEN, unpublished data) and 'Hawaiian' pyrolite at 10 kbar (JAQUES and GREEN, 1980) compared with primitive MORB glasses in the normative system Ol–Jd + CaTs–Qz–Di. Compositions are projected from plagioclase (An + Ab) onto the plane Ol–Di–Qz. Cotectics are as follows: (1) stippled line, ol + opx + cpx + L cotectic. (2) dashed line, ol + opx + L cotectic. (3) solid line, ol + L cotectic. (▼) is the MORB glass DSDP3–18–7–1, studied by GREEN et al., (1979) and (▽) is the low-olivine parent composition of BRYAN (1979).

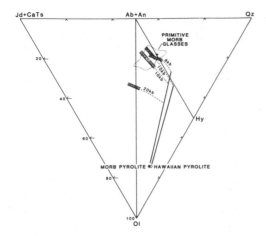

FIG. 10. Equilibrium partial melts of 'MORB' pyrolite at 8, 10 and 20 kbar and 'Hawaiian' pyrolite at 10 kbar compared with the field of primitive MORB glasses in the normative system Ol–Jd + CaTs–Qz–Di. Compositions are projected from Di onto the plane Jd + CaTs–Qz–Ol. Cotectics and symbols as for Figure 9. For purposes of clarity the primitive MORB glasses are not plotted individually as in Figure 9 but outlined by the dotted line.

olivine tholeiite liquids dominate the group but they do spread from quartz–normative compositions (5 points) to nepheline–normative compositions (2 points). The glass DSDP 3-18-7-1, studied experimentally by GREEN et al., (1979) as a primitive MORB lies centrally within the MORB field.

GREEN et al. (1979, Table 4, column D) calculated a model source composition for DSDP 3-18-7-1 and we have studied the anhydrous melting relationships of this composition (MORB pyrolite) from 8-20 kbar. A 'sandwich' technique (STOLPER, 1980) was used in which a layer of DSDP 3-18-7-1 glass is placed between two layers of MORB pyrolite. The bulk compositions of these mixtures lie in the join between these two compositions (Figures 9 and 10). This approach avoids the bulk compositional changes inherent in equilibrating diverse liquids within unrelated peridotite layers (FUJII and SCARFE, 1985; TAKAHASHI and KUSHIRO, 1983; STOLPER, 1980) although these studies provide useful comparative data for our purposes.

The results are summarized in Figures 9 and 10, showing 3-phase cotectics (Ol + Opx + Cpx), followed by (Ol + Opx) cotectics until the olivine–control line is reached as the temperature increases. Attention is particularly drawn to the compositional range of liquids along the 3-phase cotectic. For example, at 20 kbar liquids in equilibrium with olivine, orthopyroxene and clinopyroxene range from nepheline–normative picrites to picrites with low

normative hypersthene contents, before clinopyroxene disappears. Examination of Figure 10 alone would suggest that MORB glasses may be explained as primary olivine tholeiite melts separating from Ol + Opx + Cpx residues at approximately 8 kbar to 13 kbar pressure. Limited olivine fractionation and variation in percent melting could possibly account for the observed compositional variation of Figure 10. However, it is insufficient to examine only one projection or plot of the many chemical variables involved. From Figure 9, it is clear that almost all MORB glasses are too rich in normative diopside to have formed as primary liquids on Ol + Opx + Cpx cotectics. However, a few exceptions lie close to Ol + Opx + Cpx cotectics at ~8 kbar and these distinctive MORB glasses have low normative olivine contents or are quartz normative. They are similar in composition to the low–olivine parent composition of BRYAN (1979) and are consistent with being close to primary liquids or derived from ~8 kbar primary olivine tholeiite magmas via limited olivine fractionation.

Examination of Figure 9 suggests that liquids formed by small degrees of melting at 10–12 kbar, might fractionate small amounts of olivine to reach MORB glass compositions. In fact, this mechanism does not work as liquids at small degrees of melting have TiO$_2$ contents, Na/Ca ratio, K$_2$O contents and P$_2$O$_5$ contents that are too high for MORBs.

The data, as illustrated in Figures 9 and 10, but confirmed by other oxide, element ratio and normative plots, show that most MORB glasses are derivative by olivine factionation from primary picritic magmas (i.e., liquids with >20% normative olivine), segregating from residual olivine, orthopyroxene and clinopyroxene at pressures of 15–20 kbar. Melting conditions are essentially anhydrous although small quantities of reduced and oxidised volatiles may be present (BYERS et al., 1983; MATHEZ, 1984). It is inferred from Figure 10 particularly, that we do not see liquids formed by very high degrees of melting along the olivine + orthopyroxene cotectic and we believe that this is unlikely in a single stage melting process as such liquids would require >30–35% melting and melt segregation apparently occurs prior to this (GREEN and RINGWOOD, 1967; MYSEN and KUSHIRO, 1977; JAQUES and GREEN 1980).

Liquids at 10 kbar from the Hawaiian pyrolite melting study (JAQUES and GREEN, 1980) are also plotted in Figure 10. The pattern of liquid variation with progressively higher degrees of melting is similar to that for MORB pyrolite but the Ol + Opx + Cpx cotectic is at much higher normative diopside contents. Although the cotectic overlaps the MORB field in this projection the liquids have >3 weight percent TiO_2, >0.5 weight percent K_2O and Na/Ca ratios unlike MORB. The Hawaiian pyrolite composition cannot produce MORB liquids in terms of major and minor elements and vice versa—different source compositions are required.

Source for Troodos Upper Pillow Lavas

CAMERON (1985) has subdivided the Troodos Upper Pillow Lavas into three chemical groups with trace element and isotopic abundances requiring chemical differences between the source compositions for each group. DUNCAN and GREEN (1986) studied a parental Group III composition (Arakapas area) experimentally with particular attention to matching of observed phenocryst assemblages with phases crystallized under controlled pressure and temperature conditions. Conditions of magma segregation from residual harzburgite were determined as 1350°C, 8–9 kbar and the magma contained ~0.5–1 percent weight H_2O.

In Figure 11, we plot the Troodos Group III pa-

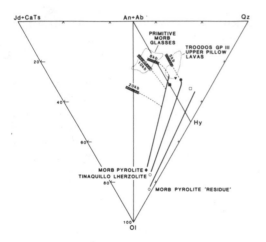

FIG. 11. Comparison of primitive MORB glasses and Troodos Group III upper pillow lavas (CAMERON, 1985) with equilibrium partial melts from 'MORB' pyrolite at 8, 10 and 20 kbar and Tinaquillo lherzolite at 5 kbar (JAQUES and GREEN, 1980; GREEN et al., unpublished data), in the normative system Ol–Jd + CaTs–Qz–Di, projected from Di onto the plane Jd + CaTs–Qz–Ol. Cotectics same as for Figure 9. Symbols are as follows: (■) calculated parental composition for the Troodos Group III upper pillow lavas studied by DUNCAN and GREEN (1986). (▼) refractory primitive magma composition identified from glass inclusions in magnesian (Fo94) olivine phenocrysts from Tonga (FALLOON and GREEN, 1986b). (●) Marianas fore–arc lava (JENNER, 1983). (□) Cape Vogel, boninite parental magma composition calculated by WALKER and CAMERON (1983). (▽) primitive MORB glass DSDP3–18–7–1 studied experimentally by GREEN et al. (1979).

rental lava composition with derivative glasses and aphyric basalts to compare with the area of MORB glasses and the range of liquids derived from MORB pyrolite. The Troodos Group III parental liquid lies marginally outside (i.e., is more refractory than) the olivine–control line from MORB pyrolite. In terms of phase relations a liquid close to Troodos Group III parental magma could be formed by melting of MORB pyrolite at the point of elimination of orthopyroxene from the residual phases, i.e., a dunite residue. However, minor and trace element contents of the MORB–pyrolite derived melt do not match these of Troodos Upper Pillow Lava—a relation readily apparent in the first plagioclase to crystallize (~An91 in Troodos UPL and ~An78 in DSDP 3–18–7–1).

If we compare the melting relations of Tinaquillo lherzolite (Figure 11), its more refractory character means that the olivine/orthopyroxene cotectic does not end at but continues through Troodos UPL composition (cf. Figure 1). In phase relations, trace element and minor element abundances, the source peridotite for Troodos UPL must be more refractory than MORB pyrolite—we infer that it is a residual lherzolite composition in which melting is enhanced by addition of a very small amount of water (0.5–1 percent weight H_2O in the melt fraction) sufficient to depress the melting temperature by about 30–50°C.

Source for north Tongan arc high magnesium lavas

In 1984, the Japanese research vessel 'Natsushima' dredged an extraordinary variety of high magnesium lavas from water depths in excess of 2000 m at the northern termination of the Tongan arc (FALLOON, 1985; FALLOON and GREEN, 1986a). In detail, the samples show evidence for decoupling of incompatible elements from major plus compatible elements. For example, rocks with similar major element chemistry have REE patterns varying from strongly LREE enriched to strongly LREE depleted. There is also unequivocal evidence for magma mixing involving several magnesian end-member liquids, differing principally in their degree of depletion in the more incompatible elements and in ratios such as Na/Ca. One sample contained Fo94 olivine phenocrysts and within these unusually magnesian olivine crystals occur large glass inclusions with $CaO/Na_2O > 20$ and magnesian quartz tholeiite compositions (FALLOON and GREEN, 1986b). Correction for quench outgrowth on the olivine host enabled calculation of the original liquid (Figure 11). In other samples from the same and other dredge sites, extremely calcic plagioclase (up to An_{100}) megacrysts occur and are compatible with crystallization from a liquid matching the composition of the glass inclusions in olivine. The role of water in depressing the liquidus and fluxing melting of refractory lherzolite composition, is as yet uncertain. However, the composition of the glass inclusion is such that the original magma is a possible partial melt of very refractory lherzolite compositions, such as Tinaquillo lherzolite, with magma segregation at ~5 kbar (anhydrous) or a slightly higher pressure if OH^- solution expands the olivine phase volume.

Source for boninite or high magnesia andesites

In young West Pacific Arcs, distinctive very high SiO_2 (>56 weight percent, high magnesium, low calcium lavas characterized by rare olivine and abundant enstatite or multiply twinned clinoenstatite (inverted from protoenstatite) have been found both outcropping in emergent islands and dredged from the sea floor (DALLWITZ *et al.,* 1966; KURODA *et al.,* 1978; CRAWFORD *et al.,* 1981; UMINO, 1985; DIETRICH *et al.,* 1978). Strongly porphyritic character, accumulation of phenocrysts and growth of quench phases makes determination of liquid compositions very difficult. Experimental studies (TATSUMI, 1981; JENNER and GREEN, unpublished) have established conditions of multiple olivine and orthopyroxene saturation and the latter studies lead us to a preferred petrogenetic model in which liquids contain ~2 weight percent water and segregate from residual olivine and orthopyroxene at ~2–5 kbar, $T \sim$ 1250–1300°C.

In Figure 11, plots of possible parental magma compositions lie on the refractory (olivine–enstatite) side of the olivine–control line for peridotite such as MORB pyrolite or Tinaquillo lherzolite. Liquids such as boninites cannot be derived from lherzolites as calcic as those and require very refractory harzburgite source rocks (*cf.* Figure 1). Potential source rocks in terms of (Jd + CaTs)–Ol–Qz–Di include the more refractory compositions of the lherzolite-harzburgite suites of Figure 8. However, examination of trace element and minor element contents of boninites reveals great variability (JENNER 1981, HICKEY and FREY 1982, CAMERON *et al.,* 1983) from incompatible elements enriched to V–shaped REE patterns suggesting addition of a minor enriched component to a previously strongly depleted peridotite. A similar mechanism is required to explain the non–refractory character of the plagioclase (which appears very late in the crystallization sequence and lacks a high Ca/Na ratio). The geo-

chemistry and petrology of boninites suggests volatile-induced (H_2O–rich) melting of very refractory residual harzburgite. There is evidence for significant addition of incompatible elements either with or prior to the addition of H_2O–rich fluid.

CONCLUSIONS

The genesis of primary magmas in the earth's mantle is controlled by

a) Pressure, temperature variation, *i.e.,* mantle geotherms, which differ in different tectonic settings;

b) The presence and composition of a fluid phase, principally within the system C–H–O; and

c) The bulk compositions of source peridotite.

Experimental studies at high pressure and temperature constrain the effects of C–H–O fluid emphasizing the importance of water–rich and water–methane fluids at oxygen fugacities of IW to IW + 1 to 3 log units. The experimental studies also show that different source peridotite compositions are required for primary magmas from different tectonic settings. These different source compositions can be simply related within a framework in which the source compositions for Hawaiian (Hot–spot) and Mid–Ocean Ridge tholeiitic picrites to olivine tholeiites are regarded as 'enriched' and 'depleted' variants of the mantle composition within the asthenosphere. The process of enrichment and depletion is attributed to upward migration of a very small (~1%) melt fraction of olivine melilitite or olivine nephelinite character within the asthenosphere.

Source rocks for back–arc basin tholeiites, island arc tholeiites, high–Mg tholeiites such as Troodos Upper Pillow Lavas or from the Tongan fore–arc, and particularly for boninite parent magmas, are highly refractory lherzolites to harzburgites. The roles of such compositions as source rocks for magmas requires second and probably third stage melting, in which residues from extraction of earlier magmatic events are re–melted in different *P, T* and tectonic settings.

In such multiple melting episodes, a key role is played by C–H–O fluids in fluxing the melting, and in the case of boninite liquids, in introducing incompatible elements to the source rock prior to or during the melting process. Although the subducted slab has long been regarded as a source for H_2O–rich fluids by dehydration of minerals within the upper parts of the slab, in this paper we suggest that an additional source of water, particularly at deeper levels of the subduction process, comes from redox interaction of reduced $CH_4 + H_2$ fluids from the deep mantle with the oxidized (FMQ) mineral assemblages of the subducted slab.

Particularly favourable environments for multistage melting, even of extremely refractory harzburgite compositions within the subducted slab, occur at the intersections of subduction zones with transform faults, *e.g.,* northern end of Tongan Trench, southern end of Marianas Trench.

Acknowledgements—We acknowledge the assistance of K. Harris (High Pressure Laboratory), W. Jablonski (microprobe laboratory), P. Robinson (geochemical laboratory) and other technical staff of the Geology Department, University of Tasmania in the research programs summarized in this paper. Mrs. J. Pongratz and Mrs. J. Beattie are thanked for drafting and manuscript preparation. The research is supported by the Australian Research Grants Scheme and research funds from the University of Tasmania.

REFERENCES

ARCULUS R. J., DAWSON J. B., MITCHELL R. H., GUST D. A. and HOLMES R. D. (1984) Oxidation state of the upper mantle recorded by megacryst ilmenite in kimberlite and type A and B spinel lherzolite. *Contrib. Mineral. Petrol.* **85**, 85–94.

BASALTIC VOLCANISM STUDY PROJECT (1981) *Basaltic Volcanism on the Terrestrial Planets.* 1286 pp. Pergamon Press.

BOYD F. R. (1973) A pyroxene geotherm. *Geochim. Cosmochim. Acta* **37**, 2533–254.

BOYD F. R. and NIXON P. H. (1975) Origins of ultramafic rocks from some kimberites of northern Lesotho and the Monastery Mine, South Africa. *Phys. Chem. Earth* **9**, 431–454.

BREY G. P. and GREEN D. H. (1975) The role of CO_2 in the genesis of olivine melilitite. *Contrib. Mineral. Petrol.* **49**, 93–103.

BREY G. P. and GREEN D. H. (1976) Solubility of CO_2 in olivine melilitite at high pressures and role of CO_2 in the Earth's upper mantle. *Contrib. Mineral. Petrol.* **55**, 217–230.

BREY G. P., BRICE W. R., ELLIS D. J., GREEN D. H., HARRIS K. L. and RYABCHIKOV I. D. (1983) Pyroxene–carbonate reactions in the upper mantle. *Earth Planet. Sci. Lett.* **62**, 63–74.

BRYAN W. B. (1979) Regional variation and Petrogenesis of Basalt Glasses from the FAMOUS Area, Mid–Atlantic Ridge. *J. Petrol.* **20**, 293–325.

BYERS C. D., MUENOW D. W. and GARCIA M. D. (1983) Volatiles in basalts and andesites from the Galapagos Spreading Center, 85° to 86°W. *Geochim. Cosmochim. Acta* **47**, 1551–1558.

CAMERON W. E. (1985) Petrology and origin of primitive lavas from the Troodos ophiolite, Cyprus. *Contrib. Mineral. Petrol.* **89**, 239–255.

CAMERON W. E., McCULLOCH M. R. and WALKER D. A. (1983) Boninite petrogenesis: chemical and Nd–Sr isotopic constraints. *Earth Planet. Sci. Lett.* **65**, 75–89.

CRAWFORD A. J., BECCALUVA A. L. and SERRI G. (1981) Tectonomagmatic evolution of the West Philippine-Mariana region and the origin of boninites. *Earth Planet. Sci. Lett.* **54**, 346–356.

DALLWITZ W. B., GREEN D. H. and THOMPSON J. E. (1966) Clinoenstatite in a volcanic rock from the Cape Vogel area, Papua. *J. Petrol.* **7**, 375–403.

DIETRICH V., EMMERMAN R., OBERHANSLI R. and PUCHELT H. (1978) Geochemistry of basaltic and gabbroic rocks from the West Mariana Basin and the Mariana Trench. *Earth Planet. Sci. Lett.* **39**, 127–144.

DUNCAN R. A. (1981) Hotspots in the Southern Oceans—An absolute frame of reference for motion of the Gondwana Continents. *Tectonophys.* **74**, 29–42.

DUNCAN R. A. and GREEN D. H. (1986) The genesis of refractory melts in the formation of oceanic crust. *Contrib. Mineral. Petrol.,* (Submitted).

EGGLER D. H. (1975) CO_2 as a volatile component of the mantle: The system Mg_2SiO_4–SiO_2–H_2O–CO_2. *Phys. Chem. Earth* **9**, 869–882.

EGGLER D. H. (1978) The effect of CO_2 upon partial melting of peridotite in the system Na_2O–CaO–Al_2O_3–MgO–H_2O–CO_2 to 35 kb, with an analysis of melting in a peridotite–H_2O–CO_2 system. *Amer. J. Sci.* **278**, 305–343.

EGGLER D. H. (1983) Upper mantle oxidation state: Evidence from olivine–orthopyroxene–ilmenite assemblages. *Geophys. Res. Lett.* **10**, 365–368.

EGGLER D. H. and BAKER D. R. (1982) Reduced volatiles in the system C–O–H: implications to meltings, fluid formation, and diamond genesis. In *High-Pressure Research* in *Geophysics,* (eds. S. AKIMOTO and M. MANGHNANI), pp. 237–250 Center for Academic Publications Japan, Tokyo.

FALLOON T. J. (1985) Preliminary petrography and geochemistry of igneous rocks from the Northern Tonga Ridge and adjacent Lau Basin. In *A Marine Geological and Geophysical Survey of the Northern Tonga Ridge and Adjacent Lau Basin,* (eds. E. HONZA, K. B. LEWIS *et al.*), pp. 63–71 Mineral Resources of Tonga Field Report No. 1, Ministry of Lands, Survey and National Resources, Kingdom of Tonga.

FALLOON T. J. and GREEN D. H. (1986a) Dredged igneous rocks from the Northern Termination of the Tofua Magmatic arc, Tonga and adjacent Lau Basin. *Australian Journal of Earth Sciences,* (Submitted).

FALLOON T. J. and GREEN D. H. (1986b) Glass inclusions in magnesian olivine phenocrysts from Tonga: Evidence for highly refractory parental magmas in the Tongan Arc. *Earth Planet. Sci. Lett.,* (In press).

FOLEY, S. F. (1986) The genesis of lamproitic magmas in a reduced fluorine–rich mantle. *Int. Kimb. Conf. Perth, Extended Abstracts 4th,* (In press).

FOLEY S. F., TAYLOR W. R. and GREEN D. H. (1986) The role of fluorine and oxygen fugacity in the genesis of the ultrapotassic rocks. *Contrib. Mineral. Petrol.,* (In press).

FREY F. A. and GREEN D. H. (1974) The mineralogy, geochemistry and origin of lherzolite inclusions in victorian basinites. *Geochim. Cosmochim. Acta* **38**, 1023–1059.

FREY F. A., GREEN D. H. and ROY D. S. (1978) Integrated models of basalt petrogenesis: a study of quartz tholeiites to olivine melilitites from south eastern Australia utilizing geochemical and experimental petrological data. *J. Petrol.* **19**, 463–513.

FREY F. S., SUEN J. C. and STOCKMAN H. W. (1985) The Ronda high temperature peridotite: Geochemistry and petrogenesis. *Geochim. Cosmochim. Acta* **49**, 2469–2491.

FROST B. R. (1979) Mineral equilibria involving mixed–

volatiles in a C–O–H fluid phase: the stabilities of graphite and siderite. *Amer. J. Sci.* **279**, 1033–1059.

FUJII T. and SCARFE C. M. (1985) Composition of liquids coexisting with spinel lherzolite at 10 kbar and the genesis of MORBS. *Contrib. Mineral. Petrol.* **90**, 18–28.

GILL J. B. (1981) *Orogenic Andesites and Plate Tectonics.* 390 pp. Springer-Verlag.

GOLD T. and SOTER S. (1980) The Deep–Earth–Gas hypothesis. *Sci. Amer.* **242**, 130–137.

GOLD T. and SOTER S. (1983) Methane and seismicity: a reply. *EOS* **64**, 663.

GREEN D. H. (1971) Composition of basaltic magmas as indicators of conditions of origin: application to oceanic volcanism. *Phil. Trans. Roy. Soc. London* **A268**, 707–725.

GREEN D. H. (1972) Magmatic activity as the major process in the chemical evolution of the earth's crust and mantle. *Tectonophys.* **13**, 47–71.

GREEN D. H. (1973) Experimental melting studies on a model upper mantle composition at high pressure under water–saturated and water–undersaturated conditions. *Earth Planet. Sci. Lett.* **19**, 37–53.

GREEN D. H. (1976) Experimental testing of equilibrium partial melting of peridotite under water–saturated, high-pressure conditions. *Can. Mineral.* **14**, 255–268.

GREEN D. H. and RINGWOOD A. E. (1967) The genesis of basaltic magmas. *Contrib. Mineral. Petrol.* **15**, 103–190.

GREEN D. H. and LIEBERMANN R. C. (1976) Phase equilibria and elastic properties of the pyrolite model for the oceanic upper mantle. *Tectonophys.* **32**, 61–92.

GREEN D. H., HIBBERSON W. O. and JAQUES A. L. (1979) Petrogenesis of mid–ocean ridge basalts. In *The Earth: Its Origin, Structure and Evolution,* (ed. M. W. McELHINNY), pp. 265–299 Academic Press.

GREEN D. H., TAYLOR W. R. and FOLEY S. F. (1986) The Earth's upper mantle as a source for volatiles. *Geol. Soc. Aust. Special Pub. 12,* (In press).

GUPTA A. K., TAYLOR W. R. and GREEN D. H. (1986) Experimental study of the system forsterite–nepheline–quartz at variable temperatures under 28 kb pressure, (Submitted).

HAGGERTY S. E. (1978) The redox state of planetary basalts. *Geophys. Res. Lett.* **5**, 443–446.

HICKEY R. L. and FREY F. A. (1982) Geochemical characteristics of boninite series volcanics: implications for their source. *Geochim. Cosmochim. Acta* **46**, 2099–2115.

HOFFMAN A. W. and WHITE W. M. (1982) Mantle plumes from ancient oceanic crust. *Earth Planet. Sci. Lett.* **57**, 421–436.

HOLLOWAY J. R. (1973) The system pargasite–H_2O–CO_2: a model for melting of a hydrous mineral with a mixed-volatile fluid—I. Experimental results to 8 kbar. *Geochim. Cosmochim. Acta* **37**, 651–666.

JAQUES A. L. and GREEN D. H. (1980) Anhydrous melting of peridotite at 0–15 kb pressure and the genesis of tholeiitic basalts. *Contrib. Mineral. Petrol.* **73**, 287–310.

JENNER G. A. (1981) Geochemistry of high–Mg andesites from Cape Vogel, Papua New Guinea. *Chem. Geol.* **33**, 307–332.

JENNER G. A. (1983) Petrogenesis of high–Mg andesites: An experimental and geochemical study with emphasis on high–Mg andesite from Cape Vogel, P.N.G. Ph.D. Dissertation, Univ. of Tasmania.

KURAT G., PALME H., SPETTEL B., BADDENHAUSEN H., HOFMEISTER H., PALME C. and WANKE H. (1980)

Geochemistry of ultramafic xenoliths from Kapfenstein, Austria: evidence for a variety of upper mantle processes. *Geochim. Cosmochim. Acta* **44**, 45–60.

KURODA N., SHIRAKI K. and URANO H. (1978) Boninite as a possible calc-alkaline primary magma. *Bull. Volcanol.* **41**, 563–575.

KUSHIRO I. (1968) Melting of a peridotite nodule at high pressures and high water pressures. *J. Geophys. Res.* **73**, 6023–6029.

KUSHIRO I. (1972) Effect of water on the compositions of magmas formed at high pressures. *J. Petrol.* **13**, 311–334.

MATHEZ E. A. (1984) Influence of degassing on oxidation states of basaltic magmas. *Nature* **310**, 371–375.

MATTIOLI G. S. and WOOD B. J. (1986) Upper mantle f_{O_2} recorded by spinel-lherzolites (abstr.) *EOS* **67**, 375.

MCGETCHIN T. R., SILVER L. T. and CHODOS A. A. (1970) Titanoclinohumite: a possible mineralogical site for water in the upper mantle. *J. Geophys. Res.* **75**, 255–259.

MENZIES M. (1983) Mantle ultramafic xenoliths in alkaline magmas: evidence for mantle heterogeneity modified by magmatic activity. In *Continental Basalts and Mantle Xenoliths,* (eds. C. J. HAWKESWORTH and M. J. NORRY), pp. 92–110 Shiva Publ., Cheshire, England.

MORGAN W. J. (1972) Plate motions and deep mantle convection. *Geol. Soc. Amer. Mem.* **132**, 7–22.

MYSEN B. O. and BOETTCHER A. L. (1975) Melting of a hydrous mantle I. Phase relations of natural peridotite at high pressures and temperatures with controlled activities of water, carbon dioxide and hydrogen. *J. Petrol.* **16**, 520–548.

MYSEN B. O. and KUSHIRO I. (1977) Compositional variations of coexisting phases with degree of melting of peridotite in the upper mantle. *Amer. Mineral.* **62**, 843–865.

MYSEN B. O., VIRGO D., HARRISON W. J. and SCARFE C. M. (1980) Solubility mechanisms of H_2O in silicate melts at high pressures and temperatures: A Raman spectroscopic study. *Amer. Mineral.* **65**, 900–914.

NICKEL K. G. and GREEN D. H. (1984) The nature of the upper–most mantle beneath Victoria, Australia as deduced from ultramafic xenoliths. In *Kimberlites II: The Mantle and Crust-Mantle Relationships,* (ed. J. KORNPROBST), pp. 161–178 Elsevier.

NICKEL K. G. and GREEN D. H. (1985) Empirical geothermobarometry for garnet peridotites and implications for the nature of the lithosphere, kimberlites and diamonds. *Earth Planet. Sci. Lett.* **73**, 158–170.

RINGWOOD A. E. (1966) The chemical composition and origin of the earth. In *Advances in Earth Science,* (ed. P. M. HURLEY), pp. 287–356 M.I.T. Press, Cambridge.

RINGWOOD A. E. (1974) The petrological evolution of island arc systems. *J. Geol. Soc. London.* **130**, 183–204.

RINGWOOD A. E. (1986) Dynamics of subducted lithosphere and implications for basalt genesis. *Terra Cognita* **6**, 67–77.

RYABCHIKOV I. D., GREEN D. H., WALL V. J. and BREY G. P. (1981) The oxidation state of carbon in the low-velocity zone. *Geochem. International* **18**, 148–158.

SATO M. (1978) Oxygen fugacity of basaltic magmas and the role of gas–forming elements. *Geophys. Res. Lett.* **5**, 447–449.

SATO M. and VALENZA M. (1980) Oxygen fugacities of the layered series of the Skaergaard Instrusion, East Greenland. *Amer. J. Sci.* **20A**, 134–158.

SCHAIRER J. F. and YODER H. S. JR. (1961) Crystallization

in the system nepheline–forsterite–silica at one atmosphere pressure. *Carnegie Inst. Wash. Yearb.* **60,** 141–144.

STOLPER E. (1980) A Phase Diagram for Mid–ocean ridge basalts: Preliminary results and implications for petrogenesis. *Contrib. Mineral. Petrol.* **74,** 13–27.

TAKAHASHI E. and KUSHIRO I. (1983) Melting of a dry peridotite at high pressures and basalt magma genesis. *Amer. Mineral.* **68,** 859–879.

TATSUMI Y. (1981) Melting experiments on a high-magnesian andesite. *Earth Planet. Sci. Lett.* **54,** 357–365.

TAYLOR W. R. (1985) The role of C–O–H fluids in upper mantle processes: a theoretical, experimental and spectroscopic study. Ph.D. Dissertation, Univ. of Tasmania.

TAYLOR W. R. (1986) A reappraisal of the nature of fluids included by diamond—a window to deep–seated mantle fluids and redox conditions. *Geol. Soc. Aust. Special Publ.* **12,** (In press).

TAYLOR W. R. and GREEN D. H. (1987) The petrogenic role of methane: Effects on liquidus phase relations and the solubility of reduced C–H–O volatiles. In *Magmatic Processes: Physicochemical Principles,* (ed. B. O. MYSEN), pp. 121–138 The Geochemical Society Spec. Publ. No. 1.

UMINO S. (1985) Volcanic geology of Chichijima, the Bonin Islands. *J. Geol. Soc. Japan* **91,** 505–523.

WALKER D. A. and CAMERON W. E. (1983) Boninite primary magmas: evidence from the Cape Vogel Peninsula, P.N.G. *Contrib. Mineral. Petrol.* **83,** 150–158.

WEDEPOHL K. H. (1985) Origin of the Tertiary basaltic volcanism in the northern Hessian Depression. *Contrib. Mineral. Petrol.* **89,** 122–145.

WELHAN J. A. and CRAIG H. (1979) Methane and hydrogen in East Pacific Rise hydrothermal fluids. *Geophys. Res. Lett.* **6,** 829–831.

WELHAN J. A. and CRAIG H. (1983) Methane, hydrogen and helium in hydrothermal fluids at 21°N on the East Pacific Rise. In *Hydrothermal Processes at Seafloor Spreading Centres,* (eds. P. A. RONA, K. BOSTROM, L. LAUBIER and K. L. SMITH), pp. 391–409 Plenum Press, New York.

WENDLANDT R. F. and MYSEN B. O. (1980) Melting phase relations of natural peridotite + CO_2 as a function of degree of melting at 15 and 30 kbar. *Amer. Mineral.* **65,** 37–44.

WINDOM K. E. and BOETTCHER A. L. (1981) Phase relations for the joins jadeite–enstatite and jadeite–forsterite at 28 kb and their bearing on basalt genesis. *Amer. J. Sci.* **281,** 335–351.

WOERMANN E. and ROSENHAUER M. (1985) Fluid phases and the redox state of the Earth's mantle: extrapolations based on experimental, phase–theoretical and petrological data. *Fortschr. Mineral.* **63,** 263–349.

WYLLIE P. J. (1978) Mantle fluid compositions buffered in peridotite–CO_2–H_2O by carbonates, amphibole, phlogopite. *J. Geology* **86,** 687–713.

WYLLIE P. J. (1979) Magmas and volatile components. *Amer. Mineral.* **65,** 469–500.

WYLLIE P. J. (1980) The origin of kimberlites. *J. Geophys. Res.* **85,** 6902–6910.

YODER H. S. JR. and TILLEY C. E. (1962) Origin of basalt magmas: an experimental study of natural and synthetic rock systems. *J. Petrol.* **3,** 342–532.

Part C.
Continental Margin Processes

Magmatic Processes: Physicochemical Principles
© The Geochemical Society, Special Publication No. 1, 1987
Editor B. O. Mysen

Continental crust subducted to depths near 100 km: Implications for magma and fluid genesis in collision zones*

W. SCHREYER

Institut für Mineralogie, Ruhr-Universitat, D-4630 Bochum, F.R.G.

H.-J. MASSONNE

Institut für Mineralogie, Ruhr-Universitat, D-4630 Bochum, F.R.G.

and

C. CHOPIN

Laboratoire de Géologie, Ecole Normale Superieure, F-75005 Paris, France

Abstract—The metasedimentary pyrope–coesite–kyanite–talc–phengite–jadeite rocks of the Dora Maira Massif, Western Alps, were formed under metamorphic conditions of about 30 kbar, 700°–800°C and thus indicate a depth of subduction of some 100 km at low average geothermal gradients (7–8°C/km) along a possible geotherm of 40 mW/m². These pressure-temperature conditions lie above the eutectic melting curve for wet alkali granite and overlap with the melting curve of jadeite + coesite in the presence of excess H_2O. While there are no obvious signs of partial melting in the pyrope–coesite rocks, occasional interlayered bands of jadeite–kyanite–almandine–SiO_2 rock could possibly have gone through a liquid state. Because of thermal barriers such as phengite–talc or phengite–pyrope in the K_2O-MgO-Al_2O_3-SiO_2-H_2O system, it is unlikely that the pyrope-coesite rock itself represents a restite composition after extraction of a trachyte–like eutectic liquid. It is clear, however, that rocks with the hydrous mineralogy of the pyrope–coesite rock will have to melt partially when, subsequent to subduction, higher and more normal average geothermal gradients are being established at these mantle depths.

Experimental studies in the model system K_2O-MgO-Al_2O_3-SiO_2-H_2O indicate that the assemblage K feldspar–phlogopite becomes unstable, in the presence of excess water, in the pressure-temperature range 15–20 kbar and 400°–700°C and forms very MgSi–rich phengite and quartz plus a (supercritical?) hydrous, KMg–rich fluid with an estimated oxide ratio near K_2O:$3MgO$:$7SiO_2$. On this basis it can be predicted that common rocks of the upper continental crust such as granites and acid gneisses will, upon deep subduction, develop KMg–rich fluids at the expense of their biotite–K feldspar parageneses. When interacting with neighboring ultramafic mantle rocks these highly reactive fluids can cause the commonly observed mantle metasomatism producing phlogopite and K–richterite at the expense of olivine and clinopyroxene. Very K–rich igneous rocks such as those of the lamproite family, which are believed to have formed from partial melts developed in highly metasomatized mantle sources, could thus exhibit close spatial relationships with zones of preceding crustal subduction. In at least five cases (Spain, Colorado Plateau, Western Alps, Corsica, Karakorum) such relations exist. Moreover, the trace element and isotopic signature of lamproites is best explained by contamination of normal mantle with material from continental crust prior to magma formation. Magma genesis in continent/continent collision zones is severely influenced, or even governed, by the fluid and melt production within slabs of the continental crust that are subducted to considerable mantle depths.

INTRODUCTION

UNTIL RECENTLY the metamorphism of sediments, both continental and oceanic, and of other rocks of the continental crust such as granites, has been regarded as an essentially internal process within the crust of the earth attaining normally a maximum thickness of some 30–40 km. Thus, disregarding the materials of the deeper "mountain roots" below young fold belts, the maximum pressures of metamorphism endured by such rocks were believed to lie near 10 kbar, and indeed all the classical petrogenetic grids of metamorphism end at this limiting pressure (*e.g.*, WINKLER, 1974; ERNST, 1983).

Petrologic studies of metasediments from the Alpine/Mediterranean area (*e.g.*, CHOPIN and SCHREYER, 1983; SCHLIESTEDT, 1986), and especially the rediscovery and detailed examination of the exceptional pyrope–coesite rock of the Dora Maira Massif in the Western Alps (CHOPIN, 1984) have indicated, however, that this limit to depth and pressure of the metamorphism of crustal rocks

---

* Dedicated to H. S. Yoder, Jr.

cannot be generally true. As a result, a newly modified and considerably expanded view of geodynamic processes involving continental crustal materials is developing. This view takes into account subduction of crustal rocks during continent/continent collisions to at least 100 km depths and perhaps even more (SCHREYER, 1985; CHOPIN, 1986a).

Within the new conceptual framework of very deep crustal subduction the problem of (partial) melting of crustal rocks at mantle depths and its possible bearing on magma genesis as a whole will have to be included. To open, or reopen, the discussion on this point using earlier experimental data is a major goal of this paper. In addition, new experimental results suggesting the evolution of highly potassic fluids from crustal rocks at mantle depths, that may cause the metasomatic changes within the ultrabasic mantle materials as known from xenoliths within kimberlites are evaluated.

THE DORA MAIRA PYROPE–COESITE ROCK

The petrology and origin of this—thus far—unique metamorphic rock was described by CHOPIN (1984), with a few additions by SCHREYER (1985). In the present context, only a brief summary can be presented, but it must be clear that the problem is not exhausted and that additional, still more detailed work will undoubtedly yield further insights.

Although the outcrops of fresh, non-retrograde pyrope–coesite rocks are limited to dimensions in the 100 m range, it is clear that the total area affected by this type of extreme metamorphism is at least 5 × 10 km (CHOPIN, 1986a). There are coarse varieties of the rock with pyrope single crystals up to 20 cm in diameter, that were initially identified as conglomerates (see VIALON, 1966), but there are also sugary quartzites with pale, pink pyropes of only centimeter size. It is in these smaller garnet crystals that inclusions of relic coesite were identified (CHOPIN, 1984). Because the typical feathery quartz aggregates formed after coesite were also found in portions of garnet that are open to the quartz matrix, Chopin concludes that all free SiO_2 of the inclusions as well as the quartz matrix had once been coesite.

In addition to pyrope and coesite (quartz) the rock contains kyanite, phengite, and talc as major constituents. Jadeite (with some 25 mol percent diopside component) is occasionally enriched in layers and irregular trails. Rutile is a common accessory mineral. The coexistence of jadeite and talc implies the instability of a partially CaMg-substituted glaucophane due to the water-conserving reacton glaucophane = jadeite + talc. However, Ca-bearing glaucophane with excess Mg is found as

relic inclusions in some large pyrope crystals (CHOPIN, 1986b). Additional inclusions within the large pyrope crystals are Mg-chlorite, the new MgAlTi-silicate mineral ellenbergerite (CHOPIN et al., 1986) and, locally enriched, relics of large crystals of the Mg-tourmaline dravite (see SCHREYER, 1985, Figure 14). The resulting compatibility relations are shown, in a simplified way, in the right-side portion of the AKF-plot of Figure 1. Due to the presence of small amounts of FeO in addition to MgO in the Dora Maira rock, four rather than three solid AKF-phases may coexist with coesite.

The overall chemistry of the Dora Maira rocks as deduced from their mineral contents is best explained by a sedimentary protolith such as an evaporitic clay, which had already been cited by SCHREYER (1977) to derive the talc–kyanite rocks (whiteschists), which are found elsewhere in the world in metamorphic sequences subjected to relatively high pressures. Because the whiteschist assemblage is related to the pyrope–coesite rock by the simple reaction

$$talc + kyanite = pyrope + coesite + H_2O, \quad (1)$$

it is believed that the Dora Maira material represents a higher-grade metamorphic equivalent of whiteschists.

The pressure-critical minerals and mineral assemblages making up the Dora Maira rock indicate, more or less independently, metamorphic conditions lying within the coesite stability field. Based on the assemblage pyrope–SiO_2–talc–phengite CHOPIN (1984) has derived a minimum pressure of

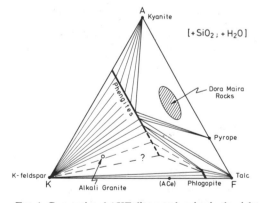

FIG. 1. Conventional AKF-diagram showing in the right portion the compatibility relations as found in the pyrope–coesite rocks of the Dora Maira Massif. The projection point for the composition of alkali granite was taken from WINKLER (1974). ACe stands for Al–celadonite, the theoretical end member of the phengite series with the formula $KMgAl[Si_4O_{10}](OH)_2$. The excess SiO_2-phase is coesite.

28 kbar and temperatures of 700°–800°C. For jadeite + kyanite, the high–pressure equivalent of paragonite, CHOPIN (1984) estimated a minimum pressure of 26 kbar. Preliminary experimental work on the stability field of the Ti-end member of the new mineral ellenbergerite (CHOPIN *et al.,* 1986) by SCHREYER, BALLER, and CHOPIN (unpublished data) indicate minimum water pressures for this phase to lie in the order of 25–30 kbar and maximum temperatures just below 800°C. This is in general agreement with the theoretical deductions by CHOPIN (1986b). Application of a new phengite barometer, for the limiting assemblage (see Figure 1) phengite + kyanite + a Mg-silicate + quartz/coesite (MASSONNE and SCHREYER, 1985) results in a water pressure of 35 kbar for the Dora Maira rock (see MASSONNE and SCHREYER, 1986b).

Conservative estimates of metamorphic conditions of the Dora Maira pyrope–coesite rock are shown in Figure 2. They are indicative of a thermal regime characterized by a very low geothermal gradient which—linearly extrapolated—is on the order of 7°/km. Such low gradients are prerequisite for blueschist facies metamorphism within subduction zones, although this type of metamorphism takes place at considerably lower pressures and temper-

atures (compare SCHREYER, 1985, Figure 19). Thus one may conclude that the Dora Maira metamorphism operating at some 100 kilometers depth is the higher grade equivalent of the blueschist facies. It is interesting to note, however, that the Dora Maira pressure-temperature conditions also fall close to the 40 mW/m² *continental* geotherm calculated by POLLACK and CHAPMAN (1977), along which all the low– as well as high–temperature lherzolite xenoliths from the Udachnaya, Siberia, kimberlite analyzed by BOYD (1984) lie.

PARTIAL MELTING IN THE DORA MAIRA ROCKS?

With the metamorphic conditions of the Dora Maira rocks located at some 30 kbar, 700°–800°C (Figure 2), the general question arises as to the possible melting behavior of crustal rocks under these conditions, and—specifically—whether or not there is any evidence of partial melting in the Dora Maira rocks. Because granites are believed to be those rocks of the continental crust that melt at the lowest temperatures, their behavior under mantle pressures is also of interest.

Pioneering experimental work on granite melting relations at pressures up to 35 kbar was performed

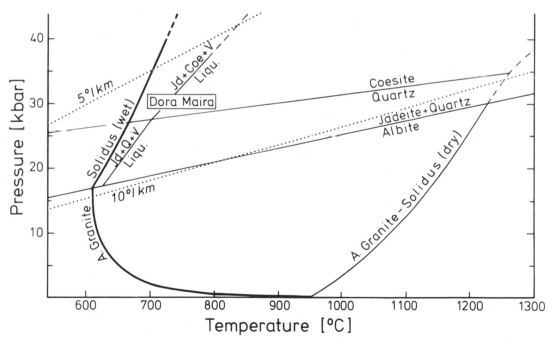

FIG. 2. Melting behavior of alkali granite as a function of pressure as simplified from the diagrams of HUANG and WYLLIE (1975, Figures 2–3). The box named Dora Maira indicates the metamorphic conditions of the pyrope–coesite rock discussed in the text. Dotted lines represent constant average geothermal gradients as labelled. Abbreviations: Coe = coesite; Jd = jadeite; Liqu. = hydrous melt; V = hydrous vapor.

during the last decade by the Wyllie school (*e.g.,* STERN and WYLLIE, 1973; HUANG and WYLLIE, 1973, 1975). Some salient features taken from the 1975 paper are reproduced in Figure 2. It can be seen from this figure that the Dora Maira pressure–temperature conditions are just above the solidus temperature of an alkali granite, provided that water pressure equalled total pressure. Because partial melting of the Dora Maira rock was not observed, CHOPIN (1984) assumed water activities well below 1.0.

An important aspect of the granite melting relations at high pressures is that granite is no longer of eutectic composition, because the eutectic melt moves toward albite composition with increasing pressure (HUANG and WYLLIE, 1975). Thus, a normal granite with a composition close to the low-pressure, crustal eutectic would—upon melting under the Dora Maira pressures—produce a more "trachytic" liquid with quartz remaining as a solid phase. Under these provisions the question arises as to whether or not the very SiO_2–rich pyrope–coesite rock of Dora Maira may in itself represent a *restite* after partial melting. A thorough examination of this possibility clearly requires more experimental melting data than presently available. Thus only a rather superficial treatment of this problem can be offered here.

The AKF-plot of Figure 1 shows the projection point of alkali granite, which lies rather close to the K (=K feldspar) corner of the diagram. Presumably, a "trachytic" liquid in the sense of HUANG and WYLLIE (1975), produced as a eutectic melt at the Dora Maira metamorphic conditions, would project into the same general area. Depending on the extension of the phengite solid solution (MASSONNE and SCHREYER, 1986a,b), it would fall either into the 3–phase field K feldspar + phlogopite + phengite, or into the 2–phase region K feldspar + phengite, and is thus clearly separated from the compositional range of the Dora Maira pyrope–coesite rocks. The phengite solid solution line, as well as the tie lines phengite–pyrope and phengite–talc, are stable during the high–pressure metamorphism and would act as thermal barriers between the pyrope–coesite rocks and the eutectic liquid. Hence, the two compositions cannot coexist; the solid assemblage cannot be the restite of "trachytic" eutectic melting. However, it must be borne in mind that this conclusion applies strictly only to compositions in the pure AKF-system K_2O-MgO-Al_2O_3-SiO_2-H_2O. Additional components such as Na_2O and FeO, which are present in subordinate amounts in the pyrope–coesite rocks, may change the situation,

especially if they had been abundant initially in a hypothetical starting material.

The possibility was also considered that at least some of the numerous polymineralic inclusions within the pyrope megacrysts once represented liquid. However, this was ruled out because these inclusions are nearly free of K_2O and Na_2O, or at least strongly depleted in these elements relative to the matrix of the rock. In the Dora Maira country rocks that are more Fe–rich, no compelling textural evidence for partial melting has so far been found (CHOPIN, 1986a).

The data in Figure 2 also indicate that the estimated metamorphic conditions at Dora Maira overlap with the eutectic melting of the assemblage jadeite + coesite in the presence of excess water after HUANG and WYLLIE (1975). This may be of interest regarding the origin of the bluish jadeite–kyanite–almandine quartzite which forms "rare, decimeter-thick layers (or veins?)" within the pyrope–coesite rock (CHOPIN, 1984). Figure 3 presents a close-up view of the mutual contact of the two rock types, which—unfortunately—does not allow discrimination between igneous intrusion of the jadeite rock into the pyrope rock and subsolidus interbanding of the two different lithologies. Cross-cutting relationships were not observed elsewhere either. Nevertheless, it is possible that the jadeite–quartz rock had been partially liquid, but has subsequently recrystallized, together with the pyrope rock, to form quartz from coesite. The Fe–rich garnet composition as opposed to that of the country rock may be in favor of this melt hypothesis, but the crucial problem concerning the availability of excess water cannot be solved.

In summary, the specific question "Was there any partial melting in the Dora Maira rocks during the high–pressure metamorphism?" cannot be answered with present knowledge. More important, however, is the result from Figure 2 that the temperatures during this metamorphism were sufficiently high to produce melts from units of appropriate bulk composition. As mentioned in the introduction, just the existence of the Dora Maira pyrope–coesite rock implies that rocks of the continental crust may be subducted to depths of about 100 km. Both SCHREYER (1985) and CHOPIN (1986a) have emphasized that the geodynamic conditions required for such deep subduction are still much easier to understand than those necessary for the retrieval of the high–pressure rocks without significant mineralogical changes back to the surface. Therefore, the chances are indeed very high that the subducted continental materials remain at

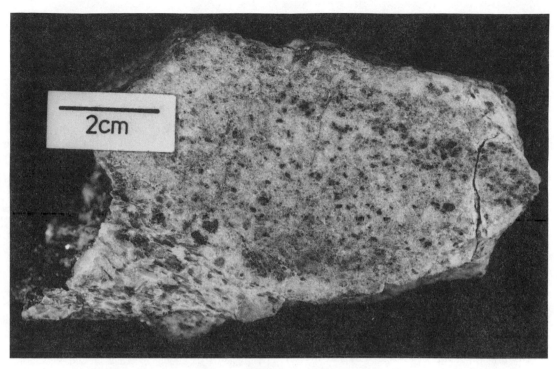

FIG. 3. Concordant contact between schistose, phengite–bearing pyrope rock (bottom) and a more massive layer of jadeite–kyanite–almandine quartzite (top). Parigi, Italy.

mantle depths sufficiently long for higher, more normal geotherms to develop (see Figure 2), which would then inevitably lead to increasing proportions of melt in these materials. These crust-derived, acid melts could then react with neighboring mantle peridotite, that is they "may experience hybridization" and produce phlogopite–bearing pyroxenities, similarly as proposed by SEKINE and WYLLIE (1982, 1985) for magmas above subducted oceanic crust.

At any rate, any model of magma genesis in continent/continent collision zones must necessarily take into account the contribution by melting of subducted continental crust. However, as we hope to show in the two final sections of this paper, there may be an additional contribution of material from continental rocks by virtue of a mechanism other than eutectic melting.

INSTABILITY OF THE K FELDSPAR–PHLOGOPITE ASSEMBLAGE AT HIGH FLUID PRESSURES

In their reports on experimental studies of phengite synthesis and stability in the system K_2O-MgO-Al_2O_3-SiO_2-H_2O MASSONNE and SCHREYER (1986a,b) emphasized that, for a fixed pressure and temperature, a critical phengite that is useful as a geobarometer must coexist with the limiting assemblage K feldspar–phlogopite–quartz. This is shown, in a simplified fashion, by the AKF diagram of Figure 1. In the course of determining the pressure–temperature stability range of this limiting assemblage the surprising discovery was made that it becomes unstable at the comparatively moderate water pressure of 15–20 kbar. Specifically, the tie line between phlogopite and K feldspar is broken, in the presence of excess water, by a reaction to form the solid phases phengite and quartz. Because the theoretical endmember of the dioctahedral mica series Al–celadonite, $KMgAl[Si_4O_{10}](OH)_2$, which lies on the join K feldspar–ideal phlogopite, $KMg_3[AlSi_3O_{10}](OH)_2$, (see Figure 1), cannot be synthesized (MASSONE and SCHREYER, 1986a), the phengite formed by the above reaction must contain less MgSi but nevertheless represent the critical composition. In order to account for the bulk composition along the K feldspar–phlogopite join, an additional phase must be present which, however, is not observed in the X-ray diffractograms of the quenched run products. For stoichiometric reasons this phase must be very poor in Al, or free from it, and consist essentially of K and Mg together with additional amounts of SiO_2 and H_2O. Earlier in-

vestigations of the system K_2O-MgO-SiO_2-H_2O by
SEIFERT and SCHREYER (1966) at the much lower
pressure of 1 kbar have indicated that hydrous ternary KMg silicate melts may appear in this system
at surprisingly low temperatures ($<500°C$). It can
be predicted that, with increasing fluid pressures,
these melts will dissolve more and more H_2O and
become low-viscosity, perhaps supercritical fluids
with rather variable K:Mg:Si ratios. Therefore, it is
most likely that the "missing components" in the
experiments of MASSONE and SCHREYER (1986b)
are contained, under run conditions, in such a K,
Mg–rich fluid. Indeed, the water observed after
opening the quenched run capsules was often found
to yield a precipitate when drying up.

In Figure 4 the AKF plot has been extended to
include the Al-free border system, in which the assumed K, Mg–rich fluids lie. It can be seen that
instead of the assemblage K feldspar + phlogopite
the limiting phengite coexists with a series of fluid
phases that are presumed to be confined to the system K_2O-MgO-SiO_2-H_2O. According to MASSONNE
(1986) the fluid in equilibrium with phengite for
the bulk composition of Al–celadonite (see above)
plus excess water has an estimated composition
given on an anhydrous basis as $K_2O \cdot 3MgO \cdot 7SiO_2$.
In increasingly more potassic bulk compositions K
feldspar appears first as a second solid phase that

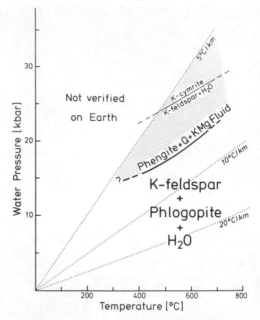

FIG. 5. Water pressure/temperature plot showing the
approximate stability field of the assemblage K feldspar–
phlogopite and the location of its high-pressure breakdown
to phengite + quartz + KMg–rich fluid (see Figure 4). The
hydration curve of K feldspar to form K–cymrite,
$KAlSi_3O_8 \cdot H_2O$, is taken from HUANG and WYLLIE
(1975). Dotted lines represent constant geothermal gradients as labelled. Thus the shaded field, in which the
KMg–rich fluid exists, can only be reached in areas with
a low thermal flux such active subduction zones.

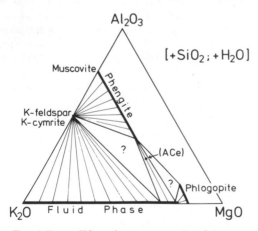

FIG. 4. Excess-SiO_2 and excess water plot of the system
K_2O-MgO-Al_2O_3-SiO_2-H_2O showing compatability relations of solid and fluid phases at fluid pressures above 20
kbar and temperatures of about 450°–650°C. ACe is Al-
celadonite as in Figure 1, K–cymrite is the phase
$KAlSi_3O_8 \cdot H_2O$. For relevant reaction curves see Figure
5. To simplify the projection all solid solutions are shown
as binary. The most aluminous phlogopite coexisting with
phengite is less aluminous than ideal phlogopite,
$KMg_3[AlSi_3O_{10}](OH)_2$ according to the data of MASSONE
and SCHREYER (1986b). The compositional range of the
KMg–rich fluid phase is hypothetical.

coexists with a KMg–fluid of invariant K/Mg-ratio,
and then phengite disappears with only K feldspar
+ fluids remaining.

Figure 5 shows the location of the univariant
curve for the reaction of K feldspar and phlogopite
in the presence of excess water as determined by
MASSONNE (1986) in the temperature range 450°–
650°C. In the case of water undersaturation, K
feldspar + phlogopite were found to coexist with
phengite + KMg–fluid towards higher pressures. At
still higher total pressures K feldspar itself reacts
with H_2O to form the solid K feldspar hydrate phase
named K–cymrite here (Figure 5). This phase is the
potassium analogue of the mineral cymrite,
$BaAl_2Si_2O_8 \cdot H_2O$, and has the composition
$KAlSi_3O_8 \cdot H_2O$. SEKI and KENNEDY (1964) first
synthesized this phase, but their reaction curve had
to be modified on the basis of more recent work.
(HUANG and WYLLIE, 1975; MASSONNE, unpublished data). Thus, at high fluid pressures on the
order of 30 kbar, phengites near the critical composition may coexist with KMg–fluid, coesite, and
K–cymrite (compare Figure 4).

CONSEQUENCES FOR MANTLE METASOMATISM AND RELATED LAMPROITIC MAGMAS

As a result of the phase relations depicted in Figures 4–5 a petrogenetic mechanism evolves that may be of considerable interest regarding the interaction of deeply subducted crustal rocks with mantle materials and the genesis of ultrapotassic lavas. The paragenesis of K feldspar with trioctahedral potassic micas (phlogopite, biotite) is widespread within the continental crust, because it occurs in such common upper crustal rocks as granites and many acid gneisses. Therefore, during deep subduction of crustal materials in collision zones, along relatively low geotherms (6–8°C/km), the reaction curve shown in Figure 5 for pure Mg-micas will be transgressed at depths of some 50–70 km, and—in the presence of a hydrous gas phase that may be formed during progressive subduction zone metamorphism—the KMg–rich fluid should develop. Presumably, the amount of fluid forming would be governed by the amount of interstitial H_2O gas present initially, and there is also an unknown influence of other components hitherto neglected such as Fe-oxides, TiO_2, and Na_2O.

Because of high reactivity of the KMg–fluids there is virtually no chance that they can ever reach the earth's surface, although their enrichment in K and Mg is reminiscent of lamproitic and perhaps K–rich kimberlitic igneous rocks. However, these rocks also contain other components such as Al_2O_3 and CaO, and all estimates and determinations of melting temperatures of lamproitic rocks yield much higher values ($\sim800°$–1000°C; see WAGNER and VELDE, 1986) than the temperatures of fluid formation (450°–650°C, Figure 5). Particularly interesting relations arise when one envisions that the KMg–fluids produced from subducted crustal rocks react with neighboring ultramafic mantle materials.

Petrological observations on mantle xenoliths from kimberlites have provided compelling evidence for mantle metasomatism (*e.g.,* DAWSON and SMITH, 1977; JONES *et al.,* 1982; BAILEY, 1982). Among the newly formed minerals that replace olivine and clinopyroxene the phases phlogopite, and K–richterite, ideally $KNaCaMg_5[Si_8O_{22}](OH)_2$, predominate. Although there is agreement that this metasomatism is due to infiltration of peridotites and harzburgites by fluids, the sources and origins of these fluids are still under debate (RYABCHIKOV *et al.,* 1982; WYLLIE and SEKINE, 1982; SCHNEIDER and EGGLER, 1986). It is clear from the present study that the KMg–fluids emerging from subducted crustal rocks containing biotite and K feldspar would represent almost ideal agents for the

mantle metasomatism observed. The additional enrichment of the metasomatites in Ba, Ti, and other large–ion–lithophile–elements may be explained by their former presence in the biotite that reacted to form the metasomatizing fluid.

If mantle metasomatism was indeed, or at least partly, caused by fluids originating from subducted crustal rocks, a spatial relationship between the metasomatites in the mantle and relics of crustal rocks could be expected at least in some cases. An example may be provided by the xenolith suites from South African kimberlite pipes which comprise, on one hand, metasomatites and the MARID (mica–amphibole–rutile–ilmenite–diopside) "pegmatites" (DAWSON and SMITH, 1977) and, on the other hand, rocks that can only be explained through crustal origin. Regarding the latter ones, xenoliths of kyanite–bearing grospydites from the Roberts Victor Mine (SMYTH and HATTON, 1977) are particularly interesting, because they contain K feldspar and coesite as well. We wish to emphasize, however, that—to our knowledge—in none of the mantle xenoliths the assemblage K feldspar + phlogopite was found.

There seems to be general consent that the K–rich magmas of the lamproite family are products of partial melting of a large–ion–lithophile–element enriched mantle source such as the metasomatites discussed above (see JAQUES *et al.,* 1984; NIXON *et al.,* 1984; WAGNER and VELDE, 1986). The high temperatures of the lamproite magmas cited before require high temperatures of magma genesis in the mantle. Therefore, we conclude that the production of these melts occurred considerably later than the metasomatism which—if effected by the mechanism proposed here—was a low–temperature process along the initially low subduction zone geotherm. The lamproite melts may have formed when more normal, post–subduction geotherms of the mantle were established, or in fact even much later during a subsequent thermal event.

Disregarding the variable time difference between subduction-induced mantle metasomatism and lamproite melt formation, the mechanism proposed here should imply a close *spatial relationship* between lamproite occurrences and underlying former collision zones. Although this problem has not been tested in any systematic way, there are indications from the regional geology of some lamproite occurrences that such relationships may indeed exist.

At least some of the lamproitic and related rocks of the Western United States, including minettes and kimberlites, may overlie subduction zones, because metamorphic rocks typically formed along subduction geotherms are in fact found as xenoliths within them (HELMSTAEDT and SCHULZE, 1977). Still more importantly, such xenoliths contain phengites that exhibit the highest MgSi-contents found in nature (SMITH and ZIENTEK, 1979; HARLEY and GREEN, 1981) and thus represent critical phengites that may have coexisted, at some stage, with the KMg–fluid discussed. Similar relationships hold for the lamproites of southeastern Spain (WAGNER and VELDE, 1986), for which most models of the tectonic evolution involve young subduction (see NELSON et al., 1986). Indeed, in the clearly subduction-zone metamorphosed crustal slab of the Sesia-Lanzo zone in the Western Alps, some 100 km north of the Dora Maira Massif, "K–rich lamprophyre" dykes (DAL PIAZ et al., 1979) of Oligocene and thus postsubduction age are found, which have compositions practically identical to lamproites. DAL PIAZ et al. (1979) emphasize the similarity of their "lamprophyres" to postcollisional lamprophyres of Karakorum (VITERBO and ZANETTIN, 1959) as well as to a minette intruded into the previously subducted schists of Corsica (VELDE, 1967; WAGNER and VELDE, 1986). Thus there are at least five regions in the world, where a spatial relationship between lamproite occurrences and former collision zones exists. For the diamond-bearing lamproites of the Kimberley region, Western Australia, no such relationship can be identified at present, because the 20 million year old intrusions occur near a mobile belt of Proterozoic age (JAQUES et al., 1984).

The sequence of events as proposed by the present model, that is 1) deep subduction of crustal rocks, 2) mantle metasomatism, 3) genesis of K–rich magmas, is also supported by geochemical data on lamproites, especially those from Spain. HERTOGEN et al. (1985) have successfully modeled the isotopic and peculiar trace element patterns of the mantle source region of these lamproites by contamination of normal mantle with material released from subducted continental sediments. Similarly, NELSON et al. (1986) conclude that the metasomatic component added to the mantle below southeastern Spain prior to lamproite formation "has the Sr, Nd, and Pb isotopic characteristics of continental crust or sediments derived from continental crust."

To summarize the implications of deep subduction of continental crust into the upper mantle in continent/continent collision zones it is clear that both the development of melts and of low–temperature fluids within the subducted crustal slabs must have profound effects on the genesis of magmas in these areas, and may even govern them.

REFERENCES

BAILEY D. K. (1982) Mantle metasomatism—continuing chemical change within the Earth. Nature 296, 525–530.

BOYD F. R. (1984) Siberian geotherm based on lherzolite xenoliths from the Udachnaya kimberlite, USSR. Geology 12, 528–530.

CHOPIN C. (1984) Coesite and pure pyrope in high-grade blueschists of the Western Alps: a first record and some consequences. Contrib. Mineral. Petrol. 86, 107–118.

CHOPIN C. (1986a) Very-high-pressure metamorphism in the Western Alps: implications for subduction of continent l crust. Philos. Trans. Roy. Soc. London (In press).

CHOPIN C. (1986b) Phase relationships of ellenbergerite, a new high-pressure Mg–Al–Ti–silicate in pyrope–coesite–quartzite from the Western Alps. Geol. Soc. Amer. Mem. 164, 31–42.

CHOPIN C. and SCHREYER W. (1983) Magnesiocarpholite and magnesiochloritoid: two index minerals of pelitic blueschists and their preliminary phase relations in the model system MgO-Al₂O₃-SiO₂-H₂O. Amer. J. Sci. 283-A, 72–96.

CHOPIN C., KLASKA R., MEDENBACH O. and DRON D. (1986) Ellenbergerite, a new high-pressure Mg–Al–(Ti, Zr)–silicate with a novel structure based on face–sharing octahedra. Contrib. Mineral. Petrol. 92, 322–330.

DAL PIAZ G. V., VENTURELLI G. and SCOLARI A. (1979) Calc-alkaline to ultrapotassic postcollisional volcanic activity in the internal northwestern Alps. Mem. Sci. Geol. Univ. di Padova 32, 16 pp.

DAWSON J. B. and SMITH J. V. (1977) The MARID (mica-amphibole–rutile–ilmenite–diopside) suite of xenoliths in kimberlite. Geochim. Cosmochim. Acta 41, 309–333.

ERNST W. G. (1983) Phanerozoic continental accretion and the metamorphic evolution of northern and central California. Tectonophys. 100, 287–320.

HARLEY S. L. and GREEN D. H. (1981) Petrogenesis of eclogite inclusions in the Moses Rock dyke, Utah, U.S.A. Tschermaks Mineral. Petrogr. Mitt. 28, 131–155.

HELMSTAEDT H. and SCHULZE D. J. (1977) Type A–Type C eclogite transition in a xenolith from the Moses Rock diatreme—further evidence for the presence of metamorphosed ophiolites beneath the Colorado Plateau. Extended Abstr. Second Internat. Kimberlite Conf., New Mexico.

HERTOGEN J., LOPEZ-RUIZ J., RODRIGUEZ-BADIOLA E., DEMAIFFE D. and WEIS D. (1985) A mantle-sediment mixing model for the petrogenesis of a ultrapotassic lamproite from S.E. Spain. EOS 66, 1114.

HUANG W. L. and WYLLIE P. J. (1973) Melting relations of muscovite–granite to 35 kbar as a model for fusion of metamorphosed subducted oceanic sediments. Contrib. Mineral. Petrol. 42, 1–14.

HUANG W. L. and WYLLIE P. J. (1975) Melting reactions in the system NaAlSi₃O₈-KAlSi₃O₈-SiO₂ to 35 kilobars, dry and with excess water. J. Geol. 83, 737–748.

JAQUES A. L., LEWIS J. D., SMITH C. B., GREGORY G. P., FERGUSON J., CHAPPELL B. W. and McCULLOCH M. T. (1984) The diamond–bearing ultrapotassic (lamproitic) rocks of the West Kimberley Region, Western Australia. In Kimberlites. I: Kimberlites and Related Rocks, (ed. J. KORNPROBST), pp. 225–254 Elsevier.

JONES A. P., SMITH J. V. and DAWSON J. B. (1982) Mantle

metasomatism in 14 veined peridotites from Bultfontein Mine, South Africa. *J. Geol.* **90**, 435–453.

MASSONNE H.-J. (1986) Breakdown of K-feldspar + phlogopite to phengite + K, Mg–rich silicate melt under the metamorphic conditions of a subduction zone. *Symp. Experim. Mineral. Geochem. Nancy Abstr.*, 97–98.

MASSONNE H.-J. and SCHREYER W. (1985) Phengite barometry in assemblages with kyanite, Mg–rich silicates, and a SiO₂ phase. *Terra Cogn.* **5**, 432.

MASSONNE H.-J. and SCHREYER W. (1986a) High-pressure syntheses and X-ray properties of white micas in the system K₂O-MgO-Al₂O₃-SiO₂-H₂O. *Neues Jahrb. Mineral. (Abhand)* **153**, 177–215.

MASSONNE H.-J. and SCHREYER W. (1986b) Phengite geobarometry based on the limiting assemblage with K feldspar, phlogopite, and quartz. *Contrib. Mineral. Petrol.* (In press).

NELSON D. R., MCCULLOCH M. T. and SUN S.-S. (1986) The origins of ultrapotassic rocks as inferred from Sr, Nd and Pb isotopes. *Geochim. Cosmochim. Acta* **50**, 231–245.

NIXON P. H., THIRLWALL, M. F., BUCKLEY F. and DAVIES C. J. (1984) Spanish and Western Australian lamproites: aspects of whole rock geochemistry. In: *Kimberlites. I: Kimberlites and Related Rocks* (ed. J. KORNPROBST), pp. 285–296 Elsevier.

POLLACK H. N. and CHAPMAN D. S. (1977) On the regional variation of heat flow, geotherms and lithospheric thickness. *Tectonophys.* **38**, 279–296.

RYABCHIKOV I. D., SCHREYER W. and ABRAHAM K. (1982) Compositions of aqueous fluids in equilibrium with pyroxenes and olivines at mantle pressures and temperatures. *Contrib. Mineral. Petrol.* **79**, 80–84.

SCHLIESTEDT M. (1986) Eclogite–blueschist relationships as evidenced by mineral equilibria in the high-pressure metabasic rocks of Sifnos (Cycladic Islands), Greece. *J. Petrol.* (In press).

SCHNEIDER M. E. and EGGLER D. H. (1986) Fluids in equilibrium with peridotite minerals: Implications for mantle metasomatism. *Geochim. Cosmochim. Acta* **50**, 711–724.

SCHREYER W. (1977) Whiteschist: their compositions and pressure–temperature regimes based on experimental, field, and petrographic evidence. *Tectonophys.* **43**, 127–144.

SCHREYER W. (1985) Metamorphism of crustal rocks at mantle depths: High-pressure minerals and mineral assemblages in metapelites. *Fortschr. Mineral.* **63**, 227–261.

SEIFERT F. and SCHREYER W. (1966) Fluide Phasen im System K₂O-MgO-SiO₂-H₂O und ihre mögliche Bedeutung für die Entstehung ultrabasischer Gesteine. *Ber. Bunsenges. Phys. Chem.* **70**, 1045–1050.

SEKI Y. and KENNEDY G. C. (1964) The breakdown of potassium feldspar, KAlSi₃O₈, at high temperatures and pressures. *Amer. Mineral.* **49**, 1688–1706.

SEKINE T. and WYLLIE P. J. (1982) The system granite–peridotite–H₂O at 30 kbar, with applications to hybridization in subduction zone magmatism. *Contrib. Mineral. Petrol.* **81**, 190–202.

SEKINE T. and WYLLIE P. J. (1985) Hybridization of magmas above subducted oceanic crust. *Geol Zb.-Geol. Carpathica* (Bratislava) **36**, 259–268.

SMITH D. and ZIENTEK M. (1979) Mineral chemistry and zoning in eclogite inclusions from Colorado Plateau diatremes. *Contrib. Mineral. Petrol.* **69**, 119–131.

SMYTH J. R. and HATTON (1977) A coesite–sanidine grospydite from the Roberts Victor kimberlite. *Earth Planet. Sci. Lett.* **34**, 284–290.

STERN C. R. and WYLLIE P. J. (1973) Water-saturated and under-saturated melting relations of a granite to 35 kilobar. *Earth Planet. Sci. Lett.* **18**, 163–167.

VELDE D. (1967) Sur un lamprophyre hyperalcalin potassique: la minette de Sisco (Corse). *Bull. Soc. Fr. Mineral. Cristallogr.* **90**, 214–223.

VIALON P. (1966) Etude géologique du massif cristallin Dora-Maira, Alpes cottiennes internes, Italie. Thèse d'état, Univ. Grenoble.

VITERBO C. and ZANETTIN B. (1959) I filoni lamprofirici dell'alto Baltoro (Karakorum). *Mem. Acc. Patavina Sc. Lett. Arti* **71**, 39 pp.

WAGNER C. and VELDE D. (1986) The mineralogy of K–richterite-bearing lamproites. *Amer. Mineral.* **71**, 17–37.

WINKLER H. G. F. (1974) *Petrogenesis of Metamorphic Rock.* 3d edition. 320 pp. Springer-Verlag.

WYLLIE P. J. and SEKINE T. (1982) The formation of mantle phlogopite in subduction zone hybridization. *Contrib. Mineral. Petrol.* **79**, 375–380.

Magmatic Processes: Physicochemical Principles
© The Geochemical Society, Special Publication No. 1, 1987
Editor B. O. Mysen

A petrological model of the mantle wedge and lower crust in the Japanese island arcs

IKUO KUSHIRO

Geological Institute, University of Tokyo, Hongo, Tokyo, and Geophysical Laboratory,
Carnegie Institution of Washington, Washington, D.C. 20008, U.S.A.

Abstract—A petrological model of the mantle wedge and the lower crust in the Japanese Island arcs has been developed from deep-seated inclusions, melting relations of arc basalts, and the phase relations of peridotite combined with seismic wave velocities in the mantle wedge and laboratory measurements of velocities of peridotite at high pressures and temperatures. It is suggested that the amount of melt in the mantle wedge source region is approximately 2 volume percent below the volcanic zone and increases in ascending diapirs. The maximum amount of water in most parts of the mantle wedge is 0.2 weight percent. The highest temperature region in the mantle wedge would exist at a depth of 70–80 km beneath the volcanic front and slightly deeper near the coast of the Japan Sea and has the lowest density. A density reversal below the volcanic zone would cause the ascent of diapirs in the mantle wedge. Magmas would segregate at a level where the diapirs stop ascending. This depth would be determined by the peridotite solidus in the absence of excess vapor phase, and increase from about 35 km to about 60 km from the volcanic front to the Japan Sea coast.

Lower crustal materials in the Japanese Islands inferred from the deep-seated inclusions in basalts and tuffs are hornblende-gabbro, granulite, and amphibolite. In NE Honshu, at least near the coast of Japan Sea, the original rocks for these metamorphic rocks are mostly hornblende gabbros formed by accumulation from hydrous basaltic magmas, and subsequently metamorphosed with decreasing temperature toward the ambient geotherm. In SW Honshu, the original rocks are both gabbros and sedimentary rocks enriched in silica and alumina, as inferred from the chemical analyses of the inclusions and the presence of aluminous phases such as kyanite and corundum in some of the inclusions. The geotherm estimated by the pyroxene geothermometry passes 850°C at about 25 km in NE Honshu and at about 35 km in SW Honshu. The crust of the volcanic zone in NE Honshu would have been thickened essentially by the intrusion of basalt magmas from the mantle wedge, whereas that of the SW Honshu may have been formed by both magmatic and sedimentary or accretionary processes. By the latter process, sedimentary rocks could be brought to the base of the crust. The crust of the island arcs can thus be classified as igneous crust and sedimentary or accretionary crust. There is also a mixed crust consisting of both the rocks as major constituents, such as at SW Honshu. I-type and S-type granites may be formed in the igneous crust and sedimentary or accretionary crust, respectively. "Paired granite belts" may be formed in mature island arcs.

INTRODUCTION

ISLAND ARCS and active continental margins associated with the subduction of oceanic lithospheres are the place of active magmatic processes and formation of the earth's crust. Although the amount of magmas produced in these regions is much less than that in the mid-oceanic ridges, the process taking place in the subduction zone are important for the formation and development of the continental-type crust (*e.g.,* KAY, 1985).

The Japanese island arcs consist of five different segments of arcs with different geophysical and geological characteristic; they are southern Kurile, NE Japan, Izu-Bonin, SW Japan and Ryukyu arcs (Figure 1). For example, the angle of subduction is steeper in the Izu-Bonin and Ryukyu arcs than in the NE Japan and southern Kurile arcs, and Quaternary calc-alkalic andesite is common in the NE Japan and southern Kurile arcs, whereas it is less common in the Izu-Bonin arc. Heat flow data are

different between the NE and SW Japan arcs. In addition, most of these arcs have a double arc structure consisting of an outer arc (presently non-volcanic) and an inner arc (presently volcanic) (MI-YASHIRO, 1974). Such variations in the Japanese island arcs may reflect the complexity of the processes in the development of the island arcs. Detailed studies of the Japanese island arcs will, therefore, be important for understanding the nature and origin of arcs in general.

A number of geological, geophysical and petrological studies have been made in the Japanese Island arcs, and these arcs are among the most extensively studied arcs in the world. However, the materials, structure and evolution of the lower crust and the upper mantle in the Japanese island arcs are still not well understood. In the present study this problem is revisited on the basis of a synthesis of recent experimental and petrological studies together with previous studies, especially those by KUNO (1959, 1960), SUGIMURA (1960), AOKI

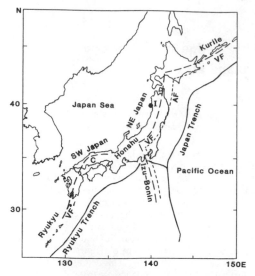

FIG. 1. Japanese Island arcs. Abbreviations: VF, volcanic front; AF, aseismic front; I, Ichinomegata; C, Chugoku district.

(1971), MIYASHIRO (1974), ISHIHARA (1977) and SAKUYAMA (1979). The conditions of magma generation and a petrological model of the mantle wedge beneath the NE Japan arc are discussed first. This is an extension of the work by TATSUMI *et al.* (1983); here, the physical properties of the mantle wedge are discussed in more detail. Secondly, the lower crustal materials and the geotherms in the NE and SW Japan arcs are compared based on deep-seated inclusions in volcanic rocks. Finally, the process of formation of the lower crustal materials and the evolution of the crust of the Japanese arcs are briefly discussed.

CONDITIONS OF BASALT MAGMA GENERATION

KUNO (1959) first proposed a model for generation of basalt magmas in the Japanese Islands and their adjacent areas based on his petrological observations on basalts and the deep seismicity studied by Wadati and his colleagues (*e.g.,* WADATI and IWAI, 1954). According to this model, basalt magmas are generated along the deep seismic zone (Wadati-Benioff zone); tholeiitic basalt magma, which is erupted near the volcanic front, is generated at depths shallower than 200 km, whereas alkali basalt magma, which is erupted near the Japan Sea coast, at depths greater than 200 km. Later, KUNO (1960) proposed that high-alumina basalt, which is erupted in a zone between the tholeiite and alkali basalt zones, is generated at depths around 200 km.

The depths of generation of basalt magmas proposed by Kuno were found to be too deep when experimental data on natural and synthetic basalt systems became available (*e.g.,* GREEN and RINGWOOD, 1967; KUSHIRO, 1968). Based on experimental results, GREEN *et al.* (1967) proposed a model for the generation of basalt magmas in island arcs; tholeiitic magma is segregated from the source upper mantle materials at depths shallower than 30 km, high-alumina basalt magma, at 30–40 km, and alkali olivine basalt magma, at 50–60 km. The chemical compositions of the basalts which GREEN *et al.* (1967) used for this estimation are, however, not appropriate for the primary basalt magmas in the Japanese Islands. Recently, TATSUMI *et al.* (1983) estimated the chemical compositions of the primary magmas for tholeiite, high-alumina basalt, and alkali olivine basalt in the NE Japan arc using the compositions of the least fractionated basalts of these three basalt types and maximum fractionation model. The most magnesian olivine tholeiite found in the Ryozen district, NE Honshu (KOTOKU, 1986) has a composition very close to that of the primary olivine tholeiite magma estimated by TATSUMI *et al.* (1983) as shown in Table 1. This olivine tholeiite contains about 10 volume percent olivine phenocrysts and microphenocrysts in a fine-grained groundmass; however, their compositions (Fo_{90}) in

Table 1. Chemical compositions of olivine tholeiite from Ryozen district and estimated primary olivine tholeiite magma

	Olivine tholeiite, Ryozen district, NE Honshu*	Primary olivine tholeiite magma†
SiO_2	49.03	49.71
TiO_2	0.64	0.74
Al_2O_3	13.95	14.97
Fe_2O_3	2.36	—
FeO	6.83	10.57††
MnO	0.15	0.14
MgO	12.38	13.03
CaO	10.75	9.00
Na_2O	1.93	1.56
K_2O	0.27	0.28
$H_2O(-)$	0.60	—
$H_2O(+)$	1.30	—
P_2O_5	0.07	—
Cr_2O_3	0.12	—
Total	100.38	100.00
Fe/(Fe + Mg)	0.289	0.313

* Described by KOTOKU (1986); this sample (IK85060803) was collected by the author and analyzed by H. Haramura.
† TATSUMI *et al.* (1983).
†† Total iron as FeO.

the core and the bulk composition of the rock indicate that the olivine phenocrysts are not cumulates.

TATSUMI et al. (1983) made a series of experiments on the estimated compositions of the three primary basalt magmas under both anhydrous and hydrous conditions to determine the pressure-temperature conditions where these magmas can be in equilibrium with olivine and orthopyroxene with or without clinopyroxene. The results indicate that the primary tholeiite (olivine tholeiite) magma can be in equilibrium with olivine and orthopyroxene at about 11 kbar and 1320°C under anydrous conditions, the primary high-alumina basalt magma can be in equilibrium with olivine, orthopyroxene and clinopyroxene at about 15 kbar and 1340°C under anhydrous conditions and 17 kbar and 1325°C in the presence of 1.5 weight percent H_2O, and the primary alkali olivine basalt magma can be in equilibrium with the above three minerals at about 17 kbar and 1360°C under anhydrous conditions and 23 kbar and 1320°C in the presence of 3 weight percent H_2O. These H_2O contents applied are maximum values estimated by SAKUYAMA (1979). The primary tholeiite magma would contain less than 1 weight percent H_2O.

In the present study, melting experiments have been made on the olivine tholeiite from Ryozen district. The results, shown in Appendix, indicate that both olivine ($Fo_{89}Fa_{11}$) and orthopyroxene ($En_{89}Fs_{11}$) crystallize on the liquidus at about 1315°C at 11 kbar. The results are very close to those obtained by TATSUMI et al. (1983) on the primary olivine tholeiite composition; the temperature of olivine-orthopyroxene cotectic is only about 5° lower in the present experiments.

Because of high temperature of equilibration with peridotites (1300°C) at relatively shallow depths, it was suggested that the primary basalt magmas are initially formed within hot diapirs that have ascended from the deeper regions in the mantle wedge (SAKUYAMA, 1983a; TATSUMI et al., 1983; KUSHIRO, 1983). A similar model was suggested by GREEN et al. (1967). If the diapirs (or magmas) ascend adiabatically, the temperature of the source region must be higher than that during the final equilibration; the temperature at a depth of 80 km is estimated to be about 1400°C or a little higher (TATSUMI et al., 1983).

GEOTHERMS AND PHASE RELATIONS IN THE MANTLE WEDGE

As discussed above, the temperature of the source region for primary olivine tholeiite magma is esti-

mated to be about 1400°C or a little higher at a depth of about 80 km. Because the depth of the top of the subducting slab below the volcanic front in NE Japan is about 100 km (e.g., YOSHII, 1972, 1979; HASEGAWA et al., 1978; HASEMI et al., 1984), the zone of highest temperature near the volcanic front must be shallower than 100 km, and is most likely 70–80 km deep. The temperature at the Moho is estimated at about 850°C near the coast of the Japan Sea in NE Honshu where the depth of the Moho is about 25 km, as discussed later, and is probably 950–1000°C near the volcanic front where the depth of the Moho is about 35 km. A possible geotherm near the volcanic front (VF) in NE Honshu is drawn based on these estimated temperatures (Figure 2). The geotherm near the coast of the Japan Sea is also constrained by the temperatures of the source regions for the primary high-alumina basalt and alkali olivine basalt magmas. The geotherm near the coast of the Japan Sea is slightly lower than that near the volcanic front. The temperature along the subducting slab is from TOKSOZ et al. (1971).

The phase relations for the mantle wedge in NE Honshu constructed from the experimental data are consistent with the mineral assemblages of peridotite inclusions in the tuff of Ichinomegata crater at the coast of Japan Sea, NE Honshu. The most abundant deep-seated inclusions at Ichinomegata are spinel-lherzolite. Many of them contain amphibole and some show evidence of partial melting such as the presence of interstitial glass and fine-grained crystal aggregates. These inclusions indicate

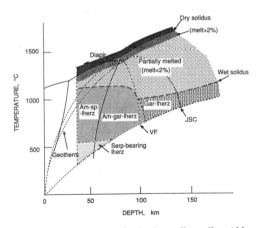

FIG. 2. Phase relations for hydrous lherzolite. Abbreviations: am, amphibole; gar, garnet; lherz, lherzolite; serp, serpentine; sp, spinel; VF, volcanic front; JSC, Japan Sea coast. Wet solidus with about 0.2 weight percent H_2O is from GREEN (1973), dry solidus from TAKAHASHI and KUSHIRO (1983), boundary between spinel- and garnet-lherzolite from O'HARA et al. (1971). See text for geotherms and partially melted region.

168					I. Kushiro

that H_2O exists in the mantle wedge in NE Honshu. The amount of $H_2O(+)$ in the Ichinomegata peridotite inclusions ranges from 0.09 to 0.44 weight percent with an average of 0.25 weight percent (KUNO and AOKI, 1970). These values, however, do not necessarily represent the H_2O content in the upper mantle. As shown below, this water content can be estimated indirectly from seismic wave velocities combined with the results of laboratory measurements of elastic wave velocities at high pressures and temperatures. The amount of H_2O thus estimated is less than 0.2 weight percent in most part of the mantle wedge. The phase relations for lherzolite under water-undersaturated conditions such as those with about 0.2 weight percent H_2O determined by GREEN (1973) would, therefore, be applicable. In Figure 2, the mineral assemblages for lherzolite are shown on the basis of the experimental results of GREEN (1973) for the stability field of amphibole-bearing lherzolite and the hydrous solidus, in addition to those of KUSHIRO and YODER (1966), YODER (1967), MACGREGOR (1974), O'HARA et al., (1971) and OBATA (1976) for those of spinel- and garnet-lherzolites. The water-saturated solidii of some lherzolites are 100–150° lower depending on bulk composition according to the results by MYSEN and BOETTCHER (1975). The anhydrous solidus of lherzolite is based on the experimental results by TAKAHASHI and KUSHIRO (1983). Above the hydrous solidus, lherzolite containing H_2O should be partially melted. The amount of melt increases only slightly with increasing temperature above the hydrous solidus; however, near the anhydrous solidus, the amount of melt would increase rapidly. According to the experimental results of MYSEN and KUSHIRO (1977) on partial melting of a lherzolite in the presence of about 1.9 weight percent H_2O,* the amount of melt increases rapidly near the anhydrous solidus; from 5 to about 20% in a 50° temperature interval just below the anhydrous solidus. Because the amount of H_2O in the mantle wedge is about $1/10$ of that in the experiments by MYSEN and KUSHIRO (1977), the amount of melt formed would be also much smaller. However, the amount of melt must increase rapidly (>2 weight percent) within a 50° temper-

ature interval immediately below the anhydrous solidus of lherzolite.

From the geotherms estimated for the volcanic front and the coast of the Japan Sea and the phase relations for lherzolite, a petrologic model in the mantle wedge of the NE Japan arc has been constructed (Figure 3). Beneath the volcanic front, amphibole-bearing lherzolite may be present in the uppermost part of the mantle, which is underlain by a region of partially melted lherzolite. The amount of melt in this region increases with increasing depth to 70–80 km or the highest temperature region and then decreases rapidly.

The partially melted spinel-lherzolite is underlain by a partially melted garnet-lherzolite at depths greater than about 80 km. Just above the subducted slab, there may be a thin zone of unmelted garnet-lherzolite. The thickness of this unmelted zone, however, depends on the temperatures of the hydrous solidus of lherzolite and the bottom of the mantle wedge and, because of the uncertainties of these temperatures, the thickness of this zone is not certain at present.

As mentioned above, the geotherm near the coast of the Japan Sea would be less steep than that near the volcanic front. In consequence, the amphibole-bearing lherzolite should have a wider stability range and the initiation of partial melting should be deeper. It is also expected that the zone of high degree of partial melting (melt >2%) becomes thinner toward the Japan Sea, although the partially melted region becomes thicker.

In the partially melted region where the amount of melt is less than 2%, the melt may not segregate but may stay along the grain boundaries. The source region of magmas probably contains more melt; that is, it is the region of higher degree of melting (>2%) as shown in Figure 3. As discussed below, this zone would also be the source of the diapirs. The presence of a sharp volcanic front requires termination of the zone of high degree of melting at the volcanic front (Figure 3).

The region of partial melting may extend to the aseismic front which is the ocean-side limit of the non-seismic region of the mantle wedge. The upper limit of partial melting may correspond to the depths at which diapirs stop ascending or the rate of their ascent become much less and magmas segregate from the diapirs. The depth of magma segregation increases from 35 to 60 km from the volcanic front to the coast of the Japan Sea in NE Japan (TATSUMI et al., 1983). The region of partial melting should have low Q values. The low-Q region in the mantle wedge beneath the NE Japan arc (e.g., YOSHII, 1979), though not well defined, is consis-

---

* The amount of H_2O is probably lower. This amount is the total ignition loss including H_2O and CO_2. MYSEN and KUSHIRO (1977) assumed that this ignition loss was all due to H_2O. The composition of the melt formed, however, is slightly more silica-deficient than that formed under volatile-free conditions, suggesting the presence of CO_2 in the starting material.

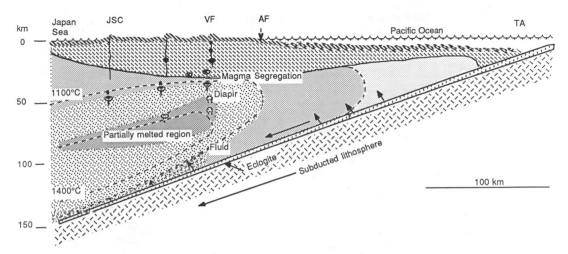

FIG. 3. A schematic cross section of the mantle wedge beneath the NE Japan arc. Abbreviations: TA, trench axis; others as in Figs. 1 and 2. Isotherms and depths of segregation of magmas from diapirs are from TATSUMI *et al.* (1983).

tent with the region of partial melting shown in Figure 3.

THE VELOCITY STRUCTURE OF SEISMIC WAVES AND THE AMOUNT OF MELT IN THE MANTLE WEDGE

The velocity structure in the mantle wedge beneath the NE Japan arc can be determined from the basis of the above petrological model and from the velocity measurements of peridotites at high pressures and temperatures by MURASE and KUSHIRO (1979) and MURASE and FUKUYAMA (1980). MURASE and KUSHIRO (1979) demonstrated that the compressional wave velocity of a spinel-lherzolite at 10 kbar decreases from 7.8 km/sec at 1000°C to 7.4 km/sec at 1225°C and sharply decreases to 7.0 km/sec at 1270°C when the amount of melt is about 3 volume percent. MURASE and FUKUYAMA (1980) measured the shear wave velocity of the same peridotite at 5 and 10 kbar in the temperature range between 1000 and 1400°C. They showed that the velocity decreases from 4.5 km/sec at 1100°C to 4.4 km/sec at 1200°C and to 4.0 km/sec at 1270°C at 10 kbar. The relation between the change of the compressional wave velocity and the degree of melting at 10 kbar is close to that at 5 kbar, although the absolute velocity is about 3% higher at 10 kbar than at 5 kbar. Assuming that the relation between the change of velocity and the degree of melting at 10 kbar can be applied to higher pressures (~30 kbar), the velocity structure for compressional wave in the shallow portion of the mantle wedge beneath the NE Japan arc can be constructed (Figure 4). The effect of pressure on the

velocity obtained at 5 and 10 kbar was extrapolated to estimate the velocities at depths to 100 km. The velocity immediately below the Moho is estimated to be 7.6–7.8 km/sec in most part of the NE Japan arc (NE Honshu), which agrees fairly well with the actual velocities obtained (*e.g.,* YOSHII, 1979). The so-called "anomalous mantle" with relatively low velocity (as low as 7.5 km/sec), which lies just below the Moho in NE Japan, does not require phases with low velocities such as melt and plagioclase, but may be just due to relatively high temperatures in this zone. The velocity in the partially melted region with melt greater than 2% is smaller than 7.3 km/sec.

The three dimensional velocity structures in the mantle wedge beneath NE Honshu have been determined by HASEMI *et al.* (1984) and HASEGAWA *et al.* (1985) using the method of Aki and Lee (1976). The velocity structure in Figure 4 is compared with that obtained recently by HASEGAWA *et al.* (1985) who derived the velocity structures for both P and S waves. The results on the compressional wave velocity obtained by them at latitude of about 39°N (Figure 5) is quite consistent at least qualitatively with the velocity structure obtained by the experimental-petrological method. The velocity structures at different latitudes are more or less similar to that at 39°N. The rate of velocity decrease is, however, slightly smaller in the seismic model compared to the petrologic model; the region with velocity decrease greater than 4% in the seismic model corresponds to that with more than 5% decrease in the petrological model. This suggests that the melt in the mantle wedge assumed in the pet-

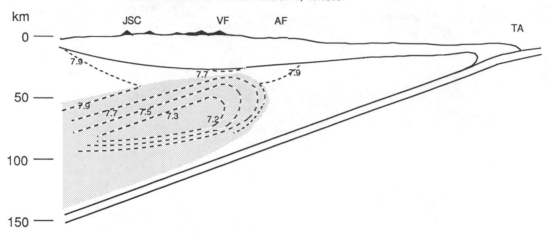

FIG. 4. Velocity structure for compressional wave in the mantle wedge beneath the NE Japan arc
obtained by experimental petrological methods.

rologic model is slightly overestimated; about 1.5% instead of 2% melt would be most consistent with the velocity structure obtained by the seismic observations. Also, the depth of the low-velocity region is about 10 km deeper in the petrologic model than that in the three-dimensional velocity model. A similar velocity structure has been shown in the northern Kurile area by BOLDYREV (1985). In this case, the velocity decrease is about 6% beneath the volcanic zone at depths between 50 and 100 km, and 2% melt may explain such a velocity decrease. However, the velocity measurements and the esti-

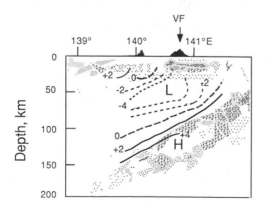

FIG. 5. Velocity structure for compressional wave in the mantle wedge beneath the NE Japan arc at 39°N obtained by the method of AKI and LEE (1976) using local earthquakes (HASEGAWA *et al.,* 1985). Dashed lines are isopleths for compressional wave velocity, and numbers indicate change of the compressional wave velocity (in percent) from the reference isopleths denoted as 0%. L and H denote the regions of low and high velocities, respectively. Dots indicate foci of local earthquakes.

mation of such a small amount of melt include considerable uncertainties, and therefore, the above estimation of the amount of melt also includes significant uncertainties (but probably less than ±1% melt).

The three-dimensional velocity structure for shear waves is similar to that for compressional waves, although the slowest region is located slightly to the ocean-side of the volcanic front. The maximum drop of the velocity is 6%, and the amount of melt for such a decrease estimated from the velocity measurements in the laboratory (MURASE and FUKUYAMA, 1980) is about 2% when the pressure effect is taken into account.

The maximum H_2O content in the mantle wedge can be estimated from the amount of melt and the solubility of H_2O in the melt. The solubility of H_2O in basaltic melts ranges from 10 to 20 weight percent in the pressure range between 10 and 20 kbar at 1000–1200°C. If the melt is saturated with H_2O and the amount of melt is 1.5 volume percent (~1.3 weight percent), the amount of H_2O in most parts of the mantle wedge is 0.07–0.13 weight percent. However, there is a possibility that local enrichment of H_2O may occur in some part of the mantle wedge.

DENSITIES OF MELT AND PARTIALLY MELTED MANTLE WEDGE

The densities of melts of basaltic compositions at pressures to 15 kbar can be obtained by a simple equation derived from the results of the density measurements of basaltic melts at high pressures with the falling/floating sphere method (FUJII and KUSHIRO, 1977; KUSHIRO, 1982, 1983). Within the limited pressure range, the equation is linear with a coefficient of 0.014 g/cm³ kbar (KUSHIRO, 1986a). This is consistent with the compressibility of silicate melts obtained by shock wave measurements on a synthetic silicate melt (RIGDEN et al., 1985). By using this equation and the method of BOTTINGA and WEILL (1970) for estimating the density of silicate melts at 1 atm, the densities of melts at shallow levels in the upper mantle can be calculated. The densities of anhydrous partial melts formed in the upper mantle thus calculated increases from 2.78 g/cm³ at 8 kbar to 2.92 g/cm³ at 15 kbar along the anhydrous solidus of peridotite (KUSHIRO, 1986a). Extrapolating the above equation to higher pressures, the density of partial melts is calculated to be 2.96 g/cm³ at 20 kbar. The effect of H₂O on the density of silicate melts can be evaluated by the partial molar volume of H₂O given by BURNHAM and DAVIS (1969) and experimental data (KUSHIRO, 1986b); for example, 1 weight percent H₂O reduces the density of a basaltic melt by 0.05 g/cm³ at 10 kbar.

The density variations of the mantle wedge beneath the NE Japan arc have been calculated based on the geotherms at the volcanic front and near the coast of Japan Sea. The compressibility and thermal expansivity of olivine and pyroxene (CLARK, 1966) have been used to calculate the densities of unmelted lherzolite and the crystalline portion of the partially melted lherzolite in the mantle wedge. The thermal expansivity of olivine above 1200°C used for the calculation is the extrapolated value. Figure 6 shows the results of these calculations.

At the volcanic front, the density of unmelted lherzolite is 3.23 g/cm³ just below the Moho, decreases slightly to 3.215–3.22 g/cm³ at 80–100 km where the temperature is highest (i.e., ~1400°C), and increases considerably with increasing depth below the minimum. In the presence of melt, the density decrease is more pronounced; if the maximum amount of melt is 3% at the highest temperature (>2% in the region of high-degree of partial melting), the density at the minimum is about 3.20 g/cm³. Such a density reversal, though it is not large, may cause a density instability and trigger the ascent of diapirs. Once diapirs start ascending adiabatically, their bulk density decreases, although the temperature of the diapirs decreases. TATSUMI et al. (1983) calculated the rate of temperature decrease of a diapir, taking into account the heat of fusion of basalt at high pressures measured by FUKUYAMA (1985b). The density decrease of a diapir based on their calculation is shown in Figure 6. A diapir with a temperature of 1320°C at a depth of about 40 km, which contains about 20 weight percent melt as es-

timated on the basis of the experimental results by MYSEN and KUSHIRO (1977) has a density of about 3.14 g/cm³.

Near the coast of the Japan Sea, the density of unmelted lherzolite is essentially constant to the depth of about 100 km and increases with increasing depth below 100 km. The amount of melt in this region is less than that at the volcanic front, so that the effect of melt on the density variation is smaller. The density reversal is, therefore, very small and ascent of diapirs may be less frequent and less intense than near the volcanic front. In addition, the rate of increase in the amount of melt in the diapirs is also smaller because the adiabatic ascending path of the diapirs from the region of highest temperature is close to the anhydrous solidus of peridotite, so that the amount of melt remains relatively small. This may explain the fact that the volcanic activity and the volume of the erupted materials decrease from the volcanic front toward the coast of Japan Sea across the NE Japan arc.

PETROLOGY OF THE LOWER CRUST IN NE AND SW JAPAN ARCS

Deep-seated inclusions are found in alkali basalts and tuffs in many localities in the Japanese Islands. Among these localities, Ichinomegata at the coast of the Japan Sea, NE Honshu (Figure 1), is best known, because the inclusions are abundant, fresh and variable in rock types and have been studied in detail by several investigators (KUNO, 1967; KUNO and AOKI, 1970; AOKI, 1971; TAKAHASHI, 1980, 1986; TANAKA and AOKI, 1981; FUKUYAMA, 1985a). Ichinomegata is one of the three tuff cones formed by an explosive eruption about 10,000 years ago and is one of the rare localities for deep-seated inclusions in island arcs. Most other localities in Japan are not in the typical island arc geological setting, but rather are in the back arc basin settings.

The inclusions in the Ichinomegata tuff cone consist of ultramafic rocks such as spinel-lherzolite with or without amphibole and plagioclase, harzburgite, wehrlite, and pyroxenite, and mafic rocks such as gabbro, granulite, amphibolite, and low grade metamorphic rocks. Most of the latter rocks are of crustal origin, and have not been studied in detail. AOKI (1971) analyzed minerals and bulk rocks of some gabbros and amphibolite, and TANAKA and AOKI (1981) analyzed trace elements (REE and Ba) in some gabbroic inclusions. Recently, FUKUYAMA (1985a) analyzed 57 gabbroic inclusions by the XRF method. In the present study the author re-examined under the microscope more than 200 mafic inclusions collected by H. Kuno,

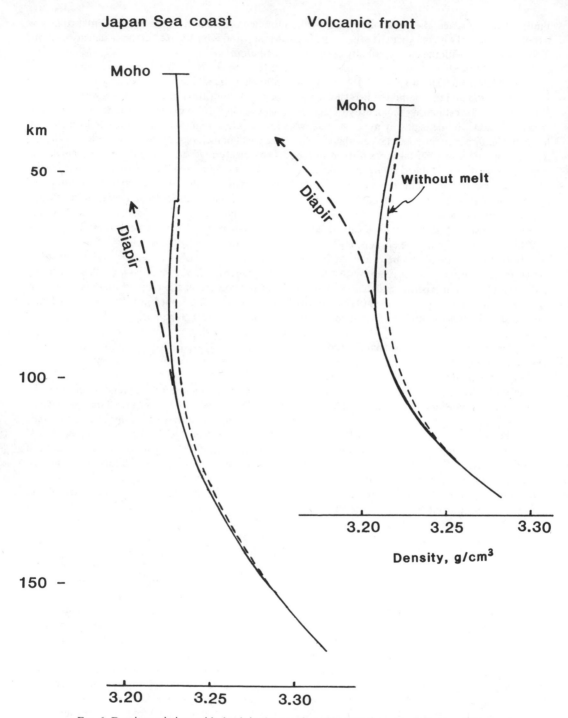

FIG. 6. Density variations with depth in the mantle wedge near the volcanic front and the coast of the Japan Sea and those of the ascending diapirs. The density has been calculated on the basis of the model shown in Figure 3. Dashed lines are the density variations without melt.

K. Ozawa and T. Fujii. Minerals in some of them have been analyzed with the electron microprobe. According to AOKI (1971), FUKUYAMA (1985a) and this work, the mineral assemblages of the mafic inclusions are classified into the following different types; (1) ol + cpx + opx + hb + pl + mt, (2) ol

+ cpx + hb + pl + mt, (3) cpx + opx + hb + pl + mt, (4) cpx + hb + pl + mt, (5) opx + hb + pl + mt, and (6) hb + pl + mt (Abbreviations: cpx, clinopyroxene; hb, hornblende; mt, magnetite; ol, olivine; opx, orthopyroxene; pl, plagioclase). Apatite is present in most of these inclusions as an accessory mineral. Green spinel exists in some them, and is especially abundant in the assemblages (1) and (2) as a reaction product between olivine and plagioclase. Some of these inclusions show igneous textures and are gabbros but others show recrystallized textures and are metamorphic rocks. The gabbroic inclusions are in general very similar to the inclusions found in the Lesser Antilles lavas (e.g., WAGER, 1962; LEWIS, 1973; ARCULUS and WILLIS, 1980); in particular the assemblages (1), (2) and (3) are identical to those described in St. Kitts, Dominica, and St. Vincent. Amphibole-bearing inclusions have been also reported in calc-alkaline volcanoes on Adak Island in the Aleutian arc (CONRAD and KAY, 1985; KAY et al., 1986; DEBARI et al., 1986), and are considered to be cumulates from calc-alkaline magmas. Phenocryst assemblages in calc-alkaline andesites in Shirouma-Oike, Myoko and other volcanoes in central Honshu (SAKU-YAMA, 1978; 1981; and 1983b) are the same as the above six assemblages. Most of the metamorphic mafic inclusions have been described as amphibolite (AOKI, 1971; TAKAHASHI, 1978; FUKUYAMA, 1985a), and granulites have not been clearly mentioned among the Ichinomegata mafic inclusions, except IRVING (1974) who pointed the presence of granulite in these inclusions. In the present work, it was found that about 20% of the metamorphic inclusions are granulite consisting of aluminous clinopyroxene and orthopyroxene, plagioclase, amphibole with or without green spinel and magnetite. They show granular textures. Pyroxene crystals are in contact with one another with about 120° angle, indicating recrystallization to textural equilibrium. The granulites have mineral assemblages (1), (3) and (5); however, some are almost free of amphibole. Most of amphibole crystals contain numerous fine-grained inclusions of pyroxene and spinel and appear to be partially broken down. Some mafic inclusions contain olivine, which is often surrounded or replaced by pyroxene-spinel symplectite, especially near plagioclase crystals. The analyses of pyroxene (K. OZAWA and K. AOKI, personal communication, 1985) indicate that clinopyroxene ($\sim Ca_{49}Mg_{40}Fe_{11}$) and orthopyroxene ($\sim Ca_{1.1}Mg_{76.0}Fe_{22.9}$) contain up to 7.2 and 5.2 weight percent Al_2O_3, respectively. They are homogeneous in single rock specimens. Temperatures estimated by pyroxene geothermometry (WOOD

and BANNO, 1973; WELLS, 1977) range from 830° to 870° depending on the inclusions. These temperatures are within the temperature range of the granulite facies. More detailed descriptions will be presented by Ozawa (in preparation).

The granulite inclusions with the orthopyroxene-clinopyroxene-spinel assemblage must have been formed or equilibrated on the higher-pressure or lower-temperature side of the reaction olivine + calcic plagioclase $\rightleftharpoons$ orthopyroxene + clinopyroxene + spinel. The pressure-temperature conditions for this reaction has been determined experimentally by KUSHIRO and YODER (1966), YODER (1967), GREEN and HIBBERSON (1970), EMSLIE (1970) and HERZBERG (1978). According to these studies the reaction takes place at 8–9 kbar at 1100°C and 7–8 kbar at 800°C for basaltic compositions (Figure 7). The inclusions containing relic olivine surrounded by spinel-pyroxene symplectite must have been cooled from a higher temperature to lower temperatures crossing the curve for this reaction. The absence of garnet gives a further constraint on the pressure; two-pyroxene granulite transforms to garnet granulite at about 10 kbar at 1100°C and about 9 kbar at 900°C (IRVING, 1974).

The lower crust of the Ichinomegata or NE Honshu has been considered to consist of amphibolite and hornblende gabbros (AOKI, 1971; TAKAHASHI, 1978, 1986). However, it is most likely that the lowest part of the crust in this region consists of granulite, although it may belong to the hornblende subfacies.

Some mafic inclusions have a greenschist facies assemblage consisting of chlorite, epidote, actinolitic amphibole and albite. They often show gabbroic textures; large elongated feldspar crystals and aggregates of mafic minerals, the form of which appears to be subhedral pyroxene or hornblende in gabbros. Such inclusions may have been originally gabbros which recrystallized under greenschist facies conditions.

FUKUYAMA (1985a) analyzed 57 mafic inclusions (mostly gabbroic inclusions) and found that their chemical compositions lie essentially between those of amphibole and plagioclase; they were interpreted to be cumulates from hydrous basaltic magmas. The chemical compositions of the metamorphic mafic inclusions (AOKI, 1971; FUKUYAMA, 1985a) are included in the compositional range of the gabbroic inclusions and they cannot be distinguished from each other.

It is strongly suggested that the metamorphic mafic inclusions were originally hornblende gabbro and they were later metamorphosed to granulite, amphibolite and greenschist depending on the depth

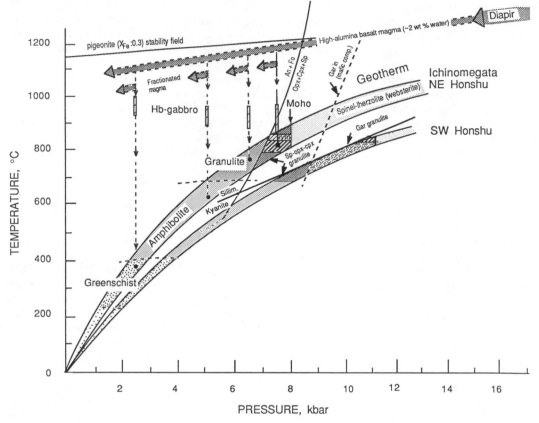

FIG. 7. Geotherms and rock types in the crusts at Ichinomegata in NE Japan and in central Chugoku district in SW Japan. The boundary curve for An + Fo ⇌ Opx + Cpx + SP is from KUSHIRO and YODER (1966), OBATA (1976), and HERZBERG (1978), the sillimanite-kyanite transition curve is from HOLDAWAY (1971), and the garnet-in curve for basaltic composition from IRVING (1974).

of metamorphism. The final equilibration temperature is assumed to lie on the geotherm. KAY and KAY (1985) and KAY et al. (1986) suggested that amphibole-free pyroxene granulites in the Aleutian arc could be remnants from the original oceanic crust on which the Aleutian volcanic islands were formed. Although this possibility cannot be entirely discarded in the NE Japan arc, the presence of amphibole with various degrees of breakdown and the bulk chemical compositions strongly suggest that the granulites were originally hornblende gabbros formed from arc magmas.

The geotherm has been estimated on the basis of the equilibrium temperature of granulite and the fact that there are many plagioclase-bearing lherzolite inclusions. The equilibration temperature of spinel- and plagioclase-lherzolites ranges from 800° to 1000°C (TAKAHASHI, 1980). Because these lherzolite inclusions are most probably upper mantle materials, the upper most mantle beneath Ichinomegata would have been once in the stability field

of olivine and plagioclase. On the other hand, as mentioned above, spinel-two pyroxene granulites were probably materials of the lower part of the crust. This assemblage is stable on the higher pressure side of the above reaction where olivine + plagioclase assemblage is unstable. Such an apparent inconsistency may be explained by the compositional effect on the reaction; the reaction would take place at slightly higher pressures for ultramafic (more magnesian) compositions than for mafic compositions. Another possibility is that the upper mantle is more frequently heated locally by uprising hot diapirs or magmas than the lower crust, so that it becomes located to the higher temperature side of the reaction curve. In any case the Moho must be close to the boundary between the stability fields of olivine + calcic plagioclase and spinel + orthopyroxene + clinopyroxene assemblages. The equilibrium temperature of the spinel-two pyroxene granulites is 830°–870°C. The pressure range of the boundary within this temperature range is 7–8 kbar or a depth range of 20–25 km, which is not inconsistent with the depth of the Moho in this region estimated from seismic observations (YOSHII, 1979).

A possible geotherm at Ichinomegata is drawn in Figure 7 assuming that the spinel-two pyroxene granulite lies near the base of the crust, the condition of which is very close to that of the reaction of olivine and plagioclase. The geotherm should also extend to the source region of the basalt magmas (*i.e.,* ~1400°C at depth of about 100 km). This geotherm would be applicable to that near the coast of the Japan Sea in the NE Japan arc where the heat flow data, the distribution of volcanoes and the chemical compositions of volcanic rocks are similar to those near Ichinomegata. Near the volcanic front, however, the geotherm may be steeper. As mentioned before, the temperature of the source region for basalt magmas is higher near the volcanic front than near the coast of the Japan Sea when the temperature at the same depth is compared. This indicates that the temperatures of basalt magmas (at least those which formed gabbroic inclusions) were higher than those of basaltic magmas formed near the coast of Japan Sea. Other evidence which might support the higher geotherm near the volcanic front is that large volumes Quaternary felsic magmas have been erupted mostly near the volcanic front. At least some of such felsic magmas would have been formed by melting of the lower crust. Heat supply from the mantle must be significantly larger for the generation of the large volume of felsic magmas and in consequence, the geotherm would have become steeper.

In SW Honshu the geotherm is lower than that at Ichinomegata. Deep-seated inclusions in basaltic rocks in the Chugoku district, SW Honshu (Figure 1), have been studied by several investigators (MURAKAMI, 1954, 1975; KURASAWA *et al.,* 1961; KOYAGUCHI and FUJII, 1981). In the present study, inclusions of crustal rocks in alkali basalts in the central Chugoku district collected by M. Iwamori and the author have been examined. Among these inclusions, pyroxene gabbros and granulites are most abundant. Granulites are mostly quartz-bearing and often contain aluminous phases such as spinel, corundum, kyanite and garnet (MURAKAMI, 1954; KOYAGUCHI and FUJII, 1981; IWAMORI, 1986). No sillimanite has been found among the inclusions examined. Primary amphibole is absent in most of the inclusions. The presence of kyanite in granulites provides a constraint on the temperature of equilibration of these rocks. In Figure 7, the sillimanite-kyanite transition curve by HOLDAWAY (1971) is given. The temperature of a garnet-bearing granulite has been estimated by KOYAGUCHI and FUJII (1981) at about 820°C from plagioclase geothermometry. If kyanite-bearing granulite was equilibrated at a similar temperature, the geotherm must be very close to but at the lower temperature side of the kyanite-sillimanite transition curve, as shown in Figure 7. This geotherm, which is late Pliocene, is significantly lower than the Quaternary geotherms at Ichinomegata and that near the volcanic front in the NE Japan arc. In the central Chugoku district, necks and minor lava flows of alkali basalts of late Pliocene age are distributed widely, but no Quaternary volcanoes are present and the heat flow is low compared to the northern part of the Chugoku district where several Quaternary volcanoes are present (UYEDA and HORAI, 1964). Eruption of a small amount of alkali basalt magmas is a feature often observed in back-arc basin and continental regions. As shown in Figure 7, the depth of the Moho must be deeper than 30 km in the central Chugoku district; it is significantly deeper than that at Ichinomegata. This is consistent with the thickness of the crust in the SW and NE Japan arcs determined from seismic wave velocities (*e.g.,* KANAMORI, 1963).

The geotherms in NE and SW Honshu are similar to the pressure-temperature conditions for some of the metamorthic facies series by MIYASHIRO (1961). The geotherm in NE Honshu especially that near the volcanic front is close to but slightly to the lower temperature side of the pressure-temperature path for the low pressure/temperature type metamorphic facies series represented by metamorphism in the Abukuma plateau studied by MIYASHIRO (1958) and SHIDO (1958). On the other hand, the geotherm in SW Honshu is close to the intermediate pressure/temperature type represented by metamorphism in the Grampian Highlands, Scotland studied by BARROW (1893), TILLEY (1925) and CHINNER (1960) among others. Such similarities of the geotherms might be applicable to the tectonic and geological settings of these old metamorphic terranes.

CHEMICAL COMPOSITIONS OF THE LOWER CRUSTAL MATERIALS

The chemical compositions of the inclusions of the lower crustal rocks in the NE and SW Honshu are compared. The analyses of the gabbroic inclusions at Ichinomegata have been made by KUNO (1967), AOKI (1971) and FUKUYAMA (1985a) with the conventional wet chemical analysis and the XRF methods. The average chemical composition by wet chemical analysis of the mafic inclusions including gabbroic rocks, granulite and amphibolite is given in Table 2. The composition is low in SiO_2 and alkalies and high in Al_2O_3 and CaO, and is not close to those of any basalt. The analyses are enriched in plagioclase and hornblende components. However, none of these analyses shows normative corundum. Most of the analyses actually plot in the region between calcic plagioclase and hornblende (FUKUYAMA, 1985a), and TANAKA and AOKI (1981) showed that two of three gabbro inclusions analyzed have positive Eu anomalies. It is probable that most of the mafic inclusions were formed by the accumulation of hornblende and calcic plagioclase crystallized from hydrous basaltic magmas. Under lower crustal conditions, calcic plagioclase floats in anhydrous basaltic magmas (KUSHIRO and FUJII, 1977); however, the density of hydrous basaltic magmas with an H_2O content of 1.5 weight percent is less than that of calcic plagioclase with An > 60 (KUSHIRO, 1986a). Cumulates of hornblende gabbro can, therefore, be formed from hydrous basaltic magmas containing 1.5 weight percent H_2O even at the base of the crust. The chemical composition of the residual liquids after subtraction of the average composition of the mafic inclusions from the primary high-alumina basalt is calc-alkalic andesite. It is certainly possible that calc-alkalic andesite magma can be formed from high-alumina basalt magma by removal of amphibole, magnetite and plagioclase with or without clinopyroxene.

The chemical compositions of 18 gabbroic and granulitic inclusions in basalts in SW Honshu have been analyzed by MURAKAMI (1975). Their average chemical composition is given in Table 2. It is higher in SiO_2 and alkali and poorer in CaO and MgO than that of the inclusions from Ichinomegata. Six

Table 2. Average chemical composition of deep-seated crustal rocks in NE and SW Honshu

	NE Honshu* (29)	SW Honshu† (18)
SiO_2	43.99	50.98
TiO_2	0.92	0.89
Al_2O_3	18.06	18.28
Fe_2O_3	4.62	4.87
FeO	6.38	4.43
MnO	0.16	0.22
MgO	10.91	5.73
CaO	13.18	10.74
Na_2O	1.40	3.10
K_2O	0.22	0.67
P_2O_5	0.16	0.09
Total	100.00	100.00

* KUNO (1967), AOKI (1970) and FUKUYAMA (1985a). XRF analyses are not included.

† MURAKAMI (1975). Numbers in parentheses are the numbers of analyses averaged.

of the granulite inclusions are, however, high in Al_2O_3 relative to CaO and alkalies and contain significant amounts of normative corundum in their norm. As mentioned above, some granulite inclusions, which were not analyzed, contain aluminous phases such as garnet, kyanite and corundum. These inclusions and especially those containing latter two minerals would also have normative corundum. The presence or absence of alumina-excess inclusions is a distinct difference between the inclusions at Ichinomegata and those in the SW Honshu.

Formation of excess-alumina rocks by magmatic processes is rather difficult. Partial melting of peridotite or gabbro does not generate alumina-excess liquids at least under anhydrous conditions in the pressure range 10–30 kbar. Under hydrous conditions, however, it is possible to produce alumina-excess liquids by fractional crystallization when the reaction plagioclase + orthopyroxene $\rightleftharpoons$ clinopyroxene + liquid takes place (KUSHIRO and YODER, 1972; ELLIS and THOMPSON, 1986). Alumina-excess calc-alkalic andesite and dacite may have been formed by such a reaction. KUNO (1950) observed a fractional crystallization trend toward normative corundum compositions in the hypersthenic rock series ($\approx$ calc-alkalic rock series) in the Izu-Hakone region. The amount of normative corundum in the calc-alkalic andesite and dacite is, however, a few percent at most. On the other hand, the amount of normative corundum in the granulites mentioned above is generally larger. In addition, most of the inclusions in the SW Honshu do not contain primary hydrous minerals and may have been formed under nearly anhydrous conditions. Some of them are high in SiO_2 and the compositions are rather similar to those of pelitic sediments. It is most probable that aluminous sediments were metamorphosed to granulite under the lower crustal conditions. MURAKAMI (1975) mentioned this possibility based on the presence of rounded zircon crystals in some of the granulite inclusions. It is, thus, most probable that the lower crust of SW Honshu consists of gabbroic and metamorphosed sedimentary rocks. Some of the gabbroic rocks have a two-pyroxene spinel assemblage, and were probably recrystallized under the granulite facies conditions, similar to those in the Ichinomegata region.

A MODEL FOR THE FORMATION OF LOWER CRUSTAL ROCKS

As mentioned above, most of the mafic inclusions at Ichinomegata were originally hornblende gabbro formed by accumulation from hydrous basaltic magmas. The original magma of the cumulates is probably hydrous high-alumina basalt magmas because the Quaternary basalts erupted near Ichinomegata are high-alumina basalts. The condition of generation of the high-alumina basalt magma is about 40 km deep and 1350°C in the presence of a small amount of H_2O (1–2 weight percent). As mentioned above, the density of hydrous tholeiitic magmas with H_2O contents greater than 1.5 weight percent is smaller than that of calcic plagioclase ($>An_{65}$) at 8 kbar (KUSHIRO, 1986a), indicating that calcic plagioclase sinks in basaltic magmas even near the base of the crust. Based on thermodynamic calculations, POWELL (1978) suggested that the gabbroic inclusions in St. Vincent, Dominica and St. Kitts where mineral assemblages are the same as those of the Ichinomegata gabbroic inclusions were formed at pressures between 6 and 8 kbar (20–30 km). The cumulates containing calcic plagioclase would have been formed at various depths within the crust in the temperature range 900–1000°C. Although some cumulates were preserved, others were crystallized to metamorphic rocks with different metamorphic facies upon cooling toward the stable geotherm. These processes are illustrated in Figure 7.

The crust in NE Honshu must have been thickened essentially by the input of basalt magmas from the upper mantle, as illustrated in Figure 8. The basalt magmas were fractionated to produce cumulates of gabbros and residual magmas of andesite, dacite and rhyolite compositions. Apparently no sedimentary rocks were involved in the development of the major part of the crust. Only in the shallow parts or near the surface, sedimentary rocks are involved, but they are products of recycling processes near the surface and would not contribute to the overall thickening of the crust in general. Remelting of the lower part of the crust may have produced granitic magmas as inferred from the intrusions of Tertiary granites in the NE Honshu. Even if it occurred, however, it should not contribute to the thickening of the crust. Near the coast of the Japan Sea, magmas of high-alumina basalt or alkali olivine basalt with higher H_2O contents are supplied from the upper mantle into the crust, whereas near the volcanic front, olivine tholeiite magmas with lower H_2O contents are supplied. The cumulates formed within the crust must be also different; more hornblende gabbros near the coast of Japan Sea and more pyroxene gabbro near the volcanic front. In the Hakone volcano, for example, some andesitic breccias contain many pyroxene gabbro inclusions. They often include inverted pigeonite which has never been reported in hornblende gabbros at Ichinomegata and other areas near the coast of Japan Sea. Apparently, the magmas from which pyroxene gabbros crystallized were higher in temperature and contain less H_2O.

DEVELOPMENT OF "IGNEOUS CRUST"

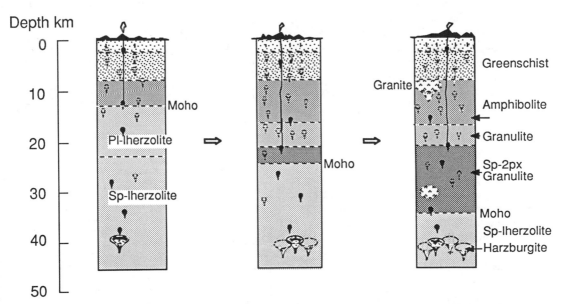

FIG. 8. Schematic pictures illustrating the development of 'igneous crust' in the volcanic zone of NE Japan and island arcs in general.

On the other hand, the deep-seated inclusions in SW Honshu would be partly of sedimentary origin as discussed above. Emplacement of the sedimentary rocks near the base of the crust must involve large scale tectonic movements. One possible mechanism is subduction of the oceanic plate, by which the sedimentary rocks as well as the volcanic rocks are accreted to the crust of SW Honshu. In the southern parts of SW Honshu (*i.e.,* southern Shikoku and Kyushu), the Shimanto group strata ranging in age from late Jurassic to early Tertiary are widely distributed. They are mainly sedimentary rocks with some basaltic rocks and are considered to have been accreted in association with subduction processes (*e.g.,* SUZUKI and HADA, 1979).

The evidence which may support the involvement of sedimentary rocks in the deep crust of the southern regions of SW Japan arc is that the Tertiary ilmenite-series granites, which are similar but not identical to the S-type granites (CHAPPEL and WHITE, 1974) or peraluminous granites, are distributed in these regions. On the other hand, the Tertiary magnetite-series granites, which are similar to the I-type granite, are distributed in the northern part of SW Honshu (ISHIHARA, 1977; TAKAHASHI *et al.,* 1980). It is suggested that these ilmenite-series granites were formed by remelting of the rocks of sedimentary origin or by reaction of basaltic magmas with these rocks in the lower part of the crust in this region (TAKAHASHI, 1986). On the other hand, magnetite-series granites were probably formed by remelting of igneous rocks or by fractional crystallization of basaltic magmas without significant interaction with sedimentary rocks.

In the island arcs where both magmatic activity and accretionary process take place, granite magmas of two different types (S and I types or ilmenite and magnetite series) would be generated; the I-type granites are formed mainly in the Quaternary volcanic zone and the S-type granites in the presently non-volcanic zone closer to the trench. ISHIHARA (1980) pointed out that the magnetite-series granites are generally distributed in the back arc basin side, whereas the ilmenite-series granites in the ocean side in island arcs, forming a double granite belt. The double granite or paired granite belts may be formed in matured island arcs within a limited time period. The paired granite belts appear to exist not only in the island arcs, but also in some parts in the continents such as SE Australia and the eastern part of China. Some of those might represent ancient island arc systems.

Acknowledgments—The author thanks K. Ozawa, K. Aoki and M. Iwamori for providing mafic inclusions from Ichinomegata and central Chugoku district and microprobe analyses of minerals in some of the inclusions. He is indebted to C. T. Herzberg and R. W. Kay for critical reading of the manuscript and R. J. Arculus, J. Gill, H. Nagahara, K. Ozawa, Y. Tatsumi and S. Uyeda for discussion.

REFERENCES

AKI K. and LEE W. H. K. (1976) Determination of three-dimensional velocity anomalies under a seismic array using first P arrival times from local earthquakes 1. A homogeneous initial model. *J. Geophys. Res.* **81,** 4381–4399.

AOKI K. (1970) Petrology of mafic inclusions from Ichinome-gata, Japan. *Contrib. Mineral. Petrol.* **30,** 314–331.

ARCULUS R. J. and WILLS K. J. A. (1980) The petrology of plutonic blocks and inclusions from the Lesser Antilles Island arc. *J. Petrol.* **21,** 743–799.

BARROW G. (1893) On an intrusion of muscovite-biotite gneiss in the southeastern Highlands of Scotland, and its accompanying metamorphism. *Quart. J. Geol. Soc. London* **44,** 350–358.

BOTTINGA Y. and WEIL D. F. (1970) Densities of liquid silicate systems calculated from partial molar volumes of oxide components. *Amer. J. Sci.* **269,** 169–182.

BOLDYREV S. A. (1985) Mantle heterogeneities within active margins of the world oceans and their seismological characteristics. *Tectonophys.* **112,** 255–276.

BURNHAM C. W. and DAVIS N. F. (1969) Partial molar volume of water in albite melts (abstract). *Trans. Amer. Geophys. Union* **50,** 338.

CHAPPELL B. W. and WHITE A. J. R. (1974) Two contrasting granite types. *Pacific Geol.* **8,** 173–174.

CHINNER G. A. (1960) Pelitic gneisses with varying ferrous/ferric ratios from Glen Clova, Augus, Scotland. *J. Petrol.* **1,** 178–217.

CLARK S. P. (1966) Handbook of Physical Constants. *Geol. Soc. Amer. Mem.* **97.**

CONRAD W. and KAY R. W. (1984) Ultramafic and mafic inclusions from Adak Island: Crystallization history and implications for the nature of primary magmas and crustal evolution in the Aleutian arc. *J. Petrol.* **25,** 88–125.

DEBARI S. M., KAY S. M. and KAY R. W. (1986) Ultramafic xenoliths from Adagdak volcano, Adak, Aleutian Islands, Alaska: Deformed igneous cumulates from the Moho of an island arc. *J. Geol.* (In press).

ELLIS D. J. and THOMPSON A. B. (1986) Subsolidus and partial melting reactions in the quartz-excess CaO + MgO + Al$_2$O$_3$ + SiO$_2$ + H$_2$O system under water-excess and water-deficient conditions to 10 kb: some implications for the origin of peraluminous melt from mafic rocks. *J. Petrol.* **27,** 91–121.

EMSLIE R. E. (1970) Liquidus relations and subsolidus reactions in some plagioclase-bearing systems. *Carnegie Inst. Wash. Yearb.* **69,** 148–155.

FUJII T. and KUSHIRO I. (1977) Density, viscosity, and compressibility of basaltic liquid at high pressures. *Carnegie Inst. Wash. Yearb.* **76,** 419–424.

FUKUYAMA H. (1985a) Gabbroic inclusions of Ichinome-gata tuff cone: Bulk chemical composition. *J. Fac. Sci. Univ. Tokyo, Sec. II* **22,** 67–80.

FUKUYAMA H. (1985b) Heat of fusion of basaltic magma. *Earth Planet. Sci. Lett.* **73,** 407–414.

GREEN D. H. (1973) Experimental melting studies on a model upper mantle composition of high pressure under water-saturated and water-undersaturated conditions. *Earth Planet. Sci. Lett.* **10,** 37–45.

GREEN D. H. and HIBBERSON W. O. (1970) The stability of plagioclase in peridotite at high pressure. *Lithos* **3**, 209–221.

GREEN D. H. and RINGWOOD A. E. (1967) The genesis of basaltic magmas. *Contrib. Mineral. Petrol.* **15**, 103–190.

GREEN T. H., GREEN D. H. and RINGWOOD A. E. (1967) The origin of high-alumina basalts and their relationships to quartz tholeiites and alkali basalts. *Earth Planet. Sci. Lett.* **2**, 41–51.

HASEGAWA A., UMINO N. and TAKAGI A. (1978) Double-planned deep seismic zone and upper mantle structure in the northeastern Japan Arc. *Geophys. J. Roy. Astron. Soc.* **54**, 281–296.

HASEGAWA A., OBARA K. and TAKAGI A. (1985) P and S wave low-velocity zones existing under active volcanoes (in Japanese). *Rep. Nat. Diast. Res.* 102–110.

HASEMI A. H., ISHII H. and TAKAGI A. (1984) Fine structure beneath the Tohoku district, northeastern Japan arc, as derived by an inversion of P-wave arrival times from local earthquakes. *Tectonphys.* **101**, 245–265.

HERZBERG C. T. (1978) The bearing of phase equilibria in simple and complex systems on the origin and evolution of some well-documented garnet-websterite. *Contrib. Mineral. Petrol.* **66**, 375–382.

HOLDAWAY M. J. (1971) Stability of andalusite and the aluminium silicate phase diagram. *Amer. J. Sci.* **271**, 97–131.

IRVING A. J. (1974) Geochemical and high pressure experimental studies of garnet pyroxenite and pyroxene granulite xenoliths from Delegate basaltic pipes, Australia. *J. Petrol.* **15**, 1–40.

ISHIHARA S. (1977) The magnetite-series and ilmenite-series granitic rocks. *Mining Geol.* **27**, 293–305.

ISHIHARA S. (1980) Granite and rhyolite (in Japanese). In *Chikyukagaku (Earth Sciences)*, vol. 15, Geology of Japan, (eds. K. KANMARA, M. HASHIMOTO and T. MATSUDA) 105–141. Iwanami, Tokyo.

IWAMORI M. (1986) Petrograhgy of alkali basalts of Kibi plateau, southwest Japan. B.Sc. Thesis, Univ. Tokyo.

KANAMORI H. (1963) Study on the crust-mantle structure in Japan. Parts 1, 2 and 3. *Bull. Earthq. Res. Inst.* **41**, 743–749, 761–779 and 801–818.

KAY R. W. (1985) Island arc processes relevant to crustal and mantle evolution. *Tectonophys.* **112**, 1–15.

KAY R. W., RUBENSTONE J. L., WASSERBURG G. J. and KAY S. M. (1986) Aleutian terranes from Nd isotopes. *Nature*, (In press).

KAY S. M. and KAY R. W. (1985) Role of crystal cumulates and the oceanic crust in the formation of the lower crust of the Aleutian arc. *Geology* **13**, 461–464.

KOTOKU M. (1986) Geology and petrology of Ryozen volcanic rocks. M.Sc. Thesis, Univ. Tokyo.

KOYAGUCHI T. and FUJII T. (1981) Garnet-bearing inclusions in basalt in Hato-zima, SW Honshu. *J. Geol. Soc. Japan* **87**, 489–492.

KUNO H. (1950) Petrology of Hakone volcano and adjacent areas, Japan, *Bull. Geol. Soc. Amer.* **61**, 957–1020.

KUNO H. (1959) Origin of Cenozoic petrographic provinces of Japan and surrounding areas. *Bull. Volcanol., Sec. II* **20**, 37–76.

KUNO H. (1960) High-alumina basalt. *J. Petrol.* **1**, 125–145.

KUNO H. (1967) Mafic and ultramafic nodules from Ichinomegata, Japan. In *Ultramafic and Related Rocks*, (ed. P. J. WYLLIE), 337–342. Wiley, New York.

KUNO H. and AOKI K. (1970) Chemistry of ultramafic nodules and their bearing on the origin of basaltic magmas. *Phys. Earth Planet. Interior* **3**, 273–301.

KURASAWA H., NOZAWA T. and TAKAHASHI K. (1961) Gneissose xenoliths in basalt in the northern part of Yamaguchi Prefecture (in Japanese). *J. Geol. Soc. Japan* **67**, 184–185.

KUSHIRO I. (1968) Compositions of magmas fromed by partial zone melting of the earth's upper mantle. *J. Geophys. Res.* **73**, 619–634.

KUSHIRO I. (1982) Density of tholeiite and alkali basalt magmas at high pressures. *Carnegie Inst. Wash. Yearb.* **81**, 305–309.

KUSHIRO I. (1983) On the lateral variations in chemical composition and volume of Quaternary volcanic rocks across Japanese arcs. *J. Volcanol. Geotherm. Res.* **18**, 435–447.

KUSHIRO I. (1986a) Density of basalt magmas at high pressures and its petrological application. In *Physical Chemistry of Magmas*, (eds. L. L. PERCHUK and I. KUSHIRO), Springer-Verlag, Berlin. (In press).

KUSHIRO I. (1986b) Viscosity of partial melts in the upper mantle. *J. Geophys. Res.* **91**, 9343–9350.

KUSHIRO I. and YODER H. S. (1966) Anorthite-forsterite and anorthite-enstatite reactions and their bearing on the basalt-eclogite transformation. *J. Petrol.* **7**, 337–362.

KUSHIRO I. and YODER H. S. (1972) Origin of calc-alkalic peraluminous andesite and dacite. *Carnegie Inst. Wash. Yearb.* **71**, 411–413.

KUSHIRO I. and FUJII T. (1977) Floatation of plagioclase in magma at high pressures and its bearing on the origin of anorthosite. *Proc. Japan Acad., Ser. B*, **53**, 262–266.

LEWIS J. F. (1973) Petrology of the ejected plutonic blocks of the Soufriere volcano, St. Vincent, West Indies. *J. Petrol.* **14**, 81–112.

MACGREGOR I. D. (1974) The system $MgO-Al_2O_3-SiO_2$: solubility of Al_2O_3 in enstatite for spinel and garnet peridotite compositions. *Amer. Mineral.* **59**, 110–119.

MIYASHIRO A. (1958) Regional metamorphism of the Gosaisyo-Takanuki district in the central Abukuma Plateau. *J. Fac. Sci. Univ. Tokyo, Sec. 2*, **11**, 219–272.

MIYASHIRO A. (1961) Evolution of metamorphic belts. *J. Petrol.* **2**, 277–311.

MIYASHIRO A. (1974) Volcanic rock series in island arcs and active continental margins. *Amer. J. Sci.* **274**, 321–355.

MURAKAMI N. (1954) Gneissose gabbroic xenoliths in basalt in the northern part of Yamaguchi Prefecture (in Japanese). *Chigaku-Kenkyu* **7**, 98–103.

MURAKAMI N. (1975) High-grade metamorphic inclusions in Cenozoic volcanic rocks from west Sanin, southwest Japan. *J. Japan. Assoc. Min. Petrol. Econ. Geol.* **70**, 424–439.

MURASE T. and KUSHIRO I. (1979) Compressional wave velocity in partially molten peridotite at high pressures. *Carnegie Inst. Wash. Yearb.* **78**, 559–562.

MURASE T. and FUKUYAMA H. (1980) Shear wave velocity in partially molten peridotite at high pressures. *Carnegie Inst. Wash. Yearb.* **79**, 307–311.

MYSEN B. O. and BOETTCHER A. L. (1975) Melting in a hydrous mantle, I. Phase relations of natural peridotite at high pressures and temperatures with controlled activities of water, carbon dioxide and hydrogen. *J. Petrol.* **16**, 520–548.

MYSEN B. O. and KUSHIRO I. (1977) Compositional variations of coexisting phases with degree of melting of peridotite in the upper mantle. *Amer. Mineral.* **62**, 843–865.

OBATA M. (1976) The solubility of Al_2O_3 in orthopyroxenes in spinel and plagioclase peridotites and spinel pyroxenite. *Amer. Mineral.* **61,** 804–816.

O'HARA M. J., RICHARDSON S. W. and WILSON G. (1971) Garnet-peridotite stability and occurrence in crust and mantle. *Contrib. Mineral. Petrol.* **32,** 48–68.

POWELL M. (1978) Crystallization conditions of low-pressure cumulate nodules from the Lesser Antilles island arcs. *Earth Planet. Sci. Lett.* **39,** 162–172.

RIGDEN S. M., AHRENS T. J. and STOLPER E. M. (1984) Densities of liquid silicates at high pressures. *Science* **226,** 1071–1074.

SAKUYAMA M. (1978) Evidence of magma mixing: petrological study of Shirouma-Oike calc-alkaline andesite volcano, Japan. *J. Volcanol. Geotherm. Res.* **5,** 179–208.

SAKUYAMA M. (1979) Lateral variation of H_2O contents in Quaternary magmas of north-eastern Japan. *Earth Planet. Sci. Lett.* **43,** 103–111.

SAKUYAMA M. (1981) Petrological study of Myoko and Kurohime volcanoes, Japan: crystallization sequence and evidence of magma mixing. *J. Petrol.* **22,** 553–583.

SAKUYAMA M. (1983a) Petrology of arc volcanic rocks and their origin by mantle diapir. *J. Volcanol. Geotherm. Res.* **18,** 297–320.

SAKUYAMA M. (1983b) Phenocryst assemblages and H_2O content in circum-Pacific arc magmas. In *Geodynamics of the Western Pacific and Indonesian Regions.* Geodynamics Ser. Vol. 11, 143–158. Amer. Geophys. Union.

SHIDO F. (1958) Plutonic and metamorphic rocks of the Nakoso and Iritono districts in the central Abukuma Plateau. *J. Fac. Sci. Univ. Tokyo, Sec. 2,* **11,** 131–217.

SUGIMURA A. (1960) Zonal arrangements of some geophysical and petrological features in Japan and its environs. *J. Fac. Sci. Univ. Tokyo, Sec. 2,* **12,** 133–153.

SUZUKI T. and HADA S. (1979) Cretaceous tectonic melange of the Shimanto belt in Shikoku, Japan. *J. Geol. Soc. Japan* **85,** 467–479.

TAKAHASHI E. (1978) Petrologic model of the crust and upper mantle of the Japanese island arcs. *Bull. Volcanol.* **41,** 529–547.

TAKAHASHI E. (1980) Thermal history of lherzolite xenoliths I. Petrology of lherzolite xenoliths from Ichinomegata crater, Oga peninsula, northeast Japan. *Geochim. Cosmochim. Acta* **44,** 1643–1658.

TAKAHASHI E. (1986) Genesis of calc-alkali andesite magma in a hydrous mantle-crust boundary: Petrology of lherzolite xenoliths from the Ichinomegata crater, Oga peninsula, northeast Japan, Part II. *J. Volcanol. Geotherm. Res.* **29** (In press).

TAKAHASHI E. and KUSHIRO I. (1983) Melting of a dry peridotite at high pressures and basalt magma genesis. *Amer. Mineral.* **68,** 859–879.

TAKAHASHI M. (1986) Island arc magmatism before and after the opening of Japan Sea (in Japanese). *Kagaku (Science)* **56,** 103–111.

TAKAHASHI M., ARAMAKI S. and ISHIHARA S. (1980) In Granitic magmatism and related mineralization, *Mining Geol., Spec. Issue, No.* **8,** 13–28.

TANAKA T. and AOKI K. (1981) Petrogenic implications of REE and Ba on mafic and ultramafic inclusions from Ichinomegata, Japan. *J. Geol.* **89,** 369–390.

TATSUMI Y., SAKUYAMA M., FUKUYAMA H. and KUSHIRO I. (1983) Generation of arc basalt magmas and thermal structure of the mantle wedge in subduction zones. *J. Geophys. Res.* **88,** 5815–5825.

TOKSOZ M. N., MINEAR J. W. and JULIAN B. R. (1971) Temperature field and geophysical effects of a downgoing slab. *J. Geophys. Res.* **76,** 1113–1138.

TILLEY C. E. (1925) A preliminary survey of metamorphic zones in the southern Highlands. *Quart. J. Geol. Soc. London* **81,** 100–112.

UYEDA S. and HORAI K. (1964) Terrestrial heat flow in Japan. *J. Geophys. Res.* **69,** 2121–2141.

WADATI K. and IWAI Y. (1954) The minute investigation of seismicity in Japan. *Geophys. Mag.* **25,** 167–173.

WAGER L. R. (1962) Igneous cumulates from the 1902 eruption of Soufriere, St. Vincent. *Bull. Volcanol.* **24,** 93–99.

WELLS P. R. A. (1977) Pyroxene thermometry in simple and complex systems. *Contrib. Mineral. Petrol.* **62,** 129–139.

WOOD B. J. and BANNO S. (1973) Garnet-orthopyroxene and orthopyroxene-clinopyroxene relationships in simple and complex systems. *Contrib. Mineral. Petrol.* **42,** 109–124.

YODER H. S. (1967) Spilites and serpentinites. *Carnegie Inst. Wash. Yearb.* **65,** 269–279.

YOSHII T. (1972) Feature of the upper mantle around Japan as inferred from gravity anomalies. *J. Phys. Earth* **20,** 23–34.

YOSHII T. (1979) A detailed cross-section of the deep seismic zone beneath northeastern Honshu, Japan. *Tectonophys.* **55,** 349–360.

APPENDIX

The melting experiments on the olivine tholeiite (IK85060803) from Ryozen district, NE Honshu have been made at pressures between 9.5 and 12.5 kbar and in the temperature range between 1290° and 1325°C under anhydrous conditions, using a solid-media, piston-cylinder apparatus. The piston-out method was employed. Temperature was measured with $Pt/Pt_{90}Rh_{10}$ thermocouples. Finely ground powder (<5 μm) of the rock was used as

Table 3. Results of runs on Ryozen olivine tholeiite

Pressure (kbar)	Temperature (°C)	Time (min.)	Results
9.5	1315	45	Ol + Gl
9.5	1305	60	Ol + Opx + Gl
9.5	1300	60	Ol + Opx + Gl
9.5	1290	150	Ol + Opx + Gl
10.5	1325	40	Gl
10.5	1315	60	Ol + Gl
11	1315	80	Ol + Opx + Gl
11	1315	420*	Ol + Opx + Gl
11	1306	60	Ol + Opx + Cpx(r) + Gl
11	1300	60	Ol(r) + Opx + Cpx + Gl
11	1292	60	Ol(r) + Opx + Cpx + Gl
11.5	1316	60	Opx + Gl
12.5	1325	60	Opx + Cpx(?) + Gl
12.5	1315	60	Opx + Cpx + Gl

Abbreviations: Cpx, clinopyroxene; Gl, glass; Ol, olivine; Opx, orthopyroxene; r, rare.

* Thus run was made in a graphite capsule which was sealed in a Pt capsule.

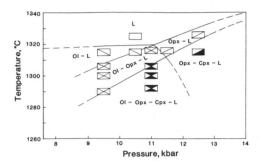

FIG. 9. Melting phase relations of olivine tholeiite (IK85060803) from Ryozen district, NE Honshu. The chemical composition is given in Table 1. Abbreviations: L, liquid; Ol, olivine; Opx, orthopyroxene; Cpx, clinopyroxene.

the starting material. Graphite capsule was used for all the runs. One critical run was made with a graphite capsule which was sealed in a Pt capsule for a longer duration. The results are shown in Table 3 and plotted in Figure 9. At pressures lower than 10.5 kbar, olivine is on the liquidus and above 11.5 kbar orthopyroxene is the liquidus phase. At 11 kbar both olivine (Fo_{89}) and orthopyroxene ($En_{89}Fs_{11}$) crystallize almost simultaneously on the liquidus at 1315°C. This run was made for both 80 and 420 minutes. The results of these runs were the same, except that the longer run showed growth of both olivine and orthopyroxene. At 11 kbar and 1306°C, olivine (Fo_{88}), orthopyroxene ($En_{88}Fs_{12}$) and clinopyroxene crystallize. The results indicate that the melt of this olivine tholeiite is in equilibrium with olivine and orthopyroxene at 11 kbar under anhydrous conditions. It is also indicated that the composition is close to an olivine-orthopyroxene-clinopyroxene cotectic under this condition.

Magmatic Processes: Physicochemical Principles
© The Geochemical Society, Special Publication No. 1, 1987
Editor B. O. Mysen

Impact of hornblende crystallization for the genesis of calc-alkalic andesites

KENZO YAGI

Department of Geology and Mineralogy, Hokkaido University, Sapporo 060, Japan

and

HISASHI TAKESHITA

Department of Earth Science, Fukuoka University of Education, Minakata, Fukuoka, 811-41 Japan

Abstract—Hornblende gabbros, including hornblendite or hornblende megacrysts, often enclosed in calc-alkalic andesites, are common along the Japan Sea coast of northeastern Japan. The andesites, in turn, are closely associated with high–alumina basalts.

Hornblende gabbros consist of plagioclase, hornblende, clinopyroxene, olivine, orthopyroxene, magnetite and apatite. The first two minerals are essential constituents, with modal plagioclase contents ranging from 15 to 55 volume percent and hornblende from 30–60 volume percent,—rarely reaching 90 volume percent in hornblendite. Magnetite is an intercumulus phase, with a modal abundance reaching 7 volume percent. These phases occur as cumulus minerals, poikilitically enclosing clinopyroxene and magnetite. Most hornblende is pargasitic with a small portion of tschermakite. Clinopyroxene is high in Al_2O_3, being consistent with a suggestion that clinopyroxene crystallized at high pressure.

Experimental studies in the system basalt-water indicate that hornblende and plagioclase will crystallize from a wet magma in the pressure range 7–10 kbar at temperatures near 1000°C. At these pressures and temperatures hornblende and plagioclase will sink to form cumulate hornblende and plagioclase.

It is suggested that calc-alkalic andesite magma may be formed by separation of about 40 percent of a hornblende gabbro mineral assemblage from a proposed hydrous, parental high–alumina basalt magma near the base of the crust in the island arcs. Extensive hornblende gabbros, or metamorphic rocks derived from them, may, therefore, be expected to underlie calc-alkalic andesites. Such a model is consistent with the structure of the lower crust in northeastern Japan from seismological observations.

INTRODUCTION

THE GENESIS of calc-alkalic andesite, commonly occurring in island arcs and along continental margins, is one of the most important problems in igneous petrology. YODER and TILLEY (1962), in an experimental study of liquidus phase relations of tholeiitic and related basalts at high water pressure, found that hornblende has a wide pressure-temperature stability field, and sometimes is stable near the liquidus. This study revived the importance of hornblende fractionation [originally postulated by BOWEN (1928)] in the evolution of igneous rocks. Based mainly on experimental studies in hydrous synthetic and natural basalt systems, hornblende fractionation has been considered an important factor in the genesis of calc-alkalic andesites (e.g., GREEN and RINGWOOD, 1968; HOLLOWAY and BURNHAM, 1972; ALLEN et al., 1975; ALLEN and BOETTCHER, 1978; CAWTHORN and O'HARA, 1976).

Hornblende gabbro inclusions, or hornblende megacrysts in calc-alkalic andesite and high–alumina basalt, are common in the Fossa Magna region in central Japan (TAKESHITA, 1974, 1975). Hornblende gabbro closely associated with calc-alkalic andesite has also been reported from numerous other localities in Japan and in other island arcs (e.g., SATO et al., 1975; SHIMAZU et al., 1979; KUNO and AOKI, 1970; ARCULUS and WILLS, 1980). The purpose of this report is to describe the petrology and geochemistry of these rocks, and to discuss possible origins of calc-alkalic andesites in light of these data.

OCCURRENCE OF HORNBLENDE GABBRO INCLUSIONS

In the Green Tuff region (Figure 1) there are several localities along the Japan Sea coast of northeastern Japan where hornblende gabbros, including hornblendite and hornblende megacrysts, are closely associated with calc-alkalic andesites.

(1) Shigarami

Felsic volcanic rocks, andesites of both the calc-alkalic and tholeiitic series, and high–alumina basalt of Miocene to early Pleistocene age occur in this district (TAKESHITA, 1974, 1975).

Andesites. Calc-alkalic augite-hypersthene andesites are predominant in the Shigarami district.

S   Shigarami
U   Umikawa
Y   Yoneyama
I   Ichinomegata
O   Oshima-Ōshima

~ Green Tuff regions

N

Fossa Magna

0   500 km

FIG. 1. Localities of hornblende gabbroic inclusions in northeastern Japan.

Their SiO_2 content ranges from 50 to 59 weight percent. Plagioclase phenocrysts range from labradorite to bytownite, sometimes to anorthite. Augite and hypersthene are abundant and hornblende is sometimes present. Magnetite also occurs as phenocrysts. The groundmass, pilotaxitic to hyalophilitic in texture, consists of andesine-labradorite laths and prismatic crystals of hypersthene and rare clinopyroxene. In the leucocratic portions, anhedral anorthoclase and quartz poikilitically enclose plagioclase laths, pyroxene rods and magnetite grains.

Tholeiitic augite-hypersthene andesites are less common. The phenocrysts in these rocks are the same as in calc-alkalic andesite, but the groundmass is composed of plagioclase laths and pigeonite. Hypersthene, when present, is always sandwiched between pigeonite or subcalcic augite (YAGI and YAGI, 1958).

High-alumina basalt. The phenocrysts are plagioclase, olivine, augite and rare orthopyroxene and hornblende. The plagioclase is bytownite with anorthite cores, surrounded by labradorite-andesine rims. Augite is abundant in calcium-rich basalts, but is rare in magnesium-rich ones. The groundmass consists of labradorite laths, augite, subcalcic augite or pigeonite together with small amounts of olivine without reaction rims. Magnetite is confined to the groundmass.

Hornblende gabbro inclusions. The calc-alkalic andesites carry inclusions of hornblende gabbro or hornblendite as well as hornblende megacrysts (YAMAZAKI *et al.,* 1966; TAKESHITA and OJI, 1968;

TAKESHITA, 1974). The texture of the hornblende gabbro is medium (0.5–1.0 mm) to coarse–grained (2.0–7.0 mm grain size), and is essentially a plagioclase ($An_{85}–An_{95}$) cumulate (WAGER *et al.,* 1960). The plagioclase occurs as rectangular, euhedral, chemically homogeneous crystals, rarely containing globular magnetite grains. Long prismatic hornblende shows subhedral form against plagioclase. Hornblende-rich gabbro, containing more than 50 volume percent hornblende, is, therefore, regarded as plagioclase-hornblende cumulate in which euhedral to subhedral hornblende poikilitically encloses clinopyroxene. Most magnetite grains are subhedral to anhedral, occurring as an intercumulus phase.

The modal abundance of hornblende ranges from 30 to 60 volume percent in hornblende gabbro and may reach 95 volume percent in hornblendite. In contrast to phenocrystic hornblende, the hornblende in the gabbros is generally homogeneous with only a weak zoning in some subhedral grains. Anorthite and hornblende megacrysts in the andesites are homogeneous in composition and structure except for a thin labradorite rim on plagioclase.

(2) Yoneyama and Umikawa districts

The Yoneyama district to the northeast, and the Umikawa district to the northwest of the Shirigama district (Figure 1) have mostly Pliocene volcanics, ranging from olivine-augite high–alumina basalt to pyroxene andesites and biotite–hornblende andesites for the calc-alkalic series. The pyroxene andesites are the most common (SATO *et al.,* 1975; SHIMAZU *et al.,* 1979).

In the Yoneyama district, olivine-augite or hornblende-bearing olivine-augite high alumina basalt occurs. Although such basalts have not been found in the Umikawa district, basaltic andesite is present. Calc-alkalic andesite is common in both districts.

Hornblende gabbroic inclusions. The calc-alkalic andesites in the Yoneyama and Umikawa districts often contain hornblende gabbro inclusions as well as hornblende and diopside megacrysts, and rarely anorthite, up to 8 cm in size. The gabbro consists principally of hornblende (6–47 volume percent) and plagioclase, with varying amounts of augite, olivine, magnetite and rare hypersthene. As in the gabbros from the Shigarami district, fine-grained patches of basalt occur in hornblende gabbro from the Umikawa district (SHIMAZU *et al.,* 1979).

(3) Ichinomegata

Ichinomegata in northeastern Honshu (Figure 1) is well known for the occurrence of various ultramafic and mafic inclusions. These inclusions are garnet lherzolite, spinel lherzolite, websterite, amphibolite, hornblende gabbro, hornblendite and various crustal rocks (KUNO and AOKI, 1970; AOKI, 1971; TANAKA and AOKI, 1981; AOKI and FUJI-MAKI, 1982; FUKUYAMA, 1985; TAKAHASHI, 1986). AOKI (1971) considered that hornblende gabbro, hornblendite, and amphibolite form the greater part of the lower crust in this region.

Among mafic inclusions, hornblende gabbro with 30–60 volume percent hornblende, is the most common. Modal plagioclase contents are usually less than 10 percent in hornblendite, but reaches 50 volume percent in the gabbro. Cumulus olivine and clinopyroxene, surrounded by intercumulus, poikilitic hornblende, are described by AOKI (1971) in one type of hornblendite. This hornblendite resembles the coarse–grained, hornblende-rich gabbros from the Shigarami district. The mineral assemblages of the hornblende gabbro at Ichinomegata are the same as in the other districts.

(4) Oshima-Oshima Volcano

Oshima-Oshima is an insular volcano off the Japan Sea coast of Hokkaido (Figure 1) with about 30 percent calc-alkalic andesite and about 70 percent alkalic basalt among the lavas and pyroclastics. The alkalic basalts are augite-olivine basalts with phenocrysts of plagioclase, olivine and augite in a groundmass of plagioclase, anorthoclase, fine-grained clinopyroxene, olivine and magnetite. Calc-alkalic andesites comprise olivine-pyroxene andesite, hornblende-pyroxene andesite and biotite-hornblende bearing pyroxene andesite. Both andesites and basalts are intimately associated; for example, in the 1741–1742 eruption pyroclastics changed from calc-alkalic andesite pumice to olivine basalt scoria.

Most andesites and some basalts contain ultramfic and mafic inclusions such as dunite, wehrlite, olivine-clinopyroxenite, hornblendite and hornblende gabbro. Hornblende gabbro is predominant and consists mainly of hornblende and plagioclase, with variable amounts of olivine, clinopyroxene, orthopyroxene, magnetite and apatite.

PETROCHEMISTRY OF VOLCANIC ROCKS

(1) High–alumina basalts

Bulk chemical compositions of several high–alumina basalts from the Shigarami and Yoneyama districts (SATO et al., 1975), and alkali basalt from Oshima-Oshima (YAMAMOTO, 1984) show that the Al_2O_3 contents are always high, especially in the

Yoneyama basalts. There may also be an increase in K_2O/Na_2O from the volcanic front toward the Japan Sea coast (KUNO, 1960; KAWANO et al., 1961).

(2) Calc-alkalic andesites

The calc-alkalic andesites from Shigarami (TAKESHITA, 1975), Yoneyama (SATO et al., 1975) and Oshima-Oshima (YAMAMOTO, 1984) are characterized by their higher Al_2O_3 and alkalies compared with aphanitic andesite of the hypersthenic rock series (KUNO, 1950) and are also rather poor in MgO and CaO.

(3) Hornblende gabbros

The hornblende gabbros from the Shigarami (TAKESHITA, 1975), Yoneyama (SHIMAZU et al., 1979), Umikawa (SHIMAZU et al., 1979) and Ichinomegata (AOKI, 1971) districts exhibit generally low silica contents, ranging from 35 to 47 weight percent, whereas $FeO^*/(FeO^* + MgO)$ is fairly high and ranges from 0.335 to 0.796. Normative nepheline is always present, except in an orthopyroxene-clinopyroxene hornblende gabbro from Umikawa (SHIMAZU et al., 1979). In one case from Ichinomegata (AOKI, 1971), there is leucite in the norm. Diopside and magnetite contents are always high. These features are similar to those of hornblende gabbro from the Lesser Antilles (LEWIS, 1973a, b; ARCULUS and WILLS, 1980).

It is noted that in the $FeO^*-MgO-(Na_2O + K_2O)$ diagram (Figure 2) hornblende gabbros associated with the calc-alkalic andesites in the Shigarami district fall near the extension of the tie-lines connecting calc-alkaline andesites with high–alumina basalts. These gabbros do fall within the tholeiite field.

Hornblende. All hornblende analyses (TAKESHITA, 1975; YAMAZAKI et al., 1966; SHIMAZU et al., 1979; ONUKI, 1965; YAMAMOTO, 1984) are compositionally similar with low SiO_2 and high concentrations of Al_2O_3 and Fe_2O_3. The TiO_2 content is low (less than 2.5 weight percent).

The hornblende compositions are plotted in the $Al^{IV}-(Al^{VI} + Fe^{3+} + Ti)$ and $Al^{IV}-(Na + K)$ diagrams (Figure 3) used by DEER et al. (1963). The pargasite contents range from 40 to 90 percent, and tschermakite from 10 to 30 percent, with less than 10 percent of the edenite molecule.

Clinopyroxene, present as an accessory mineral, has generally high Al_2O_3 and low Na_2O and TiO_2, and thus exhibit high concentrations of the tschermak's molecule (4–12 mol percent). The Ca-tschermak's molecule has a wide high-pressure sta-

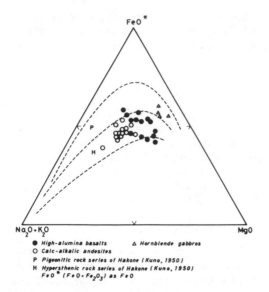

FIG. 2. Plots of high–alumina basalts, calc-alkalic andesites and hornblende gabbros in the MgO–FeO^*–$(Na_2O + K_2O)$ diagram. Data from KUNO (1950), SATO et al. (1975), YAMAMOTO (1984), TAKESHITA (1974, 1975), SHIMAZU et al. (1979), AOKI (1971).

bility field (HAYS, 1966) with increasing tschermak's molecule content with increasing pressure (KUSHIRO and YODER, 1966; KHANUKHOVA et al., 1976). Therefore, it is probable that clinopyroxene crystallization took place at high pressure.

Magnetite consists mainly of FeO and Fe_2O_3,

with variable amounts of Al_2O_3, TiO_2, MnO and MgO. An electron microprobe analysis of an intercumulus magnetite showed (weight percent); TiO_2 = 7.59, Al_2O_3 = 9.00, Cr_2O_3 = 0.16, FeO^* = 70.10, MnO = 0.33, and MgO = 7.53 (T. UENO, personal communication).

The Fe/Mg ratios among hornblende, clinopyroxene and orthopyroxene in the gabbros of the Shigarami district are displayed in Figure 4 with hornblende showing higher Fe/Mg than the co–existing minerals. Similar relations among hornblende, clinopyroxene, orthopyroxene, olivine and glasses have been observed in experimental run products from basalt-H_2O systems (HOLLOWAY and BURNHAM, 1972; ALLEN et al., 1975).

GENESIS OF HORNBLENDE GABBROS

It is inferred from the textural and compositional relations that the crystallization sequence in the gabbros was olivine, clinopyroxene, plagioclase, hornblende and magnetite. It is noted that hornblende is not an early crystallizing phase. In view of the high water contents required to stabilize hornblende on or near the liquidus of mafic magmas (e.g., YODER and TILLEY, 1962; HOLLOWAY and BURNHAM, 1972; ALLEN et al., 1975; ALLEN and BOETTCHER, 1978), it is suggested that the parental magma for the calc-alkalic andesites did not have very high water contents. Thus, it is suggested that the magma from which hornblende crystallized had

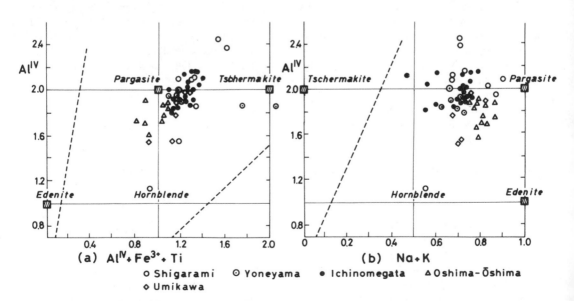

FIG. 3. Hornblende compositions in Al^{IV}-$(Al^{VI} + Fe^{3+} + Ti)$ and Al^{IV}-$(Na + K)$ diagrams. Data from TAKESHITA (1975), AOKI (1971), YAMAZAKI et al. (1966), SHIMAZU et al. (1979), ONUKI (1965), YAMAMOTO (1984). Symbols as in Figure 4.

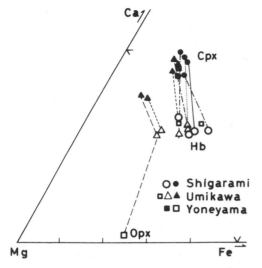

FIG. 4. Clinopyroxene, hornblende and orthopyroxene compositions from the Shigarami, Umikawa and Yoenyama districts. Data sources as in Figure 3.

undergone some fractionation of anhydrous phases (point B in Figure 5). Most likely, the early crystallizing phases were olivine and anorthite (see also discussion below) to shift the liquid composition along the line from cpx-pl toward point B at which point hornblende began crystallizing and driving the liquid from B toward c (Figure 5).

The difference between hornblende gabbro and hornblendite may be related to whether or not plagioclase sinks in the magma. FUJII and KUSHIRO (1977) found that plagioclase of any composition will float in anhydrous basaltic melt at pressures higher than about 5 kbar, whereas in the pressure range 1–5 kbar, plagioclase more calcic than An_{90} will sink. Recently, KUSHIRO (1987) found that plagioclase more calcic than An_{60} would sink in a basaltic melt with 1.5 weight percent H_2O at pressures corresponding to those at the base of the crust. Therefore, plagioclase may sink in the basalt magma together with hornblende and magnetite, sometimes also clinopyroxene, resulting in the formation of hornblende gabbro. If plagioclase does not sink, hornblendite, instead of hornblende gabbro, will be formed, giving rise to gradational facies between the two types of rocks.

From the available experimental data and mineral assemblages in the gabbros, AOKI (1971) suggested that hornblende gabbro and hornblendite of Ichinomegata were formed at about 7–9 kbar near 1000°C, corresponding to a depth of 25–30 km, whereas SHIMAZU et al. (1979) estimated the pressure and temperature of crystallization of gabbros

with the mineral assemblage olivine–clinopyroxene-plagioclase-hornblende from the Umikawa district to be about 10 kbar and 1000°C, respectively (about 30 km depth). In both models, the basalt magma was hydrated, with total pressure greater than water pressure. It appears, therefore, that the crystallization of these gabbroic inclusions occurred near the base of the crust in northeastern Japan.

ORIGIN OF CALC-ALKALIC ANDESITE MAGMA

The relationship of hornblende fractionation in high–alumina basalt magma to the formation of calc-alkalic andesite will be considered next,—first for the Shigarami district. The composition of a suggested parental high–alumina basalt magma B (No. 1; Table 1), a representative calc-alkalic andesite (No. 3, sample 66862; Table 1), hornblende gabbro HG (No. 2; Table 1), clinopyroxene (No. 7; Table 1) and hornblende (No. 8; Table 1) from the Shigarami district are shown.

Plagioclase and hornblende are present in the gabbro inclusions in nearly equal proportions, and the proportions of inclusions in the magma are estimated to be less than 50 percent of the total amount of material. We choose 40 percent for the amount to be subtracted from the parental magma. Liquid compositions a, b and c (Figure 5) can be

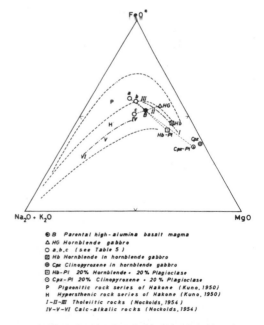

FIG. 5. High–alumina basalt, hornblende gabbro, hornblende, clinopyroxene and residual compositions (from Table 1).

Table 1. Comparison of compositions of residua after separation of hornblende-free cumulate, hornblende-bearing cumulate and hornblende gabbro from the high–alumina basalt magma

$$B(100\%) - Cpx(20\%) - Pl\,An_{90}(20\%) \rightarrow a(60\%)$$

$$B(100\%) - Hb(20\%) - Pl\,An_{90}(20\%) \rightarrow b(60\%)$$

$$B(100\%) - HG(40\%) \qquad\qquad \rightarrow c(60\%)$$

Number	1	2	3	4	5	6		7	8
Symbol	B	HG		a	b	c		Cpx	Hb
Weight percent									
SiO_2	48.57	43.40	54.31	49.93	51.69	52.06		48.26	43.22
TiO_2	0.94	1.19	0.93	1.30	1.04	0.77		0.82	1.58
Al_2O_3	20.42	21.25	19.15	20.41	18.29	19.88		6.22	12.59
Fe_2O_3	3.95	4.74	3.91	—	—	3.43		—	—
FeO	5.62	6.12	3.58	12.84*	10.87*	5.29		7.63*	13.55*
MnO	0.16	0.16	0.11	0.27	0.21	0.16		—	0.10
MgO	5.07	6.60	3.62	3.91	3.85	4.05		13.63	13.90
CaO	11.64	14.20	9.42	5.68	9.33	9.94		23.18	12.14
Na_2O	2.37	1.65	2.77	3.59	2.96	2.85		0.26	2.16
K_2O	0.97	0.38	1.94	1.63	1.38	1.36		—	0.76
P_2O_5	0.24	0.31	0.25	0.40	0.31	0.19		—	—
Total	99.95	100.00	99.99	100.00	99.96	99.98		100.00	100.00
MgO	28.8	34.7	23.4	17.8	20.2	24.3		63.3	45.8
FeO*	52.2	54.6	46.0	58.4	57.0	50.4		35.5	44.6
$Na_2 + K_2O$	19.0	10.7	30.5	23.8	22.8	25.3		1.2	9.6

Norm Compositions							Atomic Ratios		
Q	—	—	7.31			3.04		$O = 6$	$O = 23$
Or	5.73	2.25	11.49			8.06	Si	1.802	6.258
Ab	20.12	10.82	23.46			24.10	Al^{IV}	0.198	1.742
An	42.12	49.39	33.94			37.36	Al^{IV}	0.076	0.403
Ne	—	1.68	—			—	Ti	0.023	0.172
Di { Wo	5.95	8.05	4.72			4.55	Fe^{3+}	—	—
Di { En	3.81	5.51	3.44			2.72	Fe^{2+}	0.238	1.635
Di { Fs	1.71	1.89	0.83			1.58	Mn	—	0.012
Hy { En	7.67	—	5.60			7.41	Mg	0.764	3.019
Hy { Fs	3.50	—	1.18			4.30	Ca	0.928	1.884
Ol { Fo	0.82	7.69	—			—	Na	0.019	0.606
Ol { Fa	0.42	2.90	—			—	K	—	0.141
Mt	5.73	6.87	5.67			4.96			
Il	1.78	2.26	1.76			1.46			
Ap	0.57	0.67	0.61			0.44			

* Total iron represented as FeO

1 (B): Average of four analyses of lime-rich high–alumina basalts from Shigarami (TAKESHITA, 1975)

2 (HG): Average of four analyses of hornblende gabbros from Shigarami (TAKESHITA, 1975; SHIMAZU et al., 1979)

3: Calc-alkalic andesite, No. 66862 from Shigarami (TAKESHITA, 1975)

7 (Cpx): Average of three analyses of clinopyroxene in clinopyroxene hornblende gabbro from Shigarami (SHIMAZU et al., 1979)

8 (Hb): Average of five analyses of hornblende in the hornblende gabbro from Shigarami (SHIMAZU et al., 1979)

obtained by subtracting 40 percent of clinopyroxene (20 percent) + plagioclase (An_{90}; 20 percent), hornblende (20 percent) + plagioclase (20 percent) and hornblende gabbro (40 percent), respectively. In Figure 5, a plots in the tholeiite field, c in the calc-alkalic field and b on the boundary between those two fields. The results indicate that hornblende fractionation from high–alumina basalt will tend to shift the residual liquid compositions along the calc-alkalic trend. Magnetite, which may reach modal abundances of 7 percent in some gabbros, can also play an important role in the fractionation processes (OSBORN, 1959). Magnetite fractionation alone, however, drives the residual liquid away from the FeO* apex toward the MgO–($Na_2O + K_2O$) side. This trend differs from the calc-alkalic trend represented by the shift of liquid compositions from b to c in Figure 5. Therefore, silicate minerals, *e.g.,* hornblende, must crystallize together with magnetite in order for the residual liquids to follow the calc-alkalic trend.

TAKESHITA and OJI (1968) concluded that calc-alkalic andesites were formed as residua after separation of hornblende gabbro from high alumina basalt. In the Oshima-Oshima volcano, YAMAMOTO *et al.* (1977) and YAMAMOTO (1984) suggested that amphibole fractionation (hornblende and hornblende gabbro) in basalt magma is an important factor for the genesis of calc-alkalic andesites. SHIMAZU *et al.* (1979), however, recognized implicitly the same possibility for the Umikawa district, though conclusive evidence was lacking.

Diverse opinions have been proposed for the Ichinomegata district. KATSUI *et al.* (1979), from the observation that the host rocks of mafic and ultramafic inclusions are pumiceous calc-alkalic andesite, were lead to the conclusion that calc-alkalic andesite may have been derived primarily by partial melting of upper mantle peridotite under hydrous conditions as suggested by KUSHIRO (1969, 1972, 1974), MYSEN and BOETTCHER (1975) and YODER (1969). AOKI and FUJIMAKI (1982) ruled out the possibility of production of andesite magma through amphibole fractionation from basalt magma on the basis of similar REE patterns and concentrations in the basalts and calc-alkalic andesites. Instead, they suggested that andesite and basalt in the Ichinomegata district could have been formed by nearly the same degree of partial melting of mantle peridotite at depths of 40–60 km with different activities of water.

TAKAHASHI (1986) proposed a two-stage melting process. Basalt magma was generated by partial melting in an ascending diapir. The heat supplied by the invading basalt magma might cause partial melting of hornblende gabbros or amphibolites near the crust/mantle boundary to form calc-alkalic andesite. Although most investigators have interpreted the hornblende gabbros or hornblende-rich rocks as cumulates from hydrous basaltic magma, TAKAHASHI (1986) suggested that they could be residues after extraction of calc-alkalic magma from basaltic materials.

In our calculations (Table 1; see also Figure 5) the production of calc-alkalic andesite magma from parental high–alumina basalt would leave about 40 percent hornblende gabbro. Whether or not such large amounts of hornblende gabbro exist in the deep portion of the Japanese crust remains a subject of discussion. AOKI (1971) considered that hornblende gabbro and related mafic rocks form the main part of the lower crust from which the inclusion in the Ichinomegata extrusives were derived. In the Paleogene volcanic rocks of northern Caucasus, ZAKARIADZE and LORDKIPANDZE (1971) showed that calc-alkalic andesites and dacites are underlain by extensive hornblende gabbros or related rocks in the lower crust. From the prevalent occurrence of inclusions of hornblende gabbro, amphibolite and granulite rocks in the Cenozoic volcanic rocks from the west San-in region in southwestern Japan, MURAMAKI (1975) concluded that these rocks form the lower crust in that part of Japan.

CONCLUSIONS

(1). Inclusions of hornblende gabbro or hornblendite frequently observed in the calc-alkalic andesite and associated rocks along the Japan Sea coast in northeastern Japan are considered to represent the cumulates from the parental hydrated basalt magma.

(2). Calc-alkalic andesite magma was formed by separation of hornblende gabbro or hornblendite from the parental high–alumina basalt.

(3). The presence of large amounts of hornblende gabbro and related mafic rocks in the lower crust in northeastern Japan is consistent with crustal models derived from seismic data.

Acknowledgements—We are grateful to Professor I. Kushiro for valuable discussions and to Dr. E. Takahashi for his courtesy in allowing us to see his unpublished paper. We also thank Ms. S. Kumano for drafting the figures and to Dr. T. Ueno for his chemical analysis of magnetite.

REFERENCES

ALLEN J. C., BOETTCHER A. L. and MARLAND G. (1975) Amphiboles in andesites and basalts. I. Stability as a function of P-T-f_{O_2}. *Amer. Mineral.* **60**, 1068–1085.

ALLEN J. C. and BOETTCHER A. L. (1978) Amphiboles in andesites and basalts. II. Stability as a function of P-T-f_{H_2O}-f_{O_2}. *Amer. Mineral.* **63**, 1074–1087.

AOKI K. (1971) Petrology of mafic inclusions from Ichinomegata, Japan. *Contrib. Mineral. Petrol.* **30**, 314–331.

AOKI K. and FUJIMAKI H. (1982) Petrology and geochemistry of calc-alkaline andesite of presumed upper mantle origin from Ichinomegata, Japan. *Amer. Mineral.* **67**, 1–13.

ARCULUS A. J. and WILLS K. J. A. (1980) The petrology of plutonic blocks and inclusions from Lesser Antilles island arc. *J. Petrol.* **21**, 743–799.

BOWEN N. L. (1928) *The Evolution of the Igneous Rocks.* 334 pp. Princeton University Press.

CAWTHORN R. G. and O'HARA M. J. (1976) Amphibole fractionation in calc-alkaline magma genesis. *Amer. J. Sci.* **276**, 309–329.

DEER W. A., HOWIE R. A. and ZUSSMAN A. (1963) *Rock-Forming Minerals,* Vol. 2, 370 pp. Longman.

FUJII T. and KUSHIRO I. (1977) Density, viscosity and compressibility of basaltic liquid at high pressure. *Carnegie Inst. Wash. Yearb.* **76**, 419–424.

FUKUYAMA H. (1985) Gabrroic inclusions of Ichinomegata tuff cone: Bulk chemical composition. *J. Fac. Sci. Univ. Tokyo, Sect. II* **22**, 67–80.

GREEN T. H. and RINGWOOD A. E. (1968) Genesis of the calc-alkaline rock-suite. *Contr. Mineral. Petrol.* **18**, 105–162.

HAYS J. F. (1966) Lima-alumina-silica. *Carnegie Inst. Wash. Yearb.* **65**, 234–239.

HOLLOWAY J. R. and BURNHAM C. W. (1972) Melting relations of basalt with equilibrium water pressure less than total pressure. *J. Petrol.* **13**, 1–29.

KATSUI Y., YAMAMOTO M., NEMOTO S. and NIIDA K. (1979) Genesis of calc-alkalic andesites from Oshima-Oshima and Ichinomegata volcanoes, North Japan. *J. Fac. Sci. Hokkaido, Ser. IV* **19**, 157–168.

KAWANO Y., YAGI K. and AOKI K. (1961) Petrography and petrochemistry of the volcanic rocks of Quaternary volcanoes of northeastern Japan. Sci. Rep. Tohuku Univ. Ser. 3, No. 7, 1–47.

KHANUKHOVA L. T., ZHARIKOV V. A., ISHBULATOV R. A. and LITVIN YU. A. (1976) Pyroxene solid solutions in the system, $NaAlSi_2O_6$-$CaAl_2SiO_6$-SiO_2 at 35 kbars and 1200°C. *Akad. Sci. USSR, Dokl.* **231**, 140–142.

KUNO H. (1950) Petrology of the Hakone volcano and adjacent areas, Japan. *Bull. Geol. Soc. Amer.* **61**, 957–1020.

KUNO H. (1960) High–alumina basalt. *J. Petrol.* **1**, 121–145.

KUNO H. and AOKI K. (1970) Chemistry of ultramafic nodules and their bearing on the origin of basaltic magmas. *Phys. Earth Planet. Int.* **3**, 273–301.

KUSHIRO I. (1969) Discussion of the paper "The origin of basaltic and nephelinitic magmas in the earth's mantle". *Tectonophys.* **7**, 427–436.

KUSHIRO I. (1972) Effect of water on the composition of magmas formed at high pressures. *J. Petrol.* **13**, 311–334.

KUSHIRO I. (1974) Melting of hydrous upper mantle and possible generation of andesitic magma: An approach from synthetic systems. *Earth Planet. Sci. Lett.* **22**, 294–299.

KUSHIRO I. (1987) A petrological model of the upper mantle and lower crust in the Japanese island arcs. In *Magmatic Processes: Physicochemical Principles,* (ed. B. O. MYSEN), The Geochemical Society Spec. Publ. No. 1, pp. 165–181.

KUSHIRO I. and YODER H. S. JR. (1966) Anorthite-for-sterite and anorthite-enstatite reactions and their bearing on the basalt-eclogite transformation. *J. Petrol.* **7**, 337–362.

LEWIS J. F. (1973a) Petrology of ejected plutonic blocks of the Soufriere Volcano, St. Vincent, West Indies. *J. Petrol.* **14**, 81–112.

LEWIS J. F. (1973b) Mineralogy of ejected plutonic blocks of Soufriere Volcano, St. Vincent: Olivine, pyroxene, amphibole and magnetite paragenesis. *Contrib. Mineral. Petrol.* **38**, 197–220.

MURAMAKI N. (1975) High-grade metamorphic inclusions in Cenozoic volcanic rocks from West San-in, Southwest Japan. *J. Japan. Assoc. Mineral. Petr. Econ. Geol.* **70**, 424–439.

MYSEN B. O. and BOETTCHER A. L. (1975) Melting of a hydrous mantle. II. Geochemistry of crystals and liquids formed by anatexis of mantle peridotite at high pressures and temperatures as a function of controlled activities of water, hydrogen and carbon dioxide. *J. Petrol.* **16**, 549–590.

NOCKOLDS S. R. (1954) Average chemical compositions of some igneous rocks. *Bull. Geol. Soc. Amer.* **65**, 1007–1032.

ONUKI H. (1965) Petrochemical research on the Horoman and Miyamori ultramafic intrusives, northern Japan. Sci. Rep. Tohuku Univ. Ser. III, No. 9, 217–276.

OSBORN E. F. (1959) Role of oxygen pressure in the crystallization and differentiation of basaltic magma. *Amer. J. Sci.* **257**, 609–647.

SATO T. and YONEYAMA RESEARCH GROUP (1975) Volcanic rocks in the Yoneyama district, Niigata Prefecture, Japan. *Earth Sci.* **29**, 211–229.

SHIMAZU M., YANO T. and TAJIMA M. (1979) Gabbroic inclusions in the calc-alkali andesites of the Fossa Magna region, Central Japan. Sci. Rept. Niigata Univ., Ser. E., No. 5, 63–85.

TAKAHASHI E. (1986) Genesis of calc-alkalic andesite magma in a hydrous mantle-crust boundary. *J. Volcan. Geotherm. Res.* (In press).

TAKESHITA H. (1974) Petrological studies on the volcanic rocks of the northern Fossa Magna region, Nagano Prefecture, Japan I. *Pac. Geol.* **7**, 65–95.

TAKESHITA H. (1975) Petrological studies on the volcanic rocks of the northern Fossa Magna region, Nagano Prefecture, Japan I. *Pac. Geol.* **10**, 1–32.

TAKESHITA H. and OJI Y. (1968) Hornblende gabbroic inclusions in calc-alkaline andesites from the northern district of Nagano Prefecture, I. and II. *J. Jap. Assoc. Mineral. Petr. Econ. Geol.* **60**, 1–26; 57–73.

TANAKA T. and AOKI K. (1981) Petrogenetic implications of REE and Ba data on mafic and ultramafic inclusions from Ichinomegata, Japan. *J. Geol.* **89**, 369–390.

WAGER L. R., BROWN G. M. and WADSWORTH W. J. (1960) Types of igneous cumulates. *J. Petrol.* **1**, 73–85.

YAGI T. and YAGI K. (1958) *Geology of the Kami-Minochi Region.* 480 pp. Kokin Shoin, Tokyo.

YAMAMOTO M. (1984) Origin of calc-alkaline andesite from Oshima-Oshima Volcano, North Japan. *J. Fac. Sci. Hokkaido Univ. Ser. IV* **21**, 77–131.

YAMAMOTO M., KATSUI Y. and NIIDA K. (1977) Petrology of volcanic rocks and ultramafic inclusions of Oshima-Oshima Volcano. *Bull. Volcanol. Soc. Japan Ser. II* **22**, 241–248.

YAMAZAKI T., ONUKI H. and TIBA T. (1966) Significance of hornblende gabbroic inclusions in calc-alkalic rocks. *J. Jap. Assoc. Mineral. Petr. Econ. Geol.* **55**, 87–103.

YODER H. S. JR. (1969) Calc-alkalic andesites: Experimental data bearing on their assumed characteristics. *Bull. Dept. Geol. Mineral. Res. Oregon 65* (ed. A. R. MCBIRNEY) pp. 43–63.

YODER H. S. JR. and TILLEY C. E. (1962) Origin of basalt magmas: An experimental study of natural and synthetic rock systems. *J. Petrol.* **3**, 342–532.

ZAKARIADZE K. and LORDKIPANIDZE M. (1971) Differentiation of basalt magma in deep crustal chambers as evidenced by cognate inclusions of hornblendite and related rocks (abstr.). *General Meeting Int. Union Geodesy Geophys. Moscow, USSR, 1971, 1-2-103.*

Magmatic Processes: Physicochemical Principles
© The Geochemical Society, Special Publication No. 1, 1987
Editor B. O. Mysen

Mafic meta–igneous arc rocks of apparent komatiitic affinities, Sawyers Bar area, central Klamath Mountains, northern California*

W. G. ERNST

Department of Earth and Space Sciences, Institute of Geophysics and Planetary Physics,
University of California, Los Angeles, California 90024, U.S.A.

Abstract—Metamorphosed mafic igneous rocks of probable Permo–Triassic age, informally designated the Yellow Dog metavolcanics, apparently overly fine–grained metasedimentary strata in the eastern portion of the Western Paleozoic and Triassic Belt. These rocks have been recrystallized under biotite zone greenschist facies pressure-temperature conditions. Bulk–rock XRF analyses of the metavolcanics document the existence of a mafic-ultramafic trend. Fourteen of 55 analyzed protoliths appear to have been komatiitic or picritic basalts. The mean bulk–rock chemistry of these 14 specimens in oxide weight percent is: 49.0 SiO_2; 10.8 Al_2O_3; 14.8 MgO; 10.0 CaO; 1.4 Na_2O; 0.7 K_2O; 11.3 Fe_2O_3*; 1.2 TiO_2; 0.2 MnO; and 0.2 P_2O_5. Cr and Ni contents average 660 and 280 ppm respectively. Most Yellow Dog greenstones represent magnesian tholeiites, but, for the chemical parameters plotted, appear to be chemically intergradational with the consanguineous, apparently aphanitic, komatiitic(?) units. Associated coarser grained hypabyssal dikes and sills tend to be more evolved as a group, and exhibit transitions towards calc–alkaline basalts; evidently more leisurely ascent towards the surface allowed shallow-level crystal fractionation of the intrusive units. All analyzed rocks interpreted as komatiitic display hypersthene in the norm, and only one of the non–komatiitic samples carries significant normative nepheline (4 weight percent). Electron microprobe analyses of relict clinopyroxene phenocrysts from three mafic meta–igneous but non–komatiitic Yellow Dog rocks document a tholeiitic rather than an alkalic trend.

Twelve bulk–rock INAA demonstrate that, unlike most Archean ultramafic lavas, komatiitic(?) members of the Yellow Dog suite are strongly fractionated in rare earth elements and are LREE enriched, whereas some of the more normal mafic units display nearly flat, less evolved patterns; two magma series may therefore be present in the Yellow Dog metavolcanics. Six bulk–rock oxygen analyses show that the metavolcanics ($\delta^{18}O$ = 9.6–15.3) are substantially enriched in ^{18}O relative to Archean komatiites and Phanerozoic MORBs, probably as a consequence of partial re–equilibration with large quantities of magmatic water or seawater; isotopic fractionation is thought to have occurred at temperatures on the order of 100°C. However, major element + REE chemistries and mineralogies of the greenstones do not reflect metasomatism comparable to the extent of oxygen isotopic exchange.

Lack of an alkali basalt suite, occurrence of distal metaclastic sedimentary units intricately inter-layered with the Yellow Dog metavolcanics and apparently conformably underlying them, and ul-tramafic affinities of some of the greenstones argue for an origin by partial melting of relatively undepleted mantle ± eclogite at depths approaching 200 km and high temperatures beneath a primitive continental margin, or island arc. Eliminated by the petrochemistry are back–arc basin, intraplate, and rifted plate tectonic settings. Partial fusion of deep–seated portions of a garnet-bearing (REE-enriched?) protolith beneath this sector of the Californian convergent margin evidently was attended by abnormally elevated temperatures during Permo–Triassic subduction, volcanism and accretion, possibly due to oceanic-ridge or hot–spot descent. Subsequent interaction with magmatic water (or seawater) disturbed $^{18}O/^{16}O$ ratios in the analyzed rocks, but seems not to have involved widespread major element migration within the volcanic arc during regional metamorphism.

INTRODUCTION

THE SO-CALLED Western Paleozoic and Triassic belt (WTrPz) of the Klamath Mountains consists of four major lithotectonic units, serially juxtaposed along east-dipping thrust faults, and variably intruded by calc-alkaline plutons of approximately Nevadan age (IRWIN, 1960, 1966, 1985). Regional relationships in northern and central California, after JENNINGS (1977), are illustrated in Figure 1. The lithotectonic

units contain fossils of late Paleozoic and early Me-sozoic age; some, especially the older macrofossils, appear to occur in olistolithic blocks (IRWIN et al., 1978). The structurally highest unit, on the east, is the Stuart Fork terrane, which bears the effects of a Late Triassic blueschist-type metamorphism (HOTZ, 1973; HOTZ et al., 1977; BORNS, 1984; GOODGE, 1985). The three successively more oceanward units are the North Fork, Hayfork, and Rattlesnake Creek terranes, respectively (IRWIN, 1972). Similar to the inboard Stuart Fork, all were pervasively deformed and overprinted by a green-schist facies metamorphism during final assembly

---

* Institute of Geophysics and Planetary Physics Publi-cation No. 2901.

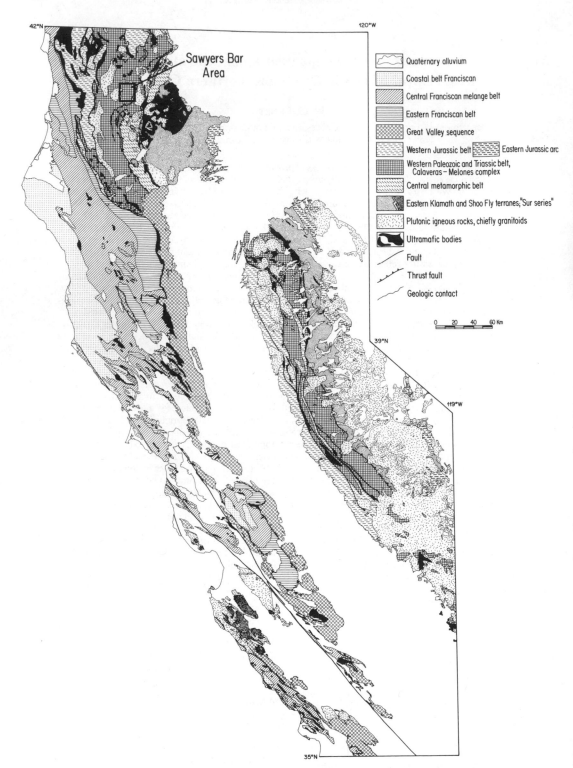

FIG. 1. Geologic setting and lithotectonic belts of northern and central California (JENNINGS, 1977; ERNST, 1983). Note the apparent extension of Klamath terranes into the northern Sierra Nevada foothills (DAVIS, 1969; HIETANEN, 1981). Location of the Sawyers Bar area (Figure 2) is also indicated.

of these units (the Siskiyou event of COLEMAN *et al.*, 1987; see also BARNES *et al.*, 1986), probably preceeding and accompanying emplacement of the Nevadan and slightly older granitoids (LANPHERE *et al.*, 1968; WRIGHT, 1982; HARPER and WRIGHT, 1984; HILL, 1985; COTKIN *et al.*, 1985).

Premetamorphic lithologies of the WTrPz apparently included widespread ophiolitic lenses + associated cherts, as well as abundant, locally melanged argillites, minor limestones and interlayered mafic volcanics. Although ophiolitic metagabbros, metabasalts, serpentinized peridotites and metacherts are of oceanic origin (IRWIN, 1985), the provenance of voluminous terrigineous debris was more probably a nearby calc–alkaline arc (WRIGHT, 1982). The Stuart Fork terrane is interpreted as a section of oceanic crust and distal fan deposits carried down a subduction zone in the vicinity of the eastern Klamath region approximately 220 m.y. (BORNS, 1980; ERNST, 1984; GOODGE, 1985). Its juxtaposition with—and the origins of—the North Fork, Hayfork, and Rattlesnake Creek terranes are less clear, however (*e.g.*, BURCHFIEL and DAVIS, 1981; IRWIN, 1981; WRIGHT, 1982; ANDO *et al.*, 1983; MORTIMER, 1985).

Do these WTrPz terranes constitute a single, evolving, imbricated lithotectonic unit, or several unrelated, far-travelled entities? Are the original rocks parts of a landward arc, a continental margin, or portions of an oceanic plateau? Is the metamorphism coeval throughout the three western units, and is it characteristic of an island arc, a transpression boundary, a subduction zone, a back–arc basin, or yet another plate tectonic environment? In an attempt to address some of these questions, the author has been conducting field, petrographic, and geochemical investigations in the Sawyers Bar area of the central Klamath Mountains since 1979.

GENERAL GEOLOGY OF THE SAWYERS BAR AREA

The mapped area includes small portions of the Stuart Fork and possibly Hayfork terranes, but evidently consists chiefly of North Fork rocks. Local geologic relationships are shown in Figure 2. A small segment of the Stuart Fork thrust sheet, intruded by the Russian Peak pluton, is situated in the southeast corner of the map area (GOODGE, 1986); the English Peak Granodiorite crops out along the western margin of the studied area and appears to have intruded chiefly the eastern portion of the Hayfork belt (WRIGHT, 1982; IRWIN, 1985; COLEMAN *et al.*, 1987). Both granitoids were emplaced during mid Jurassic time subsequent to, or at the

terminal stages of, the WTrPz regional dynamothermal event.

The major portion of the mapped region, however, consists of massive to weakly foliated greenstones (mafic metavolcanics), and less voluminous interlayered metaclastic schists; the metavolcanics lie on strike with, and directly north of North Fork rocks termed ophiolitic by ANDO *et al.* (1983); the formational age of this suite is apparently Permo–Triassic. SNOKE *et al.* (1982) drew attention to a widely distributed group of basalts, basaltic andesites, and andesites of the western Klamaths and western Sierra Nevada, and correlated these volcanics (of which the Sawyers Bar greenstones represent a small part) with a series of coextensive ultramafic to dioritic intrusive complexes. BLAKE *et al.* (1982) recognized the North Fork metavolcanics as a separate exotic unit—the so-called Salmon River terrane; others (MORTIMER, 1985; DONATO, 1987) have included mafic meta–igneous rocks directly north of the Sawyers Bar area in the eastern Hayfork subterrane. Recognizing their on–strike correlation with ophiolitic rocks mapped by ANDO *et al.* (1983), the present author tentatively assigns the Sawyers Bar greenstones to the North Fork terrane of IRWIN (1972).

Primary(?) compositional layering and indistinct foliation in the greenstones and better layering in the metaclastics coincide in attitude, strike NS to N30E, and dip steeply, mostly to the east. Although the greenstones are relatively massive, rare pillows, providing unequivocal top indicators, were recognized in the northeastern part of the map area near locality 127M (Figure 2). Based on this occurrence and the stratigraphic coherence displayed by the several mapped lithologies, the region is provisionally interpreted as consisting of a major greenstone synform on the east, and progressively westward, very gently south–plunging, overturned, west-verging metasediment-cored antiforms and greenstone-cored synforms. Tectonic discontinuities between metasedimentary and metavolcanic lithologic packages were not recognized, although a few small lenses of serpentinite are present in the northern part of the mapped area. Protoliths for the metasedimentary rocks consist of pelitic units, argillites, micrographywackes, calcareous quartzofeldspathics, rare limey units, and minor cherty layers—in aggregate probably representing distal turbiditic (continental rise?) deposits.

In general, the greenstone units appear stratigraphically to overlie the fine–grained metaclastics, and dominate the exposed section in the Sawyers Bar area. Depositional interlayering and tight folding are evident from the intricately interdigitated

General Geology of the Sawyers Bar Area, Central Klamath Mountains

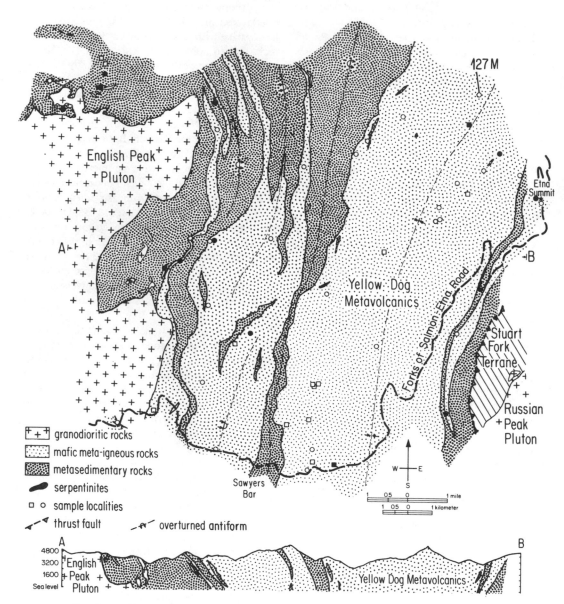

FIG. 2. Provisional, interpretive structure and general geology of the Sawyers Bar area, showing sample localities for metamorphosed mafic igneous rocks. Coarser grained hypabyssal rocks are indicated by square symbols; fine–grained, massive extrusives by circles. Filled symbols denote ultramafic bulk–rock compositions, open symbols mafic bulk–rock chemistry. Locality 127M is the site of unambiguous pillows, with tops facing east.

outcrop patterns of the metasedimentary and mafic meta–igneous units (Figure 2). Although the meta-volcanic units comprise chiefly massive flows, minor cross–cutting and concordant tabular bodies—most of which display coarser grained, relict diabasic texture—attest to the presence of hypabyssal meta-morphosed mafic intrusives and slowly-cooled flows distributed within the volcanic pile. The entire greenstone assemblage, including the associated dikes and sills, is here referred to informally as the Yellow Dog metavolcanics, after the prominent peak northeast of Sawyers Bar. The association with distal turbidites, and the fact that the extrusive section apparently overlies the sediments argue for construction in an island–arc setting. As will be shown later, the predominantly basaltic volcanics also include more evolved, calc–alkaline type units, supporting the hypothesis of formation in a primitive arc.

Except for the preservation of diabasic texture in some of the coarser grained intrusives, pillows in a few of the flows, and rare relict cpx in several of the Yellow Dog (chiefly hypabyssal) rocks, recrystallization has obliterated most original features in both meta–igneous and metasedimentary units. Petrographic studies demonstrate that the investigated rocks have been recrystallized under physical conditions of the upper greenschist facies (DONATO et al., 1982; MORTIMER, 1985; COLEMAN et al.,

1987). Biotite is an important phase in metasedimentary (especially pelitic and quartzofeldspathic) units, and is a widespread accessory in the mafic metavolcanics. The typical assemblage of the latter is albite + chlorite + actinolitic hornblende + epidote + sphene. Average modes for 297 pre-Nevadan metamorphic rocks examined petrographically are listed in Table 1. Low-variance, pressure-temperature diagnostic assemblages are rare in the studied area and geothermometric/geobarometric investigations are not yet completed. Judging from the experimental pressure-temperature location of the greenschist–epidote amphibolite facies boundary (LIOU et al., 1974; APTED and LIOU, 1983), and occurrences of andalusite directly to the north (DONATO, 1987), provisional estimated physical conditions for greenschist facies metamorphism in the Sawyers Bar area are 400° ± 75°C, 3 ± 1 kbar.

Megascopically, a few of the metavolcanic rocks which are situated adjacent to the English Peak and Russian Peak plutons are dark green, and were

Table 1. Average modes of pre-Nevadan North Fork/Hayfork metamorphic lithologics, Sawyers Bar area*

Mineral / Rock type (no. of samples)	Siliceous (22)	Pelitic (33)	Carbonaceous (18)	Quartzo-feldspathic (44)	Serpentinite (6)	Mafic volcanic (21)	More felsic volcanic (118)	Hypabyssal (35)
Quartz	79	57	54	38	—	2	2	5
Plagioclase	1	2	2	17	—	17	26	34
Pyroxene	—	—	—	1[a]	1[b]	—	2[b]	3[b]
Amphibole	—	—	1	12	14	43	36	31
Biotite	4	13	2	6	—	3	1	2
White mica	7	14	14	9	—	2	1	2
Chlorite	2	8	2	5	3	16	9	8
Talc	—	—	—	—	31	—	—	tr
Serpentine	—	—	—	—	40	—	—	—
Epidote	tr	tr	1	1	—	6	16[c]	8
Garnet	tr	tr	—	tr	—	—	—	—
Carbonate	tr	tr	—	5	6	3	2	2
Carbonaceous matter	3	2	19	2	—	—	—	—
Opaques	4	4	5	3	5	4	2	2
Sphene ± rutile	—	tr	—	1	—	4	3	3

a = diopside in marble layers; b = relict igneous pyroxene; c = tr prehnite, stilbite.
* Modes visually estimated.

mapped in the field as mafic metavolcanics. Most of the greenstones possess pale, apple–green or light–gray colors, however, and were originally regarded as more felsic metavolcanics and intrusive equivalents. Modes presented in Table 1 demonstrate that both rock types are, in fact, quite mafic in composition, although the lighter–colored metavolcanics and meta-hypabyssals contain greater proportions of albite + epidote and less amphibole + chlorite than the darker metavolcanics. The conclusion that all the metavolcanics are mafic is supported by chemical analyses now to be discussed.

MAJOR ELEMENT CHEMISTRY OF MAFIC META–IGNEOUS ROCKS

X-ray fluorescence (XRF) bulk–rock analyses of 17 hypabyssal and 38 extrusive mafic metavolcanics from the Sawyers Bar area are presented in Table 2. Analytical methods were described by CORTE-SOGNO et al. (1977). Locations of analyzed Sawyers Bar samples are indicated on Figure 2. Only samples which appeared fresh and generally free from veining were collected in the field. After petrographic examination, the least altered of these were chosen for bulk–rock and mineral analyses. CIPW normative minerals were computed for the rock analyses listed in Table 2, assuming a $Fe^{2+}/(Fe^{3+} + Fe^{2+})$ ratio of 0.80, as is reasonable for fresh basalts. Only one rock (sample no. 1M) contains normative

nepheline—4 weight percent; two others, 288M and 372M, are critically undersaturated. The remaining 52 specimens are hypersthene normative.

Bulk XRF data indicate that the darker metavolcanics (Table 2a) in general are slightly more titaniferous than the lighter green units (Table 2b) with which they are regionally associated. The modal abundance of Ti–rich oxides + sphene, and slightly more voluminous hornblende evidently account for the more ferromagnesian aspect of the dark green units. However, the latter do not occur as stratigraphically distinct flows. Accordingly, they are not distinguished from the lighter–colored mafic metavolcanics in Figures 2–9. Most of the coarser grained, recrystallized hypabyssal rocks appear to be slightly more silicic and alkalic and lower in iron + magnesium than the meta–extrusives (compare Table 2c with 2a,b). Bulk chemical variations are plotted for all the Yellow Dog meta–igneous rocks in Figures 3–8. Intrusive and extrusive protoliths are shown separately. As will now be discussed, fourteen of the analyzed rocks—eleven extrusives and three hypabyssals—possess compositional affinities with picritic or komatiitic basalts. Several others (e.g., sample 81M) are transitional in chemical properties between those exhibiting ultramafic affinities and the more normal basaltic meta–igneous rocks. The highly magnesian character of the komatiitic(?) rocks was only recognized after XRF analyses, hence field collection and petrographic

Table 2a. Bulk–rock major element chemistry, mafic metavolcanics, Sawyers Bar area

Weight percent	1M	11M†	38M†	46M	67M†	89M†	91M†	92M†	153M	170M	174M†	201M	207M	Ave₁₃
SiO_2	43.75	46.40	49.70	51.90	52.00	46.80	45.20	48.80	51.40	51.80	50.15	49.35	48.35	48.89 ± 2.69
Al_2O_3	14.30	12.40	10.50	14.60	10.70	10.00	10.60	10.80	11.10	13.95	10.30	13.70	13.84	12.06 ± 1.76
MgO	10.30	14.10	14.30	10.70	14.10	12.25	14.00	15.00	9.60	9.40	16.10	9.20	7.40	12.03 ± 2.74
CaO	6.10	10.25	7.50	11.20	9.60	13.25	12.70	13.40	6.40	6.40	9.20	6.30	10.80	9.47 ± 2.73
Na_2O	3.40	1.70	2.00	2.30	1.35	0.91	0.91	1.20	3.50	2.00	1.50	2.90	1.12	1.91 ± 0.89
K_2O	1.45	0.52	0.90	0.32	1.02	0.34	0.70	0.10	0.36	0.71	0.58	0.16	1.50	0.67 ± 0.45
Fe_2O_3‡	15.90	11.70	12.30	8.30	10.00	13.50	12.30	9.00	14.70	13.10	11.20	14.60	13.30	12.30 ± 2.25
TiO_2	4.00	2.20	2.17	0.20	0.65	2.38	2.64	0.11	2.20	2.20	0.50	3.10	2.80	1.93 ± 1.20
MnO	0.22	0.17	0.19	0.17	0.18	0.15	0.18	0.18	0.17	0.15	0.18	0.18	0.17	0.18 ± 0.02
P_2O_5	0.28	0.32	0.28	0.09	0.14	0.29	0.43	0.01	0.15	0.30	0.15	0.38	0.32	0.24 ± 0.12
Total	99.90	99.97	99.99	99.85	99.91	100.02	99.89	99.71	99.63	99.96	99.98	99.96	99.70	
ppm														
Cr	60	485	425	75	981	403	549	644	97	136	642	113	176	368 ± 287
Cu	94	97	48	55	80	33	26	24	42	22	31	74	89	55 ± 28
Ni	43	374	299	83	219	310	359	131	42	66	153	46	68	169 ± 127
Rb	14	5	28	10	34	2	17	1	0	21	10	1	27	13 ± 12
Sr	508	635	326	450	261	421	954	202	100	360	216	379	440	404 ± 218
Zn	113	274	111	42	81	133	114	44	77	79	61	77	81	99 ± 59
Zr	148	217	254	0	73	242	271	0	116	206	87	245	208	159 ± 95

† Indicates komatiitic meta–igneous rock.
Analyst: G. Stummer, UCLA.

Table 2b. Bulk–rock major element chemistry, more felsic metavolcanics, Sawyers Bar area

Weight percent	5M	6M	9M†	80M	84M	85M	86M	103M	115M	117M	127M	146M
SiO_2	50.10	52.50	48.00	51.50	51.50	49.80	50.60	50.20	51.20	51.35	53.90	51.80
Al_2O_3	12.80	11.50	12.77	11.60	12.60	12.90	12.00	12.65	12.60	13.30	11.80	12.60
MgO	9.77	9.23	14.90	11.56	10.50	11.15	12.10	10.85	10.50	9.30	10.70	11.70
CaO	9.97	11.50	9.40	9.90	10.20	11.15	8.20	11.50	7.94	10.10	9.45	9.30
Na_2O	2.75	2.70	1.70	2.90	2.75	1.90	2.90	2.40	3.20	3.02	2.60	2.60
K_2O	0.69	0.15	0.30	0.30	0.21	1.00	0.42	0.11	0.34	0.08	0.14	0.15
$Fe_2O_3^{\ddagger}$	12.10	10.75	10.80	10.70	10.70	10.80	12.10	10.75	12.45	11.35	10.10	10.40
TiO_2	1.22	1.17	0.82	1.00	1.02	0.75	1.13	1.00	1.09	0.97	0.76	0.93
MnO	0.18	0.18	0.17	0.17	0.16	0.17	0.19	0.17	0.24	0.14	0.14	0.14
P_2O_5	0.10	0.09	0.09	0.08	0.09	0.08	0.10	0.11	0.08	0.07	0.07	0.07
Total	99.76	99.84	99.83	99.78	99.79	99.79	99.83	99.81	99.70	99.73	99.71	99.75
ppm												
Cr	224	169	506	212	209	367	318	231	138	129	230	245
Cu	70	79	68	121	86	115	113	56	55	45	83	35
Ni	100	133	269	105	97	181	173	67	55	41	84	99
Rb	13	2	20	7	6	28	15	5	13	3	4	2
Sr	265	169	109	139	85	101	66	274	135	90	76	183
Zn	91	79	197	76	70	72	93	38	110	54	46	40
Zr	73	60	68	59	86	69	93	83	70	74	59	80

Weight percent	161M	215M	237M	245M	249M	272M	288M	312M	320M†	354M†	372M
SiO_2	50.75	51.00	48.30	52.30	55.30	50.20	50.50	50.20	49.20	46.05	48.55
Al_2O_3	13.25	14.20	14.30	14.20	13.00	14.80	15.80	13.45	9.50	11.83	13.40
MgO	11.30	9.15	10.30	8.50	8.50	10.60	6.25	10.60	18.60	13.12	7.10
CaO	8.96	9.40	12.00	10.50	8.00	7.60	13.10	8.32	8.10	9.33	12.10
Na_2O	3.00	2.90	2.60	2.80	2.50	3.40	3.10	2.80	1.50	1.54	3.20
K_2O	0.23	0.35	0.11	0.59	2.10	0.15	0.60	0.30	0.62	0.42	0.20
$Fe_2O_3^{\ddagger}$	11.00	11.60	10.70	9.80	9.00	12.00	9.00	12.50	11.00	14.80	12.85
TiO_2	0.88	1.06	1.12	0.89	0.75	1.01	1.16	1.24	0.79	2.07	1.95
MnO	0.17	0.17	0.17	0.16	0.14	0.18	0.27	0.15	0.18	0.17	0.14
P_2O_5	0.08	0.08	0.10	0.07	0.24	0.07	0.21	0.11	0.16	0.36	0.25
Total	99.67	99.96	99.79	98.29	99.53	100.01	100.03	99.73	99.68	99.82	99.81
ppm											
Cr	178	118	270	185	240	300	62	117	1000	445	118
Cu	54	59	75	105	63	95	20	62	64	38	30
Ni	66	48	135	75	40	110	40	55	400	345	53
Rb	2	6	13	8	40	10	0	6	6	10	0
Sr	113	206	110	110	340	160	150	200	115	200	290
Zn	45	54	70	66	68	80	81	85	88	105	84
Zr	66	87	60	38	90	52	92	95	80	160	140

Weight percent	374M	375M†	Ave (25 analyses)
SiO_2	51.20	48.30	50.57 ± 1.93
Al_2O_3	14.00	9.00	12.79 ± 1.49
MgO	8.80	18.18	10.93 ± 2.88
CaO	9.30	10.00	9.82 ± 1.46
Na_2O	2.60	1.35	2.59 ± 0.56
K_2O	0.67	0.39	0.46 ± 0.44
$Fe_2O_3^{\ddagger}$	11.75	11.11	11.20 ± 1.24
TiO_2	1.08	0.75	1.06 ± 0.32
MnO	0.14	0.18	0.17 ± 0.03
P_2O_5	0.10	0.20	0.12 ± 0.07
Total	99.72	99.65	
ppm			
Cr	205	1000	288 ± 238
Cu	70	22	67 ± 28
Ni	72	500	134 ± 120
Rb	0	0	9 ± 9
Sr	190	150	161 ± 72
Zn	77	86	78 ± 31
Zr	114	105	82 ± 27

Table 2c. Bulk–rock major element chemistry, hypabyssal meta–igneous rocks, Sawyers Bar area

Weight percent	54M	57M	65M	69M	81M	124M†	212M†	220M	221M	222M	235M
SiO_2	54.40	52.50	54.00	56.20	49.10	52.30	48.80	51.80	58.70	55.90	49.55
Al_2O_3	13.35	14.50	13.25	14.25	12.30	11.00	10.90	16.95	15.90	13.80	15.30
MgO	8.50	11.50	9.40	7.30	12.15	14.00	16.40	7.30	5.00	7.20	9.50
CaO	8.10	8.95	9.40	6.70	9.60	8.75	8.50	9.00	5.30	5.80	10.20
Na_2O	1.75	2.00	2.12	2.70	2.90	2.00	1.60	2.80	3.20	3.50	2.90
K_2O	1.92	1.40	1.72	1.73	0.76	1.15	1.10	1.00	2.80	1.35	0.20
$Fe_2O_3^*$	10.30	8.25	8.60	9.57	11.40	9.60	11.10	9.60	7.65	10.90	10.90
TiO_2	1.10	0.53	0.85	1.02	1.22	0.68	0.85	0.82	0.65	0.86	1.07
MnO	0.18	0.14	0.17	0.17	0.18	0.15	0.17	0.19	0.13	0.16	0.16
P_2O_5	0.26	0.08	0.21	0.25	0.22	0.13	0.17	0.29	0.25	0.23	0.10
Total	99.96	99.95	99.81	99.98	99.97	99.87	99.73	99.86	99.69	99.78	99.88
ppm											
Cr	126	232	227	265	641	715	696	56	36	115	260
Cu	37	49	85	50	88	26	41	35	74	29	80
Ni	22	169	60	50	208	206	242	21	21	10	108
Rb	86	55	35	50	10	27	26	22	72	36	14
Sr	500	327	330	380	306	93	222	768	660	404	100
Zn	108	81	86	90	76	56	57	69	68	77	76
Zr	165	49	80	85	71	75	90	120	235	130	45

Weight percent	241M	285M	321M	329M	347M	351M†	Ave (17 analyses)
SiO_2	53.90	56.40	55.50	54.46	53.47	54.10	53.59 ± 2.71
Al_2O_3	15.90	15.40	15.25	13.90	14.35	11.20	13.97 ± 1.80
MgO	5.30	5.90	5.50	7.77	9.52	12.56	9.11 ± 3.27
CaO	6.20	8.30	7.55	8.35	8.32	9.88	8.17 ± 1.43
Na_2O	2.70	3.15	2.80	2.73	2.22	0.58	2.45 ± 0.72
K_2O	2.10	1.05	2.42	1.53	0.97	1.05	1.43 ± 0.64
$Fe_2O_3^*$	12.20	8.50	9.50	10.11	9.95	9.52	9.86 ± 1.20
TiO_2	1.03	0.85	0.87	0.78	0.76	0.78	0.87 ± 0.18
MnO	0.18	0.15	0.15	0.18	0.17	0.18	0.17 ± 0.02
P_2O_5	0.41	0.27	0.28	0.26	0.13	0.25	0.22 ± 0.08
Total	99.92	100.12	99.91	100.20	99.95	100.23	
ppm							
Cr	50	134	75	170	325	680	283 ± 236
Cu	70	570	55	70	65	50	87 ± 126
Ni	19	70	35	65	60	160	90 ± 77
Rb	40	40	35	27	16	28	36 ± 20
Sr	540	490	450	760	245	210	399 ± 204
Zn	105	75	90	90	85	82	81 ± 14
Zr	150	160	160	145	92	130	117 ± 50

selection for detailed chemical investigation, based on freshness of specimens, is not regarded as having introduced a sample bias. Collecting in the northeastern corner of the mapped area, DONATO (1985) also reported the presence of several dikes and sills characterized by high MgO contents.

Although alkali metasomatism may have affected these units during metamorphism, Figure 3 demonstrates that protoliths of analyzed metavolcanic rocks from the Sawyers Bar area belong chiefly to tholeiitic and high–alumina basalt suites. Of the 55 analyses, only one (non–komatiitic sample 1M) appears to be distinctly alkalic, as is also clear from the presence of four weight percent normative nepheline required by this bulk composition. The nearly complete absence of alkali basaltic protoliths is also supported by clinopyroxene analyses to be presented farther on. The specimens regarded as possibly komatiitic have low total alkalis and moderate silica contents. Clearly, Si–, Na–, K–metasomatism have not strongly modified the rock bulk compositions during metamorphism, except probably that of sample 1M.

An AFM diagram is illustrated in Figure 4. Analyzed Yellow Dog greenstones in general plot along the Skaergaard trend, but do not exhibit marked

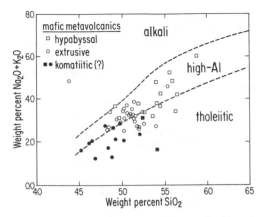

FIG. 3. Total alkalis versus silica variations for mafic meta–igneous bulk–rock analyses, Sawyers Bar area (data of Table 2). The tholeiitic, high–alumina and alkali basalt fields for unaltered volcanics are shown after KUNO (1960, 1966). As with Figures 4 and 8, field boundaries are valid only if oxides plotted have not been remobilized subsequent to igneous solidification.

iron enrichment. Most hypabyssal units are alkali enriched, and show chemical characteristics transitional between tholeiitic and calc–alkaline differentiation trends. In contrast, the komatiitic(?) basalts and a few hypabyssal rocks are impoverished in alkalis, and tend towards more magnesian compositions compared to the rest of the extrusive greenstones.

Komatiitic lavas are characterized by high CaO/Al_2O_3 weight percent ratios, as well as by elevated MgO contents. VILJOEN and VILJOEN (1969) defined komatiites as having CaO/Al_2O_3 ratios ex-

ceeding unit value, although others (*e.g.,* ARNDT and NISBET, 1982; XU and CHEN, 1984) have indicated that slightly lower ratios, 0.8 or even 0.7, may still qualify highly magnesian aphyric lavas as komatiitic. High MgO contents are diagnostic of ultramafic lavas. Hybrid picritic rocks (*i.e.,* the solidification products of mafic melts + cumulate olivine crystals) and boninites are also characterized by high magnesia, although the CaO/Al_2O_3 values for such crystal + liquid assemblages are much lower than unity. (However, as will be shown in Table 6, some island–arc picrites appear to be chemically rather similar to komatiites.) The ratio of CaO/Al_2O_3 is plotted against MgO in Figure 5 for the Yellow Dog metavolcanics. Clearly, the fourteen analyzed magnesian samples as a group have komatiitic(?) chemistries. The lack of pseudomorphs after olivine and the probable aphanitic nature of the precursor lavas (as indicated by a lack of textural relics) also suggest a komatiitic rather than a picritic protolith.

Elevated nickel and chromium contents typify olivine–, chrome diopside– and spinel–rich rocks. Liquids derived through high degrees of partial fusion of such solid precursor assemblages, therefore, tend to be enriched in these transition elements, whereas olivine–cumulate basalts should exhibit high Ni but more nearly normal Cr contents relative to MORB. (Of course, olivine + chrome–diopside cumulate hybrids, ankaramites, would also exhibit high Cr and Ni values, but ankaramites are typically alkalic and titaniferous.) The Yellow Dog magnesian metabasalts shown in Figure 6, a Ni–Cr plot, are compositionally distinct from—but transitional with—the associated mafic metavolcanics of more normal chemistry.

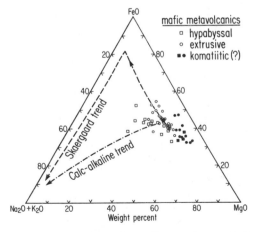

FIG. 4. AFM diagram for mafic meta–igneous bulk–rock analyses, Sawyers Bar area (data of Table 2). The Skaergaard and calc–alkaline trends of igneous differentiation are shown after WAGER and DEER (1939) and DALY (1933), respectively.

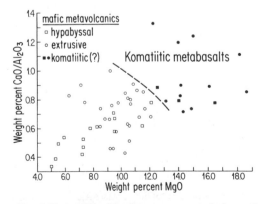

FIG. 5. Ratio of lime/alumina versus magnesia for mafic meta–igneous bulk–rock analyses, Sawyers Bar area (data of Table 2). The dashed line separating metamafics of komatiitic(?) affinities from more normal meta–igneous rocks is arbitrary.

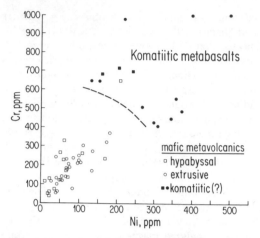

FIG. 6. Proportions of chromium versus nickel for mafic meta–igneous bulk–rock analyses, Sawyers Bar area (data of Table 2). The dashed line separating metamafics of komatiitic(?) affinities (+ hypabyssal sample 81M) from more normal meta–igneous rocks is arbitrary.

More evolved melts in general have higher titanium contents, and display notable iron enrichment. The Yellow Dog greenstone bulk–rock analyses are chemographically portrayed in Figure 7 in terms of weight percent TiO_2 versus mol fraction Fe (*i.e.*, Fe*/(Mg + Fe*), or X_{Fe}). The metamorphosed hypabyssal rocks show nearly constant TiO_2 contents, regardless of X_{Fe}. Among the metamorphosed extrusives, for any given X_{Fe}, rocks of presumed komatiitic affinity are slightly more titaniferous than the associated normal greenstones; in general, although a range in X_{Fe} exists, the former are less fractionated than the latter. If the analyzed non-alkalic komatiitic(?) lavas were cumulate (olivine + clinopyroxene)–enriched hybrids, the titania

contents would be expected to be lower than parental basalts rather than higher, as is observed.

For metamorphosed rocks—and in particular, for mafic metavolcanics—metasomatism may alter the chemical proportions of the more mobile constituents. JENSEN (1976) pointed out that the most stable, inert components (*e.g.*, KORZHINSKII, 1959) should provide the best indication of original lithologies, and on this basis instituted a triangular diagram, the apices of which are the cations Mg, (Fe* + Ti), and Al. Jensen subdivided this ternary to provide separate fields for the well–recognized calc–alkaline, tholeiitic and komatiitic magma types. According to this diagram, presented as Figure 8, the Yellow Dog metavolcanics are komatiitic

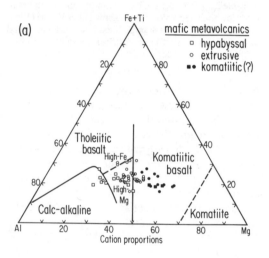

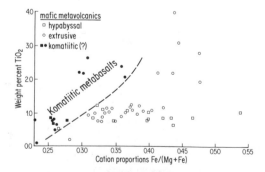

FIG. 7. Proportions of titania versus X_{Fe} for mafic meta–igneous bulk–rock analyses, Sawyers Bar area (data of Table 2). The dashed line separating metamafics of komatiitic(?) affinities + hypabyssal specimen 57M (and hypabyssal sample 81M situated on the line) from more normal meta–igneous rocks is arbitrary.

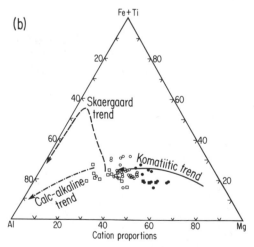

FIG. 8. Metal proportions of Mg, (Fe* + Ti) and Al for mafic meta–igneous bulk–rock analyses, Sawyers Bar area (data of Table 2). The fields for the various magma types (a), and crystal-fractionation trends (b) are illustrated after JENSEN (1976).

basalts and high–Mg tholeiites, with an important portion of the hypabyssals exhibiting chemical tendencies transitional towards calc–alkaline basalts. They appear to represent an intergradational, consanguineous suite. The relatively high proportion of komatiitic samples (14 of 55 analyzed specimens) supports the contention that these rocks are products of a widely operating process, and may represent true liquids rather than crystal + melt hybrids.

The low–alkali, low–silica nature of the Sawyers Bar mafic meta–igneous suite, as well as systematic magmatic chemical differentiation trends, leads to the suggestion that major element metasomatism played only a minor role during the subsequent metamorphism.

RARE EARTH ELEMENT CHEMISTRY OF MAFIC META–IGNEOUS ROCKS

Bulk–rock instrumental neutron activation analyses (INAA) of rare earth elements (REE) for twelve Yellow Dog metavolcanics are presented in Table 3, along with chondrite normalizing values. Analytical data were provided employing routine techniques (*e.g.,* BAEDECKER, 1976; KALLEMEYN and WASSON, 1981). Rare earth patterns are plotted in Figure 9.

The six komatiitic(?) greenstones form a very well–defined group characterized by marked REE enrichments relative to chondrites, and high values of the LREE/HREE ratio (Figure 9a). In contrast, although two of the six Yellow Dog greenstones of more basaltic major element chemistry (sample numbers 207M and 32IM) exhibit patterns quite similar to the units of ultramafic affinity, the other four possess somewhat variable but unfractionated patterns, much less enriched in rare earths (Figure 9b). One of the latter (sample number 46M) possesses a distinct positive Eu anomaly, suggesting the possibility of cumulate plagioclase in this protolith.

Uniformity of the komatiitic(?) metavolcanic REE compositions (and two of six analyzed metabasaltic units) indicates the crustal introduction of a chemically homogeneous magma type. Elevated LREE concentrations and fractionated rare earth element patterns suggest the likelihood of derivation from a garnet-bearing mantle protolith, but argue against high degrees of partial fusion of the source rock. Strong REE enrichment relative to chondrites allows the conjecture that, at depths within the stable non-subducted lithosphere, the postulated garnet lherzolite source had been previously enriched in REE, especially LREE, possibly by mantle metasomatism associated with descent of the downgoing, devolatilizing slab (*e.g.,* MYSEN, 1979). Alternatively, small amounts of strongly LREE–enriched, siliceous liquid produced by partial fusion of eclogitic material within the subducted plate (see APTED, 1981), when mixed with mantle–derived subsiliceous melt containing near-chondritic REE concentrations, might account for the observed major element and REE chemistry of the Yellow Dog komatiitic metavolcanics. Yet another possibility might involve reaction with crustal materials, but it is hard to imagine how the ultramafic affinities of the Yellow Dog metakomatiites could be retained during assimilation of sialic components. However, Late Archean magnesian metabasalts from western Australia have been shown to contain older zircon populations, and bulk–rock chemistries are strongly LREE enriched, indicating that contamination of felsic crust has played a role in the petrogenesis of some greenstone belts (COMPSTON *et al.,* 1986; ARNDT and JENNER, 1986).

The diverse REE patterns which characterize the more normal basaltic metavolcanics may be a consequence of: (1) two or more mafic magma types present in the terrane; (2) exchange with mantle wall rocks accompanying melt ascent and emplacement; (3) variable degrees of crustal contamination;

Table 3. Bulk–rock rare earth element chemistry, mafic metavolcanics, Sawyers Bar area

ppm	Normalizing values	11M*	46M	57M	89M*	117M	207M	212M*	215M	320M*	321M	351M*	375M*
La	0.315	25.9	0.3	3.9	25.7	3.7	28.0	11.0	3.5	10.2	20.0	15.5	14.8
Ce	0.813	55	2	11	56	14	60	25	10	23	41	32	35
Nd	0.597	27	<3	6	27	7	29	12	7	12	22	15	17
Sm	0.192	6.05	0.42	1.53	6.38	2.23	5.40	2.91	2.35	2.98	5.08	3.93	3.48
Eu	0.0722	2.01	0.43	0.60	2.77	0.90	1.78	0.95	0.90	0.93	1.38	1.13	1.04
Tb	0.049	0.8	0.1	0.2	1.0	0.5	0.9	0.6	0.5	0.3	0.6	0.7	0.4
Yb	0.209	1.78	0.60	1.27	1.31	2.63	1.95	1.75	2.55	1.60	2.10	1.85	1.37
Lu	0.0323	0.28	0.10	0.21	0.18	0.39	0.27	0.26	0.40	0.25	0.35	0.27	0.19

* Indicates komatiitic meta–igneous rock.
Analyst: X-ray Assay Laboratories, Ltd., Don Mills, Ontario.

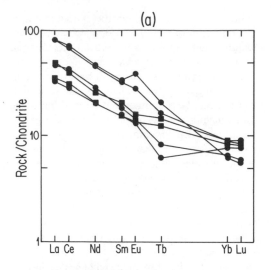

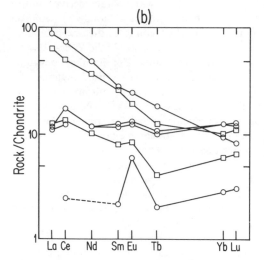

FIG. 9. Chondrite-normalized bulk–rock rare earth patterns for mafic meta–igneous rocks, Sawyers Bar area (data of Table 3): (a) rocks of komatiitic(?) affinity; (b) rocks of basaltic affinity. Symbols conform to usage of Figures 2–8.

judging by this negative textural evidence. For the porphyritic mafic meta–igneous rocks, clinopyroxenes enriched in Al, Cr, Na and Ti would be expected in rocks of ankaramitic—and other alkali basaltic—bulk compositions (*e.g.*, HUCKENHOLZ, 1965, 1966; HAWKINS and NATLAND, 1975; IMSLAND, 1984).

Electron microprobe analyses of Sawyers Bar mineral assemblages were obtained on a Cameca automated electron microprobe, employing analytical and data-reduction techniques described previously (ERNST, 1976). Core and rim values were obtained for two or three clinopyroxene in each of three non–komatiitic rocks; in all cases, compositional zoning was searched for but not detected. As shown in Table 4, averaged relict clinopyroxene compositions from three specimens, metavolcanic 215M, and hypabyssals 221M and 321M, exhibiting X_{Fe} values of 0.174, 0.297 and 0.303 respectively, appear to conform to a tholeiitic trend (CARMICHAEL, 1967; GARCIA, 1978). The subcalcic augites are uniformly low in aluminum, titanium, chromium and sodium. Rock sample no. 321M, although not highly magnesian, does exhibit strong LREE enrichment, like the analyzed komatiitic(?)

Table 4. Major element mineral chemistry of relict clinopyroxenes from mafic metavolcanics, Sawyers Bar area

Weight percent	215M	221M	321M
SiO_2	52.54	51.34	51.52
Al_2O_3	2.23	2.30	2.07
Cr_2O_3	0.09	0.01	0.04
TiO_2	0.38	0.50	0.62
FeO	6.58	10.53	11.36
MgO	17.57	13.96	14.67
MnO	0.18	0.31	0.34
CaO	19.45	20.11	19.17
Na_2O	0.17	0.39	0.27
K_2O	0.00	0.00	0.01
Total	99.20	99.44	100.05
cations/ 6 oxygens			
Si	1.939	1.934	1.931
Al^{IV}	0.061	0.066	0.069
Al^{VI}	0.036	0.036	0.022
Cr	0.003	0.000	0.001
Ti	0.011	0.014	0.017
Fe^{2+}	0.203	0.332	0.356
Mg	0.967	0.783	0.819
Mn	0.006	0.010	0.011
Ca	0.769	0.812	0.770
Na	0.012	0.029	0.019
K	0.000	0.000	0.000
Total	4.006	4.015	4.016

Analyst: W. G. Ernst, UCLA.

and/or (4) metasomatic alteration during WTrPz regional dynamothermal metamorphism. Too few data are available to allow the erection of a compelling hypothesis.

RELICT CLINOPYROXENE COMPOSITIONS OF MAFIC META–IGNEOUS ROCKS

Subcalcic augites are present as relict, stumpy igneous phenocrysts in a few of the Yellow Dog hypabyssals and the more felsic greenstones (Table 1), but are totally lacking in the 14 analyzed rocks which display ultramafic characteristics. Evidently protoliths of the komatiitic extrusives were aphyric, metavolcanics. These data reinforce the conclusion that the Yellow Dog metavolcanics are not of alkaline affinities, and therefore do not represent alkali basaltic or ankaramitic lavas. Except for somewhat lower Al contents, the analyzed Sawyers Bar relict clinopyroxene phenocrysts are virtually indistinguishable from subcalcic augites of Archean komatiitic flows from northeastern Ontario (ARNDT et al., 1977).

OXYGEN ISOTOPE CHEMISTRY OF MAFIC META-IGNEOUS ROCKS

Bulk–rock mass spectrometric analyses of ^{16}O and ^{18}O were obtained for six Yellow Dog metavolcanic specimens, employing routine techniques. Data are presented in Table 5.

The investigated samples, four metakomatiitic and two of more normal chemistry, are all ^{18}O–enriched ($\delta^{18}O$ = 9.6–15.3) relative to Phanerozoic MORBs as well as Archean komatiites ($\delta^{18}O$ = 6; TAYLOR, 1968; LONGSTAFFE et al., 1977; HOEFS and BINNS, 1978; BEATY and TAYLOR, 1982). Limited isotopic analyses of the Sawyers Bar greenstones prevent the demonstration of detailed relationships, but several observations, nevertheless, can be made: (1) All analyzed rocks have exchanged isotopically with large quantities of oxygen-bearing fluid under low–grade metamorphic conditions. Employing the method of ELTHON et al. (1984), which involves weighting fractionation factors for the silicates present and assumes equilibrium between aqueous fluid and the condensed assemblage, it appears that isotopic exchange took place with magmatic water (or seawater) at low temperatures, on the order of 100°C; nominal water/rock ratios indicated for this isotherm range from 0.4 (or 1.0) to 3 (or 20). (2) There is no systematic isotopic contrast between rocks of mafic versus ultramafic affinity, or between extrusives (sample numbers 11M, 89M, 215M, 375M) versus hypabyssals (sample nos. 321M, 351M). (3) An inverse correlation between chromium–, nickel– or magnesium–contents on the one hand, and $\delta^{18}O$ enhancement on the other is not evident, nor are the rare earth element–, SiO_2–, Na_2O– or K_2O–contents, or modes of minerals such as alkali feldspars and/or micas proportional to $\delta^{18}O$—as might be expected if metasomatism had modified bulk–rock concentrations in these units.

Therefore, although the Yellow Dog metavolcanics have been subjected to the throughput of large volumes of an oxygen-bearing fluid during recrystallization, and have at least partially re-equilibrated isotopically, the major element and REE chemistries of these greenstones are not principally metasomatic in origin.

PETROCHEMICAL DISCUSSION

Bulk–rock major element chemical data have been presented which demonstrate that, among the investigated mafic metavolcanic rocks of the Sawyers Bar area, fourteen are strongly magnesian, possess high CaO/Al_2O_3 ratios, and elevated concentrations of Cr and Ni; they are not notably titaniferous or iron–enriched, and are neither alkalic nor, apparently, the result of olivine ± clinopyroxene accumulation. Employing a variety of major element discrimination diagrams, these fourteen greenstones (three of which are hypabyssal) appear to represent recrystallized komatiitic(?) basalts. Unlike Archean komatiitic basalts, they are neither associated with true komatiites, nor display spinifex texture. The remaining 27 analyzed meta–extrusives are chiefly magnesian metabasalts, although a few display modest iron–enrichment. With several exceptions, the associated coarser grained dikes and sills are more evolved and are enriched in Na_2O, K_2O and SiO_2; most possess bulk chemistries gradational to calc–alkaline basalt/andesite. Some of the meta–extrusives and metahypabyssals exhibit compositions transitional between the more normal recrystallized mafic igneous rocks and the picritic/ komatiitic types.

Approximately a quarter of the analyzed Yellow Dog metavolcanics are highly magnesian [MgO (average of 14 analyses) = 14.83 ± 1.90 weight percent] in their bulk–rock chemistries. Nevertheless, lacking diagnostic textural criteria, ambiguities exist in the distinction of former komatiitic liquids from hybrid basaltic melts containing cumulate olivine ± clinopyroxene (picrites ± ankaramites), as discussed by many authors (e.g., ARNDT et al., 1977; ARNDT and NISBET, 1982; NESBITT et al., 1982).

Table 5. Bulk–rock oxygen isotope chemistry, mafic metavolcanics, Sawyers Bar area

Sample no.	$\delta^{18}O$
11M*	9.6†
89M*	13.2†
215M	10.0
321M	11.4
351M*	15.3†
375M*	11.1

* Indicates komatiitic meta–igneous rock.

† Indicates duplicate analyses on separate portions of sample analyst: Geochron Laboratories, Cambridge, Massachusetts.

The Sawyers Bar mafic meta–igneous rocks of komatiitic affinities contrast markedly with average calc–alkaline island–arc volcanic rocks (*e.g.*, EWART, 1976; TAYLOR, 1977) which are much higher in silica, alumina and alkalis, and substantially lower in magnesia and lime. Oceanic and abyssal tholeiites from the North Atlantic (*e.g.*, JAKOBSSON, 1972; WOOD, 1976; SIGURDSSON *et al.*, 1978; IRVINE, 1979) and plateau basalts from peninsular India (*e.g.*, GHOSE, 1976) all possess low CaO/Al_2O_3 ratios, low MgO, and generally high TiO_2 contents compared with the Sawyers Bar komatiitic(?) greenstones.

The latter are compared in Table 6 with modern boninites and a Cenozoic picrite from the South Pacific, a mean of picrites from the circum–Atlantic, Phanerozoic ankaramites from southernmost British Columbia and from the Atlantic realm, and Archean komatiites from Ontario and South Africa. All rock types listed are characterized by high concentrations of Cr and Ni. The average for fourteen Yellow Dog komatiitic(?) greenstones most closely corresponds to the Ontario olivine–rich pyroxenitic komatiite, except that the TiO_2 and K_2O contents of the Sawyers Bar lithology are slightly higher. The

Barberton komatiite is a good match too, although it is somewhat more siliceous. The southwest Pacific picritic basalt, surprisingly, has a CaO/Al_2O_3 ratio of unity, and a very high MgO content. Perhaps two different kinds of rock have been termed "picrite" in the literature, olivine–cumulate occurrences, such as have been described from the British Isles, and more nearly primitive Pacific arc lavas. Lavas of the former do not represent true melts, but the latter possibly might, and appear to be chemically indistinguishable from komatiitic basalts. It is to this second category that the Sawyers Bar metavolcanics evidently belong. Boninites (CRAWFORD *et al.*, 1981; HAWKINS *et al.*, 1984) are too rich in SiO_2 and impoverished in CaO (also probably they are too poor in TiO_2, MgO and $Fe_2O_3^*$) to compare satisfactorily with the Yellow Dog ultramafic rocks. Ankaramites possess bulk chemistries rather similar to picrites and komatiites, but most are markedly enriched in alkalis and iron plus titania relative to the Sawyers Bar and Archean metakomatiites (SORENSON, 1974; IMSLAND, 1984); so, too, are nephelinites and basanites (HAWKINS and NATLAND, 1975). Thus, komatiitic and (noncumulate?) picritic basalt seems to best describe the

Table 6. Bulk–rock major element chemistry, mafic–igneous rocks of ultramafic affinities

Weight percent	(1)	(2)	(3)	(4)	(5)	(6)	(7)	(8)
SiO_2	48.99 ± 2.54	57.1 ± 1.7	48.08 ± 1.35	40.93 ± 2.65	44.99	42.18 ± 2.53	48.30	52.92
Al_2O_3	10.82 ± 1.03	9.8 ± 1.6	12.24 ± 1.58	9.94 ± 2.11	6.54	8.27 ± 3.20	10.80	8.14
MgO	14.83 ± 1.90	12.5 ± 2.6	12.45 ± 1.13	15.05 ± 3.05	28.10	23.29 ± 4.96	13.90	11.81
CaO	9.99 ± 1.86	5.41 ± 1.64	9.36 ± 1.68	12.87 ± 1.28	6.56	5.99 ± 2.28	10.00	11.53
Na_2O	1.42 ± 0.41	1.70 ± 0.75	1.97 ± 0.24	2.43 ± 0.93	0.92	1.56 ± 0.89	2.24	2.29
K_2O	0.71 ± 0.34	0.60 ± 0.29	1.66 ± 0.53	0.86 ± 0.49	0.57	0.53 ± 0.35	0.00	0.04
$Fe_2O_3^*$	11.28 ± 1.59	9.0 ± 1.0	14.62 ± 3.65	13.53 ± 0.40	10.77	14.42 ± 1.43	10.61	12.00
TiO_2	1.24 ± 0.84	0.20 ± 0.08	0.81 ± 0.07	3.29 ± 0.59	0.33	1.19 ± 0.54	0.62	0.50
MnO	0.17 ± 0.01	0.17 ± 0.02	0.19 ± 0.01	0.16 ± 0.06	0.19	0.22 ± 0.06	0.18	0.19
P_2O_5	0.21 ± 0.11	0.03 ± 0.01	0.35 ± 0.15	0.65 ± 0.41	0.08	0.20 ± 0.11	—	0.67
ppm								
Cr	655 ± 210	1061 ± 641	664 ± 106	—	1152	—	1460	904
Cu	46 ± 23	—	118 ± 47	—	—	—	—	—
Ni	283 ± 106	231 ± 134	212 ± 67	—	590	—	647	138
Rb	14 ± 12	—	39 ± 12	—	—	—	1	1
Sr	294 ± 237	99 ± 36	542 ± 79	—	—	—	41	33
Zn	106 ± 62	—	97 ± 15	—	—	—	—	—
Zr	132 ± 83	39 ± 26	62 ± 11	—	—	—	—	44

(1) Sawyers Bar area, average of 14 samples: 9M, 11M, 38M, 67M, 89M, 91M, 92M, 124M, 174M, 212M, 320M, 351M, 354M and 375M (Table 2).

(2) Average of four representative aphyric modern boninites from Papua New Guinea, Mariana Trench, Bonin Islands and New Caledonia (CRAWFORD and CAMERON, 1985, Table 3).

(3) Average of four ankaramites, Ominica Crystalline Belt, southern British Columbia (BEDDOE-STEPHENS, 1982, Table 1).

(4) Average of four ankaramites, Atlantic Ocean Islands (BORLEY, 1974, Table 3A, B).

(5) Picritic basalt no. NG 143/2, from Solomon Islands (STANTON and BELL, 1969).

(6) Average of nine picritic lavas and sills, mostly from the British Isles (DREVER and JOHNSTON, 1967, Table 3.1).

(7) Olivine-rich pyroxenitic komatiite lava no. P9-132, northeastern Ontario (ARNDT *et al.*, 1977).

(8) Komatiitic basalt no. 331/780 from Barberton, South Africa (NESBITT, *et al.*, 1979).

protolith for the now-metamorphosed Yellow Dog mafic igneous series. The average Al_2O_3/TiO_2 value for 14 Sawyers Bar metakomatiites(?), 8.7, likewise, compares favorably with Barberton analogues (SUN and NESBITT, 1978). The clinopyroxene analyses presented in Table 4 are also in accord with this comparison.

It is of interest to note that analyses of a picritic flow and a dike in the Lower Devonian Copley Greenstone from the Redding section of the Eastern Klamath plate (Figure 1), presented by LAPIERRE et al. (1985), have somewhat lower CaO/Al_2O_3 ratios than the Yellow Dog metavolcanics; the Copley picrites also possess much higher Al_2O_3/TiO_2 ratios, 23 ± 5. As described by LAPIERRE et al., these metamorphosed, immature arc volcanics evidently represent low-K tholeiitic melts enriched by cumulate olivine, and are not chemically analogous to the Sawyers Bar greenstones.

The marked LREE enrichment of Yellow Dog metavolcanics of ultramafic affinities does not compare favorably with typical Archean komatiites (e.g., see Basaltic Volcanism Study Project, 1981, pages 21–23; ARNDT and NESBITT, 1982; ARNDT, 1986) or with Phanerozoic ridge basalts such as the Oman Ophiolite (PALLISTER and KNIGHT, 1981) inasmuch as both komatiites and MORBs are characterized by flat or LREE-depleted patterns.* The REE enrichments and high LREE/HREE ratios of the Sawyers Bar magnesian greenstones also exceed those of modern back–arc basin basalts (HAWKINS and MELCHIOR, 1985). Yellow Dog REE patterns are rather comparable, however, to some mid–plate alkali basalts (CHEN and FREY, 1985) and the more potassic rifted oceanic ankaramites (MAALOE et al., 1986) and island–arc lavas (GILL, 1981; POLI et al., 1984); none of these series appears to be closely related genetically to the Yellow Dog greenstones, however. Perhaps, as stated previously, partial fusion of previously REE–enriched garnitiferous protolith (lherzolite ± eclogite) was responsible for the generation of the Sawyers Bar komatiitic(?) suite.

TECTONIC SPECULATIONS

In what sort of geologic environment was this mafic to komatiitic complex formed? SNOKE et al. (1982) suggested that petrogenesis of such volcanics

and related peridotitic to dioritic plutonic suites in the western Klamaths and western Sierra Nevada was related to repeated fracturing and magmatism in a rifted oceanic arc. Absence of an alkali basalt suite, however, seems to preclude a rifted plate tectonic setting. The occurrence of distal clastic metasediments, apparently lying conformably beneath the greenstone pile, argues for a convergent plate tectonic environment, and against a mid-oceanic ridge origin for the mafic–komatiitic protolith. Transitions of the slower cooled, more chemically fractionated hypabyssals towards calc–alkaline compositions suggest that the Sawyers Bar greenstones were produced during the early stages of construction of a continental margin or an island arc, with the dikes and sills reflecting low–pressure crystal fractionation attending later injection into the volcanic pile. This origin is rather similar to that proposed for many boninites (KUSHIRO, 1972; HAWKINS et al., 1984).

Major element bulk–rock chemistry is compatible with the idea that the Yellow Dog suite is komatiitic in its affinities. However, this hypothesis is at variance with the REE bulk–rock chemistry unless the preexisting mantle source had been a rare earth–enriched garnet lherzolite, or a mixture of peridotite and eclogite, and was subjected to only low degrees of partial melting. How REE enrichment of the mantle lithosphere beneath the arc took place is unclear, but could have been a manifestation of metasomatism involving aqueous fluids or eclogitic partial melting and invasion during Permo–Triassic underflow of the paleopacific plate.

The high degrees of partial fusion required to produce komatiitic melts at moderate mantle depths involve unusually high temperatures (GREEN, 1975, 1981). TAKAHASHI (1986) and HERZBERG et al. (1986) have recently demonstrated, however, that only very small degrees of partial melting would be necessary to generate picritic, komatiitic and lherzolitic liquids from mantle peridotite at pressures approaching 60 kbar—thus somewhat easing the problem of obtaining LREE–enriched komatiitic chemistries. In any case, such pressure-temperature conditions are not characteristic of shallow portions of divergent or transform plate boundaries, but could have attended the generation of magmas at profound depths (e.g., 180–210 km) beneath an evolving continental margin or island arc.

The subduction of a spreading center (e.g., UYEDA and MIYASHIRO, 1974) or hot spot is one way in which a high heat–flow regime could have been emplaced beneath the WTrPz during its constructional stage. Ridge or hot–spot destruction would represent an uncommon but episodic phenomenon in the history of the circum–Pacific, thus

---

* An unusual LREE–enriched Archean komatiite from Ontario has recently been reported by STONE (1986), but this flow is iron rich and HREE poor compared to the Sawyers Bar ultramafic greenstones. Some hangingwall basalts from Kambalda, western Australia similarly display LREE–enriched patterns, but these relationships are ascribed to melt contamination by assimilation (CHAUVEL et al., 1985).

explaining the relative rarity of mafic–komatiitic lithologies in the Phanerozoic rock record (for another example, see ECHEVERRIA, 1982). The source of Yellow Dog volcanism thus appears to have been within a convergent plate setting, with low degrees of partial melting of REE-enriched garnetiferous mantle (± eclogite) taking place at depths near 200 km, followed by emplacement into a primitive island arc or continental margin.

Acknowledgments—Field and analytical studies in the Klamath Mountains have been supported in part by the U.S. Geological Survey, the UCLA Research Committee, and the National Science Foundation, most recently through grant EAR83-12702. I thank N. T. Arndt and Gautam Sen for suggestions and R. G. Coleman, M. M. Donato, Donald Elthon, W. P. Irwin and Paul Warren for helpful reviews.

REFERENCES

ANDO C. J., IRWIN W. P., JONES D. L. and SALEEBY J. B. (1983) The ophiolitic North Fork terrane in the Salmon River region, central Klamath Mountains, California. *Bull. Geol. Soc. Amer.* **94**, 236–252.

APTED M. J. (1981) Rare earth element systematics of hydrous liquids from partial melting of basaltic eclogite: a re-evaluation. *Earth Planet. Sci. Lett.* **52**, 172–182.

APTED M. J. and LIOU J. G. (1983) Phase relations among greenschist, epidote amphibolite and amphibolite in a basaltic system. *Amer. J. Sci.* **283A**, 328–354.

ARNDT N. T. (1986) Differentiation of komatiite flows. *J. Petrol.* **27**, 279–301.

ARNDT N. T. and NESBITT R. W. (1982) Geochemistry of Munro Township basalts. In *Komatiites,* (eds. N. T. ARNDT and E. T. NISBET), pp. 309–329 George Allen and Unwin.

ARNDT N. T. and NISBET E. T. (eds.) (1982) *Komatiites.* George Allen and Unwin. 526 pp.

ARNDT N. T. and JENNER G. A. (1986) Crustally contaminated komatiites and basalts from Kambalda, western Australia. *Chem. Geol.* **56**.

ARNDT N. T., NALDRETT A. J. and PYKE D. R. (1977) Komatiitic and iron–rich tholeiitic lavas of Munro Township, northeast Ontario. *J. Petrol.* **18**, 319–369.

BAEDECKER P. A. (1976) SPECTRA: Computer reduction of gamma–ray spectroscopic data for neutron activation analysis. In *Advances in Obsidian Glass Studies: Archaeological and Geochemical Perspectives,* (ed. R. E. TAYLOR), pp. 334–349 Noyes Press.

BARNES C. G., RICE J. M. and GRIBBLE R. F. (1986) Tilted plutons in the Klamath Mountains of California and Oregon. *J. Geophys. Res.* **91**, 6059–6071.

BASALTIC VOLCANISM STUDY PROJECT (1981) *Basaltic Volcanism on the Terrestrial Planets.* 1286 pp. Pergamon Press.

BEATY D. W. and TAYLOR H. P. JR. (1982) The oxygen isotope geochemistry of komatiites: evidence for water–rock interaction. In *Komatiites,* (eds. N. T. ARNDT and E. T. NISBET), pp. 267–280 George Allen and Unwin.

BEDDOE-STEPHENS B. (1982) The petrology of the Rossland volcanic rocks, southern British Columbia. *Bull. Geol. Soc. Amer.* **93**, 585–594.

BLAKE M. C. JR., HOWELL D. G. and JONES D. L. (1982) Preliminary tectonostratigraphic terranes map of California. U.S. Geol. Surv. Open-file Rep. No. 82-593.

BORLEY G. D. (1974) Oceanic Islands. In *The Alkaline Rocks,* (ed. H. SORENSEN), pp. 311–330 John Wiley & Sons.

BORNS D. J. (1980) Blueschist metamorphism of the Yreka-Fort Jones area, Klamath Mountains, northern California. Ph.D. Diss., Univ. of Washington.

BORNS D. J. (1984) Ecologites in the Stuart Fork terrane, Klamath Mountains, California. *Geol. Soc. Amer. Abstr. Prog.* **16**, 271.

BURCHFIEL B. C. and DAVIS G. A. (1981) Triassic and Jurassic tectonic evolution of the Klamath Mountains-Sierra Nevada geologic terrane. In *The Geotectonic Development of California,* (ed. W. G. ERNST), pp. 50–70 Prentice-Hall.

CARMICHAEL I. S. E. (1967) The mineralogy of Thingmuli, a tertiary volcano in eastern Iceland. *Amer. Mineral.* **52**, 1815–1841.

CHAUVEL C., DUPRÉ B. and JENNER G. A. (1985) The Sm–Nd age of Kambalda volcanics is 500 Ma too old! *Earth Planet. Sci. Lett.* **74**, 315–324.

CHEN C. Y. and FREY F. A. (1985) Trace element and isotopic geochemistry of lavas from Haleakala Volcano, east Maui, Hawaii: implications for the origin of Hawaiian basalts. *J. Geophys. Res.* **90**, 8743–8768.

COLEMAN R. G., MORTIMER N., DONATO M. M., MANNING C. E. and HILL L. B. (1987) Tectonic and regional metamorphic framework of the Klamath Mountains and adjacent Coast Ranges, California and Oregon. In *Metamorphism and Crustal Evolution of the Western United States,* (ed. W. G. ERNST), Prentice-Hall. (In press).

COMPSTON W., WILLIAMS I. S., CAMPBELL I. H. and GRESHAM J. J. (1986) Zircon xenocrysts from the Kambalda volcanics: age constraints and direct evidence for older continental crust below the Kambalda–Norseman greenstones. *Earth Planet. Sci. Lett.* **76**, 299–311.

CORTESOGNO L., ERNST W. G., GALLI M., MESSIGA B., PEDEMONTE G. M. and PICCARDO G. G. (1977) Chemical petrology of eclogitic lenses in serpentinite, Gruppo di Voltri, Ligurian Alps. *J. Geol.* **85**, 225–277.

COTKIN S. J., MEDARIS L. G. JR. and KISTLER R. W. (1985) Rb–Sr systematics of the Russian Peak pluton, northern California. *Trans. Amer. Geophys. Union* **66**, 1118.

CRAWFORD A. J. and CAMERON W. E. (1985) Petrology and geochemistry of Cambrian boninites and low–Ti andesites from Heathcote, Victoria. *Contr. Mineral. Petrol.* **91**, 93–104.

CRAWFORD A. J., BECCALUVA L. and SERRI G. (1981) Tectonic–magmatic evolution of the west Philippine-Mariana region and the origin of boninites. *Earth Planet. Sci. Lett.* **54**, 346–356.

DALY R. A. (1933) *Igneous Rocks and the Depths of the Earth.* 598 pp. McGraw-Hill.

DAVIS G. A. (1969) Tectonic correlations, Klamath Mountains and western Sierra Nevada, California. *Bull. Geol. Soc. Amer.* **80**, 1095–1108.

DONATO M. M. (1987) Evolution of an ophiolitic tectonic melange, Marble Mountains, northern California Klamath Mountains. *Bull. Geol. Soc. Amer.* (In press).

DONATO M. M. (1985) Metamorphic and structural evolution of an ophiolitic tectonic melange, Marble Mountains, northern California. Ph.D. Diss., Stanford Univ.

DONATO M. M., BARNES C. G., COLEMAN R. G., ERNST W. G. and KAYS M. A. (1982) Geologic map of the Marble Mountain wilderness area, Siskiyou County, California. U.S. Geol. Survey map MF 1452A, scale 1:48,000.

DREVER H. I. and JOHNSTON R. (1967) The ultrabasic facies in some sills and sheets. In *Ultramafic and Related Rocks,* (ed. P. J. WYLLIE), pp. 51–63 John Wiley & Sons.

ECHEVERRIA L. M. (1982) Komatiites from Gorgona Island, Columbia. In *Komatiites,* (eds. N. T. ARNDT and E. T. NISBET), pp. 199–209 George Allen and Unwin.

ELTHON D., LAWRENCE J. R., HANSON R. E. and STERN C. (1984) Modelling of oxygen isotope data from the Sarmiento ophiolite complex, Chile. In *Ophiolites and Oceanic Lithosphere,* (eds. T. G. GASS, S. J. LIPPARD and A. W. SHELTON), pp. 185–197, Pub. No. 13, Geol. Soc. London.

ERNST W. G. (1976) Mineral chemistry of eclogites and related rocks from the Voltric Group, western Liguria, Italy. *Schweiz. Mineral. Petrogr. Mitt.* **56,** 293–343.

ERNST W. G. (1983) Phanerozoic continental accretion and the metamorphic evolution of northern and central California. *Tectonophys.* **100,** 287–320.

ERNST W. G. (1984) California blueschists, subduction, and the significance of tectonostratigraphic terranes. *Geology* **12,** 436–440.

EWART A. (1976) Mineralogy and chemistry of modern orogenic lavas—some statistics and implications. *Earth Planet. Sci. Lett.* **31,** 417–432.

GARCIA M. O. (1978) Criteria for the identification of ancient volcanic ages. *Earth Sci. Rev.* **14,** 147–165.

GHOSE M. C. (1976) Composition and origin of Deccan basalts. *Lithos* **9,** 65–73.

GILL J. B. (1981) *Orogenic Andesites and Plate Tectonics.* 390 pp. Springer–Verlag.

GOODGE J. W. (1985) Widespread blueschist assemblages in the Stuart Fork terrane, central Klamath Mountains, northern California. *Geol. Soc. Amer. Abstr. Prog.* **17,** 357.

GOODGE J. W. (1986) Relations of Stuart Fork and North Fork terranes in the central Klamath Mountains, northern California. *Geol. Soc. Amer. Abstr. Prog.* **18,** 109.

GREEN D. H. (1975) Genesis of Archean peridotitic magmas and constraints on Archean geothermal gradients and tectonics. *Geology* **3,** 15–18.

GREEN D. H. (1981) Petrogenesis of Archaean ultramafic magmas of implications for Archaean tectonics. In *Precambrian Plate Tectonics,* (ed. A. KRONER), pp. 469–489 Elsevier.

HARPER G. D. and WRIGHT J. E. (1984) Middle to Late Jurassic tectonic evolution of the Klamath Mountains, California–Oregon. *Tectonics* **3,** 759–772.

HAWKINS J. W. and NATLAND J. H. (1975) Nephelinites and basanites of the Samoan linear volcanic chain: their possible tectonic significance. *Earth Planet. Sci. Lett.* **24,** 427–439.

HAWKINS J. W. and MELCHIOR J. T. (1985) Petrology of Mariana Trough and Lau Basin basalts. *J. Geophys. Res.* **90,** 11,431–11,468.

HAWKINS J. W., BLOOMER S. H., EVANS C. A. and MELCHIOR J. T. (1984) Evolution of intra–oceanic arc–trends system. *Tectonophys.* **102,** 175–205.

HERZBERG C. T., HASEBE K. and SAWAMOTO H. (1986) Origin of high Mg komatiites: constraints from melting experiments to 8 GPa (abstr.). *Trans. Amer. Geophys. Union* **67,** 408.

HIETANEN A. (1981) Petrologic and structural studies in the northwestern Sierra Nevada, California. *U.S. Geol. Surv. Prof. Paper 1225A–C,* 59 pp.

HILL L. B. (1985) Metamorphic, deformational, and temporal constraints on terrane assembly, northern Klamath

Mountain, California. In *Tectonostratigraphic Terranes of the Circum–Pacific Region,* (ed. D. G. HOWELL), pp. 173–186 Circum–Pacific Council Energy Mineral Resources, Earth Sci. Ser. No. 1, Houston.

HOEFS J. and BINNS R. A. (1978) Oxygen isotope compositions in Archean rocks from Western Australia, with special reference to komatiites. U.S. Geol. Surv. Open-file Rep. No. 78-701, 180–182.

HOTZ P. E. (1973) Blueschist metamorphism in the Yreka-Fort Jones area, Klamath Mountains, California. *J. Res. U.S. Geol. Surv.* **1,** 53–61.

HOTZ P. E., LANPHERE M. A. and SWANSON D. A. (1977) Triassic blueschists from northern California and north-central Oregon. *Geology* **5,** 659–663.

HUCKENHOLZ H. G. (1965) Der petrogenetische Werdegang der Klinopyroxene in den tertiaren Vulkaniten der Hocheifel, I. *Beitr. Mineral. Petrogr. Mitt.* **11,** 138–195.

HUCKENHOLTZ H. G. (1966) Der petrogenetische Werdegang der Klinopyroxene in den tertiaren Vulkaniten der Hocheifel, III. *Beitr. Mineral. Petrogr. Mitt.* **12,** 73–95.

IMSLAND P. (1984) Petrology, mineralogy and evolution of the Jan Mayen magmas system. *Visindafelag Islendinga, Rit* **43,** 332.

IRVINE T. N. (1979) Rocks whose composition is determined by crystal accumulation and sorting. In *The Evolution of the Igneous Rocks,* (ed. H. S. YODER, JR.), pp. 245–306 Princeton Univ. Press.

IRWIN W. P. (1960) Geologic reconnaissance of the northern Coast Ranges and the southern Klamath Mountains, California, with a summary of the mineral resources. *Bull. Calif. Div. Mines Geol.* **179,** 80.

IRWIN W. P. (1966) Geology of the Klamath Mountains Province. In *Geology of Northern California,* (ed. E. H. BAILEY), pp. 19–38. *Bull. Calif. Div. Mines Geol.* **190,** 508.

IRWIN W. P. (1972) Terranes of the western Paleozoic and Triassic belt in the southern Klamath Mountains, California. *U.S. Geol. Surv. Prof. Paper 800-C,* 103–111.

IRWIN W. P. (1981) Tectonic accretion of the Klamath Mountains. In *The Geotectonic Development of California,* (ed. W. G. ERNST), pp. 29–49 Prentice-Hall.

IRWIN W. P. (1985) Age and tectonics of plutonic belts in accreted terranes of the Klamath Mountains, California and Oregon. In *Tectonostratigraphic Terranes of the Circum-Pacific Region,* (ed. D. G. HOWELL), pp. 187–199 Circum–Pacific Council Energy Mineral Resources, Earth Sci. Ser. No. 1, Houston.

IRWIN W. P., JONES D. L. and KAPLAN T. A. (1978) Radiolarians from pre-Nevadan rocks of the Klamath Mountains, California and Oregon. In *Mesozoic Paleography of the Western United States* (ed. D. G. HOWELL and K. A. McDOUGALL), pp. 303–310 Pacific Sec., Soc. Econ. Paleont. Mineral., Los Angeles.

JAKOBSSON S. P. (1972) Chemistry and distribution patterns of recent basaltic rocks in Iceland. *Lithos* **5,** 365–386.

JENNINGS C. W. (1977) Geologic Map of California. Calif. Div. Mines Geol., scale 1:750,000.

JENSEN L. S. (1976) A new cation plot for classifying subalkaline volcanic rocks. *Ont. Div. Mines, Misc. Paper 66,* 22 pp.

KALLEMEYN G. W. and WASSON J. T. (1981) The compositional classification of chondrites—I. The carbonaceous chondrite groups. *Geochim. Cosmochim. Acta* **45,** 1217–1230.

KORZHINSKII D. S. (1959) *Physicochemical Basis of the*

Analysis of the Paragenesis of Minerals. 142 pp. Consultants Bur., New York.

KUNO H. (1960) High–alumina basalt. *J. Petrol.* **1,** 121–145.

KUNO H. (1966) Lateral variation of basaltic magma type across continental margins and island arcs. *Bull. Volcanol.* **29,** 195–202.

KUSHIRO I. (1972) Effect of water on the composition of magmas formed at high pressures. *J. Petrol.* **13,** 11–334.

LANPHERE M. A., IRWIN W. P. and HOLTZ P. E. (1968) Isotopic age of the Nevadan orogeny and older plutonic and metamorphic events in the Klamath Mountains, California. *Bull. Geol. Soc. Amer.* **79,** 1027–1052.

LAPIERRE H., ALBAREDE F., ALBERS J., CABANIS B. and COULON C. (1985) Early Devonian volcanism in the eastern Klamath Mountains, California: evidence for an immature island arc. *Can. J. Earth Sci.* **22,** 214–227.

LIOU J. G., KUNIYOSHI S. and ITO K. (1974) Experimental studies of the phase relations between greenschist and amphibolite in a basaltic system. *Amer. J. Sci.* **274,** 613–632.

LONGSTAFFE F. J., MCNUTT R. J. and SCHWARTZ H. P. (1977) Geochemistry of Archean rocks from the Lake Despair area, Ontario: a preliminary report. *Geol. Surv. Can. Paper 77–1A,* 169–178.

MAALOE S., SORENSEN I. and HERTOGEN J. (1986) The trachybasaltic suite of Jan Mayen. *J. Petrol.* **27,** 439–466.

MORTIMER N. (1985) Structural and metamorphic aspects of Middle Jurassic terrane juxtaposition, northeastern Klamath Mountains, California. In *Tectonostratigraphic Terranes of the Circum–Pacific Region,* (ed. D. G. HOWELL), pp. 201–214 Circum–Pacific Council Energy Mineral Resources, Earth Sci. Ser. No. 1, Houston.

MYSEN B. O. (1979) Trace-element partitioning between garnet peridotite minerals and water-rich vapor: experimental data from 5 to 30 kbar. *Amer. Mineral.* **64,** 274–287.

NESBITT R. W., SUN S. S. and PURVIS A. C. (1979) Komatiites: geochemistry and genesis. *Can. Mineral.* **17,** 165–186.

NESBITT R. W., JAHN B-M. and PURVIS A. C. (1982) Komatiites: an early Precambrian phenomenon. *J. Volcanol. Geotherm. Res.* **14,** 31–45.

PALLISTER J. S. and KNIGHT R. J. (1981) Rare-earth element geochemistry of the Semail Ophiolite near Ibra, Oman. *J. Geophys. Res.* **86,** 2673–2697.

POLI G., FREY F. A. and FERRARA G. (1984) Geochemical characteristics of the south Tuscany (Italy) volcanic province: constraints on lava petrogenesis. *Chem. Geol.* **43,** 203–221.

SIGURDSSON H., SCHILLING J. G. and MEYER P. S. (1978) Skagi and Langjokull volcanic zones in Iceland: petrology and structure. *J. Geophys. Res.* **83,** 3971–3982.

SNOKE A. W., SHARP W. D., WRIGHT J. E. and SALEEBY J. P. (1982) Significance of mid–Mesozoic peridotitic to dioritic intrusive complexes, Klamath Mountains—western Sierra Nevada, California. *Geology* **10,** 160–166.

SORENSEN H. (1974) *The Alkaline Rocks.* 622 pp. John Wiley & Sons.

STANTON R. L. and BELL J. D. (1969) Volcanic and associated rocks of the New Georgia Group, British Islands Protectorate. *Overseas Geol. Miner. Res. (G.B.)* **10,** 113–145.

STONE W. E. (1986) An unusual Fe-rich Al-depleted komatiite from the Abitibi greenstone belt. *Geol. Assoc. Canada Abstr. Prog.* **11,** 132.

SUN S. S. and NESBITT R. W. (1978) Geochemical regularities and genetic significance of ophiolitic basalts. *Geology* **6,** 689–693.

TAKAHASHI E. (1986) Melting of a dry peridotite KLB–1 up to 14 GPa: implications on the origin of peridotite upper mantle. *J. Geophys. Res.* **91,** 9367–9382.

TAYLOR H. P. JR. (1968) The ^{18}O geochemistry of igneous rocks. *Contrib. Mineral. Petrol.* **19,** 1–71.

TAYLOR S. R. (1977) Island arc models and the composition of the continental crust. *Maurice Ewing Series, Monograph 1,* 325–335 Amer. Geophys. Union.

UYEDA S. and MIYASHIRO A. (1974) Plate tectonics and the Japanese Islands: a synthesis. *Bull. Geol. Soc. Amer.* **85,** 1159–1170.

VILJOEN M. J. and VILJOEN R. P. (1969) Evidence for the existence of a mobile extrusive peridotitic magma from the Komati formation of the Onverwacht group. *Geol. Soc. S. Afr. Spec. Publ. 2, Upper Mantle Projects,* 87–112.

WAGER L. R. and DEER W. A. (1939) Geological investigations in east Greenland, part III, The petrology of the Skaergaard intrusion, Kangerdlugssuaq, east Greenland. *Medd. om Gronland* **105,** No. 4, 352 pp.

WOOD D. A. (1976) Spatial and temporal variation in the trace element geochemistry of the Eastern Iceland flood basalt succession. *J. Geophys. Res.* **81,** 4353–4360.

WRIGHT J. E. (1982) Permo–Triassic accretionary subduction complex, southwestern Klamath Mountains, northern California. *J. Geophys. Res.* **87,** 3805–3818.

XU G. R. and CHEN H. J. (1984) A preliminary study of komatiites in Anshan–Benxi–Fushun region, northeast China. *Geochemistry* **8,** 128–141.

Magmatic Processes: Physicochemical Principles
© The Geochemical Society, Special Publication No. 1, 1987
Editor B. O. Mysen

Studies of volcanic series related to the origin of some marginal sea floors

Leonid L. Perchuk

Institute of Experimental Mineralogy, USSR Academy of Sciences,
Chernogolovka, Moscow District, 142432, U.S.S.R.

Abstract—The evolution of basalt series reflects the thermal regime and dynamics of mantle diapirism as well as interaction between ultramafic magma and crustal rocks. The pressure–temperature regime of the origin and fractionation for two types of basalt series has been determined with the following semi-empirical equations:

$$T(°C) = 1047 + 44X - 13.77X^2 + Y(314.97 + 1941.55X - 630.69X^2) + 4P(\text{kbar}), \qquad (1)$$

where X is the mol fraction of forsterite in olivine and Y, the abundance of nonbridging oxygen associated with Fe and Mg in the melt. The pressure may be estimated from;

$$P = 40.57 + 21.33 \ln (a_K^L/a_{Na}^L) + 3.794 \ln (a_K^L/a_{Na}^L)^2, \qquad (2)$$

which was derived by least–squares regression from the bulk compositions of 23 nondifferentiated magmatic liquids, from experimental data, and partial molar volumes of the components involved. The results are compared with pressure estimates for ultramafic inclusions in alkali basalts.

The first group of basalts includes tholeiite and alkali basalt that are typical of growing igneous crust at both convergent and divergent plate margins. The evolution of these rocks show typical fractional crystallization behavior with gradual enrichment of alkalies and silica in the magma. This series evolves rapidly with the tholeiitic members equilibrating at 35–40 km depth and the alkali basalt members at depths between 60 and 100 km in the upper mantle.

The second volcanic series shows an evolution from acid to mafic members over a long time period ($\sim$60 m.y.). The pressures and temperatures required to produce the initial magmas for this group are 4–10 kbar and $\leq$1470°C, respectively. This rock series is typical of marginal sea floors, and may reflect interaction (thermal erosion and magma mixing) between molten diapirs of ultramafic composition with crustal rocks. This process may include magmatic replacement of continental crust by newly–formed oceanic crust at an average rate of $\sim$0.6 cm/year. It is proposed that this process is typical for active continental margins as exemplified by the western Pacific sea floor (*e.g.*, the Philippine, Japan and Okhotsk sea floors). A possible mechanism for simultaneous origin of deep–sea depressions and island arcs is also discussed.

INTRODUCTION AND FORMULATION OF THE PROBLEM

Both nondifferentiated and well–differentiated basalts series are widespread in the crust of the Earth. Figure 1 illustrates the relationship between alkalies and silica for several basalt series that are typical of continents, island arcs and oceanic islands. These series do not represent the total diversity of the volcanic series known, but are the most typical. In particular, the most acidic series and high–alkali groups will not be discussed here. The former shows a maximum alkali content, whereas the latter exhibits a silica maximum on the $(Na_2O + K_2O)$ − SiO_2 differentiation curves (*e.g.*, Cox *et al.*, 1979; Perchuk and Frolova, 1980).

The paths in Figure 1 may be viewed as projections of cotectic minima in basalt systems onto the plane silica + alkalies. The position of each projection depends principally on bulk composition and pressure. These parameters are not well known. The bulk composition of a series (*i.e.*, composition of the initial basalt magma) is difficult to calculate be-

cause of lack of knowledge about the proportions of the products differentiated. Pressure estimates are not always available due to lack of calibrated petrological barometers. Nevertheless, the values of some of these parameters may be estimated.

Nondifferentiated basalts have been suggested as analogues of "primary" magmas (Perchuk and Frolova, 1979; Perchuk, 1984a). The principal nondifferentiated basalts are alkali basalt and tholeiite (see Figure 2). These magmas may undergo fractionation and subsequent formation of volcanic series. At low pressures (shallow depths), tholeiitic magmas are produced. As the depth increases, the silica content and degree of partial melting of the mantle material decrease, whereas the amount of alkalies may increase (*e.g.*, Kushiro, 1983). Bulk compositions of 23 nondifferentiated and slightly differentiated volcanic series are listed in Table 1 with additional data on Al–(Fe, Mg)–Ca variation shown in Figure 2.

The dots in Figure 2C are distributed within a path limited by clear lines of a negative slope. These

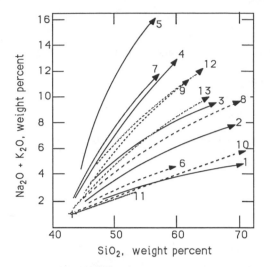

FIG. 1. Curves fitted by least squares regression of alkalies and silica for basalt series developed in the crust of continents (1–5), island arcs (6–10) and oceanic islands (11–13). 1, 2 = Karoo, South Africa (WALKER and POLDEVAART, 1949); 3 = volcanic suite of Aden and Little Aden (COX and MALLICK, 1970); 4 = Ethiopa trap formation (MOHR, 1960; SCHEINMANN, 1968); 5 = high alkali basalts from Central Africa (SCHAIMANN, 1968; PERCHUK and VAGANOV, 1977); 6, 7 = Little Kuriles (FROLOVA et al., 1985; PERCHUK and FROLOVA, 1982); 8 = Shidra suite, Japan (KUNO, 1968); 9 = Morotu subvolcanics, Sakhalin (YAGI, 1953); 10 = South Sandwich Islands (FROLOVA and RUDNIK, 1974); 11, 12 = Hawaii (MACDONALD and KATSURA, 1964; MURATA, 1970; WINCHELL, 1947; THOMPSON and TILLEY, 1969); 13 = Galapagos Islands (MCBIRNEY and WILLIAMS, 1969; PERCHUK and FROLOVA, 1979). The symbol denotes coordinates of intersection of the curves (an average of the Earth's mantle after PERCHUK and VAGANOV, 1980). Numbers indicate the series discussed in the text.

lines can be described with the following two equations:

$$\text{Alkalies (weight percent)} = 44.35 - 0.84 SiO_2;$$
$$r^2 = 0.954, \quad (1)$$

and

$$\text{Alkalies (weight percent)} = 39.06 - 0.68 SiO_2;$$
$$r^2 = 0.910. \quad (2)$$

In order to understand the regularities of the magmatism in marginal sea floors, three basins (Okhotsk, Japan and Philippine sea floors) of different geodynamic history have been compared. The northernmost of these, the Okhotsk sea floor, is composed ($\sim$80 percent) of typical continental crust (Precambrian gneisses, Jurassic granites, etc.). Its thickness is about 50 percent of that of the adjacent continents. The crustal thickness in the South Okhotsk depression is 11–15 km, including a 4 km thick sediment layer (Neogene basalts and terrigenous material have been dredged from the floor). The basement of the Okhotsk sea floor (the Okhotsk dome) is covered by Upper Cretaceous (113–133 m.y.) and Paleogene (45–55 m.y.) rhyolites, andesites and basalts. An inverse volcanic sequence appears typical for this region (GNIBIDENKO and KHVEDCHUK, 1984). A regular change in volcanism from acidic to basic over the last 50–60 m.y. (Figures 3 and 4; see also Table 2) has also been suggested for the Japan sea floor (KONOVALOV, 1984; PERCHUK, 1985; PERCHUK et al., 1985).

Basalts (10–15 m.y.; see also Table 2) with high Na, K, Ti and P contents and variable Al/Mg occur

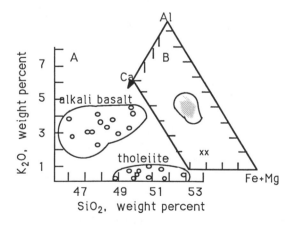

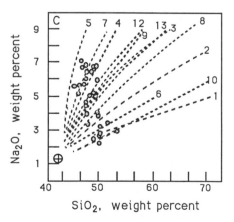

FIG. 2. Average values of petrochemical parameters for 23 nondifferentiated and slightly differentiated basalt complexes formed in continents, oceans and island arcs (see also Table 1). Symbols and trend numbers in Figure 2C as in Figure 1. Asterisks in Figure 2B indicate the average Yakutian kimberlite and composition of the upper mantle (PERCHUK and VAGANOV, 1980). Dashed lines from Figure 1.

Table 1. Some petrochemical parameters for nondifferentiated and slightly differentiated basalt complexes (after PERCHUK and FROLOVA, 1979; KEPHEZHINSKAS, 1979)

No.	Province	Weight percent		Mol ratio		
		SiO_2	$Na_2O + K_2O$	K/Na	ψ*	a_{SiO_2}**
	Mid-Atlantic Ridge					
1	Rift valley, 30°N	50.11	2.30	0.057	0.086	0.463
2	Rift valley, 35°N	50.35	3.35	0.122	0.122	0.469
3	Red Sea Rift	50.09	2.67	0.036	0.099	0.464
4	Hess depression and the Galapagos Rift, Pacific	48.96	2.55	0.045	0.099	0.453
	Varton Depression, Indian Ocean					
5	Tholeiites	51.06	2.58	0.067	0.067	0.472
6	Alkali basalt	51.10	4.75	0.194	0.166	0.475
7	Jan Mayen Island, North Atlantic	46.69	5.59	0.539	0.196	0.429
	Primorjye, Far East, USSR					
8	Shufan Plateau	53.24	2.90	0.099	0.101	0.494
9	Shkotov Plateau	51.61	3.49	0.230	0.199	0.477
10	Berezovskaya Formation: Ural	49.72	4.93	0.143	0.181	0.458
11	South Australia and Victoria	47.75	5.75	0.319	0.207	0.437
	Eastern Siberia, USSR					
15	Irkut River	47.75	4.47	0.351	0.161	0.440
13	East Sayan	48.04	3.47	0.309	0.147	0.442
16	Tuva	49.21	5.24	0.517	0.175	0.455
	Mongolia					
12	Jida	48.44	6.58	0.369	0.231	0.446
14	SW Mongolia	46.02	5.18	0.382	0.231	0.458
17	Potassium–rich formation	47.96	6.79	0.434	0.237	0.434
18	Sodium–rich formation	45.99	5.60	0.314	0.210	0.419
21	Hangay	48.90	6.82	0.516	0.230	0.447
23	Dariganga	48.21	5.92	0.315	0.202	0.458
19	Eastern Pacific Elevation	49.74	3.91	0.188	0.142	0.402
20	Kunlun and NW Tibet	47.60	7.19	0.955	0.233	0.437
22	Northern China	48.21	5.92	0.432	0.206	0.443

* $\psi = (Na + K)/Si$
** Ideal mixing: $a_{SiO_2} \sim X_{SiO_2}$

in the Tsushima basin and on the submarine islands of Ulyndo and Chukto, whereas younger tholeiites (3–10 m.y.) are typical of the main Japan sea floor depressions (Figure 4). These tholeiites differ, however, from Mid–Ocean Ridge Basalts (MORB) by their high MgO content and higher K/Na (PERCHUK et al., 1985). Almost all Japan Sea floor depressions are characterized (RODNIKOV et al., 1982) by high heat flow (2.75–2.92 × 10⁻⁶ cal/cm² sec), seismic inhomogeneity, and the absence of a granitic layer.

Tholeiitic basalts are widely distributed in the Philippine sea floor (SHARASKIN, 1984; PEIVE et al., 1980). Its western portion, i.e., the Philippine depression, shows an east–west orientation of paleomagnetic and tectonic lineaments, whereas in the eastern portion of this sea floor these lineaments

are oriented in north–south directions. The eastern portion of the Philippine sea floor is also younger and has a complex morphology and tectonic setting; three submarine ridges may be relict island arcs. The geological structure of all three ridges is similar to submarine ridges on the Japan sea floor. They comprise calc–alkaline volcanic rocks as well as ophiolites, greenschists and amphibolites of Jurassic to Miocene age. In contrast to the Japan sea floor, boninites occur regularly in the volcanic sequences from ophiolite formations of the Philippine sea floor.

All marginal sea floors in southeast Asia show an inverse correlation between crustal thickness and the square root of their age (HAYES, 1984). An inter–arc spreading mechanism has been suggested to ex-

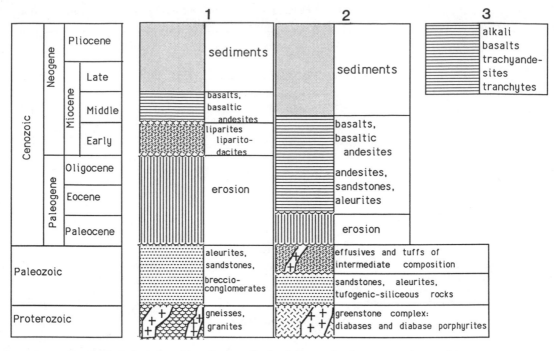

FIG. 3. Typical stratigraphic columns reflecting inverse volcanic sequences in morphostructures in the Japan sea floor (after KONOVALOV, 1984; PERCHUK *et al.*, 1985). 1 = sea floor margin; 2 = submarine ridges and islands (submarine arcs); 3 = depressions.

plain the origin of all marginal sea floors in this region (KARIG, 1974, 1975; HAYES, 1984; SHARASKIN, 1984).

Typical continental crust (gneisses, amphibolites, granites, etc.) is exposed in the basement of some island arcs. For example, the crustal thickness beneath the Ryukyu arc is about 20 km including a 10 km thick granitic layer. Its crust resembles the crust of the Okhotsk sea floor. The basement of the eastern Philippine sea floor, however, is composed principally of ophiolites, greenschists and amphibolites (MEIJER, 1976; PEIVE *et al.*, 1980).

Basalts resembling MORBs are the main constituents of the Philippine sea floor. However, some petrochemical variation is observed with differences between younger basalts and older typical oceanic tholeiites. For example, in the western Mariana depression, the basalts exhibit a "boninite tendency", with relatively high concentrations of SiO_2, MgO, Al_2O_3, Ba, Sr, Th and Ni (PEIVE *et al.*, 1980; SHARASKIN, 1984; KARPENKO *et al.*, 1984).

In order to explain the magmatic activity in marginal sea floors, three interrelated problems must be addressed. These are (1) the temperature and pressure of formation of primary (nondifferentiated) mantle–derived magmas, (2) the origin of boninites and related rocks, and (3) the origin of back–arc

basins and island arcs adjacent to oceanic trenches. For example, in order to support the idea of magmatic replacement of the Earth's crust by high–temperature, ultramafic magmas, it is necessary (a) to estimate the pressure–temperature constraints by the mantle on diapirism and magmatism in the mantle and the crust, (b) to find volcanic equivalents of the proposed mixed magmas, and (c) to interpret how back–arc depressions are related to the growth of igneous crust in island arcs. In view of the limited field documentation, available experimental data and the absence of adequate numerical models of the proposed processes, I will discuss only a petrological hypothesis for the origin of some marginal sea depressions.

BASALT SERIES: PRESSURE–TEMPERATURE CONTROL BASED ON GEOTHERMOMETRY AND GEOBAROMETRY

Xenoliths in alkali basalts

Mafic and ultramafic inclusions are found in many oceanic and continental alkali basalts. Because these inclusions are xenoliths, geothermometers and geobarometers provide only estimates of the minimum pressure–temperature conditions of the basalt magma generation. Mineral compositions

SiO₂, weight percent

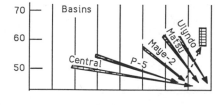

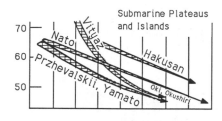

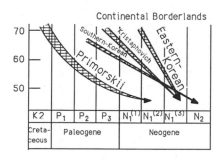

FIG 4 Time–composition paths illustrating inverse sequence of volcanism in the Japan sea floor. Arrows summarize petrochemical data for each area. Data sources as in Figure 3; see also text for additional information.

of xenoliths in alkali basalts are in places notably different from those in kimberlites, and the two–pyroxene thermometer calibrated originally by DAVIS and BOYD (1966) at 30 kbar cannot be used. Instead, for the high–temperature range I have used experimental data (AKELLA and BOYD, 1973, 1974; GREEN and RINGWOOD, 1966; HENSEN, 1973), and for the low–temperature range, estimates for the same equilibria from PERCHUK (1967, 1968, 1969, 1977c) have been employed in order to improve on the two–pyroxene thermometer (PERCHUK, 1977a). The empirical form of a modified version of this thermometer is (PERCHUK, 1977a) (all symbols in this and subsequent equations are defined in Table 3):

$$T(\text{K}) = 10^3/(0.4305 - 0.1651 \ln K_{\text{FeO}} + 1.1071 X_{\text{Mg}}$$
$$- 1.0525 X_{\text{Mg}}^2 - 0.0304 X_{\text{Mg}}^2 \ln K_{\text{FeO}}), \quad (3)$$

Table 3. Symbols used in formulae and equations

$N_{\text{Al}_2\text{O}_3}^{\text{Opx}}$	Weight percent Al_2O_3 in orthopyroxene
X_i^θ	mol fraction of component, i, in phase θ
a_i^θ	activity of component, i, in phase θ
ΔG	Gibbs free energy change in a reaction
K_D	distribution coefficient
K_i	the partition coefficient of component i between two phases α and β; $K_i - X_i^\alpha/X_i^\beta$
P	pressure, bar (kbar)
T, K	temperature, K
$T\,°\text{C}$	temperature, °C
R	gas constant, 1.987 cal/K mol
ΔV	volume change in a reaction
V_m^i	partial molar volume of component i
V_m^e	excess molar volume of component i

Abbreviations used in the text and figures

Ab	Albite
Ak	Akermanite
Am	Amphibole
An	Anorthite
Bi	Biotite
CaTs	Ca–tschermakite
Cpx	Clinopyroxene
Di	Diopside
En	Enstatite
Fo	Forsterite
Gr	Garnet
Ks	Kalsilite
Ne	Nepheline
Ol	Olivine
Opx	Orthopyroxene
Py	Pyrope
Qz	Quartz
Usp	Ulvospinel

Systems

KFQ	Ks–Fo–Qz
NFQ	Ne–Fo–Qz
CMAS	CaO–MgO–Al₂O₃–SiO₂

Table 2. Age of the Japan Sea submarine plateaus (K–Ar method) (after KONOVALOV, 1984).

Plateau	Rocks	Age (m.y.)
Vityaz	Basaltic andesite	16
	Andesitic dacite	32
	Rhyodacite	24
	Rhyodacite	53.2
Yamoto	Basalt	35, 34, 32, 23, 22
	Basaltic andesite	46, 34.5, 34, 32
	Andesite	49, 54, 61.1
	Dacite	46
Korean Sea floor margin	Basalt	15.5
	Trachyrhyolite	17
	Rhyodacite tuff	23.4
	Tuff of acid composition	53.2
Bogorov	Basalt	52.7, 17.9, 6.5
	Rhyolite	49.5

Table 4. Estimates of temperature and pressure from mineral equilibria in ultramafic xenoliths from alkali basalts.

Source of Analyses	Sample no.	Mineral Analyzed						Cpx + Opx + Gr + Bi		Opx + Gr	
		Sp	Opx	Cpx	Gr	Ol	—	T, °C	P, kbar	T, °C	P, kbar
Salt Lake Crater, Honolulu, Hawaii											
PERCHUK (1984)	SL-3	+	+	+	+	+		1050	20.6	970	13
	SL-4	+	+	+	+	+		1060	21	1020	18
	SL-5	+	+	+	+	+	Bi	1000	20	1075	11
	SL-2	+	+	+	+	+	Usp	995	20	1050	11
WHITE (1966)	140.1	—	+	+	+	—		1020	16	1115	22
	138.2	—	—	+	+	+	Am	1020	—	—	—
	137.1	+	—	—	—	—		1057	21	1151	21
YODER and TILLEY (1962)	6611.8	—	+	+	+	—		1115	20.5	990	12
Shavarin Tsaram Volcano, Mongolia											
AGAFONOV et al. (1975)	A-1	—	+	+	+	—	Bi	1150	22	—	—
		—	+	+	+	—		1175	21.9	1150	19
KEPEZHINSKAS (1979)	155	—	+	+	+	—	1150	21.2	1080	21	
	157	—	—	—	—	—	1155	20	1085	20	
	158	—	+	—	+	+	—	—	1065	17	
RYABCHIKOV et al. (1983)	14	+	2*	4	5	+	Bi	1158	21.3	1147	20
	14	+	+	+	5	+	6	1075	21.3	1147	20
	11	+	10	13	+	+	Bi	1163	22.2	1187	21
	11	+	12	14	+	+	Bi	1158	21.7	1180	32
	11	+	8	13	+	+	Bi	1145	22.0	1138	20
	11	+	9	14	—	—	Bi	1160	21.6	1175	21
	11	+	+	+	16	+	17	1190	21.5	1215	21.5
Lashaine Volcano, Tanzania											
DAWSON et al. (1970)	BD 738	—	+	+	+	—	Bi	1166	54	1112	39
	BD 738	—	+	+	+	+		1189	54	—	—
	BD 730	—	+	+	+	+		1063	36	—	—
REID et al. (1975)	775	—	+	nd	—	—		—	—	1100	39
	797	—	+	nd	+	—	Bi	—	—	1250	43
	747P	—	+	+	—	—		1231	58.6	—	—

Sample										
796	39	1080	−	−	−	−	+	nd	+	−
796	35	980	−	−	−	−	+	nd	+	−
740	39	1101	64	1207	−	−	+	+	+	−
794	38	1139	56	1174	−	−	+	+	+	−
794	−	−	45	1170	−	−	−	+	+	−
776A	39	1079	48	1182	−	−	+	+	+	−
782	48	1305	−	−	−	−	+	−	+	−
730	36	1107	46	1129	−	−	+	+	+	−
773A	−	−	52.6	1182	−	−	−	c	+	−
773A	−	−	49.3	1150	−	−	−	r	+	−
773B	−	−	50	1164	−	−	−	c	+	−
773B	−	−	57	1233	−	−	−	r	+	−
772	−	−	48.6	1074	−	−	−	−	c	−
772	−	−	48.3	1075	−	−	−	−	r	−

* Analysis number in original reference.
(+) Mineral is present and analyzed.
(−) Mineral not analyzed.
r Rim.
c Core.

where,

$$X_{Mg} = 0.5(X_{Mg}^{Opx} + X_{Mg}^{Cpx}), \qquad (4)$$

$$X_{Mg}^{Opx(Cpx)} = Mg/(Mg + Fe + Mn), \qquad (5)$$

and

$$K_{FeO} = \frac{FeO \text{ (weight percent) in Cpx}}{FeO \text{ (weight percent) in Opx}}. \qquad (6)$$

Linear regression of ln K_{Fe} versus $1/T$ (at X_{Mg} constant) from the experimental data results in r^2 = 0.96–0.98. The experimental data, themselves, are reproduced with an uncertainty of ±30°C with Equation (3) (PERCHUK, 1977b).

The following reaction;

$$x(Mg, Fe)_2CaAl_2Si_3O_{12} + y(Mg, Fe)SiO_3$$

$$= xCa(Mg, Fe)Si_2O_6 + (x+y)(Mg, Fe) \cdot SiO_3$$

$$\cdot xAl_2O_3, \qquad (7)$$

has been studied (PERCHUK, 1977b) to solve the derivative;

$$\left(\frac{\partial \ln a_{Al_2O_3}^{Opx}}{\partial P}\right)_T = \frac{(V_{Al_2O_3}^{Opx})^e}{RT}, \qquad (8)$$

for orthopyroxene from deep–seated ultramafic xenoliths. Depending on the concentration of Al_2O_3 in orthopyroxene, the following two equations to estimate pressure have been obtained (PERCHUK, 1977b: PERCHUK and VAGANOV, 1980) for

(1) $N_{Al_2O_3}^{Opx} \leq 4.0$ weight percent:

$$P(kbar) = \{15 + [26.5 - 0.045(T - 273)]$$

$$\times (\ln N_{Al_2O_3}^{Opx} - 3.778 + 3150/T)\} \quad (\pm 2\,kbar), \qquad (9)$$

(2) $N_{Al_2O_3}^{Opx} \geq 4.0$ weight percent:

$$P(kbar) = \{20 + [27.53 - 0.028(T - 273)]$$

$$\times (\ln N_{Al_2O_3}^{Opx} - 4.175 + 3077/T)\} \quad (\pm 0.5\,kbar), \qquad (10)$$

where the temperature is obtained from Equation (3).

Table 4 lists pressure–temperature estimates based on mineral equilibria in garnet–bearing xenoliths in lavas from both continents and oceanic islands. Some estimates have been obtained with amphibole–garnet (PERCHUK, 1967) and biotite–garnet thermometers (PERCHUK and LAVRENT'EVA, 1983). For clinopyroxene-free samples, the orthopyroxene–garnet thermometer (PERCHUK et al., 1985), recently revised by ARANOVICH and KOSYAKOVA (1986), has been used.

The liquidii of basalts generated in the mantle and the crust

Since ROEDER and EMSLIE (1970), the liquidus temperature of basalt saturated with olivine at 1 bar pressure could be estimated. The influence of alkalies on the liquid–olivine equilibria has not been studied experimentally. The activity of an olivine component in liquids as a function of the activity coefficient for Na has been derived from systems involving fayalite, albite and nepheline (PERCHUK and VAGANOV, 1977). However, those equations are valid only for a relatively narrow range of liquidus temperatures and bulk compositions. Recently, GHIORSO and KELEMEN (1987) summarized data and completed their version of the super-liquidus phase relations of magmatic rocks.

Because of the high liquidus temperatures of komatiite, basaltic komatiite and boninite, thermodynamic treatment of olivine–liquid equilibria for temperatures in the range 1350–1650°C is necessary. Figure 5 has been constructed from a thermodynamic treatment of all available phase diagrams involving olivine on the liquidus at 1 bar (see APPENDIX for sources, and also references in PERCHUK and VAGANOV, 1977; PERCHUK, 1984a; LEEMAN, 1978). The diagram makes use of the oxygen model for silicate melts (see review by BOTTINGA *et al.,* 1970):

$$O^0 + O^{2-} = 2O^-, \qquad (11)$$

where O^0 is bridging oxygen (*i.e.,* Si–O–Si bridge), O^{2-} is free oxygen (*i.e.,* Me–O–Me, where Me is a metal cation not in tetrahedral coordination) and O^- is nonbridging oxygen (*i.e.,* Si–O–Me). The abscissa in Figure 5 indicates the fraction, Y, of oxygen (nonbridging) associated with Fe^{2+} and Mg^{2+} in the melt. Least squares regression of Y versus temperature (153 data points) yields the expression (see APPENDIX; Table A1, for data sources)

$$T(°C) = 1089 + 1634Y; \quad r^2 = 0.988. \qquad (12)$$

Equation (12) is valid for all systems in the composition range $Y = 0$–0.5.

For volatile–free systems with potassium contents ≥ 5 weight percent, the liquidus temperatures have been estimated ($\pm 25°C$) with the equation:

$$T(°C) = 1047 + 44X - 13.77X^2$$
$$+ Y(314.97 + 1941.55X - 630.69X^2), \qquad (13)$$

where,

$$X = X_{Mg}^{Ol} = 3.33X_{Mg}^{rock}/(1 + 2.33X_{Mg}^{rock}),$$

$$X_{Mg}^{rock} = Mg/(Mg + Fe + Mn),$$

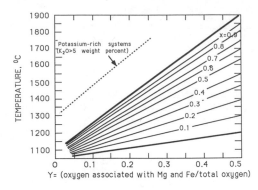

FIG. 5. Temperature–dependence of the structure-chemical parameter, Y, for Fe–Mg systems with olivine on the liquidus at $P = 1$ bar. Dashed line is calculated from the data of SCHAIRER (1954) and WENDLANDT and EGGLER (1980) for the system Fo–Ks–Qz. Other data sources are listed in the APPENDIX.

and

$$Y = \frac{\text{oxygen associated with Mg and Fe}}{\text{total oxygen}}.$$

At $K_2O \geq 5$ weight percent, the activity coefficient for MgO increases rapidly, and olivine appears on the liquidus at a temperature about 210°C higher than that predicted from Equation (13). This effect is so pronounced that the activity coefficient of MgO becomes practically independent of the potassium content in the system up to $K_2O = 16$–17 weight percent. For iron–free, potassium–rich systems, the liquidus temperature may be expressed as a linear function of Y (see Figure 5 and APPENDIX; Table A2):

$$T(°C) = 1294 + 1690.84Y; \quad r^2 = 0.955;$$
$$\text{accuracy} = \pm 30°C. \qquad (14)$$

This inference is of great importance when using Equations (13) and (14) as petrological thermometers. The effect of Na_2O is not yet known.

Equation (13) was deduced for $P = 1$ bar. From the experimental data of DAVIS and ENGLAND (1963), the pressure–dependence of the forsterite liquidus temperature may be approximated with the equation;

$$T(°C) = 1898 + 4.77P(kbar), \qquad (15)$$

whereas for fayalite (HSU, 1967), the expression is:

$$T(°C) = 1205 + 4.85P(kbar). \qquad (16)$$

The temperatures of the melting curves for basalt compositions with olivine on the liquidus (typically up to pressures near 10 kbar) show a relatively constant pressure derivative, $(\partial T/\partial P) = 4.08°C/kbar$

(PERCHUK, 1983; see also APPENDIX; Table A3). It is suggested, therefore, that in the pressure range where olivine is on the liquidus, the pressure dependence of the liquidus temperature can be approximated with the expression:

$$T_P(°C) = T_{1\,bar}(°C) + 4P(kbar). \qquad (17)$$

There are two unknown parameters in Equation (17)—pressure and temperature. Methods are required to calculate the values of these variables. One method was proposed by PERCHUK (1984b) and PERCHUK et al. (1982). The approach is based on the observation (KUSHIRO, 1975; PERCHUK, 1973; p. 269), also postulated by KORZHINSKII (1959), that for volatile–free silicate systems the shift in liquidus boundaries with pressure is correlated with the radius of additional cations in the system. The data in Figure 2C may indicate a decrease of silica activity in the parent magma as the alkali activity increases. This observation is supported by thermodynamic calculations (NICHOLLS and CARMICHAEL, 1972; NICHOLLS et al., 1971).

Silicate melts are not ideal mixtures (e.g., CARMICHAEL et al., 1977; PERCHUK and VAGANOV, 1977; NAVROTSKY, 1980; GHIORSO et al., 1983). NICHOLLS (1977) and DEPAOLO (1979) have shown, however, that for simple silicate systems such as MgO–SiO$_2$, the following inference is valid:

$$\left(\frac{\partial \ln a_{SiO_2}}{\partial T}\right)_P \simeq 0. \qquad (18)$$

Some, but not all, silicate melts may, therefore, be approximated as ideal solutions. The validity of this inference is also supported by constant a_{SiO_2} along isobaric liquidus boundaries between forsterite and enstatite in the systems forsterite–kalsilite–silica (Fo–Ks–Qz) and forsterite–nepheline–silica (Fo–Ne–Qz) at high pressure (KUSHIRO, 1968, 1980) and at atmospheric pressure (SCHAIRER, 1954). From the data reported by KUSHIRO (1968, 1980), it is inferred that:

Fo–Ks–Qz at 30 kbar $a_{SiO_2}^L \simeq 0.47$;   and

Fo–Ne–Qz at 30 kbar $a_{SiO_2}^L \simeq 0.49$.

At other pressures, the SiO$_2$ activity varies, and the higher the alkali content, the greater the effect of pressure on the shift of the enstatite–forsterite (En–Fo) liquidus boundaries. The change in SiO$_2$ mol fraction with pressure at constant Na$_2$O and K$_2$O ($X_{alkalies}^L = 0.12$) can be described with the linear equations for the systems

(1) Fo–Ks–Qz:

$$-\ln K_{SiO_2}^L = 0.39 - 0.0123P(kbar), \qquad (19)$$

(2) Fo–Ne–Qz:

$$-\ln K_{SiO_2}^L = 0.49 - 0.0167P(kbar). \qquad (20)$$

In such a case the activity coefficient of silica is considered independent of pressure and the partial molar volume of SiO$_2$, $(V_{SiO_2}^m)^L$ can be estimated. By taking into account Equation (18), integration of

$$RT \ln a_{SiO_2}^L = \int_0^P (V_{SiO_2}^m)^L dP, \qquad (21)$$

at $X_{alkalies}^L = 0.12$ and $T = 1500°C$ gives the following average value for both the system Fo–Ne–Qz and Fo–Ks–Qz:

$$(V_{SiO_2}^m)^L = V_{SiO_2} - V_{SiO_2}^0 = -0.051\,cal/bar. \qquad (22)$$

This value is much larger than that calculated with Equation (5) of DEPAOLO (1979) for the alkali–free system MgO–SiO$_2$ under the assumption that $\ln a_{SiO_2}^L$ is a linear function of pressure:

$$(V_{SiO_2}^m)^L = V_{SiO_2} - V_{SiO_2}^0 = -0.012\,cal/bar, \qquad (23)$$

where in Equations (22) and (23), V_{SiO_2} is the partial molar volume of SiO$_2$ in the melt (BOTTINGA and WEILL, 1970) and $V_{SiO_2}^0$ is the molar volume of molten SiO$_2$ at the same temperature. Thus, Maxwell's relation for the alkali–bearing systems,

$$\left(\frac{\partial \ln a_{SiO_2}^L}{\partial P}\right) = -1.23 \times 10^{-5}, \qquad (24)$$

leads to the conclusion that with increasing pressure, addition of alkalies to volatile–free systems results in a large decrease of silica activity and forces the En–Fo boundary to shift toward alkali–rich compositions. There are also significant deviations from ideality in alkali–bearing silicate melts. Nonideal behavior of these silicate solutions has been demonstrated at 1 bar pressure both in terms of the random network model (PERCHUK and VAGANOV, 1977, 1980; LEEMAN, 1978) and the Temkin model of silicate melt structure (HERZBERG, 1979). The use of ESIN's (1946) polymer theory–based model allows us to avoid excess enthalpy of mixing.

In order to calibrate the pressure–dependence of the silica activity, isobaric cross–sections from the system MgO–SiO$_2$–Al$_2$O$_3$–(Na, K)$_2$O should be studied in detail. Only a limited number of experimental data is, however, available for thermodynamic treatment of the systems Fo–Ks–Qz and Fo–Ne–Qz (KUSHIRO, 1965, 1980; WENDLANDT and EGGLER, 1980; MODRESKI and BOETTCHER, 1973; SCHAIRER, 1954; IRVINE, 1976). As was shown above, the alkali/silica is a good indicator of the pressure effect on the peritectic in Fe–Mg silicate systems. This effect can be roughly calibrated under the assumption of statistically equal association of free oxygen with Na and K in the silicate melt, which, in turn, can be described with the internal distribution coefficient:

$$K_D = \left(\frac{a_{K_2O}^L}{a_{SiO_2}^L}\right)_{FoKsQz} \left(\frac{a_{SiO_2}^L}{a_{Na_2O}^L}\right)_{FoNeQz} \simeq 1.0. \quad (25)$$

This assumption is based on experimental data on sodium and potassium distribution in haplogranite melt–chloride melt, haplophonolite melt–chloride melt and haplophonolite melt–chloride + water fluid (ZYRIANOV, 1987; ZYRIANOV and PERCHUK, 1978; PERCHUK and LINDSLEY, 1982). The following equilibria have been studied under the pressure–temperature conditions of interest:

(i) $NaAlSi_3O_8 \cdot 3SiO_2$ (melt) + KCl (melt)

$$= NaCl (melt) + KAlSi_3O_8 \cdot 3SiO_2 (melt), \quad (26)$$

$$1100°C, 6 \text{ bar}, K_D = 0.78$$

$$(n = 10 \text{ data points}; \sigma = \pm0.102)$$

(ii) $NaAlSiO_4 \cdot SiO_2$ (melt) + KCl (melt)

$$= NaCl (melt) + KAlSiO_4 \cdot SiO_2 (melt), \quad (27)$$

$$1300°C, 6 \text{ bar}, K_D = 1.24$$

$$(n = 7 \text{ data points}; \sigma = \pm0.177)$$

(iii) $NaAlSiO_4 \cdot SiO_2$ (melt) + $KCl \cdot nH_2O$ (fluid)

$$= NaCl \cdot nH_2O (fluid) + KAlSiO_4 \cdot SiO_2 (melt), \quad (28)$$

$$1100°C, 5 \text{ kbar}, K_D = 1.146$$

$$(n = 10 \text{ data points}; \sigma = \pm0.207)$$

where;

$$K_D = \left(\frac{X_K}{1 - X_K}\right)_{silicate} \left(\frac{1 - X_K}{X_K}\right)_{chloride}, \quad (29)$$

where X_K is the mol fraction of potassium.

Equilibria (i) and (ii) were studied in sealed quartz tubes at high temperature with a calculated internal pressure of approximately 6 bar. Equilibrium (iii) was studied by using sealed platinum containers at 5 kbar in an internally–heated, gas–medium apparatus. Silicate glasses in the system Ne–Ks–Qz were prepared from oxides, melted at 1200°C and examined with the electron microprobe for chemical homogeneity. Experimental charges were similarly studied. The K/Na of coexisting glasses, salts and fluids was determined by flame photometry.

The K_D equals 0.78 for the silica–rich composition, (i), whereas for the silica–poor compositions,

(ii) and (iii), $K_D \simeq 1.2$. To a first approximation, these data permit the approximation of ideal mixing of Na and K in silicate melt systems.

By assuming that K_D in Equation (25) is independent of pressure and temperature, the pressure–dependence of $\ln (a_{Na_2O}^L/a_{SiO_2}^L)$ for the systems Fo–Ne–Qz and Fo–Ks–Qz, respectively, has been calculated for pressures corresponding to constant silica activity. From Equation (24) the position of the Fo–En boundary in the system $Na_2O-K_2O-MgO-SiO_2-Al_2O_3$ has been calculated (Figure 6). The values for the calculated parameters are similar to those given in Table 1 and Figure 2C for natural magmatic liquids. By taking into account Equation (18), the lines in Figure 6 can be described with the following expressions:

$$P(kbar) = 51.3(\Psi/a_{SiO_2}^L - 0.017), \quad (30)$$

where,

$$\Psi = \frac{a_{(Na,K)_2O}^L}{a_{SiO_2}^L}, \quad (31)$$

to yield the expression:

$$a_{SiO_2}^L = \sqrt{\frac{a_{(Na,K)_2O}^L}{0.0195P - 0.017}}. \quad (32)$$

Equations (30) and (32) show the dependence of the liquidus pressure on the alkali content of a melt in equilibrium with olivine and orthopyroxene.

The equations above are analytical forms of the regularities in liquidus phase relations predicted by KUNO (1960). Despite the fairly large uncertainty (±30% compared with the input data), Equations (30) and (32) could be used to calculate the last

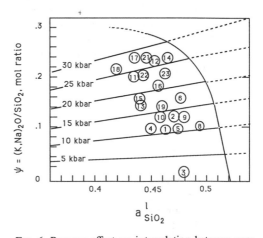

FIG. 6. Pressure effect on interrelation between composition parameters in the volatile–free system MgO–(Na, K)$_2$O–SiO$_2$–Al$_2$O$_3$–SiO$_2$. Solid line: Theoretically calculated shift of the incongruent melting point as a result of pressure. Numbered circles correspond to the numbers in Table 1. For other details, see text.

term in Equation (17), deduced for volatile–free conditions.

The value of K_D in Equation (25) is based on data collected for simple systems. In order to approximate a mixing model for natural magmatic compositions, the following internal oxygen exchange equilibrium in the melts (equilibrated with a solid residue in the mantle) should be considered:

$$[Si-O^0]_{FoKsQz} + [Na-O^{2-}]$$

$$= [Si-O^0]_{FoNeQz} + [K-O^{2-}]. \quad (33)$$

In this expression, $[Si-O^0]_{FoKsQz}$ and $[Si-O^0]_{FoNeQz}$ are activities of bridging oxygen in the systems Fo–Ks–Qz and Fo–Ne–Qz, respectively, whereas [K–O^{2-}] and [Na–O^{2-}] are activities of free oxygen associated with potassium and sodium in the melts. Because it is assumed that the activities of bridging oxygen in Equation (33) are the same, the equilibrium constant is:

$$K_{(33)} = \frac{[Na-O^{2-}]}{[K-O^{2-}]} = \exp\left(\frac{\Delta G_{(33)}}{RT}\right)_P. \quad (34)$$

The volume change of reaction (33) can be calculated from partial molar volumes of K_2O and Na_2O in silicate melts (BOTTINGA and WEILL, 1970; STEBBINS *et al.*, 1984). The value of $\Delta V_{(33)}$, calculated per atom of K and Na, is 0.204 cal/bar at 1400°C, which results in:

$$\left(\frac{\partial \ln K_{(17)}}{\partial P}\right)_{T=1673} = -6.13 \times 10^{-5}. \quad (35)$$

This value is quite similar to the value that can be calculated, to a first approximation, by comparing K/Na of volcanic rocks. For example, in mid–ocean ridge basalts, $(K_2O/Na_2O) = 0.056$; whereas in kimberlite this ratio is near 5.54 (see Table 1). If it assumed that the mid–ocean ridge basalts were formed at pressures near 10 kbar, kimberlite would be formed at $P = 75–80$ kbar (PERCHUK, 1977b). By assuming linear relations between $\ln K_D$ and pressure and that basalt and kimberlite were formed at the same temperature, the following value of the derivative has been calculated:

$$\left(\frac{\partial \ln K_{(17)}}{\partial P}\right)_T = -6.23 \times 10^{-5}. \quad (36)$$

The value from Equation (36) is surprisingly similar to that of Equation (35). Consequently, in addition to Equation (30), we can use this large effect of pressure on initial magma composition as a petrological barometer. By using 25 data points from Table 1, the following expression has been derived:

$$P(kbar) = 40.57 + 21.33\psi + 3.794\psi^2, \quad (37)$$

where $\psi = \ln (K/Na) \simeq 1/K_{(33)}$ in basalt (molar ratio). The pressures obtained with Equations (30) and (37) are in accord to within ±4.3 kbar.

The least–squares fitted lines in Figures 1 and 2C intersect at $SiO_2 = 42.5 \pm 1.2$ weight percent. In the concentration range between 42.5 and 54 weight percent SiO_2, each trend could be approximated by a straight line. From Equations (30) and (37), the following empirical expression has been obtained:

$$P(kbar) = 9.13 + 20.76\alpha, \quad (38)$$

where

$$\alpha = (Na_2O + K_2O - 1.25)/(SiO_2 - 42.5). \quad (39)$$

Equation (38) is useful for pressure estimates of the most mafic members of differentiated basalt series as discussed further below.

The deduced equations imply an empirical correlation between the alkalinity of a cotectic basalt and degree of melting of the mantle material. This effect is several orders of magnitude greater than any other variation. Equations (30), (36) and (37) were deduced on the basis of phase relations in portions of the system $Na_2O-K_2O-MgO-Al_2O_3-SiO_2$. It is, therefore, possible that their application to calculate the pressure of formation of magmatic liquids in the mantle may be restricted. Any significant variation in bulk compositions of the mantle–derived basalt will affect the value of the calculated pressure. It is likely, though, that these equations will provide useful estimates of relative pressures of formation of basaltic magmas with different alkalinity.

Pressure estimates from Equations (30) and (37) can be compared with results from geobarometry from xenoliths from alkali basalts and kimberlites (Tables 4 and 5). Because the xenoliths are not genetically related to the host magmas, the pressures estimated from their mineral assemblages are minimal values insofar as the magma is concerned. Average compositions from Kilauea and Mauna Kea are also included in Table 5, where the average depth of their magma chambers is known from seismic data. Two volcanic samples from Mongolia are also included (KEPENZHINSKAS, 1979), known as sodium–rich and potassium–rich basalts. The first group is characterized by spinel–bearing xenoliths (Solhitin Volcano), whereas the second group is characterized by magma–types found in the Shavarin Volcano.

As can be seen from the data in Table 5, the pressure estimates obtained with the three different methods are in fair agreement. As expected, the

Table 5. Chemical analyses (recalculated to 100 weight percent) and some compositional parameters of volcanic rocks containing deep-seated ultramafic xenoliths

Country	USA, Hawaii			Mongolia		Tanzania
Volcano	Kilauea (1)	Mauna Kea (2)	Salt Lake, Oahu (3)	Solhitin (4)	Tsaram (5)	(6)
SiO_2	49.82	47.48	40.82	49.61	50.86	41.39
TiO_2	2.57	2.68	2.54	3.12	2.47	2.49
Al_2O_3	12.87	14.17	11.66	9.98	15.99	6.13
Fe_2O_3	2.06	4.05	2.52	5.75	2.37	9.67
FeO	9.51	8.38	12.41	7.43	7.57	4.82
MgO	10.09	9.59	13.02	9.62	5.86	18.52
MnO	0.17	0.18	0.20	0.20	0.14	0.18
CaO	10.22	9.83	13.50	9.21	6.05	12.85
Na_2O	2.20	2.71	2.62	3.91	4.84	2.07
K_2O	0.50	0.92	0.71	1.15	3.84	1.04
P_2O_5	—	—	—	—	—	0.80
$Na_2O + K_2O$	2.70	3.64	3.33	5.06	8.68	3.11
α	0.198	0.479	−1.24	0.535	0.89	−1.69
a_{SiO_2}	0.462	0.439	0.374	0.458	0.463	0.377
X_{Fo}^{Ol}	0.838	0.823	0.839	0.817	0.779	0.889
T °C at 1 atm	1265	1258	1406	1265	1193	1405
P, kbar*	12	17	20	15	20	17–20
(Na, K)/Si	0.098	0.136	0.146	0.82	0.280	0.129
P, kbar†	12	17	21	21	32	18
ln(K/Na)	−1.48	−1.07	−1.224	−1.22	−0.232	−0.688
P, kbar‡	14	17	15	16	28	19
T, °C at P	1316	1326	1407	1340	1313	1485

* Pressure estimated with geobarometers or seismic data.
† Pressure estimated with Equation (30).
‡ Pressure estimated with Equation (37).

pressures from the ultramafic xenoliths are slightly lower than the pressure values obtained with either Equation (30) or (37).

APPROXIMATE PRESSURE–TEMPERATURE CONDITIONS OF INITIAL MAGMA FRACTIONATION

The equations derived above may be used to estimate the approximate depth of magma chambers and their thermal regime during fractionation. A few examples of such estimates will be provided.

Two volcanic series (alkali basalt and tholeiite) are known in the Little Kurile arc north of Hokkaido. The island arc is bounded by the South Okhotsk depression and Kurile–Kamchatka trench to the northwest and southeast, respectively. The arc comprises two subparallel island chains divided by a basin. The frontal arc consists of three islands— Zeleny, Polonsky and Panfilov composed of volcanics with common trachybasaltic sills. Radiometric dating of the rocks yields ages in the range 50–68 m.y. The volcanic rocks are younger than the host flysch (FROLOVA et al., 1985) and occur

as thick layered sills (olivine–bearing gabbro–norites, teshenites, monzonites and syenites) and as small bodies and dikes composed of trachybasalt and trachydolerite. These rocks make up one alkali basalt series denoted 7 in Figure 1. The depth of the initial magma chamber is estimated [by Equation (37)] to be near 70 km (Figure 7). The calculated temperature interval of fractionation is about 150°C [from Equation (17)].

Tholeiite series from the Tyatya volcano (Kunashir Island in the inner portion of the Little Kurile arc) was generated at a depth of about 40 km (see Figure 7). In spite of the fact that the calculated temperature interval of fractionation was only about 80°C [from Equation (17)], the series is well differentiated, possibly in a magma chamber located directly beneath the island arc crust. This conclusion is consistent with the reconstruction of the lithosphere beneath Kunashir Island based on recent seismic data (ZLOBIN et al., 1982; FROLOVA et al., 1985) suggesting a magma chamber near the crust–mantle boundary at 40–60 km depth. Similar volcanism is typical of oceanic islands such as Hawaii

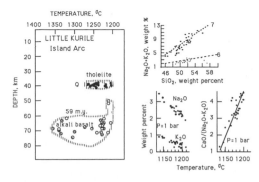

FIG. 7. Approximate depth and temperature regime of differentiation of the Paleogene (59 m.y.) alkali basalts and Holocene (Q) tholeiites from the Little Kuriles. Left-hand diagram shows temperatures of magma generation calculated with Equation (17) for appropriate pressures from Equation (37). The temperature regime of differentiation of tholeiite basalts is calculated with Equation (17).

(WINCHELL, 1947; MACDONALD and KATSURA, 1964) and continents.

The origin of boninite, characteristic of some ancient and present–day island arcs, is considered next. It has been suggested that boninite and komatiite series exhibit kindred petrology, geochemistry and very high temperature melt inclusions. Furthermore, in some greenstone belts, both series are located within ophiolite formations (CAMERON et al., 1979; NESBITT et al., 1979). According to existing views (PEIVE et al., 1980; SHARASKIN, 1980; MEIJER, 1980, 1983; CAMERON et al., 1979, 1983; SOBOLEV et al., 1986) and also suggested from experimental data (KUSHIRO, 1972; GREEN, 1976; TATSUMI, 1987), boninite could be a product of partial melting of ultramafic upper mantle material at high water pressures. However, recent studies on melt inclusions in olivine and orthopyroxene show low contents (1–3 weight percent) of primary water (SOBOLEV et al., 1986; DANUSHEVSKII et al., 1986; SOBOLEV and DANUSHEVSKII, 1986) with homogenization temperatures in the range 1090–1430°C. Earlier, WALKER and CAMERON (1983) suggested that parent boninite magma in equilibrium with Fo_{94} most likely would have been generated at 1250–1300°C at $P < 10$ kbar and with relatively low water contents (2–3 weight percent).

Compositions of boninites from some western Pacific and Australian island arcs are plotted in Figure 8. The stratigraphic position of boninites in ophiolite formations suggests fractionation of the initial magma to produce a differentiated series beginning with olivine–rich, enstatite–bearing products and ending with olivine–free, high–silica and low–alkali melts. Figure 8 shows, however, a regular change in boninite composition with respect to Mg,

Al and Si, whereas there is no correlation between silica and alkalies. Boninite occupies the field between ultramafics and dacite–andesite in Figure 8. Similar relationships have been demonstrated by SAKUYAMA (1979, 1981) for some calc–alkaline series in the Japanese island arcs. This relationship suggests that perhaps boninite might be a product of interaction between ultramafic (harzburgitic or komatiitic?) magma and crustal rocks, where the latter were melted before mixing with the postulated, mantle–derived ultramafic magma. The lack of correlation of alkalies with silica might be the result of such a process. In other words, this possible interaction should lead to thermal erosion of the Earth's crust and magma mixing rather than forming these rock series through fractional crystallization. In order to support this possibility, we have to show (1) that boninite and related rocks have a shallow origin, (2) very high temperature of magma mixing, (3) absence of correlation between alkalies and silica, and (4) mixed sources of compatible and incompatible elements. Evidence relating to the first three requirements is shown in Figure 9.

Let us consider the abundance of compatible and incompatible trace elements in boninites and related rocks. Boninite series rocks are characterized by the following geochemical features (HICKEY and FREY, 1982; KARPENKO et al., 1984; SHARASKIN, 1984; ARMSTRONG and NIXON, 1981; DEPAOLO and WASSERBURG, 1977):

(1) $54 < SiO_2 < 59$ weight percent at $X_{Mg} > 0.5$ with Mg, Ni, Co and Cr contents typical of primary mantle–derived magmas.

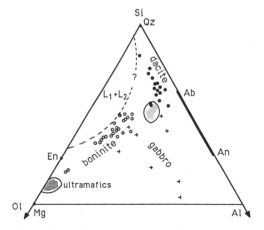

FIG. 8. Compositional variations in boninite series (open circles) ophiolites and calc–alkaline rocks from Bonin Island, the Mariana Trench (Philippine Sea) and Cape Vogel, Papua, New Guinea (WALKER and CAMERON, 1983; PEIVE et al., 1980). Dashed lines = likely immiscibility gap. Small dots = composition of glasses from boninites.

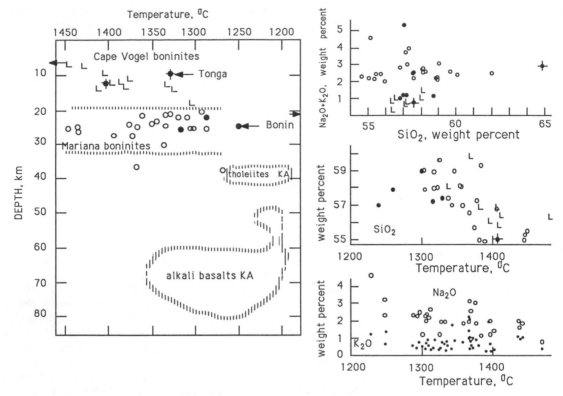

FIG. 9. Temperature regime of boninite series formation in the Earth's lower crust from contemporary island arcs calculated under the assumption of volatile–free conditions with Equations (17) and (30). Note the absence of silica versus alkali correlation in the initial magma in the temperature range 1220–1460°C. Symbols as in Figure 8. KA = Kurile arc (from Figure 7).

(2) Abundance of alkalies, Ba and Sr typical of island arc rocks.

(3) Variations in $^{143}Nd/^{144}Nd$ between those of chondrite and MORBs. The $^{143}Nd/^{144}Nd$ shows a good inverse correlation with Zr/Ti and $^{87}Sr/^{86}Sr$ and positive correlation with Sm/Nd and La/Sm.

(4) High and variable Zr/Ti and variable HREE/LREE.

(5) Correlation between Ti and Th as well as between $^{143}Nd/^{144}Nd$ and the abundance ratios, Sm/Nd and Ti/Zr.

(6) A wide variation of Ti/Zr and La/Yb.

Many models have been suggested to explain this variable geochemistry of boninites (*e.g.*, metasomatic alteration of the mantle before its partial melting, partial melting of a subducted, lithospheric slab). The geochemical data from ophiolite complexes (involving boninites) suggest, however, that mixing of primary magma with crustal material might be possible in the course of their formation. On the basis of REE data (Figure 10), drawn mainly from data for rocks in the western Pacific island arcs, heterogeneous sources of boninites might be

suggested. KARPENKO *et al.* (1984) also suggested two sources (at the minimum) to explain the isotopic and rare earth element data from boninites, although ultramafic magma–crustal rock mixing was not postulated in their model.

The pressure–temperature range for generation and fractionation of basaltic liquids in the Earth's crust and mantle is summarized in Figure 11. Alkali basalts are produced by ~5 percent melting of ultramafic rocks in the mantle at depths down to about 100 km, whereas boninites and related rocks (including some tholeiite complexes; *e.g.*, Bushveld and Karoo) were generated at shallower depth.

HYPOTHESIS FOR THE ORIGIN OF THEIR TERNARY GEOSTRUCTURES: MARGINAL SEA FLOORS–ISLAND ARCS–TRENCHES

From the evidence discussed in the preceding section, it is possible to conclude that the eastern and southeastern margins of Asia have been gradually destroyed over the last 50–60 m.y. and converted into oceanic crust. The appearance of typical sea floor instead of continental crust as well as

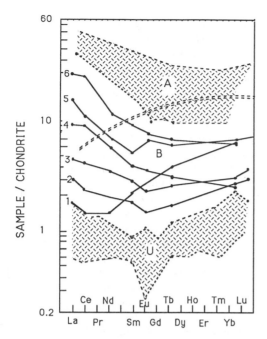

FIG. 10. Distribution of rare earth elements from rocks from western Pacific island arcs (after KARPENKO *et al.*, 1984; SHARASKIN *et al.*, 1983; ISHIZAKA and CARLSON, 1983; FROLOVA *et al.*, 1985). A = andesites and basalts in the northern Kuriles; high–magnesian andesites from southwest Japan fall in the same region. U = ultrabasites from the Mariana fore–arc. B = boninites from Bonin Islands (2), the Mariana Trench (3), Cape Vogel, Papua, New Guinea (4), Tonga Trench (5), and New Caledonia (6). Data from Cyprus boninites are given for comparison. See text for data sources.

thickening of the island arc crust have been observed in many areas. Possible mechanisms to explain these observations have been widely discussed, and two groups of possible mechanisms have been proposed. Back–arc spreading was suggested by, for example, KARIG (1974, 1975), HAYES (1984), SHARASKIN (1984) and others. A second mechanism has been proposed by ARTYUSHKOV (1979), BELOUSOV (1981, 1982) and PERCHUK (1984). This group of mechanisms is of a different nature and includes rock density inversions (BELOUSOV, 1981, 1982), eclogitization of the lower crust (ARTYUSHKOV, 1979) and magmatic replacement as a result of interaction of mantle–derived, high–temperature magma with crustal rocks (PERCHUK, 1985). Supporting evidence (see also discussion above) for the latter hypothesis includes:

(1) Gradual change in bulk composition of volcanic rocks from rhyolite and alkali basalt to trachybasalt and tholeiite during the last 60 m.y. in the Japan Sea. The volume of basic magma erupted increased significantly in the late Neogene.

(2) Increase in intensity of transformation of continental into oceanic crust, *i.e.* from the Okhotsk sea floor through the Japan Sea toward the Philippine sea floor. The process is accompanied either by crustal thinning beneath the sea floors or by thickening of the island arc igneous crust (KUSHIRO, 1985; TATSUMI, 1987; PEIVE *et al.*, 1980; SHARASKIN, 1984). This may reflect different stages of the same process.

(3) Alkali basalts that are diagnostic of the early (rift volcanism) and later stages (post–tholeiite volcanism) of continental destruction.

(4) Lateral correlation of alkalies and incompatible trace element contents with degree of partial melting of the mantle [across Japan and the Kuriles (KUSHIRO, 1983; TATSUMI, 1987)].

(5) Magma mixing beneath island arcs (SAKU-YAMA, 1979, 1981).

(6) Fluctuations in differentiation paths in terms of alkalies and silica with simultaneous strong correlation between Mg, Al and Si in boninite series; high temperature (up to 1450°C) and low pressure (less than 10 kbar) for the generation of boninites

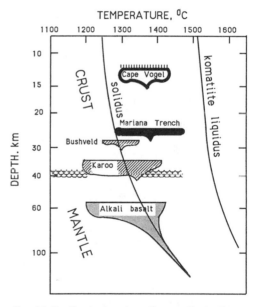

FIG. 11. Depth—temperature diagram illustrating possible regimes of basalt magma generation in the crust and the mantle. The contoured arcs show depth of each magma chamber and the temperature range of differentiation (not necessarily at a given pressure). Bushveld Complex—analytical data from CAWTHORN and DAVIES (1983). Cape Vogel and Mariana Trench are represented by boninite series. Karoo Province data are from WALKER and POL-DEVAART (1949) to represent tholeiite and high–alumina basalt series. Alkali basalt series from numerous regions in the world. Wavy line represents the crust–mantle boundary beneath continental shields.

and fractionation of magma with 1–3 weight percent H_2O (WALKER and CAMERON, 1983; SOBOLEV *et al.*, 1986).

(7) Predominance of alkali basalt and tholeiite among differentiated and nondifferentiated volcanic complexes.

(8) Relatively high heat flow at low seismic and volcanic activity in the depressions of marginal sea floors and inverse relationships in island arcs (HAYES, 1984; SHARASKIN, 1984; RODNIKOV *et al.*, 1980).

(9) Distribution of major and trace elements and their isotopes in boninites and related rocks.

By taking into account these observations together with the discussion above, the following hypothesis is being proposed. Mantle magmatism is caused by decompression melting of a plume, enriched in incompatible trace elements, that ascends from the core–mantle boundary (*e.g.*, ARTUSHKOV, 1979; RAMBERG, 1972). Because the rate of heat dissipation by conduction is several orders of magnitude lower than by thermal convection, the plume mechanism may be the only mechanism responsible for mantle diapirism (*e.g.*, RAMBERG, 1972). The temperature of the plume ($\sim$500 km in diameter at the boundary between the lower and upper mantle) could have reached 1900–2000°C where, however, only a small amount of melting would occur ($P = 130$–150 kbar). At a depth of about 300 km, the first komatiite composition melts could be produced (NESBITT *et al.*, 1979). The amount of komatiitic magma depends on the thermal environment, *i.e.* heat distribution in the upper part of the diapir. The relationships between the temperature and crystallization of the magma and the thermal conditions in a given volume of the mantle determine the composition of the diapiric front and the extent of its interaction with the surrounding rocks. The appearance of komatiitic liquid might lead to accumulation of volatiles in the melt and, as a result, to an increase in the ascent rate. Some of this melt may segregate and rise as an intrusion in the mantle. As the pressure and temperature decrease, differentiation of the komatiitic magma might lead to crystallization in the form of layered intrusions. If tectonic conditions allow, a small amount of the komatiitic liquid might appear on the Earth's surface along deep–seated faults. This feature is illustrated with the well–known ophiolitic complex in Cyprus.

The next event in the diapiric ascent is illustrated by the appearance of late Cretaceous alkali basaltic volcanism in rift zones in island arcs. For example, pressure–temperature and age determinations of

alkali basalt magma generation in the Little Kurile island arcs (Figure 7) reflect the partial melting of the upper mantle 50–60 m.y. ago in the front of the diapir at depths of 60–100 km at a temperature near 1400°C. Fluid and melt inclusions in glasses and olivine phenocrysts in alkali basalts reflect enrichments in the initial magma of CO_2, H_2, N_2, CH_4, Ar, CO and H_2O (LETNIKOV *et al.*, 1977). These data show some differentiation processes near the solidus of a diapiric material with increasing abundance of volatiles in the frontal portions of the body (at 100–200 km or shallower). This enrichment might result in a lower temperature of partial melting at the depth corresponding to alkali basalt liquid formation as compared with a volatile–free environment. Deep–seated faults favor fast ascent of the low–density magma thus produced, and alkali basaltic volcanism is likely to be the first stage of the magmatic process observed on the surface.

The stratigraphy of the Japan sea floor shows erosion of sedimentary and igneous rocks during the Paleogene (VASILKOVSKII *et al.*, 1978; see also Figure 3). This period corresponds to the time of formation of an anticline structure observed in the modern Japan sea floor crust. Penetration of mantle–derived fluids into this structure might lead to partial melting of the crustal rocks and, as a result, to the development of acid magmatism in this area. The change in bulk composition of crustal–derived magmas (see Figure 4) correlates with the ascent of the diapir and its partial melt followed by complete melting resulting from the decompression (RAMBERG, 1972).

Thus, the ascent of the diapir results in an increase in heat flow upward and partial melting of both the crust and the mantle. This process results in the development of liparito–dacitic magmas followed by the basaltic andesite series shown in Figures 3 and 4. Komatiite could also be a product of lherzolitic or harzburgitic magma assimilated by crustal material when the completely molten diapir ascended toward the continental crust. Melting and mixing processes in the frontal part of the diapir would lead to cooling and subsequent fractionation. When the diapir reaches the crust–mantle interface, a completely molten diapir could react with the crust to produce mixed magmas whose compositions vary from komatiite through boninite to dacite. A peak in volcanic activity occurred in the Japan Sea area 3–5 m.y. ago when deep depressions formed in the sea floor (RODNIKOV *et al.*, 1980; PERCHUK *et al.*, 1985; see also Figure 3). In the Okhotsk sea floor, this stage was essentially limited to tholeiitic volcanism (accompanied by the formation of the South Okhotsk depression). In the

Philippine sea floor, the maximum intensity of this late stage was manifested by tholeiitic volcanism and tectonic activity.

From isotopic age determinations, an ascent rate from 350–400 km to the surface over the last ~60 m.y. can be estimated (0.4–0.6 cm/year). This rate is consistent with RAMBERG's (1972) prediction. Thermal convection cells must exist adjacent to the diapir, with a relatively high rate of horizontal displacement of the mixed magma toward cooler zones. This process is accompanied by fractionation. As the temperature decreases, isolated magma chambers may be formed. The largest extent of fractionation of the mixed magma might lead to magmatic activity and growth of igneous crust in island arcs and continents (PERCHUK et al., 1985).

Deep–sea depressions are caused by magmatic replacement of the crust (melting, magma mixing and transport of material toward relatively low–temperature zones beneath the island arcs) and increasing density of the material beneath the sea floors. A decrease in crustal thickness beneath the sea floor is accompanied by increased thickness in the island arc provided by the thermal gradient discussed above. This gradient may be a function of the geometry of the subducted slab. With a very steep descent angle, the thermal gradient is low and horizontal displacement is slow, resulting in only slow growth of the igneous crust in a volcanic arc. This situation could have existed beneath the Philippine sea floor, where island arc growth at early stages of diapir–crust interaction was replaced by mantle–derived ultramafic magma during the latest stages. A sharp decrease in heat flow, and displacement of isotherms to greater depth, lead to the disappearance of the magma flow beneath the sea floor. The asthenospheric surface beneath the Philippine sea floor, at present, varies between 20 km beneath the Mariana trench and fore–arc and 80 km beneath the Philippine basin, between the Palau–Kyushu ridge and Ryukyu (A. G. RODNIKOV, personal communication, 1986).

According to the suggested hypothesis, fractionation and crystallization of diapiric material in separate convection cells produce numerous layered mafic–ultramafic bodies in the upper mantle. The body size depends on the relationships between the thermal gradient in a given volume of mantle and the solidus temperature of komatiite.

This hypothesis suggests a simultaneous origin of marginal sea floors and island arcs and also suggests replacement of early island arc volcanics by mantle–derived diapiric material. This hypothesis can explain, for example, the evolution of volcanism from acid to basic accompanied by a decrease of sea floor crustal thickness. The hypothesis is not at variance with well–known models for calc–alkaline magma generation in subduction zones, and can also shed some light on some unresolved problems related to this mechanism. In contrast to models developed by Japanese petrologists, for example (e.g., SAKUYAMA, 1983; KUSHIRO, 1985; TATSUMI, 1987), this hypothesis emphasizes interaction of high–temperature ultramafic magma with crustal rocks and the appearance of hybrid melts whose subsequent fractionation could lead to development of the basalt series in frontal arcs.

It is suggested that similar processes were inherent in many periods in the Earth's evolution. For example, the most suitable conditions for interaction processes existed in the Precambrian because of the higher geothermal gradients in the crust and upper mantle.

Acknowledgements—First of all, I wish to thank Bjorn Mysen who invited me to contribute to both the symposium and this Volume. His assistance, criticism and patience are highly appreciated. I would also like to thank T. I. Frolova for fruitful two–year discussions of my ideas on the subject. I am indebted to Ikuo Kushiro for numerous discussions concerning the origin of island arc crust. This manuscript was greatly improved by critical reviews by Ikuo Kushiro, Mark Ghiorso, Peter Kelemen and one anonymous reviewer whom I thank very much for their helpful comments made in a constructive spirit. With a great pleasure I wish to thank Mrs. G. G. Gonchar for her assistance during translation of the paper into English. Finally, I wish to express gratitude to Mrs. E. V. Zabotina for patiently typing the manuscript.

Editorial Comment—Several of the reviewers objected strongly to the logic of and support for important conclusions in this paper. These objections led to some changes in the revised manuscript. It was then subjected to extensive editorial changes in regard to style and phraseology. I am indebted to I. Kushiro, R. W. Luth, D. Virgo and H. S. Yoder, Jr., for their assistance in this process.

REFERENCES

AKELLA J. and BOYD F. R. (1973) Effect of pressure on the composition of coexisting pyroxenes and garnets in the system $CaSiO_3$–$FeSiO_3$–$MgSiO_3$–$CaAlTi_2O_6$. *Carnegie Inst. Wash. Yearb.* **72**, 523–526.

AKELLA J. and BOYD F. R. (1974) Petrogenetic grid for garnet peridotites. *Carnegie Inst. Wash. Yearb.* **73**, 269–285.

ARANOVICH L. YA. and KOSYAKOVA N. A. (1986) Garnet–orthopyroxene geobarometer: Thermodynamics and implication. *Geokhimiya* **8**, 1181–1202.

ARTYUSHKOV A. (1979) *Geodynamics,* 324 pp. Nauka Press, Moscow.

BELOUSOV V. V. (1981) Some problems of structure and conditions of development of transitional zones between continents and oceans. *Geotectonika* **3**, 3–24.

BELOUSOV V. V. (1982) *Transitional Zones between Continents and Oceans.* 210 pp. Nedra Press, Moscow.

BOTTINGA Y. and WEILL D. F. (1970) Densities of liquid silicate systems calculated from partial molar volumes of oxide components. *Amer. J. Sci.* **269,** 169–182.

BOTTINGA Y., WEILL D. F. and RICHET P. (1980) Thermodynamic modelling of silicate melts. In *Thermodynamics of Minerals and Melts,* (eds. R. C. NEWTON, A. NAVROTSKY and B. J. WOOD), pp. 207–247 Springer-Verlag.

CAMERON W. E., NISBET E. G. and DIETRICH V. J. (1979) Boninites, komatiites and ophiolitic basalts. *Nature* **280,** 550–553.

CAMERON W. E., MCCULLOCH M. T. and WALKER D. A. (1983) Boninite petrogenesis: Chemical and Nd–Sr isotopic constraints. *Earth Planet. Sci. Lett.* **65,** 75–89.

CARMICHAEL I. S. E., NICHOLLS J., SPERA F. J., WOOD B. J. and NELSON S. A. (1977) High temperature properties of silicate liquids: Applications to the equilibration and ascent of basic magma. *Phil. Trans. Roy. Soc. London* **286A,** 373–431.

CAWTHORN R. G. and DAVIES G. (1983) Experimental data at 3 kbar pressure on parental magma of the Bushveld Complex. *Contrib. Mineral. Petrol.* **83,** 128–135.

COX K. G. and MALLICK D. J. (1970) The peralkaline volcanic suite of Aden and Little Aden, South Arabia. *J. Petrol.* **11,** 433–463.

COX K. G., BELL J. D. and PANKHURST R. J. (1979) *The Interpretation of Igneous Rocks.* Allen and Unwin.

DANUSHEVSKII L. V., SOBOLEV A. V. and KONONKOVA N. N. (1986) Initial melt of the high titanium boninite series from Tonga trench. In *Geochemistry of Magmatic Rocks, Abstr. Seminar XII,* Institute of Geochemistry, Moscow.

DAVIS B. C. T. and BOYD F. R. (1966) The join $Mg_2Si_2O_6$–$CaMgSi_2O_6$ at 30 kb pressure and its application to pyroxenes from kimberlites. *J. Geophys. Res.* **71,** 3567–3576.

DAVIS B. C. T. and ENGLAND J. L. (1963) Melting of forsterite, Mg_2SiO_4, at pressures up to 47 kilobars. *Carnegie Inst. Wash. Yearb.* **62,** 119–121.

DAWSON J. B., POWELL D. G. and REID A. M. (1970) Ultrabasic xenoliths and lava from Lashaine Volcano, Northern Tanzania. *J. Petrol.* **11,** 519–548.

DEPAOLO D. J. (1979) Estimation of the depth of origin of basaic magmas: A modified thermodynamic approach and a comparison with experimental melting studies. *Contrib. Mineral. Petrol.* **69,** 265–278.

DEPAOLO D. J. and WASSERBURG G. J. (1977) The sources of island arcs as indicated by Nd and Sr isotopic studies. *Geophys. Res. Lett.* **4,** 465–468.

ESIN O. A. (1946) *Elektrolitischeskaya Priroda Zhidkich Shlakov.* Izd. Doma Tekhniki Ural'skogo Industr. Inst., Sverdlovsk.

FROLOVA T. I., BURIKOVA I. A. and GUSCHIN A. V. (1985) *Origin of the Island Arc Volcanic Series,* 230 pp. Nedra Press, Moscow.

FROLOVA T. I. and RUDNIK G. B. (1974) An anorthositic tendency of differentiation in volcanic rocks of early stages in evolution of island arcs (an example from the South Sandwich Islands). Vestnik Moscow University, *Geology,* 20–26.

GHIORSO M. S. and KELEMEN P. B. (1987) Evaluating reaction stoichiometry evolving under generalized thermodynamic constraints: Examples comparing isothermal and isenthalpic assimilation. In *Magmatic Processes: Physicochemical Principles,* (ed. B. O. MYSEN). The Geochemical Special Publ. No. 1., pp. 319–358.

GHIORSO M. S., CARMICHAEL I. S. E., RIVERS M. L. and SACK R. O. (1983) The Gibbs free energy of mixing of natural silicate liquids: An expanded regular solution approximation for the calculation of magmatic intensive variables. *Contrib. Mineral. Petrol.* **84,** 107–145.

GNIBIDENKO G. S. and KHVEDCHUK I. I. (1984) The main features of geology for the Okhotsk sea floor. In *27th Intern. Congress Moscow, R.S.F.S.R.,* Vol. 23, pp. 19–41 VNU Science Press, Utrecht, Netherlands.

GREEN D. H. (1976) Experimental testing of the "equilibrium" partial melting of peridotite under water–saturated, high–pressure conditions. *Can. Mineral.* **14,** 255–268.

GREEN D. H. and RINGWOOD A. E. (1966) Origin of the calc–alkaline igneous rock suite. In *High Pressure Experimental Investigation into the Nature of the Mohorovicic Discontinuity.* Australian National University, Canberra.

HAYES D. E. (1984) Marginal seas of Southeast Asia: Their geophysical characteristics and structure. In *27th Intern. Congress Moscow, R.S.F.S.R.,* Vol. 23, pp. 123–154 VNU Science Press, Utrecht, Netherlands.

HENSEN B. J. (1973) Pyroxenes and garnets as geothermometers and geobarometers. *Carnegie Inst. Wash. Yearb.* **72,** 527–534.

HERZBERG C. T. (1979) The solubility of olivine in basaltic liquids: An ionic model. *Geochim. Cosmochim. Acta* **43,** 1241–1253.

HICKEY R. L. and FREY F. A. (1982) Geochemical characteristics of boninite series volcanics: Implications for their source. *Geochim. Cosmochim. Acta* **46,** 2099–2117.

HSU L. C. (1967) Melting of fayalite up to 40 kilobars. *J. Geophys. Res.* **72,** 4235–4244.

IRVINE T. N. (1976) Metastable liquid immiscibility and MgO–FeO–SiO_2 fractionation patterns in the systems Mg_2SiO_4–Fe_2SiO_4–$CaAl_2Si_2O_8$–$KAlSi_3O_8$–SiO_2. *Carnegie Inst. Wash. Yearb.* **75,** 597–611.

ISHIZAKA K. and CRALSON R. W. (1983) Nd–Sr systematics of the Setouchi volcanic rocks, southwest Japan: A clue to the origin of orogenic andesite. *Earth Planet. Sci. Lett.* **54,** 327–340.

KARIG D. E. (1974) Evolution of arc systems in the Western Pacific. *Ann. Rev. Earth Planet. Sci.* **2,** 51–75.

KARIG D. E. (1975) Basin genesis in the Philippine sea. *Init. Rep. DSDP 31,* pp. 857–879. U.S. Government Printing Office.

KARPENKO S. F., SHARASKIN A. JA., BALASHOV JU. A., LYALIKOV A. V. and SPIRIDONOV V. G. (1984) The isotopic and geochemical criteria of boninite genesis. *Geokhimiya* **7,** 958–970.

KEPEZHINSKAS V. V. (1979) *Quaternary Alkali Basalts of Mongolia and their Deep–Seated Inclusions.* 311 pp. Nauka Press, Moscow.

KONOVALOV YU. I. (1984) Volcanism of the Japan Sea Floor. Ph.D. Dissertation, Moscow University.

KORZHINSKII D. S. (1959) Acid–base interaction of components in the dry silicate melts and direction of cotectic lines. *Akad. Nauk. USSR, Dokl.* 128.

KUNO H. (1960) High–alumina basalts. *J. Petrol.* **1,** 121–145.

KUNO H. (1968) Differentiation of basalt magmas. In *Basalts,* (eds. H. H. HESS and A. POLDEVAART), Vol. 2, pp. 623–688. Interscience.

KUSHIRO I. (1968) Composition of magmas formed by partial zone melting in the Earth's upper mantle. *J. Geophys. Res.* **73,** 619–634.

Kushiro I. (1972) Effect of H_2O on the composition of magmas at high pressures. *J. Petrol.* **13**, 311–334.

Kushiro I. (1975) On the nature of silicate melt and its significance in magma genesis: Regularities in the shift of the liquidus boundaries involving olivine, pyroxene and silica minerals. *Amer. J. Sci.* **275**, 411–431.

Kushiro I. (1980) Changes with pressure of degree of partial melting and K_2O content of liquids in the system Mg_2SiO_4–$KAlSiO_4$–SiO_2. *Carnegie Inst. Wash. Yearb.* **79**, 267–271.

Kushiro I. (1983) On the laterial variations in chemical composition and volume of Quaternary volcanic rocks across Japanese arcs. *J. Volcanol. Geotherm. Res.* **18**, 435–447.

Kushiro I. (1985) A petrological model of the mantle wedge and crust of the Japanese Islands. In *Correlation of Metamorphic Processes and Geodynamic Regimes, Initial Meeting of IGCP, Project 235, Extended Abstracts*, pp. 16–23.

Leeman W. P. (1978) Distribution of Mg^{2+} between olivine and silicate melt structure. *Geochim. Cosmochim. Acta* **42**, 789–801.

Letnikov F. A., Karpov I. K., Kisiles A. H. and Shkandriy B. O. (1977) *Fluid Regime in the Earth's Crust and Upper Mantle*, 213 pp. Nauka Press, Moscow.

MacDonald G. A. and Katsura T. (1964) Chemical composition of Hawaiian lavas. *J. Petrol.* **5**, 81–133.

McBirney A. R. and Williams H. (1969) Geology and Petrology of Galapagos Islands. *Geol. Soc. Amer. Mem. 118.*

Meijer A. (1976) Pb and Sr isotopic data bearing on the origin of volcanic rocks from the Mariana island arc system. *Bull. Geol. Soc. Amer.* **87**, 1358–1369.

Meijer A. (1980) Primitive arc volcanism and a boninite series: Examples from western Pacific island arcs. In *Amer. Geophys. Union Mon. 23*, (ed. D. E. Hayes), pp. 269–282.

Meijer A. (1983) The origin of low-K rhyolites from Mariana frontal arc. *Contrib. Mineral. Petrol.* **88**, 45–51.

Modreski P. J. and Boettcher A. L. (1973) Phase relationships of phlogopite in the system K_2O–MgO–CaO–Al_2O_3–SiO_2–H_2O to 35 kilobars: A better model for micas in the interior of the Earth. *Amer. J. Sci.* **273**, 385–414.

Mohr P. A. (1960) *The Geology of Ethiopia.* Univ. College, Addis–Ababa.

Murata K. J. (1970) Tholeiitic basalt magmatism of Kilauea and Mauna Loa volcanoes of Hawaii. *Naturwissenschaften* **57**, 108–113.

Navrotsky A. (1980) Thermodynamics of mixing in silicate glasses and melts. In *Thermodynamics of Minerals and Melts*, (eds. R. C. Newton, A. Navrotsky and B. J. Wood), pp. 189–206 Springer-Verlag.

Nesbitt R. W., Sun S.-S. and Purvis A. C. (1979) Komatiites: Geochemistry and genesis. *Can. Mineral.* **17**, 165–186.

Nicholls J. (1977) The activities of components in natural silicate melts. In *Thermodynamics in Geology*, (ed. D. G. Fraser), pp. 327–347 Reidel, Dordrecht, Holland.

Nicholls J., Carmichael I. S. E. and Stormer J. C. (1971) Silicate activity and P^{total} in igneous rocks. *Contrib. Mineral. Petrol.* **33**, 1–20.

Nicholls J. and Carmichael I. S. E. (1972) Equilibrium temperature and pressure of various lava types with spinel and garnet peridotite. *Amer. Mineral.* **57**, 941–959.

Peive A. V., Coleman R. G. and Bogdanov N. A. (1980)

Geology of the Philippine Sea Floor. 259 pp. Nauka Press, Moscow.

Perchuk L. L. (1967) Analysis of thermodynamic conditions of mineral equilibria in the amphibole–garnet rocks. *Izv. Akad. Nauk. SSSR* **3**, 57–83.

Perchuk L. L. (1968) Pyroxene–garnet equilibrium and the depth of facies of eclogites. *Int. Geol. Rev.* **10**, 280–318.

Perchuk L. L. (1969) The effect of temperature and pressure on the equilibrium of natural iron–magnesium minerals. *Int. Geol. Rev.* **11**, 875–901.

Perchuk L. L. (1973) *Thermodynamic Regime of Depth of Petrogenesis.* 310 pp. Nauka Press, Moscow.

Perchuk L. L. (1977a) Modification of the two–pyroxene geothermometer for deep–seated peridotites. *Akad. Nauk. USSR, Dokl.* **233**, 456–459.

Perchuk L. L. (1977b) Proxene barometer and "pyroxene geotherm". *Akad. Nauk. USSR, Dokl.* **233**, 1196–1199.

Perchuk L. L. (1977c) Thermodynamic control of metamorphic processes. In *Energetics of Geological Processes*, (eds. S. K. Saxena and S. Bhattacharji), pp. 285–352 Springer.

Perchuk L. L. (1983) Pressure dependence of liquidus temperatures for the dry basalt systems. *Akad. Nauk. USSR, Dokl.* **271**, 702–705.

Perchuk L. L. (1984a) Theoretical consideration of basalt series: Temperature control. *Geol. Zb.—Geol. Carpathica* **35**, 329–354.

Perchuk L. L. (1984b) The pressure and H_2O/CO_2 control of generation and evolution of basalt series. *Geol. Zb.—Geol. Carpathica* **35**, 463–487.

Perchuk L. L. (1985) Petrological aspects of formation of marginal sea depressions. *Akad. Nauk. SSSR, Dokl.* **280**, 178–182.

Perchuk L. L. and Vaganov V. I. (1977) Temperature regime of formation of continental volcanic series. Acad. Sci. USSR, Special Issue, Moscow.

Perchuk L. L. and Frolova T. I. (1979) Causes of the volcanic series variation. *Izv. Akad. Nauk. SSSR* **9**, 28–44.

Perchuk L. L. and Frolova T. I. (1980) Influence of fluid composition on the evolution of basalt series. *Akad. Nauk. USSR, Dokl.* **253**, 1456–1439.

Perchuk L. L. and Vaganov V. I. (1980) Petrochemical and thermodynamic evidence of the origin of kimberlites. *Contrib. Mineral. Petrol.* **72**, 219–228.

Perchuk L. L. and Lindsley D. H. (1982) Fluid–magma interaction at high pressure–temperature conditions. *Adv. Earth Planet. Sci.* **12**, 251–259.

Perchuk L. L. and Lavrent'eva I. V. (1983) Experimental investigation of exchange equilibria in the system cordierite–garnet–biotite. *Adv. Phys. Geochem.* **3**, 199–239.

Perchuk L. L., Aranovich L. Ya. and Kosyakova N. A. (1982) Thermodynamic models of the origin and evolution of basalt magmas. *Vest. Moskov. Univ. Ser. Geol.* **4**, 3–25.

Perchuk L. L., Frolova T. I. and Konovalov Yu. I. (1985) Role of volcanic processes in the creation of geostructure of the Japan sea floor. *Contrib. Phys. Chem. Petrol.*, **13**, 81–105.

Ramberg H. (1972) Mantle diapirism and its tectonic and magma–genetic consequences. *Phys. Earth Planet. Int.* **5**, 45–60.

Reid A. M., Donaldson C. H., Brown R. W., Ridley W. I. and Dawson J. B. (1975) Mineral chemistry of peridotite xenoliths from the Lashaine Volcano. *Phys. Chem. Earth* **9**, 525–543.

RODNIKOV A. G., FAINANOV A. G., ERMAKOV B. V. and KATO T. (1982) Geotraverse across Suikhote-Alin'-the Sea of Japan-the Honshu Island-the Pacific. USSR Geophys. Committee, Spec. Issue, Moscow.

ROEDER P. L. and EMSLIE R. F. (1970) Olivine-liquid equilibrium. *Contrib. Mineral. Petrol.* **29**, 275–289.

RYABCHIKOV I. D., KOVALENKO V. I., IONOV D. A. and SOLOVOVA I. P. (1983) Thermodynamic parameters of mineral equilibria in garnet-spinel lherzolites of Mongolia. *Geokhimiya* **7**, 567–578.

SAKUYAMA M. (1979) Evidence of magma mixing: Petrological study of Shirouma-Oike calc-alkaline andesite. *J. Volcan. Geotherm. Res.* **5**, 179–208.

SAKUYAMA M. (1981) Petrological study of the Myoko and Kurohime volcanoes, Japan: Crystallization sequence and evidence of magma mixing. *J. Petrol.* **22**, 553–583.

SAKUYAMA M. (1983) Petrology of arc volcanic rocks and their origin by mantle diapir. *J. Volcan. Geotherm. Res.* **18**, 297–320.

SCHAIRER J. F. (1954) The system K_2O–MgO–Al_2O_3–SiO_2. I. Results of quenching experiments on four joins in the tetrahedron cordierite-forsterite-leucite-silica and on the join cordierite-mullite-potash feldspar. *J. Amer. Ceram. Soc.* **37**, 501–533.

SCHAIMANN JU. M. (1968) *Contribution to Abyssal Petrology*, 320 pp. Nedra Press, Moscow.

SHARASKIN A. YA. (1984) Structure and tectono-magmatic origin of the Philippine sea floor. In Intern. Geol. Congr. Symp. Rep. 6, pp. 44–58 Nauka Press, Moscow.

SHARASKIN A. YA., PUSTCHIN I. K., ZLOBIN S. K. and KOLESOV G. M. (1983) Two ophiolite sequences from the basement of the northern Tonga arc. *Ofioliti* **8**, 411–430.

SOBOLEV A. V. and DANUSHEVSKII L. V. (1986) Evidence of magmatic origin of H_2O and determination of its content in the residual boninite melts. *Akad. Nauk. USSR, Dokl.* **288**, 262–265.

SOBOLEV A. V., TSAMERUAN O. P., DMITRIEV L. V. and KONONKOVA N. N. (1986) Water-bearing komatiites as a new type of the komatiitic melts and the origin and ultramafic lavas from the Troodos Complex, Cyprus. *Akad. Nauk. USSR, Dokl.* **286**, 422–425.

STEBBINS J. F., CARMICHAEL I. S. E. and MORET L. K. (1984) Heat capacities and entropies of silicate liquids and glasses. *Contrib. Mineral. Petrol.* **86**, 131–148.

TATSUMI I. (1987) The origin of subduction zone magmas based on experimental petrology. In *Physical Chemistry of Magmas*, (eds. I. KUSHIRO and L. L. PERCHUK), Springer. (In press).

THOMPSON R. N. and TILLEY C. E. (1969) Melting and crystallization relations of Kilauean basalt of Hawaii. The lavas of the 1959–60 eruption. *Earth Planet. Sci. Lett.* **5**, 469–477.

VASILKOVSKII N. V., BEZVERKMHII V. L. and DERKACHEV A. N. (1978) *The Basic Features of the Geology of the Japan Sea Floor*, 263 pp. Nauka Press, Moscow.

WALKER D. A. and CAMERON W. R. (1983) Boninite primary magmas: Evidence from the Cape Vogel peninsula, PNG. *Contrib. Mineral. Petrol.* **83**, 150–158.

WALKER F. and POLDEVAART A. (1949) Karoo tholeiites of the Union of South Africa. *Bull. Geol. Soc. Amer.* **60**, 591–706.

WENDLANDT R. F. and EGGLER D. H. (1980) The origins of potassic magmas, 1. Melting relations in the systems $KAlSiO_4$–Mg_2SiO_4–SiO_2 and $KAlSiO_4$–$MgSiO_3$–CO_2 to 30 kbars. *Amer. J. Sci.* **280**, 385–420.

WHITE R. W. (1966) Ultramafic inclusions in basaltic rocks from Hawaii. *Contrib. Mineral. Petrol.* **12**, 245–314.

WINCHELL H. (1947) Honolulu series, Oahu, Hawaii. *Bull. Geol. Soc. Amer.* **58**, 1–48.

YAGI K. (1953) Petrological studies of the alkalic rocks of the Morotu district, Sakhalin. *Bull. Geol. Soc. Amer.* **64**, 769–810.

YODER H. S. and TILLEY C. E. (1962) Origin of basalt magmas: An experimental study of natural and synthetic rock systems. *J. Petrol.* **3**, 342–532.

ZLOBIN T. K., FEDORENKO V. I., PETROV A. V. and NEMCHENKO G. S. (1982) Lithosphere structure beneath Kunashir (the Kurile Islands) based on seismic data. *Pac. Geol.* **1**, 92–100.

ZYRIANOV V. N. (1987) Interaction of phonolitic melt with fluid and the origin of pseudoleucite. *Contrib. Phys. Chem. Petrol.*, (In press).

ZYRIANOV V. N. and PERCHUK L. L. (1978) Origin of the sodium and potassium magmas enriched by silica. *Akad. Nauk. USSR, Dokl.* **242**, 187–189.

APPENDIX

Table A1. Sources of data used for evaluation of Equation (13)

A. System CaO–MgO–Al_2O_3–SiO_2 at atmospheric pressure		B. Alkali–bearing systems at atmospheric pressure	
Fo–Di–SiO_2	BOWEN (1914); KUSHIRO (1972a)	Fo–Ne–Di	SCHAIRER and YODER (1960a,b); YODER and KUSHIRO (1972)
MgO–SiO_2	BOWEN and ANDERSEN (1914)	En–Ab–Di	SCHAIRER and MORIMOTO (1959)
Fo–An–SiO_2	ANDERSEN (1915); BIRD (1971)	Fo–Ab–Di	WATSON (1977); HART et al. (1976)
MgO–Al_2O_3–SiO_2	RANKIN and MERWIN (1918)	Di–Ak–Ne–CaTs	YAGI and ONUMA (1969)
Fo–Di–An	OSBORN and TAIT (1952)	Di–Ab–An–NiO	LINDSTROM (1976)
En–Di–An	HYTONEN and SCHAIRER (1961)	K_2O–Al_2O_3–MgO–SiO_2	SCHAIRER (1954); IRVINE (1976); WENDLANDT and EGGLER (1980); KUSHIRO (1980)
CaO–MgO–Al_2O_3–SiO_2	MACGREGOR (1965); PRESNALL et al. (1973)		
FeO–MgO–SiO_2	BOWEN and SCHAIRER (1935)		

Table A1. (Continued)

C. Synthetic systems at high pressure

Fo–Di–Py	DAVIS and SCHAIRER (1965)
CaO–MgO–Al$_2$O$_3$– SiO$_2$	PRESNALL *et al.* (1973)
CaO–MgO–Al$_2$O$_3$– SiO$_2$–Na$_2$O	KUSHIRO (1972b)
Fo–Di–SiO$_2$	KUSHIRO (1969)
MgO–TiO$_2$–SiO$_2$	MACGREGOR (1969)

Table A2. Data used for linear regression of Y versus temperature [Equation (14)] for the system Fo–Ks–Qz (weight percent) for K$_2$O > 5 weight percent (SCHAIRER, 1954; WENDLANDT and EGGLER, 1980)

Fo	Ks	Qz	Y	$T_{\text{experimental}}$	$T_{\text{theoretical}}$
15	31.8	53.2	0.071	1450	1413
7	36.9	56.1	0.031	1400	1349
2	42.6	55.4	0.010	1200	1311
8	48.3	43.7	0.039	1400	1359
2	45.5	52.2	0.010	1300	1311
21	57	22	0.107	1490	1474
30	51	19	0.179	1560	1553
35	47	18	0.204	1640	1637
51	36	13	0.260	1700	1732

Table A3. Chemical compositions of basalts studied at dry conditions between 1 and 10 kbar

Sample no.	(1) A-90	(2) TF-38	(3) KRB 2	(4) NT-233	(5) NM-51	(6) 527-1-1
SiO$_2$	49.73	44.31	47.47	49.25	47.93	48.60
TiO$_2$	2.10	1.58	2.43	0.79	1.34	0.61
Cr$_2$O$_3$	n.d.	n.d.	n.d.	0.20	—	0.06
Al$_2$O$_3$	15.82	13.89	17.15	13.64	16.75	16.30
Fe$_2$O$_3$	9.57*	10.47*	2.26	—	—	0.30
FeO	—	—	7.22	9.77†	11.40†	8.40
MgO	6.41	13.78	8.76	17.61	7.59	10.20
CaO	9.30	10.52	8.03	9.58	9.33	12.30
Na$_2$O	3.45	2.49	3.67	0.89	2.95	1.90
K$_2$O	1.34	0.73	1.89	0.06	0.19	0.07
H2O†	n.d.	0.98	1.69	n.d.	—	0.08
H2O$^-$	n.d.	—	0.22	n.d.	—	0.13
P$_2$O$_5$	n.d.	0.17	0.66	n.d.	—	0.06
Total	99.05	99.09	100.38	99.93	100.00	99.16
dT/dP °C/kbar	3.8	4.0	4.5	4.7	4.0	3.5

(1)–(2): data from PERCHUK (1983); (3): TAKAHASHI (1979); (4): ELTHON and SCARFE (1979); (5): COHEN *et al.* (1967); (6): BENDER *et al.* (1978).

* All iron as Fe$_2$O$_3$

† All iron as FeO

REFERENCES TO APPENDIX

ANDERSEN O. (1915) The system anorthite–forsterite–silica. *Amer. J. Sci.* **39**, 407–454.

BENDER L. F., HODGES F. N. and BENCE A. E. (1978) Petrogenesis of basalts from the Project FAMOUS area: Experimental study from 0 to 15 kbar. *Earth Planet. Sci. Lett.* **41**, 277–302.

BIRD M. L. (1971) Distribution of trace elements in olivines and pyroxenes—An experimental study. Ph.D. Thesis, Univ. Missouri.

BOWEN N. L. (1914) The ternary system diopside–forsterite–silica. *Amer. J. Sci.* **37**, 207–264.

BOWEN N. L. and ANDERSEN O. (1914) The binary system MgO–SiO$_2$. *Amer. J. Sci.* **37**, 487–500.

BOWEN N. L. and SCHAIRER J. F. (1935) The system MgO–FeO–SiO$_2$. *Amer. J. Sci.* **29**, 151–217.

COHEN L., ITO K. and KENNEDY G. C. (1967) Melting and phase relations in an anhydrous basalt to 40 kilobars. *Amer. J. Sci.* **265**, 519–539.

DAVIS B. C. T. and SCHAIRER J. F. (1965) Melting relations in the join diopside–forsterite–pyrope at 40 kilobars and at 1 atmosphere. *Carnegie Inst. Wash. Yearb.* **64**, 123–126.

HART S. R., DAVIS K. E., WATSON E. B. and KUSHIRO I. (1976) Partitioning of nickel between olivine and silicate liquids. *Geol. Soc. Amer. Abstr. Progr.* **8**, 906.

HYTONEN K. and SCHAIRER J. F. (1961) The plane enstatite–anorthite–diopside and its relation to basalts. *Carnegie Inst. Wash. Yearb.* **60**, 124–141.

KUSHIRO I. (1969) The system forsterite–diopside–silica with and without water at high pressures. *Amer. J. Sci.* **267-A**, 269–294.

KUSHIRO I. (1972a) Determination of liquidus relations in synthetic silicate systems with electron probe analysis: The system forsterite–diopside–siliva at 1 atmosphere. *Amer. Mineral.* **57**, 1260–1271.

KUSHIRO I. (1972b) The effect of water on the composition of magmas formed at high pressure. *J. Petrol.* **13**, 311–334.

LINDSTROM D. J. (1976) Experimental study of the partitioning of the transition metals between clinopyroxene and coexisting silicate liquids. Ph.D. Thesis Univ. Oregon.

MacGREGOR I. D. (1965) Stability fields of spinel and garnet peridotites in the synthetic system MgO–CaO–Al_2O_3–SiO_2. *Carnegie Inst. Wash. Yearb.* **64**, 125–134.

MacGREGOR I. D. (1969) The system MgO–TiO_2–SiO_2 and its bearing on the distribution of TiO_2 in basalts. *Amer. J. Sci.* **267-A**, 342–363.

OSBORN E. F. and TAIT D. B. (1952) The system diopside–forsterite–anorthite. *Amer. J. Sci.* **250-A**, 413–433.

PRESNALL D. C., O'DONNELL T. H. and BRENNER N. L. (1973) Cusps on the solidus curves as controls for primary magma composition: A mechanism for producing oceanic tholeiites of uniform composition. *Geol. Soc. Amer. Abstr. Progr.* **5**, 771–772.

RANKIN G. A. and MERWIN H. W. (1918) The system MgO–Al_2O_3–SiO_2. *Amer. J. Sci.* **45**, 301–325.

SCHAIRER J. F. (1954) The system K_2O–MgO–Al_2O_3–SiO_2: I. Results of quenching experiments of four joins in the tetrahedron cordierite–forsterite–leucite–silica and on the join cordierite–mullite–potash feldspar. *J. Amer. Ceram. Soc.* **37**, 501–533.

SCHAIRER J. F. and MORIMOTO N. (1959) The system forsterite–diopside–silica–albite. *Carnegie Inst. Wash. Yearb.* **58**, 113–118.

SCHAIRER J. F. and YODER H. S. JR. (1960a) The system forsterite–nepheline–silica. *Carnegie Inst. Wash. Yearb.* **59**, 70–71.

SCHAIRER J. F. and YODER H. S. JR. (1960b) Crystallization in the system nepheline–forsterite–silica at one atmosphere pressure. *Carnegie Inst. Wash. Yearb.* **59**, 141–144.

TAKAHASHI E. (1980) Melting relations of an alkali–olivine basalt at 30 kbar, and their bearing on the origin of alkali basalt magmas. *Carnegie Inst. Wash. Yearb.* **79**, 271–276.

WATSON E. B. (1977) Partitioning of manganese between forsterite and silicate liquid. *Geochim. Cosmochim. Acta* **41**, 1363–1374.

YAGI K. and ONUMA K. (1969) An experimental study on the role of titanium in alkalic basalts in light of the system diopside–akermanite–nepheline–$CaTiAl_2O_6$. *Amer. J. Sci.* **267-A**, 509–549.

YODER H. S. and KUSHIRO I. (1972) Composition of residual liquids in the nepheline–diopside system. *Carnegie Inst. Wash. Yearb.* **71**, 413–415.

Part D.
Magma Ascent, Emplacement and Eruption

Magmatic Processes: Physicochemical Principles
© The Geochemical Society, Special Publication No. 1, 1987
Editor B. O. Mysen

The significance of observations at active volcanoes: A review and annotated bibliography of studies at Kilauea and Mount St. Helens

THOMAS L. WRIGHT

U.S. Geological Survey, Hawaiian Volcano Observatory, P.O. Box 51, Hawaii National Park, Hawaii 96718, U.S.A.

and

DONALD A. SWANSON

U.S. Geological Survey, Cascades Volcano Observatory, 5400 MacArthur Boulevard,
Vancouver, Washington 98661, U.S.A.

Abstract—Study of active volcanoes yields information of much broader significance than to only the discipline of volcanology. Some applications are 1) interpretation of lava-flow structures, stratigraphic complexities, and petrologic relations in older volcanic units; 2) interpretation of bulk properties of the mantle and constraints on partial melting and deep magma transport; 3) interpretation of geophysical characteristics of potentially active volcanic systems; 4) direct determination of physical properties of molten and solidified basalt, and of intensive variables (*e.g.,* oxygen fugacity and temperature) accompanying cooling and crystallization; 5) quantitative assessment of crystal fractionation and magma mixing, 6) tests of theoretical and experimental geochemical, geophysical, and rheologic models of volcanic behavior; and 7) confirmation in nature of laboratory experiments related to crystallization in igneous systems.

The critical factors that make real-time study of volcanic activity valuable are that the location and timing of events are known, and that molten rock and gases are available for direct observation and sampling for subsequent study. Observations made over a period of time make it possible to calculate rates of magma transport, storage, and crystallization, as well as to quantitatively determine elastic and inelastic deformation and the build up and decay of stress within the active volcanic system.

Discussion of these topics is keyed to an annotated bibliography from which quantitative information on properties and processes may be obtained. Emphasis is on Hawaii's active basaltic volcanoes for which the most information is available. Additional references are made to research at Mount St. Helens, one of the first real-time studies of an active volcano of dacitic composition.

INTRODUCTION

SCIENTISTS have long been attracted by active volcanoes and have made observations and measurements that have increased our understanding of how volcanic systems behave. Much of this information has been published in the volcanologic literature and disseminated within the small community of volcanologists. The thesis of this paper is that study of active volcanoes provides important constraints on geological processes to disciplines as diverse as geology, seismology, geochemistry, experimental petrology, and rock mechanics.

Continuous monitoring of volcanoes enables quantitative measurement of the rates at which certain volcanic processes take place (*5a, b*). Such processes range from accumulation and release of strain in rocks to rates of magma transport and their effect on igneous differentiation. We discuss these applications below, beginning with surface observations and progressing to inferences regarding deeper processes. The text is supported by an annotated bibliography for Kilauea with additional references to recent work at Mount St. Helens. We have not attempted to cover studies of other active volcanoes. For Kilauea the reference list includes 1) review articles which themselves contain useful bibliographic coverage, 2) papers that contain quantitative determinations of physical parameters or quantitative demonstrations of specific volcanic processes, and 3) the most recent articles. A primary reference for recent research at Kilauea is *U.S. Geological Survey Professional Paper 1350,* published on the occasion of the 75th anniversary of the founding of the Hawaiian Volcano Observatory in January of 1987 (*4a01*). A primary reference for the catastrophic events at Mount St. Helens in 1980 is *U.S. Geological Survey Professional Paper 1250* (*4a02*).

OBSERVATION OF ERUPTIONS

Observation of volcanic eruptions is a key to understanding the origin of volcanic units preserved in the geologic record. A useful distinction here is between catastrophic and milder volcanic events. For catastrophic eruptions (*e.g.,* Crater Lake, the

Bishop tuff, or flood basalts of the Columbia Plateau) a nearly complete record of eruption may be preserved in the deposits themselves. Observations of explosive eruptions such as that of Mount St. Helens in 1980 are difficult to make but serve to limit the time frame in which various kinds of deposits—such as surges, pyroclastic flows, and airfall ejecta—are formed (*4a02*). Eruptions at Kilauea, in contrast, leave a very confused stratigraphic record that is virtually impossible to decipher after the fact (*3a01–04; 3b05, 06*). Continuous monitoring of eruptions as they take place is the best means of specifying what constitutes a single eruption, or of determining the sequence of events that accompany construction of a large volcanic edifice or lava field that may be preserved in the geologic record. One application of real-time observation of volcanic activity to older rocks involves the origin of thin, dense, streaky layers of basalt exposed locally on the Columbia Plateau. They were identified as collapsed pahoehoe from their similarity to layers that were observed to form during the Mauna Ulu eruption of Kilauea (see Figure 4 in *3b05*). The occurrence of collapsed pahoehoe on the Columbia Plateau was then used as evidence for the proximity of source vents. Another example is how observed formation of the debris avalanche at Mount St. Helens on May 18, 1980, elucidated the process and enabled proper reconstruction of similar past events at many volcanoes in the world (*6a01*).

Volcanic eruptions are also a laboratory for studying the complex behavior of natural silicate melts, consisting of liquid, one or more crystalline phases, and a volatile phase. Laboratory simulations and theoretical study of fountain dynamics and flowing lava can be tested by direct observation. Field observation of the transition from pahoehoe to a'a (*3b04*) can likewise be used to constrain theoretical or experimental derivation of the rheologic properties of basalt. Finally, deuteric alteration and weathering may be quantified by study of lavas of known age (*3a06;* last paragraph of following section.)

KILAUEA LAVA LAKES

In contrast to study of eruption dynamics and flowing lava, the filling of circular pit craters at Kilauea by basaltic lava created natural crucibles for study, by core drilling, of the cooling and crystallization of basaltic magma at low pressure and under static conditions (*1a–1d*). The rates of cooling and crystallization of basaltic magma have been measured, and direct measurements have been made

of properties such as density, viscosity, and oxygen fugacity at different stages in the cooling history (*1c01b, 1d02, 1d03*). These observations combined with estimates of primary volatile contents (*2d*) can be used in the laboratory to simulate real conditions in basaltic systems under study by experimental petrologists. It is encouraging that the pioneering experimental work of N. L. Bowen, H. S. Yoder, C. E. Tilley, and others at the Geophysical Laboratory on crystallization and differentiation of basaltic magma has found solid confirmation in the observations made on the natural lava lakes (see, for example the temperatures of crystallization of different phases in Kilauea Iki lava lake presented in *1b09*).

Several processes of crystal-liquid fractionation have been observed in the lava lakes, and the chemistry of a liquid line of descent has been directly specified. The question of whether high-SiO_2 rhyolite can be derived from a crystallizing basalt is answered affirmatively (*1b01*). Residual glass of rhyolitic composition is present in Alae and Makaopuhi, comprising 6 per cent by weight of the starting basalt. In the larger Kilauea Iki lava lake, liquids of rhyolitic composition were segregated and entered fractures; deep in the lake, however, slower cooling resulted in complete crystallization with no residual glass remaining. Thus we have placed quantitative limits on the possibility of deriving rhyolite from a basaltic liquid that should lead to better understanding of the rhyolite-basalt association in large volcanic fields.

Study of Alae and Makaopuhi revealed differentiation processes that could be directly applied to the origin of differentiated liquids erupted from Kilauea's rift zones. The deeper and more olivine-rich lake, Kilauea Iki, underwent different differentiation processes, including diapiric transfer of one melt through another without appreciable mixing (*1b01, 02*), that can be applied to the origin of chemical and mineralogical layering in shallow mafic intrusions.

Another significant benefit from lava lake studies is the geophysical determination of the relative proportions of liquid and solid. Geoelectrical measurements made in Kilauea Iki (*1b06*) generally agreed with the petrologists' definition of "melt" and "crust" (see *1a01; 1b01*). Seismic measurements generally underestimated the amount of "melt," presumably due to differences between the seismic attenuation of shear waves expected for passage through a single phase liquid and those observed for a liquid which contained a large population of suspended olivine crystals (*1b07*). These results have important implications with regard to geophysical prospecting for magma, either deep (*e.g.*, hot dry rock) or shallow (*e.g.*, beneath active geothermal systems).

The lava lakes give data on the initial oxidation state of basaltic magma and on the effects of deuteric oxidation during cooling. Minimum weight-percent ratios of ferric to ferrous iron are about 0.12, in equilibrium with directly measured oxygen fugacities slightly above the QFM buffer (*1d01, 03*). High oxygen fugacities that developed in Makaopuhi lava lake (*1d01, 03*) resulted only in a streaky hematitic alteration of mafic silicates. No hydrous mineraloids, such as iddingsite, have been observed. The occurrence of iddingsite in older Hawaiian flows can thus be ascribed to low temperature alteration long after initial cooling and solidification. Kilauea Iki lava lake, whose upper crust has a well developed hydrothermal system at temperatures lower than 100°C, shows increasing deuteric alteration at a given depth in successive drillings of the lake. Eventually we should be able to specify the alteration history, mineralogically and chemically, as a reference for the interpretation of alteration in older ponded basaltic lava flows.

INTERPRETATION OF SUB-VOLCANIC PROCESSES

Kilauea is an ideal laboratory for study of subsurface volcanic processes because of its accessibility and frequent activity. Major events—eruptions and earthquakes—occur with a frequency that permits repeated testing of hypotheses of volcanic and seismic behavior. Kilauea's shallow plumbing system has been intensely studied (*2a–e*) with the result that we can specify depth and size of a primary magma-storage reservoir (*2a01–05; 2b01, 02*), measure rates of magma transport from storage to eruption (*2a07, 08; 2b17, 18*), identify locations of secondary storage (*2a05; 2b08, 09; 2c05; 2e04, 05*), and evaluate seismicity and ground deformation as they relate to accumulation and release of strain in brittle rocks surrounding the magma reservoir and active rift systems (*2a06; 2b14–16; 2e01, 02*). The record of the chemistry and petrography of all eruptions since 1952 has resulted in quantitative interpretation of crystal-liquid fractionation and high-temperature magma mixing in the rift zones (*2c05–08*). These processes are not restricted to Kilauea. Numerous papers on the mid-ocean ridge basalt system and, more recently, on the Krafla system in Iceland, have described processes similar to those first documented for Kilauea.

The concept that volcanic rocks provide a window into the earth's mantle is not new. Nevertheless most geophysical and geochemical models of mantle processes do not use the constraints and insights gained from study of active volcanic areas. The process of magma generation in the mantle often leads to eruption of lava on the earth's surface. At Kilauea a key observation connecting shallow and deep processes is the magma supply rate, the rate at which basaltic magma is supplied to shallow storage (*2f01, 02; 2b04*). Kilauea's magma supply rate has been estimated at 0.1 km^3 per year. From storage the magma may either be intruded or extruded; the net result of endogenous and exogenous growth (*2f02*) and the isostatically controlled subsidence of the volcanic pile into the crust (*3a05*) is what determines the rate of growth of an Hawaiian volcano above the sea floor.

Several other critical observations constrain melt generation in the Hawaiian hot spot:

1. More than one volcano can be active at a given time.
2. Simultaneously active volcanoes are chemically identifiable (*2c01,02*).
3. Deeper storage areas in the mantle have not been identified (*2a03*).
4. Magma is rapidly resupplied from depth following partial draining of the summit reservoir during summit (*2b04*) or rift (*2b08, 09*) activity.

Observations 3 and 4, combined with knowledge of the size of the shallow reservoir and the estimated constant magma-supply rate, require a geologically short time between melting in the mantle and eruption at the surface; current estimates are in the range of a few decades to at most 100 years. Thus each of us is a witness, well within one's lifetime, to the entire process of melting, magma transfer to shallow storage, and eruption or intrusion of a particular batch of magma. The short times involved make it likely that melting occurs at a relatively shallow depth, but the question of the ultimate depth of the source material brought up to be melted remains unresolved. The fundamental parameters for melting in the mantle, for example, degree and depth of partial melting and the chemistry and mineralogy of the bulk mantle, all may be better addressed in the context of the magma supply rate and the chemistry of erupted lavas associated with an active volcanic system. Again we emphasize the importance of knowing the magma supply rate. To the extent that the supply rate reflects the melting rate within the Hawaiian hot spot, the degree of partial melting is constrained by consideration of the total volume of mantle available to be melted per unit time, the physical dimensions of magma transport systems in both the asthenosphere and lithosphere, and the rates at which melts can be homogenized to produce a broadly uniform composition. New methodologies in seismology (*e.g.*, tomographic mapping), following on earlier work (*2a02, 03*), may be important to apply to the roots of volcanic areas (*e.g.*, the Hawaiian hot spot) to elucidate further the distribution of melt in the uppermost mantle. Likewise, theoretical and experimental modeling of magma transport mechanisms (*e.g.*, rates of collection of magma in the asthenosphere from an originally solid crystal framework; fracture mechanisms in a rigid lithosphere) may be constrained by knowledge of magma-supply rates for specific volcanic systems.

A final application of real-time volcanic studies to deeper processes is the determination of volatile contents and rates of outgassing of the earth's mantle. These are critically dependent on determinations of the concentrations and saturation pressures of volatiles in magmas, as well as quantitative determination of the rates of outgassing of volatiles from an active volcano. Kilauea has provided both kinds of data. Determination of volatile concentrations in rapidly quenched volcanic glasses, both subaerial (glass inclusions in phenocrysts—*2d09*) and submarine (pillow rinds—*2d07, 08*), has given estimates of saturation concentrations for CO_2, S, and H_2O). Study of Kilauea gases emitted from fumaroles and active vents (*2d04–06*) as well as measurement of volcanic plumes both between and during eruptions (*2d01–04*) yield limiting values for volatile content (*2d05*) and outgassing rate (*2d06, 10*) of the mantle.

COMPARISON OF MOUNT ST. HELENS WITH KILAUEA

Volcanic monitoring by the U.S. Geological Survey is currently conducted at two very different active volcanic centers (*5a, b*). Kilauea is a basaltic volcano in an oceanic intraplate tectonic setting; Mount St. Helens is a dacitic volcano in a continental margin setting. Similar instrumentation for study of seismicity and ground deformation is used on both volcanoes. The contrasts in eruptive style, magma supply rate, shape of volcanic edifice, types of volcanic deposits, and chemistry and petrology of erupted magmas are obvious. What are perhaps more interesting are the similarities in magma storage and the constancy of magma supply rate as inferred from instrumental monitoring.

Seismic evidence obtained prior to and during the eruption of May 18, 1980 suggests the presence of a magma reservoir 7–8 km below the surface (*4a05*). This depth correlates exactly with the depth of storage and crystallization inferred on the basis of experimental work on pumice erupted on May 18 (*4a04*). Seismic studies since 1980, however, have failed to detect a magma body beneath Mount St. Helens, although the volcano has been in episodic eruption during this time. This failure might relate to the resolution scale of existing techniques, but by analogy with the Kilauea Iki experiment (*1b07*) the absence of a clear seismic definition of magma might also reflect a high crystal content in the reservoir that feeds the 50-percent-crystalline dacite into the dome.

The magma supply rate at Mount St. Helens is estimated from study of post-1980 dome growth to be about one order of magnitude less than that estimated for Kilauea. It is, like Kilauea, remarkably constant over the short period of time in which measurements have been made (*4a06*).

Acknowledgements—This paper grew out of an afternoon workshop moderated by one of us (TLW) entitled "Study of Active Volcanism: Constraints on Petrologic and Geophysical Models of Dynamic Earth Processes" held at the symposium honoring Hat Yoder. We are grateful to the following persons who also participated in the workshop: Fred Anderson, Rosalind Helz, Peter Lipman, Christina Neal, Michael Ryan, and George Ulrich. The present paper covers the overview and not the substance of the workshop. We appreciate the cooperation of Bjorn Mysen in offering a forum in Hawaii and in this volume to present some of our thoughts on the importance of studies at active volcanoes.

Finally, it is appropriate to recognize the contribution that the Geophysical Laboratory and Hatten S. Yoder himself have made to understanding Hawaiian volcanic processes. From the earliest visits of Day, Allen, and Shepherd to collect gas at Halemaumau to the weighty synthesis of basalt genesis by Yoder and Tilley, based in part on experiments using Kilauea tholeiite, there has been an important scientific interchange between the Carnegie Institution's Geophysical Laboratory and the U.S. Geological Survey's Hawaiian Volcano Observatory. We dedicate this paper to that continued association and hope that a similar association can be established between the Geophysical Laboratory and the fledgling Cascades Volcano Observatory.

The manuscript was reviewed and improved by Paul Greenland, Charlie Bacon, Peter Lipman, and Bjorn Mysen. Fred Anderson provided additional insight into the significance of volatile studies at Kilauea.

ANNOTATED REFERENCES

1. KILAUEA LAVA LAKES

a. Overviews of lava lake studies   **1a**
WRIGHT T. L., PECK D. L. and SHAW H. R. (1976) Kilauea lava lakes, natural laboratories for the study of the cool-

ing, crystallization, and differentiation of basaltic magma. In *The Geophysics of the Pacific Ocean Basin and Its Margin.* Amer. Geophys. Union Monograph 19, p. 375–392. **1a01**
A summary of data and interpretations from earlier studies of Kilauea lava lakes.

WRIGHT T. L. and HELZ R. T. (1987) Recent advances in Hawaiian petrology and geochemistry: *U.S. Geol. Survey Prof. Paper 1350,* Chap. 23. **1a02**
Gives an updated summary of differentiation processes observed in the Kilauea lava lakes.

b. Kilauea Iki. **1b**
Eruption of November–December 1959. Drilled in 1960–61, 1967, 1975, 1976, 1979, and 1981. Depth = 365 feet.

HELZ R. T. (1987) Studies of Kilauea Iki lava lake: an overview. In *Magmatic Processes: Physicochemical Principles,* (ed. B. O. MYSEN), The Geochemical Society Spec. Publ. 1, pp. 241–258. **1b01**
A valuable reference to geophysical and petrological studies of the largest, and only extant, Kilauea lava lake.

HELZ R. T. Diapiric transfer of melt in Kilauea Iki lava lake: a rapid and efficient process of igneous differentiation. *Bull. Geol. Soc. Amer.* (In press) **1b02**
This paper contains a wealth of information on igneous processes of potential importance to the crystallization of olivine-rich basaltic intrusions.

HELZ R. T. (1987) Character of olivine in lavas of the 1959 eruption of Kilauea Volcano, Hawaii, and its bearing on eruption dynamics. *U.S. Geol. Survey Prof. Paper 1350,* Chap. 25. **1b03**

EATON J. P., RICHTER D. H. and KRIVOY H. L. (1987) Cycling of magma between the summit reservoir and Kilauea Iki lava lake during the 1959 eruptions of Kilauea Volcano, Hawaii. *U.S. Geol. Survey Prof. Paper 1350,* Chap. 48. **1b04**
These papers provide insight into the unusual character of the 1959 eruption (see also 2c04). They document the appearance of new magma during the early part of the eruption, and Helz presents evidence that this magma came directly from the mantle. The 1959–60 eruption of Kilauea Volcano, Hawaii.

MURATA K. J. and RICHTER D. H. (1966) Chemistry of the lavas. In *U.S. Geol. Survey Prof. Paper 537,* Chap. A. **1b05a**

RICHTER D. H. and MOORE J. G. (1966) Petrology of the Kilauea Iki lava lake. In *U.S. Geol. Survey Prof. Paper 537,* Chap. B. **1b05b**

MURATA K. J. (1966) An acid fumarolic gas from Kilauea Iki. In *U.S. Geol. Survey Prof. Paper 537,* Chap. C. **1b05c**

RICHTER D. H. and MURATA K. J. (1966) Petrography of the lavas. In *U.S. Geol. Survey Prof. Paper 537,* Chap. D. **1b05d**

RICHTER D. H., EATON J. P., MURATA K. J., AULT W. U. and KRIVOY H. L. (1966) Chronologic narrative. In *U.S. Geol. Survey Prof. Paper 537,* Chap. E. **1b05e**
This classic reference is the first detailed study of a Kilauea eruption using modern seismic and geodetic techniques, and documents the formation of the first modern lava lake in a pit crater.

ANDERSON L. A. (1987) The geoelectric character of the Kilauea Iki lava lake crust, Hawaii. *U.S. Geol. Survey Prof. Paper 1350,* Chap. 50. **1b06**

CHOUET B. and AKI K. (1981) Seismic structure and seismicity of the cooling lava lake of Kilauea Iki, Hawaii. *J. Volcan. Geotherm. Res.* 9, 41–56. **1b07**
These papers summarize the structure of the lava lake as studied by geophysical methods. The electrical structure agrees with that obtained by drilling and with petrologic definitions of

"melt" and "crust." The seismic structure is biased toward lower amounts of "melt", presumably because the olivine-rich crystal mush transmits shear waves, an important observation relative to seismic prospecting for buried bodies of melt.

ZABLOCKI C. J. and TILLING R. I. (1976) Field measurements of apparent Curie temperatures in a cooling basaltic lava lake, Kilauea Iki, Hawaii. *Geophys. Res. Lett.* 3, 487–490. **1b08**
Shows agreement of curie temperatures measured in the field with laboratory measurements on drill core from other lava lakes.

HELZ R. T. and THORNBER C. T. Geothermometry of Kilauea Iki lava lake. *Bull. Volcanol.* (submitted) **1b09**
This paper shows between lava temperature and the CaO and MgO contents of natural glass. The geothermometer as constructed is applicable to most Kilauea lavas.

c. Alae. **1c**
Eruption of August 1963. Depth = 50 feet. Drilled until lake reached ambient temperature in 1967. Covered by new lava in February 1969. The data for Alae lava lake are basic to studies of Hawaiian olivine-poor tholeiite.

Solidification of Alae lava lake, Hawaii. **1c01**

PECK D. L. and KINOSHITA W. T. (1976) The eruption of August 1963 and the formation of Alae lava lake. In *U.S. Geol. Survey Prof. Paper 935,* Chap. A. **1c01a**

PECK D. L. (1978) Cooling and vesiculation of Alae lava lake, Hawaii. In *U.S. Geol. Survey Prof. Paper 935,* Chap. B. **1c01b**
Chapter B gives (1) important measured and derived physical data for Hawaiian basalt, including density of solidified and molten basalt, vesicularity and conductivity, and (2) cooling history, with an evaluation of additional physical properties by fitting theoretical cooling models to the observed data.

WRIGHT T. L. and PECK D. L. (1978) Crystallization and differentiation of the Alae magma, Alae lava lake, Hawaii. In *U.S. Geol. Survey Prof. Paper 935,* Chap. C. **1c01c**
Chapter C contains a direct determination of the liquid line of descent for tholeiitic basalt, the order and temperature of crystallization of different phases, and the process in which low-temperature basaltic liquids are segregated.

SKINNER B. J. and PECK D. L. (1969) An immiscible sulfide melt from Hawaii. *Econ. Geol., Monograph No. 4,* 310–322. **1c02**
Identification of the temperature of formation, composition, and mineralogy of primary sulfide segregations. These may become altered and disseminated with further cooling or lava movement, thus obscuring their origin in older lava.

d. Makaopuhi. **1d**
Eruption of March 1965. Depth = 270 feet. Drilled until covered by new lava in February 1969.

WRIGHT T. L. and OKAMURA R. T. (1977) Cooling and crystallization of tholeiitic basalt, 1965 Makaopuhi lava lake, Hawaii. *U.S. Geol. Survey Prof. Paper 1004,* 78 p. **1d01**
Reports all results of the four-year study of cooling, crystallization, and differentiation. Processes of convection-driven high temperature fractionation and differentiation by filter-pressing at low temperatures also apply to olivine-poor magma bodies stored in the Kilauea rift zones.

SHAW H. R., WRIGHT T. L., PECK D. L. and OKAMURA R. T. (1968) The viscosity of basaltic magma: an analysis of field measurements in Makaopuhi lava lake, Hawaii. *Amer. J. Sci.* 266, 225–264. **1d02**
First direct measurement of viscosity in a cooling lava pond. Demonstrated the non-newtonian behavior of the melt.

SATO M. and WRIGHT T. L. (1966) Oxygen fugacities directly measured in magmatic gases. *Science* 153, 113–1105. **1d03**

First direct measurement of oxygen fugacity-temperature profiles through the upper crust of a cooling basalt flow. Updated in *1d01*.

2. REAL-TIME STUDY OF MAGMA TRANSPORT AND STORAGE

a. Seismic   **2a**

KOYANAGI R. Y., UNGER J. D., ENDO E. T. and OKAMURA A. T. (1976) Shallow earthquakes associated with inflation episodes at the summit of Kilauea Volcano, Hawaii. *Bull. Volcanol.* **39**, 621–631.   **2a01**
First detailed seismic portrayal of summit reservoir, using distribution pattern of shallow earthquakes.

THURBER C. H. (1984) Seismic Detection of the Summit Magma Complex of Kilauea Volcano, Hawaii. *Science* **223**, 165–167.   **2a02**
A different seismic definition of summit reservoir using inversion of local earthquake P-wave arrival times.

ELLSWORTH W. L. and KOYANAGI R. Y. (1977) Three-dimensional crust and mantle structure of Kilauea Volcano, Hawaii. *J. Geophys. Res.* **82**, 5379–5394. **2a03**

THURBER C. H. (1987) Seismic structure and tectonics of Kilauea Volcano, Hawaii. *U.S. Geol. Survey Prof. Paper 1350*, Chap. 38.   **2a04**
These papers apply 3-dimensional seismic methods to the subvolcanic structure. Paper *2a03* gives evidence for the absence of deep areas of magma storage.

RYAN M. P., KOYANAGI R. Y. and FISKE R. S. (1981) Modeling the three-dimensional structure of macroscopic magma transport systems: application to Kilauea Volcano, Hawaii. *J. Geophys. Res.* **86**, 7111–7129.   **2a05**
Presents seismic data in three dimensions to illustrate the plumbing system of Kilauea.

RYAN M. P. The Mechanics and Three-dimensional internal structure of active magmatic systems: Kilauea volcano, Hawaii. *J. Geophys. Res.* (In press)   **2a06**
Continuation of the work above, adding finite-element modeling of summit and rift-zone responses to intrusion of magma.

KLEIN F. W., KOYANAGI R. Y., NAKATA J. S. and TANIGAWA W. R. (1987) The seismicity of Kilauea's magma system, Hawaii. *U.S. Geol. Survey Prof. Paper 1350*, Chap. 43.   **2a07**
A comprehensive study of the relationship between seismic activity and magmatic intrusion. Data are given on both lateral and vertical rates of intrusion.

KARPIN T. L. and THURBER C. H. The relationship between earthquake swarms and magma transport. Kilauea Volcano. *Pure Appl. Geophys.* (In press)   **2a08**
Additional definition of dike propagation using analysis of earthquake focal mechanisms.

KOYANAGI R. Y., CHOUET B. and AKI K. (1987) The origin of volcanic tremor in Hawaii. *U.S. Geol. Survey Prof. Paper 1350*, Chap. 45.   **2a09**
A comprehensive summary of the occurrence of volcanic tremor and its relationship to magma transport. Assessment of the relative importance of aseismic magma transport and the identification of common source regions for different active Hawaiian volcanoes.

KOYANAGI R. Y., TANIGAWA W. R. and NAKATA J. S. Seismicity associated with the eruption of Kilauea from January 1983 to July 1984. In The Pu'u O'o eruption of Kilauea volcano, Hawaii: the first 1½ years, (ed. E. W. WOLFE), *U.S. Geol. Survey Prof. Paper 1463*, (In press)   **2a10**
A fundamental study of seismicity associated with a long-lived eruption of Kilauea. *2a07, 2a09*, and this paper form a basic reference for study of volcanic seismicity.

b. Ground deformation   **2b**

EATON J. P. and MURATA K. J. (1960) How volcanoes grow. *Science* **132**, 925–938.   **2b01**

EATON J. P. (1962) Crustal structure and volcanism in Hawaii. *Amer. Geophys. Union Monograph 6*, 13–29.   **2b02**
Classic papers defining the Kilauea summit reservoir from the modeling of ground deformation during inflation-deflation cycles.

FISKE R. S. and KINOSHITA W. T. (1969) Inflation of Kilauea Volcano prior to its 1967–1968 eruption. *Science* **165**, 341–349.   **2b03**

KINOSHITA W. T., KOYANAGI R. Y., WRIGHT T. L. and FISKE R. S. (1969) Kilauea Volcano: the 1967–1968 summit eruption. *Science* **166**, 459–468.   **2b04**
These companion papers illustrate the complex structure of the Kilauea summit reservoir during a long inflation cycle and the response of the volcano to a major summit eruption.

DECKER R. W., KOYANAGI R. Y., DVORAK J. J., LOCKWOOD J. P., OKAMURA A. T., YAMASHITA K. M. and TANIGAWA W. R. (1983) Seismicity and surface deformation of Mauna Loa volcano, Hawaii. *EOS* **64**, 545–547.   **2b05**

LOCKWOOD J. P., BANKS N. G., ENGLISH T. T., GREENLAND L. P., JACKSON D. B., JOHNSON D. J., KOYANAGI R. Y., MCGEE K. A., OKAMURA A. T. and RHODES J. M. (1985) The 1984 eruption of Mauna Loa volcano, Hawaii. *EOS* **66**, 169–171.   **2b06**

LOCKWOOD J. P., DVORAK J. J., ENGLISH T. T., KOYANAGI R. Y., OKAMURA A. T., SUMMERS M. L. and TANIGAWA W. R. (1987) Mauna Loa volcano, Hawaii. A decade of intrusive and extrusive activity. *U.S. Geol. Survey Prof. Paper 1350*, Chap. 19.   **2b07**
These papers establish the similarity of Mauna Loa's magma reservoir and feeding system to that of Kilauea, in the first fully documented account of a Mauna Loa eruption cycle.

JACKSON D. B., SWANSON D. A., KOYANAGI R. Y. and WRIGHT T. L. (1975) The August and October 1968 east rift eruptions of Kilauea Volcano, Hawaii. *U.S. Geol. Survey Prof. Paper 890*, 33 p.   **2b08**

SWANSON D. A., JACKSON D. B., KOYANAGI R. Y. and WRIGHT T. L. (1976) The February 1969 east rift eruption of Kilauea Volcano, Hawaii. *U.S. Geol. Survey Prof. Paper 891*, 30 p.   **2b09**
These two papers are the most complete summaries of typical short rift eruptions that combine extrusive and intrusive processes, the latter revealed largely by measurement of ground deformation.

DUFFIELD W. A., CHRISTIANSEN R. L., KOYANAGI R. Y. and PETERSON D. W. (1982) Storage, migration, and eruption of magma at Kilauea Volcano, Hawaii, 1971–1972. *J. Volcanol. Geotherm. Res.* **13**, 273–307. **2b10**
The first documentation of intrusion above the Kilauea summit reservoir. This activity resulted in the eruption at a later time (1974—see *2c07*) of slightly fractionated lava at Kilauea summit, the first documented occurrence in historic time.

DIETERICH J. H. and DECKER R. W. (1975) Finite element modeling of surface deformation associated with volcanism. *J. Geophys. Res.* **80**, 4094–4102.   **2b11**
First modeling of surface deformation using a dike rather than a point source geometry.

RYAN M. P., BLEVINS J. Y. K., OKAMURA A. T. and KOYANAGI R. Y. (1983) Magma reservoir subsidence mechanics, theoretical summary and application to Kilauea Volcano, Hawaii. *J. Geophys. Res.* **88**, 4147–4181.   **2b12**
Presents a novel approach to modeling the ground deformation over the reservoir using emplacement of sill-like bodies instead of a point source (*2b01, 02*) or vertical dikes (*2b11*).

DVORAK J., OKAMURA A. and DIETERICH J. H. (1983) Analysis of surface deformation data, Kilauea Volcano, Hawaii October 1966 to September 1970. *J. Geophys. Res.* **88**, 9295–9304.   **2b13**
Models deformation data for a variety of source geometries, concluding that a point-source is adequate. Documents inelastic deformation at Kilauea summit (See also *2e03*).

SWANSON D. A., DUFFIELD W. A. and FISKE R. S. (1976) Displacement of the south flank of Kilauea Volcano: the result of forceful intrusion of magma into the rift zones. *U.S. Geol. Survey Prof. Paper 963*, 39 p.   **2b14**
Shows geometry of deformation related to diking, as determined by geodetic surveys throughout the century. Documents that deformation follows and is probably caused by intrusion.

LIPMAN P. W., LOCKWOOD J. P., OKAMURA R. T., SWANSON D. A. and YAMASHITA K. M. (1985) Ground deformation associated with the 1975 magnitude 7.2 earthquake and resulting changes in activity of Kilauea Volcano, Hawaii. *U.S. Geol. Survey Prof. Paper 1276*, 45 p.   **2b15**
Details nature of deformation event resulting from diking in east rift zone and documents how this deformation increased the storage capacity of the rift zone for several years.

DVORAK J. J., OKAMURA A. T., ENGLISH T. T., KOYANAGI R. Y., NAKATA J. S., SAKO M. K., TANIGAWA W. T. and YAMASHITA K. M. (1986) Mechanical response of the south flank of Kilauea Volcano, Hawaii, to intrusive events along the rift systems: *Tectonophysics* **124**, 193–209.   **2b16**
Further investigation of the relationship between intrusive activity and seismicity and ground movement on Kilauea's south flank.

OKAMURA A. T., DVORAK J. J., KOYANAGI R. Y. and TANIGAWA W. R. Surface deformation during dike propagation: the 1983 east rift eruption of Kilauea Volcano, Hawaii. In The Pu'u O'o eruption of Kilauea Volcano, Hawaii: the first 1½ years, (ed. E. W. WOLFE), *U.S. Geol. Survey Prof. Paper 1463*,(In press)   **2b17**
The first quantitative modeling of dike emplacement at Kilauea using data from continuously recording electronic tiltmeters.

DVORAK J. J. and OKAMURA A. T. (1984) Variations in tilt rate and harmonic tremor amplitude during the January–August 1983 east rift eruptions of Kilauea Volcano, Hawaii. *J. Volcanol. Geotherm. Res.* **25**, 249–258.   **2b18**
A novel approach to estimating magma-transport rates from ground deformation and seismic measurements. Comparison with observed extrusion rates permits an estimate of the ratio of material erupted to that left underground as intrusions.

c. Petrology   **2c**

WRIGHT T. L. (1971) Chemistry of Kilauea and Mauna Loa in space and time. *U.S. Geol. Survey Prof. Paper 735*, 40 p.   **2c01**
Summarizes high-quality major-oxide analyses for all olivine-controlled historical eruptions of Kilauea through 1968, all historical eruptions of Mauna Loa through 1950, and selected prehistorical eruptions from both volcanoes. Interpretations made in this paper are largely superseded by later work (*e.g.*, *2c02, 09, 10*).

TILLING R. I., WRIGHT T. L. and MILLARD H. T. JR. (1987) Trace-element chemistry of Kilauea and Mauna Loa lava in space and time: A reconnaisance. *U.S. Geol. Survey Prof. Paper 1350*, Chap. 24.   **2c02**
A sequel to *2c01*, presenting trace-element data for the same samples, and showing that both Kilauea and Mauna Loa show long-term chemical trends.

RHODES J. M. (1983) Homogeneity of lava flows: Chemical data for historic Mauna Loan eruptions. *J. Geophys. Res.* **88**, A869–879.   **2c03**

Demonstrates the homogeneity of different Mauna Loa eruptions after correcting for olivine-controlled chemical variation.

WRIGHT T. L. (1973) Magma mixing as illustrated by the 1959 eruption, Kilauea Volcano, Hawaii. *Bull. Geol. Soc. Amer.* **84**, 849–858.   **2c04**
A well-documented example of an eruption fed from two sources.

WRIGHT T. L. and FISKE R. S. (1971) Origin of the differentiated and hybrid lavas of Kilauea Volcano. *Hawaii. J. Petrol.* **12**, 1–65.   **2c05**

WRIGHT T. L., SWANSON D. A. and DUFFIELD W. A. (1975) Chemical composition of Kilauea east-rift lava, 1968–1971. *J. Petrol.* **16**, 110–133.   **2c06**

WRIGHT T. L. and TILLING R. I. (1980) Chemical variation in Kilauea eruptions, 1971–1974. In *The Jackson Volume*, (ed. A. IRVING), *Amer. J. Sci.* **280-A**, pt. 2, 777–793.   **2c07**
These three papers constitute a comprehensive treatment of crystal-liquid fractionation and high-temperature mixing of Kilauea magmas, derived by analysis of the complex record of eruption chemistry from 1952–1974.

GARCIA M. O. and WOLFE E. W. Petrology of the lava from the Pu'u O'o eruption of Kilauea Volcano, Hawaii: episodes 1–20. In The Pu'u O'o eruption of Kilauea Volcano, Hawaii: the first 1½ years, (ed. E. W. WOLFE), *U.S. Geol. Survey Prof. Paper 1463*, (In press)   **2c08**
Provides alternative interpretations to those in *2c05–07* to explain the chemical variation in Kilauea's most recent eruption.

WRIGHT T. L. (1984) Origin of Hawaiian tholeiite: a metasomatic model. *J. Geophys. Res.* **89**, 3233–3252.   **2c09**
Provides critical data obtained from real time study of recent Kilauea activity that constrain geochemical and geophysical models of magma generation.

HOFMANN A. W., FEIGENSON M. D. and RACZEC I. (1984) Case studies on the origin of basalt: III. Petrogenesis of the Mauna Ulu eruption, Kilauea, 1969–1971. *Contrib. Mineral. Petrol.* **88**, 24–35.   **2c10**
Precise determinations of incompatible trace elements and Sr isotopes for part of the data set for which major-oxide chemistry is presented in *2c06*. Model for origin of Kilauea magma differs substantially from that presented in *2c09*. The importance of working with samples documented as to time and place of eruption is acknowledged in both papers.

d. Geochemistry of magmatic volatiles   **2d**

GREENLAND L. P., ROSE W. I. and STOKES J. B. (1985) An estimate of gas emissions and magmatic gas content from Kilauea Volcano. *Geochim. Cosmochim. Acta* **49**, 125–129.   **2d01**

GERLACH T. M. and GRAEBER E. J. (1985) The volatile budget of Kilauea Volcano. *Nature* **313**, 273–277.   **2d02**

GERLACH T. M. (1980) Evaluation of volcanic gas analyses from Kilauea Volcano. *J. Volcanol. Geotherm. Res.* **7**, 295–317.   **2d03**

GREENLAND L. P. (1987) Composition of Hawaiian eruptive gases. *U.S. Geol. Survey Prof. Paper 1350*, Chap. 28.   **2d04**
Together these papers address the chemical composition of volcanic gas in equilibrium with Kilauea magma 1) as it arrives from the mantle, 2) after degassing in shallow storage at pressures of 1–2 kbar, and 3) in solidified lava after degassing during eruption. Oxygen fugacities derived from gas collections are similar to those directly measured in the Kilauea lava lakes.

GREENLAND L. P. Estimated mantle content of volatiles from basaltic compositions: *Bull. Volcanol.* (In press)   **2d05**
Uses data from real-time gas collections to estimate volatile contents of the magma source.

GREENLAND L. P., OKAMURA A. D. and STOKES J. B. (1987) Constraints on the mechanics of eruption of PU'u O'o, In The Pu'u O'o eruption of Kilauea Volcano, Hawaii: the first 1½ years, (ed. W. E. WOLFE). *U.S. Geol. Survey Prof. Paper.*   **2d06**
This paper extends studies of gas chemistry to address subjects such as lava fountain dynamics, dimensions of eruptive conduits, size of immediate source areas, and magma supply rates.

MOORE J. G. (1965) Petrology of deep-sea basalt near Hawaii. *Amer. J. Sci.* **263**, 40–52.   **2d07**
First determination of water saturation values in naturally quenched basaltic glass.

MOORE J. G. and FABBI B. P. (1971) An estimate of the juvenile sulfur content of basalt. *Contrib. Mineral. Petrol.* **33**, 118–127.   **2d08**
First determination of sulfur saturation values in naturally quenched basaltic glass, and an estimate of sulfur loss during eruption.

HARRIS D. M. and ANDERSON A. T. (1983) Concentrations, sources, and losses of H_2O, CO_2, and S in Kilauea basalt. *Geochim. Cosmochim. Acta* **47**, 1139–1150.   **2d09**
Analyses of volatiles in glass inclusions in olivines from the 1959 eruption of Kilauea agree with saturation values determined from pillow glasses (*2d07, 08*) and the estimates of restored equilibrium compositions of Kilauea gases (*2d03*).

GERLACH T. M. Exsolution of H_2O, CO_2, and S during eruptive episodes at Kilauea Volcano, Hawaii: *J. Geophys. Res.* (In press)   **2d10**
Quantitative modeling of the equilibrium pressures at which different volatile species are exsolved.

e. Other geophysical studies   **2e**

RYAN M. P. (1987) The elasticity and contractancy of Hawaiian olivine tholeiite. *U.S. Geol. Survey Prof. Paper 1350*, Chap. 52.   **2e01**
This study combines laboratory measurements and theory related to rock mechanics, and real-time deformation and seismic data, to derive a model showing that magma reservoirs bear a fixed relationship to the surface altitude and size of the summit caldera on Hawaiian shield volcanoes.

RYAN M. (1987) Neutral bouyancy and the mechanical evolution of magmatic systems. In *Magmatic Processes: Physicochemical Principles*, (ed. B. O. MYSEN), The Geochemical Society Spec. Publ. 1, pp. 259–288. **2e02**
A comprehensive treatment of the relationship between the location of magma storage reservoirs and the mechanical properties of the surrounding basaltic shield.

JOHNSON D. J. (1987) Elastic and inelastic magma storage at Kilauea Volcano, Hawaii. *U.S. Geol. Survey Prof. Paper 1350*, Chap. 47.   **2e03**
Applies gravity data obtained during the period of episodic east rift eruption to refine interpretations of the summit reservoir to reconcile vertical and horizontal displacement data. The analysis complements those obtained by modeling deformation data with conventional source geometries.

JACKSON D. B. and KAUAHIKAUA J. (1987) Regional self—potential anomalies at Kilauea Volcano, Hawaii. *U.S. Geol. Survey Prof. Paper 1350*, Chap. 40.   **2e04**

JACKSON D. B. Geoelectric observations: September 1982 summit eruption and the first year of the 3 January 1983 middle east rift eruption. In The Pu'u O'o eruption of Kilauea Volcano, Hawaii. The first 1½ years, (ed. E. W. WOLFE), *U.S. Geol. Survey Prof. Paper 1463*, (In press)   **2e05**

JACKSON D. B., KAUAHIKAUA J. and ZABLOCKI C. J. (1985) Resistivity monitoring of an active volcano using the controlled-source electro-magnetic technique Kilauea, Hawaii. *J. Geophys. Res.* **90**, 12545–12555.   **2e06**

KAUAHIKAUA J., JACKSON D. B. and ZABLOCKI C. J. (1986) The subsurface resistivity structure of Kilauea Volcano, Hawaii. *J. Geophys. Res.* **91**, 8267–8284.   **2e07**

ZABLOCKI C. J. (1976) Mapping thermal anomalies on an active volcano by the self-potential method, Kilauea, Hawaii. In *2nd U.N. Symposium on the Development and Use of Geothermal Resources, San Francisco, California, May 1975, Proc. 2*, 1299–1309.   **2e08**

ZABLOCKI C. J. (1978) Applications of the VLF induction method for studying some volcanic processes of Kilauea Volcano, Hawaii. *J. Volcanol. Geotherm. Res.* **3**, 155–195.   **2e09**
These papers document the electrical structure of Kilauea and discuss dike impoundment of high-standing water tables and aseismic magmatic intrusion without accompanying ground deformation.

HARDEE H. C. (1987) Heat and mass transport in the east rift magma conduit of Kilauea Volcano, Hawaii. *U.S. Geol. Survey Prof. Paper 1350*, Chap. 54.   **2e10**
Derives the physical state of basaltic magma from modeling flow in active volcanic conduits using real-time monitoring data.

RYAN M. and SAMMIS C. G. (1981) The glass transition in basalt. *J. Geophys. Res.* **86**, 9519–9535.   **2e11**
A generalized rheologic model for the subsolidus portions of basaltic lava lakes, and a general model for the origin and development of columnar jointing in basalt, based on the combination of in-situ field measurements and laboratory measurements at high temperature.

f. Estimates of magma supply rate   **2f**

SWANSON D. A. (1972) Magma supply rate at Kilaueas Volcano, 1952–1971. *Science* **175**, 169–170.   **2f01**
First documentation of constant magma supply rate (about 9×10^6 m³/mo) during long-lived eruptions of Kilauea.

DZURISIN D., KOYANAGI R. Y. and ENGLISH T. T. (1984) Magma supply and storage at Kilauea Volcano, Hawaii, 1956–1983. *J. Volcanol. Geotherm. Res.* **21**, 177–206.   **2f02**
Attempts to calculate minimum supply rates on the basis of tilt changes, and to partition the magma into that erupted and that stored in the two rift zones.

3. REAL-TIME STUDY OF VOLCANIC ERUPTIONS: APPLICATION TO OLDER VOLCANIC ROCKS

a. Construction of volcanic edifices and lava fields   **3a**

SWANSON D. A., DUFFIELD W. A., JACKSON D. B. and PETERSON D. W. (1979) Chronological narrative of the 1969–1971 Mauna Ulu eruption of Kilauea Volcano, Hawaii. *U.S. Geol. Survey Prof. Paper 1056*, 55 p.   **3a01**
A detailed account of the growth of a satellitic lava shield formed of dominantly pahoehoe lava.

TILLING R. I., CHRISTIANSEN R. L., DUFFIELD W. A., ENDO E. T., HOLCOMB R. T., KOYANAGI R. Y., PETERSON D. W. and UNGER J. D. (1987) The 1972–1974 Mauna Ulu eruption, Kilauea Volcano, Hawaii: an example of "quasi-steady state" magma transfer. *U.S. Geol. Survey Prof. Paper 1350*, Chap. 16.   **3a02**
Completes the study of the growth of the Mauna Ulu shield.

WOLFE E. W., NEAL C. A., BANKS N. G. and DUGGAN T. J. Geologic observations and chronology of eruptive events during the first 20 episodes of the Pu'u O'o eruption, January 3, 1983, through June 8, 1984. In The Pu'u O'o eruption of Kilauea Volcano, Hawaii: the first 1½ years, (editor E. W. WOLFE), *U.S. Geol. Survey Prof. Paper.* (In press)   **3a03**

A detailed narrative account of the growth of the Pu'u O'o cone and associated a'a lava flows.

WOLFE E. W., GARCIA M. O., JACKSON D. B., KOYANAGI R. Y., NEAL C. A. and OKAMURA A. T. (1987) The Pu'u O'o eruption of Kilauea Volcano, episodes 1–20, January 1983 to June 1984. *U.S. Geol. Survey Prof. Paper 1350*, Chap. 17.   **3a04**
A synthesis of geological observations made during the first 1½ years of the current Kilauea east rift eruption.

MOORE J. G. (1987) Subsidence of the Hawaiian Ridge. *U.S. Geol. Survey Prof. Paper 1350*, Chap. 2.   **3a05**
Emphasizes the high rates of subsidence of Hawaiian volcanoes and the implications of this for volumes and growth rates of Hawaiian shields.

MOORE J. G., FORNARI D. J. and CLAGUE D. A. (1985) Basalts from the 1877 submarine eruption of Mauna Loa, Hawaii: New data on the variation of palagonitization rate with temperature. *U.S. Geol. Survey Bull. 1663*, 11 p.   **3a06**
Shows that vesicularity of flows can provide information on depth of cooling of ancient lava.

b. Lava flow dynamics   **3b**

NEAL C. A., DUGGAN J. J., WOLFE E. W. and BRANDT E. L. Lava samples, temperatures, and compositions, Pu'u O'o eruption of Kilauea Volcano, Hawaii, episodes 1–20, January 3, 1983–June 8, 1984. In The Pu'u O'o eruption of Kilauea Volcano, Hawaii: the first 1½ years, (ed. E. W. WOLFE), *U.S. Geol. Survey Prof. Paper 1463*, (In press)   **3b01**
An excellent summary of field methods used to obtain quantiative data on erupted lava.

LIPMAN P. W. and BANKS N. G. (1987) Aa flow dynamics, 1984 eruption of Mauna Loa Volcano, Hawaii. *U.S. Geol. Survey Prof. Paper 1350*, Chap. 57.   **3b02**

LIPMAN P. W., BANKS N. G. and RHODES J. M. (1985) Degassing-induced crystallization of basaltic magmas and effects of lava rheology. *Nature* 317, 604–607.   **3b03**
These two papers analyze behavior of lava flows during a typical Mauna Loa eruption. Data on "aging" of the feeding channel, undercooling, and volatile loss.

PETERSON D. W. and TILLING R. I. (1980) Transition of basaltic lava from pahoehoe to aa, Kilauea Volcano, Hawaii: field observations and key factors. *J. Volcanol. Geotherm. Res.* 7, 271–293.   **3b04**
A comprehensive treatment of the pahoehoe to a'a transition in terms of the combined effects of viscosity and rate of shear strain. Chemical composition, temperature, and volatile content, considered independently, are found to be unimportant in determining whether lava is a'a or pahoehoe.

SWANSON D. A. (1973) Pahoehoe flows from the 1969–1971 Mauna Ulu eruption, Kilauea Volcano, Hawaii. *Bull. Geol. Soc. Amer.* 84, 615–626.   **3b05**
Describes different kinds of pahoehoe flows and their preservation in sections of the lava pile exposed by collapse. First explanation of the origin of dense pahoehoe far from the vent as being supplied in lava degassed during flow in tubes.

SWANSON D. A., DUFFIELD W. A., JACKSON D. B. and PETERSON D. W. (1972) The complex filling of Alae Crater, Kilauea Volcano, Hawaii. *Bull. Volcanol. 36*, pt. 1, 105–126.   **3b06**
Discusses the difficulties in deriving eruptive history from stratigraphic sections of older volcanic rocks.

SWANSON D. A. and FABBI B. P. (1973) Loss of volatiles during fountaining and flowage of basaltic lava at Kilauea Volcano, Hawaii. *J. Res. U.S. Geol. Survey 1*, 649–658.   **3b07**
Points out the use of analyses of volatiles in cooled lava to indicate distance from and direction to vent.

MOORE J. G., PHILLIPS R. L., GRIGG R. W., PETERSON

D. W. and SWANSON D. A. (1973) Flow of lava into the sea, 1969–1971, Kilauea Volcano, Hawaii. *Bull. Geol. Soc. Amer.* 84, 537–546.   **3b08**

MOORE J. G. (1975) Mechanism of formation of pillow lava. *Amer. Sci.* 63, 269–277.   **3b09**
These two papers given an account of the formation of basaltic pillows from direct observation of lava flowing into the ocean, the first such observations ever made.

PECK D. L. and MINAKAMI T. (1968) The formation of columnar joints in the upper part of Kilauean lava lakes, Hawaii. *Bull. Geol. Soc. Amer.* 79, 1151–1166.   **3b10**
The first real-time study of joint formation, including determination of maximum temperature at which joints propagate (1,000°C), the direction of propagation, and the effects of rainfall.

DUFFIELD W. A. (1972) A naturally occurring model of global plate tectonics. *J. Geophys. Res.* 77, 1543–2555.   **3b11**
A fascinating paper which treats an active lava lake surface as a scaled analogue of mantle-crust processes occurring at much higher viscosities and much slower rates.

4. SELECTED BIBLIOGRAPHY TO ENABLE COMPARISON OF MECHANICAL BEHAVIOR OF MOUNT ST. HELENS WITH THAT OF KILAUEA   4a

U.S. Geol. Survey (1987) Hawaiian Volcanism, (eds. R. W. DECKER, T. L. WRIGHT, and P. H. STAUFFER), *U.S. Geol. Survey Prof. Paper 1350*, Chap. 1–63.   **4a01**
A modern reference to studies of active Hawaiian volcanoes. Individual chapters are annotated throughout this reference list.

U. S. Geol. Survey (1981) The 1980 eruptions of Mount St. Helens, Washington, (eds. P. W. LIPMAN and D. R. MULLINEAUX), *U.S. Geol. Survey Prof. Paper 1250*, 844 p.   **4a02**
A comprehensive summary of the May, 1980 catastrophic eruption of Mount St. Helens. Articles include eyewitness accounts, geophysical monitoring before, during, and after, detailed accounts of the volcanic deposits formed, and environmental effects of the 1980 eruption.

CAREY S. and SIGURDSSON H. (1985) The May 18, 1980, eruption of Mount St. Helens. 2. Modeling of dynamics of the Plinian phase. *J. Geophys. Res.* 90, 2948–2958.   **4a03**
Challenging model relating eruption dynamics to decompression and vesiculation at a depth of about 4.5 km.

RUTHERFORD M.J., SIGURDSSON H., CAREY S. and DAVIS A. (1985) The May 18, 1980, eruption of Mount St. Helens. 1. Melt composition and experimental phase equilibria. *J. Geophys. Res.* 90, 2929–2947.   **4a04**
Provides experimental evidence consistent with seismic evidence (4a05) for a magma reservoir at 7–8 km depth.

SCANDONE ROBERTO and MALONE S. D. (1985) Magma supply, magma discharge and readjustment of the feeding system of Mount St. Helens during 1980. *J. Volcanol. Geotherm. Res.* 23, 239–262.   **4a05**
Provocative discussion of effects of varying supply and discharge rates on nature of eruptions. Presents seismic evidence for existence of magma reservoir whose top is about 7–8 km deep.

SWANSON D. A., DZURISIN D., HOLCOMB R. T., IWATSUBO E. Y., CHADWICK W. W. JR., CASADEVALL T. J., EVERT J. W. and HELIKER C. C. Growth of the lava dome at Mount St. Helens, Washington (USA), 1981–83. In The emplacement of silicic domes and lava flows, (ed. J. H. FINK). *Geol. Soc. Amer. Spec. Paper 212*. (In press)   **4a06**
Provides volume data documenting relatively constant rate of magma supply during eruption.

VOIGHT B., JANDA R. J., GLICKEN H. and DOUGLASS
P. M. (1983) Nature and mechanics of the Mount St.
Helens rockslide-avalanche of 18 May 1980. *Geotechnique* **33**, 243–273.   **4a07**
   Includes good discussion of development of bulge resulting
from intrusion of magma into the cone.

EICHELBERGER J. C. and HAYES D. B. (1982) Magmatic
model for the Mount St. Helens blast of May 18, 1980.
J. Geophys. Res. **87**, 7727–7738.   **4a08**

KIEFFER S. W. (1981a) Blast dynamics at Mount St. Helens
on 18 May, 1980. *Nature* **291**, 568–570.   **4a09**

KIEFFER S. W. (1981b) Fluid dynamics of the May 18
blast at Mount St. Helens. *U.S. Geol. Survey Prof. Paper
1250*, 379–400.   **4a10**
   These three papers discuss the relative roles of magmatic volatiles and vaporized groundwater in driving the lethal blast. Eichelberger and Hayes favor magmatic gas as the driving agent,
whereas Kieffer stresses the role of flashing water in the hydrothermal system. These papers demonstrate how even with close
observaton, transient events can have controversial origins.

5. GENERAL REFERENCES ON THE MONITORING OF ACTIVE VOLCANOES

BRANTLEY S. and TOPINKA L. (eds.) (1984) Volcanic
studies at the U.S. Geological Survey's David A. Johnston Cascades Volcano Observatory, Vancouver,
Washington. *Earthquake Information Bull. 16*, 44–
122.   **5a**

HELIKER C. C., GRIGGS J. D., TAKAHASHI T. J. and
WRIGHT T. L. (1986) Volcano monitoring at the U.S.
Geological Survey's Hawaiian Volcano Observatory.
Earthquakes and Volcanoes **1**, 1–70.   **5b**
   Comprehensive summaries of visual and instrumental mon-
itoring of two currently active volcanic areas.

6. OTHER REFERENCES

SIEBERT L. (1984) Large volcanic debris avalanches:
Characteristics of source areas, deposits, and associated
eruption. *J. Volcanol. Geotherm. Res.* **22**, 163–197.
                                                              6a01

Magmatic Processes: Physicochemical Principles
© The Geochemical Society, Special Publication No. 1, 1987
Editor B. O. Mysen

Differentiation behavior of Kilauea Iki lava lake, Kilauea Volcano, Hawaii: An overview of past and current work

ROSALIND TUTHILL HELZ

U.S. Geological Survey, Reston, VA 22092, U.S.A.

Abstract—The 1959 eruption of Kilauea Volcano, which produced the only picritic lava at Kilauea's summit in historic time, formed Kilauea Iki lava lake, a large pond of lava over 100 m deep. Both this unusual eruption and the lava lake have been the subject of many studies.

Petrologic investigations of subsolidus and partially molten drill core samples from Kilauea Iki carried out to date include studies of zoning or re-equilibration patterns, or both, in olivines and chromites, and determination of the quenching temperature of the core using a variety of geothermometers. In addition, detailed study of interstitial glass compositions shows that (1) the basaltic liquids in the lake are capable of fractionating to rhyolitic liquid compositions and (2) the succession of liquid compositions found is significantly affected by the amount of olivine present. At low olivine contents, glasses are relatively Fe-rich and low in SiO_2, Al_2O_3, alkalies and P_2O_5, whereas glasses in olivine–rich rocks show the opposite effects. The reason for the effect is that as olivine re-equilibrates with the liquid, it takes up FeO, so the liquids at a given MgO content become progressively lower in FeO and higher in SiO_2, Al_2O_3 the alkalies and P_2O_5.

Processes occurring within Kilauea Iki lava lake include (1) redistribution of olivine phenocrysts by gravitative settling; (2) migraton and loss of vapor bubbles (vesicles); (3) redistribution of olivine + augite during convection; (4) migration of low-density liquid from the bottom of the lake to the top via diapiric melt transfer; (5) formation of pipe-like vuggy olivine–rich bodies by a second, lower-temperature diapiric process; (6) formation of segregation veins, sill-like internal differentiates of ferrobasaltic composition; and (7) formation of minor veinlets of andesitic to rhyolitic composition by flowage of melt into cracks that form as the partially molten rock cools and fractures. Quantification of the physical parameters controlling these magmatic processes may eventually make it possible to apply results from Kilauea Iki to other, larger mafic magma chambers.

INTRODUCTION

SEVERAL TIMES in recent decades, lava erupted at Kilauea Volcano has ponded in one or another of the pit craters that occur along the upper east rift zone of Kilauea volcano (Figure 1). The resulting lava lakes are natural laboratories for the study of the processes by which basaltic lava cools, crystallizes and differentiates. Historic lava lakes investigated periodically include the 1963 Alae lava lake (PECK *et al.,* 1966; WRIGHT and PECK, 1978) and the 1965 Makaopuhi lava lake (WRIGHT *et al.,* 1968; WRIGHT and OKAMURA, 1977). In addition, the prehistoric Makaopuhi lava lake was investigated by J. G. Moore and B. W. Evans (MOORE and EVANS, 1967; EVANS and MOORE, 1968). An earlier summary of lava lake studies was given by WRIGHT *et al.* (1976).

Kilauea Iki lava lake, formed by the 1959 eruption of Kilauea is the biggest of the historic lava lakes and by far the most thoroughly studied. Unlike the 1963 Alae and 1965 Makaopuhi lakes, which have been buried by the 1969–1974 lavas from Mauna Ulu (SWANSON *et al.,* 1972, 1979; TILLING *et al.,* 1987), Kilauea Iki has not been the scene of any more recent eruptive activity, and so has remained accessible from 1959 to the present (1986).

Its size and accessibility, and the unusual character of the eruption which fed it have made it an object of interest to petrologists and geophysicists for over two decades. The purpose of this paper is to review previous work and to present some of the work in progress on both the 1959 eruption and lava lake.

THE 1959 ERUPTION

The spectacular 1959 summit eruption of Kilauea Volcano was a benchmark in modern volcanology because it was the first major Kilauean eruption for which most of the array of modern volcano-monitoring techniques were used. Descriptions of the relationship between earthquake activity, summit deformation, and the 1959 and 1960 eruptions of Kilauea, by EATON (1962) and EATON and MURATA (1960), gave an unprecedentedly detailed picture of Kilauea's magmatic plumbing. More recently, EATON *et al.* (1987) have interpreted the tilt records in greater detail, showing the pattern of exchange of magma between Kilauea Iki lava lake and the subsurface reservoir during the 1959 eruption, and documenting the arrival of new magma into the system through the first half of the eruption.

A day-by-day account of the 1959 eruption is given by RICHTER *et al.* (1970). It lasted from November 14 to December 20, formed a prominent cinder cone and pumice blanket as well as Kilauea Iki lava lake, and included seventeen phases of high fountaining and lava output, separated by periods of relative quiescence. Temperatures as high as 1215°C, the highest in Kilauea's history, were re-

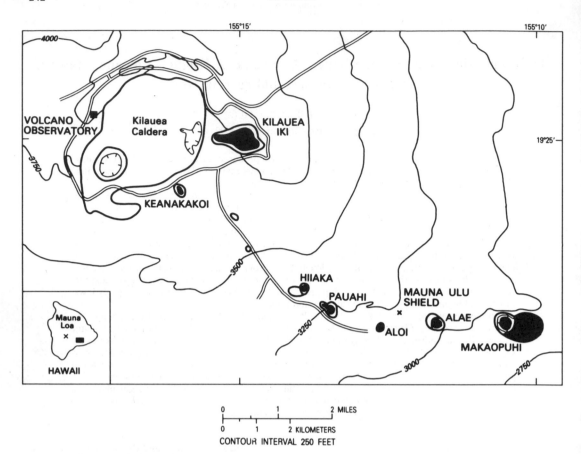

FIG. 1. Index map of the summit area of Kilauea Volcano. Pit craters containing ponded flows (= lava lakes) are labelled by name. Of these, only Kilauea Iki (formed in 1959), Alae (1963) and Makaopuhi (1965) have been studied in any detail. The prehistoric Makaopuhi lava lake, shown in the stippled pattern, has also been studied. The craters of Aloi, Alae and the west pit of Makaopuhi are now completely filled by lavas of the 1969–74 Mauna Ulu shield.

corded, and fountain heights of up to 1900 feet (580 m), the highest of any historic summit eruption, were observed. Repeated crustal foundering and growth of new crust continued for three days after the eruption ended; it was December 23, 1959 before the crust stabilized and studies of the lava lake could begin (RICHTER et al., 1970; AULT et al., 1961).

Early work on the 1959 lavas included studies of their chemistry (MURATA and RICHTER, 1966a; MACDONALD and KATSURA, 1961) and of their petrology (RICHTER and MURATA, 1966). Trace-element data for a suite of 1959 samples, presented by GUNN (1971), showed clearly the effects of the varying olivine content of the eruption pumices. In addition, the gases emitted during and shortly after the eruption were analyzed (HEALD et al., 1963; MURATA, 1966). Processes documented as occurring during the eruption include olivine settling (MURATA and RICHTER, 1966b) and magma mixing (WRIGHT, 1973). WADGE (1981) analyzed the variation in magma-discharge rate as the 1959 eruption proceeded and concluded that it was compatible with progressive degassing of a closed system, leading to higher effusion rates as the magma became progressively cooler and less gas–rich, and hence more viscous and non-Newtonian.

The 1959 lava was picritic in character, the only picritic lava erupted at or near Kilauea's summit in historic times. WRIGHT (1973) estimated the average MgO content of the eruption pumices to be 15.43 percent by weight. As would be expected from their chemistry, the eruption samples contain abundant (17–20 weight percent, on the average) olivine phenocrysts, plus minor chromite, in a matrix of vesicular brown glass. The chromite, which occurs almost exclusively as inclusions in olivine, was analyzed by EVANS and WRIGHT (1972). Compositional data on olivine and glass compositions were reported by LEEMAN and SCHEIDEGGER (1977) and by HARRIS and ANDERSON (1983). The latter study also presents data on the H_2O, CO_2, and S content of glasses included in olivine phenocrysts.

More recently, the olivine phenocryst population has been re-examined and classified on the basis of crystal morphology, characteristic inclusions, etc. by SCHWINDINGER et al. (1983), SCHWINDINGER (1986), and by HELZ (1983, 1987a), who also presents analytical data for olivines and glasses in the pumice samples and for olivines in the lake. The results on the pumice samples corroborate the inferences of EATON et al. (1987) that new material was entering the erupting system through phase 9 of the eruption. In addition, HELZ (1987a) documents the occurrence in the lava lake of dunitic inclusions having deformed or annealed textures, or both, like those of mantle-derived xenoliths (see *e.g.*, KIRBY and GREEN, 1980), and of rare olivine megacrysts (up to 2 cm long). The presence of these materials, plus the unusually high eruption temperatures and fountain heights observed during the 1959 eruption, have been explained by a model in which one of the two mixing components of the eruption (the S-1 component of WRIGHT, 1973) is hypothesized to have come directly from the mantle, without being stored for a significant time in Kilauea's shallow summit reservoir. In this model, the juvenile component originated at 45–60 km depth, where it was liberated from the mantle during the August 14–19 earthquake swarm (EATON, 1962). It reached the surface November 14. It thus had an average ascent velocity of 0.6–0.8 cm/sec, just sufficient to offset the settling velocity of, and hence to entrain, the largest (1 × 2 cm) olivine grains and dunitic aggregates found in the 1959 lavas.

PREVIOUS LAVA LAKE STUDIES

In early 1960, shortly after the crust stabilized, a network of leveling nails (shown in Figure 2a) was established and drilling of Kilauea Iki lava lake began in April 1960. The lake was drilled in 1960, 1961, and 1962, by Hawaii Volcano Observatory (HVO) staff and also by workers from Lawrence Radiation Laboratories. The resulting drill core was described and analyzed by RICHTER and MOORE (1966). Downhole temperature measurements showed temperatures of 1060–1106°C at the crust-melt interface, as defined by drilling (RAWSON, 1960; AULT *et al.*, 1961, 1962; RICHTER and MOORE, 1966). Gas from one of the early drill holes, was collected and analyzed by HEALD *et al.* (1963), and determined to be dominantly $SO_2 + H_2O$.

The lake was drilled by the U.S. Geological Survey (U.S.G.S.) in 1967 and 1975, whereas drillings in 1976, 1979, and 1981 were cooperative efforts between the U.S. Geological Survey (U.S.G.S.) and Sandia National Laboratories. Figures 2a and 2b show the location in plan view and in cross–section of the 1967–1981 drill holes. Observations on this drilling experience, including discussions of (1) the significance of the crust-melt interface, (2) backfilling of the holes by melt and crystal-liquid mushes, and (3) degassing of the lake with time are presented in WRIGHT *et al.* (1976) and HELZ and WRIGHT (1983). Petrographic logs of all of the core are given in HELZ *et al.* (1984) and HELZ and WRIGHT (1983).

The open drill holes have been used for a variety of downhole geophysical studies and measurements. ZABLOCKI and TILLING (1976) determined the Curie point of the cooling basalt in situ. Sandia personnel conducted a number of heat-extraction and other experiments in the 1981 drill holes (HARDEE *et al.*, 1981). In addition, since 1967, temperature profiles have been measured in most of the drill holes. Most of these profiles are still unpublished, but examples can be found in ZABLOCKI and TILLING (1976; profiles from holes KI75-1 and KI75-2), in HERMANCE and COLP (1982; profile from hole KI76-1), and in HARDEE *et al.* (1981, profile from KI81-1). The latter profile is anomalous in shape, probably because another hole (KI81-5) was being drilled only 5 m away at the time the profile was taken. Generalized thermal modeling of the lake has been done in one dimension (PECK *et al.*, 1977) and two dimensions (RYAN, 1979). The only attempt to date to work with and interpret the shapes of actual temperature profiles is that of HARDEE (1980).

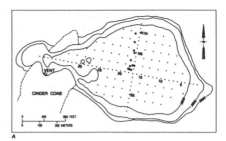

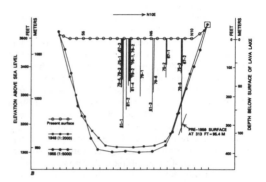

FIG. 2. (A) Plan view of the post-1959 surface of Kilauea Iki lava lake. The large dots indicate the location of drilling sites occupied from 1967 to 1981. The network of levelling stations is shown by the small dots. Individual levelling stations along the principal N–S line of stations have been labelled in Figures 2a and 2b to facilitate comparison. (B) Cross–section of Kilauea Iki lava lake, taken along the N–S line of closely-spaced levelling stations shown in Figure 2a. The present surface of the lava lake and two pre-eruption profiles are shown. The actual location of the lake bottom below hole KI79-5 is significantly different from either, as indicated; possible causes are discussed in HELZ (1980). Vertical exaggeration is 4:1. The drill holes, most of which lie along a line 100 feet west of the section given in Figure 2b, are shown projected onto this cross–section. Several of the drill hole locations have been drilled more than once, in order to sample the same section of crust as cooling proceeded. These clusters of closely spaced drill holes are shown schematically in this figure; the spacing between holes within clusters is not to scale.

During the 1970's the lake was the object of many geophysical investigations. CHOUET (1979) and RYAN (1979) studied the seismic activity associated with crack formation in the lake. ZABLOCKI (1976), FLANIGAN and ZABLOCKI (1977) and SMITH *et al.* (1977) used a variety of electromagnetic sounding methods to locate the boundaries of the lens of melt as it existed in the mid-seventies. A summary of the geoelectric structure of the lava lake is given by ANDERSON (1987). CHOUET and AKI (1981) used both active and passive seismic techniques to characterize the lake. HERMANCE and COLP (1982) attempted a synthesis of these various geophysical studies of the lava lake. The lateral extent of the lens of melt, as defined electrically and seismically, agreed well with that inferred from the drilling results and the leveling data (HVO unpublished data). The thickness of the lens of melt was less well defined, with estimates ranging from <10 m (based on seismic data) to >30 m (from the electromagnetic sounding experiments). Both the drilling results (HELZ et al., 1984) and early petrographic reports (HELZ, 1980) suggested the greater thickness was correct, and this was subsequently confirmed by the results of the 1981 drilling (HELZ and WRIGHT, 1983). The lens of melt, where "melt" is defined as a crystal-liquid mush too fluid to drill into, was 30–40 m thick in 1976. The crystallinity or vesicle content, or both, of the lens of melt at that time were nevertheless sufficiently high for the molten lens to transmit shear waves, as reported by CHOUET and AKI (1981). This result suggests that the delineation of partially molten zones in the crust or upper mantle by seismic methods will be a very difficult problem.

Most other studies of the lake have focused on the drill core, using the observed bulk compositions, mineral assemblages and phase compositions to delineate the differentiation history of the lava lake, a task which is still in progress. HELZ (1980) presented a summary of the gross petrographic variations of the 1967–1979 core, including a detailed discussion of the pattern of occurrence of segregation veins in the lake. LUTH and GERLACH (1980) and LUTH et al. (1981) characterized the phase compositions and assemblages of selected 1979 and 1981 core samples. In addition, HELZ and THORNBER (1981) determined the 1-atmosphere melting relations of two core samples. Analysis of the resulting melts showed that CaO and MgO in the melt varied linearly with temperature, suggesting that the concentration of these oxides in glass found in quenched partially molten drill core could be used as empirical geothermometers, to assign quenching temperatures to all glassy core from the lake. Lastly, JOHNSON (1980) determined the porosity and permeability of a suite of samples from hole KI76-1.

CURRENT WORK

Repeated drilling of Kilauea Iki lava lake (in 1960, 1961, 1962, 1975, 1976, 1979 and 1981) has provided a suite of drill cores that record in detail the crystallization and differentiation of this small, self-roofed magma chamber. Throughout most of the period 1960–1981, drilling was stopped at the base of the upper crust, that is, wherever the drillers encountered material too fluid to drill into. Most holes, therefore, sample only the upper crust, the exceptions being hole KI79-5, which passed through the entire lake near its north edge (see Figure 2B) and several of the 1981 holes (KI81-1, 2 and 3) which passed the thermal maximum in the lake (see Figure 2B; also HELZ and THORNBER, 1987). Core recovery rates in 1960–62 were quite low, and varied from 70% to 98% for the 1967–1975 holes (HELZ et al., 1984). In all subsequent drillings (1976–1981), use of diamond drill bits resulted in 100% recovery of core in all holes (HELZ and WRIGHT, 1983; HELZ et al., 1984).

The drill core is very fresh, except for local oxidation at temperatures of 100–950°C. Much of the core was partially molten prior to being quenched by the water used to cool the bit. Only core quenched from just above the incoming of the first oxide phase (ilmenite or ferro-pseudobrookite) shows any quench-phase crystallization, in the form of black fuzzy rims on plagioclase or augite.

The available core has been extensively sampled (>400 samples) for petrography and chemical work, and the bulk compositions of 143 samples have been determined gravimetrically, by H. R. Kirschenbaum and J. W. Marinenko, of the U.S.G.S. The trace element characteristics of the lava lake are being investigated by M. M. Lindstrom, who has analyzed a suite of 55 of these same, analyzed samples, by using neutron activation techniques. The whole-rock analyses will not be summarized here, but some of the following discussion draws on this extensive body of largely unpublished data. In all of the work summarized below, including that done by workers other than the present author, this same group of samples has been used when possible, to facilitate comparison among the various studies.

Crystalline assemblages and phase compositions

Examination of the drill core has led to recognition of three texturally had chemically distinct rock types in the lake (not counting the complicated textures, produced by annealing and small-scale filterpressing, seen in foundered crust). Most of the lake is olivine-phyric basalt, with normal igneous textures, referred to in this paper and others (e.g., HELZ, 1980) as "matrix rock." (Photomicrographs of typical matrix rock are shown in HELZ, 1980, 1987a,b). The principal macroscopic variation in the matrix rock is variation in the amount of phenocrystic olivine, which ranges from 1–3 volume percent (MANGAN and HELZ, 1985) to >40 volume percent.

In addition, the upper crust of the lake, especially the interval between 18 and 56 m below the surface, contains 6–10% by volume segregation veins. These veins are coarse-grained, diabase-textured micropegmatites (see photomicrographs in RICHTER and MOORE, 1966; HELZ, 1987b), and are irregular to sill-like in form. They occur in all the Hawaiian lava lakes (WRIGHT et al., 1976), but are particularly numerous and well-developed in Kilauea Iki (HELZ, 1980). Chemically, the segregation veins are olivine-poor ferrobasalt. Their range of composition in Kilauea Iki does not overlap that of the olivine-phyric matrix rock from which they are derived (see HELZ, 1987b; also Figure 3).

The third rock type consists of the vertical olivine–rich bodies (vorbs) described and illustrated in HELZ (1980, 1987b). These irregular pipe-like bodies, found principally between 40 and 58 m below the lake surface, extend vertically up the side of the core, and are locally traceable for a meter or more. They are enriched in iron-rich olivine and contain an excess of differentiated, segregation-vein-like liquid relative to normal olivine–bearing matrix rock of similar MgO content (HELZ, 1987b). Much of the textural contrast between these cross-cutting bodies and the adjacent matrix results from the fact that the groundmass phases (olivine + augite + plagioclase) in the vorb mostly occur adjacent to olivine, so that the main pools of melt are free of crystals. This textural rearrangement has been interpreted as resulting from differential vertical flow of crystals, liquid and vesicles with the vorb (HELZ, 1980).

These textural variants are not mineralogically distinct, the bulk compositional differences between them being accommodated by variations in the amount of the different phases present. In the matrix rock and vorbs, olivine is the only phenocryst phase. The groundmass of the olivine-phyric basalt consists of olivine, tiny (0.1 mm or less) augite grains and lathy plagioclase, with or without Fe–Ti oxides, pigeonite or hypersthene, apatite, sulfide and glass depending on the quenching temperature of the core. Cristobalite rosettes occur in vesicles or diktytaxitic cavities, especially in olivine–poor holocrystalline samples; they have not been observed to coexist with glass in any olivine–bearing sample. No discrete alkali feldspar phase has been observed.

The segregation veins are aphyric. They contain only minor, late–crystallizing olivine and always contain pigeonite rather than hypersthene, but are not otherwise mineralogically distinct from the matrix rock. Rarely, cristobalite has been observed crystallizing from glass in segregation veins quenched from just above the solidus.

The amount and petrographic and chemical character of the olivine phenocrysts vary, as described in HELZ (1980, 1987a). The olivine phenocrysts (typically 0.5–8 mm long) characteristically contain inclusions of chromite. The composition of the chromite and its compositional changes with time from 1959 until quenched during drilling have been documented by SCOWEN (1986; see also SCOWEN et al., 1986). She found that chromite continued to re–equilibrate with the host olivine and with the liquid outside the olivine, probably by diffusion of ions through the olivine. With the passage of time since the eruption, Mg, Al and Cr in chromite decreased, while Ti, Fe^{2+} and Fe^{3+} increased, the effect being greater for core from progressively deeper in the lake. The results have implications for the study of olivine-chromite assemblages in other mafic and ultramafic bodies.

The pyroxenes in Kilauea Iki are similar in composition to pyroxenes from other Hawaiian lavas (see analyses in MURATA and RICHTER, 1966a; also in WRIGHT and PECK, 1978). They are of interest because of the opportunity to determine rather precisely (1) the bulk compositional controls on the occurrence of pigeonite vs hypersthene, both of which occur in the lake; (2) the composition of all coexisting phases, including that of the melt at the point where the low–Ca pyroxene appears (see e.g., the glass in Table 1, Column 5); and (3) the stoichiometry of the two reactons at that point.

Another noteworthy feature of the pyroxenes is the tendency of hypersthene to occur as oikocrysts. The monoclinic pyroxenes rarely exhibit this habit: only one augite oikocryst has been found in the lake to date. In contrast, most hypersthene in the lake is oikocrystic, occurring in plates up to a few millimeters in length. The hypersthene encloses plagioclase laths, the grain size and random orientation of which are identical to those of plagioclase laths outside the oikocrysts. The augite and groundmass olivine, which occur uniformly throughout the matrix, are almost completely absent within the oikocrysts. The resorption of olivine by hypersthene is not surprising; the reaction relationship olivine + liquid → hypersthene is familiar to all petrologists. There is not a recognized reaction relationship between augite and hypersthene, however, and yet the augite (which first crystallized at ≥1170°C) has been resorbed by the oikocrysts, which begin to grow at ~1090°C. Why and how this resorption takes place is not yet clear. These observations suggest, however, that the inclusion population in pyroxene oikocrysts in layered intrusives is not a reliable indicator of the presence or absence of cumulus minerals other than plagioclase.

Plagioclase in the lake varies in composition from $An_{78.2}Ab_{21.2}Or_{0.6}$ to $An_{8.7}Ab_{69.8}Or_{21.5}$, though this range is not found within a single sample. The range of plagioclase compositions found within each sample increases from 5 mol percent An at 1140°C to 30–35 mol percent An in near-solidus samples. Maximum zoning observed within single grains is typically 10 mol percent An in plagioclase in olivine-phyric rock; in segregation veins and other, more differentiated samples, An content may vary by as much as 25% within a single crystal. The plagioclase zonation patterns in the lava lake are of interest because of their bearing on the development and preservation of zoning in plagioclase in other mafic rocks.

The phase assemblages exhibited by the Fe–Ti oxides are somewhat unusual because of the existence of three oxide phases (ilmenite, ferropseu-

Table 1. Representative electron microprobe analyses of interstitial glasses from Kilauea Iki lava lake.

Sample no.	1	2	3	4	5	6	7	8	9
	KI81-1-224.4	KI75-1-145.1	KI75-1-143.8	KI75-1-141.8	KI81-1-299.9	KI67-3-83.8	KI79-1-167.6	KI75-1-130	KI79-5-193.3
No. points analyzed	14	7	19	13	6	7	18	18	10
SiO_2	51.4	50.7	51.8	51.7	54.2	55.0	60.6	71.4	68.9
TiO_2	3.89	4.14	5.08	5.40	3.98	3.11	2.17	1.00	1.52
Al_2O_3	13.5	13.0	12.6	12.1	13.2	12.6	14.0	13.8	14.6
Cr_2O_3	.02	.01	.03	.00	.00	.00	.00	.00	.00
FeO^a	10.3	11.8	12.1	12.7	10.6	13.2	7.51	2.71	2.45
MnO	.12	.14	.14	.18	.14	.16	.06	.02	.01
MgO	6.26	5.88	4.99	4.52	4.33	3.18	2.21	.42	.59
CaO	10.1	9.84	8.82	8.48	7.67	6.98	4.53	.94	1.01
Na_2O	2.87	2.57	2.67	3.12	3.36	2.71	3.99	3.65	3.77
K_2O	.78	.88	1.03	1.06	1.66	1.62	3.05	5.59	5.85
P_2O_5	.31	.40	.43	.39	.62	.71	1.45	.12	.22
Sum	99.6	99.3	99.7	99.6	99.8	99.2	99.6	99.7	98.9
Comments:	Glass at thermal maximum	Glass below crust/melt interface	Glass in typical matrix	Glass just above oxide-in	Glass just below opx-in	Glass below pigeonite-in, at magnetite-in	Glass in typical matrix	Last glass, olivine basalt (MgO = 10%)	Last glass, olivine-rich basalt (MgO = 21%)
Chemical distinction:	Most Mg-rich, low-Fe melt series	Most Mg-rich, mid-Fe melt series	Mid-Fe melt series	Most Ti-rich, mid-Fe melt series	Low-Fe melt series	Most Fe-rich, high-Fe melt series	Most P-rich, mid-Fe melt series	Most Si-rich, Mid-Fe melt series	Most Si-rich, low-Fe melt series
Quenching temperature:									
T_{MgO} (°C)[b]	1141	1134	1114	1102	1099	1076	1053	—	—
T_{CaO} (°C)[b]	1138	1129	1114	1110	1096	1080	1046	986	990

[a] Total iron as FeO.

[b] T_{MgO} indicates a temperature estimate based on MgO content of the glass; T_{CaO} indicates a temperature estimate based on CaO content.

dobrookite and magnetite) in a variety of assemblages and crystallization sequences. In olivine–poor core, the first oxide to crystallize is ilmenite (at 1105°–1110°C) followed by magnetite (at 1080°–1090°C). As the olivine content of the rock increases, however, the Fe/Ti ratio of the final, near-solidus oxide assemblage decreases, passing from ilmenite + magnetite, to ilmenite-dominant assemblages with minor ferropseudobrookite and magnetite, to ferropseudobrookite-dominant assemblages where groundmass spinel is absent, and only a trace of ilmenite is found.

The magnetite-ilmenite assemblage is a well-known geothermometer (BUDDINGTON and LINDS-LEY, 1964) and has been used as such for the lake samples. HELZ and THORNBER (1987) compared the results of oxide geothermometry with glass geothermometry and downhole temperature measurements, and concluded that oxide and glass geothermometry gave essentially identical results, over the temperature range (980°–1090°C) where glass, magnetite and ilmenite coexist.

Melt compositions

Among the phases present in the lake, the quenched melt, or glass, present in the core varies the most in composition. The range of melt compositions present and their significance has been determined via (1) melting experiments on selected core samples (HELZ and THORNBER, 1981, 1987) and (2) electron microprobe analysis of glasses in situ in drill core and bulk chemical analysis of naturally produced liquid segregations and of artificial melt segregations ("oozes") that flowed into drill holes and were subsequently recovered. The latter data were presented briefly in HELZ (1984) and will be discussed in more detail here.

Melting experiments, at 1 atmosphere and at f_{O_2} conditions of the nickel-bunsenite buffer, have been performed on sample KI75-1-143.8 (MgO = 12.15 weight percent) and on KI67-3-83.8 (MgO = 7.54 weight percent). Details of the experimental procedures are given in HELZ and THORNBER (1987). The resulting phase relations are given in Table 2. These results agree closely with those of THOMPSON and TILLEY (1969), who used 1959 eruption samples, rather than core from the lava lake, as starting materials.

The most useful feature of these experimental results is that the MgO and CaO contents of the melts were observed to vary linearly with temperature, as follows: the MgO content of glasses coexisting with olivine is linear from 1050° to 1250°C, whereas the CaO content of glasses coexisting with olivine [or (orthopyroxene) + augite + plagioclase]

Table 2. Temperatures (°C) of first occurrence of phases in 1-atmosphere melting experiments (from HELZ and THORNBER, 1987).

Phase	K167-3-83.8	K175-1-143.8
Olivine	1169	1260
Augite	1160 ± 10	1172 ± 8
Plagioclase	1160 ± 10	1163 ± 2
Ilmenite or pseudobrookite	1100 ± 10	1106 ± 6
Magnetite	1080 ± 10	1090 ± 10

is linear from 1000° to 1160°C. These results have therefore been used to assign quenching temperatures to every piece of olivine–bearing glassy core analyzed (see Table 1, for example). The shapes of the resulting profiles generally resemble those predicted by theoretical thermal modeling; the minor deviations from predicted shapes are also interpretable (see discussion in HELZ and THORNBER, 1987).

Figure 3 shows all bulk analyses and electron microprobe analyses from the 1967–1981 drill core for which MgO < 11 weight percent. Most analyses of the olivine-phyric samples lie at higher bulk MgO contents. They were omitted to better display the full range of compositions seen in the interstitial glasses. In addition, representative analyses of a variety of interstitial glasses are presented in Tables 1 and 3.

Inspection of Figure 3 shows the resulting suite of microprobe analyses to be internally coherent and consistent with the wet chemical analyses of segregation veins and oozes. The analyses also provide a very complete determination of the line of descent of the liquid in olivine tholeiite. The smoothness and tightness of the variation of most oxides testify to the internal coherence of the data set. The fact that compositions of high-temperature (1100°–1140°C) glasses largely overlap wet-chemical analyses of oozes and segregation veins (samples with MgO = 2.61 to 5.96 weight percent, see also Figure 8, below), suggests that the glass compositions are of acceptable accuracy. Given that, and the geothermometer scale implicit in the glass compositions (shown along the top of the individual plots in Figure 3), one can determine the sequence of crystallization of phase in the lake, and the temperature at which they first appear, by inspection, as follows:

(1) The first silicate phase to crystallize is olivine. The liquidus temperature of individual samples varies from 1170°C (at MgO = 7.5 weight percent) to ⩾1200°C, depending on the olivine content of

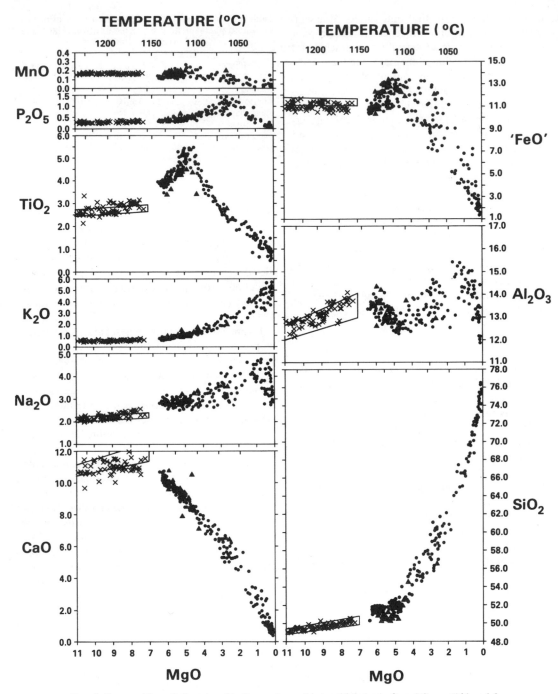

FIG. 3. Composition of all analyzed bulk samples and interstitial glasses from Kilauea Iki lava lake with MgO ≤ 11 weight percent, plotted against their MgO content. All quantities are in weight percent, with FeO = total iron as FeO. Wet chemical analyses of olivine–phyric rocks are indicated by crosses; those of segregation veins and oozes are shown by solid triangles. Electron microprobe analyses of interstitial glass are shown as small black dots. The temperature scale at the top gives the quenching temperature of each glass, corresponding to its MgO content, based on the calibration of HELZ and THORNBER (1987). Rectangular field at MgO = 7–11 weight percent outlines the field of composition of low-MgO eruption samples.

Table 3. Additional glasses from Kilauea Iki laval lake.

Sample no.	1 KI81-3-185.5	2 KI81-3-185.5	3 KI79-1-158	4 KI79-1-158	5 KI79-1-158	6 KI79-1-158	7 KI79-1-158
No. points analyzed	8	10	2	2	2	2	14
SiO_2	65.2	62.9	43.1	70.7	40.1	72.1	73.5
TiO_2	1.90	1.70	6.80	1.30	6.48	1.04	.59
Al_2O_3	14.4	12.9	4.56	11.1	4.11	10.9	12.7
Cr_2O_3	.00	.00	.00	.00	.00	.00	.00
FeO	5.05	9.28	26.6	5.56	29.9	5.51	2.69
MnO	.07	.12	.32	.04	.40	.06	.02
MgO	1.39	1.62	5.24	.94	3.99	.60	.19
CaO	2.87	4.60	9.81	1.97	9.59	1.68	.92
Na_2O	4.22	3.57	1.35	(2.66)[a]	1.32	(2.76)[a]	4.24
K_2O	3.73	2.46	.62	4.22	.48	3.65	4.41
P_2O_5	.58	.80	1.94	.12	3.05	.12	.08
Sum	99.4	100.0	100.3	98.6	99.4	98.4	99.4

Comments: Glass in olivine-phyric matrix (cols 1–2) / Glass in segregation vein (col 2). Coexisting immiscible liquids included in plagioclase in a segregation vein (cols 3–6). Glass outside plagioclase, same thin section (col 7).

Both in same thin section, but not coexisting

[a] Total iron as FeO.
[b] Parentheses indicate probable loss of Na_2O during analysis.

the rock. [The most magnesian glass found in the eruption samples was quenched from 1216°C (HELZ, 1981a), corresponding to MgO = 10 weight percent. Samples with higher MgO contents were not erupted as liquids.]

(2) Augite begins to crystallize between 1175° and 1160°C (see CaO *vs* MgO, Figure 3). No drill core this hot has been recovered, so the CaO peak is not defined very precisely.

(3) Plagioclase begins to crystallize between 1165° and 1145°C (Al_2O_3 *vs* MgO). Again, no core was recovered from this temperature range, so the Al_2O_3 peak is located only approximately.

(4) A TiO_2-rich Fe–Ti oxide (either ilmenite or ferropseudobrookite) begins to crystallize at 1110 ± 5°C (see TiO_2 *vs* MgO). The TiO_2 peak is very sharp in glasses from any individual drill hole; the slight broadening of the peak in Figure 3 probably results from the fact that the two different oxide phases come in at slightly different temperatures. Note that the peak TiO_2 content (~5.45 weight percent, see column 4, Table 1) is the same regardless of which oxide is crystallizing.

(5) A broad peak in FeO reflects the first appearance of hypersthene or pigeonite, ± magnetite

at 1090 ± 10°C, as well as the incoming of the first Fe–Ti oxide at 1110 ± 5°C.

(6) Apatite begins to crystallize at 1065 ± 10°C (see P_2O_5 *vs* MgO).

As for the other major oxides, MnO follows TiO_2, whereas SiO_2 and K_2O increase monotonically in the liquid. This increase reflects the absence of K-feldspar and the scarcity of coexisting cristobalite + glass, noted above. The Na_2O variation is problematical. It may drop off slightly at extremely low MgO contents, but most of the decline seen in Figure 3 is more likely the result of loss of Na_2O under the electron microprobe beam.

This liquid descent line does not reflect either perfect equilibrium or perfect fractional crystallization. Rather it is a hybrid, especially in the high-temperature olivine-phyric samples. Olivine re-equilibrates with the liquid (HELZ, 1987a) as do chromite (SCOWEN, 1986) and the Fe–Ti oxides (HELZ, unpublished data; HELZ and THORNBER, 1987). At the other extreme, plagioclase is markedly zoned, as noted above. The pyroxenes are intermediate between olivine and plagioclase in the degree of compositional variability exhibited.

As temperature decreases in the olivine-phyric samples, at some point all the remaining glass becomes isolated from the olivine, and the liquid compositions produced by subsequent crystallization approach a perfect fractional crystallization trend. The temperature at which this occurs lies between 1050° and 1080°C. The characteristic endpoint liquids in olivine-phyric rock (see, for example columns 7 and 8 in Table 1) are thus produced by extreme in situ fractional crystallization, and are not in equilibrium with olivine.

In the segregation veins, by contrast, the pyroxenes as well as plagioclase exhibit strong zoning, so that liquids in the veins should approach a perfect fractional crystallization trend from the start. This kind of hybrid fractionation trend should not be peculiar to the lava lake, but may occur in other magma chambers as well.

It is obvious from Figure 3 that the concentration of some oxides inthe melt is controlled quite closely by the crystalline assemblage. Thus olivine ± low-Ca pyroxene controls MgO, ilmenite or ferropseudobrookite controls TiO_2, and the assemblage olivine + augite + plagioclase constrains CaO. Other oxides, notably FeO and Al_2O_3, appear to be much more variable.

The wide spread in FeO contents results from the re-equilibration of olivine with the liquid as crystallization proceeds (HELZ, 1984). As the olivine grows more Fe-rich as a result of this in-situ re-equilibration, it takes iron from the adjacent interstitial melt. The more olivine present initially, the more severely depleted in iron the liquid becomes,

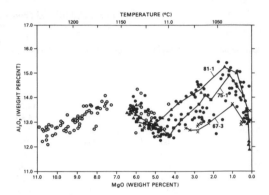

FIG. 5. Al_2O_3 vs MgO, for the same samples as in Figure 3. Symbols for samples from holes KI67-3, KI75-1 and KI81-1 as in Figure 4.

at a given MgO content. Core samples from the 1967 drill holes, which bottomed in the olivine-depleted zone, thus all contain interstitial glasses that are relatively Fe-rich (see *e.g.*, Column 6, Table 1). Samples from the partially molten zones of KI79-5 and from the 1981 drill holes, which come from the most olivine-rich parts of the lake, contain glasses that are very Fe-depleted. Core samples from the 1975, 1976 and the other 1979 holes contains glasses of intermediate iron content. This effect is clearly displayed in Figure 4, where liquid descent lines for three individual holes [KI67-3, KI75-1 and KI81-1, from the southern-most cluster in the center of the lake (Figure 2a,b)] are shown separately.

The effect on FeO in the glass of varying olivine content can also be seen wherever a segregation vein cuts an olivine-rich host. An extreme example of this is shown by the analyses in Columns 1 and 2 of Table 3, for two melts from a single thin section, quenched from 1025°–1040°C. The glass in the segregation vein contains 9.28 weight percent FeO, while that in the adjacent wall rock, a few millimeters away, contains 5.05 weight percent FeO. This difference exists because the melt composition is controlled by *local* bulk composition, even though a significant amount of melt is present.

The variation in FeO in the melt imposed by re-equilibration of olivine is compensated for by increases in all the other, relatively unbuffered, oxides (SiO_2, Al_2O_3, Na_2O, K_2O and P_2O_5). The apparent scatter in Al_2O_3 vs MgO (Figure 5) thus resolves into a set of nested curves, tightly defined for any one drill hole, but offset from each other so as to produce a wide smear when all data are superimposed.

Figure 6 shows SiO_2 vs MgO variation for different drill holes: at a given MgO content and hence

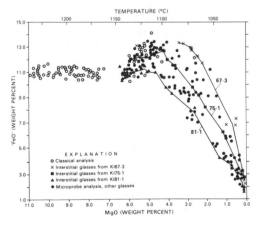

FIG. 4. FeO vs MgO, for the same samples as in Figure 3. Glasses from three drill holes are distinguished, illustrating the succession of liquid compositions for samples from within individual holes. Temperature scale as in Figure 3.

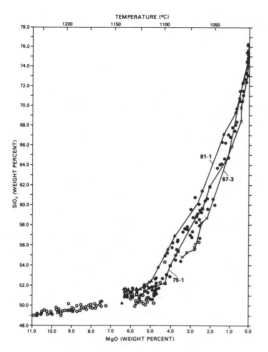

FIG. 6. SiO₂ vs MgO, for the same samples as in figure 3. Symbols for samples from holes KI67-3, KI75-1 and KI81-1 as in Figure 4.

temperature, the melts in the most olivine-rich rocks are the richest in SiO_2 and alkalies, as well as Al_2O_3. These effects, like the FeO effect, can also be seen in melts from the segregation vein/matrix pairs within a single thin section (see again columns 1 and 2, Table 3). The *maximum* SiO_2 enrichment reached by the interstitial glass in olivine-phyric rock decreases as olivine content increases, of course; this can be seen by comparing columns 8 and 9 in Table 1.

The uppermost levels of SiO_2 enrichment seen in Figures 3 and 6 are attained by rest liquids in the segregation veins (see, *e.g.*, column 7, Table 3). As segregation veins are all olivine–poor, the various liquid descent lines tend to converge in this range. Because the liquid that makes up the bulk vein was derived in part from the adjacent rock, analysis of the interstitial liquids within the segregation vein show that they too reflect the olivine content of the adjacent host rock. Accordingly, peak SiO_2 contents (≥ 76 weight percent SiO_2) are found in the residual liquids in segregation veins in olivine–rich host rock. Thus, initial differences in liquid composition, established by re–equilibration of olivine at 1100–1150°C propagate through all subsequent crystallization/fractionation events, to the last drop of liquid.

These varying liquid descent lines affect the nature of the later crystalline phases, of course. Orthopyroxene is the low–Ca pyroxene in olivine–rich samples, whereas pigeonite is found in olivine–poor core and in the segregation veins. The effects on the Fe–Ti oxide assemblages have already been noted above.

One feature of these glass data worth noting is the temperature at which compositional divergence starts. In the olivine-phyric rock, it clearly begins at 1125°–1140°C. Liquids in segregation veins start to diverge from those in the host at temperatures of 1070°C in most core, by 1080°–85°C in some of the deepest veins, where the compositional contrast with the host rock is largest. In both cases, the melt does not occur in isolated pools, but remains continuous. At the higher temperatures, the crystal network is barely self-supporting, and yet melt compositions start to reflect the local bulk compositional environment. This observation implies that interstitial liquids in other bodies, such as layered intrusives, may also start to differ between adjacent layers at relatively high melt contents, unless circumstances favor circulation of such melts out into the main chamber. This behavior also raises the possibility that one could get more than one group of lavas, significantly different in composition, from a single magma chamber, not because of varying source chemistry or degree of contamination, but by tapping different parts of a chamber undergoing crystal-liquid fractionation.

Liquid immiscibility is not a significant petrogenetic process in Kilauea Iki lava lake. It occurs only at a very late stage of crystallization, and then principally in melts included within plagioclase. Plagioclases in segregation veins, which are coarse grained and may contain large melt inclusions, locally contain spectacular examples of coexisting liquids, however. Two pairs of such coexisting liquids are given in Table 3 (Columns 3–4, 5–6). The segregation vein in which they occur was quenched from 985°C. The extreme compositions of the mafic liquids are unlike any known basalt in Hawaii.

PROCESSES ACTIVE IN KILAUEA IKI

One of the principal objectives of the Hawaiian lava lake studies has been to characterize, as completely as possible, the physical processes responsible for producing the chemical variations observed in the lakes. A summary of the processes active in Kilauea Iki, Alae and Makaopuni was given in WRIGHT *et al.* (1976). An updated summary is given in WRIGHT and HELZ (1987). What follows is an

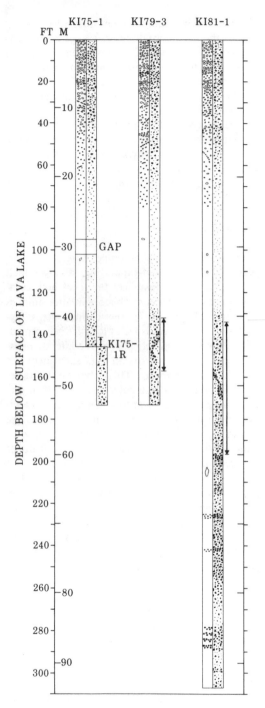

expanded version of the latter summary, as it applies to Kilauea Iki lava lake.

Olivine settling and compaction

Gravitative settling of coarser phenocrystic olivines has occurred in Kilauea Iki, producing a zone between 10 and 40 m depth in which olivine content is <3% and olivines >3 mm^2 are essentially absent (MANGAN and HELZ, 1985). The process did not become efficient until 1963 (HELZ, 1980). The final distribution of olivines in the central part of the lake is illustrated semi–quantitatively in Figure 7. Some size sorting is observed, so that average grain size and amount of phenocrysts are positively correlated. On plots of SiO_2 vs MgO and Al_2O_3 vs MgO (Figure 3) it can be seen that most of the olivine–bearing samples define a trend parallel to the original field of compositions of the eruption pumices, a field that can be extrapolated to an olivine composition of Fo_{86}. Thus olivine redistribution was achieved while the olivine phenocrysts retained their original compositions (average Fo_{86}), that is, before significant cooling of the lake had occurred (HELZ, 1987b). This conclusion implies that olivine redistribution took place at $T \geq 1180°C$ (HELZ, 1987b; HELZ and THORNBER, 1987).

It has been observed (HELZ, 1980; WRIGHT *et al.*, 1976) that settling of olivine in this and other lava lakes was not nearly as efficient as would be predicted from Stokes' law calculations. The decrease in olivine size and content observed between 10 and 20 m depth correlates with a decrease in vesicularity in the same range, suggesting that early settling of olivine was inhibited by the presence of vesicles (HELZ, 1980). No actual flotation of olivine crystals by individual vesicles seems to be involved (MANGAN, unpublished data). Rather the idea is that the presence of vesicles gives the melt the properties of a Bingham plastic rather than a Newtonian fluid, as was inferred by SHAW *et al.* (1968) for the 1965 Makaopuhi lava lake. If so, the gradual decrease in olivine size and content with depth in the 10–20 m range would reflect a gradual decrease in the yield strength of the Bingham-body melt as vesicle content decreased. M. Mangan has quantified the size and distribution of olivine and vesicles in the core, in order to verify this model (MANGAN and HELZ, 1985).

FIG. 7. Schematic representation of the distribution of vesicles (left-hand columns) and olivine phenocrysts (right-hand columns) with depth, for three drill holes from the center of Kilauea Iki lava lake. Hole KI75-1R is an extension of hole KI75-1, drilled in December 1978 (HELZ and WRIGHT, 1983). Symbols convey qualitative information on the size, shape and clustering patterns of vesicles and olivine crystals, as well as on their relative amounts. Arrows beside the columns indicate the depth range in which vorbs are particularly well-developed in each drill hole. Very deep vesicular layers at 85–90 m in KI81-1 are filled with melt: this interval is all foundered crust (HELZ and WRIGHT, 1983).

A related problem is the diffuseness of the cumulate zone, in which olivine content never exceeds 50% and is usually <40%, with the olivines rarely touching each other. HELZ (1980) suggested that migrating olivines carried with them a stagnant boundary layer of melt which slowed their settling and inhibited compaction. Such "skins" of melt have not been observed in lake samples, but some olivine crystals in the eruption pumices do exhibit boundary layers, detectable because of their contrasting composition (HELZ, 1987a).

Vesicle movement and retention

The distribution of vesicles in the lake, shown semiquantitatively in Figure 7, has been investigated quantitatively by M. Mangan (MANGAN and HELZ, 1985). The vesicles are of interest because they demonstrate the existence of a gas phase in the lake, and allow us to monitor the degassing history of the lake. There is some evidence (HELZ and WRIGHT, 1983) that loss of original gases was complete before 1979. Nevertheless, bubble-driven processes continued to be active in the lake, and late-stage vesicle accumulation at the final position of the thermal maximum was observed in the 1981 drill holes (see Figure 7, vesicles at 70–75 m depth). This and other lines of evidence suggest that the lake may have picked up meteoric water, probably through cracks in the sides and bottom of the lake.

The other reason for studying vesicles is that they are extremely mobile in the lake because of their low density, and their ability to coalesce and to deform as they pass through a crystal framework. Hence they interact with other processes, triggering some and inhibiting others. Olivine settling may have been greatly slowed by the flux of vesicles rising from below. Large–scale convective overturn of the lake appears to have been virtually eliminated, in part because of vesicle gradients. The vesicle contents involved are mostly <10% (MANGAN and HELZ, 1985). Evaluating the effects of such moderate vesicle contents on the various processes that occur in the lake should be relevant to processes in other subvolcanic magma chambers as well.

Convective redistribution of crystals

Within the olivine–depleted zone, small (1–2 mm) clots of augite microphenocrysts are present in core samples from near the edge of the lake, and largely absent from samples from the middle. This differential concentration of augite raises the CaO content of the core from 10.5 weight percent in the middle to 11.6 weight percent CaO at the edge. The small olivine phenocrysts present show a similar pattern. The implied lateral redistribution of crystals seems most readiy explicable by convection within the olivine–depleted zone. Flow differentiation by convection was a major process in the 1965 Makaopuhi lava lake (WRIGHT and OKAMURA, 1977), and affected plagioclase distribution as well as olivine + augite in that lake. In Kilauea Iki, by contrast, flow differentiation of this sort was a minor process, confined to temperatures above the incoming of plagioclase and below the incoming of augite and probably confined to the most olivine–depleted part of the lake (20–40 m) (HELZ, 1987b).

Diapiric melt transfer

"Diapiric melt transfer" refers to a process whereby relatively low-density melt from within the olivine–rich mush at the base of the lens of melt extricates itself from that mush, rises through the main lens of melt, to pool and mingle with the melt at the base of the upper crust. This process, documented in HELZ (1986, 1987b), involves a compositionally unique melt present at 1145°–1160°C, at and just below the incoming of plagioclase. Not coincidentally, this melt is lower in density than any other along the liquid descent line at temperatures >1100°C.

The process has been hypothesized to occur because it was observed (HELZ, 1980, 1987b) that core from the lake shows strong chemical zonation with depth, in which the upper part of the lake is enriched in TiO_2 and incompatible elements, and depleted in FeO and CaO, whereas deeper parts of the lake show the opposite signature. Details of the pattern of chemical variation, and the very small grain size of augite and plagioclase in the lake, make it very unlikely that this zonation was produced by large-scale downward movement of olivine + augite + plagioclase crystals (HELZ, 1987b). Earlier, with only the upper, enriched zone available for analysis, HELZ (1980) tentatively ascribed the enrichment to filterpressing of liquid formed during partial remelting of foundered crust, deep in the lake. Recovery and analysis of deeper core has led to our present recognition, first that this process does not require foundered crust, and second that it has affected the chemistry of the lake at virtually all levels. The only more pervasive process was olivine settling.

The melt extraction process is very efficient. The depleted zone at 56–78 m has lost 21–42% liquid. Some aspects of it are still not well understood, however. For example, we do not know the minimum value of the density inversion in the melt column required for the process to occur. From knowledge of the rate at which the lake has crystallized, gained from repeated drilling of the lake, we can infer that diapiric melt transfer started in 1960 and ceased around 1971.

Formation of vertical olivine–rich bodies

The core between 40 and 58 m depth contains the curious vertical olivine–rich bodies (vorbs) described above. They are hypothesized to form as streams of vesicles ± melt rise from within the lower crust, interact with the olivine–rich core of the lake and emplace themselves in the base of the upper crust. The vesicles are hypothesized to entrain first melt, then coarse olivines, from the lower part of the lake. As the plumes rise into the cooler material they begin to segregate, the vesicles + melt eventually detaching themselves from the trail of coarse olivines. These olivine–rich bodies contain excess iron-rich olivine (Fo_{77-79}) and differentiated liquid, relative to the adjacent host rock, so they are active at lower temperatures than is the diapiric transfer process (HELZ, 1987b). Analogous vuggy, vertical bodies also occur in the interval between 22–40 m, but they are more difficult to spot, as they lack coarse olivines (HELZ, 1980). The melt– and vesicle–rich top of the vorbs, where present, looks like a segregation vein, which led HELZ (1980) to suggest that segregation veins might be related to these bodies in some way.

Formation of segregation veins

Segregation veins, as discussed earlier, are coarse-grained diabasic sills formed by processes internal to the lake. Such veins are more abundant in Kilauea Iki than in the other lava lakes. The most important subgroup forms an apparently continuous set of sills, concentrated between 18 and 56 m depth, which make up 5–10% by volume of the upper crust of the lake (HELZ, 1980). Bulk compositions of the segregation veins correspond to liquids produced at 1135°–1105°C, estimated by using the calibration of HELZ and THORNBER (1987), as shown in Figure 8. In Kilauea Iki, they are more magnesian and presumably form at higher temperatures than segregation veins in Alae, Makaopuhi, and the prehistoric Makaopuhi lava lakes. The process by which they form is some sort of filterpressing. Temperature profiles through partially molten crust surrounding segregation veins have anomalous shapes above the veins (HELZ and THORNBER, 1987) suggesting that the melt in the segregation vein was hotter than the overlying crust and hence

was derived from deeper in the lake. Samples with the chemical signature indicative of loss of segregation-vein liquid are found at almost all levels in the lake, except for the 56–78 m interval. This clearly implies that some of the material in any given segregation vein may be derived from the crust near the vein. The presence of such samples at depths below 78 m, however, suggests that some of the material now present as segregation veins was transported, via the vorbs, from below 78 m to the 18–56 m depth range, (see HELZ, 1987b for a discussion of the variation of core chemistry with depth) though the full details of this process remain to be worked out.

Formation of late differentiates

Locally the core contains late fractures or large vesicles partly or completely filled by highly differentiated liquids, with liquidus temperatures of 1060°–1000°C. Figure 8 shows the compositions of some of these minor late differentiates plotted on a magnesia variation diagram, along with the bulk compositions of segregation veins and oozes (MgO = 4–6 weight percent) and the olivine-phyric matrix rock (MgO ≥ 7.5 weight percent). The fractures are produced by brittle failure of the lake as it shrinks during crystallization and cooling. The compositional gap between about 4.2 and 3.0 weight percent MgO is probably real. This gap corresponds to the temperature range over which the segregations have already formed within the crust, but are not yet rigid enough to break. So long as the crust contains a few slushy layers, stress cannot build to the breaking point. Once the veins have "set," pervasive fracturing becomes possible. It is probably not coincidental that the highest-temperature late differentiates have the composition of the liquid recovered from a half-crystallized, barely drillable segregation (c.f., Figure 8 and the ooze analysis in Table 3, Col. 5, HELZ, 1980).

The bulk compositions of these tiny bodies are andesitic to rhyolitic (SiO_2 = 56–70 weight percent; MgO = 3.0–0.5 weight percent). Their existence demonstrates the mobility of interstitial liquid even when very little (<10 percent) melt is present and confining pressures are low (HELZ, 1984). The driving mechanism is inferred to be gas filterpressing (ANDERSON et al., 1984). In deeper, larger intrusives, the process of segregation might be different, but the following observations are still relevant. First, basalt can fractionate to produce dacitic and rhyolitic liquids. Second, when partially molten rock fractures, the interstitial melt can move, even at very high melt viscosities.

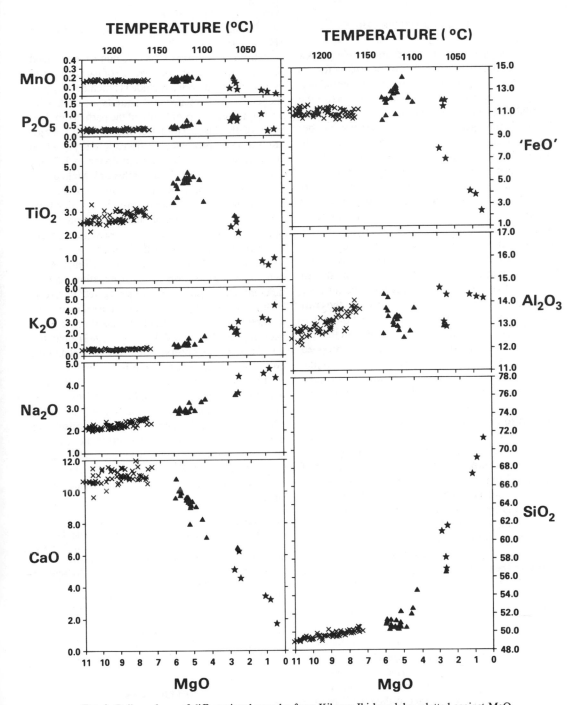

FIG. 8. Bulk analyses of differentiated samples from Kilauea Iki lava lake, plotted against MgO. All quantities in weight percent. Wet chemical analyses of olivine-phyric rock (crosses) and of segregation veins and oozes (triangle) as in Figure 3. Bulk compositions shown by stars were obtained by microprobe analyses of fused glasses and/or by rastered-beam analyses of the differentiated veinlets in polished section. Where both methods were used they gave comparable results.

SUMMARY COMMENTS

This overview of studies, past and present, of Kilauea Iki lava lake should give some feeling for the range of studies possible and for their potential petrologic applications. The work does not exhaust the possibilities. Other aspects of the lava lake's behavior (trace element geochemistry and partitioning, detailed thermal modeling, for example) are still in the earliest stages or have not been done at all.

The most important feature of Kilauea Iki is that it is not a fossil system. In deciphering the history of the lava lake, for example, it has been invaluable to have drill core that records many different stages of the lake's crystallization history. The main reason we can state with confidence that the diapiric melt transfer process began in 1960 and ended about 1971, or that the vorbs began to form by at least 1967 and ceased in 1979 is precisely because we know directly from the drilling results how the crust grew with time, and hence how the position of particular isotherms varied with time. The fact that the melt diapirism and the vorbs started and stopped independently in turn confirms that these two diapiric processes operated independently of each other, and hence that from 1967–1971, there were two different populations of diapirs migrating from bottom to top within the lens of melt.

The simple availability of any partially molten core is extremely important because it enables us to "capture" processes in progress. Recognition of the vorbs, in particular of their textures and significance, would have been much more difficult if only subsolidus core were available for study. Similarly, the fact that we have drilled into segregation veins at all stages of their formation (HELZ, 1980; HELZ and WRIGHT, 1983) tells us where and at what temperature they form.

Another valuable set of constraints has been provided by the detailed field observations on the 1959 eruption, and by the availability of samples collected day by day as the eruption proceeded. Information on the composition and petrography of the material fed into the lake, plus knowledge of the sequence in which it was erupted, permits us to conclude unequivocally, for example, that the present distribution of olivine phenocrysts in the lake is not inherited from the eruption. Furthermore, the observation that olivine-phyric samples lie on Fo_{86} control lines in certain projections, just as the eruption pumices do, constrains the time and temperature at which olivine settling took place. The latter information is one of several pieces of data that rule out three-phase crystal settling of olivine + augite + plagioclase in the lake, which in turn led to the

recognition of the process of diapiric melt transfer (HELZ, 1987b).

Petrologic features of the lake that have emerged as being critical to the occurrence of particular processes discussed above include (1) melt density gradients (2) the presence and abundance of vesicles and (3) the exact crystallinity of the partially molten parts of the lake. The latter has exerted a major control on all processes. As noted in HELZ (1987b) the presence of 17–20% initial olivine in Kilauea Iki vs <5% in the 1965 and prehistoric Makaopuhi lava lakes was sufficient to curtail lake-wide convection, with flow differentiation, as a major fractionation process. Also, because of the higher crystallinity in Kilauea Iki, other processes (such as segregation-vein formation) occur at higher temperature in Kilauea Iki than in the other lakes; hence the composition of the segregation veins is different. Drilling behavior is also affected: the crust-melt interface temperature in Kilauea Iki varies as a function of olivine content and has almost always been higher (HELZ and WRIGHT, 1983) than the 1075°C typically observed in Alae and Makaopuhi (WRIGHT et al., 1976). There is no unique crust-melt interface temperature, therefore. What is important is the percent crystallinity, and to a lesser extent, the shape of the crystals present (HELZ, 1980). Quantification of the physical processes and the *physical* parameters that control them in the lava lake, may eventually make it possible to use our knowledge of the lava lakes to decipher magmatic processes within other, larger magma chambers.

Acknowledgements—Study of Kilauea Iki has required the cooperation and support of many people, of whom I would particularly like to acknowledge T. L. Wright, R. T. Okamura, and R. I. Tilling. This paper was presented as part of a workshop on "Study of Active Volcanism: Constraints on Petrologic and Geophysical Models of Dynamic Earth Processes," held at the symposium honoring Hatten S. Yoder, Jr. In preparing the paper I chose to focus on the differentiation processes active in the lake and, in particular, on the succession of liquids present along the line(s) of descent of olivine tholeiitic liquid, in honor of Dr. Yoder's many contributions to our knowledge of the differentiation of basalts.

REFERENCES

ANDERSON A. T., JR., SWIHART G. H., ARTOLI G. and GEIGER C. A. (1984) Segregation vesicles, gas filter-pressing, and igneous differentiation. *J. Geol.* **92,** 55–72.

ANDERSON L. A. (1987) The geoelectric character of the Kilauea lava lake crust, Hawaii. *U.S. Geol. Survey Prof. Paper 1350,* Chap. 50. (In press).

AULT W. U., EATON J. P. and RICHTER D. H. (1961) Lava temperatures in the 1959 Kilauean eruption and cooling lake. *Geol. Soc. Amer. Bull.* **72,** 791–794.

AULT W. U., RICHTER D. H. and STEWART D. B. (1962) A temperature measurement probe into the melt of the Kilauea Iki lava lake in Hawaii. *J. Geophys. Res.* **67,** 2809–2912.

BUDDINGTON A. F. and LINDSLEY D. H. (1964) Iron-titanium oxide minerals and synthetic equivalents. *J. Petrol.* **5,** 310–357.

CHOUET B. (1979) Sources of seismic events in the cooling lava lake of Kilauea Iki, Hawaii. *J. Geophys. Res.* **84,** 2315–2330.

CHOUET B. and AKI K. (1981) Seismic structure and seismicity of the cooling lava lake of Kilauea Iki, Hawaii. *J. Volcanol. Geotherm. Res.* **9,** 41–56.

EATON J. P. (1962) Crustal structure and volcanism in Hawaii. *Amer. Geophys. Union Monograph* **6,** 13–29.

EATON J. P. and MURATA K. J. (1960) How volcanoes grow. *Science* **132,** 925–938.

EATON J. P., RICHTER D. H. and KRIVOY H. L. (1987) Cycling of magma between the summit reservoir and Kilauea Iki lava lake during the 1959 eruption of Kilauea volcano, Hawaii. *U.S. Geol. Survey Prof. Paper 1350,* Chap. 48. (In press).

EVANS B. W. and MOORE J. G. (1968) Mineralogy as a function of depth in the prehistoric Makaopuhi tholeiitic lava lake, Hawaii. *Contrib. Mineral. Petrol.* **17,** 85–115.

EVANS B. W. and WRIGHT T. L. (1972) Composition of liquidus chromite from the 1959 (Kilauea Iki) and 1965 (Makaopuhi) eruptions of Kilauea volcano, Hawaii. *Amer. Mineral.* **57,** 217–230.

FLANIGAN V. J. and ZABLOCKI C. J. (1977) Mapping the lateral boundaries of a cooling basaltic lava lake, Kilauea Iki, Hawaii. *U.S. Geol. Survey Open–File Report 77–94,* 21 pp.

GUNN B. M. (1971) Trace element partition during olivine fractionation of Hawaiian basalts. *Chem. Geol.* **8,** 1–13.

HARDEE H. C. (1980) Solidification in Kilauea Iki lava lake. *J. Volcanol. Geothermal Res.* **7,** 211–223.

HARDEE H. C., DUNN J. C., HILLS R. G. and WARD R. W. (1981) Probing the melt zone of Kilauea Iki lava lake, Kilauea volcano, Hawaii. *Geophys. Res. Lett.* **8,** 1211–1214.

HARRIS D. M. and ANDERSON A. T., JR. (1983) Concentrations, sources and losses of H_2O, CO_2, and S in Kilauean basalt. *Geochim. Cosmochim. Acta* **47,** 1139–1150.

HEALD E. F., NAUGHTON J. J. and BARNES I. L., JR. (1963) The chemistry of volcanic gases 2. Use of equilibrium calculations in the interpretation of volcanic gas samples. *J. Geophys. Res.* **68,** 545–557.

HELZ R. T. (1980) Crystallization history of Kilauea Iki lava lake as seen in drill core recovered in 1967–1979. *Bull. Volcanol.* **43–4,** 675–701.

HELZ R. T. (1983) Diverse olivine population in lavas of the 1959 eruption of Kilauea volcano, Hawaii. *EOS* **64,** 900.

HELZ R. T. (1984) In situ fractionation of olivine tholeiite: Kilauea Iki lava lake, Hawaii. *Geol. Soc. Amer. Abstr. Prog.* **16,** 536–7.

HELZ R. T. (1986) Diapiric transfer of melt in Kilauea Iki lava lake, Hawaii. *Geol. Assoc. Can. Abstr. Prog.* **11,** 80.

HELZ R. T. (1987a) Character of olivines in lavas of the 1959 eruption of Kilauea Volcano and its bearing on eruption dynamics. *U.S. Geol. Survey Prof. Paper 1350,* Chap. 25. (In press).

HELZ R. T. (1987b) Diapiric transfer of melt in Kilauea

Iki lava lake: A quick, efficient process of igneous differentiation. (In review).

HELZ R. T. and THORNBER C. R. (1981) Geothermometry of Kilauea Iki lava lake. *EOS* **62,** 1073.

HELZ R. T. and WRIGHT T. L. (1983) Drilling report and core logs for the 1981 drilling of Kilauea Iki lava lake (Kilauea Volcano, Hawaii) with comparative notes on earlier (1967–1979) drilling experiences. *U.S. Geol. Survey Open–File Report 83-326,* 66 pp.

HELZ R. T. and THORNBER C. R. (1987) Geothermometry of Kilauea Iki lava lake. (In preparation).

HELZ R. T., BANKS N. G., CASADEVALL T. J., FISKE R. S. and MOORE R. B. (1984) A catalogue of drill core recovered from Kilauea Iki lava lake from 1967–1979. *U.S. Geol. Survey Open-File Report 84-484,* 72 pp.

HERMANCE J. F. and COLP J. L. (1982) Kilauea Iki lava lake: Geophysical constraints on its present (1980) physical state. *J. Volcanol. Geotherm. Res.* **13,** 31–61.

JOHNSON G. R. (1980) Porosity and density of Kilauea volcano basalts, Hawaii. *U.S. Geol. Survey Prof. Paper 1123-B,* 6.

KIRBY S. H. and GREEN H. W. II (1980) Dunite xenoliths from Hualalai volcano: Evidence for mantle diapiric flow beneath the island of Hawaii. *Amer. J. Sci.* **280-A,** 550–575.

LEEMAN W. P. and SCHEIDEGGER K. F. (1977) Olivine/liquid distribution coefficients and a test for crystal-liquid equilibrium. *Earth Planet. Sci. Lett.* **35,** 247–257.

LUTH W. C. and GERLACH T. M. (1980) Composition and proportions of major phases in the 1959 Kilauea Iki lava lake in December 1978. *EOS* **61,** 1142.

LUTH W. C., GERLACH T. M. and EICHELBERGER J. C. (1981) Kilauea Iki lava lake: April 1981. *EOS* **62,** 1073.

MANGAN M. T. and HELZ R. T. (1985) Vesicle and phenocryst distribution in Kilauea Iki lava lake, Hawaii. *EOS* **66,** 1133.

MACDONALD G. A. and KATSURA T. (1961) Variations in the lava of the 1959 eruption in Kilauea Iki. *Pacific Sci.* **15,** 358–369.

MOORE J. G. and EVANS B. W. (1967) The role of olivine in the crystallization of the prehistoric Makaopuhi tholeiitic lava lake, Hawaii. *Contrib. Mineral. Petrol.* **15,** 202–213.

MURATA K. J. (1966) An acid fumarolic gas from Kilauea Iki. *U.S. Geol. Survey Prof. Paper 537-C,* 6 pp.

MURATA K. J. and RICHTER D. H. (1966a) Chemistry of the lavas of the 1959–60 eruption of Kilauea volcano, Hawaii. *U.S. Geol. Survey Prof. Paper 537-A,* 26 pp.

MURATA K. J. and RICHTER D. H. (1966b) The settling of olivine in Kilauean magma as shown by lavas of the 1959 eruption. *Amer. J. Sci.* **264,** 194–203.

PECK D. L., WRIGHT T. L. and MOORE J. G. (1966) Crystallization of tholeiitic basalt lava in Alae lava lake, Hawaii. *Bull. Volcanol.* **29,** 629–656.

PECK D. L., HAMILTON M. S. and SHAW H. R. (1977) Numerical analysis of lava lake cooling models: Part II. Application to Alae lava lake, Hawaii. *Amer. J. Sci.* **277,** 415–437.

PECK D. L., WRIGHT T. L. and DECKER R. W. (1979) The lava lakes of Kilauea. *Scientific Amer.* **241,** 114–128.

RAWSON D. E. (1960) Drilling into molten lava in the Kilauea Iki Volcanic Crater, Hawaii. *Nature* **188,** 930–931.

RICHTER D. H. and MOORE J. G. (1966) Petrology of the Kilauea Iki lava lake, Hawaii. *U.S. Geolog. Survey Prof. Paper 537-B,* 26 pp.

258   R. T. Helz

RICHTER D. H. and MURATA K. J. (1966) Petrography of the lavas of the 1959–60 eruption of Kilauea volcano, Hawaii. *U.S. Geol. Survey Prof. Paper 537-D*, 12 pp.

RICHTER D. H., EATON J. P., MURATA K. J., AULT W. U. and KRIVOY H. L. (1970) Chronological narrative of the 1959–60 eruption of Kilauea volcano, Hawaii. *U.S. Geol. Survey Prof. Paper 537-E*, 73 pp.

RYAN M. P. (1979) High-temperature mechanical properties of basalt, Ph.D. Thesis, The Penn. State Univ.

SCHWINDINGER K. R. (1986) Petrogenesis of olivine aggregates from the 1959 eruption of Kilauea Iki: synneusis and magma mixing. Ph.D. Thesis, Univ. of Chicago.

SCHWINDINGER K. R., ROPPO P. S. and ANDERSON A. T., JR. (1983) Synneusis of olivine phenocrysts into glomeroporphyritic clusters. *Geol. Soc. Amer. Abst. Prog.* **15**, 682.

SCOWEN P. A. H. (1986) Re-equilibration of chromite from Kilauea Iki lava lake, Hawaii. M.Sc. Thesis, Queens University, Kingston, Ontario.

SCOWEN P. A. H., ROEDER P. L. and HELZ R. T. (1986) Re-equilibration of chromite from Kilauea Iki lava lake, Hawaii. *Geol. Assoc. Can. Abstr. Prog.* **11**, 125.

SHAW H. R., WRIGHT T. L., PECK D. L. and OKAMURA R. (1968) The viscosity of basaltic magma: an analysis of field measurements in Makaopuhi lava lake, Hawaii. *Amer. J. Sci.* **266**, 225–264.

SMITH B. D., ZABLOCKI C. J., FRISCHKNECHT F. and FLANIGAN V. J. (1977) Summary of results from electromagnetic and galvanic soundings on Kilauea Iki lava lake, Hawaii. *U.S. Geol. Survey Open-File Report 77-59*, 27 pp.

SWANSON D. A., DUFFIELD W. A., JACKSON D. B. and PETERSON D. W. (1972) The complex filling of Alae crater, Kilauea volcano. *Bull. Volcanol.* **36**, 105–126.

SWANSON D. A., DUFFIELD W. A., JACKSON D. B. and PETERSON D. W. (1979) Chronological narrative of the 1967–71 Mauna Ulu eruption of Kilauea Volcano, Hawaii. *U.S. Geol. Survey Prof. Paper 1056*, 55 pp.

THOMPSON R. N. and TILLEY C. E. (1969) Melting and crystallization relations of Kilauean basalts of Hawaii:

The lavas of the 1959–60 Kilauea eruption. *Earth Planet. Sci. Lett.* **5**, 469–477.

TILLING R. I., CHRISTIANSEN R. L., DUFFIELD W. A., ENDO E. T., HOLCOMB R. T., KOYANAGI R. Y., PETERSON D. W. and UNGER J. D. (1987) The 1972–1974 Mauna Ulu eruption, Kilauea Volcano: An example of quasi-steady-state magma transfer. *U.S. Geol. Survey Prof. Paper 1350*, Chap. 16. (In press).

WADGE G. (1981) The variation of magma discharge during basaltic eruptions. *J. Volcanol. Geotherm. Res.* **11**, 139–168.

WRIGHT T. L. (1973) Magma mixing as illustrated by the 1959 eruption, Kilauea volcano, Hawaii. *Geol. Soc. Amer. Bull.* **84**, 849–858.

WRIGHT T. L. and OKAMURA R. T. (1977) Cooling and crystallization of tholeiitic basalt, 1965 Makaopuhi lava lake, Hawaii. *U.S. Geol. Survey Prof. Paper 1004*, 78 pp.

WRIGHT T. L. and PECK D. L. (1978) Crystallization and differentiation of the Alae magma, Alae lava lake, Hawaii. *U.S. Geol. Survey Prof. Paper 935-C*, 20 pp.

WRIGHT T. L. and HELZ R. T. (1987) Recent advances in Hawaiian petrology and geochemistry. *U.S. Geol. Survey Prof. Paper 1350*, Chap. 23. (In press).

WRIGHT T. L., KINOSHITA W. T. and PECK D. L. (1968) March 1965 eruption of Kilauea volcano and the formation of Makaopuhi lava lake. *J. Geophys. Res.* **73**, 3181–3205.

WRIGHT T. L., PECK D. L. and SHAW H. R. (1976) Kilauea lava lakes: natural laboratories for study of cooling, crystallization and differentiation of basaltic magma. In *The Geophysics of the Pacific Ocean Basin and its Margin*. Amer. Geophys. Union Monograph 19, 375–392.

ZABLOCKI C. J. (1976) Some electrical and magnetic studies of Kilauea Iki lava lake, Hawaii. *U.S. Geol. Survey Open-File Report 76-304*, 19 pp.

ZABLOCKI C. J. and TILLING R. I. (1976) Field measurements of apparent Curie temperatures in a cooling basaltic lava lake, Kilauea Iki, Hawaii. *Geophys. Res. Lett.* **3**, 487–490.

Magmatic Processes: Physicochemical Principles
© The Geochemical Society, Special Publication No. 1, 1987
Editor B. O. Mysen

Neutral buoyancy and the mechanical evolution of magmatic systems

MICHAEL P. RYAN

Branch of Igneous and Geothermal Processes, 959 National Center,
U.S. Geological Survey, Reston, Virginia 22092 U.S.A.

What once sprung from
the Earth sinks back
into the Earth.

Lucretius (99–55 B.C.)

Abstract—Regions of neutral buoyancy are produced by the crossover in the *in-situ* densities of magmatic fluids and the countryrock of subcaldera magma reservoirs and rift systems. Beneath this horizon, magmatic parcels ascend under positive buoyancy forces, where above it, they descend under the influence of negative buoyancy. In Hawaii, the region of neutral buoyancy is coincident with the location of the summit reservoirs of Kilauea and Mauna Loa volcanoes. The horizon of neutral buoyancy, coupled with the contractancy mechanism which produces it, thus provides for the long–term stability (the prolonged existence) of subcaldera magma reservoirs, and their rift systems. As Hawaiian volcanoes evolve, they carry their contractancy profile and region of neutral buoyancy upward with them. Thus the evolutionary progression from seamount, through a subaerial immature shield, to a mature volcano, is one characterized by the progressive elevation of the summit reservoir complex and rift zones—a process that leaves beneath a wake of high velocity mafic and ultramafic rocks within the core region of the shield. In Iceland, the upper levels of the depth range for subcaldera magma storage and lateral injection in the active rift zones of the Krafla central volcano are coincident with the horizon of neutral buoyancy. Intrusion dynamics at Krafla, analogous to that of Kilauea's rift zones, demonstrate that the depth interval 2 to 4 km contains the parabolic nose of the expanding magmatic fracture front—symmetrically moving along the horizon of neutral buoyancy. Along the East Pacific Rise at 21°N latitude, and near the Siqueiros Fracture Zone, regions of marked reduction in elastic wave velocity at depth intervals of 2.5–4.5 km and 2.0–3.5 km, respectively, are inferred to contain partial melt and are consistent with the location of a horizon of neutral buoyancy. This suggests that neutral buoyancy, and the related changes in *in-situ* density produced through crustal contractancy, play a fundamental role in the inception, dynamics and evolution of magma reservoirs and their intrusive complexes within the Earth's oceanic spreading centers. Moreover, inspection of the magma and countryrock densities relevant to regions of calc-alkaline volcanism above subduction zones, and at select centers of granitic–rhyolitic complexes in continental interiors, suggest that their upper levels have a depth extent regulated by the neutral buoyancy principle.

INTRODUCTION

MAGMATIC FLUIDS, generated at depth in the Earth's mantle often come to rest at surprisingly shallow levels within the crust. It is, of course, remarkable that so long an odyssey should be interrupted just short of subaerial or submarine eruption. But such interruptions in the form of shallow crustal intrusive events are widespread, long-lived, and have played an enduring and important role in the evolution of the Earth's crust.

Why does magma come to rest at shallow levels in the crust? That is, why do shallow magma reservoirs exist? What are the mechanical controls that regulate the growth and evolution of subcaldera magma reservoirs and their associated rift systems? How do these controls operate within the context of the short-term dynamics of a volcanic system and its longer term evolutionary trends? Are there important mechanical and dynamic processes whose controls are, in part, independent of the deeper geodynamic regimes that support and feed them? In this paper, these questions will be addressed by presenting a series of relations that exist between the space-time behavior of shallow magmatic storage and transport regimes, pressure-dependent physical properties of the mafic and ultramafic rocks that surround them, and the magma contained within.

As active Hawaiian volcanoes grow upward, their summit magma reservoirs rise progressively. Such a pattern was suggested by JACKSON (1968) and HILL (1969) to explain the roughly cylindrical core of ultramafic cumulates inferred on the basis of high compressional wave velocities within active and extinct Hawaiian volcanoes. Gravity surveys (KINOSHITA, 1965; KINOSHITA and OKAMURA, 1965; Hawaii Institute of Geophysics, 1965) have demonstrated that this pattern of high density volcanic cores exists throughout the Hawaiian archipelago, and must, therefore, represent a fundamental process in the construction of Hawaiian shields. Recently, DECKER et al. (1983) have geodetically confirmed this trend of upward reservoir growth, by demonstrating that the relative position of Mauna Loa's (elevation: 4,169 meters above mean sea level (AMSL)) summit reservoir is approximately the same as its diminutive younger neighbor, Kilauea (elevation: 1,240 meters AMSL) (Figure 1). That is, for each volcano, the magma reservoir roof lies at ≈ 2 km depth, and the depth inferred for the central region of pressure lies near 3 km respectively, beneath their summit calderas. The reasons for this relationship were not, however, suggested by these authors (DECKER et al., 1983), and although confirmed, it has remained puzzling why such a systematic pattern should occur.

Since the recognition that Icelandic crustal accretion is intimately bound up in the plate margin accretionary process itself (BODVARSSON and WALKER, 1964; PÁLMASON and SAEMUNDSSON, 1974) it has been determined—partly through the study of the dissected Tertiary volcanoes in the eastern fjörds (WALKER, 1959, 1963)—that the emplacement of intrusives progressively stretches the Icelandic crust. Confirmation of this crustal stretching, and its role in the plate accretion process, has been documented through geodetic studies of the current magmatic episode at the Krafla central volcano (BJÖRNSSON et al. 1979; TRYGGVASON 1980, 1984, 1986). Similarly, whereas the geodetic signature of crustal evolution and plate accretion has been firmly established, the mechanical controls have not been made clear.

DEFINITIONS

Contractancy may be defined as the progressive reduction in macroscopic and microscopic pore space as a function of increases in depth and confining pressure. A central element in the definition is the progressive sealing shut—or contraction—of fractures, joints and irregular inter-flow porosity with depth. A synthesis of the structural geology of active rift systems and their *in-situ* compressional wave velocity structure from seismic refraction experiments, suggests that the large-scale vertical fractures form an important component of closable shallow porosity. It may be thought of as an anelastic volume decrease under applied loading, and is distinguished from dilantancy—the anelastic increase in volume under differential stresses. The importance of the mechanism of contractancy lies in its ability to change significantly the *in-situ* density of the uppermost fractured veneer of rock above active magma reservoirs and rift systems, and in so doing, produces horizons of neutral buoyancy and magma stabilization. Within a geotechnical and engineering mechanics context, the term has an established usage fully consistent with the physical mechanism under discussion here: the progressive closure of dilatant cracks under increasing normal stress

components. For example, GOODMAN and DUBOIS (1972) and GOODMAN (1976) discuss the closure history of cracks as functions of normal and shear stress components, present constitutive relations for jointed media, and outline the relative roles of contractancy and dilatancy in fractured media deformation. The term contractancy thus has an established usage, a succinctness that obviates the need for a more lengthy description, and the ability to clearly convey a physical mechanism.

Neutral buoyancy is produced by an equity in the effective, large-scale *in-situ* density of magma and the host countryrock. As such, the net driving force for vertical motion has been removed, and the magmatic structure or region has been stabilized. No restrictions are placed on the geologic context within which neutral buoyancy can be achieved; that is, within magma chambers, multiply-connected magma reservoirs, dikes, sills or structures of other geometries. Implicit in the definition is the concept of an integrated or net equity between the fluid components of an active magma reservoir and the far–field host countryrock. It is emphasized that the concept focuses on the long–term stability of a magmatic region or major structural subdivision of a magmatic system. In addition, it provides a control on the geometry, location and dynamics of dike emplacement that are laterally-directed from high-level reservoirs.

The horizon of neutral buoyancy is that depth interval within which the magma density and the aggregate country rock density are equal. It is expected to have a narrow vertical extent, and wide lateral extent. In three dimensions, it is not a plane, but rather a layer having its own topography, as produced by local heterogenieties in country rock composition and local variations in porosity and structure. The wide lateral extent of the horizon of neutral buoyancy is a primary control on the depth of magma injection: it corresponds to the locus of the parabolic magma fracture front during the dike formation process. It's lateral extent corresponds, therefore, to the maximum lateral reach of magmatic injection within an active rift zone. In this fashion, local and regional horizons of neutral buoyancy may be integrated, to form surfaces of neutral buoyancy with regional, plate or global dimensions. It may be contrasted with the region of neutral buoyancy, which, as used in this paper, has a finite depth extent and a lateral extent that is restricted to the dimensions of subcaldera magma storage.

THE PHENOMENOLOGY OF ROCK CONTRACTANCY

A characteristic feature of the pressure dependence of compressional wave velocities in the mafic and ultramafic rocks of Hawaii and Iceland, is the nonlinear increase in V_p with increasing confining pressure. Figures 2, 3, and 4 illustrate this behavior to 1000 MPa. In general, V_p and its pressure dependence, $\partial V_p/\partial P$, show two types of behavior: a low pressure, non-linear dependence, ranging from 0 to 200 MPa, and a second, generally linear regime that starts at 200 MPa, and extends to higher pressures. Since the work of BIRCH (1960), studies of dry holocrystalline rocks (TODD and SIMMONS, 1972), as well as fluid-saturated porous lithologies (KING, 1966), have attributed this change in wave velocity to the progressive closure of microfractures within the rock matrix. This microfracture closure—or contractancy—produces the early rapid rise in V_p and V_s characteristic of the low pressure environment.

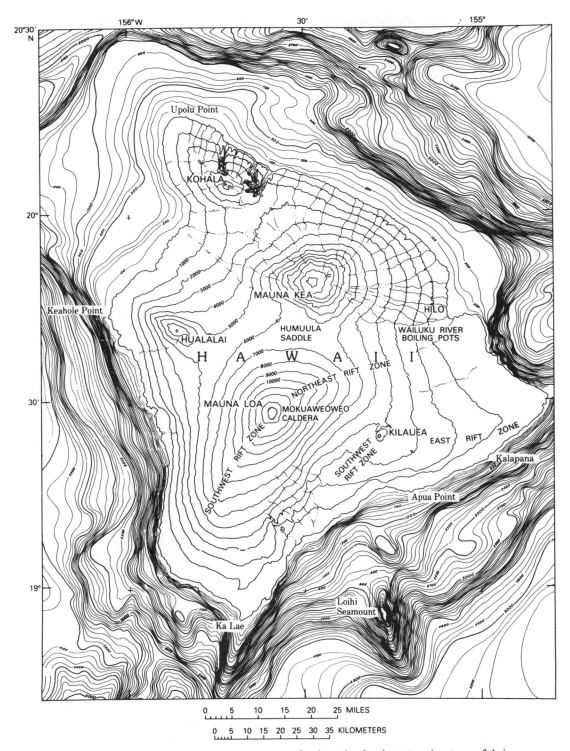

FIG. 1. The island of Hawaii, and three centers of active volcanism in progressive stages of their life cycle: Loihi Seamount, Kilauea and Mauna Loa. The evolutionary progression of their subcaldera magma reservoirs is illustrated in vertical cross-section in Figure 11. Marine bathymetry is based on the compilation of CHASE *et al.* (1980), and contours are in meters. Topographic contours above mean sea level are in feet.

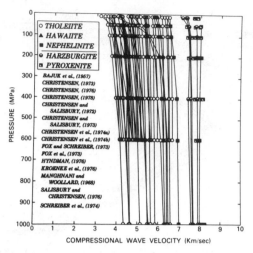

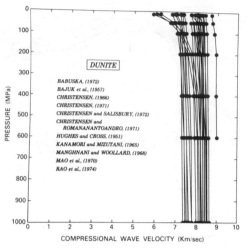

FIG. 2. The pressure dependence of compressional wave velocities (V_p) in the tholeiitic, alkalic and nephelinitic basalt of Hawaiian shield volcanoes, and in the rocks of their ultramafic upper mantle foundations. Particularly noteworthy is the non-linear behavior of V_p at confining pressures below 200 MPa.

FIG. 4. The pressure dependence of compressional wave velocity (V_p) in dunite. For olivine-rich cumulates that may floor shallow storage compartments in subcaldera reservoirs, or plate the walls of deep rift zone conduits, the non-linear sections of the compressional wave curves are expected to apply.

The linear change in wave velocity above 200 MPa, is produced by the compressibility of the minerals within the matrix (BIRCH, 1960). Additional evidence of the important changes in the mechanical and transport properties of crystalline igneous rocks as confining pressures change in the near-surface environment are (see Figure 5): (1) order of magnitude increases in fluid permeability over the range 200 to 0.1 MPa, (BRACE, 1972); (2) increases in compressibility over 800 to 0.1 MPa (BRACE, 1972); (3)

order of magnitude increases in d.c. electrical conductivity for fluid-saturated rocks over the range 300 to 0.1 MPa; (BRACE and ORANGE, 1968a,b); and (4) corresponding decreases in aggregate thermal conductivity over the range 100 MPa to 0.1 MPa (BRACE, 1972). These confining pressure ranges have depth equivalents that extend into the magma storage regions of the active volcanoes in Hawaii, in Iceland and in the Earth's mid–ocean rift systems. Thus, these mechanical and transport properties will all exhibit the characteristic non-linear behavior induced by the low confining pressures associated with the Earth's surface, in countryrock that surrounds subcaldera and central reservoirs and their radiating rift systems. In Figure 5, those properties that characterize a mechanical response to applied loading (V_p, V_s, β, K and μ) are referred to as aggregate mechanical properties, whereas the transport of heat ($\bar{K}$) and pore fluids (K_f) are referred to as aggregate transport properties.

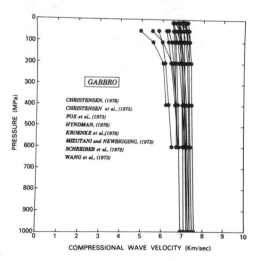

FIG. 3. The pressure dependence of compressional wave velocity in gabbro. Such a confining pressure dependence is expected to characterize the intact gabbroic masses in rift zone and central region cores. Changes in the pressure derivative of V_p ($\partial V_p/\partial P$) above and below 200 MPa, are suggestive of the progressive closure of microcracks within the upper levels of the volcanic shield.

CONTRACTANCY WITHIN VOLCANIC CENTERS

The superstructures of active Hawaiian and Icelandic volcanoes contain evidence of dilatant fractures—void spaces that are inferred to extend to depths approaching 2 km. This evidence includes the following:

(1) gaping cracks in rift systems and summit calderas, which can have crack opening displacements from 10's of centimeters to meters (*e.g.*, the "Great Crack" of Kilauea's southwest rift zone and the dilatant fracture network of the Námafjall-Krafla–Gjástykki rift system);

(2) exceptionally low compressional wave velocities within the upper two kilometers (≈ 2.5 km/

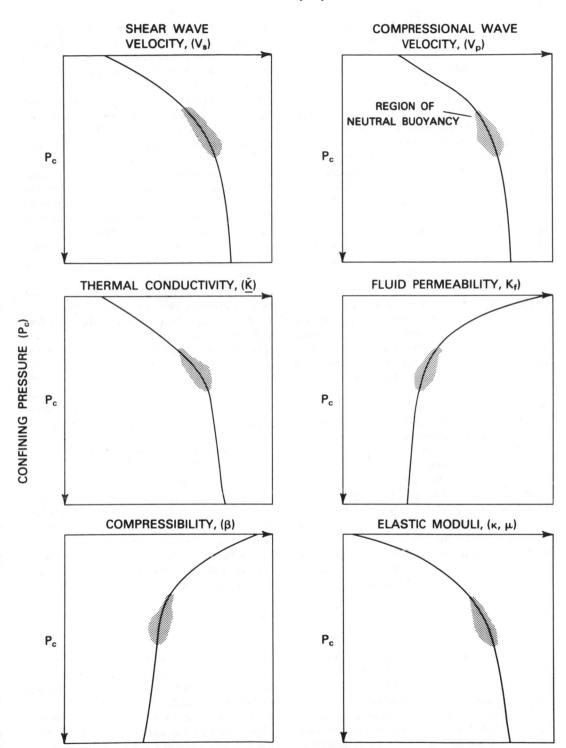

FIG. 5. The synoptic pressure dependence of the mechanical (V_s, V_p, β, K, and μ) and transport ($\bar{K}$ and K_f) properties of mafic and ultramafic rocks under the low to moderate confining pressures appropriate to shallow magma storage in oceanic shield volcanoes, central volcanoes and their associated rift zones, and in mid–ocean spreading centers. These properties have been schematically summarized, and are based on the studies of BRACE (1971a, 1972), BRACE and ORANGE (1968a,b), the combined compilations of Figures 2, 3 and 4, as well as WALSH (1965, 1981) and WALSH and DECKER (1966).

sec for Hawaii (ZUCCA *et al.,* 1982) and ≈2.0–2.5 km/sec for Iceland (FLOVENZ, 1980; FLOVENZ *et al.,* 1985).

(3) exceptionally low *in-situ* densities within the upper two kilometers; (ZUCCA *et al.,* 1982; PÁL-MASON, 1963, 1971).

(4) the common observation of massive lava drainback at the conclusion of an eruptive phase, as well as the observations of concurrent eruption and drainback, as exemplified by the 1983 activity at Pu'u O'o, in Kilauea's east rift zone, and completely analogous activity in the Gjástykki rift system north of the Krafla caldera;

(5) the long-observed discrepancy between the vertical and horizontal displacement fits to geodetic model predictions, suggesting considerable horizontal inelastic behavior at shallow depths (see RYAN *et al.,* 1983, pp. 4150–4152).

The depth intervals inferred for Kilauea's summit magma reservoir have been compiled in Figure 6a. These constraints are based on (a) three–dimensional deformation modeling (RYAN *et al.,* 1983); (B) three-dimensional studies of the seismicity associated with the hydraulic inflation of the summit reservoir (RYAN *et al.,* 1981); (c) simply-connected models of the reservoir as sources of pressure or displacement (*e.g.,* MOGI, 1958); and (d) inversions of summit tilt, trilateration and displacement data (*e.g.,* DVORAK *et al.,* 1983). The total depth range is ≈2 to 9 km, including the reservoir "roots." Uncertainties in the geodetic and seismic resolution of depth intervals typically span the range 500 m to 1000 m. Because 15 of the 16 studies of these depth ranges are included within the 2 to 4 km interval, this is the most frequently retrieved interval for subcaldera magma storage. On purely geometric grounds, however, the interval 2.2 to 3.5 km, for example, also encompasses this same 15-study consensus. As such, it explicitly acknowledges the uncertainties in both the seismic and the geodetic resolution. The aseismic region beneath Kilauea caldera extends to a depth of 6 to 7 km, thus suggesting that the geodetic studies tend to 'see' the upper levels of the reservoir. The seismic surveys must therefore be taken into greater account when evaluating the total depth range available for storage.

Within the Hawaiian and Icelandic crust, we expect to find the fracture porosity saturated with water. It is relevant then to consider the contractancy of basalt with both dry and H_2O–saturated porosity and its pressure dependence. This dependence is illustrated in Figure 6b for basalt (CHRISTENSEN and SALISBURY, 1975). The characteristic $\partial V_p/\partial p$—non-linear for pressures less than 200 MPa—is well developed, especially for the "dry" curve where the fluid-accessible fracture space is occupied by air. Both the dry and H_2O–saturated curves show an additional effect: there is a pressure interval above which (in the direction of increasing pressure) the pore space accessible to the mobile second phase (air or H_2O) has been squeezed out. For basalt, gabbro, and dunite, this pressure window comprises a transition region centered at 200–300 MPa. This region of transition coincides with the convergence of the fluid-saturated and dry branches, and is herein referred to as the contractancy transition zone. The zone separates the figure into two subregions: a shallow region where the dominant mechanism is the compression of fractures and pore fluids, and a deeper region dominated by the compression of the crystal structures in the component mineral phases, as well as intercrystalline glass.

A comparison of Figures 6a and 6b makes clear the 1:1 correlation between the depth range associated with magmatic storage, and the equivalent confining pressure-depth level required for the elimination of fluid-accessible fracture porosity in basalt. This transition, from the compression of pore fluids and fracture space above 9 km, to the compression of minerals below 10 km, is illustrated by the schematic volume elements in Figure 6b, and denoted by a stippled region in both figures.

CONTRACTANCY AND SUBCALDERA MAGMA STORAGE

If the bulk compression of the volcanic system is dominated—at shallow levels—by the progressive elimination of macrofracture, joint, microfracture and vesicle porosity due to increasing confining pressure, then we should expect to see correlation between the *in-situ* density of the volcanic superstructure and increases in confining pressure. Moreover, the systematics of the mechanical and transport properties with depth, as exemplified by changes in the compressional and shear wave velocities ($\partial V_p/\partial Z$, $\partial V_s/\partial Z$), seismic attenuation ($\partial Q/\partial Z$), bulk compressibility ($\partial \beta/\partial Z$), bulk thermal conductivity ($\partial \bar{K}/\partial Z$), *in-situ* fluid permeability (gas, water, magma) ($\partial K_f/\partial Z$), and electrical resistivity ($\partial \Omega/\partial Z$), as well as the predictions of volumetric strain for jointed media (RYAN, 1985; 1987), demonstrate that this dependence will be non-linear.

Studies of the three–dimensional distribution of seismicity at Kilauea provide an additional means of evaluating the depth distribution of magma associated with subcaldera storage (RYAN *et al.,* 1981). These results may be summarized in vertical profiles taken beneath the caldera as well as beneath Halemaumau (RYAN *et al.,* 1983), (Figure 7a). From

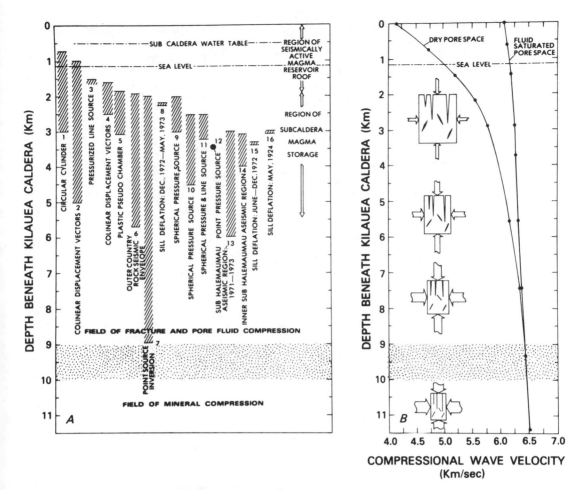

FIG. 6a, b. (A) The depth range of subcaldera magma storage for Kilauea, as inferred from geodetic and seismic surveys. The total range for a given survey is given by the vertical cross-ruled band. The identity of the source model geometry and the theoretical inversion or forward modeling procedure is given beneath each inferred magmatic interval. Respective source references are: (1) DIETERICH and DECKER (1975); (2) DUFFIELD et al. (1982); (3) WALSH and DECKER (1971); (4) MOGI (1958); (5 and 7) JACKSON et al. (1975); (6) FISKE and KINOSHITA (1969); (8) DAVIS et al. (1974); (9) RYAN et al. (1981); (10) DVORAK et al. (1983); (11, 14, 15 and 16) RYAN et al. (1983); (12) EATON (1962); and (13) KOYANAGI et al. (1976). (B) The compressional wave velocity for dry and seawater saturated basalt as a function of depth-equivalent confining pressure. The interval 0–9 km corresponds to the confining pressure required for the closure of fluid accessible pore space in the absence of an applied pore pressure. Block diagrams schematically illustrate the progressive closure of the macroscopic and microscopic fracture networks as confining pressures are progressively increased. The convergence of the dry and H_2O–saturated V_p branches at a 9–10 depth equivalent, separates the upper field of fracture and pore fluid compression from the lower field of mineral compression. The stippled band is the contractancy transition zone. V_p data from CHRISTENSEN (1974) and CHRISTENSEN and SALISBURY (1975). Figure modified from RYAN (1987).

0 to 2 km depth, intense seismicity is produced by the flexure of the storage complex roof; from 2 to 7 km, an increasingly aseismic region attests to relatively high fluid-to-rock ratios, with the interval 4 to 7 km being very aseismic; at depths in excess of 7 km, the intensity of hydraulic fracturing rises, suggesting the active vertical transport of magma within the dikes that make up the primary conduit.

In Figure 7b, the *in-situ* densities of Kilauea and Mauna Loa have been compiled from the gravity inversion studies of ZUCCA et al. (1982). These correspond to both summit and flank profiles for each volcanic shield. For both volcanoes, the density profiles increase in a non-linear manner with depth from 0 to 7 km. This band generally corresponds with the non-linear portion of the experimentally-

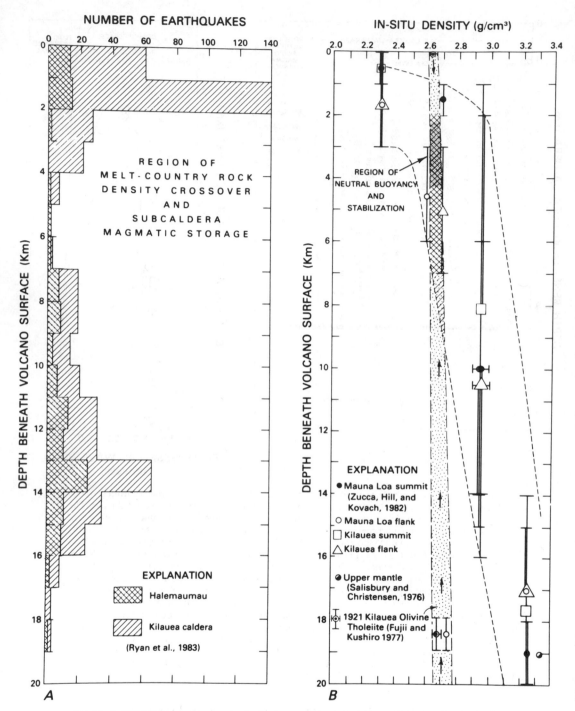

FIG. 7a, b. (A) The representative depth distribution of the rate of earthquake occurrence for two prismatic volumes of Kilauea volcano. Both volumes extend to 20 km beneath the caldera floor; however, one volume is laterally extensive, and contains the entire caldera (single cross-ruled pattern). The second volume is focused on Halemaumau crater within the caldera, and it's substructure (double cross-ruling). In both cases, an aseismic region extends from 4 to 7 km, suggesting differentially high magma to countryrock ratios within the summit storage reservoir. For the immediate substructure of Halemaumau, this region extends from 2 to 7 km. The interval 2 to 4 km is the most frequently retrieved storage depth from geodetic studies, as illustrated in Figure 6a. (B) *In-situ* density values and their depth ranges as derived from the gravity inversion study of the summit and flanks of Kilauea based contractancy curves for basalt and gabbro (Figures 2, 3). In addition, the band corresponds closely with the 0–9 km depth range necessary to achieve a convergence in the air-saturated and H_2O–saturated branches of the V_p vs. confining pressure ($P_{conf.}$) profile in Figure 6b.

Moreover, there is additional congruence between the non-linear variation in contractancy-related material properties associated with the Earth's surface: the systematics of the *in-situ* change in compressional wave velocity for Hawaii based on the seismic refraction surveys of HILL (1969); and the compilations of WARD and GREGERSEN (1973), CROSSON and KOYANAGI (1979), and KLEIN (1981). These velocity profiles are assembled in RYAN (1987) and are compared and contrasted with the $\partial V_p/\partial Z$ for basalt, gabbro, peridotite, and dunite, which represent the lithologies of the volcanic shields, oceanic crust, upper mantle and refractory path regions of the tholeiitic suite of magmas. These results suggest that several physical states and derivative properties—*in-situ* densities, *in-situ* compressional wave velocities, laboratory-derived V_p and V_s, and field studies of gravity are all responding to the non-linear contractancy associated with the Earth's surface; that is, the progressive elimination of void space on all scales with depth. Figure 8, based on the seismic refraction survey of ZUCCA *et al.* (1982), summarizes the two-dimensional velocity structure for the traverse from Mauna Loa to western offshore Hawaii. Attention is drawn to the very low V_p values of the highly porous surficial veneer and to the velocity values suggestive of gabbro and gabbro + dunite mixtures within the core of Mauna Loa. This last region is labeled 'intrusive complex' in the figure.

NEUTRAL BUOYANCY AND MAGMA STABILIZATION

The density of the 1921 Kilauea olivine tholeiite has been measured over the pressure range 0.1 MPa to 1500 MPa, at temperatures of 1250 to 1400°C

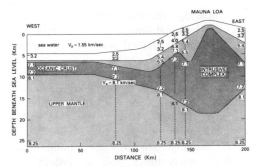

FIG. 8. Seismic velocity structure of the island of Hawaii along a profile linking the Mauna Loa substructure with west and offshore Hawaii. Compressional wave velocities (in km/sec) along six vertical profiles, as inferred from refraction survey travel time data, are the numbers linked by dashed lines. Attention is drawn to the high velocity core region of Mauna Loa's intrusive complex, and to the very low velocities deduced from the surficial veneer of fracture basalt and hyaloclastite. Modified from ZUCCA *et al.* (1982).

by FUJII and KUSHIRO (1977). These results have been incorporated in Figure 7b. Attention is drawn to the cross-over in melt and countryrock density—a cross-over produced by the low *in-situ* density of the heavily fractured veneer that makes up the upper 2 to 3 km of the volcanoe's superstructure. *Within the core of the cross-over region lies a subregion of density equilibration between the melt and countryrock. This is referred to as the region of neutral buoyancy and magma stabilization.*

Magma arriving in Kilauea's summit reservoir has been estimated to contain about 0.32 weight percent H_2O (GREENLAND *et al.*, 1985), in broad accord with MOORE's (1965) estimate of 0.45 ± 0.15 weight percent H_2O for deep sea Hawaiian basalts. KUSHIRO (personal communication, 1986) has measured the density change produced by the addition of 1.08 weight percent H_2O to a basaltic melt, and finds a decrease of 0.05 g/cm^3 compared with the anhydrous liquid. This represents an upper limit for the expected effect of dissolved H_2O. For all

and Mauna Loa volcanoes. The vertical extent of the bars delimit the depth range for *in-situ* densities, whereas the central circles, squares and triangles correspond to vertical profiles beneath Mauna Loa's or Kilauea's summit or flank. The combined results suggest a non-linear *in-situ* density vs. depth profile. Superposed on the countryrock density profile is the pressure (depth) dependence of density for the 1921 Kilauea olivine tholeiite (stippled band), from the falling sphere densitometry experiments of FUJII and KUSHIRO (1977). An *in-situ* cross-over region occurs in the 1 to 7 km depth interval. Within the heart of this interval is a subregion of inferred neutral buoyancy and magma stabilization, that has a 1:1 correspondence with the 2 to 4 km depth interval associated with summit magma storage—for both Kilauea and Mauna Loa volcanoes. Upward-directed arrows in the lower portion of the stippled band suggest magma motion directions in response to positive buoyancy. The downward-directed arrow at top of band suggests the motion direction for negative buoyancy. *In-situ* density values are from the study of ZUCCA *et al.* (1982). Figure modified from RYAN (1987).

conceivable H_2O contents for these tholeiitic melts, therefore, the small density perturbation would be well within the stippled melt density band of Figure 7b.

What is the relationship between the region of neutral buoyancy, and the depth range associated with the summit magma reservoirs of active Hawaiian volcanoes? For Kilauea, this range has been constrained to occupy the region from 2 to 9 km, with the subrange 2 to 4 km representing the consensus of nearly thirty years of geodetic study (MOGI, 1958; EATON, 1962; FISKE and KINOSHITA, 1969; WALSH and DECKER, 1971; DIETERICH and DECKER, 1975; JACKSON et al., 1975; SWANSON et al., 1976; DUFFIELD et al., 1982; RYAN et al., 1983; DVORAK et al., 1983; DAVIS et al., 1974; and DAVIS, 1986). With respect to the in-situ density cross-over region of Figure 7b, the core of neutral buoyancy occupies the 2.5 to 4.5 km depth levels. This suggests a remarkable correspondence between the depths at which magma is in approximate mechanical equilibrium with its surroundings and the region independently demonstrated to comprise Kilauea's subcaldera magma storage region. A further comparison of the overall 2 to 7 km interval (Figure 7) shows additional congruence: there is a 1:1 correlation between the region associated with subcaldera magma storage, and the region of magma-country rock equilibrium inferred from the equality of in-situ melt-rock densities.

THE MECHANICS AND DYNAMICS OF NEUTRAL BUOYANCY

Consider the steady, low velocity influx of melt, rising vertically into a magma reservoir, and having an initial density ρ_0, that is different from the ambient density, ρ_a. Due to the density contrast, the fluid exiting the conduits that supply the base of the magma reservoir is subject to buoyancy forces.

If the environmental density progressively decreases upwards—in either a linear or non-linear pattern—the ambient medium is referred to as 'stably stratified' (FOX, 1970; HIMASEKHAR and JALURIA, 1982; HIRST, 1971). Rising melt batches then undergo a continuous reduction in their buoyancy flux, until it eventually approaches zero. This region is the 'point of neutral buoyancy'. Above this point is the region of negative buoyancy ($\rho_{melt} > \rho_{environment}$), and the flow decelerates markedly, where its velocity falls to zero at the maximum height of rise (JALURIA, 1986). After reaching the peak height, fluid parcels descend, spreading outwards towards a laterally-extensive horizon of neutral buoyancy. Along this horizon, they further mix

with the ambient fluid (RODI, 1982; HUPPERT et al., 1986).

In the discussion above it has been assumed that the melt has a density that interacts with the far-field host rock environment of the volcanic shield or the near-field 'host fluid' environment of the reservoir itself. Melt density, however, will exhibit a temperature and time dependence during fractional crystallization, which will lead to a set of embedded systems: the large-scale system of the reservoir as a whole, as it interacts with the surroundings on a scale of several kilometers, and an embedded, smaller-scale subsystem described by the dynamics of individual melt parcels and the ambient fluid that forms their immediate environment.

With the data of BENDER et al. (1978), MCBIRNEY and AOKI (1968), SHIBATA et al. (1979), and WALKER et al. (1979) from 0.1 MPa experiments and residual glasses of basalts from oceanic islands and mid-ocean ridge basalts (MORB), STOLPER and WALKER (1980) recognized that the fractional crystallization of a picritic melt leads to progressive changes in the density of the remaining liquid. Olivine removal (with or without concurrent clinopyroxene (cpx) crystallization) reduces the density of the residual melt. The extreme range was a reduction of ≈ 0.1 g/cm^3; however, HUPPERT and SPARKS (1980) suggested a decrease of ≈ 0.2 g/cm^3 for the picritic basalt-to-MORB minimum point. Subsequent comparisons of measured 'fractionation densities' and densities calculated by a weighted average mixing procedure (SPARKS and HUPPERT, 1984), suggest that, while numerically small, the effect is real.

At the magma reservoir margins and beneath the rift system axes, the horizon of neutral buoyancy is coincident with the peak in the available magma driving pressure—that pressure required to do the work of rock fracture at the propagating dike formation front. This occurs regardless of the depth of origin of the magma, variations in magma density, or the stratigraphic details of where in the overlying section the final density balance is achieved.

Consider the differences in pressure between a lithostatic and magmatic column as a function of magma density and with provision for a non-linear distribution of countryrock densities in the near-surface. The driving pressure, P_d, at a given height above its source, h, is (e.g., JOHNSON and POLLARD, 1973):

$$P_d = P_m - P_L, \tag{1}$$

$$P_d = P_m - \rho_{CR} G \bar{H}, \tag{2}$$

$$P_d = (P_s - \gamma_m h) - \rho_{CR} G \bar{H}, \qquad (3)$$

where P_m is the magma pressure, P_L is the lithostatic load, γ_m is the unit weight of the magma, and P_s is the lithostatic pressure in the source region. The lithostatic load is

$$P_L = \rho_{CR} G \bar{H}, \qquad (4)$$

where $\bar{H}$ is the depth beneath the upper reference surface, and ρ_{CR} is the countryrock density. Curves of P_d vs. depth (e.g., JOHNSON and POLLARD, 1973, p. 284; FEDOTOV, 1977, p. 677) have a characteristic parabolic profile for conditions of increasing pressure (P_L) with depth if and only if the near surface veneer has an in-situ density lower than the magma. The nose of the driving pressure parabola then lies at the depth at which $\rho_{magma} \cong \rho_{countryrock}$, that is, the neutral buoyancy surface. It is interesting to note that in terms of the mechanical and transport properties of a subcaldera magma reservoir and of active rift systems the peak in magma driving pressure coincides with that depth at which rock moduli and strength-related parameters (e.g., V_p and V_s) begin a marked fall-off (Figures 2–5 and BRACE, 1971b; BYERLEE, 1968), whereas fluid-accessible permeability (k_f) begins a corresponding marked increase. This is the region of neutral buoyancy. It is then no surprise that rupture of the magma reservoir walls should occur over and over again at the same preferred depth, as discussed below.

How does the region of neutral buoyancy control the magma injection-dike formation process, and what is its mechanical signature? Earthquake hypocenters produced by the hydraulic fracturing associated with magma injection and dike formation in rift zones provide a means of answering these questions. Hypocenters have been determined for Kilauea's Southwest and East Rift Zones. By plotting their locations on a depth-time diagram (Figure 9), temporal variations in the top and the keel of a growing dike may be inferred. In addition, the relationship of the region of neutral buoyancy to the point of dike initiation, the centerline axis of dike symmetry, the depth position of the advancing dike snout and the final resting place of the leading magma fracture front may all be determined from hypocenter data.

Figures 9a through 9f present a series of depth-time dike propagation envelopes for both of Kilauea's rift zones. They show the following general characteristics: at the moment of summit reservoir rupture and rift zone intrusion, reservoir breaching occurs by failure of the confining wall rock at the depth of the region of neutral buoyancy. This region is centered at the 3 km depth level, beneath the caldera floor. Rift zone intrusion occurs by the lateral propagation of a magma-filled fracture, along the horizon of neutral buoyancy. Propagation continues until the fluid pressure falls below that required to meet and exceed the in-situ tensile strength of host rock at the advancing magma fracture front.

In detail, three modes of neutral buoyancy-controlled dike formation have been recognized. (1) Slow initial leakage with a gradually enlarging magma fracture front as illustrated by results shown for Kilauea's east rift zone in Figure 9a. Initiated along the horizon of neutral buoyancy, the growing dike has a top that rises progressively towards the surface, while the keel simultaneously descends. The intrusion may end abruptly (Figure 9a), forming a cross-sectional depth-time signature that resembles a tomahawk head, with the blade facing down-rift. (2) Abrupt initial leakage with a rapidly enlarging magma fracture front. Here, the top of the dike grows rapidly towards the surface, while the keel simultaneously (and rapidly) descends (Figure 9b). Subsequent inferred pressure reductions gradually narrow the advancing fracture front, and inferred positive and negative buoyancy forces return parcels of magma below and above the horizon of neutral buoyancy, to positions of mechanical equilibrium. The resting place of the dike snout after growth has been completed is thus along the horizon of neutral buoyancy. The depth-time signature therefore resembles an up-rift facing tomahawk (Figure 9b). (3) Harmonic oscillations of the top and keel of the growing dike (Figures 9c–9f). Here, the intrusion is again initiated within the region of neutral buoyancy, and the top and bottom of the extending dike antithetically rises and falls together as time progresses. The resulting depth-time signature resembles a knife blade with doubly-serrated edges: the valleys and hills on each "edge" of the dike face each other on either side of an axial plane of bilateral symmetry. For Kilauea, this type of harmonic neutral buoyancy-controlled dike growth occurs within both rift zones. Amplitudes of the rising and falling tops and keels vary from ≈ 2 to 3 km (Figures 9a and b) to ≈ 6 km (Figure 9d). Periods of the harmonic fracture process vary from ≈ 6 hrs. (Figure 9c), to 78 hrs. (Figure 9f). In all cases (types (1), (2) and (3)), the horizon of advance corresponds to the depth horizon of neutral buoyancy. Above and below this level, magma rises and falls in vertical expansion and contraction pulses, corresponding to the momentary buildup of inferred pressure differentials within the momentarily stalled dike.

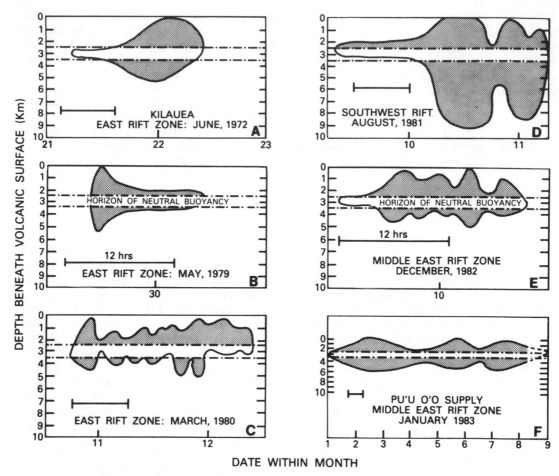

FIGS. 9a–9f. The time variation in the vertical extent of magma during rift zone intrusion at Kilauea volcano, Hawaii. These envelopes, in depth-time plots, enclose the cross-sectional upper and lower limits of the seismicity associated with the magmatic fracture and dike formation process, and have been constructed from the compilation in KLEIN et al. (1987). The horizontal stippled band corresponds to the horizon of neutral buoyancy, where the in-situ $\rho_{melt} \approx \rho_{countryrock}$. In the frames, the relevant time periods are: (9a) June 21–23, 1972; (9b) May 29–30, 1979; (9c) March 10–12, 1980; (9d) August 9–11, 1981; (9e) December 9–10, 1982; and (9f) January 1–8, 1983. The datum for the depth is the Earth's surface above the seismic swarm produced by rising magma pressure. Particularly striking throughout this sequence of frames, is the marked symmetry of the vertical extent of magma about the horizon of neutral buoyancy. In all cases, the horizontal bar represents 12 hours.

Achieving the requisite pressure to exceed the critical stress intensity factor (K_{Ic}) required for dike advance, then reduces this inferred pressure level, extends the dike, and produces a "necking" (narrowing) in the vertical extent of the magma fracture front. The cycle is then repeated.

Figure 10 illustrates the three–dimensional internal structure of Kilauea with respect to primary magma transport routes and the horizon of neutral buoyancy. The model has been constructed by integrating seismic constraints (RYAN et al., 1981; RYAN, 1987) with the continuum mechanics of the magma transport process, combined with the results of geodetic surveys and the structural and eruptive geology of the volcanic shield.

After ascent through the primary conduit, magma stored within the summit reservoir is periodically injected laterally into the rift zones, in dike formation episodes. This may take the form of discrete dikes whose rear is connected to the reservoir structure, or the transfer of magma batches, to still-molten regions of rift zone storage, as occurred during the Pu'u O'o eruptive sequence. Because of the tendency for magma to be both stored and transferred

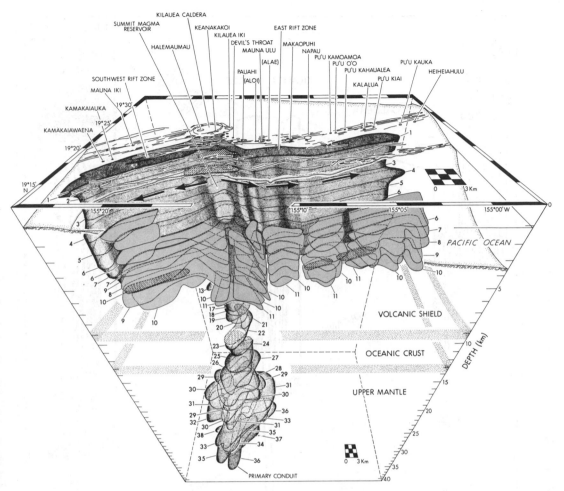

FIG. 10. The three–dimensional internal structure of Kilauea volcano, in a northward-directed view. The medium stippled pattern denotes the substructure of the southwest and east rift zones, that radiate from the summit magma reservoir, as well as the primary conduit, rising from a depth of 40 km into the base of the reservoir. The reservoir appears as a lightly-shaded central region with an overall depth extent of 2 to 7 km, whose upper levels lie behind the tectonic blocks of the Koae Fracture System (cross-hatched pattern) from this perspective. The rift zones lie behind the seismically active tectonic blocks of the south flank (medium grey pattern). The periodic high level injection of magma into the rift zones occurs along the horizon of neutral buoyancy (arrowed pathways) and is associated with the lateral formation of dikes at the ≈3 km local reference level beneath the volcanic surface. Figure modified from RYAN (1987).

within the horizon of neutral buoyancy, magma mixing in active volcanic systems is expected to occur preferentially within this horizon.

NEUTRAL BUOYANCY AND THE EVOLUTION OF HAWAIIAN SHIELD VOLCANOES

The zero confining pressure datum for the contractancy profile of a subaerial volcano is, of course, positioned at the Earth's surface. For submarine volcanoes, this will correspond to sea level, because the weight of the superjacent column of seawater will contribute to the compressive stress exerted across vertically-oriented planes within the seamount, and in the oceanic crust beneath. During the growth of a Hawaiian shield, the dilatant fracture network associated with the Earth's surface (free surface reference level) produces a contraction profile. For subaerial volcanoes, the profile's datum is the caldera floor, or the rift zone crest; that is, that portion of the Earth's surface directly above the intrusive complex of interest. An increase in depth

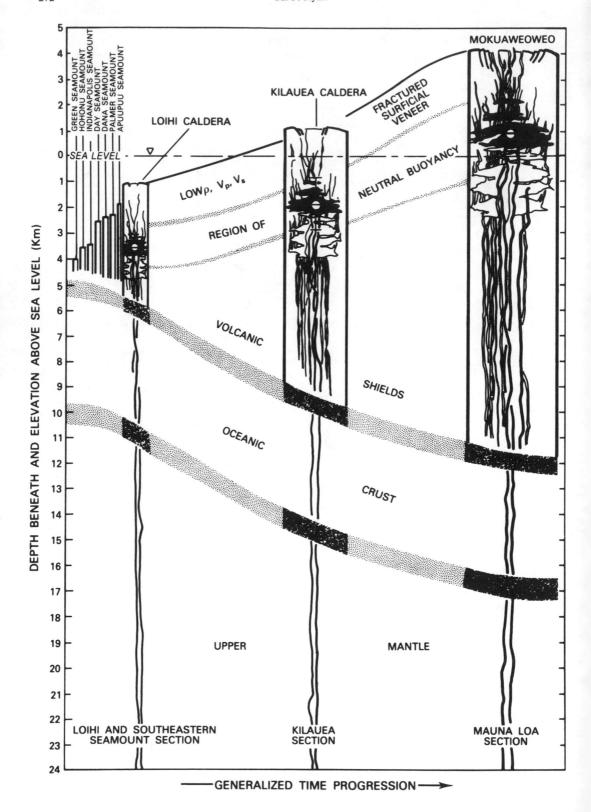

produces an increase in confining pressure—first squeezing shut the dilatant and permeable fracture space, then, beneath the contractancy transition zone—compressing the minerals of the rock matrix. *It is therefore the depth beneath the free-surface reference level that regulates the amount of intrinsic dilatant fracture permeability in shallow portions of volcanic systems. Raising the free surface, therefore, progressively raises the contractancy transition region within the system.* In simpler terms, as a volcano grows upward, it carries its contractancy profile with it.

Inside a given magma storage complex, the region of neutral buoyancy will regulate the positioning of the upper levels of the stable reservoir, whereas the lower basal levels will be ordered by the final squeezing shut of the fracture permeability associated with passage through the contractancy transition zone. *In an almost a literal sense,* Hawaiian magma reservoirs lift themselves by their own "boot straps": *the addition of eruptive products to the top of the system, progressively raises all the critical reference levels within the system—the zero confining pressure datum, the horizon of neutral buoyancy, and the contractancy transition zone. The summit storage reservoir and its associated rift systems must now elevate themselves to achieve mechanical equilibrium within the shield.* Thus, the life cycle of Hawaiian magmatism is controlled by an evolutionary track along which the fluid regions of active storage and differentiation are continuously squeezed upwards. Only a reduction or curtailment of the supply at depth can slow or stop this evolutionary process. Figure 11 illustrates the evolution and upward growth of a Hawaiian volcanic shield, from the initial stage of submarine activity, through subaerial growth to relative inactivity in the waning stages of subaerial development. It represents the vertical progression illustrated topographically in Figure 1. In the terms of Davisian geomorphology, these three evolutionary stages would be submarine infancy, subaerial maturity and subaerial old age, respectively.

NEUTRAL BUOYANCY AND ICELANDIC MAGMATISM

The remarkable correlations between regions of intra-shield magma transport and storage for Hawaii and the relevant physical properties of basic and ultrabasic rocks and melts suggests that we look beyond Hawaii for similar correlations. Encouraged by the essential simplicity of the rift zone intrusion process, we next turn our attention to the north of Iceland and the current eruptive episode at the Krafla central volcano and its associated rift systems (Figure 12).

The depths of subcaldera magma storage at Krafla have been studied through three general approaches: (1) seismic shear wave attenuation patterns (EINARSSON, 1978); (2) inversion modeling of caldera displacement and tilt patterns (BJÖRNSSON et al., 1979; TRYGGVASON, 1981, 1986; MARQUART and JACOBY, 1985); and (3) forward modeling of magma withdrawal from sills combined with a collateral study of exhumed intrusives in the Tertiary central volcano complexes of eastern Iceland and adjunct seismic constraints (RYAN, in preparation). Figure 13 summarizes the depth intervals suggested for Krafla magma beneath the central caldera. The total range is 2.5 to 7.0 km, with the sub-interval 2.5 to 4.0 km representing that range common to all nine studies. As in the Hawaiian context, the geodetic surveys emphasize the upper levels of the storage region. In a companion plot, Figure 14 presents a depth comparison for Krafla's rift systems, where the total interval is 1.0 to 5.75 km, and the subinterval 1.0 to 3.5 km is the most frequently retrieved depth range for dike formation.

The compressional wave velocity profile for the Icelandic crust is a nonlinear but continuous in-

---

FIG. 11. The nature of the evolutionary track during the growth and development of oceanic volcanoes. The volcanic shield-oceanic crust interface is stippled and is isostatically depressed under the volcanic load. As a volcano grows and evolves, it carries its contractancy profile with it. Concomitant changes along this evolutionary track include: (1) a progressive upward migration in the transition zone that separates the fields of fracture and pore fluid compression above, from mineral compression below (not illustrated); (2) a progressive upward migration of the zone of neutral buoyancy (circle with horizontal dash); and (3) a progressive elevation of the fractured surficial veneer, with its characteristic non-linear mechanical and transport properties. This provides the mechanical rationale for the evolutionary climb of subcaldera magma reservoirs in the progression from submarine activity (infancy) (A), to the waning stages of development (subaerial maturity) (C). The crustal structure and the volcanic shield-oceanic crust and oceanic crust-upper mantle transition zones have been constrained by the studies of HILL (1969), ZUCCA and HILL (1980) and ZUCCA et al. (1982). Active magma reservoirs and their dike systems are illustrated in black. Inactive reservoirs and their dikes are white. Only the most recent generation of inactive reservoirs has been shown. Seamounts adjacent to the island of Hawaii have been ranked by their summit depth below sea level.

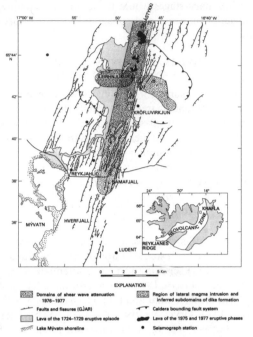

FIG. 12. The juncture of the caldera and the rift systems associated with the Krafla central volcano, northeast Iceland. Magma accumulates beneath the caldera floor in the region of neutral buoyancy prior to the initiation of intrusive and eruptive activity. Intrusion south-southwestward into the Námafjall rift system or north-northeastward into the Gjástykki rift system is guided by the topography of the horizon of neutral buoyancy—inferred to lie at a local depth of 2 to 4 km beneath the volcano's surface. It is also postulated to descend slightly towards the north and south, mimicking at depth the general surface topography of the volcano. Domains of shear wave attenuation are from EINARSSON (1978) and occur beneath the caldera. The region of lateral magma intrusion and inferred dike formation is based on the seismic and geodetic surveys of EINARSSON and BRANDSDOTTIR (1980), TRYGGVASON (1984, 1986) and BJÖRNSSON et al. (1979). (Modified after the structural sketch map of K. Saemundsson, in BJÖRNSSON et al., 1979.)

crease in V_p with depth (FLOVENZ, 1980). The lowest velocities are above the water table, in the near-surface of the neovolcanic zone (≈ 2.0–2.5 km/sec), and reach ≈ 6.5 km/sec at a depth of about 6 km (FLOVENZ et al., 1985). Part of this non-linear increase is ascribed to the progressive closure of fractures, whereas part is ascribed to secondary minerals and the progressive metamorphism of the crust (CHRISTENSEN and WILKENS, 1982). In regions of active volcanism, we should expect to see a constant competition between the hydrothermal alteration and secondary mineralization processes that tend to seal fractures, and the pervasive cracking process associated with magmatic movement and dike em-

placement—creating new fracture networks and dilating preexisting cracks. In this report, the transition from non-linear to linear increases in V_p, occurring at a depth of 6 km within the volcanic zone, is taken to represent the region of transition from fracture and pore fluid compression to deeper mineral compression (FLOVENZ, written communication, 1986). It corresponds closely with the lowest levels of inferred magmatic storage beneath Krafla caldera (Fig. 13), and lies at or beneath the inferred keels of rift zone dikes as illustrated in Figures 14 and 15.

Stimulated by the strong similarities in the depth ranges for magma within the upper levels of the Hawaiian and Icelandic volcanic systems, we now inquire about the existence of regimes of neutral buoyancy in Iceland. A summary of the geologic nature and density of Icelandic crustal layers is given in Table 1. Particularly noteworthy are inferred densities in the range 2.1 to 2.5 g/cm^3, that characterize the uppermost ≈ 1000 m. Lithologically,

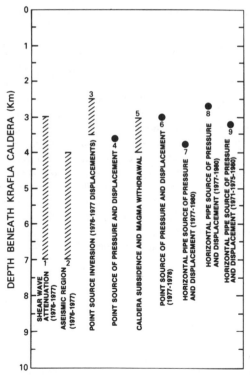

FIG. 13. Depth of subcaldera magma storage at the Krafla central volcano, northeast Iceland. The data and inversion modeling references are (1 and 2) EINARSSON (1978); (3) BJÖRNSSON et al. (1979); (4) TRYGGVASON (1981, 1986); (5) M. P. RYAN, unpublished numerical results (1983); (6) JOHNSEN et al. (1980); (7, 8 and 9) MARQUART and JACOBY (1985). Dots represent discrete inferred depths.

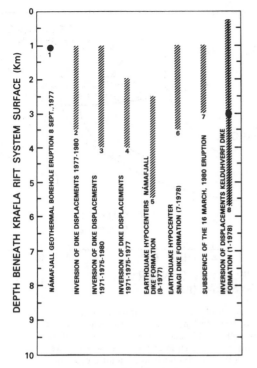

FIG. 14. Depth and inferred depth ranges of rift system dike formation during lateral magma injection pulses at the Krafla central volcano. The data and inversion modeling references are: (1) BJÖRNSSON and SIGURDSSON (1978); LARSEN *et al.* (1979); (2, 3 and 4) MARQUART and JACOBY (1985); (5 and 6) BRANDSDÓTTIR and EINARSSON (1979); (7) TRYGGVASON (1980); and (8) POLLARD *et al.* 1983).

this is a heavily fractured layer of basalts, hyaloclastic breccias, and tuffs. Beneath lie the progressively altered basalts of layer 1, with densities of about 2.60 g/cm³, and an average estimated thickness of ≈ 1000 m. Still deeper are flood basalts that have been locally intruded with gabbroic and granophyric masses. Here the *in-situ* density is ≈ 2.65 g/cm³, and the average thickness is estimated at ≈ 2100 m. This comprises crustal layer 2. Recalling that the experimentally determined density of tholeiitic melt above its liquidus is 2.62 g/cm³ (FUJII and KUSHIRO, 1977), two conclusions may now be drawn: (1) a region of neutral buoyancy exists within the heavily fractured neovolcanic zone, and is constrained to lie at a depth of 2 to 4 km; and (2) the presence of masses of gabbro and granophyre suggest that the upper levels of Icelandic magma reservoirs occur within layer 2; that is, these intrusive complexes record the fossil zone of neutral buoyancy, and their emplacement was regulated by it. WALKER (1974, 1975) also recognized the equity in density between magma and high levels of the

Icelandic crust, and suggested that sheet intrusions may, in part, be controlled by this balance.

A summary of the dynamics and kinematics of the current tectonic episode at the Krafla central volcano illustrates the role of neutral buoyancy within a spreading center context. Following the Mývatn Fires (Mývatnseldar) rifting and eruptive episode of 1724–1729, the Krafla region lay quiet until December, 1975. Eruptive and intrusive activity then resumed in a series of 9 eruptions and 12 intrusions that injected magma both north and south of the central caldera (Figure 12). During these periods of lateral rift zone magma injection, the volcanic surface surrounding Leirhnjúkur—a horst-like hyaloclastite ridge studding the topographic high of Krafla's caldera floor—subsided as much as 230 cm, in response to withdrawn magma volumes of 130×10^6 m³ (JOHNSEN *et al.*, 1980; BJÖRNSSON, 1985). The breaching of the reservoir wall rock at depths of 2 to 4 km periodically released batches of magma during dike-forming intrusive episodes. The seismicity associated with these intrusions, when viewed as a time progression during the course of hydraulic crack propagation in vertical cross-section, allows one to keep track of the evolving geometry of the magmatic fracture front, and, by inference, the evolution of the lateral dike formation process. Figure 15 illustrates such an intrusion sequence.

THE GLOBAL ROLE

Like a staircase to the mid-Atlantic ridge, Iceland's Reykjanes ridge descends to become one element of the planet's largest magmatic system and principal volcanic feature: the active centers of sea floor spreading. It is natural, then, to wonder if the relations we have discovered for Hawaii and Iceland have counterparts within these larger rift zones that form the active constructional plate margins (Figure 16). Firm constraints of the kind obtained at Mauna Loa, Kilauea and Krafla are not available and must await the development of spreading center observatories. As a working hypothesis, however, it is suggested that spreading center magma reservoirs and rift systems have a relative depth extent for shallow magmatic storage comparable to that of Iceland, that they are characterized by a pervasive region of neutral buoyancy beneath the median (axial) valley, and that this neutral buoyancy horizon regulates the manner in which magma is stored at shallow depths and is subsequently injected in oceanic rift systems. If such a hypothesis is correct, the existence of neutral buoyancy must have had a profound role in the mechanics and kine-

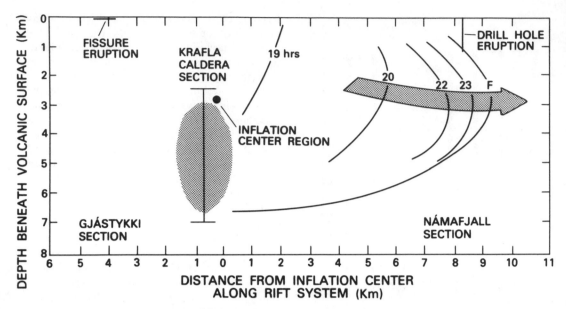

FIG. 15. Cross sectional profiles of the sequential position of the magmatic fracture front during rift zone intrusion from the Krafla central volcano, northeast Iceland. Numbers on the profiles are the times (in hours) following the onset of the intrusive phase. 'F' denotes the final position of the magmatic fracture front. The large shaded arrow delimits the time-depth-distance pathway followed by the parabolic apex of the intrusion as it moved along the inferred horizon of neutral buoyancy. The vertical bar that penetrates the aseismic region (shaded) beneath the Krafla caldera, delimits the maximum depth range of subcaldera magma storage. Parabolic intrusion profiles have been inferred by the progression of seismic hypocenter shifts within the rift zone. The inflation center is in essential coincidence with the inferred point of neutral buoyancy. The intrusion sequence culminated in a geothermal borehole eruption on 8 September 1977. Modified from BRANDSDÓTTIR and EINARSSON (1979).

matics of the plate accretionary process, as well as the manner in which shallow magmatic evolution has occurred as our planet has cooled and differentiated. We now marshall evidence supporting this hypothesis.

Direct observations by submersibles, and indirect photo reconnaissance by towed camera sleds, have revealed that the first order structural features of oceanic rift zone axial valley floors bear a striking resemblance to the eruptive and tectonic structures of subaerial basaltic rift zones. These features include:

(a) axial graben characteristically bounded by normal faulting (e.g., HEKINIAN et al., 1985);

(b) normal fault escarpments arranged in a terrace or staircase fashion away from the axial rift;

(c) shear normal fault escarpments of near vertical orientation that marginally bound flat valley floor blocks (e.g., W. F. B. RYAN, 1985);

(d) dilatant fractures with crack-opening-displacements in the range 0.5 to 15 m (e.g., HEKINIAN et al., 1985; THOMPSON et al., 1985);

(e) lava drainback structures attesting to large dilatant fracture volumes at depth, and drainback "bathtub rings", that record the incremental crack dilatations at depth which episodically lower the level of ponded lava lakes (e.g., BALLARD et al., 1979; RENARD et al., 1985).

(f) hydrothermal venting associated with dilatant fracture networks (e.g., RENARD et al., 1985);

(g) gabbroic dike-like intrusives and dike "walls" exposed in three dimensions and in cross-section (e.g., HEKINIAN et al., 1985; HEY et al., 1985);

(h) thin, sheeted basalt flows (e.g., BALLARD et al., 1979); and

(i) the common development of en echelon fractures throughout the valley floor. These may be eruptive or non-eruptive (e.g., HEKINIAN et al., 1985).

The continued imprint of these surface structures may be seen in cross-sectional seismic velocity profiles taken by refraction surveys across and along oceanic rift systems. Two surveys of the East Pacific Rise illustrate this structure. REID et al. (1977) conducted a seismic refraction survey at 21°N latitude on a central portion of the rise segment bounded

Table 1. Approximate lithologic and density relationships in the Icelandic crust

depth thickness (m)	Layer number	Refs.	Geologic environment	Density ρ (g/cm^3)
0–1000	0	1 2 3 4 5	Relatively fresh and heavily fractured basalt of the surficial layer containing dilatant fractures. Hyaloclastic tuffs and breccias. Thickness: 0–1 km within the active volcanic zones.	2.1–2.5
T: 500–2000 m; average $T \approx 1000$ m	1	1 2 5	The Tertiary and Quaternary basaltic sequence. Progressively altered with increasing depth. On either side of the active volcanic zones. Beneath layer of minor intercalated tuff layers in the neovolcanic zones.	2.60
T: 1000–3000 m aver. $T \approx 2100$ m	2	1 2 5	Flood basalts locally intruded with basic and acidic rock masses.	2.65
$T \approx 4000$–5000 m	3	1 2 5	Basaltic. Equivalent to the oceanic layer. Heavily intruded with gabbroic dikes. Increasingly metamorphosed. Locally elevated beneath central volcanoes.	2.9
	4		Upper Mantle	3.1

[1] PÁLMASON (1963).
[2] PÁLMASON (1971).
[3] BJÖRNSSON et al. (1972).
[4] TÓMASSON and KRISTMANNSDÓTTIR (1972).
[5] PÁLMASON and SAEMUNDSSON (1974).

by the Tamayo Fracture Zone to the north and the Rivera Fracture Zone to the south. The fractured and porous upper levels were characterized by a compressional wave gradient of ≈1.6 km/sec at the sea floor to ≈6.7 km/sec at a depth of 2.5 km. From 2.5 to 4.5 km, the low velocity of 4.0 to 4.7 km/sec suggested the presence of partial melt (a completely molten tholeiitic region is characterized by $V_p \approx 2.6$ km/sec). Beneath 4.5 km, V_p increased again to 6.5 km/sec, and then to 7.8 km/sec by a depth of 10 km. It is noteworthy that the interval 2 to 4.5 km brackets the region of neutral buoyancy inferred for the East Pacific Rise from in-situ density-velocity estimates, with allowance for the superjacent water column load. ORCUTT et al. (1976) conducted a refraction survey along the axis of the East Pacific Rise near the Siqueiros Fracture Zone. Near the rise crest, a low velocity zone was deduced over the depth interval 2.0 to 3.4 km, where V_p ≈ 4.8 km/sec. The roof of the low velocity zone (LVZ) is characterized by a gradient from ≈5 km/sec near the sea floor, to 6.7 km/sec at 2 km. Beneath the LVZ is a gradient from 6.2 km/sec (at 3.4 km), to 7.7 km/sec at a depth of 6 km. As at 21°N, the interval 2 to 3.4 km embraces the estimated position of the horizon of neutral buoyancy. The low velocity zone inferred by ORCUTT et al. (1976) is completely confined to a position within the 10 km-wide central

rise axis. Thus, at two relatively fast-spreading ridge segments, evidence in the form of a low velocity zone centered near 3 km depth suggests the presence of magma, and this is, in turn, consistent with the inferred position of the horizon of neutral buoyancy. For the relatively slow-spreading mid-Atlantic ridge segments, however, seismic refraction surveys have not produced clear evidence for magmatic storage. For example, the 3-component instrument surveys of FOWLER and MATTHEWS (1974) and FOWLER (1976, 1978) at the FAMOUS site (37°N latitude) as well as 45°N latitude, have demonstrated that no special attenuation was produced in shear waves beneath the axial valley.

How does neutral buoyancy work within the context of ridge crest volcanism and the tectonics of plate separation? Rising magma, driven by positive buoyancy, is stabilized at shallow depths (2–4 km) beneath the median valley of an active ridge. With respect to the valley axis along the segment strike, this is expected to be centrally located. The surface expression of the central magma reservoir beneath may include (a) a topographic high, (b) a horst, (c) well-developed hydrothermal venting, or (d) major collapse structures that are sub-circular in plan. Within the core of the reservoir will be the point of neutral buoyancy. During replenishment, rising parcels of melt may intertially overshoot this

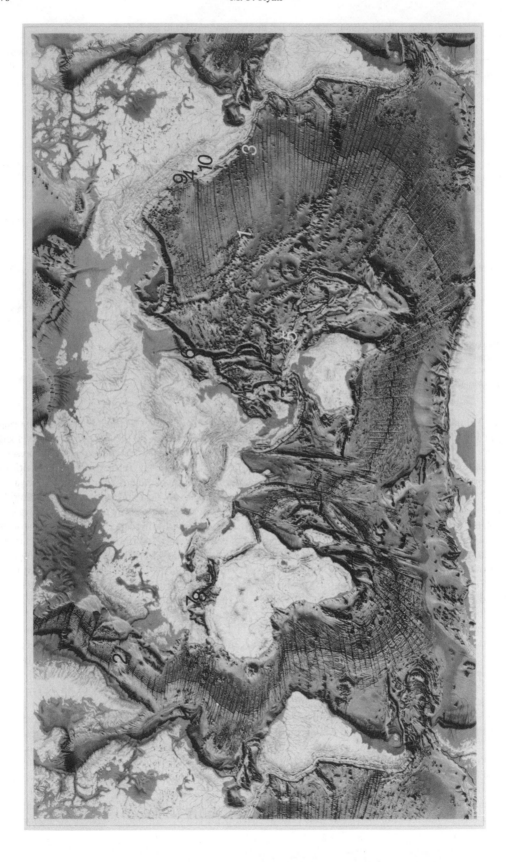

point of neutral buoyancy, and be subsequently driven down by negative buoyancy forces, spreading outward and approaching the horizon of neutral buoyancy. The continued influx of magma raises the fluid pressure level to the *in-situ* tensile strength of the reservoir ceiling and walls. For regions mantling the reservoir, fluid pressures above lithostatic will lead to sill inflation above the complex, and the topographic highs developed on the surface of the mid–ocean ridge—as well as occasional central horsts—may be accentuated by these integrated episodes of sill emplacement.

Rupture of the reservoir walls is expected to occur preferentially along the strike of the median valley axis and at a depth of 2 to 3 km beneath the valley floor. This rupture will be close to or essentially coincident with the intersection of the horizon of neutral buoyancy and the reservoir wall. For the East Pacific Rise, the work of ORCUTT *et al.* (1976), suggests a depth of ≈ 3 km. *During rupture, magmatic injection occurs laterally. That is, the newly forming dike has a distinct upper and lower edge (keel), advances with a parabolic snout, and is connected to the central reservoir only at its rear. The lateral dike-forming injection process is controlled by the horizon of neutral buoyancy.* This is control achieved through a rough, integrated balance of positive and negative buoyancy forces as the injection event proceeds: magma at the top of the dike is driven down by the low density heavily fractured countryrock above, whereas magma along the keel of the dike is prevented from descending (or may be driven up) by the relatively high *in-situ* densities associated with the gabbroic and ultramafic lithologies beneath. The result is one of lateral dike emplacement with a median height coincident with a subhorizontal plane of neutral buoyancy (RYAN, 1987). Dike widths will be controlled by the elastic strain energy stored in the host rock at the moment of intrusion, a quantity proportional to the magnitude of the minimum compressive stress (σ_3) and the host rock elastic moduli (K, μ, E, ν). Injection will continue until overpressures within the magma chamber are reduced.

A point to be emphasized is the inherent efficiency in the lateral dike formation process—a direct result of the neutral buoyancy principle. Because of the negative and positive buoyancy forces along the top and the bottom of the dike, respectively, the lateral injection process is self–regulating. An approximately constant dike thickness means that for a given volume of magma released from the reservoir, dike height is conserved, because both the top and the bottom of the intrusion attempt to reach the horizon of neutral buoyancy. The result is the attainment of remarkable dike lengths over short time periods. In this manner, a centrally located magma reservoir can activate very long segments in the rise of a mid–ocean rift system. Hence the role of neutral buoyancy in the genesis and tectonics of oceanic crust plays a dual role: a mechanical rationale for the geometry and dynamics of dike formation within spreading centers, and a means of maximizing the efficiency of the lateral injection process along the rift zone axis—providing the hydraulic force required for lithospheric plate separation beneath the median axial valley. This conclusion is broadly consistent with the three–dimensional framework for oceanic ridge segments suggested by WHITEHEAD *et al.* (1984), and SCHOUTEN *et al.* (1985). It goes beyond it, however, in providing a mechanical rationale for the stability and dynamics of the mid-segment magma reservoirs and their associated rift zone intrusive complexes.

Figure 17 illustrates the role of the horizon of neutral buoyancy within the context of an oceanic spreading center. Magma produced during the ascent of an upper mantle diapir, collects within a central region of neutral buoyancy. Lateral injection along the horizon of neutral buoyancy occurs periodically, dissipating the fluid pressure within the reservoir and leading to dike formation episodes. This produces the endogenous growth and stretching of the newly-formed oceanic crust, and the development of axial graben.

THE MAGMATIC ENVIRONMENTS OF SUBDUCTION AND THE CONTINENTAL INTERIOR

As a series of concluding observations, the depth extents of subcaldera magma storage have been compiled from an extended range of volcanic cen-

FIG. 16. Generalized map of the world's ocean floor, showing the principal features of the active oceanic ridges, plate margins, and loci of oceanic rift zones and dike intrusion. Regions whose evolution is influenced by the neutral buoyancy principle are: (1) Hawaii; (2) Iceland; (3) East Pacific Rise; (4) Crater Lake; and (5) Rabaul Caldera, Papua, New Guinea. Additional centers hypothesized to have high-level magmatic storage and/or intrusion influenced by neutral buoyancy are the Earth's oceanic rift systems; (6) Usu Volcano, Hokkaido, Japan; (7) Phlegraean Fields, Italy; (8) Mt. Etna, Sicily; (9) Mt. St. Helens, Washington; and (10) Long Valley, California. Map modified from the Mercator projection of HEEZEN and THARP (1977).

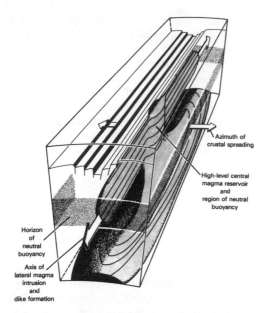

FIG. 17. Prismatic volume element of a spreading center ridge axis, illustrating the role of the horizon of neutral buoyancy (stippled horizon) in the origin and development of oceanic crust. The laterally-injected snouts of the magmatic fracture front, emanate from the central region of neutral buoyancy and move towards the front and rear of the reference volume. Mean sea level corresponds to the upper surface of the volume.

ters where geodetic and seismic monitoring have provided constraints. These include island arc, continental margin and continental interior settings, as well as inter-plate and intra-plate oceanic volcanoes. Figure 18 summarizes this depth distribution, and suggests the interval 1.5 to 5 km as a range that includes the depth region of inferred stored magma in 20 of the 21 plotted ranges. An exception occurs at Sakurajima Volcano, Japan, where an early analysis (MOGI, 1958), suggested a depth of 9–10 km.

Taken collectively, these observations embrace a wide spectrum of magma types, countryrock lithologies, and regional to local tectonic settings. To evaluate critically the proposition that these depth ranges represent the shallow equilibrium position of magma as influenced by neutral buoyancy, we require data on the in-situ density of the relevant magma types, as well as the vertical density profile appropriate to the countryrock of the volcanic edifices and their crustal foundations. Complete data sets of the type desired are not presently available for each of the centers listed in Figure 18; however, initial estimates at select centers may be made.

Modeling of the residual Bouguer gravity profiles for Cascade Range volcanoes has been undertaken

by WILLIAMS and FINN (1986). After corrections for terraine effects, two- and three-dimensional modeling (CADY and SWEENEY, 1980; CORDELL and HENDERSON, 1968) of the Bouguer profiles produced by embedded polygons of variable density, have allowed inferences to be drawn about the sub-volcanic density distributions in cross-section. At Medicine Lake, Crater Lake, Newberry, Mount Hood, Mount Shasta and Mount St. Helens, the inferred densities of the volcanic edifices span the range 2.1–2.3 g/cm^3. For Mount St. Helens, this is compatible with the edifice density of 2.3 g/cm^3 used by JACHENS et al. (1981) in an interpretation of gravity changes during March–May, 1980.

High pressure falling-sphere densitometry experiments for Crater Lake andesite (KUSHIRO, 1978) have indicated densities of 2.57 ± 0.03 g/cm^3 at 1000 MPa and 1175°C, and 2.69 ± 0.03 g/cm^3 at 1250 MPa and 1200°C. For the low confining pressures appropriate to the near surface, the melt density (at 2.86 to 2.98 weight percent H$_2$O) may be expected to be somewhat less than Kushiro's (1978) measurement of 2.47 g/cm^3, evaluated at 500 MPa. Thus, on the basis of these observations, combined with the in-situ aggregate density of the volcanic edifice, Cascade magma may be expected to achieve mechanical equilibrium with its surroundings at shallow levels within and beneath the volcanic centers. Such an inference is then consistent with the concept of a horizon of neutral buoyancy beneath the Cascade system.

Additional evidence comes from Papua, New Guinea, and the recent unrest at Rabaul Caldera. Gravimetric monitoring of the caldera uplift associated with the intra-caldera seismicity of 1983–1984, has been conducted by McKEE et al. (1984), who have observed (p. 406–407):

"These (gravimetric) changes can be fully explained by the measured uplift if a density of about 1.9 g/cm^3 is assumed in the . . . near surface lithology within the caldera. The apparent absence of a gravity anomaly associated with the inferred magma body beneath the entrance to Great Harbour is possible due to lack of density contrast between the intruded magma and its surroundings. This suggests that the combined effects of hydrostatic head and buoyancy have allowed the magma to rise to a level at which its density matches its surroundings. At the estimated depth of the magma body (about 2 km), a density of 2.5 g/cm^3 can be derived . . . for the crustal rocks from their seismic velocity of about 4.5 km/sec. This value of density would be appropriate for a hydrous intermediate magma."

Thus, a synthesis of available data supports the hypothesis that the equilibrium resting place for the upper regions of subcaldera magma storage is coincident with the region of neutral buoyancy for the respective volcanic center.

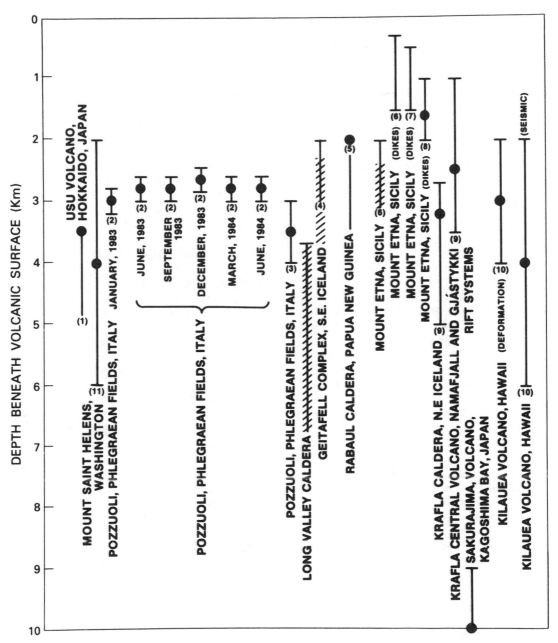

FIG. 18. The depth ranges for shallow magmatic storage in island arc, continental interior, interplate and intraplate oceanic settings. The compositional range spans the interval olivine tholeiite to rhyolite. Data sources for the respective magmatic centers are: (1) Usu volcano, Hokkaido, Japan; YOKOYAMA (1985). (2 and 3) Pozzuoli, Phlegraean fields, Italy, BERRINO et al. (1984), and ORTIZ et al. (1984); (4) Geitafell complex, southeast Iceland, FRIDLEIFSSON (1983); (5) Rabaul Caldera, Papua New Guinea, McKEE et al. (1984); (6, 7 and 8) Mount Etna, Sicily, SANDERSON et al. (1983), SANDERSON (1982), MURRAY and GUEST (1982); (9a) Krafla Caldera, northeast Iceland, this report, Figure 13 and references therein; (9b) Krafla central volcano, Námafjall and Gjástykki rift systems, this report, Figure 14 and references therein; (10a) Kilauea volcano, Hawaii—deformation-based studies, this report, Figure 6a, and references therein; (10b) Kilauea volcano, Hawaii—seismic studies, this report, Figure 6a and references therein; (11) Mount Saint Helens, Washington, ENDO et al. (1981); (12) Sakurajima volcano, Kagoshima Bay, Japan; MOGI (1958). Long Valley, California, ELBRING and RUNDLE (1987), HILL (1976), HILL et al. (1985), KISSLING et al. (1984), MURPHY et al. (1985), RUNDLE et al. (1986), RYALL and RYALL (1981), SANDERS (1983, 1984), and STEEPLES and IYER (1976).

Volcanic eruptions within a neutral buoyancy context require episodes of continued volumetric displacement by new magma batches, high velocity inertial overshooting of the horizon of neutral buoyancy, the onset of differential vesiculation, or shallow and abrupt tectonic kneading. It is likely that many eruptions involve a subtle interplay of each of these factors. In a related consideration, high-momentum alkalic and kimberlitic fluids would be expected to experience significant deceleration in regions where their bulk density matched that of their surroundings.

SUMMARY RELATIONSHIPS

The three principal aspects of shallow magma reservoirs and their rift systems may be understood and integrated within the neutral buoyancy hypothesis.

(1) *Existence.* Beneath the heavily fractured veneer of active basaltic volcanoes, a crossover is produced between the *in-situ* densities of the edifice and tholeiitic melt. In Hawaii, the crossover occurs throughout the 2–7 km depth range, with the upper sections of this interval (≈ 2 to ≈ 4 km) representing a region of inferred neutral buoyancy. The horizon of neutral buoyancy lies at ≈ 3 km local depth. For Kilauea and Mauna Loa, the region of neutral buoyancy has a 1:1 coincidence with the geodetically and seismically determined subcaldera magma reservoirs. Beneath the region, magma ascends under the influence of positive buoyancy forces, whereas above it, magma may descend by negative buoyancy. Within this region, magma is in approximate mechanical equilibrium with its surroundings. Thus, the neutral buoyancy of tholeiitic melt provides for the long–term stability and existence of magma reservoirs and their rift systems.

Intercomparisons between the depths of shallow emplacement and the *in-situ* density of their surroundings, suggests that the equilibrium position for the upper levels of magma in dioritic-andesitic centers and in granitic–rhyolitic complexes is also influenced by the neutral buoyancy principle.

(2) *Structure and function.* A 1:1 correspondence exists between the total depth range inferred for subcaldera magma storage and the depth extent required to eliminate fluid-accessible pore space in the absence of an applied pore pressure. In Hawaii, this characterizes Kilauea and Mauna Loa volcanoes, and in Iceland, it describes the Námafjall-Krafla–Gjástykki central volcano and rift systems.

The contractancy transition zone separates the shallow field of fracture and pore fluid compression from the deeper field of mineral compression, and

corresponds to a depth equivalent of ≈ 200–300 MPa. This zone is coincident with the root section of magma reservoirs and the deepest keels of the active rift zone dike complexes. This general relationship also characterizes Kilauea, Mauna Loa, and Krafla.

For both Hawaiian and Icelandic rift systems, the lateral dike injection process follows the horizon of neutral buoyancy, where $\rho_{countryrock} \approx \rho_{melt}$, with the leading edge of dike advance tracking this horizon. Neutral buoyancy thus stabilizes dikes within their rifts, such that dike keels experience $\rho_{countryrock} - \rho_{melt} > 0$ (positive buoyancy), whereas dike tops are described by $\rho_{countryrock} - \rho_{melt} < 0$ (negative buoyancy). Therefore, the density contrasts associated with the Earth's near-surface produce a mechanism of "self regulation" in the dike injection process. It is Nature's way of maximizing dike lengths (per fixed magma volume) by minimizing dike height. It is this self–regulation that permits these intrusions to achieve such remarkable lengths in both Hawaiian and Icelandic rift systems. In a geometric context, the neutral buoyancy horizon regulates the dike height, the stored elastic strain energy in the surrounding countryrock regulates dike widths, and the available volume of stored magma within the reservoir regulates the total length of the newly formed dike. Rock contractancy produces the horizon of neutral buoyancy that orders this self–regulation process, and controls the details of the three–dimensional topography of neutral buoyancy regions at depth. The core of the region of neutral buoyancy is expected to be the favored locus of reservoir magma mixing, as well as the preferred depth for leakage into laterally-extensive rift systems.

(3) *Evolution.* As Hawaiian volcanoes increase in elevation and mature, their subcaldera reservoirs and acitve rifts undergo a progressive and systematic increase in elevation. This describes the progression from submarine infancy (*e.g.,* Loihi) to subaerial maturity (*e.g.,* Mauna Loa). The mechanical hallmarks of this evolutionary track involve the systematic and simultaneous elevation of the low density fractured veneer, the horizon of neutral buoyancy and the deeper transition region that separates the field of fracture and pore fluid compression (above 9 km local depth) from the field of mineral compression (below 10-km). In the wake of these rising reservoirs lie the older abandoned subsolidus reservoirs whose cores are characterized by high elastic wave velocities.

Within the oceanic crust, the process of lateral magma injection along the axis of rift segments, is expected to follow the horizon of neutral buoyancy.

A corollary of the neutral buoyancy-self regulation principle, is that the magma injection-dike formation contribution to plate separation is maximized along the segment strike—a consequence of the volume efficient magma partitioning process.

Thus constructed, the neutral buoyancy hypothesis brings new unity and order to a wide range of experimental, theoretical and observational data. It integrates the existence, the three–dimensional structure, the large and small-scale material properties, the short-term dynamics and the long–term evolutionary behavior of shallow portions of magmatic systems. While generally accepted separately, they have heretofore remained unrelated and uncorrelated.

Acknowledgments—The Conference convenors and the Geophysical Laboratory of the Carnegie Institution of Washington provided the forum for presenting major portions of this work. Ruthie Robertson prepared the typescript. Reginald Okamura of the Hawaiian Volcano Observatory provided logistical support, as did Axel Björnsson of Orkustofnun (The National Energy Authority of Iceland) and the staff of Kröfluvirkjun. Logistical support, and field conferences with Kristján Saemundson, Hjörtur Tryggvason and Eysteinn Tryggvason were most helpful. Field conferences in eastern and southeastern Iceland with Gudmundur Ómar Fridleifsson, Helgi Torfason and Karl Grönvold were much appreciated. The Nordic Volcanological Institute and Gudmundur Sigvaldasson are thanked for their hospitality. David P. Hill and Richard S. Williams, Jr. reviewed an early version of the manuscript, and suggested several changes for its improvement. Additional review comments by Axel Björnsson, Olafur Flovenz, Claude T. Herzberg, Bjorn O. Mysen, Frank J. Spera and Edward W. Wolfe led to further improvements. Fred Klein and John Rundle kindly provided preprints of recent work. Figures were prepared with the assistance of Lendell Keaton, Marcia Prins and M. Sandra Nelson. Resources available to M. P. Ryan under a Grove Karl Gilbert Fellowship of the Geologic Division of the U.S. Geological Survey provided partial support. Additional support was provided by the U.S. Dept. of Energy, under Interagency Agreement No. DE-AI01-86CE31002.

REFERENCES

BABUSKA V. (1972) Elasticity and anisotropy of dunite and bronzitite. *J. Geophys. Res.* **77**, 6955–6965.

BAJUK E. I., VALAROVICH M. P., KLIMA K., PROS Z. and VANEK J. (1957) Velocity of longitudinal waves in eclogite and ultrabasic rocks under pressures to 4 kilobars. *Stud. in Geophys. and Geod.* **11**, 271.

BALLARD R. D., HOLCOMB R. T. and VAN ANDEL TJ. H. (1979) The Galapagos Rift at 86°W; 3. Sheet flows, collapse pits, and lava lakes of the rift valley. *J. Geophys. Res.* **84**, 5407–5422.

BENDER J. F., HODGES F. N. and BENCE A. E. (1978) Petrogenesis of basalts from the project FAMOUS area: experimental study from 0 to 15 Kbars. *Earth Planet. Sci. Lett.* **41**, 277–302.

BERRINO G., CORRADO G., LUONGO G. and TORO B. (1984) Ground deformation and gravity changes accompanying the 1982 Pozzuoli uplift. *Bull. Volcanol.* **47-2**, 187–200.

BIRCH F. (1960) Velocity of compressional waves in rocks to 10 kilobars, Part I. *J. Geophys. Res.* **65**, 1083–1102.

BJÖRNSSON A. (1985) Dynamics of crustal rifting in NE Iceland. *J. Geophys. Res.* **90**, 10,151–10,162.

BJÖRNSSON A. and SIGURDSSON O. (1978) Hraungos úr borholu i Bjarnarflagi. *Náttúru Fraedingurinn*, 19–23.

BJÖRNSSON S., ANÓRSSON S. and TÓMASSON J. (1972) Economic evaluation of the Reykjanes thermal brine area. *Geothermics*, **2**, Special issue 2, 1640–1650.

BJÖRNSSON A., JOHNSEN G., SIGURDSSON S., THORBERGSSON G. and TRYGGVASON E. (1979) Rifting of the plate boundary in north Iceland 1975–1978. *J. Geophys. Res.* **84**, 3029–3038.

BODVARSSON G. and WALKER G. P. L. (1964) Crustal drift in Iceland. *Geophys. J. Roy. Astron. Soc.* **8**, 285–300.

BRACE W. F. (1971a) Resistivity of saturated crustal rocks to 40 km. based on laboratory measurements. In *The Structure and Properties of the Earth's Crust*, American Geophysical Union Monograph 14, pp. 243–255.

BRACE W. F. (1971b) Micromechanics in rock systems. In *Solid Mechanics and Engineering Design*, (ed. M. TE'ENI), PART I, pp. 1878–2004. John Wiley–Interscience, London.

BRACE W. F. (1972) Pore pressure in geophysics. In *Flow and Fracture of Rocks*, Geophys. Monograph, Vol. 16, pp. 265–274 John Wiley.

BRACE W. F. and ORANGE A. S. (1968a) Electrical resistivity changes in saturated rocks during fracture and frictional sliding. *J. Geophys. Res.* **73**, 1433–1445.

BRACE W. F. and ORANGE A. S. (1968b) Further studies of the effect of pressure on electrical resistivity of rocks. *J. Geophys. Res.* **73**, 5407–5420.

BYERLEE J. D. (1968) The brittle-ductile transition in rocks. *J. Geophys. Res.* **73**, 4741–4750.

BRANDSDÓTTIR B. and EINARSSON P. (1979) Seismic activity associated with the September 1977 deflation of the Krafla central volcano in northeastern Iceland. *J. Volcanol. Geothermal. Res.* **6**, 197–212.

CADY J. W. and SWEENEY R. E. (1980) Program ZHDPOT for 2½ dimensional gravity and magnetic modeling: SEG suppl. to CADY, J. W., Calculation of gravity and magnetic anomalies of finite length right rectangular prisms. *Geophysics* **45**, 1507–1512.

CHASE T., MILLER C. P., SEEKINS B. A., NORMARK W. R., GUTMACHER C. E., WILDE P. and YOUNG J. D. (1980) Topography of the Southern Hawaiian Islands. *U.S. Geol. Survey Open File Map 81-120.*

CHRISTENSEN N. I. (1966) Elasticity of ultrabasic rocks. *J. Geophys. Res.* **71**, 5921–5931.

CHRISTENSEN N. I. (1971) Fabric, seismic anisotropy and tectonic history of the Twin Sisters dunite. *Bull. Geol. Soc. Amer.* **82**, 1681–1694.

CHRISTENSEN N. I. (1973) Compressional and shear wave velocities in basaltic rocks, DSDP leg 16. In *Initial Reports of the Deep Sea Drilling Project*, (eds. T. H. VAN ANDEL, G. R. HEATH and others), Vol. 16, pp. 647 U.S. Government Printing Office.

CHRISTENSEN N. I. (1974) The petrologic nature of the lower oceanic crust and upper mantle. In *Geodynamics of Iceland and the North Atlantic Area*, (ed. KRISTJANSSON), pp. 165–176 D. Reidel Publ. Co. Dordrecht, Netherlands.

CHRISTENSEN N. I. (1976) Seismic velocities, densities and

elastic constants of basalts from DSDP leg 35. In *Initial Reports of the Deep Sea Drilling Project*, (eds. C. D. HOLLISTER, C. CRADDOCK *et al.*), Vol. 35, pp. 335–337 U.S. Government Printing Office.

CHRISTENSEN N. I. (1978) Ophiolites, seismic velocities and oceanic crustal structure. *Tectonophys.* **47**, 131–157.

CHRISTENSEN N. I. and RAMANANANTOANDRO R. (1971) Elastic moduli and anisotropy of dunite to 10 kilobars. *J. Geophys. Res.* **76**, 4003–4010.

CHRISTENSEN N. I. and SALISBURY M. H. (1972) Sea floor spreading, progressive alteration of layer two basalts, and associated changes in seismic velocities. *Earth Planet. Sci. Lett.* **15**, 367–375.

CHRISTENSEN N. I. and SALISBURY M. H. (1973) Velocities, elastic moduli and weathering-age relations for Pacific layer 2 basalts. *Earth Planet. Sci. Lett.* **19**, 461–470.

CHRISTENSEN N. I. and SALISBURY M. H. (1975) Structure and constitution of the lower oceanic crust. *Rev. of Geophysics Space Physics* **13**, 57–86.

CHRISTENSEN N. I. and WILKENS R. H. (1982) Seismic properties, density, and composition of the Icelandic crust near Reydarfjördur. *J. Geophys. Res.* **87**, 6389–6395.

CHRISTENSEN N. I., CARLSON R. L., SALISBURY M. H. and FOUNTAIN D. M. (1974a) Velocities and elastic moduli of volcanic and sedimentary rocks recovered on DSDP leg 25. In *Initial Reports of the Deep Sea Drilling Project,* (eds. E. S. W. SIMPSON, R. SCHLICH *et al.*), Vol. 25, pp. 357–360 U.S. Government Printing Office.

CHRISTENSEN N. I., SALISBURY M. H., FOUNTAIN D. M. and CARLSON R. L. (1974b) Velocities of compressional and shear waves in DSDP leg 27 basalts. In *Initial Reports of the Deep Sea Drilling Project,* (eds. J. J. VEEVERS and J. R. HEIRTZLER), Vol. 27, pp. 445–449 U.S. Government Printing Office.

CHRISTENSEN N. I., CARLSON R. L., SALISBURY M. H. and FOUNTAIN D. M. (1975) Elastic wave velocities in volcanic and plutonic rocks recovered on DSDP leg 31. In *Initial Reports of the Deep Sea Drilling Project,* (eds. D. E. KRAIG, J. C. INGLE *et al.*), Vol. 31, pp. 607–609 U.S. Government Printing Office.

CORDELL L. and HENDERSON R. (1968) Iterative three-dimensional solution of gravity data using a digital computer. *Geophysics* **33**, 596–601.

CROSSON R. S. and KOYANAGI R. Y. (1979) Seismic velocity structure below the island of Hawaii from local earthquake data. *J. Geophys. Res.* **84**, 2331–2342.

DAVIS P. M., HASTIE L. M. and STACEY F. D. (1974) Stresses within an active volcano—with particular reference to Kilauea. *Tectonophys.* **22**, 355–362.

DAVIS P. M. (1986) Surface deformation due to inflation of an arbitrarily oriented triaxial ellipsoidal cavity in an elastic half-space, with reference to Kilauea Volcano, Hawaii. *J. Geophys. Res.* **91**, 7429–7438.

DECKER R. W., KOYANAGI R. Y., DVORAK J. J., LOCKWOOD, J. P., OKAMURA A. T., YAMASHITA K. M. and TANIGAWA W. R. (1983) Seismicity and surface deformation of Mauna Loa volcano, Hawaii. *EOS* **64**, 545–547.

DIETERICH J. H. and DECKER R. W. (1975) Finite element modeling of surface deformation associated with volcanism. *J. Geophys. Res.* **80**, 4094–4102.

DUFFIELD W. A., CHRISTENSEN R. L., KOYANAGI R. Y. and PETERSON D. W. (1982) Storage, migration and

eruption of magma at Kilauea volcano, Hawaii, 1971–1972. *J. Volcanol. Geothermal. Res.* **13**, 273–307.

DVORAK J., OKAMURA A. and DIETERICH J. H. (1983) Analysis of surface deformation data, Kilauea volcano, Hawaii: October 1966 to September 1970. *J. Geophys. Res.* **88**, 9295–9304.

EATON J. P. (1962) Crustal structure and volcanism in Hawaii. In *Crust of the Pacific Basin*, Amer. Geophys. Union Monograph 6, pp. 13–29.

EINARSSON P. (1978) S-wave shadows in the Krafla caldera in NE-Iceland, evidence for a magma chamber in the crust. *Bull. Volcanol.* **41–3**, 1–9.

EINARSSON P. and BRANDSDÓTTIR B. (1980) Seismological evidence for lateral magma intrusion during the July 1978 deflation of the Krafla volcano in NE-Iceland. *J. Geophys.* **47**, 160–165.

ELBRING G. J. and RUNDLE J. B. (1987) Analysis of borehole seismograms from Long Valley, California: implications for caldera structure. *J. Geophys. Res.* (In press)

ENDO E. T., MALONE S. D., NOSON L. L. and WEAVER S. C. (1981) Locations, magnitudes and statistics of the March 20–May 18 earthquake sequence. In The 1980 eruptions of Mount St. Helens, Washington (eds. P. W. LIPMAN and D. R. MULLINEAUX). *U.S. Geol. Surv. Prof. Paper 1250.*

FEDOTOV S. A. (1977) Mechanism of deep-seated magmatic activity below island-arc volcanoes and similar structures. *Int. Geol. Rev.* **19**, 671–680.

FISKE R. S. and KINOSHITA W. T. (1969) Inflation of Kilauea volcano prior to its 1967–1968 eruption. *Science* **165**, 341–349.

FLOVENZ O. G. (1980) Seismic structure of the Icelandic crust above layer three and the relation between body wave velocity and the alteration of the basaltic crust. *J. Geophys.* **47**, 211–220.

FLOVENZ O. G., GEORGSSON L. S. and ARNASON K. (1985) Resistivity structure of the upper crust in Iceland. *J. Geophys. Res.* **90**, 10,136–10,150.

FOX D. G. (1970) Forced plume in a stratified fluid. *J. of Geophys. Res.* **75**, 6818–6835.

FOX P. J. and SCHREIBER E. (1973) Compressional wave velocities in basalt and dolerite samples recovered during leg 15. In *Initial Reports of the Deep Sea Drilling Project,* (eds. N. T. EDGAR, J. B. SAUNDERS *et al.*), Vol. 15, pp. 1013–1016. U.S. Government Printing Office.

FOX P. J., SCHREIBER E. and PETERSON J. J. (1973) The geology of the Oceanic crust: Compressional wave velocities of oceanic rocks. *J. Geophys. Res.* **78**, 5155–5172.

FOWLER C. M. R. (1976) Crustal structure of the Mid-Atlantic ridge crest at 37°N. *Geophys. J. Roy. Astron. Soc.* **47**, 459–591.

FOWLER C. M. R. (1978) The Mid-Atlantic ridge: structure at 45°N. *Geophys. J. Roy. Astron. Soc.* **54**, 167–183.

FOWLER C. M. R. and MATTHEWS D. H. (1974) Seismic refraction experiment on the Mid-Atlantic Ridge in the FAMOUS area. *Nature* **249**, 752–754.

FRIDLEIFSSON G. O. (1983) The geology and the alteration history of the Geitafell central volcano, southeast Iceland. Ph.D. Thesis, Grant Institute of Geology, University of Edinburgh.

FUJII T. and KUSHIRO I. (1977) Density, viscosity, and compressibility of basaltic liquid at high pressures. *Carnegie Inst. Wash. Yearb.* **76**, 419–424.

GOODMAN R. E. (1976) Methods of geological engineering

in discontinuous rocks. 472 p. West Publishing Co. St. Paul.

GOODMAN R. E. and DUBOIS J. (1972) Duplication of dilatancy in analysis of jointed rocks. Jour. of the Soil Mechanics and Foundations Division, *Proceedings of the American Society of Civil Engineers 98,* 399–422.

GREENLAND L. P., ROSE W. I. and STOKES J. B. (1985) An estimate of gas emissions and magmatic gas content from Kilauea volcano. *Geochem. Cosmochim. Acta* **49,** 125–129.

HAWAII INSTITUTE OF GEOPHYSICS (1965) Data from gravity surveys over the Hawaiian archipelago and other Pacific islands. *Hawaii Inst. Geophys. Rept. 65-4.* 10 tables.

HEEZEN B. C. and THARP M. (1977) (Map of) the World's Ocean Floor. Mercator Projection. Scale: 1:23,230,300. United States Navy, Office of Naval Research.

HEKINIAN R., AUZENDE J. M., FRANCHETEAU J., GENTE P., RYAN W. B. F. and KAPPEL E. S. (1985) Offset spreading centers near 12°53′N on the east Pacific rise: submersible observations and composition of the volcanics. *Marine Geophys. Res.* **7,** 359–377.

HEY R., SINTON J. M., ATWATER T. M., CHRISTIE D. M., JOHNSON H. P., KLEINROCK M. C., MACDONALD K. C., MILLER S. P., NEAL C. A., SEARLE R. C., SLEEP N. H. and YONOVER R. N. (1985) Alvin investigation of an oceanic propagating rift system, Galapagos 95.5°W. *EOS* **66,** 1091.

HILL D. P. (1969) Crustal structure of the island of Hawaii from seismic–refraction measurements. *Bull. Seis. Soc. Amer.* **59,** 101–130.

HILL D. P. (1976) Structure of Long Valley caldera, California, from a seismic refraction experiment. *J. Geophys. Res.* **81,** 745–753.

HILL D. P., KISSLING E., LUETGERT J. H. and KRADOLFER V. (1985) Constraints on the upper crustal structure of Long Valley-Mono Craters volcanic complex of eastern California, from seismic refraction measurements. *J. Geophys. Res.* **90,** 11,135–11,150.

HIMASEKHAR K. and JALURIA Y. (1982) Laminar buoyancy-induced axisymmetric free boundary flows in a thermally stratified medium. *Int. J. Heat and Mass Transfer* **25,** 213–221.

HIRST E. (1971) Buoyant jets discharged to quiescent stratified ambients. *J. Geophys. Res.* **76,** 7375–7384.

HUGHES D. S. and CROSS J. H. (1951) Elastic wave velocities in rocks at high pressures and temperatures. *Geophysics* **16,** 577–593.

HUPPERT H. E. and SPARKS R. S. J. (1980) Restrictions on the compositions of mid–ocean ridge basalts: a fluid dynamical investigation. *Nature* **286,** 46–48.

HUPPERT H. E., SPARKS R. S. J., WHITEHEAD J. A. and HALLWORTH M. A. (1986) Replenishment of magma chambers by light inputs. *J. Geophys. Res.* **91,** 6113–6122.

HYNDMAN R. D. (1976) Seismic velocity measurements of basement rocks from DSDP leg 37. In *Initial Reports of the Deep Sea Drilling Project,* (eds. F. AUMENTO and W. G. MELSON), Vol. 37, pp. 373–387 U.S. Government Printing Office.

JACHENS R. C., SPYDELL R. D., PITTS G. S., DZURISIN D. and ROBERTS C. W. (1981) Temporal gravity variations at Mount St. Helens, March–May, 1980. In *The 1980 Eruptions of Mount St. Helens, Washington,* (eds. P. W. LIPMAN, D. R. MULLINEAUX). U.S. Geological Survey Professional Paper 1250, 193–200.

JACKSON D. B., SWANSON, D. A., KOYANAGI R. Y. and WRIGHT T. L. (1975) The August and October 1968 east rift eruptions of Kilauea volcano, Hawaii. *U.S. Geol. Surv. Prof. Paper 890,* 33 p.

JACKSON E. D. (1968) The character of the lower crust and upper mantle beneath the Hawaiian Islands. Prague, *XXIII International Geological Congress* **1,** 135–150.

JALURIA Y. (1986) Hydrodynamics of laminar buoyant jets. In *Encyclopedia of Fluid Mechanics,* Chap. 12, pp. 317–348 Gulf Publ. Co., Houston.

JOHNSON A. M. and POLLARD D. D. (1973) Mechanics of growth of some laccolithic intrusions in the Henry mountains, Utah, I. *Tectonophys.* **18,** 261–309.

JOHNSEN G. V., BJÖRNSSON A. and SIGURDSON S. (1980) Gravity and elevation changes caused by magma movement beneath the Krafla caldera, northeast Iceland. *J. Geophys. Res.* **47,** 132–140.

KANAMORI H. and MIZUTANI (1965) Ultrasonic measurements of elastic constants of rocks under high pressures. *Bull. Earthquake Res. Inst.* **43,** 173.

KING M. S. (1966) Wave velocities in rocks as a function of changes in overburden pressure and pore fluid saturants. *Geophysics* **31,** 50–73.

KINOSHITA W. T. (1965) A gravity survey of the island of Hawaii. *Pac. Sci.* **19,** 339–340.

KINOSHITA W. T. and OKAMURA R. T. (1965) A gravity survey of the island of Maui, Hawaii. *Pac. Sci.* **19,** 341–342.

KISSLING E., ELLSWORTH W. and COCKERHAM R. S. (1984) Three–dimensional structure of the Long Valley caldera, California, region by geotomography. *U.S. Geological Survey Open File Report 84-939,* 188–220.

KLEIN F. W. (1981) A linear gradient crustal model for south Hawaii. *Bull. Seis. Soc. Amer.* **71,** 1503–1510.

KLEIN F. W., KOYANAGI R. Y., NAKATA J. S. and TANIGAWA W. R. (1987) The seismicity of Kilauea's magma system. In *U.S. Geological Survey Professional Paper 1350,* (eds. R. W. DECKER, T. L. WRIGHT and P. H. STAUFFER) (In press).

KOYANAGI R. Y., UNGER J. D., ENDO E. T. and OKAMURA A. T. (1976) Shallow earthquakes associated with inflation episodes at the summit of Kilauea volcano, Hawaii. *Bull. Volcanol.* **39,** 621–631.

KROENKE L. W., MANGHNANI M. H., RAI C. S., FRYER P. and RAMANANANTOANDRO R. (1976) Elastic properties of selected ophiolitic rocks from Papua, New Guinea: nature and composition of oceanic crust and upper mantle. In *The Geophysics of the Pacific Ocean Basin and its Margins,* (eds. G. H. SUTTON, M. H. MANGHNANI and R. MOBERLY), American Geophysical Union Monograph 19, p. 407.

KUSHIRO I. (1978) Density and viscosity of hydrous calc-alkalic andesite magma at high pressures, *Carnegie Inst. Wash. Yearb.* **77,** 675–677.

LARSEN G., GRÖNVOLD K. and THORARINSSON S. (1979) Volcanic eruption through a geothermal borehole at Námafjall, Iceland. *Nature* **278,** 707–710.

MANGHNANI M. H. and WOOLLARD G. P. (1968) Elastic wave velocities in Hawaiian rocks at pressures to 10 kilobars. In *The Crust and Upper Mantle of the Pacific Area* (eds. L. KNOPOFF, C. L. DRAKE and P. J. HART), American Geophysical Union Monograph 12, pp. 501.

MAO N.-H., ITO J., HAYS J. F., DRAKE J. and BIRCH F. (1970) Composition and elastic constants of Hortonolite Dunite. *J. Geophys. Res.* **75,** 4071–4076.

MARQUART G. and JACOBY W. (1985) On the mechanism

of magma injection and plate divergency during the Krafla rifting episode in NE-Iceland. *J. Geophys. Res.* **90,** 10,178–10,192.

MCBIRNEY A. R. and AOKI K. (1968) Petrology of the island of Tahiti. *Geol. Soc. Amer. Mem.* **116,** 523–556.

MCKEE C. O., LOWENSTEIN P. L., DESAINT OURS P., TALAI B., ITIKARI I. and MORI J. J. (1984) Seismic and ground deformation crises at Rabul Caldera: prelude to an eruption? *Bull. Volcanol.* **47-2,** 397–411.

MIZUTANI H. and NEWBIGGING D. F. (1973) Elastic wave velocities of Apollo 14, 15 and 16 rocks. *Proceedings of the Fourth Lunar Science Conference 4-3,* 2601.

MOGI K. (1958) Relations between the eruptions of various volcanoes and the deformations of the ground surfaces around them. *Bull. Earthquake Res. Inst.* **36,** 99–134. University of Tokyo.

MOORE J. G. (1965) Petrology of deep sea basalt near Hawaii. *Amer. J. Sci.* **263,** 40–52.

MURPHY W. J., RENAKER E., ROBERTSON M., MARTIN A. and MALIN P. E. (1985) The 1985 mammoth wide-angle reflection surey. *EOS* **66,** 960.

MURRAY J. B. and GUEST J. E. (1982) Vertical ground deformation on Mount Etna, 1975–1980. *Bull. Geol. Soc. Amer.* **93,** 1160–1175.

ORCUTT J. A., KENNETT B. L. N. and DORMAN L. M. (1976) Structure of the east Pacific rise from an ocean bottom seismometer survey. *Geophys. J. Roy. Astron. Soc.* **45,** 305–320.

ORTIZ R., ARAÑA V., ASTIZ M. and VALENTIN A. (1984) Magnetotelluric survey in the Bradyseismic area of Phlegraean Fields. *Bull. Volcanol.* **47-2,** 239–246.

PÁLMASON G. (1963) Seismic refraction investigation of the basalt lavas in northern and eastern Iceland. *Jökull* **3,** 40–60.

PÁLMASON G. (1971) Crustal structure of Iceland from explosion seismology. *Soc. Sci. Islandica,* **RIT XL.**

PÁLMASON G. and SAEMUNDSSON K. (1974) Iceland in relation to the mid-Atlantic ridge. *Ann. Rev. of Earth and Planetary Sci.* **2.**

POLLARD D. D., DELANEY P. T., DUFFIELD W. A., ENDO E. T. and OKAMURA A. T. (1983) Surface deformation in volcanic rift zones. *Tectonophys.* **94,** 541–584.

RAO M., RAMANA Y. V. and GOGTE B. S. (1974) Dependence of compressional velocity on the mineral chemistry of eclogites. *Earth Planet. Sci. Lett.* **23,** 15–20.

REID I., ORCUTT J. A. and PROTHERO W. A. (1977) Seismic evidence for a narrow zone of partial melting underlying the east Pacific rise at 21°N. *Bull. Geol. Soc. Amer.* **88,** 678–682.

RENARD V., HEKINIAN R., FRANCHETEAU J., BALLARD R. D. and BACKER H. (1985) Submersible observations at the axis of the ultra-fast-spreading east Pacific rise (17°30′ to 21°30′S). *Earth Planet. Sci. Lett.* **75,** 339–353.

RODI W. (ed.) (1982) *Turbulent Buoyant Jets and Plumes.* 184 p. Pergamon Press, Oxford.

RUNDLE J. B., CARRIGAN C. R., HARDEE H. C. and LUTH W. C. (1986) Deep drilling to the magmatic environment in Long Valley, California. *EOS* **67,** No. 21, 490–491.

RYALL F. and RYALL A. (1981) Attenuation of P and S waves in a magma chamber in Long Valley, California. *Geophys. Res. Lett.* **8,** 557.

RYAN M. P. (1985) The contractancy mechanics of magma reservoir and rift system evolution. *EOS* **66,** 854.

RYAN M. P. (1987) The elasticity and contractancy of Hawaiian olivine tholeiite, and its role in the stability and structural evolution of sub-caldera magma reservoirs

and rift systems. In *U.S. Geological Survey Professional Paper 1350.* (eds. R. W. DECKER, T. L. WRIGHT and P. STAUFFER). (In Press).

RYAN M. P., KOYANAGI R. Y. and FISKE R. S. (1981) Modeling the three–dimensional structure of magma transport systems: application to Kilauea volcano, Hawaii. *J. Geophys. Res.* **86,** 7111–7129.

RYAN M. P., BLEVINS J. Y. K., OKAMURA A. T. and KOYANAGI R. Y. (1983) Magma reservoir subsidence mechanics: theoretical summary and application to Kilauea volcano, Hawaii. *J. Geophys. Res.* **88,** 4147–4181.

RYAN W. B. F. (1985) Back-spreading of the mid-ocean ridge axis using a model that incorporates features of shield volcanoes and seamounts. *EOS* **66,** 1091.

SALISBURY M. H. and CHRISTENSEN N. I. (1976) Sonic velocities and densities of basalts from the Nazca Plate, DSDP leg 34. In *Initial Reports of the Deep Sea Drilling Project* (eds. R. S. YEATS, S. R. HART *et al.*), Vol. 34, pp. 543–546 U.S. Government Printing Office.

SANDERS C. O. (1983) Location and configuration of magma bodies beneath Long Valley, California, determined from anamolous earthquake signals (abstr.). *EOS* **64,** (45), 890.

SANDERS C. O. (1984) Location and configuration of magma bodies beneath Long Valley, California, determined from anomalous earthquake signals. *J. Geophys. Res.* **89,** 8287–8302.

SANDERSON T. J. O. (1982) Direct gravimetric detection of magma movements at Mount Etna. *Nature* **297,** 487–490.

SANDERSON T. J. O., BERRINO G., CORVADO G. and GRIMALDI M. (1983) Ground deformation and gravity changes accompanying the March 1981 eruption of Mount Etna. *J. Volcanol. Geotherm. Res.* **16,** 299–315.

SCHOUTEN H., KLITGORD K. D. and WHITEHEAD J. A. (1985) Segmentation of mid–ocean ridges. *Nature* **317,** 225–229.

SCHREIBER E., FOX P. J. and PETERSON J. J. (1972) Compressional wave velocities in selected samples of gabbro, schist, limestone, anhydrogypsum and halite. In *Initial reports of the Deep Sea Drilling Project* (eds. W. B. F. RYAN, K. J. HSU *et al.*), Vol. 13, pp. 595–597 U.S. Government Printing Office.

SCHREIBER E., FOX P. J. and PETERSON J. J. (1974) Compressional wave velocities in samples recovered from DSDP leg 24. In *Initial reports of the Deep Sea Drilling Project,* (eds. R. L. FISHER *et al.*), Vol. 24, pp. 787–790 U.S. Government Printing Office.

SHIBATA T., DELONG S. E. and WALKER D. (1979) Abyssal tholeiites from the Oceanographer fracture zone. I. Petrology and fractionation. *Contrib. Mineral. Petrol.* **70,** 89–102.

SPARKS R. S. J. and HUPPERT H. E. (1984) Density changes during the fractional crystallization of basaltic magmas: fluid dynamic implications. *Contrib. Mineral. Petrol.* **85,** 300–309.

STEEPLES D. W. and IYER H. M. (1976) Low velocity zone under Long Valley as determined from teleseismic events. *J. Geophys. Res.* **81,** 849–860.

STOLPER E. and WALKER D. (1980) Melt density and the average composition of basalt. *Contrib. Mineral. Petrol.* **74,** 7–12.

SWANSON D. A., JACKSON D. B., KOYANAGI R. Y. and WRIGHT T. L. (1976) The February 1969 east rift eruption of Kilauea volcano, Hawaii. *U.S. Geological Survey Professional Paper 891,* 30 pp.

THOMPSON G., BRYAN W. B., BALLARD R., HAMURO K.

and MELSON W. G. (1985) Axial processes along a segment of the east Pacific rise, 10°–12°N. *Nature* **318**, 429–433.

TODD T. and SIMMONS G. (1972) Effect of pore pressure on the velocity of compressional waves in low-porosity rocks. *J. Geophys. Res.* **77**, 3731–3743.

TÓMASSON J. and KRISTMANNSDÓTTIR H. (1972) High temperature alteration minerals and thermal brines, Reykjanes, Iceland. *Contrib. Mineral. Petrol.* **36**, 123–134.

TRYGGVASON E. (1980) Observed ground deformation during the Krafla eruption of March 16, 1980. *Nordic Volcanol. Inst. Rept. 80-05,* 12 pp.

TRYGGVASON E. (1981) Pressure variations and volume of the Krafla magma reservoir. *Nordic Volcanol. Inst. Rept. 81-05,* University of Iceland, 17 pp.

TRYGGVASON E. (1984) Widening of the Krafla fissure swarm during the 1975–1981 volcano tectonic episode. *Bull. Volcanol.* **47-1**, 47–69.

TRYGGVASON E. (1986) Multiple magma reservoirs in a rift zone volcano: Ground deformation and magma transport during the September 1984 eruption of Krafla, Iceland. *J. Volcanol. Geotherm. Res.* **28**, 1–44.

WALKER G. P. L. (1959) Geology of the Reydarfjordur area, eastern Iceland. *Quart. J. Geol. Soc. London* **114**, 367–393.

WALKER G. P. L. (1963) The Breiddalur central volcano, eastern Iceland. *Quart. J. Geol. Soc. London* **119**, 29–63.

WALKER G. P. L. (1974) Eruptive mechanisms in Iceland. In *Geodynamics of Iceland and the North Atlantic Area,* (ed. L. KRISTJANSSON), pp. 189–201, D. Reidel Publ. Co., Dordrecht-Holland.

WALKER G. P. L. (1975) Intrusive sheet swarms and the identity of crustal layer 3 in Iceland. *J. Geol. Soc. London* **131**, 143–161.

WALKER D., SHIBATA T. and DELONG S. E. (1979) Abyssal tholeiites from the Oceanographer fracture zone. II. Phase equilibria and mixing. *Contrib. Mineral. Petrol.* **70**, 111–125.

WALSH J. B. (1965) The effect of cracks on the uniaxial elastic compression of rocks. *J. Geophys. Res.* **70**, 399–411.

WALSH J. B. (1981) Effect of pore pressure and confining pressure on fracture permeability (abstr.). *Int. J. Rock Mech. Min. Sci. Geomech.* **18**, 429–435.

WALSH J. B. and DECKER E. R. (1966) Effect of pressure and saturating fluid on the thermal conductivity of compact rock. *J. Geophysical Res.* **71**, 3053–3061.

WALSH J. B. and DECKER R. W. (1971) Surface deformation associated with volcanism. *J. Geophys. Res.* **76**, 3291–3302.

WANG H., TODD T., RICHTER D. and SIMMONS G. (1973) Elastic properties of plagioclase aggregates and seismic velocities. In *The Moon: Proc. of the Fourth Lunar Sc. Conf. 4-3,* 2663–2671.

WARD P. L. and GREGERSEN S. (1973) Comparison of earthquake locations determined with data from a network of stations and small tripartite arrays on Kilauea volcano, Hawaii. *Bull. Seism. Soc. Amer.* **63**, 679–711.

WHITEHEAD J. A., DICK H. J. B. and SCHOUTEN H. (1984) A mechanism for magmatic accretion under spreading centers. *Nature* **312**, 146–147.

WILLIAMS D. L. and FINN C. (1986) Analysis of gravity data in volcanic terrain and gravity anomalies and subvolcanic intrusions in the Cascade Range, U.S.A., and at other selected volcanoes. In *The Utility of Regional Gravity and Magnetic Anomaly Maps,* (ed. W. T. HINZE), Society of Exploration Geophysicists Publ.

YOKOYAMA I. (1985) Volcanic processes revealed by geophysical observations of the 1977–1982 activity of Usu volcano, Japan. *J. Geodynamics* **3**, 351–367.

ZUCCA J. J. and HILL D. P. (1980) Crustal structure of the southeast flank of Kilauea volcano, Hawaii from seismic refraction measurements. *Bull. Seis. Soc. Amer.* **70**, 1149–1159.

ZUCCA J. J., HILL D. P. and KOVACH R. L. (1982) Crustal structure of Mauna Loa volcano, Hawaii, from seismic refraction and gravity data. *Bull. Seis. Soc. Amer.* **72**, 1535–1550.

Magmatic Processes: Physicochemical Principles
© The Geochemical Society, Special Publication No. 1, 1987
Editor B. O. Mysen

Steady state double–diffusive convection in magma chambers heated from below

STEPHEN CLARK[1,2], FRANK J. SPERA[2] and DAVID A. YUEN[3]

1) Department of Geological and Geophysical Sciences, Princeton University, Princeton, NJ 08544 U.S.A.
2) Department of Geological Sciences, University of California, Santa Barbara, CA 93106 U.S.A.
3) Department of Geology, University of Minnesota, Minneapolis, MN 55455
and Minnesota Supercomputer Institute, University of Minnesota U.S.A.

Abstract—In order to deduce the vertical structure of doubly–diffusive convection cells in magma chambers, solutions to the horizontally–averaged conservation equations governing the distribution of temperature, composition and velocity in a two component (rhyolite–basalt) Newtonian melt have been obtained. Boundary conditions were chosen to model a chamber with a hot, dense (basaltic) base, overlain by cooler more silicic magma, where the influences of sidewall cooling, crystallization and melting are not considered. Parameters of the problem are: the Lewis number, a ratio of thermal to compositional diffusivities ($Le \equiv \kappa/D$); the Prandtl number, a ratio of viscosity to thermal diffusivity ($Pr \equiv \nu/\kappa$); the Rayleigh number, a ratio of thermal buoyancy to viscous forces ($Ra \equiv \alpha g d^3 \Delta T/\kappa\nu$, where α is thermal expansivity, d is the magma chamber depth and ΔT the temperature difference across the chamber); the buoyancy ratio, a ratio of compositional to thermal buoyancy ($R\rho \equiv \alpha_C \Delta C/\alpha\Delta T$, where α_C is compositional expansivity and ΔC the compositional difference across the chamber); the ratio of maximum to minimum magma viscosities (Λ) and the wavenumber of convection (k).

Steady state solutions have been obtained for values of these parameters in the range appropriate to magma chambers and include the effects of a strongly temperature–dependent viscosity. Calculations show that a critical Lewis number (Le_{crit}) separates steady single–cell convection from unsteady convection and conduction. For isoviscous convection at a wavenumber $k = \pi$ and for $Le > Le_{crit} = 6.7$ $Ra^{-0.12}R\rho^{1.67}$, single–layer convection cells characterized by thin chemical and thermal boundary layers and well–mixed interiors develop. Magma chambers lie above this critical Lewis number and, therefore, steady–state model magma chambers exhibit single cell convection. When a temperature dependent viscosity is assumed, the style of convection is not qualitatively different. Steady state values of the heat flux through the chamber roof and the flux of light component downward are given by $q = 0.4k_T\Delta T d^{-1} Ra^{0.23} Le^{0.01}R\rho^{0.01}\Lambda^{0.10}$ and $j = 0.4\kappa\Delta C d^{-1} Ra^{0.23} Le^{-0.65}R\rho^{0.01}\Lambda^{0.05}$, respectively, where k_T is the thermal conductivity. Calculated heat flow values in the range 2000 to 20 000 mW/m^2 compare favorably with measurements in active geothermal areas. Redistribution times for major elements by advective–diffusive transport are in the range 10^5 to 10^6 years. These redistribution rates indicate effective eddy diffusivities 10^5 to 10^6 times larger than Fickian chemical diffusivities. Maximum convection velocities are given by $W = 0.10\kappa d^{-1} Ra^{0.62}$, which implies maximum velocities in the range 1 to 10^2 km/year (far greater than crystal settling rates) for typical chambers. Crystal settling is therefore restricted to the chamber margins where convective velocities are much smaller. A difference in steady–state behavior is observed as the Prandtl number is lowered below twenty. It is suggested that effects of inertia can cause differences in the dynamic behaviour of double–diffusive convection.

For convection cell aspect ratios greater than about 3 (*i.e.,* cell 3 times deeper than wide) two steady state solutions are found. The solution corresponding to a high heat flux is a single layer (whole chamber) convection cell whereas the low heat flow solution consists of two vertically stacked convection cells separated by a thin diffusive interface. Similar layers are found in thermal convection experiments performed at high wavenumbers. Such layering must therefore be due to chamber geometry and not the influence of compositional buoyancy.

INTRODUCTION

NUMEROUS STRATIGRAPHIC STUDIES of individual pyroclastic flow deposits over the past 50 years have revealed the presence of discrete compositional gaps in major, minor and trace elements (*e.g.,* WILLIAMS, 1942; TSUYA, 1955; SMITH, 1960; LIPMAN, 1967; HILDRETH, 1981). For example, a 30 m thick ash flow tuff has been described from Gran Canaria, that consists of a single cooling unit with a rhyolitic base, a basaltic top and a thin (~ 2 m) mixed zone interior (SCHMINCKE, 1969; CRISP, 1984; CRISP and SPERA, 1986). Similarly, at Crater Lake, Oregon, USA, WILLIAMS (1942) described "the eruption of two magma types in rapid succession", a 66–69 weight percent SiO$_2$ dacitic and a 54–57 weight percent basic magma (see also BACON, 1983). In addition to cases where compositional gaps occur in vertical stratigraphic sections, many examples of mixed–pumice eruptions have been described (*e.g.,*

SMITH, 1979; HILDRETH, 1981). In a mixed–pumice pyroclastic flow, two or more compositionally distinct magma types are simultaneously erupted from a single vent.

The mechanism(s) for generation of compositional gaps in ash flow deposits are not clear. Recently it has been shown that significant variations in discharge rate during an eruption can produce compositional gaps in deposits even if the *in situ* (pre–eruptive) chemical gradient is linear (GREER, 1986; SPERA *et al.*, 1986a). That is, a compositional gap could be an artifact of the hydrodynamics of magma withdrawal. On the other hand, it has been suggested that layering may be the product of double–diffusive convection within magma chambers (MCBIRNEY, 1980; TURNER, 1980; RICE, 1981; HUPPERT and SPARKS, 1984). Two distinct models for magma chamber convection are illustrated in Figure 1. Double–diffusive convection exists when two scalars of differing molecular diffusivities, such as temperature and composition, contribute to the buoyancy of the fluid in opposing directions. Two types of double–diffusive convection may be distinguished. The first occurs when the fast diffusing 'component' (*i.e.*, heat) has an unstable distribution and is called the "diffusive regime". Layered convection may develop in this regime (Figure 1). The second regime occurs when the slow diffusing component has an unstable distribution. This regime is called the "finger regime" and is discussed at length by SCHMITT (1979, 1983) and PIASEK and TOOMRE (1980). In a crustal magma chamber, dense hot basic magma commonly underlies cooler silicic magma. Heat diffuses much faster than any chemical component in magmatic liquids, and, therefore, magma chambers are commonly in the "diffusive regime". The finger regime may also be relevant to magma

chamber evolution. For instance, in a closed system chamber (*i.e.*, no replenishment) where crystallization can produce an iron–enriched liquid, a relatively dense iron-rich melt can come to lie above a cold stagnant bottom boundary layer (JAUPART *et al.*, 1984; BRANDEIS and JAUPART, 1986). In this paper, attention is focused on the diffusive regime because it is most relevant to the origin of compositional layering in pyroclastic flows. Exhumed layered intrusions, such as the Stillwater Complex in Montana, USA, present difficulties in interpretation because of the drastic compositional effects imposed by magma crystallization. Pyroclastic flows, on the other hand, are melt dominated often containing only 5 to 10 volume percent phenocrysts.

Previous work on double–diffusive convection (or more generally multiple–component convection) includes analytic, experimental and numerical studies (*e.g.*, VERONIS, 1965; BAINES and GILL, 1969; PROCTOR, 1981; KNOBLOCH and PROCTOR, 1981; DA COSTA *et al.*, 1981; CHEN and TURNER, 1980; TURNER, 1980; NEWELL, 1984; VERONIS, 1968; ELDER, 1969; HUPPERT and MOORE, 1976; GOUGH and TOOMRE, 1982; MOORE *et al.*, 1983; KNOBLOCH *et al.*, 1986). The recent summaries by CHEN and JOHNSON (1984) and especially TURNER (1985) provide comprehensive surveys of multicomponent convection from both historical and modern viewpoints.

The aim of this work is to carry out numerical experiments under a set of conditions applicable to magma chambers with the hope of discovering the nature of double–diffusive convection there. Of particular relevance to volcanologists and geochemists is the question of whether or not layered convection exists in magma reservoirs from which pyroclastic flows are erupted. In the present work, attention is focused on possible steady–state layered chambers. The details of flow development will be discussed later. The nature of multicomponent convection in the steady state is the logical starting point for any future studies aimed at the transient behavior of convection in magma reservoirs. A similar approach was taken by mantle convection workers who studied steady state solutions (TURCOTTE and OXBURGH, 1967) long before time–dependent solutions (MCKENZIE *et al.*, 1974). The present paper is an amplification of the preliminary study by SPERA *et al.* (1986b).

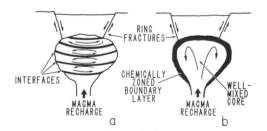

FIG. 1. Schematic depiction of (a) layered convection and (b) single–cell convection in a large volume magma chamber. In (a) distinct well–mixed and nearly isothermal regions are separated by thin diffusive interfaces characterized by steep gradients of composition and temperature. Case (b) represents single–cell convection occupying the whole magma chamber. The main body of the chamber is chemically homogeneous and isothermal. The only chemical and thermal zonation is restricted to thin boundary layers surrounding the chamber.

ANALYSIS

Mean field approximation

The equations that govern the form of the velocity, temperature and compositional fields within a convecting magma body include the conservation of mass, linear momentum, energy and composition. Magma is treated as a binary fluid composed of a silicic (light) component and a complementary mafic (dense) component. The chemical properties of the light component enter into the equations only in terms of its molecular diffusivity (D) and a thermodynamic parameter (α_C) analogous to the isothermal expansivity. The α_{SiO_2} and α_{H_2O} in intermediate composition melts are roughly 0.3 and 2 respectively. Values of α_C are listed in Table 1.

In this study the single–mode mean field method has been utilized (HERRING, 1963). This method is able to exhibit both the basic physics of the convection and also the scaling relations between quantities such as heat flux and vigor of convection. This approach basically entails the *a priori* prescription of the size and form of the convection cells. This allows horizontal averaging of quantities such as velocity, and hence the dimensionality of the problem is reduced to one dimension; depth. In comparison with two–dimensional (2–D) methods the mean field approach is considerably cheaper; because of this it is possible to perform many experiments over a range of difficult conditions. We regard the mean field method as a reconnaissance tool which enables a better interpretation of two–dimensional (2–D) numerical simulations (OLDENBURG *et al.*, 1985, HANSEN and YUEN, 1985, HANSEN, 1986). This consideration is a primary motivation for the present study.

Calculations for the case of constant viscosity convection by QUARENI and YUEN (1984) have found good agreement between mean–field and 2–

D results in both steady–state and time dependent situations. For variable viscosity thermal convection, an extensive study (QUARENI *et al.*, 1985), involving both temperature– and pressure–dependent rheologies, gave further support to the usage of the mean field for the purpose of obtaining scaling relations. For example the power–law exponents relating heat flux and the vigor of convection differ by at most 10% between mean field and 2–D methods. The recent findings of P. OLSON (1986) demonstrate the relationship between mean field, 2–D and boundary layer predictions. The essential features of the dependence of boundary–layer thicknesses upon the strength of convection is captured by both mean field and 2–D methods. More important is the finding that the 2–D results lie between those of the mean field and boundary–layer methods when heat transfer or boundary layer thicknesses are compared. Furthermore the ability of the mean field method to resolve internal interfaces is clearly demonstrated by the prediction of layered thermal convection in narrow slots. Such convection was also found in the experimental study of J. M. OLSON and ROSENBERGER (1979).

When comparing two– and three–dimensional (3–D) studies workers (LIPPS and SOMERVILLE, 1971, and MCKENZIE *et al.*, 1974) have noted discrepancies. These discrepancies result from the fact that in a full 3–D model the convective planform wavelength is dependent on the vigor of convection, as measured by the Rayleigh number. If this dependence is known, and used in 2–D and mean field simulations, good results are obtained; frequently, however, it is not known. In our numerical experiments planform wavelength is fixed *a priori*. An estimate of the error incurred by this procedure can be obtained by performing a set of numerical experiments over a range of wavelengths. It is shown later that the error introduced is *at most* a factor of two in terms of predicted heat fluxes, convection rates and boundary layer thicknesses.

The inherent limitations of the mean–field approximation are overshadowed by our immense ignorance concerning the most basic features of magma chambers. For example geological and geophysical constraints on magma chamber geometries are meager (IYER, 1984), as are data on relevant rates of heat and mass transfer along chamber–country rock contacts. Furthermore, it is an experimental fact that magma is a rheologically complex fluid (SHAW, 1969; SPERA *et al.*, 1982; MURASE *et al.*, 1985) with a constitutive relation that changes as crystals nucleate and grow. In view of this incomplete understanding, we have chosen a canonical set of magma chamber properties. In particular we study large chambers heated from below, where

Table 1. Calculated values for the coefficient of isothermal chemical expansivity*

$$\alpha_C = -\rho^{-1}\left(\frac{\partial\rho}{\partial C}\right)_T$$

	Basaltic Magma	Rhyolitic Magma
SiO_2	+0.41	+0.22
TiO_2	−0.30	−0.43
Al_2O_3	−0.05	−0.08
Fe_2O_3	−0.28	−0.39
FeO	−0.62	−0.68
MnO	−0.47	−0.54
MgO	−0.20	−0.32
CaO	−0.21	−0.33
Na_2O	+0.25	+0.01
K_2O	+0.26	+0.05
H_2O	+3.00	+2.00

* Calculated from BOTTINGA *et al.* (1982). Reference density and temperature for basalt and rhyolite are 1200°C, 2.6738 g/cm^3 and 900°C, 2.2781 g/cm^3 respectively. Values for water derived from BURNHAM and DAVIS (1971).

the influence of sidewall cooling can be neglected. The role of various sidewall boundary conditions has been previously examined (e.g., SPERA et al., 1982; LOWELL, 1985; SPERA et al., 1984; NILSON et al., 1985; NILSON and BAER, 1982).

Conservation equations

The mean field equations are derived from the two–dimensional equations of conservation of momentum, energy and composition. Following VERONIS (1968) the Boussinesq equation of motion is,

$$\left(\frac{\partial}{\partial t}+v\cdot\nabla\right)v=\frac{-1}{\rho}\nabla p+g(\alpha T+\alpha_C C)\hat{\imath}+\frac{1}{\rho}\nabla\cdot\tau, \quad (1)$$

conservation of mass is,

$$\nabla\cdot v=0, \quad (2)$$

conservation of energy is,

$$\left(\frac{\partial}{\partial t}+v\cdot\nabla\right)T=\kappa\nabla^2 T, \quad (3)$$

and the conservation of composition is,

$$\left(\frac{\partial}{\partial t}+v\cdot\nabla\right)C=D\nabla^2 C. \quad (4)$$

In these equations t is time, $v[v(u, v, w)]$ velocity, ρ the mean density, p is pressure, T is temperature, C is the concentration of the light component in the binary magma, κ is the thermal diffusivity, and D is the chemical diffusivity of the light component. $\hat{\imath}$ is the unit vector in the Z (downward vertical) direction. The viscous stress tensor, τ, for a Newtonian fluid is given by:

$$\tau_{ij}=\eta\left(\frac{\partial v_i}{\partial X_j}+\frac{\partial v_j}{\partial X_i}\right), \quad (5)$$

where the viscosity, η, is given by:

$$\eta(\bar{T})=\eta_0\exp(-A\bar{T}), \quad (6)$$

where A is a constant and η_0 is a reference viscosity. $\bar{T}$ is a dimensionless temperature and is defined shortly. The coefficients of thermal expansion and its analogous compositional counterpart are given by:

$$\alpha\equiv\frac{-1}{\rho}\left(\frac{\partial\rho}{\partial T}\right)_{C,p} \quad \text{and} \quad \alpha_C\equiv\frac{-1}{\rho}\left(\frac{\partial\rho}{\partial C}\right)_{T,p}. \quad (7)$$

Equations (1–4) are made dimensionless by choosing $\bar{Z}=Z/d$, $\bar{t}=t\kappa/d^2$, $\bar{p}=pd^2/\rho\nu\kappa$, $\bar{T}=(T-T_0)/\Delta T$ and $\bar{C}=(C-C_1)/\Delta C$ where d is the depth of the chamber, T_0, T_1, C_0 and C_1 the temperature and mass fraction of light component at the top

and bottom of the chamber, respectively and $\Delta T=T_1-T_0$ and $\Delta C=C_0-C_1$.

Following the work of HERRING (1963, 1964), GOUGH et al. (1975) and TOOMRE et al. (1977), the single-mode mean field approximation is used to simplify Equations (1) through (4). The dimensionless vertical velocity ($\bar{w}$), temperature ($\bar{T}$) and composition ($\bar{C}$) are decomposed according to:

$$\bar{w}=\tilde{W}(\bar{Z},\bar{t})f(\bar{X},\bar{Y}) \quad (8a)$$

$$\bar{T}=\tilde{T}(\bar{Z},\bar{t})+\theta(\bar{Z},\bar{t})f(\bar{X},\bar{Y}) \quad (8b)$$

$$\bar{C}=\tilde{C}(\bar{Z},\bar{t})+\phi(\bar{Z},\bar{t})f(\bar{X},\bar{Y}). \quad (8c)$$

The first term in Equations (8b) and (8c) is a non-fluctuating component which varies solely in $\bar{Z}$ and $\bar{t}$. This non-fluctuating component is the horizontal average of the quantity. In an arrangement of convection cells in which there are as many upwelling as downwelling plumes the horizontal average of $\bar{w}$ is zero and those of $\bar{T}$ and $\bar{C}$ non-zero. The second component is the fluctuating component. This component has a magnitude, which is solely a function of $\bar{Z}$ and $\bar{t}$, and which is multiplied by the periodic function $f(\bar{X}, \bar{Y})$. This function represents the convective planform and wavelength. In the case of convection cells in the form of infinitely long rolls $f(\bar{X}, \bar{Y})$ has the form $\cos k\bar{X}$. The reader is referred to SEGAL and STUART (1962) and GOUGH et al. (1975) for further discussion of the form of $f(\bar{X}, \bar{Y})$. The velocity vector, v, has the components:

$$v=\frac{1}{k^2}\frac{\partial f}{\partial X}\frac{\partial W}{\partial Z},\frac{1}{k^2}\frac{\partial f}{\partial Y}\frac{\partial W}{\partial Z},f(X,Y)W(Z,t), \quad (8d)$$

in terms of the mean field formulation. In Equation (8d) and subsequent expressions all parameters are dimensionless unless otherwise stated and tildes have been dropped from the mean quantities $\tilde{W}$, $\tilde{T}$ and $\tilde{C}$.

The equation of motion becomes:

$$\frac{1}{\text{Pr}}\left(\frac{\partial}{\partial t}-D\right)DW+\frac{F}{\text{Pr}}(2W'DW+WDW')$$

$$+\frac{2\eta'}{\eta}\left(W'''-k^2W'+\frac{\eta''}{\eta}W''+k^2W\right)$$

$$=-k^2(\text{Ra}\theta+Rc\phi), \quad (9)$$

the mean thermal and mean compositional equations are:

$$\frac{\partial T}{\partial t}+(W\theta)'=T'' \quad (10)$$

$$\frac{\partial C}{\partial t}+\text{Le}\,(W\phi)'=C'' \quad (11)$$

and the fluctuating thermal and compositional equations are:

$$\left(\frac{\partial}{\partial t} - D\right)\theta + F(2W\theta' + \theta W') = -T'W \quad (12)$$

$$\left(\frac{\partial}{\partial t} - D\right)\theta + F(2W\phi' + \phi W') = -\text{Le } C'W. \quad (13)$$

The operator D used here is defined as:

$$D \equiv \frac{\partial^2}{\partial Z^2} - k^2. \quad (14)$$

In the above equations, a prime indicates partial differentiation with respect to Z. Five dimensionless numbers occur in this set of equations including the Prandtl number (Pr), the Lewis number (Le), the thermal Rayleigh number (Ra), the compositional Rayleigh number (Rc), and the viscosity ratio (Λ) (see Table 2). The kinematic viscosity (ν_0) is defined according to $\nu_0 = \eta_0/\rho$. The Prandtl and Lewis numbers are the ratios of diffusivities and therefore depend solely on magma properties. The two Rayleigh numbers also depend on the external parameters of the problem, namely the depth of the chamber and the compositional and temperature contrasts across it. It is also convenient to define the buoyancy ratio $R\rho \equiv Rc/\text{Ra} = \alpha_C\Delta C/\alpha\Delta T$. The other parameters are k, the horizontal wavenumber and F, a planform constant, which is related to $f(X, Y)$. For rolls and rectangular planforms $F = 0$ whereas for hexagons $F = 1/\sqrt{6}$ (GOUGH et al., 1975). k is related to the cell aspect ratio, a (width to depth), by $k = \pi/a$.

Configuration and boundary conditions

The magma chamber is considered to be infinite in the horizontal direction and bounded, top and bottom, by no–slip horizontal surfaces which have fixed temperature and composition. The no–slip condition at the boundaries ($u = w = 0$) implies that both W and W' are zero. We choose to model a chamber of cool, light silicic magma underlain by hotter, denser mafic magma. These boundary conditions are:

$$W = W' = T = \theta = C - 1 = \phi$$
$$= 0 \quad \text{at} \quad Z = 0 \text{ (top)} \quad \text{and}$$
$$W = W' = T - 1 = \theta = C = \phi = 0 \quad \text{at} \quad Z = 1 \text{ (base)}$$

where C represents the mass fraction of the light silicic component. This configuration is that of classical Rayleigh–Bernard convection and as such takes no account of the effects at sidewalls.

It should be noted that the specification of T and C at the boundaries implies an unknown flux of heat and composition respectively. The boundaries of the chamber can be considered as infinite thermal and compositional reservoirs. The two implied fluxes are an output of the model and can be tested against geological observations. Alternatively, one could specify the gradients of T or C, or both, at the boundaries and solve for the distribution of T and C.

The model also requires specification of a planform and wavelength. Our experiments have been carried out for hexagonal, roll and rectangular planforms. The majority of the experiments have been performed at a wavenumber of π. The π is

Table 2. Important dimensionless numbers

Dimensionless number		Value	Magma chamber range	
Prandtl Number	Pr	ν/κ	10^4–10^8	Ratio of viscosity to thermal diffusivity
Lewis Number	Le	κ/D	10^4–10^{13}	Ratio of thermal to compositional diffusivity
Rayleigh Number	Ra	$\alpha gd^3\Delta T/\kappa\nu$	10^9–10^{17}	Ratio of thermal buoyancy to viscous forces
Compositional Rayleigh Number	Rc	$\alpha_C gd^3\Delta C/\kappa\nu$	See $R\rho$	Ratio of compositional buoyancy to viscous forces
Buoyancy Ratio	$R\rho$	$\alpha_C\Delta C/\alpha\Delta T$	0–100	Ratio of Rc to Ra
Viscosity Contrast	Λ	e^Λ	1–10^8	Ratio of maximum to minimum magma chamber viscosities

close to the critical wave number for the onset of convection as described by linear stability analysis (BAINES and GILL, 1969; VERONIS, 1965). However, a series of experiments was conducted in which k was varied extensively.

The computer program employed an adaptive finite difference grid with up to 400 points to solve discretized versions of Equations (9–13) (PEREYRA, 1978). In cases where a single steady state solution exists this solution can be arrived at regardless of initial profiles.

RESULTS

Isoviscous, steady–state, infinite Prandtl number convection

When Equation (9) is rewritten for infinite Prandtl number, constant viscosity, steady state convection it becomes much simpler. Additionally, if we choose to model convection rolls or a rectangular convection planform, Equations (12) and (13) lose their dependence on the planform constant (F). It is noted that in the infinite Prandtl number case the momentum equation has no dependence on F. The work here was carried out at a wavenumber (k) of π. After having made these assumptions, the principle variables in the problem are the Lewis number and the compositional and thermal Rayleigh numbers. Given these parameters one solves for W, T, θ, C and ϕ as functions of Z, depth in the chamber. In this work three kinds of results can be distinguished.

(1) For low Rayleigh numbers the steady state solution to the problem is a conductive solution. In this case W, θ and ϕ are all zero, $T = Z$ and $C = 1 - Z$.

(2) A second class of results is characterized by the lack of a solution to the steady state problem. HUPPERT and MOORE (1976) noted the existence of oscillatory and aperiodic solutions to the double–diffusive convection equations for certain conditions. There have been many studies of simplified versions of the equations governing the time dependence of double–diffusive convection. These studies (KNOBLOCH and PROCTOR, 1981; DA COSTA *et al.*, 1981; MOORE *et al.*, 1983; KNOBLOCH *et al.*, 1986; GOLLUB and BENSON, 1980) have intrinsic relevance to the fluid dynamics of turbulence and chaos, and to bifurcation theory.

(3) The third class of results is that of a convective steady state. All such solutions have a velocity profile with depth that has just one maximum. These are single convection cells. Figure 2 shows the W, T, θ, C and ϕ profiles as an example of this

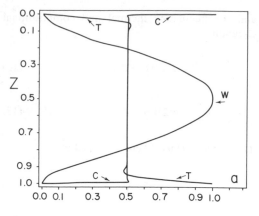

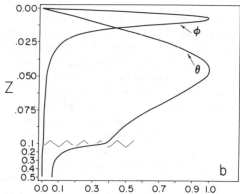

FIG. 2. Representative fields of T, C, W, θ, and ϕ for steady, single–layer, isoviscous, doubly–diffusive systems. Parameters are Le = 300, Ra = 10^7, $R\rho = 25$, $k = \pi$ and Pr $\rightarrow \infty$. (a) Mean temperature (T), mean composition (C) and normalized velocity (W/W_0). $W_0 = 1190$. Computed Nusselt numbers are $Nu_T = 12.5$ and $Nu_c = 70.9$. (b) Fluctuating temperature (θ/θ_0) and composition (ϕ/ϕ_0) for same parameters as (a). $\theta_0 = 0.167$ and $\phi_0 = 0.456$.

class of solutions. The temperature and composition profiles are both characterized by thin boundary layers and isothermal/isochemical cores in the center of the cell. From these solutions, one can calculate rates of heat transfer and chemical transport and investigate the dependence of these rates on Le, Ra and $R\rho$.

Steady state, isoviscous, infinite Prandtl number numerical experiments have been conducted over the range $1 \leq$ Le $\leq 10^6$, $10^3 \leq$ Ra $\leq 10^{10}$, $0 \leq R\rho \leq 40$ and $k = \pi$. One of the important contributions of this study is the mapping in Le–$R\rho$–Ra space of the region of steady state solutions. The results of almost 300 experiments are displayed in Figure 3. Two fields are distinguished in this figure. The uppermost field is that region of parameter space characterized by the class (3), steady convective solutions. In all cases these solutions are single–layer

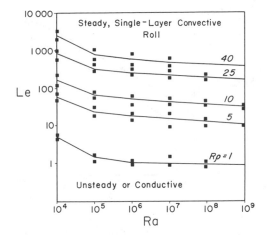

FIG. 3. Regime diagram in the Le–Ra plane separating regions of steady, single–layer, isoviscous roll convection from those characterized by either unsteady convection or conduction profiles ($k = \pi$). At given $R\rho$, Le–Ra values above the curve lead to steady, single–layer solutions characterized by thin thermal and very thin chemical boundary layers. Note that virtually all magma reservoirs with Ra $> 10^{10}$ and Le $> 10^4$ will be in the steady, single–layer mode even at quite large ratios of chemical to thermal buoyancy ($R\rho$).

convective cells. The lowermost field is made up of both the class (1), conductive, and class (2), unsteady, solutions. A critical Lewis number is defined above for which all solutions are steady convective solutions. It is important to note that this critical Lewis number decreases as the Rayleigh number increases. This is important because magma chambers typically have Rayleigh numbers in excess of 10^9. The critical Lewis number, as determined by a least–squares fit to the data, is given by:

$$Le_{crit} = 6.7 \, Ra^{-0.12} R\rho^{5/3}. \qquad (15)$$

The uncertainty in the exponents of this power–law relationship does not exceed $\pm 10\%$.

Fluxes

The vigor of convection can be measured by the rate at which heat is transferred. The Nusselt number is the ratio of the total heat flux to that which would be carried by conduction alone given the imposed ΔT. Nusselt numbers of unity indicate pure conduction. In the mean field formulation the thermal Nusselt number (Nu_T) and the analogous compositional Nusselt number (Nu_C) are calculated by:

$$Nu_T = \frac{\partial T}{\partial Z} - W\theta, \qquad (16)$$

and

$$Nu_C = -\left(\frac{\partial C}{\partial Z} - Le \, (W\theta)\right). \qquad (17)$$

Thermal Nusselt numbers have been calculated for all of the steady convective solutions [class (3)] and these data are plotted on Figure 4. Figure 4a shows the relationship between Nu_T and Ra. It is seen that at high Rayleigh numbers (*i.e.*, $> 10^6$) the relationship is linear in log–log space. When plotted against Lewis number (Figure 4b) it can be seen that for high Lewis numbers the thermal Nusselt number is almost independent of Lewis number. This relationship holds so long as Le/Le_{crit} is greater than 10. The Le_{crit} is the critical Lewis number for steady convection and given by Equation (15). Thermal Nusselt number is plotted in Figure (4c) versus $R\rho$. When $Le/Le_{crit} > 10$, thermal Nusselt number is almost independent of $R\rho$.

In conclusion, note that Nu_T has little dependence on either Le or $R\rho$. Results of multivariate linear regression are included in Table 3. A simplified equation derived from these results is:

$$Nu_T = 0.42 \, Ra^{0.23}. \qquad (18)$$

The uncertainty in the coefficients in Equation (18) and in similar parameterized expressions that follow is about 5%. Physically, this relationship implies that in multicomponent convection heat transport is not affected by the compositional buoyancy of a slow diffusing chemical species. Heat transport is affected only by the magnitude of the thermal driving force and the viscous resistance.

The relationship between heat flux and Rayleigh number can be compared with previous 2–D thermal convection studies. The power-law exponent deduced by the asymptotic solutions of ROBERTS (1979) is 0.2. From the work of LIPPS and SOMERVILLE (1971) at Pr = 200, ROBERTS (1979) calculated the value of the multiplicative coefficient to be 0.426. QUARENI and YUEN (1984) calculated a power law exponent of 0.25. The values of the exponents and coefficients for thermal convection are in reasonable agreement.

In an analogous manner compositional Nusselt numbers have been calculated for the steady convective solutions (Figure 5). Figure 5a indicates a similar relationship to Equation (18) for dependence on Ra. However, the compositional Nusselt number depends also upon the Lewis number (Figure 5b), the power law exponent being about 0.35. Compositional Nusselt number is plotted versus $R\rho$ in Figure 5c. We deduce the following relationship (see also Table 3):

$$Nu_C = 0.39 \, Ra^{0.23} \, Le^{0.35}. \qquad (19)$$

The downward flux of light silicic material depends on both the thermal driving force of the convection and on the ratio of thermal and compositional dif-

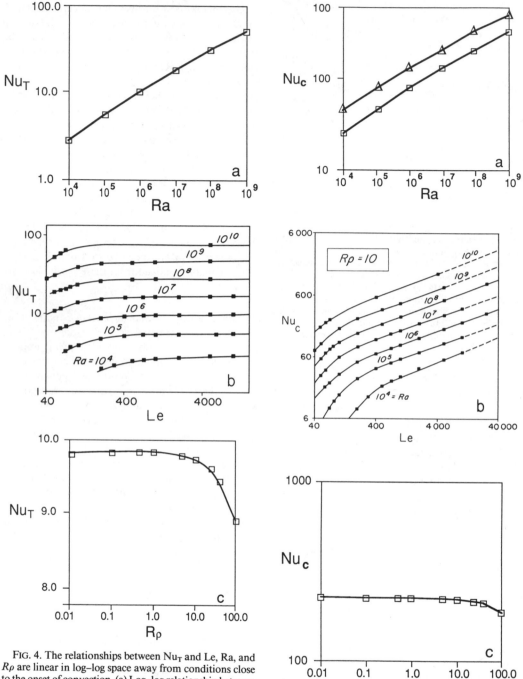

FIG. 4. The relationships between Nu_T and Le, Ra, and $R\rho$ are linear in log–log space away from conditions close to the onset of convection. (a) Log–log relationship between Nu_T and Ra with slope 0.23. Other parameters are Le = 5000, $R\rho$ = 10, $k = \pi$. (b) Asymptotic relationship between Nu_T and Le with slope zero. Other parameters are $R\rho$ = 10 and $k = \pi$. (c) Nu_T vs. $R\rho$ for Le = 10^4, Ra = 10^6 and $k = \pi$. Slope is close to zero. Dependence of Nu_T on $R\rho$ is weak for $R\rho$ less than a critical value of $R\rho$ (about 10 for the conditions of Figure 4c). Critical values of $R\rho$ depend on Ra [see Equation (15) in text] such that for Ra $\geqslant 10^9$, $(R\rho)_{crit} > 400$. Hence, in the geologically important range, the Le and $R\rho$ dependence of Nu_T is negligible.

FIG. 5. The relationships between Nu_C and Le, Ra, and $R\rho$ are linear in log–log space away from conditions close to the onset of convection. (a) Log–log relationship between Nu_C and Ra with slope 0.23. Other parameters are Le = 1000 (squares) and 5000 (triangles), $R\rho$ = 10, $k = \pi$. (b) Asymptotic relationship between Nu_C and Le with slope 0.35. Other parameters are $R\rho$ = 10 and $k = \pi$. (c) Nu_C vs. $R\rho$ for Le = 10^4, Ra = 10^6 and $k = \pi$.

Table 3. Multiple linear regression results for Equations (18) and (19).

	Numerical coefficient	Ra exponent	Le exponent	$R\rho$ exponent
Thermal Nusselt Number Equation (18)	0.417 ± 0.011	0.229 ± 0.001	0.008 ± 0.002	0.009 ± 0.002
Compositional Nusselt Number Equation (19)	0.387 ± 0.010	0.225 ± 0.001	0.345 ± 0.002	0.010 ± 0.002

Correlations include 38 runs and cover the range $10^5 \leqslant Ra \leqslant 10^9$, $20 \leqslant Le \leqslant 10^4$ and $0.01 \leqslant R\rho \leqslant 40$.

fusivities. It does not depend on the magnitude of the compositional driving force.

The parameterizations which follow, in addition to the two above, are made for the asymptotic portions of the dataset. These parameterizations therefore imply conditions far into the convective regime, i.e., (high Lewis and Rayleigh numbers) $Le/Le_{crit} > 10$ and $Ra \geqslant 10^6$.

Velocity

Figure 6 shows the relationship between maximum dimensionless velocity (W_{max}) and Rayleigh number (Ra). At conditions away from the critical parameters for class (3) solutions, the velocity is solely dependent upon thermal Rayleigh number. The following relationship has been deduced from the numerical experiments:

$$W = 0.09 \, Ra^{0.62}. \qquad (20)$$

Again comparison can be made to previous thermal convection studies. ROBERTS (1979) found a power–law relationship between Rayleigh number and velocity with an exponent of 0.6. Our results are quite close to this exponent.

Boundary layer thicknesses

The convection cells have narrow thermal and compositional boundary layers of thickness, δ_T, and, δ_C, respectively. The boundary layer thicknesses are defined by the region in which 95% of the variation in temperature or composition, in one half of the convection cell, takes place. The δ_T and δ_C vary inversely with Nu_T and Nu_C respectively. It is found that

$$\delta_T = 0.77d \, Ra^{-0.23}, \qquad (21)$$

and,

$$\delta_C = 0.77d \, Ra^{-0.23} \, Le^{-0.35}. \qquad (22)$$

Note that the ratio of δ_T/δ_C is given by:

$$\frac{\delta_T}{\delta_C} = 0.85 \, Le^{0.35}. \qquad (23)$$

Finite Prandtl number solutions

A series of numerical experiments was conducted for finite Prandtl number assuming a hexagonal convective planform. Figure 7 shows the variation in thermal Nusselt number as Prandtl number is decreased. There is no dependence of thermal Nusselt number on Prandtl number for values of $Pr > 10^2$. For $Pr < 10^2$ the thermal Nusselt number begins to rise quite sharply, indicating increased vigor of convection. Physically, this situation corresponds to an increasing importance of the inertial terms in the momentum equation. Laboratory double–diffusion experiments have been conducted for fluids with Prandtl numbers less than 10. Extrapolation of these laboratory experiments to magma conditions, for which Prandtl numbers exceed 10^2, may be problematical. Future laboratory or numerical experiments are warranted to shed light on this important point.

Effect of wavenumber

The effect of wavenumber on the style of convection has been extensively studied in this work for the conditions $Ra = 10^6$, $Le = 500$, $R\rho = 5$, Pr

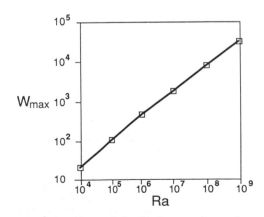

FIG. 6. Log–linear relationship between the maximum velocity, W_{max}, and Ra with slope 0.62. Parameters are $Le = 1000$, $R\rho = 10$ and $k = \pi$.

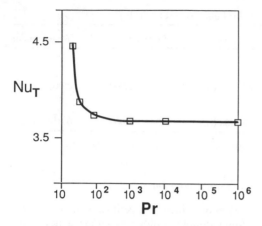

FIG. 7. Variation of thermal Nusselt number, Nu_T with Prandtl number, Pr. There is no dependence on Pr for $Pr > 10^3$. The other parameters are $Le = 100$, $R\rho = 10$, $Ra = 10^5$ and $k = \pi$ and the planform chosen is hexagonal. Solutions for $Pr < 15$ are non-convergent. Note asymptotic limit for $Pr > 100$.

$= \infty$, and $1 \leq k \leq 28$ (Figure 8). For $k \geq 10$ two branches of steady state solution are found.

The upper branch has a maximum thermal Nusselt number (Nu_{mx}) of 12.8 at wavenumber of 12. This branch of solutions is made up of single convection cells like those illustrated in Figure 2. No solution along this branch could be found for $k > 24$. The lower branch of solutions extends from $k = 10$ to $k = 20$. The solutions along this branch have thermal Nusselt numbers which are roughly half those on the upper branch. The form of these

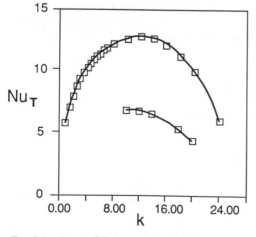

FIG. 8. Variation of thermal Nusselt number, Nu_T, with wavenumber, k, for $Le = 500$, $R\rho = 5$, $Ra = 10^6$. For $k \geq 10$, two sets of solutions can be obtained. The lower branch is a branch of steady double–layer convection cells. The upper branch is one of steady single–layer convection cells.

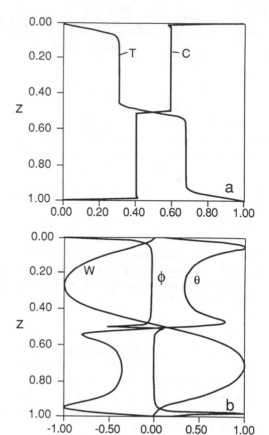

FIG. 9. Representative fields of T, C, W, θ, and ϕ for steady, two–layer, isoviscous, doubly–diffusive systems at high wavenumber. Parameters are $Le = 500$, $Ra = 10^6$, $R\rho = 5$, $k = 12$ and $Pr \rightarrow \infty$. (a) Mean temperature (T) and mean composition (C) plotted versus depth in the chamber (Z). Computed Nusselt numbers are $Nu_T = 6.793$ and $Nu_C = 79.473$. (b) Normalized velocity (W/W_0) (curve W), fluctuating temperature (θ/θ_0) (curve θ) and composition (ϕ/ϕ_0) (curve ϕ) versus depth (Z). $W_0 = 186$, $\theta_0 = 0.105$ and $\phi_0 = 0.021$.

solutions is illustrated in Figure 9. Two vertically stacked convecting layers are separated by a stationary diffusing interface.

The following remarks can be made about these results.

(1) In cases where two steady state solutions have been found for a fixed set of parameters, the initial profiles determine whether a single or double layer solution is found.

(2) The effect of wavenumber on mean field thermal convection has been studied by TOOMRE et al. (1977) and QUARENI and YUEN (1984). Both groups found only single layer solutions. In the present work, pure thermal convection runs were performed for $Ra = 10^6$, $Pr = \infty$ and $1 \leq k \leq 24$.

The steady state calculation produced single layer results over all of this range given $\theta_{init} = \phi_{init} = 0.01$ Sinπz. If, however, the double–layer θ profile from the runs described above was used as the initial guess for the thermal convection case, a fully converged two–layer solution would be found in the range $8 \leqslant k < 20$. The laboratory experiments of J. M. OLSON and ROSENBERGER (1979) reported multi-layer configurations for thermal convection in narrow enclosures. Our mean–field predictions for thermal convection are in good agreement with these experiments and are displayed in Figure 10.

(3) Solutions for $k \geqslant 20$ in the thermal and double diffusive cases along double and single layer solution branches all show non–isothermal cores in convection cells. Composition (in the double diffusive case) still remains isochemical.

(4) TOOMRE et al. (1977) note that there is no logical reason for choosing an appropriate wavelength for convection within the mean field formulation. They compared the results of physical experiments to their work and found that the Nusselt number obtained in experiments was typically less than the maximum, Nu_{mx}. This conclusion leaves two choices of wavenumber. TOOMRE et al. (1977) suggest that the lower of these two wavelengths is the appropriate one to model convection. They find that numerical modelling of higher wavelengths leads to profiles, such as in (3) above, which are not seen in laboratory experiments. Over the range $10^6 < Ra < 10^8$ they find that the value

of wavenumber (k) that compares best with experimental data lies in the range $1.5 < k < 3$.

In modelling steady state double–diffusive convection we have fewer relevant physical experiments with which to compare our data. However, as is illustrated in the above, the qualitative similarity between thermal and double–diffusive convection, with respect to wavenumber, is very strong. On this basis we feel the choice of $k = \pi$ made for the majority of steady state double–diffusive convection runs is an appropriate one.

Effect of variable viscosity

A series of numerical experiments in which viscosity varied as a function of temperature was performed. In Equation (9) this condition results in non–zero values for those terms containing spatial derivatives of viscosity. This result complicates the solution of the momentum equation. In this work we follow QUARENI et al. (1985) and make the substitution:

$$Y = \frac{d^3 W}{dZ^3} + \frac{1}{\eta}\frac{d\eta}{\partial Z}\left(\frac{d^2 W}{dZ^2} + k^2 W\right). \tag{24}$$

Then for infinite Prandtl number Equation (9) becomes:

$$\frac{dY}{dZ} + \frac{1}{\eta}\frac{d\eta}{dZ}\left(Y - 3k^2\frac{dW}{\partial Z} - 2k^2\frac{d^2 W}{\partial Z^2}\right) + k^4 W$$
$$= \frac{-k^2}{\eta}(Ra\,\theta + Rc\phi). \tag{25}$$

The conditions over which viscosity is included as a variable are $1 \leqslant R\rho \leqslant 10$, $1 \leqslant Le \leqslant 100$, $10^4 \leqslant Ra \leqslant 10^7$ and $Pr = \infty$. The parameter A [equation (6)] is positive in all numerical experiments; therefore, viscosity never exceeds the reference viscosity (η_0). The ratio of maximum and minimum viscosities in the chamber is given by $\Lambda = e^A$. The greatest value of Λ in these experiments was about 600.

Figure 11 presents a comparison between results with $\Lambda = 1$ and $\Lambda \sim 25$ for $R\rho = 5$, $Ra = 10^6$, $Le = 100$ and $k = \pi$. High Λ leads to a higher average velocity in the chamber. This conclusion is a direct consequence of the lowered viscosity as temperature increases away from the upper boundary. The velocity field also becomes asymmetric as Λ increases. Slower velocities are present in the upper cooler boundary layer. In the upper boundary layer, diffusional transport is more important, relative to advection, than it is at the lower boundary. For this reason the contrasts in T and C across the upper boundary layer are greater than they are across the lower boundary layer. The differing importance of

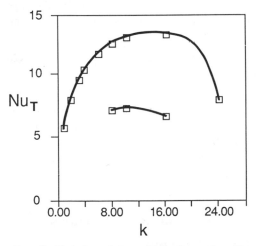

FIG. 10. Variation of thermal Nusselt number, Nu_T, with wavenumber, k, for purely thermal convection and $Ra = 10^6$. For $k \geqslant 8$ two sets of solutions can be obtained. The upper branch (single layer) can be continued to lower wavenumbers, whereas the lower branch (double layer) cannot.

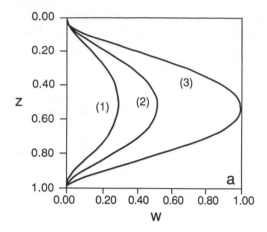

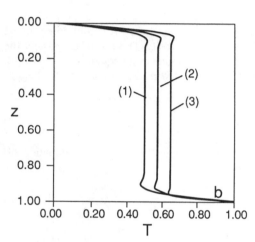

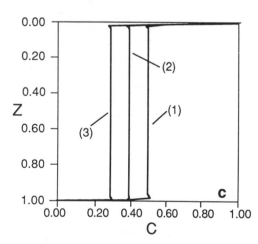

FIG. 11. Representative fields of T, C and W for steady, single–layer, double–diffusive systems with temperature dependent viscosity. Plots are made against depth (Z). Parameters are Le = 100, Ra = 10^6, $R\rho$ = 5, $k = \pi$ and Pr $\rightarrow \infty$. In these plots Λ = 0, 5 and 25 (curves 1, 2, and 3 respectively). (a) Normalized velocity (W/W_0), (b) mean temperature (T) and (c) mean composition (C). W_0 = 1292.

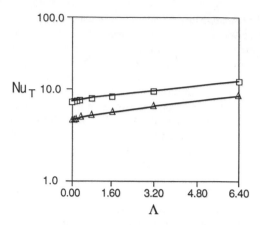

FIG. 12. Log–log relationship between Nu_T and viscosity parameter Λ. Parameterization of Nu_T with leads to the relationship $Nu_T = 0.42\ Ra^{0.23}\Lambda^{0.10}$. Graphs are for Le = 10, Ra = 10^6, $R\rho$ = 1 and $k = \pi$ (upper curve), and Le = 100, Ra = 10^5, $R\rho$ = 1 and $k = \pi$ (lower curve).

diffusion across the boundary layers accounts for T and C values in the core of the convection cell which differ from 0.5 when Λ is other than unity.

Figure 12 illustrates the dependence of Nu on Λ. Over the range of the experiments a linear fit in log–log space is obtained. Equation (18) then becomes:

$$Nu_T = 0.42\ Ra^{0.23}\Lambda^{0.10}. \qquad (26)$$

Nu_C also has a similar relationship with Λ. Equation (19) becomes:

$$Nu_C = 0.39\ Ra^{0.23}\ Le^{0.35}\Lambda^{0.05}. \qquad (27)$$

Approximate relationships describing the upper and lower thermal boundary layer thicknesses (δ_{TU} and δ_{TL}, respectively) are given by:

$$\delta_{TU} \approx 0.77\ Ra^{-0.23}\Lambda^{-0.08}, \qquad (28)$$

and,

$$\delta_{TL} \approx \delta_{TU}\Lambda^{-0.12}. \qquad (29)$$

Similarly, the upper and lower compositional boundary layer thicknesses (δ_{CU} and δ_{CL}, respectively) are given by:

$$\delta_{CU} \approx 0.77\ Ra^{-0.23}\ Le^{-0.35}\Lambda^{-0.04} \qquad (30)$$

and,

$$\delta_{CL} \approx \delta_{CU}\Lambda^{-0.22}. \qquad (31)$$

GEOLOGICAL IMPLICATIONS

Magma chamber parameters

Appropriate ranges of the governing dimensionless numbers relevant to flow in crustal and upper mantle magma reservoirs are given in Table 2. The specific chemical properties of a given component (e.g., SiO_2, H_2O, MgO, etc.) enter into both the Lewis number (Le) and buoyancy ratio ($R\rho$). Although composition has an obvious effect on viscosity, inertia is of negligible importance in most magmatic flows and the infinite Pr limit is entirely justified *a priori*. Compositional expansivities (α_C) fall between approximately 0.1 and 2 for the major components in a silicate melt; the large value of 2 is associated with H_2O component (see Table 1). Note that the range of $R\rho$ values covered by the numerical experiments corresponds well with those in nature.

Although it is generally assumed that thermal Rayleigh numbers applicable to magma chamber convection are quite large (SHAW, 1965, 1974; BARTLETT, 1969; SPERA, 1980; HARDEE, 1983; SPERA et al., 1982, 1986b), this has recently been questioned (MARSH, 1985). MARSH (1985) argued that the heat transfer through magma chambers, being limited by conductive heat transfer in the country rock, must imply low Rayleigh numbers. The following example demonstrates that the Rayleigh number is large even when heat fluxes are small (e.g., 1 HFU). The simplest way to show this is to envision the heat transfer along a vertical country–rock magma–chamber contact. A uniform heat flux assumed along the boundary is governed by heat conduction in the country rock.

A scaling analysis of the two–dimensional form of the conservation equations of heat and momentum enables one to estimate the thermal boundary layer thickness (δ_T) and the Nusselt number. The analysis gives:

$$\delta_T \sim LR^{-1/5}, \tag{32}$$

and,

$$Nu_T = \frac{qL}{\Delta T_1 k_T} \sim \frac{L}{\delta_T} \sim R^{1/5}. \tag{33}$$

In these expressions, L is the characteristic length of the chamber–country rock contact, ΔT_1 is the temperature difference between chamber interior and wall, q is the heat flux, k_T the thermal conductivity of the country rock (or magma), and R is the Rayleigh number based on the imposed (and constant) heat flux at the chamber wall. The Rayleigh number is defined according to $R = \alpha g q L 3/k\kappa\nu$. The validity of these scaling results can be demonstrated by referring to the numerical solution by SPARROW and GREGG (1956), who found:

$$Nu_T = 0.62R^{1/5}, \tag{34}$$

in the infinite Prandtl number limit. Application of these results to a magma chamber is made possible once typical parameters are assumed. For illustrative purposes set $L = 1$ km, $\alpha = 5 \times 10^{-5}$

K^{-1}, $g = 10$ m/s^2, $\kappa = 8 \times 10^{-7}$ m^2/s, $k_T = 3.35$ W m^{-1} K^{-1}, $\nu = 10^4$ m^2/s. Choosing a heat flux (q) at the wall of 1 HFU ($=1 \times 10^{-6}$ cal/cm^2 s $= 41.84$ mW/m^2) one finds:

$$R = 8 \times 10^8$$

$$\delta_T \sim 17 \text{ m}$$

$$\Delta T_1 \sim 0.2 \text{ K}.$$

Note that even for the small value of heat flux used here (1 HFU is less than the global average) the implied Rayleigh number is quite large. Even though the temperature difference across the chamber is small, the magma will be actively and vigorously convecting. As is noted below, in many geothermal areas heat flow can be one hundred times greater than the value used here; our estimate of magma chamber Rayleigh number is therefore a low one. On this basis we cannot agree with the findings of MARSH (1985).

Finally, we note that although most of numerical experiments are for Le $\leqslant 10^4$, the asymptotic dependence of Nu_C on Le enables one to estimate light–component transport rates with some degree of confidence. Similarly, the weak dependence of Nu_T on Le and $R\rho$ implies that heat fluxes may be reasonably well calculated even for species characterized by low chemical diffusivities.

The ranges here are valid for a wide variety of magma chamber conditions. The most variable is the Rayleigh number, because of its cubic dependence on length scale. Given these dimensionless numbers it is now possible to apply the results to magma chamber flows. The following calculations use parameterizations which assume a wavenumber (k) of π. This is equivalent to assuming a convection cell as deep as it is wide (i.e., aspect ratio, $a = 1$).

Flux of heat and light–silicic–component

The steady state model fixes temperature and composition at the upper and lower boundaries of the chamber. Such conditions imply fluxes through the boundaries of both heat and the light–silicic component.

Heat flux is a quantity that can be directly compared with measurements made in active geothermal areas, which are presumably underlain by active magma chambers. Unfortunately, uncertainty regarding the present–day size and shape of active magma reservoirs precludes more than a semi–quantitative comparison. For high Lewis number, the thermal Nusselt number is given by Equation (26), so that the dimensional heat flux through the roof of the chamber is given by:

$$q = 0.42 k_T \left(\frac{\Delta T}{d} \right) Ra^{0.23} \Lambda^{0.10}. \qquad (35)$$

In this equation and in Equations (36) to (39) all quantities are dimensional. For example, with thermal conductivity, $k_T = 3.35$ W m^{-1} K^{-1}, $\Delta T = 5$ K, $d = 5$ km, a magma viscosity of 10^4 Pa·s, $\Lambda = 10^2$ then the heat flux is 3500 mW/m^2 ($\sim$80 HFU where 1 HFU = 1×10^{-6} cal/cm^2 s). By allowing for latent heat effects and assuming the hydrothermal system can efficiently dissipate magmatic heat, a hypothetical magma chamber solidification time of roughly 40 000 years may be estimated.

Comparison can be made between the heat fluxes calculated and the heat loss in geothermal areas. Over large geothermal areas (1000 km^2) ELDER (1965) suggests heat flow values averaging 100 HFU (4 200 mW/m^2). More recently, SOREY (1985) has computed the heat discharge divided by caldera area for the Yellowstone, Long Valley and Valles geothermal areas. Heat flux values of 2100, 630 and 500 mW/m^2, respectively, were estimated based on present–day discharge rates of high-chloride thermal water in hot springs and seepage into rivers. It is reassuring that these measurements are in agreement with the numerical experiments.

Rates of transport of silicic–component downward can be calculated from the compositional Nusselt number relationship given by Equation (27). The compositional Nusselt number is defined according to:

$$Nu_C \equiv \frac{j_i d}{\rho D \Delta C}, \qquad (36)$$

where j_i is the flux of the ith light component, ρ the density, D the chemical diffusivity and ΔC the externally maintained difference in composition across the chamber. By combination of Equations (27) and (33), the dimensional flux of light component is determined according to:

$$j_i = \frac{0.39 \rho \kappa \Delta C}{d} Ra^{0.23} Le^{-0.65} \Lambda^{0.05}. \qquad (37)$$

With Ra = 8×10^{13}, Le = 10^4 (for water in melt), $\rho = 2500$ kg/m^3, $\kappa = 10^{-6}$ m^2/s, $\Delta C = 0.05$, $\Lambda = 100$ and $d = 5$ km, Equation (34) yields a flux of water, $j_{H_2O} = 5.0 \times 10^{-8}$ kg/m^2 s. If the cross–sectional area of the magma chamber is taken as d^2, then the mass flux corresponds to an effective mass flow rate for H_2O of 4.0×10^7 kg/year. The residence or redistribution time for H_2O within the chamber may be defined as:

$$t_{H_2O} = \frac{C_{H_2O} \rho d}{j_{H_2O}}, \qquad (38)$$

where C_{H_2O} is the average H_2O content of melt within the chamber (say $C_{H_2O} = 0.025$). For the parameters cited, $t_{H_2O} = 2.0 \times 10^5$ years. The effective rate at which H_2O is transported from top to bottom of the chamber is therefore $V_{H_2O} \sim d/t_{H_2O} \sim 2.5$ cm/year. The eddy diffusivity of H_2O corresponding to this rate is roughly D_{eddy} (H_2O) $\sim d^2/t_{H_2O} = 4.0 \times 10^{-6}$ m^2/s which is larger by a factor of 10^5 than the corresponding molecular diffusivity of H_2O of $\sim 10^{-11}$ m^2/s (DELANEY and KARSTEN, 1981; SHAW, 1974). Convection clearly plays a dominant role in the redistribution of H_2O in a magma chamber.

The mass flux can be interpreted in terms of the rate at which crystallization of anhydrous phases takes place at the top boundary of the chamber. In the above example, the implied rate of crystallization of the "upper border group" is $j_i/\rho \Delta C$ which corresponds to about 4 m per thousand years.

Convection rates

From Equation (20), the dimensional maximum convection velocity is given by:

$$W = 0.09 \frac{\kappa}{d} Ra^{0.62}. \qquad (39)$$

For Ra = 10^{12}, this corresponds to a velocity of 16 km/yr. This value is in agreement with the calculations based on boundary–layer theory made by SPERA et al. (1982). The circulation time for a magma parcel is therefore $4d/W \sim 1$ year.

It is noted that both a 5 mm crystal and a 5 cm xenolith will have settling velocities several orders of magnitude less than convective velocities in the center of the cell. This finding suggests that crystal fractionation by settling is very unlikely in melt dominated magma chambers, except at the margins where velocities are smaller. More detailed studies on the distribution of crystals in magma chambers support this finding (MARSH and MAXEY, 1985; WEINSTEIN et al., 1986).

Boundary layer thicknesses

Equations (21) and (22) give the thermal and compositional boundary–layer thicknesses, respectively. For Ra = 10^{12}, Le = 10^4 and $d = 5$ km these boundary layer thicknesses are 6 m and 0.2 m, respectively. Note that δ_C depends on the molecular diffusion coefficient of the light component. Although these calculations do predict that a continuously zoned cap of "evolved" magma will accumulate at the top of a chamber, the thickness of that zone is quite thin.

Existence of layers

Results of numerical experiments plotted in Figure 3 cover a large range of conditions. The values of Prandtl number (Pr) and buoyancy ratio ($R\rho$) in these experiments are in the range of those in magma chambers, *i.e.,* Pr = ∞ and $1 < R\rho < 50$. We are not able to model the complete range of magma chamber Lewis or Rayleigh numbers. However, it is noted that magma chambers lie at Lewis numbers well above the critical Lewis numbers that yield steady single–cell convection. Additionally, as Rayleigh number increases this critical Lewis number will drop even farther from relevant magma chamber Lewis numbers. The data comprising Figure 2 are for $\Lambda = 1$ and $k = \pi$. A primary conclusion of this paper is that given these conditions steady state magma chamber convection cells will be simple single cells. The value of Λ, although changing some of the characteristics of the cells, does not change their single–cell nature.

At values of k greater than 10 (aspect ratio ~ 3) multiple steady states exist. Two points should be noted.

(1) Single and double layer solutions are found for $k \geqslant 10$ in both multicomponent and pure thermal convection experiments. Therefore, they are not solely a phenomena of multicomponent convection.

(2) As we have studied only steady state solutions it is not possible to determine which solution (the single cell or double cell) will be produced in the evolution of a magma chamber. Both solutions should be admitted to be possible in tall, thin magma chambers.

CONCLUSIONS

This study of steady state double–diffusive convection in magma chambers was conducted for boundary conditions which prescribed fixed temperatures and compositions at the top and bottom of the chamber. Within this context, the following conclusions are reached:

(1) The mean–field approximation to the full convective equations can successfully model convection over a large range of magma chamber conditions. In particular it has been possible to model Rayleigh numbers as high as 10^{10} and Lewis numbers as high as 10^5.

(2) Isoviscous convection dominated by a wavenumber of π (as might be expected in an equidimensional magma chamber) is found to exist above a critical Lewis number. Below the critical Lewis number, unsteady and conductive solutions are found. Above the critical Lewis number, all steady-state solutions are single layer convection cells. No layered convection is found. The critical Lewis number is a function of the buoyancy ratio and the Rayleigh number. Magma chambers lie above this critical Lewis number and therefore it is suggested that steady–state magma chambers will exhibit single cell convection.

(3) Convection is characterized by thin thermal boundary layers and thinner compositional boundary layers. The cores of the convection cells are isothermal/isochemical. For the typical parameters used in the text thermal and compositional boundary layer thicknesses are 6 m and 20 cm respectively.

(4) Calculated heat fluxes for magma chambers are in agreement with measured heat fluxes in hydrothermal areas. Such fluxes are of the order of 4200 mW/m^2 (100 HFU).

(5) The boundary conditions imply fluxes of chemical species. For the parameters given in the text, the effective rate at which water is transported through a magma chamber is about 2.5 cm/yr. Evidently convection plays a dominant role in the redistribution of water in magma chambers.

(6) Characteristic convective velocities on the order of km/yr prohibit fractionation by crystal settling in both rhyolitic and basaltic magmas, except within flow along chamber margins.

(7) For hexagonal convection planforms at wavenumber $k = \pi$, the characteristics of the flow become dependent on Prandtl number (Pr) for Pr $\leqslant 100$. As Prandtl numbers for magmatic systems exceed 100, future work is needed to investigate the nature of this potentially important transition. This dependence is manifested in increased fluxes of heat and composition.

(8) When viscosity is temperature dependent the style of convection is similar to that for isoviscous convection. Velocities and boundary layer thicknesses differ quantitatively from those found in isoviscous convection.

(9) In chambers dominated by high wavenumbers ($k > 10$, such as might be expected in tall thin magma chambers) two steady–state solutions are found. The first, corresponding to a high heat flow, is a single convection cell. The second, corresponding to a low heat flow, consists of two vertically stacked convecting cells separated by a diffusive interface. Given the same parameters and a high wavenumber, both sets of solutions are found for purely thermal convection, in addition to double diffusive convection. It is possible that these double-layer convection cells could occur in tall thin magma chambers. This model, on account of its

steady–state nature, cannot discriminate between these two solutions.

Acknowledgements—This research was supported by NSF EAR85–19900, NSF EAR86–08284 to FJS and NSF EAR86–08479 to DAY. Numerical calculations were supported by the National Center for Atmospheric Research at Boulder, Colorado, the Princeton University Computer Center, the IBM Los Angeles Scientific Computing Center and the offices of the Vice Chancellor and Provost of the University of California at Santa Barbara. We thank Ulli Hansen, Peter M. Olson, Alain Trial, and Curtis Oldenburg for stimulating discussions.

REFERENCES

BACON C. R. (1983) Eruptive history of Mount Mazama and Crater Lake Caldera, Cascade Range, USA. *J. Volcanol. Geotherm. Res.* **18**, 57–115.

BAINES P. G. and GILL A. E. (1969) On thermohaline convection with linear gradients. *J. Fluid Mech.* **37**, 289–306.

BARTLETT R. W. (1969) Magma convection, temperature distribution, and differentiation. *Amer. J. Sci.* **267**, 1067–1082.

BRANDEIS G. and JAUPART C. (1986) On the interaction between convection and crystallization in cooling magma chambers. *Earth Planet. Sci. Lett.* **77**, 345–361.

BOTTINGA Y., WEILL D. F. and RICHET P. (1982) Density calculations for silicate liquids. I. Revised method for aluminosilicate compositions. *Geochim. Cosmochim. Acta* **46**, 909–917.

BURNHAM C. W. and DAVIS N. F. (1971) The role of H_2O in silicate melts, I. $P–V–T$ relations in the system $NaAlSi_3O_8–H_2O$ to 10 kilobars and 1000°C. *Amer. J. Sci.* **270**, 54–79.

CHEN C. F. and TURNER J. S. (1980) Crystallization in a double–diffusive convection system. *J. Geophys. Res.* **85**, 2573–2593.

CHEN C. F. and JOHNSON D. H. (1984) Double–diffusive convection: a report on an engineering foundation conference. *J. Fluid Mech.* **138**, 405–416.

CRISP J. A. (1984) The Mogan and Fataga formations of Gran Canaria (Canary Islands). Geochemistry, petrology and compositional zonation of the pyroclastic and lava flows; intensive thermodynamic variables within the magma chambers; and the depositional history of pyroclastic flow E/ET. Ph.D. Thesis, Princeton Univ.

CRISP J. C. and SPERA F. J. (1986) Pyroclastic flows and lavas from Tejeda volcano, Gran Canaria, Canary Islands: Mineral chemistry, intensive parameters and magma chamber evolution. *Contrib. Mineral. Petrol.* (Submitted).

DA COSTA L. N., KNOBLOCH E. and WEISS N. O. (1981) Oscillations in double–diffusive convection. *J. Fluid Mech.* **109**, 25–43.

DELANEY J. R. and KARSTEN J. L. (1981) Ion microprobe studies of water in silicate melts: concentration–dependent water diffusion in obsidian. *Earth Planet. Sci. Lett.* **52**, 191–202.

ELDER J. W. (1965) Physical processes in geothermal areas. In *Terrestrial Heat Flow*, (ed. W. H. K. LEE), Amer. Geophys. Union Monograph Series No. 8, 211–239.

ELDER J. W. (1969) Numerical experiments with thermohaline convection. *Phys. Fluids*, suppl. II, 194–197.

GOLLUB J. P. and BENSON S. V. (1980) Many routes to turbulent convection. *J. Fluid Mech.* **100**, 449–470.

GOUGH D. O., SPIEGEL E. A. and TOOMRE J. (1975) Modal equations for cellular convection. *J. Fluid Mech.* **68**, 695–719.

GOUGH D. O. and TOOMRE J. (1982) Single–mode theory of diffusive layer in thermohaline convection. *J. Fluid Mech.* **125**, 75–97.

GREER J. C. (1986) Dynamics of withdrawal from stratified magma chambers. M.S. Thesis, Arizona State Univ.

HANSEN U. S. (1986) Zeitabhangige Phaenomenon in thermische und doppelt diffusiver Konvection. Ph.D. Thesis, University of Cologne, West Germany.

HANSEN U. and YUEN D. A. (1985) Two dimensional double–diffusive convection at high Lewis numbers: applications to magma chambers. *Trans. Amer. Geophys. Union* **66**, no. 46, p. 1124.

HARDEE H. C. (1983) Convective transport in crustal magma bodies. *J. Volcanol. Geotherm. Res.* **19**, 45–72.

HERRING J. R. (1963) Investigation of problems in thermal convection. *J. Atmos. Sci.* **20**, 325–338.

HERRING J. R. (1964) Investigations of problems in thermal convection: rigid boundaries. *J. Atmos. Sci.* **21**, 277–290.

HILDRETH W. (1981) Gradients in silicic magma chambers: implications for lithospheric magmatism. *J. Geophys. Res.* **86**, 10153–10193.

HUPPERT H. E. and MOORE D. R. (1976) Non–linear double–diffusive convection. *J. Fluid Mech.* **78**, 821–854.

HUPPERT H. E. and SPARKS R. J. (1984) Double–diffusive convection due to crystallization in magmas. *Ann. Rev. Earth Planet Sci.* **12**, 11–37.

IYER H. M. (1984) Geophysical evidence for the location, shapes and sizes, and internal structures of magma chambers beneath regions of Quaternary volcanism. *Phil. Trans. Roy. Soc. London* **A310**, 473–510.

JAUPART C., BRANDEIS G. and ALLEGRE C. J. (1984) Stagnant layer at the bottom of convecting magma chambers. *Nature* **308**, 535–538.

KNOBLOCH E. and PROCTOR M. R. E. (1981) Nonlinear periodic convection in double–diffusive systems. *J. Fluid Mech.* **108**, 291–316.

KNOBLOCH E., MOORE D. R., TOOMRE J. and WEISS N. O. (1986) Transitions to chaos in two–dimensional double–diffusive convection. *J. Fluid Mech.* **166**, 409–448.

LIPMAN P. W. (1967) Mineral and chemical variations within an ash–flow sheet from Aso caldera, Southwest Japan. *Contrib. Mineral. Petrol.* **16**, 300–327.

LIPPS F. B. and SOMERVILLE R. C. J. (1971) Dynamics of variable wavelength in finite amplitude Bernard convection. *Phys. Fluids* **14**, 759–765.

LOWELL R. P. (1985) Double–diffusive convection in partially molten silicate systems: its role during magma production and in magma chambers. *J. Volcanol. Geotherm. Res.* **26**, 1–24.

MARSH B. D. (1985) Convective regime of crystallizing magmas. *Geol. Soc. Amer. Abstr. Prog.* **17**, 653.

MARSH D. B. and MAXEY M. R. (1985) On the distribution and separation of crystals in convecting magma. *J. Volcanol. Geotherm. Res.* **24**, 95–150.

McBIRNEY A. R. (1980) Mixing and unmixing of magmas. *J. Volcanol. Geotherm. Res.* **7**, 357–371.

McKENZIE D. P., ROBERTS J. M. and WEISS N. O. (1974) Convection in the earth's mantle: towards a numerical simulation. *J. Fluid Mech.* **62**, 465–538.

MOORE D. R., TOOMRE J., KNOBLOCH E. and WEISS N. O. (1983) Periodic doubling and chaos in partial differential equations for thermosolutal convection. *Nature* 303, 663–667.

MURASE T., MCBIRNEY A. R. and MELSON W. G. (1985) Viscosity of the dome of Mount St. Helens. *J. Volcanol. Geotherm. Res.* 24, 193–204.

NEWELL T. A. (1984) Characteristics of a double–diffusive interface at high density stability ratios. *J. Fluid Mech.* 149, 385–401.

NILSON R. H. and BAER M. R. (1982) Double diffusive counterbuoyant boundary layer in laminar natural convection. *Int. J. Heat Mass Transfer* 25, 285–287.

NILSON R. H., MCBIRNEY A. R. and BAKER B. M. (1985) Liquid fractionation. Part II: Fluid dynamics and quantitative implications for magmatic systems. *J. Volcanol. Geotherm. Res.* 24, 25–54.

OLDENBURG C. M., SPERA F. J., YUEN D. A. and SEWELL G. (1985) Two dimensional double–diffusive convection. *Trans. Amer. Geophys. Union* 66, no. 46, p. 1124.

OLSON P. (1986) A comparison of heat transfer laws for thermal convection at very high Rayleigh numbers. *Phys. Earth Planet. Int.* (Submitted).

OLSON J. M. and ROSENBERGER F. (1979) Convective instabilities in a closed vertical cylinder heated from below. Part 1. Monocomponent gases. *J. Fluid Mech.* 92, 609–629.

PEREYRA V. (1978) PASVA3: an adaptive finite difference FORTRAN program for first order nonlinear and ordinary boundary problems. *Lecture Notes Comp. Sci.* 76, 67–88.

PIASEK S. A. and TOOMRE J. (1980) Nonlinear evolution and structure of salt fingers. In *Marine Turbulence,* (ed. J. C. J. NIHOUL), pp. 193–219. Elsevier.

PROCTOR M. R. F. (1981) Steady subcritical thermohaline convection. *J. Fluid Mech.* 105, 507–521.

QUARENI F. and YUEN D. A. (1984) Time–dependent solutions of mean–field equations with applications for mantle convection. *Phys. Earth Planet. Int.* 36, 337–353.

QUARENI F., YUEN D. A., SEWELL G. and CHRISTENSEN U. R. (1985) High Rayleigh Number convection with strongly variable viscosity: a comparison between mean–field and two–dimensional solutions. *J. Geophys. Res.* 90, 12633–12644.

RICE A. (1981) Convective fractionation: a mechanism to provide cryptic zoning (macrosegregation), layering, crescumulates, banded tuffs and explosive volcanism in igneous processes. *J. Geophys. Res.* 86, 405–417.

ROBERTS G. O. (1979) Fast viscous Bernard convection. *Geophys. Astrophys. Fluid Dyn.* 12, 235–272.

SCHMINCKE H.–U. (1969) Ignimbrite sequence on Gran Canaria. *Bull. Volcanol.* 33, 1199–1219.

SCHMITT R. W. (1979) The growth rate of supercritical salt fingers. *Deep Sea Res.* 26A, 23–40.

SCHMITT R. W. (1983) The characteristics of salt fingers in a variety of fluid systems, including stellar interiors, liquid metals, oceans, and magmas. *Phys. Fluids* 26, 2373–2377.

SEGAL L. A. and STUART J. T. (1962) On the question of preferred mode in cellular thermal convection. *J. Fluid Mech.* 13, 289–306.

SHAW H. R. (1965) Comments on viscosity, crystal settling and convection in granitic magmas. *Amer. J. Sci.* 263, 120–152.

SHAW H. R. (1969) Rheology of basalt in the melting range. *J. Petrol.* 10, 510–535.

SHAW H. R. (1974) Diffusion in granitic liquids: part I, experimental data; part II, mass transfer in magma chambers. In *Geochemical Transport and Kinetics* (eds. A. W. HOFFMAN et al.), pp. 139–170. Carnegie Inst. Wash. Publ. 634.

SMITH R. L. (1960) Zones and zonal variations in welded ash–flows. *U.S. Geol. Survey Prof. Paper 354-F*, pp. 149–159.

SMITH R. L. (1979) Ash–flow magmatism. *Geol. Soc. Amer. Special Paper 180*, pp. 5–27.

SOREY M. L. (1985) Evolution and present state of the hydrothermal system in long valley caldera. *J. Geophys. Res.* 90, 11219–11228.

SPARROW E. M. and GREGG J. L. (1956) Laminar free convection from a vertical plate with uniform surface heat flux. *Trans. Amer. Soc. Mech. Eng.* 78, 435–440.

SPERA F. J. (1980) Thermal evolution of plutons: a parameterized approach. *Science* 207, 299–301.

SPERA F. J., YUEN D. A. and KEMP D. V. (1984) Mass transfer rates along vertical walls in magma chambers and marginal upwelling. *Nature* 310, 764–767.

SPERA F. J., YUEN D. A. and KIRSCHVINK S. J. (1982) Thermal boundary layer convection in silicic magma chambers: effects of temperature dependent rheology and implications for thermogravitational chemical fractionation. *J. Geophys. Res.* 87, 8755–8767.

SPERA F. J., YUEN D. A., GREER J. C. and SEWELL G. (1986a) Dynamics of magma withdrawal from stratified magma chambers. *Geology* 14, 723–726.

SPERA F. J., YUEN D. A., CLARK S. and HONG H.–J. (1986b) Double–diffusive convection in magma chambers: single or multiple layers? *Geophys. Res. Lett.* 13, 153–156.

TOOMRE J., GOUGH D. O. and SPIEGEL E. A. (1977) Numerical solutions of single–mode convection equations. *J. Fluid Mech.* 79, 1–31.

TSUYA H. (1955) Geological and petrological studies of Volcano Fuji. *Earth Res. Inst. Bull.* 33, 341–383. Tokyo Imp. University.

TURCOTTE D. L. and OXBURGH E. R. (1967) Finite amplitude convection cells and continental drift. *J. Fluid Mech.* 28, 29–42.

TURNER J. S. (1980) A fluid dynamical model of differentiation and layering in magma chambers. *Nature* 285, 213–215.

TURNER J. S. (1985) Multicomponent convection. *Ann. Rev. Fluid Mech.* 17, 11–44.

VERONIS G. (1965) On finite amplitude instabilities in thermohaline convection. *J. Marine Res.* 23, 1–17.

VERONIS G. (1968) Effect of a stabilizing gradient of solute on thermal convection. *J. Fluid Mech.* 34, 315–336.

WEINSTEIN S. A., YUEN D. A. and OLSON P. (1986) Evolution of crystal–settling in magma–chamber convection. *Earth Planet. Sci. Lett.,* (Submitted).

WILLIAMS H. (1942) The geology of Crater Lake National Park, Oregon, with a reconnaissance of the Cascade Range southward to Mount Shasta. *Carnegie Inst. Wash. Publ. 540,* 162 pp.

Magmatic Processes: Physicochemical Principles
© The Geochemical Society, Special Publication No. 1, 1987
Editor B. O. Mysen

Crystal sizes in intrusions of different dimensions: Constraints on the cooling regime and the crystallization kinetics

GENEVIEVE BRANDEIS and CLAUDE JAUPART

Laboratoire de Physique des Processus Magmatiques, Institut de Physique du Globe,
Université Paris, 2, place Jussieu, 75252 Paris Cedex 05, France

Abstract—Crystallization in magmas depends on the crystallization kinetics as well as on the thermal regime. A dimensional analysis is presented which allows a simple understanding of the characteristics of crystallization. Characteristic scales for the rates of nucleation and crystal growth are used, denoted by I_m and Y_m respectively. The time–scale is given by $\tau_c = (Y_m^3 I_m)^{-1/4}$, and is close to the time required for crystallization to start in supercooled magma. The crystal size scales with $(Y_m/I_m)^{1/4}$, which provides a powerful constraint on the values of the nucleation and growth rates. The influence of the form of the kinetic functions for nucleation and growth is investigated. The form of the growth function relatively unimportant, in contrast to that of the nucleation function. In natural conditions following magma emplacement in cold country rocks, temperatures are continuously changing. Local scaling laws apply, whatever the boundary conditions are, with the characteristic time and crystal size given by $\tau = (Y^3 I)^{-1/4}$ and $R = (Y/I)^{1/4}$, where Y and I are the local rates. τ is the time to achieve crystallization and R the mean crystal size in a given piece of magma, and Y and I are the rates at which crystals were nucleated and grown locally. From petrological observations and laboratory crystallization experiments, the time–scale at high undercoolings is 2×10^5 sec. This gives the characteristic time for crystallization near the margins of intrusions. The time–scale is close to 10^8 sec in equilibrium conditions prevailing in the interior of large magma chambers. These can be compared to the characteristic times for cooling by conduction and convection. Several regimes are defined, depending on the intrusion dimensions. This allows a classification based on the average crystal size which agrees with petrological observations. The detailed study of dikes, sills and igneous complexes of different dimensions will allow constraints on poorly known conditions and parameters, such as the nucleation rate at small undercoolings.

INTRODUCTION

IN NATURE, magmatic crystallization proceeds in continuously evolving conditions, with both the melt temperature and composition changing. In large magma chambers, the end result is extremely complex, and the igneous record is difficult to decipher. Most studies so far have relied on equilibrium phase diagrams (MORSE, 1980) and on various dynamical processes such as thermal and compositional convection (HESS, 1960; JACKSON, 1961; WAGER and BROWN, 1968; MORSE, 1969; IRVINE, 1974; MCBIRNEY and NOYES, 1979; KERR and TAIT, 1986). The influence of the crystallization kinetics generally has been overlooked, despite a growing number of dynamic crystallization experiments in the laboratory (GIBB, 1974; WALKER *et al.*, 1976; DONALDSON, 1979; GROVE and BENCE, 1979; LOFGREN, 1980; KIRKPATRICK *et al.*, 1981; TSUCHIYAMA, 1983; BAKER and GROVE, 1985). These studies emphasize that crystals often nucleate metastably, and that the order of appearance of different crystalline phases follows that of their respective ease of nucleation (KIRKPATRICK, 1983). They further illustrate two critical facts. One is that nucleation is usually suppressed to undercoolings of several tens of degrees. The other is that the cool-

ing rate determines which phase nucleates first. The consequence is that the internal differentiation of magma bodies may not follow equilibrium paths, which presents an obvious problem when interpreting petrological observations. MORSE (1980, p. 229) and others (KIRKPATRICK, 1983; BAKER and GROVE, 1983) have recognized the petrological implications.

To estimate cooling rates in natural conditions, a common practice is to use simple thermal models such as those by JAEGER (1968) and then to evaluate the consequences for dynamic crystallization. What is seldom recognized is that the kinetics themselves influence the cooling rate because the thermal evolution is determined not only by the heat loss mechanism but also by the crystallization rate through the release of latent heat (KIRKPATRICK, 1976; BRANDEIS *et al.*, 1984). The evolution of crystallization in a magma body depends therefore on two factors: the kinetics of crystal nucleation and growth, and the cooling regime. The first depends on local conditions (temperature) and the second on conditions in the whole magma body. Both are poorly known for natural systems, which prevents the direct interpretation of petrological observations. Only slow progress can be expected in the near future from laboratory crystallization experiments for several reasons. One is that it is difficult to study natural silicate melts with complex chemical compositions and phase relationships. Another is that such natural melts take long times to crystallize. A third reason is that true crystallization conditions are continuously changing, and determined by the coupling between cooling and latent heat release. To proceed towards quantitative understanding of igneous rocks, one can try an integrated approach where field data are used to constrain poorly known parameters and processes. For example, if the thermal regime is known, crystal sizes can be calculated using the crystallization equations (BRANDEIS et al., 1984; BRANDEIS and JAUPART, 1986a). Comparisons with field data can then be used to calibrate the crystallization kinetics. On the other hand, if the kinetics are known, the field data can be converted to temperature and hence yield information on the cooling regime. BRANDEIS and JAUPART (1986a) have used the dimensional analysis of crystallization equations for conduction cooling together with crystal size data from dikes to obtain values for the nucleation and growth rates. The analysis is valid for thin dikes. Conduction is not, however, the main mechanism of heat transport in large magma bodies (SHAW, 1965). In those, convection dominates and changes the cooling history. Because the efficiency of convection is directly related to the size of the body, constraints can be obtained by comparing magma intrusions of different dimensions. For example, dikes of various thicknesses can be studied and compared to large magma chambers.

In this paper, we attempt to define the crystallization behaviour and thermal regime of intrusions of different dimensions, using constraints from the distribution of crystal size. Recent advances on how to model dynamic crystallization and scaling laws for crystal size variations are summarized. We define and give plausible values for the characteristic crystallization time and compare it to the cooling time for magma bodies of different sizes and chemical compositions. The plan is the following. First, a brief review of crystallization kinetics is given. Then, crystallization by conduction at the margin of a dike is addressed. The influence of different expressions for the kinetic functions is investigated. Finally, the various cooling regimes are reviewed and the thickness range for intrusions discussed.

KINETIC CONTROLS ON CRYSTALLIZATION

The effects of crystallization kinetics on natural samples are well documented, most notably in lava flows and mid-ocean ridge pillow basalts where pi-

geonite, the equilibrium liquidus phase, is absent (BRYAN, 1972; BAKER and GROVE, 1985). The first hint at kinetic effects in the petrological literature came perhaps from WAGER (1959) who recognized that nucleation could control the abundance of the different mineral phases that appear in layered igneous complexes. The first specific study on silicate melts was by GIBB (1974) on Columbia River basalt, who found that the temperature at which plagioclase begins to crystallize varies as a function of cooling rate. Since then, many dynamic crystallization experiments have been made (see references above). Following common practice in metallurgy, these are carried out under two different conditions: isothermal (constant temperature) and continuous cooling (constant cooling rate). The evolution of the crystallized product is followed as a function of time, yielding TTT (time–temperature–transformation) and CT (continuous cooling-transformation) diagrams respectively (SHEWMON, 1969). There are, unfortunately, no simple relationships between the two types of experiments on the same starting material (DONALDSON, 1979; TSUCHIYAMA, 1983). Thus, it is difficult to use them for extrapolation to natural conditions which are transient and often out of the range of the laboratory.

Another approach is to study directly the crystallization process by determining the rates of nucleation and growth. Nucleation and growth are distinct phenomena obeying different rules. In silicate melts, new crystal formation results either from the presence of foreign material in the melt, such as impurities or nuclei from a distant source, and from the fortuitous formation of molecular clusters of critical size, i.e., nuclei. These are called heterogeneous and homogeneous nucleation respectively. The rates for both phenomena can be expressed as a function of temperature and undercooling following kinetic theory (JOHNSON and MEHL, 1939; TURNBULL and FISHER, 1949). The general nucleation rate function has a bell–shape, illustrated in Figure 1 and subsequently called "shape 1 function". Crystal growth, on the other hand, requires two steps: solute must be transported to the crystal surface, a process that is usually controlled by chemical diffusion, and then oriented into the crystal lattice (attachment) (KIRKPATRICK, 1975; BARONNET, 1984). The attachment kinetics are relatively well understood and the corresponding growth–rate function also has a bell–shaped curve (BARONNET, 1984), illustrated again in Figure 1. In reality, when crystals grow large and fast, diffusion of solute through the melt becomes limiting. Unfortunately, there is no simple method to treat diffusion in a crowded environment with many crystals

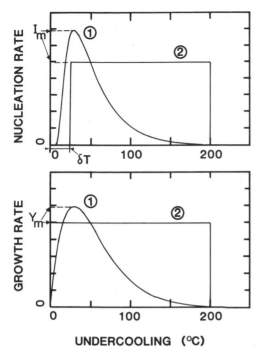

UNDERCOOLING (°C)

FIG. 1. The functions for the rates of crystal nucleation and growth, expressed as a function of undercooling. The scales are arbitrary. To define these kinetic functions completely, the scales I_m and Y_m must be specified. The bell-shaped curves correspond to standard kinetic theory (BRANDEIS *et al.*, 1984) and are termed "shape 1" functions. "Shape 2" functions are the simplest ones, and are used to investigate the influence of the shapes of the nucleation and growth curves. In each case, the nucleation and growth scales, are used for dimensional analysis. For shape 1 functions, these are the peak rates. In the shape 2 case, the scale is the constant value taken by the function. Note the delay of nucleation δT, which is the undercooling range for which nucleation is negligible.

and there is no general expression for the growth rate function. Under natural conditions, it seems that the attachment kinetics are the controlling phenomenon (KIRKPATRICK, 1977).

This brief summary probably makes it clear that there is no obvious theoretical way to specify the functions for nucleation and growth. The common practice in industrial crystallizers is to use empirical expressions derived from experimental data (RANDOLPH and LARSON, 1971, p. 110). To this end, two different steps are required. One is to specify some characteristic value for the growth and nucleation rates. These will be called the growth and nucleation scales. The other is to specify the form of the function. The bell-shaped curves of Figure 1 have been determined from standard kinetic theory (BRANDEIS *et al.*, 1984). One of the aims of this paper is to investigate the influence of the kinetic

functions on the crystallization behaviour. This will define the precision required from experimental data to allow realistic quantitative models.

Data are available on nucleation and growth rates in silicate melts. Most experiments show bell-shaped nucleation and growth curves, and in the following, we use published values to contrain the peak rates. The most complete set of measurements are those by FENN (1977) and SWANSON (1977). From these, BRANDEIS *et al.* (1984) demonstrated that the peak nucleation rates had to be between 10^{-2} and 10^2 cm^{-3} sec^{-1}. In a basaltic lava lake, KIRKPATRICK (1977) estimated values for plagioclase ranging from about 10^{-2} to 1 cm^{-3} sec^{-1}. More recently, TSUCHIYAMA (1983) obtained a value of 10^{-2} cm^{-3} sec^{-1} for diopside in the system $CaMgSi_2O_6$-$CaAl_2Si_2O_8$. The total range of possible values is, therefore, quite large, covering four orders of magnitude. Data on growth rates are more numerous and have been compiled by DOWTY (1980). From those, it appears that the growth rate decreases as the chemical system becomes more complex. For natural compositions, peak growth rates should be smaller than 10^{-7} cm sec^{-1}. In his natural crystallization experiment, the cooling lava lake, KIRKPATRICK (1977) gave values between 10^{-10} and 10^{-9} cm sec^{-1} for plagioclase. These values were obtained at small undercoolings and provide lower bounds for the peak rates. From these considerations, peak growth rates in silicate systems range from 10^{-9} to 10^{-7} cm sec^{-1}.

There is therefore quite a significant body of experimental data, showing significant differences among the various systems. One important question is to assess whether these differences lead to important variations in crystallization conditions. This question will be addressed in this paper. Once the kinetic functions for nucleation and growth are specified, it is possible to write the equation for crystallization involving the coupling between cooling and latent heat release. The first effort in the geological literature was by KIRKPATRICK (1976). However, he made a mathematical error which prevents reliable results. DOWTY (1980) investigated the test–case of a fixed rate of heat loss. He showed that latent heat release is very important and suppresses large undercoolings, bringing temperatures close to the liquidus even for rates of heat loss that are high by geological standards. His calculations give useful insights into the crystallization behaviour of silicate melts, but do not take into account the coupling with heat flow through country rocks and the continuous evolution of heat loss conditions as crystallization proceeds. Following these studies, BRANDEIS *et al.* (1984) tackled the full problem and presented solutions for conduction cooling in a variety of cases. BRANDEIS and JAUPART (1986a) later made a dimensional analysis of the same equations. In this paper, we summarize their results and investigate the consequences for the cooling regimes of magma bodies of different dimensions. In return, these considerations will allow constraints on the nucleation and growth rates.

DIMENSIONAL ANALYSIS FOR CRYSTALLIZATION

We consider a problem in one dimension only (z), perpendicular to the contact between magma and surrounding rocks. It is assumed that heat transfer is dominated by conduction, which is valid in boundary layers close to rigid boundaries (roof, floor and side-walls). The heat equation is written:

$$\partial T/\partial t = \kappa \partial^2 T/\partial z^2 + L/c_p \partial \Phi/\partial t \qquad (1)$$

where c_p is the isobaric heat capacity, κ thermal diffusivity, T temperature, t time and L the latent heat per unit mass. κ is thermal diffusivity. Φ is the crystal content per unit volume and takes values between 0 and 1. $\partial \Phi/\partial t$ depends on the rates of nucleation and growth according to the equation (KIRKPATRICK, 1976):

$$\partial \Phi/\partial t = 4\pi[1 - \Phi(z,t)]Y(t)\int_0^t I(v)$$

$$\times \left[\int_{t_v}^t Y(u)du\right]^2 dv. \qquad (2)$$

I and Y are the rates of nucleation and growth, respectively, and t_v is time of nucleation.

We consider the simplest cooling experiment. At time $t = 0$, magma with initial temperature T_L (the liquidus) is emplaced in country rocks that are colder by an amount ΔT. The initial conditions are that both magma and country rock are initially isothermal:

$$T(z,0) = T_L \quad \text{for} \quad z > 0 \quad \text{(magma)} \qquad (3a)$$

$$T(z,0) = T_L - \Delta T$$

$$\text{for} \quad z < 0 \quad \text{(country rocks)}. \qquad (3b)$$

The boundary conditions are:

$$T(+\infty, t) = T_L \qquad (3c)$$

$$T(-\infty, t) = T_L - \Delta T. \qquad (3d)$$

Note that these boundary conditions are specified at infinity. In practice, the magma body is of finite dimensions. This is not limiting for short times because conduction propagates slowly and only affects

a boundary layer that advances into uncrystallized magma. The main limitation of these equations is the conduction approximation, which will be discussed later.

The physical properties (κ, c_p) are assumed to be constant. The temperature scale is ΔT, and the scales for the rates of nucleation and growth are denoted by I_m and Y_m. A time-scale appears when Equation (2) is made dimensionless:

$$\tau_c = \{Y_m^3 I_m\}^{-1/4}. \qquad (4a)$$

Because heat transfer is by conduction, the corresponding length–scale is simply given by:

$$d_c = (\kappa \tau_c)^{1/2}. \qquad (4b)$$

The non-dimensional variables are denoted by primes:

$$t = t'\tau_c \qquad (5a)$$

$$z = z'd_c \qquad (5b)$$

$$I = I'I_m \qquad (5c)$$

$$Y = Y'Y_m \qquad (5d)$$

$$T = T'\Delta T. \qquad (5e)$$

With these, Equations (1) and (2) can be made dimensionless, which introduces a non-dimensional number called the Stefan number:

$$\sigma = \frac{L}{c_p \Delta T}. \qquad (6)$$

The Stefan number is a measure of the importance of latent heat in the temperature equation. If it is small ($\sigma \ll 1$), temperature is given by the heat equation without latent heat. In geological cases, the Stefan number is of order 1, which implies that latent heat must be taken into account.

Crystallization proceeds within a region of finite thickness called the crystallization interval, which advances into uncrystallized magma. The moving boundary between fully crystallized and crystallizing magma is called the crystallization front and is such that the crystal content Φ is equal to 0.99 (Figure 2). Its coordinate is denoted by $X(t)$. At $z = X(t)$, the temperature is $\theta(t)$. The crystallization interval is defined as the zone where $0.01 < \Phi < 0.99$ (Figure 2) and has thickness $\epsilon(t)$. Both $X(t)$ and $\epsilon(t)$ scale with the crystallization length-scale defined by Equation (4b).

It is also possible to calculate the crystal size. In dimensional variables, a unit volume of crystallized material comprises N crystals, with mean radius R:

$$N(z) = \int_0^\infty [1 - \Phi(z,t)]I(t)dt \qquad (7a)$$

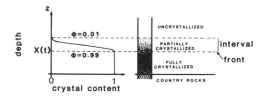

FIG. 2. Definition of the crystallization variables. The crystallization front is the moving boundary between crystallizing and fully crystallized magma. Its coordinate is denoted by $X(t)$. The region where magma is partially crystallized is called the crystallization interval, defined as the zone where $0.01 < \phi < 0.99$. Its thickness is $\epsilon(t)$.

$$R(z) = [4/3\pi N(z)]^{-1/3}. \tag{7b}$$

A size-scale appears when Equation (7) is made dimensionless (BRANDEIS and JAUPART, 1986a):

$$R_c = \{Y_m/I_m\}^{1/4}. \tag{8a}$$

The dimensionless crystal size is obtained by:

$$R' = R/R_c. \tag{8b}$$

To summarize, knowledge of characteristic scales for the rates of nucleation and growth allow the definition of a time–scale, a length–scale and a size–scale, *i.e.* of all the important crystallization parameters. An important result has already been obtained. Both the time–scale and the size–scale depend on the nucleation rate are raised to the power (¼), which shows that they are weakly sensitive to its value. The four orders of magnitude uncertainty on I_m is therefore not critical. The influence of the growth rate is slightly more important, but there are better data.

Because Equations (1) and (2) are coupled, the problem has to be solved numerically. We first derive dimensionless relationships for the different crystallization parameters. As they all depend on the kinetic rates, the influence of their shape on the crystallization behaviour will be investigated. Because both exhibit a steep increase just below the liquidus, it is worthwhile to evaluate how sensitive the results are to their shape. The influence of σ (Stefan number) is weak and only changes the values of the coefficients in the different relationships (BRANDEIS and JAUPART, 1986a).

SCALING LAWS FOR CRYSTALLIZATION

To derive the various scaling laws, "shape 1" nucleation and growth functions are used (Figure 1). These have been determined by using kinetic theory and laboratory data. For these, the scales are given by the peak values. An important parameter is the nucleation delay δT, which is the minimum un-

dercooling for the formation of nuclei. In the following, the Stefan number is taken to be 0.55, which corresponds to high temperature contrasts (BRANDEIS and JAUPART, 1986a).

The evolution of crystallization

The position of the crystallization front $X(t)$ is related to time according to a power–law (Figure 3):

$$X(t) = \lambda t^n \tag{9}$$

where the exponent, n, is close to ½, but slightly different. The ½ power–law corresponds to the ideal case of latent heat release at a fixed melting point (JAEGER, 1968) or of a binary alloy without kinetic effects (WORSTER, 1986). The crystallization kinetics do not alter significantly this simple law, due to the control by heat diffusion.

The evolution of the crystallization interval thickness $\epsilon(t)$ is shown in Figure 4. The general evolution is a slow increase with few oscillations due to the discontinuous character of the nucleation process (BRANDEIS *et al.*, 1984). For this value of σ, the data can be fitted with an error of a few percent with a law (in dimensionless variables):

$$\epsilon(t) = 0.3t^{0.31}. \tag{10}$$

As cooling proceeds, undercoolings in the crystallization interval decrease. This is illustrated by the evolution of the undercooling at the crystallization front, shown in Figure 5. The general evolution is

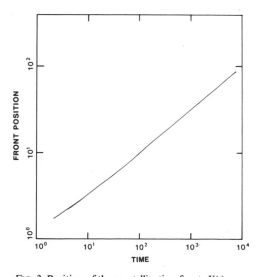

FIG. 3. Position of the crystallization front, $X(t)$, versus time in dimensionless variables for a Stefan number of 0.55. Time $t = 0$ marks the emplacement of magma in country rocks which are colder by an amount of ΔT.

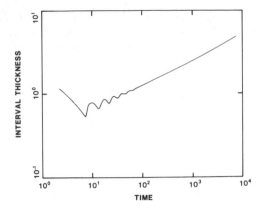

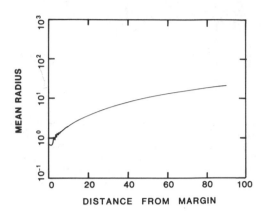

FIG. 4. Evolution of the interval thickness, $\epsilon(t)$, as a function of time in dimensionless variables for a Stefan number of 0.55.

FIG. 6. Mean crystal size as a function of distance to the margin in dimensionless variables for a Stefan number of 0.55.

a decrease towards a small value which is equal to δT (BRANDEIS and JAUPART, 1986a). There are a few oscillations due to nucleation steps. These are damped as there is a tendency to achieve an equilibrium between latent heat release and heat loss. Note that θ has not been calculated for times $t < 1$ (in dimensional variables, for times smaller than the crystallization time–scale), because crystallization has not started and hence no crystallization front can be defined. This shows that τ_c [Equation (4a)] is close to the onset time for crystallization.

The crystal size

The variation of dimensionless crystal size as a function of the distance to the margin is shown in Figure 6. Near the margin, the dimensionless crystal size is close to 1. This shows that the size–scale, R_c, is equal to the crystal size there. Two kinds of factors must be considered for extrapolation to natural

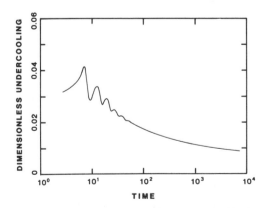

FIG. 5. Dimensionless undercooling at the crystallization front, $\theta(t)$, as a function of dimensionless time for a Stefan number of 0.55.

conditions. The first is that, as crystallization proceeds, processes other than thermal conduction may become limiting. In particular, the whole magma chamber is cooling, which implies a change in boundary condition (3c) for the temperature equation. A second factor is that most crystallization occurs in equilibrium conditions at small undercoolings. The measurements by FENN (1977) and SWANSON (1977) only allow constraints on the peak nucleation rate. At small undercoolings, a detailed understanding of how the nucleation rate varies with temperature is lacking (see the discussion on oscillatory crystallization by BRANDEIS et al., 1984). Specifically, the shape of the nucleation function must be known with precision. In the preceding calculations, the nucleation rate tends to zero continuously as θ tends towards the nucleation delay, δT. The unavoidable consequence is that the crystal size would eventually reach infinity, which is not realistic. This stresses the need for reliable data on nucleation rates at small undercoolings. The influence of the shape of the two kinetic functions is now addressed.

THE FORM OF THE KINETIC FUNCTIONS

We have so far relied on reasonable expressions for the relationships between the kinetic rates and temperature. We compare those, referenced as shape 1 functions, to the simplest ones: box–car functions, referenced as shape 2 (Figure 1). These functions have a constant value throughout the crystallization range and are zero elsewhere. They represent the limit–case of a discontinuous behavior. For the nucleation process, this is not unreasonable and approximates the effect of a finite energy barrier for the formation of one nucleus. Intuitively, these functions represent the obvious way to achieve a constant crystal size throughout the crystallization sequence. We show calculations for the crystal size for all possible combinations of shape 1 and shape 2 functions (Figure 7). We keep the same procedure for making variables dimensionless. For shape 2 functions, the scale is given by the constant value. In this section, the Stefan number, σ, is taken equal to 0, as this limiting case is simple and yet shows of the most features of the crystallization behaviour (BRANDEIS and JAUPART, 1986a).

How crystal size varies reflects the shape of the nucleation function. For a shape 2 nucleation function, the crystal size is constant after a small transient (Figure 7). The value is close to the size–scale defined by Equation (8a), with a proportionality constant between 0.5 and 0.7 depending on the shape of the growth function.

It is concluded that the shape of the nucleation function critically influences the results, contrary to that of the growth function. Nucleation is, therefore, the process governing crystallization. The main limitation of these calculations is, thus, the lack of constraints on the nucleation function. Crystal growth does play a role in determining the numerical constants in the scaling laws derived above, but does not have any effect on the crystallization behaviour. It is nucleation which is responsible for temperature oscillations as well as the tendency to maintain a constant crystal size.

DISCUSSION

Before applying this analysis to crystallization in magma bodies of different thickness, its validity in

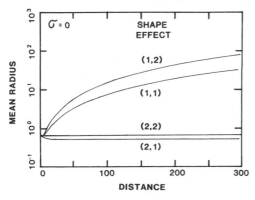

FIG. 7. Evolution of the mean crystal size as a function of the distance to the margin for several kinds of kinetic functions. The numbers in brackets represent the type of function used for nucleation and growth, respectively (shape 1 or 2, see Figure 1). The Stefan number is set equal to 0.

all cases needs assessment, as several assumptions have been made. These principally are two. First, chemical diffusion was neglected in the expressions for the kinetic rates. Second, cooling was achieved by heat conduction.

The crystallization kinetics

Crystal growth is controlled by chemical diffusion and reactions at the crystal–melt interface. The rate controlling process is diffusion at large times for a single crystal, as shown by the time variation of growth rate at fixed undercooling (LOOMIS, 1982; LASAGA, 1982). Diffusion is also important for rapidly cooled magma, for example in lava flows and chilled margins, resulting in spherulitic or dentritic crystal morphologies. In the interior of large intrusions, crystals do not exhibit these morphological instabilities and growth is assumed to be controlled by the interface attachment kinetics (CAHN, 1967; KIRKPATRICK, 1975; BARONNET, 1984). Also, there are thick accumulate layers (WAGER and BROWN, 1968). Compositional convection can occur in the porous cumulate pile (MORSE, 1969; TAIT et al., 1984; KERR and TAIT, 1986), bringing the required chemical components to achieve adcumulate growth.

Nucleation is the controlling phenomenon for crystallization. At any given depth, nucleation occurs essentially once and is followed by growing the existing nuclei (BRANDEIS et al., 1984). Crystallization occurs in a thick region which moves into uncrystallized magma. The exact growth rate function has no influence on the results (Figure 7). By considering a mean value for the growth rate, which takes into account chemical diffusion and other processes, the scaling laws can be applied successfully. The comparison between petrological observations and calculations yields values for the peak nucleation and growth rate that are close to the experimental data (BRANDEIS and JAUPART, 1986a).

An important result of this study is that the shape of the nucleation function is a critical parameter, especially at small undercoolings. The fact that the crystal size does not vary in the interior of large igneous complexes shows that the nucleation rate does not tend to zero continuously. There is probably a limiting value, which presumably corresponds to a finite energy barrier for the formation of a nucleus of critical size.

We conclude that, in order to explain both the remarkably regular variation of the crystal size at the margins of dikes (Figure 8) as well as the uniformity of crystal sizes in the interior of magma chambers, one needs a combination of shape 1 and

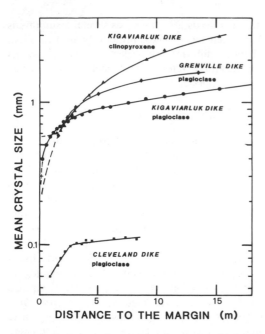

FIG. 8. Variations of crystal sizes in natural dikes (from WINKLER, 1949 for the Cleveland dike; GRAY, 1970 for the Kigaviarluk and Grenville dikes).

shape 2 functions for the nucleation rate. For scaling purposes, one must carry out two separate analyses. One is for highly transient conditions pertaining to the margins, where shape 1 functions must be used. The other is for equilibrium conditions pertaining to the interior where crystallization proceeds at small undercoolings, where a shape 2 nucleation function must be used.

Thermal regime

Two assumptions have been made in regard to the thermal regime. First, cooling is by conduction which is valid for thin dikes and sills, as well as at the bottom of large magma bodies. There, crystallization occurs mainly *in situ* (CAMPBELL, 1978; MCBIRNEY and NOYES, 1979) in stagnant layers isolated from convection (JACKSON, 1961; JAUPART *et al.*, 1984). At the bottom, there is a steep viscosity increase due both to lower temperature and higher crystallinity, and hence, any motions occurring in the chamber interior do not affect the crystallization interval. This indicates that conduction is the dominant means of heat transfer there (see also HUPPERT and SPARKS, 1980). The second assumption is made in boundary condition (3c), where we have, in fact, treated the case of an intrusion of infinite dimensions. In reality, this is not the case, which has implications investigated in the last section.

Local scaling laws

Equations (4a) and (8a) give correct values for highly transient conditions. In fact, the crystal size evolution shown in Figure 6 clearly reflects the evolution of the crystallization temperature (Figure 5). The same is true for the crystallization interval. From Equations (2) and (7), both the crystal size and the crystallization time can be defined locally, by using the instantaneous values of I and Y at the crystallization temperature, θ:

$$R = [Y(\theta)/I(\theta)]^{1/4} \qquad (11a)$$

$$\tau = [I(\theta)Y(\theta)^3]^{-1/4}. \qquad (11b)$$

By using the curve for undercoolings at the crystallization front (Figure 5) and local relationships [Equation (11)], one can reproduce to a good approximation the curves for the crystal size and the crystallization interval. Even though conditions are highly transient, the local scaling procedure gives reasonable results, and should be useful even when cooling is affected by other processes such as convection. If the mean crystal size and the growth rate are known, Equation (11a) gives the local value of the nucleation rate, independently of the thermal regime and the conditions at the other boundaries.

BRANDEIS and JAUPART (1986a) have used the data from dike margins (Figure 9) to constrain the values of the maximum kinetic rates. The nucleation and growth scales, I_m and Y_m, are found to be close to 1 cm^{-3} sec^{-1} and 10^{-7} cm sec^{-1}, respectively. For transient conditions prevailing in early stages at the margins, the characteristic time scale is 2×10^5 sec, and will be termed τ_1.

In equilibrium cooling conditions that should prevail in the interior of large magma bodies, crystallization occurs at small undercoolings and hence at smaller rates. From petrological observations, we have suggested ranges of 10^{-7} to 10^{-3} cm^{-3} sec^{-1} for the local nulceation rate and a range of 10^{-10} to 10^{-8} cm sec^{-1} for the local growth rate. These values agree with laboratory measurements (DOWTY, 1980). For those conditions, the time scale has a minimum value of 10^7 sec and is probably close to 10^8 sec. An upper bound is 10^9 sec. This time will be termed τ_3. It is large, which has important implications detailed below.

THE TIMES FOR CRYSTALLIZATION AND COOLING

A brief summary is worthwhile at this stage. In the simple cooling experiment studied above, crystallization starts after a finite time close to τ_1 at undercoolings of several tens of degrees (Figure 5).

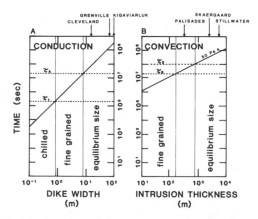

FIG. 9. Comparison between the characteristic times for cooling and crystallization in various cases. (A) Conduction cooling for dikes of different widths. The thick curve is from Equation (12) and gives the characteristic time for conduction cooling. τ_1 is the time for the onset of crystallization. Dikes which cool faster than τ_1 cannot nucleate crystals and hence are chilled. τ_2 is the time needed to achieve equilibrium cooling conditions. Dikes which cool faster than τ_2 crystallize at high undercoolings and hence develop small crystals. (B) Convection cooling in sills and magma chambers. The thick curve is from Equation (13) and gives the characteristic time for convection cooling. Intrusions which cool faster than τ_2 crystallize at high undercoolings and hence develop small crystals. τ_3 is the characteristic time for crystallization in equilibrium conditions at small undercoolings. If the cooling time is close to τ_3, crystallization has time to record the effects of convective processes operating in the intrusion interior. This implies that complex igneous structures will be found in the solidified rocks.

At later times, undercoolings in the crystallization interval decrease and equilibrium cooling conditions are approached after a time, τ_2, which is about $10^2 \times \tau_1$ (Figures 5 and 6). There is a marked difference between the cooling rates achieved in the initial transient regime, and the equilibrium rate. Note that the time needed to reach equilibrium conditions is solely determined by the crystallization kinetics. For the preferred values of the nucleation and growth scales, equilibrium conditions are reached once the thickness of crystallized magma exceeds about 5 m. This corresponds to the flattening of the crystal size curve in natural dikes (Figure 8).

In order to fit the dike data, values of 1 cm^{-3} sec^{-1} and 10^{-7} cm sec^{-1} are required for the nucleation and growth scales. These are compatible with data from laboratory experiments (see above). This shows the use of crystal size data from intrusions. As discussed above, in any volume of igneous rock, the crystal size is given by the instantaneous values of the nucleation and growth rates that prevailed at

the time it crystallized. From both the theory given is this paper and the observations, the crystal size reaches a nearly constant value close to 1 mm in the interior of dikes away from the margins. This size is also that which is observed in almost all igneous complexes (see the compilation in BRANDEIS et al., 1984). This observation suggests that most crystallization occurs under similar conditions with similar instantaneous values of the nucleation and growth rates.

Further progress in quantitative models necessitates knowledge of the nucleation and growth functions at small undercoolings. To this end, one possibility is to investigate intrusions of different dimensions. The calculations so far have relied on the assumption that the magma body is infinite. In this case, after the initial transients, uniform conditions are realized throughout the crystallization sequence, with, for instance, similar undercoolings and hence similar crystal sizes (Figures 5 and 6). In reality, a magma body is of finite dimensions and cools. The result is that it is no longer possible to assume a fixed temperature away from the crystallization interval, as was done explicitly in boundary condition (3c). This conclusion leads to a variety of cases depending on how temperature varies in the intrusion interior. For example, the limiting case is that of a very thin dike that cools very rapidly, leaving no time for crystals to nucleate, and hence chills. In the following, various possibilities for cooling conditions are discussed. The arguments are rough, and meant to illustrate basic principles and to indicate the direction for future studies. The idea is to compare the characteristic times for various phenomena. These can be defined to within maybe one order of magnitude, which is not critical because the range of natural conditions is much larger.

Conduction cooling in dikes of different widths

Conduction is the dominant mechanism of heat transport in thin dikes. For larger magma bodies, convection is important and modifies the thermal history. Because our understanding of convection in magmas is far from complete (BRANDEIS and JAUPART, 1986b), it is not easy to define the limit between conduction-dominated and convection-dominated cases. According to SHAW (1965), convection should be important in magma bodies exceeding thicknesses of a few tens of meters. For dikes that are vertical intrusions with very large aspect ratios, we consider that conduction dominate up to 100 m width. This spans the range of most dikes, including the Cleveland, Grenville and Kigaviarluk dikes. Those dikes are 16 m, 60 m and 106 m wide respectively (WINKLER, 1949; GRAY, 1970). If conduction dominates, a rough estimate for the cooling time is:

$$tc \simeq \frac{D^2}{4\kappa} \qquad (12)$$

where D is the dike width and κ thermal diffusivity, for which we take a value 7×10^{-7} m^2 sec^{-1}. This cooling time ranges from 3.6×10^3 sec for a 10 cm–wide dike to 3.6×10^9 sec for a 100 m–wide dike (Figure 9a). τ_1, the time for the onset of crystallization in supercooled magma, is about 2×10^5 sec and, therefore, falls within this range. Thus, in thin dikes, cooling occurs before crystals can nucleate, which leads to chilling. According to our estimates, this should be true up to a width of 1 m (Figure 9a), which agrees roughly with common field observations. Another point is that it takes time, τ_2, or about 2×10^7 sec, to achieve equilibrium crystallization conditions with crystal sizes of about 1 mm. This implies that dikes which cool faster than τ_2 will not experience such conditions, which means higher undercoolings and hence smaller crystal sizes. According to our rough comparison, this should correspond to dikes thinner than about 10 m (Figure 9a). Note that in the 16 m wide Cleveland dike, the crystal sizes are smaller than in the 100 m wide Kigaviarluk one (Figure 8), which agrees roughly with the prediction. The point is that studying dikes with widths in the range of 1–10 m means spanning different cooling histories implying different crystal sizes. Detailed investigations could therefore allow constraints on the crystallization kinetics for a range of undercoolings.

Convection cooling in thick sills and large magma chambers

As already discussed, convection is likely to be important in magma bodies which are thicker than a few tens of meters. In the following, we compare magma bodies with thicknesses ranging from 10 m to 10 km. One observation is that the Palisades sill, which is 330 m–thick (WALKER, 1940), has crystal sizes of less than 1 mm in its interior, somewhat smaller than values in larger intrusions (BRANDEIS et al., 1984). Let us focus on crystallization at the bottom which is controlled by conduction (see JAUPART and BRANDEIS, 1986). We have seen that it takes about 2×10^7 sec to reach equilibrium crystallization conditions. Now, the characteristic time for convective cooling is (JAUPART and BRANDEIS, 1986):

$$t_v = 15 D^2 \kappa^{-1} \text{Ra}^{-1/3} \qquad (13)$$

where Ra is the Rayleigh number defined by:

$$\text{Ra} = \frac{\rho g \alpha \Delta T D^3}{\kappa \mu}. \qquad (14)$$

D is the whole chamber thickness, μ is dynamic viscosity and α thermal expansion coefficient. This equation shows that the time for convective cooling is proportional to the intrusion thickness. Figure 9b shows its variation as a function of D for a viscosity of 50 Pa sec representative of basalts. Note that it is 6×10^5 sec for a 10 m thick sill and that it exceeds 10^8 sec for kilometer–sized chambers.

It appears, therefore, that intrusions thinner than about 300 m should cool significantly before equilibrium crystallization conditions can be reached (Figure 9b). This is like the dike case discussed above, implying higher undercoolings in the crystallization interval and hence smaller crystal sizes. One should, therefore, expect the Palisades sill to be finer grained than the large Skaergaard and Stillwater chambers, which is indeed the case. The agreement between the observations and this simple argument is not perfect: the Palisades sill lies slightly away from the boundary separating intrusions with equilibrium crystal sizes (Figure 9b). Again, the point is that making the agreement better will provide constraints on the rates of nucleation and growth. The results obtained so far are encouraging and suggest that such a study will yield meaningful information.

There is another consequence. We gave an estimate of about 10^8 sec for τ_3, the crystallization time under equilibrium conditions. This is the typical time taken for a volume of magma to become fully crystallized in the deep interior of a large magma chamber. A useful question is whether this volume of magma experiences varying temperatures due to the cooling of the intrusion. In other words, will it contain a record of the evolution of the whole magma body? As seen in Figure 9b, times for convective cooling are similar to τ_3 for most known basic and ultrabasic complexes including the Skaergaard and the Stillwater intrusions. This conclusion suggests that crystallization is slower than convective processes which occur in the interior of large magma chambers, and hence that it is able to record their effects in solidifying rocks. This also explains in a rough way why large intrusions exhibit igneous structures which are much more complex than those from small sills and dikes.

CONCLUSION

We have derived a series of scaling laws which allow a simple analysis of the crystallization behaviour. For crystallization, the most important process is nucleation, whose rate can be constrained by laboratory experiments and petrological observations. Knowledge of these kinetic rates allows calculation of the characteristic crystallization time. A simple analysis is made to compare crystallization and cooling times in intrusions of various dimensions. Different regimes are defined, which lead to observable variations in the crystal size. By using numerical values for the nucleation and growth scales, we find that chill features should be frequent in meter sized dikes. The important point is that observations of crystal sizes and morphologies in intrusions of various dimensions can be used to constrain the crystallization time, and hence the crystallization kinetics. Furthermore, the analysis allows for the treatment of different viscosity values. Hence, using data from dikes of varying chemical compositions could also prove useful.

Acknowledgements—Many of the ideas expressed in this paper developed out of conversations with Stephen Tait. We thank Hatten Yoder, Jr. for the invitation to the Hawaiian conference on the principles of magmatic processes. We appreciate the encouragement of Danielle Velde. Bjorn Mysen improved the manuscript.

REFERENCES

BAKER M. B. and GROVE T. L. (1985) Kinetic controls on pyroxene nucleation and metastable liquid lines of descent in basaltic andesite. *Amer. Mineral.* **70,** 279–287.

BARONNET A. (1984) Growth kinetics of the silicates. A review of basic concepts. *Fortschr. Mineral.* **62,** 2, 187–232.

BRANDEIS G., JAUPART C. and ALLEGRE C. J. (1984) Nucleation, crystal growth and the thermal regime of cooling magmas. *J. Geophys. Res.* **89,** 10161–10177.

BRANDEIS G. and JAUPART C. (1986a) The kinetics of nucleation and crystal growth and scaling laws for magmatic crystallization. *Contrib. Mineral. Petrol.,* (Submitted).

BRANDEIS G. and JAUPART C. (1986b) On the interaction between convection and crystallization in cooling magma chambers. *Earth Planet. Sci. Lett.* **77,** 345–361.

BRYAN W. B. (1972) Morphology of quench crystals in submarine basalts. *J. Geophys. Res.* **77,** 5812–5819.

CAHN J. W. (1967) On the morphological stability of growing crystals In *Crystal Growth,* (ed. H. S. PREISER), pp. 681–690 Pergamon Press.

CAMPBELL I. H. (1978) Some problems with the cumulus theory. *Lithos* **11,** 311–323.

DONALDSON C. H. (1979) An experimental investigation of the delay in nucleation of olivine in mafic magmas. *Contrib. Mineral. Petrol.* **69,** 21–32.

DOWTY E. (1980) Crystal growth and nucleation theory and the numerical simulation of igneous crystallization In *Physics of Magmatic Processes* (ed., R. B. HARGRAVES), pp. 419–485 Princeton Univ. Press.

FENN P. M. (1977) The nucleation and growth of alkali feldspars from hydrous melts. *Can. Mineral.* **15,** 135–161.

GIBB F. G. F. (1974) Supercooling and crystallization of plagioclase from a basaltic magma. *Mineral. Mag.* **39,** 641–653.

GRAY N. H. (1970) Crystal growth and nucleation in two large diabase dikes. *Can. J. Earth Sci.* **7,** 366–375.

GROVE T. L. and BENCE A. E. (1979) Crystallization kinetics in a multiply saturated basalt magma: An experimental study of Luna 24 ferrobasalt. *Proc. Lunar and Planetary Science Conf. 10th,* 439–479.

HESS H. H. (1960) Stillwater igenous complex, Montana: A quantitative mineralogical study. *Geol. Soc. Amer. Mem. 80,* 230 pp.

HUPPERT H. E. and SPARKS R. S. J. (1980) The fluid dynamics of a basaltic magma chamber replenished by influx of hot, dense ultrabasic magma. *Contrib. Mineral. Petrol.* **75,** 279–289.

IRVINE T. N. (1974) Petrology of the Duke Island ultramafic complex, Southeastern Alaska. *Geol. Soc. Amer. Mem. 138,* 240 pp.

JAEGER J. C. (1968) Cooling and solidification of igneous rocks, In *Basalts: The Poldervaart Treatise on Rocks of Basaltic Composition,* (eds. H. H. HESS and ARIE POLDERVAART), vol 2. pp. 503–536 New York, John Wiley & Sons Inc.

JACKSON E. D. (1961) Primary textures and mineral associations in the ultramafic zone of the Stillwater complex, Montana. *Geol. Surv. Prof. Pap. 358,* 106 pp.

JAUPART C., BRANDEIS G. and ALLEGRE C. J. (1984) Stagnant layers at the bottom of convecting magma chambers. *Nature* **308,** 535–538.

JAUPART C. and BRANDEIS G. (1986) The stagnant bottom layer of convecting magma chambers. *Earth Plan. Sci. Lett.* (In press).

JOHNSON W. A. and MEHL R. F. (1939) Reaction kinetics in processes of nucleation and growth. *Trans. Amer. Inst. Min. Metall. Pet. Eng.* **135,** 416–442.

KERR R. C. and TAIT S. R. (1986) Crystallization and compositional convection in a porous medium with application to layered intrusions. *J. Geophys. Res.* **91,** 3591–3608.

KIRKPATRICK R. J. (1975) Crystal growth from the melt: a review. *Amer. Min.* **60,** 798–814.

KIRKPATRICK R. J. (1976) Towards a kinetic model for the crystallization of magma bodies. *J. Geophys. Res.* **81,** 2565–2571.

KIRKPATRICK R. J. (1977) Nucleation and growth of plagioclase, Makaopuhi and Alae lava lakes, Kilauea volcano, Hawaii. *Bull. Geol. Soc. Amer.* **88,** 78–84.

KIRKPATRICK R. J. (1983) Theory of nucleation in silicate melts. *Amer. Mineral.* **68,** 66–77.

KIRKPATRICK R. J., KUO L. C. and MELCHIOR J. (1981) Crystal growth in incongruently melting compositions: programmed cooling experiments with diopside. *Amer. Mineral.* **66,** 223–241.

LASAGA A. C. (1982) Towards a master equation in crystal growth. *Amer. J. Sci.* **282,** 1264–1320.

LOFGREN G. E. (1980) Experimental studies on the dynamic crystallization of silicate melts, In *Physics of Magmatic Processes,* (ed., R. B. HARGRAVES), pp. 487–551 Princeton Univ. Press.

LOOMIS T. P. (1982) Numerical simulations of crystallization processes of plagioclase in complex melts: the origin of major and oscillatory zoning in plagioclase. *Contrib. Mineral. Petrol.* **81,** 219–229.

MCBIRNEY A. R. and NOYES R. M. (1979) Crystallization and layering of the Skaergaard intrusion. *J. Petrol.* **20,** 3, 487–554.

MORSE S. A. (1969) The Kiglapait layered intrusion, Labrador. *Geol. Soc. Amer. Mem., 112,* 204 pp.

MORSE S. A. (1980) *Basalts and Phase Diagrams.* 493 pp. Springer-Verlag.

RANDOLPH A. D. and LARSON M. A. (1971) *Theory of Particulate Processes.* 251 pp. Academic Press, New York.

SHAW H. R. (1965) Comments on viscosity, crystal settling and convection in granitic magmas. *Amer. J. Sci.* **263,** 120–153.

SHEWMON P. G. (1969) *Transformations in Metals.* McGraw Hill, New York.

SWANSON S. E. (1977) Relation of nucleation and crystal-growth rate to the development of granitic textures. *Amer. Mineral.* **62,** 966–978.

TAIT S. R., HUPPERT H. E. and SPARKS R. S. J. (1984) The role of compositional convection in the formation of adcumulate rocks. *Lithos,* **17,** 139–146.

TSUCHIYAMA A. (1983) Crystallization kinetics in the system $CaMgSi_2O_6-CaAl_2Si_2O_8$: the delay in nucleation of diopside and anorthite. *Amer. Mineral.* **68,** 687–698.

TURNBULL D. and FISCHER J. C. (1949) Rate of nucleation in condensed systems. *J. Chem. Phys.* **17,** 71–73.

WAGER L. R. (1959) Differing powers of crystal nucleation as a factor producing diversity in layered igneous intrusions. *Geol. Mag.* **96,** 75–80.

WAGER L. R. and BROWN G. M. (1968) *Layered Igneous Rocks.* 588 pp. Edinburgh, Oliver and Boyd.

WALKER F. (1940) Differentiation of the Palisade diabase, New Jersey. *Bull. Geol. Soc. Amer.* **51,** 1059–1106.

WALKER D., KIRKPATRICK R. J., LONGHI J. and HAYS J. F. (1976) Crystallization history of lunar picritic basalt 12002: phase equilibria and cooling-rate studies. *Bull. Geol. Soc. Amer.* **87,** 646–656.

WINKLER H. G. F. (1949) Crystallization of basaltic magma as recorded by variation of crystal size in dikes. *Mineral. Mag.* **28,** 557–574.

WORSTER M. G. (1986) Solidification of an alloy from a cooled boundary. *J. Fluid. Mech.* **167,** 481–501.

Magmatic Processes: Physicochemical Principles
© The Geochemical Society, Special Publication No. 1, 1987
Editor B. O. Mysen

Evaluating reaction stoichiometry in magmatic systems evolving under generalized thermodynamic constraints: Examples comparing isothermal and isenthalpic assimilation

MARK S. GHIORSO and PETER B. KELEMEN

Department of Geological Sciences, University of Washington, Seattle, WA 98195, U.S.A.

Abstract—Legendre transforms are utilized to construct thermodynamic potentials which are minimal in systems at thermodynamic equilibrium under conditions of fixed enthalpy, volume and chemical potential of oxygen. Numerical methods that facilitate the minimization of these potentials are described. These methods form the core of an algorithm for the calculation of reaction stoichiometry for irreversible processes in open magmatic systems proceeding under isenthalpic or isochoric constraints, or both.

As an example of the use of these methods, calculations of isothermal and isenthalpic assimilation in fractionating basaltic magma are presented. Isenthalpic calculations are most useful in considering the effects of solid–liquid reactions which are endothermic. In general, such reactions involve the assimilation of a less refractory phase in a high temperature magma, or assimilation of phases across a cotectic "valley" in the liquidus surface. The mass of crystals calculated to form during isenthalpic assimilation of pelitic rock in a magnesian MORB is six to ten times as great if the rock is assumed to be at 500°C than if the rock is at 1250°C. The examples illustrate the necessity of characterizing the temperature and phase assemblage, as well as the bulk composition, of the assimilate when considering the effect of solid–liquid reaction on the liquid line of descent.

INTRODUCTION

CHEMICAL MASS TRANSFER in magmatic systems has been simulated previously as a succession of steps in temperature, pressure or bulk composition, with each step characterized by heterogeneous equilibrium (GHIORSO, 1985; GHIORSO and CARMICHAEL, 1985). This approach has made possible the numerical simulation of magmatic processes such as equilibrium crystallization and crystal fractionation. This method has also proven useful in the investigation of the assimilation of solid phases (GHIORSO and CARMICHAEL, 1985; KELEMEN and GHIORSO, 1986). However, existing computational techniques are inadequate for calculation of reaction stoichiometry in thermodynamic systems evolving toward an equilibrium state under general system constraints. By general constraints we mean those other than temperature (T), pressure (P) and fixed bulk composition. There are two reasons for this inadequacy. Firstly, the Gibbs free energy, G, is not necessarily minimal in a thermodynamic system that is generally constrained. Thus, to calculate heterogeneous equilibrium at each step in a mass transfer calculation, proceeding under a specified set of generalized constraints, requires the generation of a thermodynamic potential which is minimal under these conditions. The second difficulty is computational. Constraints other than temperature, pressure or fixed bulk composition are usually non-linear functions of their independent variables, *i.e.,* the enthalpy, H, is a non–linear function of temperature, pressure and the number of moles of system components. Therefore, after deduction of the appropriate potential, its minimum must be computed by finding those particular values of the independent variables of the system that satisfy the non–linear constraint(s). This is a formidable numerical problem, which in magmatic systems has only been attempted in an approximate fashion (GHIORSO, 1985).

In this paper we review the theory behind the construction of potentials which are minimal in systems subject to generalized thermodynamic constraints. This theory is rendered practical for computational purposes by a modification of the numerical algorithm utilized by GHIORSO (1985) along the lines suggested by GILL *et al.* (1981). These general procedures are applied to evaluate reaction stoichiometry in magmatic systems evolving under isenthalpic or isochoric constraints, or both. In addition, a more rigorous treatment of the calculation of equilibria in thermodynamic systems open to a perfectly mobile component (KORZHINSKII, 1959; THOMPSON, 1970) is given. Numerical examples that model assimilation in magmas under both isothermal and isenthalpic conditions are provided. These examples illustrate the utility of the technique in predicting the effects of assimilation on derivative liquids.

CONSTRUCTION OF THERMODYNAMIC POTENTIALS

In this section we utilize mathematical methods for defining thermodynamic potentials which have been described by CALLEN (1961) and THOMPSON (1970). Let Ψ be an unspecified thermodynamic potential which is minimal at equilibrium. This minimum is constrained by the specification of linear and possibly non–linear relationships among the independent variables of Ψ. For example, if $\Psi = G(T, P, n_1, n_2, \ldots, n_c)$, where n_1, n_2 to n_c are variables denoting the number of moles of each of the c system components in each phase, then G is minimal at equilibrium subject to constant temperature, pressure and fixed *bulk* composition. The latter constraint does not imply that n_1 through n_c must be held constant in the processes of minimizing G, but rather that *linear* combinations of the n_1 through n_c must be constrained in order that the system bulk composition remain constant. There will always be more n_i's than compositional variables in a *multi*–phase thermodynamic system. Mathematically, finding the minimum of G is simply a case of optimizing a non–linear function subject to linear equality constraints.

Let the independent variables of Ψ be denoted by x, y, n_1, n_2, \ldots, n_c. For convenience, we will group the mol numbers of the various system components to form the elements of a vector denoted by n such that:

$$n = \begin{bmatrix} n_1 \\ n_2 \\ \vdots \\ n_c \end{bmatrix}. \tag{1}$$

Thus, we may write $\Psi = \Psi(x, y, n)$. If a new thermodynamic potential, Φ, is to be defined such that $\Phi = \Phi((\partial\Psi/\partial x)_{y,n}, y, n)$, then Φ is given by the Legendre transform (CALLEN, 1961) of Ψ:

$$\Phi\left(\left(\frac{\partial\Psi}{\partial x}\right)_{y,n}, y, n\right) = \Psi(x, y, n) - x\left(\frac{\partial\Psi}{\partial x}\right)_{y,n}. \tag{2}$$

As an example, we might consider the case treated by THOMPSON (1970) and GHIORSO (1985) of a perfectly mobile component. Imagine a magmatic system whose boundaries are open to oxygen exchange such that the chemical potential of oxygen in the system is fixed by external constraints. By denoting this chemical potential μ_{O_2}, we seek a new potential, Φ, which is minimal at equilibrium in this partially open magmatic system at constant temperature and pressure. By recognizing that μ_{O_2} is defined as $(\partial G/\partial n_{O_2})_{T,P,n^*}$, where n^* denotes all

mole numbers held constant except those of oxygen, the potential Φ can be expressed as [Equation (2)]:

$$\Phi = G - n_{O_2}\mu_{O_2}. \tag{3}$$

Potentials of the form of Equation (3), which are minimal in systems open to perfectly mobile components, are usually referred to as Korzhinskii potentials (THOMPSON, 1970).

We now utilize Equation (2) to construct a number of thermodynamic potentials which are useful in calculating reaction paths in magmatic systems.

Isenthalpic constraints

We seek the potential $\Phi(H, P, n_1, n_2, \ldots, n_c)$. From the definition of the Gibbs free energy a new potential may be defined:

$$G/T = H/T - S. \tag{4}$$

The total derivative of G/T may be deduced from Equation (4):

$$d\left(\frac{G}{T}\right) = Hd\left(\frac{1}{T}\right) + \frac{1}{T}dH - dS,$$

and given that

$$dH = TdS + VdP + \sum_{i=1}^{c}\mu_i dn_i,$$

we have:

$$d\left(\frac{G}{T}\right) = Hd\left(\frac{1}{T}\right) + \left(\frac{V}{T}\right)dP + \left(\frac{1}{T}\right)\sum_{i=1}^{c}\mu_i dn_i. \tag{5}$$

Equation (5) demonstrates that

$$\left(\frac{\partial G/T}{\partial 1/T}\right)_{P,n} = H,$$

which allows us to write, by using Equation (2):

$$\Phi(H, P, n) = \frac{G}{T} - \frac{1}{T}\left(\frac{\partial G/T}{\partial 1/T}\right)_{P,n}$$

$$= \frac{G}{T} - \frac{1}{T}H$$

$$\Phi(H, P, n) = -S. \tag{6}$$

Equation (6) establishes that equilibrium in a closed system subject to constant enthalpy and pressure is given by a maximum in entropy.

Isochoric constraints

In this case we seek the function $\Phi(T, V, n)$. The formalism of the Legendre transform is unnecessary in the case of isochoric constraints as the required function Φ is just the Helmholtz free energy, denoted by the symbol A:

$$\Phi(T, V, n) = A = G - PV. \qquad (7)$$

Combined isenthalpic and isochoric constraints

Under conditions of constant enthalpy and volume the potential Φ can be obtained from Equation (5) by recognizing that

$$\left(\frac{\partial G/T}{\partial P}\right)_{(1/T),n} = \frac{V}{T},$$

from which the Legendre transform may be constructed as:

$$\Phi(H, V, n) = \frac{G}{T} - \frac{1}{T}\left(\frac{\partial G/T}{\partial 1/T}\right)_{P,n} - P\left(\frac{\partial G/T}{\partial P}\right)_{(1/T),n}$$

$$= \frac{G}{T} - \frac{1}{T}H - P\frac{V}{T}$$

$$\Phi(H, V, n) = -S - \frac{VP}{T}. \qquad (8)$$

Equation (8) establishes the potential which is minimal under conditions of constant enthalpy, volume and fixed bulk composition.

THE MATHEMATICAL PROBLEM OF MINIMIZING Φ

In order to calculate heterogeneous equilibrium with the chemical mass transfer algorithm of GHIORSO (1985) we require a numerical method to minimize the newly defined functions, Φ, subject to the appropriate constraints. To formulate correctly the problem we must recognize initially that the fixed bulk composition constraints are always linear functions of the system components. From GHIORSO (1985) we define the rectangular matrix C such that

$$Cn = b, \qquad (9)$$

where b is a vector which describes the bulk composition of the system in terms of a minimal set of linearly independent compositional variables. As mentioned above, the length of n is greater than or equal to the length of b. Equation (9) embodies linear constraints in that the matrix C is not a function of n.

The optimization of Φ with respect to n (and possibly T and P) can be written with the aid of Equation (9) as:

minimize $\quad \Phi(x, y, n)$

such that

$$Cn - b = 0$$

$$\Delta x(n, T, P) = 0$$

and

$$\Delta y(n, T, P) = 0. \qquad (10)$$

From examination of Equation (10) it is clear that the calculation of heterogeneous equilibrium in generally constrained thermodynamic systems is essentially a problem in non–linear optimization subject to *non*–linear constraints. A numerical method for the solution of Equation (10) is provided in Appendix A.

THE ADDITION OF PERFECTLY MOBILE COMPONENTS

The necessity of modelling phase equilibria in magmatic systems open to oxygen exchange has been discussed by GHIORSO (1985), GHIORSO and CARMICHAEL (19850 and CARMICHAEL and GHIORSO (1986). This need arises out of a lack of knowledge regarding relevant oxidation–reduction couples which buffer the ferric/ferrous ratio in a crystallizing magma. For modelling purposes it is convenient to approximate the effect of these unknown homogeneous redox equilibria by specifying the chemical potential of oxygen in the system. GHIORSO (1985) formulated a solution to the minimization of the Korzhinskii potential for open system oxygen exchange by approximating the non-linear equality constraint corresponding to fixed chemical potential of oxygen in the system with an empirical linearized constraint. This formulation is adequate for many calculations, but is not an exact solution to the problem. With the algorithm discussed in Appendix A, it is possible to formulate the problem exactly.

A comprehensive expression for the chemical potential of oxygen in a magmatic liquid is provided by KILINC *et al.* (1983). Its value is a function of the fugacity of oxygen, temperature, pressure, the ferric/ferrous ratio in the melt and the number of moles of various liquid components. The Korzhinskii potential, which is minimized under equilibrium conditions in this open system, is given by (see above):

$$\Phi = G - n_{O_2}\mu_{O_2}. \qquad (11)$$

We take n_{O_2} to be the number of moles of oxygen produced or consumed in the course of crystallization; indexed by the relative increase or decrease in the ferric/ferrous ratio of the system necessary to maintain constant μ_{O_2}. The minimum in Φ is calculated subject to fixed T, P, μ_{O_2} and bulk composition with the provision of variable ferric/ferrous ratio. The optimal variables are the system components, n. Numerical methods suitable for minimizing the open system potential defined by Equation (11) are discussed in Appendix A. In Appendix B appropriate matrices and vectors needed for the minimization algorithm are derived and the results generalized to the case of the calculation of heterogeneous equilibrium in chemical systems open to oxygen and subject to isenthalpic and/or isochoric constraints.

AN ALGORITHM FOR CHEMICAL MASS TRANSFER IN MAGMATIC SYSTEMS SUBJECT TO GENERALIZED CONSTRAINTS

After having established the means of calculating heterogeneous equilibrium in arbitrarily constrained, open magmatic systems, we can now summarize a method for the calculation of chemical mass transfer in such systems. In isothermal, isobaric open magmatic systems, the appropriate Korzhinskii potential is minimized at each step in reaction progress. This calculational scheme is identical to the free energy minimization method described by GHIORSO (1985). In systems subject to isochoric or isenthalpic constraints, the initial step in calculating the evolution of the system involves determining the heterogeneous phase assemblage by minimization of the Gibbs free energy or relevant Korzhinskii potential for the system. This establishes an initial "equilibrium" volume or enthalpy which can then be used to constrain the equilibrium "path" of all future steps in reaction progress. A typical calculation might involve determining the stable solid–liquid assemblage for a magma at a given temperature and pressure, and modelling isenthalpic assimilation by adding small increments of solid to the magma, keeping track of the attendant changes in system bulk composition and increase in total system enthalpy due to the added enthalpy of the solid. At each incremental step of solid addition a new equilibrium solid–liquid phase assemblage and temperature are calculated at the new system bulk composition and total enthalpy. By choosing small enough steps, the continuous irreversible process of isenthalpic assimilation can be modelled. By examination of the compositions and proportions of solid and liquid produced at each increment of reaction progress, the stoichiometry of the assimilation reaction can be determined.

Software has been designed to implement the non–linearly constrained optimization algorithm described in Appendix A in order to allow the modelling of isenthalpic, isochoric, isothermal and isobaric reactions in magmatic systems open to oxygen. This software constitutes version 2.0 of the program SILMIN (GHIORSO, 1985). Source code is available from the first author. Thermodynamic data essential for the numerical implementation of this algorithm are summarized by GHIORSO et al. (1983) and GHIORSO (1985). Some applications of these calculations in magmatic systems are presented below.

APPLICATIONS

Assimilation of wall rock into basaltic magma has been proposed to explain a range of chemical variation trends in igneous rock series. BOWEN (1922a) explained the basic principles for thermodynamic treatment of solid–liquid reaction in silicate systems. Few subsequent studies of interaction between magma and wall rock have taken this approach, probably due to the scarcity of data on the slope of the multi–component liquidus surface, heats of fusion of solid solutions, and heats of mixing in silicate liquids. Further investigation may also have been discouraged by BOWEN's assertion that assimilation processes could not produce liquids fundamentally different from those produced by fractional crystallization alone. BOWEN (1922a) showed how, in the forsterite–anorthite–silica system determined by ANDERSON (1915), assimilation of olivine in liquids could alter the crystal and liquid line of descent. However, he believed that all such phenomena were of secondary importance, because all liquids in the ternary still possess the theoretical capacity to fractionate toward the same eutectic composition. As emphasis in igneous petrology has changed from debate over the ultimate destination of liquid fractionation to explanation of the variation between different lines of descent, the potential effect of combined assimilation and crystallization has become a subject of increasing interest.

BOWEN (1922a) showed that, for the silicate systems which had been determined experimentally by 1922, heats of mixing were negligible by comparison with heats of fusion. This observation has generally been supported by recent work. Unlike the dissolution of salts in aqueous solution, which could evolve heat without consequent crystallization, the pure dissolution of silicate minerals in magma would be endothermic. Important exceptions include the heats of fusion of the silica polymorphs β–quartz, tridymite and β–cristobalite. These enthalpies are relatively small, on the order of 30 cal/gm, in the range of estimated heats of mixing in natural silicate liquids (GHIORSO and CARMICHAEL, 1980).

In developing examples in the continuous and discontinuous reaction series (defined in BOWEN, 1922b), BOWEN (1922a,b) showed that the dissolution of refractory phases in liquids in equilibrium with less refractory phases in the same series, at constant temperature, causes crystallization of the phases with which the magma is saturated. This process maintains the composition of the magma on the liquidus surface; magma composition on the liquidus in a binary system is isothermally, isobarically invariant, whereas in a ternary system the composition may be univariant or invariant. In the examples given (BOWEN, 1922a, pp. 523–537), the mass crystallized was calculated to be larger than the mass dissolved; BOWEN (1922a) suggested that such reactions would generally be exothermic.

Similarly, BOWEN (1922a) developed examples of assimilation of less refractory phases in liquids saturated with higher–temperature phases. In the examples, it was shown that the energy required to dissolve the assimilate could not normally be provided by the heat capacity of the magma, because known heats of fusion, per gram, for silicates were two or three orders of magnitude larger than liquid heat capacities, which BOWEN (1922a) estimated at 0.2 to 0.3 cal/gm °C. The result of dissolving a low-temperature melting member of the reaction series, he reasoned, is a drop in temperature, followed by crystallization of a smaller mass of the phase(s) with which the magma is saturated.

Schematic illustration of these relationships is given in Figure 1. Note that the assimilation of a refractory phase in liquid saturated in a less refractory phase in the same reaction series, at constant temperature, need not always be exothermic. Instead,

$$\Delta H = Ma\Delta H_f^a - Mc\Delta H_f^c, \qquad (12)$$

for Ma equal to the mass assimilated, Mc equal to the mass crystallized, and ΔH_f^a and ΔH_f^c equal to the apparent heats of fusion, per gram, of the assimilate and the resultant crystals, respectively. Apparent heats of fusion are the heats of fusion of the solids at some temperature, T', below the melting temperature of the pure solid. They are calculated as:

$$\Delta H_f = \Delta H_F + \int_{T'}^{T_m} C_{P,\text{solid}} dT - \int_{T'}^{T_m} C_{P,\text{liquid}} dT, \quad (13)$$

where ΔH_f is the apparent heat of fusion at T', ΔH_F is the heat of fusion of the pure substance at its melting temperature, T_m is the melting temperature of the pure substance, and $C_{P,\text{solid}}$ and $C_{P,\text{liquid}}$ are the heat capacities of pure solid and liquid over the temperature range $T_m - T'$. The magnitude and sign

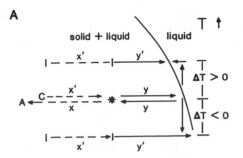

A = FORSTERITE ΔH_f^a = 200 cal/gm

C = ENSTATITE ΔH_f^c = 155 cal/gm at 1300°C

$\Delta T = [Mc\ \Delta H_f^c - Ma\ \Delta H_f^a] / M_{TOT} C_P^{TOT}$

$\Delta T = 0$ Ma/Mc = 0.78

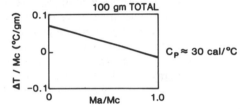

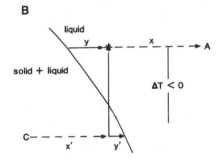

A = DIOPSIDE ΔH_f^a =151 cal/gm

C = FORSTERITE ΔH_f^c = 200 cal/gm at 1300°C

ΔT cannot be > or = 0 (?)

$\Delta T < 0$ Ma/Mc > 1.32

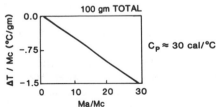

FIG. 1. (A) Schematic representation of assimilation of a refractory phase in liquid saturated with a less refractory member of the same reaction series. Resultant crystal and liquid products are shown for the cases of isothermal and isenthalpic assimilation. (B) Similar representation, for assimilation of a less refractory phase in a liquid saturated with a member of the same reaction series, or assimilation across a cotectic "valley".

of ΔH in Equation (12) may vary depending on the ratios Ma/Mc and $\Delta H_f^a/\Delta H_f^c$.

The ΔH for assimilation of a refractory phase in a magma saturated with a phase lower in the same reaction series will often be negative, but may be very close to zero. For example, take the assimilation of magnesian olivine in a liquid saturated with low-calcium pyroxene. In the $Mg_2SiO_4–SiO_2$ system, BOWEN (1922a,b) calculated Ma/Mc approximately equal to 0.2 at 1550°C. Assuming $\Delta H_f^{Fo} = 238$ cal/gm and $\Delta H_f^{En} = 186$ cal/gm (data on ΔH_F and liquid and solid heat capacities from ROBIE et al., 1978, STEBBINS et al., 1984, and GHIORSO and CARMICHAEL, 1980), $Ma/Mc = 0.2$ corresponds to an enthalpy change of -138 cal per gram clino-enstatite crystallized at 1550°C. In natural basaltic to andesitic liquids, KELEMEN (1986) and KELEMEN and GHIORSO (1986) calculate a range of Ma/Mc for the reaction olivine + liquid = orthopyroxene; a reasonable average might be 0.6. For this reaction stoichiometry, with $\Delta H_f^{Fo} \simeq 185$ cal/gm and $\Delta H_f^{En} \simeq 142$ cal/gm at 1200°C, the enthalpy change would be close to -31 cal per gram orthopyroxene produced.

During assimilation of a less refractory phase, or assimilation of material across a cotectic "valley", ΔH will always be positive. For instance, assimilation of diopside in liquid saturated with forsterite in the system forsterite–diopside–anorthite (OSBORN and TAIT, 1952) must be endothermic at constant temperature. The process of dissolving diopside will move the liquid off the forsterite liquidus surface. There will be no limit to the enthalpy change produced by this reaction until the liquid is saturated in diopside. Until then, ΔH will be positive, increasing by 151 cal per gram diopside dissolved (ΔH_f^{Di} at 1300°C from ROBIE et al., 1978; GHIORSO and CARMICHAEL, 1980; STEBBINS et al., 1984). If a reservoir of forsterite is initially present in the system, dissolution of diopside at constant temperature will lead to dissolution of forsterite as well. Note that these effects are qualitatively similar to the effects of assimilation of forsterite in liquid saturated with diopside.

If, instead of constant temperature, the system were constrained to constant enthalpy, then the temperature would drop, such that;

$$\Delta T = [Ma\Delta H_f^a - Mc\Delta H_f^c]/M^{tot}C_p^{tot}, \quad (14)$$

where M^{tot} is equal to the total mass of the system and C_p^{tot} is the heat capacity of the system (about 0.3 cal/gm °C, assuming the masses of the solid phases are negligible relative to the mass of liquid). For assimilation of a less refractory phase in the same reaction series, or assimilation across a cotec-

tic, ΔT must be less than zero. In the particular case of diopside dissolving in liquid saturated in forsterite at 1300°C, this requires Ma/Mc to be greater than about 1.3, and for assimilation of forsterite in diopside saturated liquid, Ma/Mc must be greater than 0.75.

The foregoing discussion applies to dissolution and consequent crystallization where the assimilate is at magmatic temperature prior to reaction. Where this is not the case, the energy required to heat the assimilate to magmatic temperature must also be included in the calculation of mass and energy balance at constant temperature or constant enthalpy. Again, BOWEN (1922a) treated this problem. He noted that the heat of crystallization of saturated phases can supply the energy needed to heat inclusions or wall rock, or both. Because heats of fusion are orders of magnitude greater than heat capacities in silicate systems, per gram, crystallization of a mass of liquid can raise the temperature of an equal mass of solids tens, or even hundreds, of degrees.

All these relations apply in principle to reactions in natural systems. However, in practice, such calculations have been limited by lack of quantitative understanding of such important variables as the slope of the liquidus surface in pressure-temperature-composition space. Development of a solution model for silicate liquids (GHIORSO et al., 1983; GHIORSO, 1985; GHIORSO and CARMICHAEL, 1985), implemented in the computer program SILMIN, permits investigation of solid–liquid reaction stoichiometry in natural systems (e.g., GHIORSO, 1985; KELEMEN, 1986; KELEMEN and GHIORSO, 1986).

Constant temperature may be an unreasonable constraint for assimilation of low melting phases, or assimilation of phases across a cotectic. During such exothermic reactions, magmatic temperature cannot often be externally buffered at a constant value. It is easier to understand and apply the results of calculations constrained to constant enthalpy. Isenthalpic calculations may provide a close approach to some important natural phenomena.

To illustrate the relevance of such calculations, we present the following example. In Figure 2, we have shown liquid compositions in the system forsterite–diopside–anorthite (OSBORN and TAIT, 1952), which are saturated in forsterite, and will fractionate to the forsterite–diopside cotectic. Isothermal assimilation of anorthite, on the other hand, moves the magma composition off the liquidus towards the anorthite apex, first dissolving any reservoir of forsterite in the system, until it is saturated in anorthite. Schematic, isenthalpic assimilation paths are depicted as well; given a knowledge of the heats of fusion of anorthite and forsterite, and the heat capacity of the system, combined with the lever rule, such approximate paths are simple to determine. The liquid in the isenthalpic assimilation case fractionates to the anorthite field (case 1), or to the forsterite–anorthite cotectic (case 2), with decreasing temperature. At points C1 and C2 (Figure 2), the liquid will not dissolve further anorthite added to the system; its composition and temperature will remain constant.

Figure 3 and Table 1 illustrate some of the results of a similar calculation, by using a natural system: a magnesian MORB, FAMOUS 527-1-1 (BENDER et al., 1978) plus 2 weight percent H_2O at 1 kbar, constrained to react along the FMQ oxygen buffer. This wet basalt is calculated to crystallize olivine, beginning at about 1223°C, followed by clinopyroxene at about 1165°C, and plagioclase (An_{77}) at about 1095°C. Isenthalpic assimilation of anorthite (T_{solid} = 1225°C) in this basalt, beginning at 1225°C, leads to a decrease in temperature, accompanied by crystallization of olivine beginning at about 1208°C, and plagioclase (An_{87}) at 1162°C. The calculated temperature then rises, as continued assimilation of anorthite results in the precipitation of a larger mass of increasingly calcic plagioclase. The total mass of solids produced in the temperature interval 1225 to 1165°C is much greater in the equilibrium crystallization calculation than in the case of isenthalpic assimilation of anorthite.

In order to test the notion that isenthalpic assimilation of more refractory phases in crystallizing magma might be similar in its effect to isothermal assimilation, we compared the two paths for assimilation of magnesian olivine in FAMOUS 527-1-1, dry, at 3 kbar, evolving along the FMQ oxygen

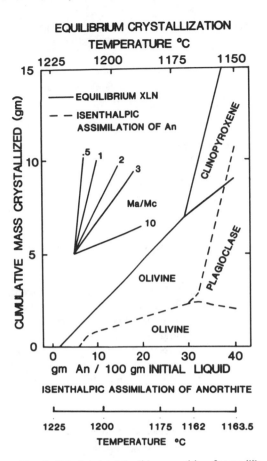

FIG. 3. Calculated total solid composition for equilibrium crystallization and isenthalpic assimilation of anorthite in a magnesian MORB (FAMOUS 527-1-1 + 2 weight percent H_2O) at 1 kbar, on the FMQ oxygen buffer. This, and all subsequent calculations depicted in Figures 4 through 7, were performed using the computer program SILMIN, described in the text. Input parameters are the initial composition of the system, temperature, pressure, oxygen fugacity, and the composition and temperature of the assimilate. Resultant liquid and solid compositions, phase proportions, and temperatures in the isenthalpic case, are calculated. Initial temperature of the liquid, and the temperature of the assimilate throughout the calculation, was set at 1225°C.

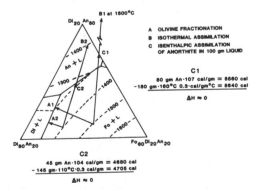

FIG. 2. (A) Paths for crystal fractionation, (B) isothermal assimilation of anorthite, (C) and isenthalpic assimilation of anorthite, for liquids (1 and 2) saturated with forsterite in the forsterite–diopside–anorthite system, determined by OSBORN and TAIT (1952).

buffer. Results are summarized in Figure 4 and Table 2. Resultant crystal and liquid compositions are almost identical in the isothermal and isenthalpic cases. In both cases, dissolution of olivine (Fo_{90}) leads to crystallization of a slightly larger mass of more iron–rich olivine, accompanied by a small percentage of plagioclase in the isothermal case. In the isenthalpic case, the calculated temperature rises 3°C during assimilation of 40 grams of Fo_{90}. Despite the rather simple reaction stoichiometry (Fo_{90} + liq = Fo_{87}, with Ma/Mc of about 0.9), this reaction

Table 1. Calculated total solid composition produced during the assimilation of anorthite in FAMOUS 527-1-1 + 2 weight percent H_2O, 1 kbar, on the FMQ oxygen buffer.

	Initial	Isothermal	Isenthalpic
Mass Assimilated/100 gm initial liquid + crystals	0	40	40
Initial Temperature of Anorthite	—	1225°C	1225°C
Total Mass Crystallized (gm)	0	0	10.72
Mass Olivine (gm)	0*	0	2.11
Mol Fraction Fo			0.82
Mass Plagioclase	0	0	8.61
Mol Fraction An			0.88
Temperature of the System	1225°C	1225°C	1164°C†

* 0.27 gm Fo_{82} @ 1220°C.

† Minimum temperature @ 1162°C, 32 gm assimilated with 2.36 gm Fo_{82} and 0.43 gm An_{87} crystallized.

has an important effect on the liquid line of descent, because the conversion of Fo_{90} to Fo_{87} fractionates FeO from the magma, effectively buffering the magnesium number (Mg#) of the derivative liquid. Similar reactions between high temperature wall rock and fractionating basalt in the upper mantle are likely. Their effects are more fully discussed by KELEMEN (1986).

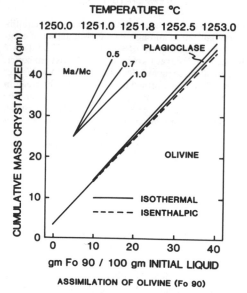

FIG. 4. Calculated total solid composition for isothermal and isenthalpic assimilation of olivine (Fo_{90}) in a magnesian MORB (FAMOUS 527-1-1), dry, at 3 kbar, on the FMQ oxygen buffer. Initial temperature of the liquid, and the temperature of the assimilate throughout the calculation, was set at 1250°C. Ma/Mc refers to the ratio of mass assimilated vs. mass crystallized.

Initial temperature of the assimilate is the most important variable in considering the potential effects of assimilation of wall rock by magmas in the crust. Although the notion of assimilation of low melting pelitic or quartzo–feldspathic metasediments, or both, in high temperature, basaltic melts has an immediate, intuitive appeal, this process is strongly endothermic and cannot be viewed simply as bulk melting and mixing of a crustal component. To illustrate this point, already made on semi–quantitative grounds by BOWEN (1922a), we have made calculations modelling the assimilation of pure albite, and a composite pelitic composition, in a magnesian MORB.

The initial liquid composition used was FA-MOUS 527-1-1, dry, at 3 kbar, constrained to react along the FMQ oxygen buffer, at an initial temperature of 1250°C. We first compare the predicted equilibrium crystallization path for this liquid to the experimental results of BENDER et al. (1978) on this composition. Results of calculations using SIL-MIN have olivine (Fo_{87}) on the liquidus at 1290°C, followed by plagioclase (An_{82}) at 1245°C, and clinopyroxene (Di_{85}) at 1235°C. Comparable values from the experiments at 1 bar are olivine ($Fo_{89.5}$) at 1268°C, and plagioclase (An_{89}) at 1235°C. At 6 kbar, oliving (Fo_{87}) appears at 1280°C, followed by plagioclase (An_{78}) at 1240°C. Clinopyroxene first crystallized between 1135 and 1150°C, at 1 bar, and between 1240 and 1250°C, at 6 kbar. Low–calcium pyroxene was not observed in any experimental runs.

Three different reaction paths were modelled: (1) isothermal assimilation, with the solid assimilate at 1250°C (a metastable solid state, at least in the case of albite); (2) isenthalpic assimilation, with the assimilate at 1250°C; and (3) isenthalpic assimilation

Table 2. Calculated total solid composition produced during the assimilation of olivine (Fo_{90}) in FAMOUS 527-1-1, at 3 kbar, on the FMQ oxygen buffer.

	Initial	Isothermal	Isenthalpic
Mass Assimilated/100 gm initial liquid + crystals	0	40	40
Initial Temperature of Olivine	1250°C	1250°C	1250°C
Total Mass Crystallized (gm)	3.44	50.55	46.44
Mass Olivine (gm)	3.44	47.71	46.44
Mol Fraction Fo	0.867	0.863	0.866
Mass Plagioclase	0	2.84	0
Mol Fraction An		0.823	
Temperature of the System	1250°C	1250°C	1253.5°C

with the assimilate at 500°C. In modelling assimilation of pelitic rock, we used a bulk composition derived by averaging the group analyses for shale and slate reported by CLARKE (1924). This composition was recast as proportions of two mineral assemblages: assemblage A (quartz–K feldspar–ilmenite–orthopyroxene–spinel–plagioclase), representative of pelitic rocks in granulite and pyroxene hornfels facies; and assemblage B (quartz–muscovite–ilmenite–biotite–garnet–plagioclase), representative of pelitic rocks in amphibolite facies. Thermodynamic data for most of these solid phases are summarized by GHIORSO et al. (1983) and constitutes part of the SILMIN data base. Data for

muscovite, phlogopite, annite, pyrope, and almandine are from HELGESON et al. (1978). Note that the phase assemblages are somewhat arbitrary. For instance, under most conditions, spinel + quartz break down to orthopyroxene + alumino–silicate. However, by use of spinel rather than sillimanite was more convenient for computational reasons. Similarly, all iron has been arbitrarily assumed to be ferrous iron. The bulk composition used and the proportions of mineral end–members are given in Table 3.

Some of the results of modelling assimilation of albite and pelitic rock are given in Figures 5, 6, and 7 and Tables 4 and 5. Figures 5 and 6 summarize

Table 3. Pelitic composition used in assimilation calculations. This composition was derived by averaging the group analyses for shale and slate reported by CLARKE (1924, pp. 552 and 631) H_2O, CO_2, SO_2, C, P_2O_5 and trace components in the analyses were ignored; other oxides normalized to 100 weight percent.

Bulk composition		Phase assemblage A		
	Weight percent		Mol percent	Weight percent
SiO_2	64.463	Hercynite	Hc_{58}	4.94
		Spinel	Sp_{42}	2.93
TiO_2	0.731	Enstatite	En_{46}	4.80
		Ferrosilite	Fs_{54}	7.33
Al_2O_3	17.107	Sanidine		21.02
		Albite	Ab_{48}	13.48
FeO^*	6.691	Anorthite	An_{52}	15.40
		β–Quartz		28.72
MgO	2.754	Ilmenite		1.39
CaO	3.105	Phase assemblage B		
Na_2O	1.593	Almandine	Alm_{61}	3.66
		Pyrope	Py_{39}	1.92
K_2O	3.556	Phlogopite	Phl_{47}	7.20
		Annite	Ann_{53}	10.19
Phase assemblage B also includes		Muscovite		14.01
1.35 weight percent H_2O in		Albite	Ab_{48}	13.48
biotite and muscovite.		Anorthite	An_{52}	15.40
		α-Quartz		32.75
		Ilmenite		1.39

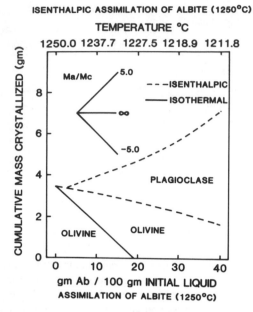

ISENTHALPIC ASSIMILATION OF ALBITE (1250°C)

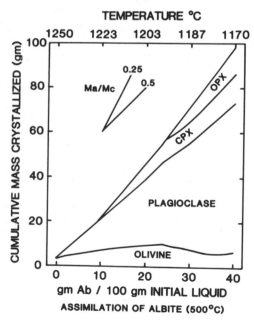

ASSIMILATION OF ALBITE (500°C)

FIG. 5. (A) Calculated total solid composition for iso-thermal and isenthalpic assimilation of albite in a mag-nesian MORB (FAMOUS 527-1-1), dry, at 3 kbar, on the FMQ oxygen buffer. Initial temperature of the liquid, and the temperature of the assimilate throughout the calcu-lation, was set at 1250°C. *Ma/Mc* refers to the ratio of mass assimilated vs. mass crystallized. (B) Input parameters as in 5(A), but with the temperature of the assimilate set at 500°C throughout the calculation.

calculated solid compositions. Note that the vertical scale (cumulative mass crystallized) is ten times as large for reaction with the assimilate at 500°C than

for reaction with the assimilate at 1250°C. Whereas isenthalpic assimilation of 40 grams of albite and pelitic rock at 1250°C have been calculated to pro-duce less than 10 grams of olivine and plagioclase, isenthalpic dissolution of the same two phase as-semblages at 500°C is predicted to result in crys-tallization of almost 100 grams of plagioclase and pyroxene. This difference results entirely from the energy requirements of heating the assimilate to magmatic temperatures. Dissolution of the high temperature solids leads to a decrease of 35 to 40°C in the temperature of the system. Dissolution of the 500°C solids leads to a decrease of 70 to 80°C.

It is also important to consider the differences between assimilation of pelitic phase assemblages A and B, both at 500°C. Phase assemblage A might be thought of as a granulite facies rock which cooled to 500°C after the peak of metamorphism. Phase assemblage B, characteristic of pelitic rocks in am-phibolite facies, includes muscovite and biotite. Equilibrium dissolution of these hydrous minerals in fluid undersaturated magma saturated with ol-ivine, pyroxenes, and plagioclase, can be thought of as a combination of three discrete steps: (1) de-hydration reactions (such as biotite + 3 muscovite = 4 Kspar + 3 spinel + 4 H_2O); (2) melting of the dehydrated solid phases; and (3) solution of the new melt, plus H_2O, in the assimilating magma. Because dehydration reactions, like melting reactions, are endothermic at crustal pressures, the apparent en-thalpy of fusion of a hydrous assemblage will gen-erally be larger than the apparent enthalpy of fusion of an equivalent anhydrous assemblage.

We reasoned that isenthalpic assimilation of a phase assemblage including muscovite and biotite would result in a larger temperature drop, per gram assimilated, and thus (at the same bulk composi-tion) more crystallization from the derivative liquid, compared to assimilation of an assemblage with Kspar and spinel replacing the hydrous phases. However, this hypothesis did not account for the actual difference in bulk compositions between phase assemblages A and B: assemblage B contains 1.35 weight percent H_2O. Addition of H_2O to the magma has the general effect of supressing crystal-lization. Our calculations illustrate this effect. Al-though assimilation of the hydrous assemblage B results in a larger temperature drop, per gram as-similated, it produces a smaller mass of crystals than assimilation of the anhydrous assemblage A.

This observation is not a general one. If the initial magma composition prior to assimilation includes a few weight percent H_2O, solution of a small amount of H_2O in the assimilate does not affect the course of reaction very much. Under these circum-

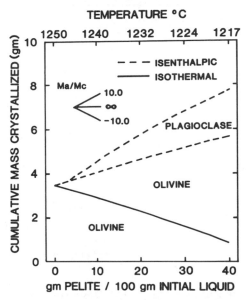

ASSIMILATION OF PELITIC ASSEMBLAGE A (1250°C)

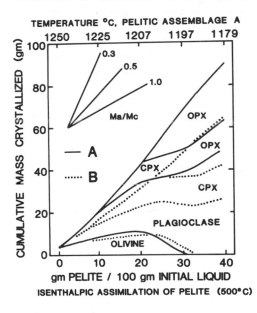

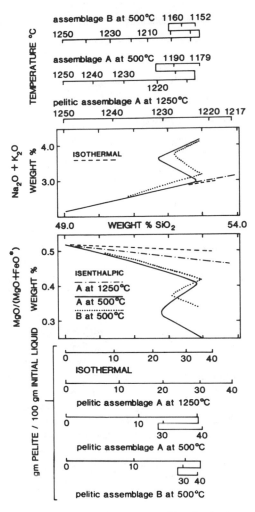

FIG. 7. Calculated liquid composition for isothermal and isenthalpic assimilation of pelitic rock in a magnesian MORB. Input parameters as for Figure 6(A) and (B).

FIG. 6. Calculated total solid composition for isothermal and isenthalpic assimilation of pelitic rock in a magnesian MORB (FAMOUS 527-1-1), dry, at 3 kbar, on the FMQ oxygen buffer. Composition and phase assemblages of the pelitic assimilate are given in Table 3. Phase assemblages A and B have identical bulk composition (except that B includes 1.35 weight percent H_2O), but A is composed of Qtz–Ilm–Kspar–Opx–Spinel–Plag, whereas B is composed of Qtz–Ilm–Musc–Bio–Garnet–Plag. Figure 6(A). Initial temperature of the liquid, and the temperature of the assimilate throughout the calculation, was set at 1250°C. Ma/Mc refers to the ratio of mass assimilated vs. mass crystallized. Figure 6(B). Input parameters as in 6(A), but with the temperature of the assimilate set at 500°C throughout the calculation.

stances, assimilation of a hydrous assemblage will result in a lower final temperature and a larger mass of crystals than assimilation of the same mass of an equivalent anhydrous assemblage.

The different reaction paths for assimilation of pelitic rock result in markedly different liquid compositions, as illustrated in Figure 7 and Table 6. Liquids produced by isenthalpic assimilation of

Table 4. Calculated total solid composition produced during the assimilation of albite in FAMOUS 527-1-1, at 3 kbar, on the FMQ oxygen buffer.

	Initial	Isothermal	Isenthalpic	
Mass Assimilated/100 gm initial liquid + crystals	0	40	40	40
Initial Temperature of Albite	—	1250°C	1250°C	500°C
Total Mass Crystallized (gm)	3.44	7.63	0	99.34
Mass Olivine (gm)	3.44	2.09	0	7.65
Mol Fraction Fo	0.87	0.82		0.74
Mass Plagioclase (gm)	0	5.54	0	67.02
Mol Fraction An		0.59		0.45
Mass Clinopyroxene (gm)	0	0	0	13.16
Mol Fraction Di				0.83
Mol Fraction Fs				0.16
Mass Orthopyroxene (gm)	0	0	0	11.51
				0.76
Temperature of the System	250°C	1250°C	1212°C	1170°C

pelitic rock at an initial temperature of 500°C diverge from those produced by assimilation of the same composition at 1250°C early in their fractionation history, becoming strongly iron–enriched and alkaline. Whereas the other two cases show continuously increasing silica contents in derivative liquids, the assimilation of "cold" pelitic rock undergoes a decrease in silica content beginning with the appearance of orthopyroxene on the liquidus, as a result of the reaction olivine + $SiO_{2(liq)}$ = orthopyroxene. Note that this reaction, which is discontinuous in the Mg_2SiO_4–SiO_2 system, is a continuous reaction occurring over a finite temperature interval in iron–bearing systems. An increase in SiO_2 concentration in the liquid resumes with the disappearance of olivine from the system.

These calculations have been done in equilibrium mode; that is, the total solid composition is in equilibrium with the liquid at every step. If the calculations were performed in fractional crystallization mode, the discontinuous reaction would be limited to a single step (in this case, defined by the addition

Table 5. Calculated total solid composition produced during assimilation of pelite in FAMOUS 527-1-1, at 3 kbar, on the FMQ oxygen buffer. Compositions of pelitic assimilates are given in Table 3.

			Isenthalpic assemblage		
	Initial	Isothermal	A	A	B
Mass Assimilated/100 gm initial liquid + crystals	0	40	40	40	40
Initial Temperature of Pelite	—	1250°C	1250°C	500°C	500°C
Total Mass Crystallized (gm)	3.44	0.83	8.23	90.66	62.99
Mass Olivine (gm)	3.44	0.83	6.08	0	0
Mol Fraction Fo	0.87	0.85	0.82		
Mass Plagioclase (gm)	0	0	2.15	49.11	26.37
Mol Fraction An			0.72	0.64	0.68
Mass Clinopyroxene (gm)	0	0	0	14.58	15.58
Mol Fraction Di				0.83	0.84
Mol Fraction Fs				0.16	0.14
Mass Orthopyroxene (gm)	0	0	0	26.97	21.04
Mol Fraction En				0.76	0.79
Temperature of the System	1250°C	1250°C	1215°C	1181°C	1152°C

Table 6. Calculations of crystal fractionation, assimilation of pelitic rock at 500°C, and assimilation of albite at 500°C, in FAMOUS 527-1-1, at 3 kbar, on the FMQ oxygen buffer. Composition and phase assemblages for pelitic assimilate are given in Table 3.

	Crystal fractionation		Isenthalpic assimilation (assimilate at 500°C)					
			1250 to 1190°C			After 40 gm assimilated		
			Pelitic			Pelitic		
	Equili-brium	Frac-tion-ation	A	B	Albite	A	B	Albite
Weight percent SiO$_2$	48.0	48.8	52.1	52.7	53.8	53.0	52.9	53.6
Weight percent MgO/(MgO + FeO*) FeO* = total Fe as FeO	0.30	0.32	0.31	0.43	0.43	0.25	0.34	0.39
Na$_2$O + K$_2$O weight percent	2.41	2.79	3.89	3.10	4.96	4.18	4.06	6.33
Gm Assimilated	0	0	34	22	28	40	40	40
Gm Crystallized	77	64	78	34	67	91	63	99
Temp °C	1190	1190	1190	1190	1190	1179	1152	1170
Gm Assimilated/°C	0	0	0.57	0.57	0.47	0.56	0.41	0.50
Mass Liquid/Initial Mass Liquid	0.21	0.35	0.55	0.87	0.60	0.48	0.76	0.40

of 2 grams of pelitic rock), and the calculated silica content of the liquid would not decrease over a long interval of assimilation. However, derivative liquids would still be far more iron–enriched and alkaline than those produced by reaction with a wall rock already at magmatic temperature.

Table 6 allows further comparison of the liquid compositions produced by assimilation of wall rock at 500°C. This comparison makes it abundantly clear that calculation of the effect of assimilation on derivative liquids demands accurate characterization of the bulk composition and the phase assemblage of the assimilate, as well as the tempera ture of both wall rock and magma. It is also instructive to compare the calculated liquid and solid lines of descent for isenthalpic assimilation with those calculated for equilibrium or fractional crystallization alone. Whereas some of the liquids produced by assimilation have almost the same Mg# as the liquids produced by crystal fractionation alone, all the liquids derived by assimilation are relatively rich in silica and alkalies.

Neither the equilibrium nor fractional crystallizatin of FAMOUS 527-1-1 were calculated to produce orthopyroxene at 1190°C. The total solid composition resulting from equilibrium crystallization at this temperature is 22 weight percent olivine (Fo$_{75}$), 23 weight percent clinopyroxene (Di$_{83}$), and 55 weight percent plagioclase (An$_{69}$). The assimilation of pelitic assemblage A leads to a predicted solid composition with 29 weight percent orthopyroxene (En$_{78}$), 16 weight percent clinopyroxene (Di$_{83}$), and 55 weight percent plagioclase (An$_{65}$) at 1190°C. Similar phase proportions were calculated for assimilation of pelitic assemblage B. This is in partial accord with BOWEN's prediction (1922a,

pp. 549–558) that the assimilation of aluminous sediment in basaltic magma would increase the modal proportion of orthopyroxene and plagioclase, and decrease the proportion of clinopyroxene, produced during crystallization.

CONCLUSIONS

Assimilation of wall rock in fractionating magma may fundamentally alter the solid and liquid line of descent, producing various distinctive magma series, all derived from a common parental liquid. Thermodynamic modelling of such processes is essential in predicting solid–liquid reaction stoichiometry, and thus the compositional effects of such reactions on derivative liquids and solids. Constant temperature assimilation of refractory solids in liquids saturated in less refractory members of the same "reaction series" will often be exothermic. Assimilation of less refractory phases in the same reaction series, and assimilation of phases across a cotectic "valley", will always be endothermic at constant temperature. Particularly for endothermic reactions, modelling of isenthalpic, rather than isothermal, conditions may yield results more generally applicable to natural phenomena. Also useful, though not explicitly presented here, are calculations based on externally controlled conditions of decreasing temperature.

The sample calculations presented here reemphasize the point made by BOWEN (1922a) that the most important variable in the energy balance of reaction between magma and wall rock, besides bulk composition, is the temperature difference between the two. Assimilation of solids with the same composition, but different initial temperatures, is calculated to produce distinctly different products. In addition, assimilation of solids with the same initial temperature and very similar bulk composition, but different phase assemblages, can also produce a variety of derivative liquids. BOWEN (1922a) argued that the theoretical eutectic "destination" of liquids modified by assimilation reactions remains unchanged. This postulate may well be correct. However, the variety of different lines of descent produced during our calculations suggests new ways of explaining the paths taken by individual magma series.

The ability to model isenthalpic crystallization in magmatic systems may also prove useful in the numerical simulation of the cooling history of magma bodies. In the past, substantial difficulty has been encountered in modelling the thermal history of "real" magmatic systems due, in part, to the necessity of knowing the temporal and spatial dependence of the enthalpy of fusion of the magma in order to calculate the evolving temperature field. These enthalpies of fusion are dependent upon the extent of crystallization and consequently, the temperature of the magma. Thus magma chamber cooling models must be solved iteratively if they involve realistic parameterizations of the enthalpies of fusion of the solid phases. With the ability to determine heterogeneous equilibrium under isenthalpic constraints, an alternative method of calculation exists. If temperature and phase proportions can be calculated given a fixed enthalpy content of the magma at any given time or point in space, then knowledge of the heat flux and the geometry of the chamber will allow simultaneous determination of the phase proportions, phase compositions and temperature field in the cooling magma body. Such calculations need not be performed recursively and are amenable to inclusion in more traditional fluid dynamical models of magma dynamics.

Acknowledgements—We are indebted to Bjørn Mysen, Alex Navrotsky and an anonymous reviewer for their thoughtful and thorough reviews. In particular, we would like to thank Alex Navrotsky for saving the general reader much agony by suggesting we "banish" the matrices to an appendix; ". . . more matter with less art . . .". The second author was supported by a David A. Johnston Memorial Fellowship and an NSF Graduate Fellowship. We would like to acknowledge material support from NSF grants EAR-8451694 and EAR-8600534.

REFERENCES

ANDERSON O. (1915) The system anorthite–forsterite–silica. *Amer. J. Sci., Ser. IV* **39**, 407–454.

BENDER J. F., HODGES F. N. and BENCE A. E. (1978) Petrogenesis of basalts from the project FAMOUS area: Experimental study from 0 to 15 kbars. *Earth Planet. Sci. Lett.* **41**, 277–302.

BOWEN N. L. (1922a) The behavior of inclusions in igneous magmas. *J. Geol.* **30**, 513–570.

BOWEN N. L. (1922b) The reaction principle in petrogenesis. *J. Geol.* **30**, 177–198.

CALLEN H. B. (1961) *Thermodynamics.* 376 pp. John Wiley and Sons.

CARMICHAEL I. S. E and GHIORSO M. S. (1986) Oxidation–reduction relations in basic magma: a case for homogeneous equilibria. *Earth Planet. Sci. Lett.* **78**, 200–210.

CLARKE F. W. (1924) Data of geochemistry: Fifth edition. *U.S. Geol. Surv. Bull.* **770**, 841 pp.

GHIORSO M. S. (1985) Chemical mass transfer in magmatic processes I. Thermodynamic relations and numerical algorithms. *Contrib. Mineral. Petrol.* **90**, 107–120.

GHIORSO M. S. and CARMICHAEL I. S. E. (1980) A regular solution model for mea-aluminous silicate liquids: Applications to geothermometry, immiscibility, and the source regions of basic magmas. *Contrib. Mineral. Petrol.* **71**, 323–342.

GHIORSO M. S. and CARMICHAEL I. S. E. (1985) Chemical mass transfer in magmatic processes II. Applications in equilibrium crystallization, fractionation and assimilation. *Contrib. Mineral. Petrol.* **90**, 121–141.

GHIORSO M. S., CARMICHAEL I. S. E., RIVERS M. L. and SACK R. O. (1983) The Gibbs Free Energy of natural silicate liquids; an expanded regular solution approximation for the calculation of magmatic intensive variables. *Contrib. Mineral. Petrol.* **84**, 107–145.

GILL P. E., MURRAY W. and WRIGHT M. H. (1981) *Practical Optimization.* 401 pp. Academic Press Inc.

HELGESON H. C., DELANY J. M., NESBITT H. W. and BIRD D. K. (1978) Summary and critique of the thermodynamic properties of rock–forming minerals. *Amer. J. Sci.* **278-A**, 1–229.

KELEMEN P. B. (1986) Assimilation of ultramafic rock in subduction–related magmatic arcs. *J. Geol.,* (In press).

KELEMEN P. B. and GHIORSO M. S. (1986) Assimilation of peridotite in zoned calc-alkaline plutonic complexes: evidence from the Big Jim Complex, Washington Cascades. *Contrib. Mineral. Petrol.* **94**, 12–28.

KILINC A., CARMICHAEL I. S. E., RIVERS M. L. and SACK R. O. (1983) The ferric–ferrous ratio of natural silicate liquids equilibrated in air. *Contrib. Mineral. Petrol.* **83**, 136–140.

KORZHINSKII D. S. (1959) *Physiochemical Basis of the Analysis of the Paragenesis of Minerals.* 142 pp. Consultants Bureau.

OSBORN E. F. and TAIT D. B. (1952) The system diopside–forsterite–anorthite. *Amer. J. Sci.,* **250-A**, 413–433.

ROBIE R. A., HEMINGWAY B. S. and FISHER J. R. (1978) Thermodynamic properties of minerals and related substances at 298.15 K and 1 bar (10^5 Pascals) pressure and at high temperature. *U.S. Geol. Surv. Bull. 1452.* 456 pp.

STEBBINS J. F., CARMICHAEL I. S. E. and MORET L. K. (1984) Heat capacities and entropies of silicate liquids and glasses. *Contrib. Mineral. Petrol.* **86**, 131–148.

THOMPSON J. B. JR. (1970) Geochemical reaction and open systems. *Geochim. Cosmochim. Acta* **34**, 529–551.

APPENDIX A NUMERICAL METHODS
FOR MINIMIZING Φ

We seek solutions (optimal values of n and possibly temperature (T) and pressure (P)) to the following problem in non–linear optimization:

$$\text{minimize} \quad \Phi(x, y, n)$$

such that

$$Cn - b = 0$$

$$\Delta x(n, T, P) = 0$$

and

$$\Delta y(n, T, P) = 0. \tag{A1}$$

We will let the vector z denote all the variables being optimized in the minimization of Φ. z may contain the elements of n as well as T and P. The last two equality constraints ($\Delta x = 0$, $\Delta y = 0$) are non–linear, so the usual minimization method of projecting with the constraint matrix and solving the unconstrained quadratic sub–problem (GHIORSO, 1985) fails. Instead, the suggestion of GILL *et al.* (1981) will be followed and the unconstrained related Lagrange function corresponding to Equation (A1) will be optimized. We denote the Lagrangian of Φ by Ξ and define it such that

$$\Xi(x, y, n, \lambda_n, \lambda_x, \lambda_y) = \Phi - \lambda_n^T(Cn - b) - \lambda_x \Delta x - \lambda_y \Delta y. \tag{A2}$$

The new variables λ_n, λ_x, and λ_y are generally referred to as Lagrange multipliers. The minimum of Ξ is obtained iteratively by seeking steps, s, away from an initial non–optimal, but feasible guess to the solution vector of Equation (A1), which we will denote z_o, and stepping along this direction while continuously maintaining the feasibility of the solution parameters. In this problem, feasible parameters refer to the set of solutions that satisfy the constraints. To define the search direction s we construct a vector c such that

$$c = \begin{bmatrix} Cn - b \\ \Delta x(n, T, P) \\ \Delta y(n, T, P) \end{bmatrix}$$

and define a matrix, A, which represents the derivative of this vector of equality constraints with respect to the optimal variables. A is commonly referred to as the Jacobian of the system of non–linear constraints and should be evaluated at the initial feasible point, z_o. Thus,

$$A = \left(\frac{dc}{dz} \right)^T \Bigg|_{z_o} = \begin{bmatrix} C^T \\ \left(\dfrac{\partial x}{\partial z} \right)^T \Big|_{z_o} \\ \left(\dfrac{\partial y}{\partial z} \right)^T \Big|_{z_o} \end{bmatrix} .$$

A matrix, Z, can be computed from the orthogonal decomposition of A (GILL *et al.*, 1981), which projects the optimal parameters of Φ into the null–space of the quality constraints. Unlike linearly constrained optimization, this null–space projection operator is only reliable "close" to the initial feasible guess z_o. An approximation, Zp, to the feasible search direction, s, is computed by solving the quadratic (second order) Newton approximation to the minimum of the Lagrangian:

$$Z^T W Z p = -Z^T g. \tag{A3}$$

The vector, g, is defined as the gradient of the original potential function [Equation (A1)] with respect to the optimal variables, z, evaluated at the initial feasible point, and the matrix W is defined as the second derivative matrix of the Lagrangian [Equation (A2)] with respect to the optimal variables and evaluated at the initial feasible point. This matrix is often referred to as the Hessian matrix of the Lagrangian function, Ξ. The Lagrange multipliers which are needed to compute the matrix, W, are estimated by solving the system of linear equations:

$$g = A^T \begin{bmatrix} \lambda_n \\ \lambda_x \\ \lambda_y \end{bmatrix} .$$

The descent direction, Zp, obtained from Equation (A3), may not be feasible because of the non–linearity of the quality constraints. Thus the new feasible guess to the solution of Equation (A1), which we will denote $z_o + s$, is found by determining the value of the steplength parameter, γ, which minimizes the potential, Φ, along the search direction, s, subject to maintaining the feasibility of the non–linear constraints; the linear constraints are implicitly maintained by the null–space operator, Z. This feasibility is guaranteed by solving, for every proposed value of γ, the system of non–linear equations

$$\Delta x(z_o + \gamma Z p + Y p_\gamma) = 0,$$

and

$$\Delta y(z_o + \gamma Z p + Y p_\gamma) = 0, \tag{A4}$$

for the elements of the vector, p_γ. The search direction, s, is defined as $\gamma Z p + Y p_\gamma$, where Y is a matrix which spans the range–space of A and p_γ is a vector whose length is equal to the total number of equality constraints. Y may be taken to be A^T (GILL et al., 1981) which emphasizes that the range–space describes the space of feasible solutions to Equation (A1). The p_γ obtained from the solution of Equation (A4) have no particular physical significance except insofar as their non–zero values indicate the sensitivity of the system of equality constraints to the optimal solution parameters about the initial feasible point.

The minimum of the potential function, Φ, is determined by solving successive quadratic approximations to the Lagrangian [Equations (A3) and (A4)] until the computed norm of the feasible step direction, s, is smaller than machine precision. This minimum is verified by examining the sufficient conditions to insure convergence. These are (1) The equality constraints are satisfied; (2) The norm of the projected gradient is approximately zero ($Z^T g \simeq 0$); and (3) The projected Hessian of the Lagrangian function ($Z_T W Z$) is positive definite.

For the convenience of the reader appropriate expressions are provided below for the quantities, g and W, in the three explicit cases of the potential, Φ, discussed in the text.

The isenthalpic case ($\Phi = -S$)

In this case the optimal variables are the system components, n and the temperature, T. The first derivative of Φ is given by the partitioned vector:

$$g = \left[\begin{array}{c} -\left(\dfrac{\partial S}{\partial n}\right)_{T,P} \\ \hline -\dfrac{C_P}{T} \end{array} \right] \tag{A5}$$

evaluated at n_o and T_o. The Jacobian of the system of equality constraints may be written:

$$A = \left[\begin{array}{c|c} C^T & 0 \\ \hline \left(\dfrac{\partial H}{\partial n}\right)^T_{T,P} & C_P \end{array} \right]. \tag{A6}$$

The elements of this matrix should be evaluated at n_o and and T_o. The second derivative of the associated Lagrangian is provided by the partitioned symmetrix matrix:

$$W = \left[\begin{array}{c|c} \left(\dfrac{\partial^2 S}{\partial n^2}\right)_{T,P} - \lambda \left(\dfrac{\partial^2 H}{\partial n^2}\right)_{T,P} & -\left(\dfrac{1}{T} + \lambda\right)\left(\dfrac{\partial C_P}{\partial n}\right)_{T,P} \\ \hline & -\left(\dfrac{1}{T} + \lambda\right)\left(\dfrac{\partial C_P}{\partial T}\right)_{n,P} + \dfrac{C_P}{T^2} \end{array} \right]. \tag{A7}$$

The elements of this matrix should be evaluated at n_o and T_o.

The isochoric case ($\Phi = A$)

For the case of constant volume, the optimal variables are the system components, n and the pressure, P. The gradient of Φ is given by the partitioned vector:

$$g = \left[\begin{array}{c} \left(\dfrac{\partial G}{\partial n}\right)_{T,P} - P\left(\dfrac{\partial V}{\partial n}\right)_{T,P} \\ \hline -P\left(\dfrac{\partial V}{\partial P}\right)_{T,n} \end{array} \right] \tag{A8}$$

evaluated at n_o and P_o. The derivative matrix of the equality constraints is provided by:

$$A = \left[\begin{array}{c|c} C^T & 0 \\ \hline \left(\dfrac{\partial V}{\partial n}\right)^T_{T,P} & \left(\dfrac{\partial V}{\partial P}\right)_{T,n} \end{array} \right]. \tag{A9}$$

This matrix should be evaluated at n_o and P_o. The Hessian matrix of the Lagrangian is given by the symmetric partitioned matrix:

$$W = \begin{bmatrix} \left(\dfrac{\partial^2 G}{\partial n^2}\right)_{T,P} - (P+\lambda)\left(\dfrac{\partial^2 V}{\partial n^2}\right)_{T,P} & -(P+\lambda)\left(\dfrac{\partial^2 V}{\partial P \partial n}\right)_{T,P} \\[2ex] & -\left(\dfrac{\partial V}{\partial P}\right)_{n,T} - (P+\lambda)\left(\dfrac{\partial^2 V}{\partial P^2}\right)_{n,T} \end{bmatrix}. \tag{A10}$$

This matrix should be evaluated at n_o and P_o.

The isenthalpic and isochoric case ($\Phi = -S - PV/T$)

In this most complicated example, the optimal variables are the system components, n, the temperature, T and the pressure, P. The gradient of Φ is given by the partitioned vector:

$$g = \begin{bmatrix} -\left(\dfrac{\partial S}{\partial n}\right)_{T,P} - P\left(\dfrac{\partial V}{\partial n}\right)_{T,P} \\[2ex] -\dfrac{C_P}{T} + \dfrac{VP}{T^2} - \dfrac{P}{T}\left(\dfrac{\partial V}{\partial T}\right)_{n,P} \\[2ex] \left(\dfrac{\partial V}{\partial T}\right)_{n,P} - \dfrac{V}{T} - \dfrac{P}{T}\left(\dfrac{\partial V}{\partial P}\right)_{T,n} \end{bmatrix} \tag{A11}$$

evaluated at the initial point n_o, T_o and P_o. The derivative matrix of equality constraints is provided by:

$$A = \begin{bmatrix} C^T & | & 0 & 0 \\[1ex] \left(\dfrac{\partial H}{\partial n}\right)^T_{T,P} & | & C_P & V - T\left(\dfrac{\partial V}{\partial T}\right)_{P,n} \\[2ex] \left(\dfrac{\partial V}{\partial n}\right)^T_{T,P} & \left(\dfrac{\partial V}{\partial T}\right)_{P,n} & \left(\dfrac{\partial V}{\partial P}\right)_{T,n} \end{bmatrix}. \tag{A12}$$

This matrix should be evaluated at n_o, T_o and P_o. The symmetric second derivative matrix of the Lagrangian is given by:

$$W = \begin{bmatrix} W_{n,n} | W_{n,T} | W_{n,P} \\ \overline{W_{T,T} | W_{T,P}} \\ \overline{W_{P,P}} \end{bmatrix} \tag{A13}$$

where

$$W_{n,n} = \left(\frac{\partial^2 S}{\partial n^2}\right)_{T,P} - \lambda_T\left(\frac{\partial^2 H}{\partial n^2}\right)_{T,P} - \left(\frac{P}{T} + \lambda_P\right)\left(\frac{\partial^2 V}{\partial n^2}\right)_{T,P} \tag{A14}$$

$$W_{n,T} = -\left(\frac{1}{T} + \lambda_T\right)\left(\frac{\partial C_P}{\partial n}\right)_{T,P} + \frac{P}{T^2}\left(\frac{\partial V}{\partial n}\right)_{T,P} - \left(\frac{P}{T} + \lambda_P\right)\left(\frac{\partial^2 V}{\partial n \partial T}\right)_{P} \tag{A15}$$

$$W_{n,P} = -\frac{1}{T}\left(\frac{\partial V}{\partial n}\right)_{T,P} + (1 + \lambda_T T)\left(\frac{\partial^2 V}{\partial n \partial T}\right)_{P} - \left(\frac{P}{T} + \lambda_P\right)\left(\frac{\partial^2 V}{\partial n \partial P}\right)_{T} \tag{A16}$$

$$W_{T,T} = \frac{C_P}{T^2} - \frac{2VP}{T^3} + \frac{2P}{T^2}\left(\frac{\partial V}{\partial T}\right)_{n,P} - \left(\frac{1}{T} + \lambda_T\right)\left(\frac{\partial C_P}{\partial T}\right)_{n,P} - \left(\frac{P}{T} + \lambda_P\right)\left(\frac{\partial^2 V}{\partial T^2}\right)_{n,P} \tag{A17}$$

$$W_{T,P} = (1 + \lambda_T T)\left(\frac{\partial^2 V}{\partial T^2}\right)_{n,P} - \frac{V}{T^2} + \left(\lambda_T - \frac{1}{T}\right)\left(\frac{\partial V}{\partial T}\right)_{n,P} + \frac{P}{T^2}\left(\frac{\partial V}{\partial P}\right)_{n,T} - \left(\frac{P}{T} + \lambda_P\right)\left(\frac{\partial^2 V}{\partial T \partial P}\right)_{n} \tag{A18}$$

and

$$W_{P,P} = (1 + \lambda_T T)\left(\frac{\partial^2 V}{\partial T \partial P}\right)_{n} - \frac{2}{T}\left(\frac{\partial V}{\partial P}\right)_{n,T} - \left(\frac{P}{T} + \lambda_P\right)\left(\frac{\partial^2 V}{\partial P^2}\right)_{n,T}. \tag{A19}$$

Equations (A14)–(A19) should be evaluated at the point n_o, T_o and P_o.

APPENDIX B HETEROGENEOUS EQUILIBRIUM IN OPEN SYSTEMS SUBJECT TO GENERALIZED THERMODYNAMIC CONSTRAINTS

The gradient necessary for the minimization of the potential Φ given by Equation (11) is:

$$g = \left(\frac{\partial G}{\partial n}\right)_{T,P} - \left(\frac{\partial n_{O_2}}{\partial n}\right)_{T,P} \mu_{O_2} - n_{O_2}\left(\frac{\partial \mu_{O_2}}{\partial n}\right)_{T,P}. \quad (B1)$$

The derivative of the constraint matrix is provided by:

$$A = \begin{bmatrix} C^{*T} \\ \left(\frac{\partial \mu_{O_2}}{\partial n}\right)_{T,P}^T \end{bmatrix} \quad (B2)$$

where C^* represents a linear bulk composition constraint matrix modified for constancy of total iron and variable ferric/ferrous ratio. The Hessian matrix of Equation (11) may be written:

$$W = \left(\frac{\partial^2 G}{\partial n^2}\right)_{T,P} - 2\left(\frac{\partial n_{O_2}}{\partial n}\right)_{T,P}\left(\frac{\partial \mu_{O_2}}{\partial n}\right)_{T,P}$$

$$- (n_{O_2} + \lambda_{O_2})\left(\frac{\partial^2 \mu_{O_2}}{\partial n^2}\right)_{T,P}. \quad (B3)$$

Equations (B1) through (B3) should be evaluated at the initial feasible point, n_0.

The potentials defined by Equations (6), (7) and (8) can be transformed to allow for the modelling of isenthalpic and isochoric processes in magmatic systems open to oxygen. We will illustrate the most complicated case. The "isenthalpic/isochoric" potential defined by Equation (8) can be modified for open systems by recognizing that:

$$\left(\frac{\partial G/T}{\partial n_{O_2}/T}\right)_{(1/T),P,n^*} = \mu_{O_2}$$

from which a new open system potential, Φ, can be defined as:

$$\Phi(H, V, n) = \frac{G}{T} - \frac{1}{T}\left(\frac{\partial G/T}{\partial 1/T}\right)_{P,n} - P\left(\frac{\partial G/T}{\partial P}\right)_{(1/T),n}$$

$$- \frac{n_{O_2}}{T}\left(\frac{\partial G/T}{\partial n_{O_2}/T}\right)_{(1/T),P,n^*}$$

$$= \frac{G}{T} - \frac{1}{T}H - P\frac{V}{T} - \frac{n_{O_2}}{T}\mu_{O_2}$$

$$\Phi(H, V, n) = -S - \frac{VP}{T} - \frac{n_{O_2}\mu_{O_2}}{T}. \quad (B4)$$

The optimal variables of this new potential Φ are n, T and P; the oxygen chemical potential constraint does not introduce new optimal variables and only modifies the relationships among the elements of n. If g and W denote the gradient and Hessian defined by Equations (A11) and (A13), then the gradient of Equation (B4) is defined by:

$$g_{(B4)} = g + \begin{bmatrix} g_{O_2} \\ g_T \\ g_P \end{bmatrix}, \quad (B5)$$

where, given that $G_{O_2} = n_{O_2}\mu_{O_2}$, we have;

$$g_{O_2} = -\frac{1}{T}\left(\frac{\partial G_{O_2}}{\partial n}\right)_{T,P}, \quad (B6)$$

$$g_T = \frac{G_{O_2}}{T^2} + \frac{S_{O_2}}{T}, \quad (B7)$$

$$g_P = -\frac{V_{O_2}}{T}. \quad (B8)$$

This gradient must be evaluated at the initial feasible point n_0, T_0, P_0. The Hessian matrix corresponding to the associated Lagrangian of Φ is given by:

$$W_{(B4)} = W + \begin{bmatrix} W_{n,n,O_2} | W_{n,T,O_2} | W_{n,P,O_2} \\ W_{T,T,O_2} W_{T,P,O_2} \\ W_{P,P,O_2} \end{bmatrix}, \quad (B9)$$

where,

$$W_{n,n,O_2} = -\left(\frac{\partial^2 G_{O_2}}{\partial n^2}\right)_{T,P} - \lambda_{O_2}\left(\frac{\partial^2 \mu_{O_2}}{\partial n^2}\right)_{T,P}, \quad (B10)$$

$$W_{n,T,O_2} = \frac{1}{T^2}\left(\frac{\partial G_{O_2}}{\partial n}\right)_{T,P} + \frac{1}{T}\left(\frac{\partial S_{O_2}}{\partial n}\right)_{T,P} + \lambda_{O_2}\left(\frac{\partial \bar{s}_{O_2}}{\partial n}\right)_{T,P}, \quad (B11)$$

$$W_{n,P,O_2} = -\frac{1}{T}\left(\frac{\partial V_{O_2}}{\partial n}\right)_{T,P} - \lambda_{O_2}\left(\frac{\partial \bar{v}_{O_2}}{\partial n}\right)_{T,P}, \quad (B12)$$

$$W_{T,T,O_2} = -\frac{2}{T^3}G_{O_2} - \frac{2}{T^2}S_{O_2} + \frac{1}{T^2}C_{P,O_2} + \lambda_{O_2}\frac{1}{T}\bar{C}_{P,O_2}, \quad (B13)$$

$$W_{T,P,O_2} = \frac{V_{O_2}}{T^2} - \frac{1}{T}\left(\frac{\partial V_{O_2}}{\partial T}\right)_{n,P} - \lambda_{O_2}\left(\frac{\partial \bar{v}_{O_2}}{\partial T}\right)_{n,P}, \quad (B14)$$

and,

$$W_{P,P,O_2} = -\frac{1}{T}\left(\frac{\partial V_{O_2}}{\partial P}\right)_{T,n} - \lambda_{O_2}\left(\frac{\partial \bar{v}_{O_2}}{\partial P}\right)_{T,n}. \quad (B15)$$

For use in the algorithm described in Appendix A the matrix defined in Equation (B9) must be evaluated at the initial feasible point.

Magmatic Processes: Physicochemical Principles
© The Geochemical Society, Special Publication No. 1, 1987
Editor B. O. Mysen

Comparison of hydrothermal systems in layered gabbros and granites, and the origin of low–^{18}O magmas*

HUGH P. TAYLOR, JR.

Division of Geological and Planetary Sciences, California Institute of Technology, Pasadena, CA 91125, U.S.A.

Abstract—The style of hydrothermal alteration in layered gabbros is very different from that in granitic plutons. Non-equilibrium ^{18}O/^{16}O effects are observed in both types of bodies, but whereas the granitic rocks with ^{18}O–exchanged feldspars commonly contain abundant chlorite, sericite, epidote, etc., most such gabbros are mineralogically virtually unaltered (*e.g.,* they contain fresh olivine). These differences imply that the bulk of the externally–derived hydrothermal fluid passes through the gabbros at temperatures of 450°–900°C, much higher than the range of 250°–450°C generally found in the granites. This contrasting behavior is in part explained by the higher solidus temperature, higher latent heat of crystallization, and generally higher melt/phenocryst ratio of the gabbro magmas. However, even more important is: (1) *The geometry of crystallization.* Cumulate gabbros crystallize from the floor upward, and the last part to crystallize is a sub-horizontal sheet of liquid near the roof. This magma sheet provides a thermally insulating lid that is impermeable to the (hydrostatic) hydrothermal convection system in the country rocks, typically producing two decoupled hydrothermal systems, a lower–temperature system at high water/rock ratios above the intrusion, and a much higher–temperature system at low water/rock ratios within the cumulate gabbro below the late–stage magma sheet. (2) *Presence of a magmatic H_2O envelope.* Granitic magmas typically contain much more H_2O than tholeiitic gabbro melts; hence, evolution of a magmatic gas phase will usually occur at the late stages of crystallization of granites. This envelope of magmatic water fills available fractures and is under lithostatic pressure. It is thus impermeable to the outer, convecting meteoric–hydrothermal fluids, which gain access to the pluton only after the magmatic H_2O has dissipated and the body has cooled.

The two major Cenozoic occurrences of low–^{18}O magmas are Iceland and the Yellowstone Plateau, Wyoming. These types of magmas are much less common than heretofore believed, and they typically seem to be developed only in extensional tectonic environments where brittle fracture of the crust allows continued replenishment of the magma reservoir from below, as well as penetration of meteoric ground waters down to great depths. Such magmas are formed by melting or assimilation, or both, of hydrothermally altered volcanic roof rocks by the underlying magma reservoir, *not* by influx of low–^{18}O ground waters directly into the melt.

INTRODUCTION

THE AIM of this paper is two–fold: (1) To point out intrinsic differences in the style of hydrothermal alteration between layered gabbros and granites, specifically trying to explain the observation that ^{18}O–depleted granites are almost invariably *mineralogically* heavily altered to chlorite, sericite, epidote, etc., whereas their counterparts, the low–^{18}O layered gabbros (that also have clearly been altered by meteoric–hydrothermal fluids), are either mineralogically *unaltered* or only weakly altered (*e.g.,* typically containing fresh olivines, for example). (2) To review some of the examples of low–^{18}O magmas formed on Earth, to discuss their possible origin, and to make a case that the mechanism of formation of low–^{18}O magmas is usually operative only in rift–zone environments, which is also where the best examples of hydrothermally exchanged, but mineralogically unaltered, low–^{18}O layered gabbros are found.

First, we briefly review the isotopic systematics displayed by the Skaergaard intrusion (the best described ^{18}O–depleted gabbro in the world), and then we compare the sub–solidus hydrothermal isotope effects observed in such low–^{18}O gabbros with those in low–^{18}O granites. Less discussion is devoted to the granites, because the hydrothermal alteration effects in these bodies are familiar to most petrologists, and the effects are produced at ordinary hydrothermal temperatures of 200°–400°C (and thus are not controversial). In contrast to the above types of effects that are characteristic of low–^{18}O igneous *rocks,* in the second part of the paper we briefly review the problems of low–^{18}O *magmas,* particularly in Iceland and Yellowstone Park, Wyoming, the two largest and best documented areas of low–^{18}O magmas in the world. We conclude with a critical evaluation of some of the pertinent facts that bear on the mechanism of formation of low–^{18}O magmas.

* Contribution No. 4347, Division of Geological and Planetary Sciences, Caltech, Pasadena, CA 91125.

Some of the discussion that appears in this paper is also included in several recent papers by CRISS and TAYLOR (1986), LARSON and TAYLOR (1986), HILDRETH *et al.* (1984), and TAYLOR (1986).

COMPARISON OF HYDROTHERMAL EFFECTS IN GABBROS AND GRANITES

Permeabilities of rocks

Permeability is the single physical parameter that is of most importance in determining hydrothermal fluid flow through rocks. The permeabilities of common geologic materials vary by a factor of about 10^{12}, from about 10^{-6} cm^2 (100 darcies) to about 10^{-18} cm^2 (10^{-7} millidarcies). A permeability of about 10^{-14} cm^2 is generally considered to be impermeable, as far as hydrothermal systems are concerned (NORTON and KNIGHT, 1977).

The permeabilities of detrital sediments increase markedly with increased grain size and with a decrease in the degree of cementation and compaction. Limestones exhibit widely ranging permeabilities, with low values in argillaceous samples and with extremely high values characterizing cavernous types. Basalt permeabilities are also highly variable, with low values being typical of ancient flows that are strongly altered or metamorphosed; high values are found in highly jointed lavas in areas of recent volcanic activity. Crystalline rocks have relatively low permeabilities except where they are fractured. Because fracture permeability is a large-scale phenomenon that may be dominated by a few, large, widely-spaced fissures, the permeabilities of crystalline rocks measured on a small scale in the laboratory are commonly several orders of magnitude lower than the true *in situ* permeabilities (*e.g.,* as measured in drill holes, BRACE, 1984).

It has become very clear through studies of ^{18}O/^{16}O effects in metamorphic terranes (*e.g.,* GARLICK and EPSTEIN, 1967; RYE *et al.,* 1976; FORESTER and TAYLOR, 1977) that many rocks undergoing ductile or plastic deformation are relatively impermeable to hydrothermal fluids. This is conceptually easy to understand, because such rocks will deform rather than fracture, thus closing up any through-going fractures and greatly reducing permeability. Examples would be salt domes and evaporite beds in sedimentary basins or relatively pure limestones and marbles in areas of strong folding or metamorphism (CRISS and TAYLOR, 1986).

There are some very peculiar notions about permeability in the literature. For example, CATHLES (1983) writes: "We conclude that intrusions hotter than ~350°C must be quite impermeable—oth-

erwise surface venting of solutions 400°C to 500°C or even 600°C or hotter would be commonplace. We conclude that significant fluid circulation does not occur in rock hotter than ~400°C." This statement seems to be contradicted by a great deal of geological and geochemical data, particularly by combined stable isotope and mineralogical data on layered gabbros (see below). However, such statements merely highlight the fact that we have at present very little *direct* knowledge of bulk permeabilities of rocks at high temperatures and pressures.

$^{18}O/^{16}O$ effects in hydrothermally altered intrusions

Figure 1 shows the steep, linear trajectories on plots of δ^{18}O feldspar vs. δ^{18}O quartz or δ^{18}O pyroxene that are a characteristic of hydrothermally altered rocks in the Earth's crust (TAYLOR and FORESTER, 1979; CRISS and TAYLOR, 1983; MAGARITZ and TAYLOR, 1986). The systematics and implications of these trajectories are discussed by CRISS *et al.* (1986) and GREGORY and TAYLOR (1981, 1986). Basically, the trajectories result from the fact that: (1) feldspar exchanges ^{18}O with hydrothermal fluids *much* faster than does the coexisting quartz or pyroxene; and (2) typically, external waters en-

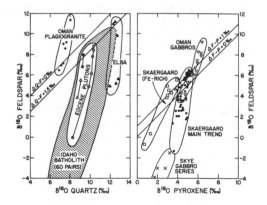

FIG. 1. The left–hand diagram is a plot of δ^{18}O–feldspar versus δ^{18}O quartz for hydrothermally altered granitic rocks from the Idaho batholith (lined pattern), individual Eocene plutons from the Idaho batholith, a Tuscan granodiorite from the island of Elba, Italy, and a plagiogranite from the Oman ophiolite complex (data from CRISS and TAYLOR, 1983; TAYLOR and TURI, 1976; and GREGORY *et al.,* 1980). The right–hand diagram is a plot of δ^{18}O–feldspar versus δ^{18}O pyroxene for hydrothermally altered rocks from the layered gabbro complexes from the Skaergaard intrusion, the island of Skye, Scotland, and the Oman ophiolite (data from TAYLOR and FORESTER, 1979; FORESTER and TAYLOR, 1977; and GREGORY and TAYLOR, 1981). In all cases, marked isotopic disequilibrium is observed in these altered rock suites.

tering a rock system will have $\delta^{18}O$ values that are either lower (meteoric H_2O, high–temperature sea water) or higher (metamorphic waters, sedimentary formation waters, or low–temperature sea water) than the H_2O that would be in equilibrium with a given igneous mineral assemblage at a particular temperature.

Because the in–flowing H_2O is not in isotopic equilibrium with the mineral assemblage, the mineral assemblage will be pulled to either higher or lower $\delta^{18}O$ values; however, because the feldspar reacts faster with the H_2O than does the quartz or pyroxene, the feldspar undergoes ^{18}O exchange at a much faster rate than the coexisting quartz or pyroxene, thereby producing the steep trajectory. The process seldom goes to completion, so the final mineral assemblage is in isotopic disequilibrium, and the disequilibrium $\delta^{18}O$ values provide a "signature" of the hydrothermal event. This isotopic signature has a distinct advantage over the simple mineralogical signature of the hydrothermal episode (*e.g.*, the presence of chlorite, sericite, etc.), because the effects are still observable even at very high temperatures where the original igneous mineral assemblage may be stable (*e.g.*, pyroxene hornfels facies). In addition, the $\delta^{18}O$ effects can provide a much better estimate of the amount of water that moved into the pluton than one can get from the mineralogical data alone. This amount of water is invariably far in excess of the small amounts required simply to make the new OH–bearing minerals. These $\delta^{18}O$ effects are thus the best and least ambiguous way to characterize a fossil hydrothermal system, *if* the external water is isotopically sufficiently different from the pluton.

Isotopic relationships in the Skaergaard intrusion

The best studied example of a fossil meteoric-hydrothermal system associated with a gabbro body is that of the 55 Ma Skaergaard intrusion (TAYLOR and EPSTEIN, 1963; TAYLOR and FORESTER, 1979; NORTON *et al.*, 1984). This 70 km^2 pluton was emplaced during the early stages of opening of the North Atlantic Ocean as a single pulse of magma of tholeiitic basalt composition. The pluton cuts across a subhorizontal unconformity separating Precambrian gneiss from an overlying, 9–km thick sequence of tholeiitic lavas (see Figure 4 below).

The $\delta^{18}O$ values in the Skaergaard intrusion decrease markedly from northwest to southeast (Figure 2). Essentially pristine $\delta^{18}O$ values ($\delta^{18}O$ plag $= 6.2 \pm 0.2$) occur in the northwest, in the lower parts of the cumulate sequence that lie stratigraphically beneath the projection of the major regional

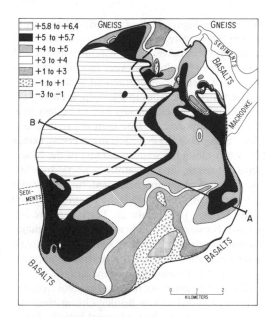

FIG. 2. Generalized distribution of $\delta^{18}O$ values of plagioclase within the Skaergaard intrusion (after TAYLOR and FORESTER, 1979). The 'normal' $\delta^{18}O$ values (+5.8 to +6.4) occur in the NW portion of the pluton, stratigraphically below the heavy dashed curve. This curve represents the projected trace on the outcrop of the unconformity between the highly permeable basalts and the gneiss. The line A–B is the position of the cross–section shown in Figures 4 and 5.

unconformity. The lowest $\delta^{18}O$ values (to -2.4) occur highest in the intrusion in the Upper Border Group; this part of the pluton is surrounded by the very permeable Early Tertiary basaltic lavas. Oxygen isotopic disequilibrium between coexisting plagioclase and pyroxene in this intrusion is clearly indicated by the steep positive slope of the data points in Figure 1.

The contrast in permeability across the regional unconformity clearly played an important role in the development of the hydrothermal $\delta^{18}O$ pattern in the Skaergaard intrusion (see Figure 4 below). Vigorous circulation in the permeable basalts allowed penetration of considerable amounts of H_2O into the solidified gabbro *above* the level of the unconformity, southeast of the heavy dashed line on Figure 2. Below this horizon, much smaller amounts of H_2O penetrated into the gabbro, because such H_2O had to travel through the much less permeable gneiss.

Hydrous minerals such as hornblende, biotite, chlorite, actinolite, and stilpnomelane are found only as rare alteration products within the gabbro matrix, or as replacement and open–space fillings along fractures, in veins, and in miarolitic cavities.

As shown in Figure 3, all of the OH–bearing minerals in the vicinity of the Skaergaard intrusion (in the gabbros themselves, in the basement gneiss, and in the plateau basalts) have very low δD values indicative of meteoric–hydrothermal exchange. The overall water/rock ratios in the deepest part of the system (the basement gneiss) had to be less than about 0.05, however, because there is a general lack of any discernible ^{18}O depletion in the gneiss, except (1) locally along the western contact of the Skaergaard intrusion, and (2) along a north–trending fracture zone containing brick–red, turbid feldspar and thoroughly chloritized mafic minerals (red gneiss $\delta D = -136$, Figure 3). These data are important because they indicate that along major fracture conduits, there *was* significant meteoric water circulation through the basement, approximately 10 km beneath the Eocene surface.

Blocks of basaltic roof rock and gabbros from the early–crystallized Upper Border Group were depleted in ^{18}O by the hydrothermal activity *before* they fell into the magma (TAYLOR and FORESTER, 1979). For example, a 6 m–wide basalt xenolith in the Middle Zone of the layered series has $\delta^{18}O = -4.0$. However, in spite of the fact that the hydrothermal system was operating for the entire

130,000–year crystallization history of the pluton (NORTON and TAYLOR, 1979), there was no measurable depletion of ^{18}O in the liquid magma. This conclusion is proved by the fact that the pyroxenes throughout the layered series have relatively normal $\delta^{18}O$ values, commonly higher than the more easily exchanged plagioclase. Thus, most of the observed ^{18}O depletion in the Skaergaard intrusion was confined to the plagioclase, and this occurred after crystallization under sub–solidus conditions.

Numerical modeling of the Skaergaard intrusion

NORTON and TAYLOR (1979) carried out a computer simulation of the Skaergaard magma–hydrothermal system, producing detailed maps of the temperature, pressure, fluid velocity, integrated fluid flux, and $\delta^{18}O$ values in rock and fluid as a function of time for a two–dimensional cross–section through the pluton. An excellent match was made between calculated $\delta^{18}O$–values and the measured $\delta^{18}O$ values in the three principal rock units, basalt, gabbro, and gneiss, as well as in xenoliths of roof rocks that are now embedded in the gabbro cumulates (compare Figures 4a and 4b). The best match was realized for a system in which the bulk rock permeabilities were 10^{-13} cm^2 for the intrusion, 10^{-11} cm^2 for basalt, and 10^{-16} cm^2 for gneiss. These represent *average* permeabilities for the lifetime of the hydrothermal system. Because of self–sealing by mineral deposition in fractures, it is likely that during the early, highest–temperature stages of hydrothermal alteration, the pluton had a somewhat higher permeability of about 10^{-12} cm^2 (NORTON *et al.*, 1984).

The thermal history calculated for the Skaergaard system by NORTON and TAYLOR (1979) showed that extensive fluid circulation was largely restricted to the permeable basalts and to regions of the pluton stratigraphically above the basalt-gneiss unconformity. During the initial 130,000–year period of crystallization, fluids circulated in all the rocks surrounding the magma body, but fluid flow paths were deflected around the impermeable magma sheet. After final crystallization of the late–stage sheet of magma, fractures could form in the gabbro, allowing the circulation system to shift toward the center of the intrusion (Figure 5).

During the initial 150,000 years, the average temperature of the intrusion was high (>700°C) and reaction rates were fast; thus, fluids flowing into the intrusion quickly equilibrated with plagioclase. By 500,000 years, the pluton had cooled to approximately ambient temperatures, and the final $\delta^{18}O$ values were 'frozen in'. Reactions between hy-

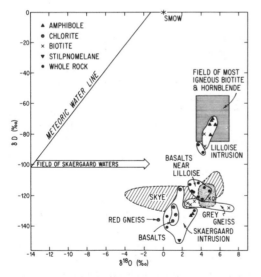

FIG. 3. Plot of δD vs. $\delta^{18}O$, showing the position of the meteoric water line and SMOW (standard mean ocean water, defined as equal to zero for both $\delta^{18}O$ and δD). The positions of the Skaergaard and Lilloise plutons and the East Greenland basalts and gneiss are indicated, as well as the general field of the Skye igneous center in NW Scotland and the field of another pluton in East Greenland (Kangerdlugssuaq, KQ), together with the 'normal' values of most primary igneous biotites and hornblendes (modified after TAYLOR and FORESTER, 1979).

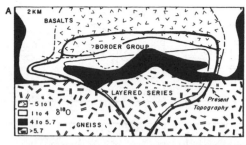

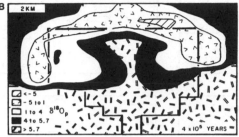

FIG. 4A. Composite geological cross-section of the Skaergaard intrusion, showing interpolated $\delta^{18}O$-values in plagioclase with respect to major rock types and topography along the restored geological section line A–B shown in Figure 2 (after data by TAYLOR and FORESTER, 1979).

FIG. 4B. Calculated $\delta^{18}O$ values of plagioclase in an idealized cross–section of the Skaergaard intrusion at 400,000 years (see Figure 5), based on the numerical modeling study of this hydrothermal system by NORTON and TAYLOR (1979). The good match between the above figures indicates that the bulk permeabilities assumed for the various rock types (gneiss = 10^{-16} cm^2, gabbro = 10^{-13} cm^2, basalts = 10^{-11} cm^2) are probably quite accurate (NORTON and TAYLOR, 1979).

drothermal fluid and the intrusion occurred over a broad range in temperature, 1000°–200°C, but 75 percent of the fluid circulated through the intrusion while its average temperature was >480°C (NORTON and TAYLOR, 1979). The relative quantities of water to rock integrated over the entire cooling history were 0.52 for the upper part of intrusion, 0.88 for the basalt, and 0.003 for the gneiss (weight units).

One of the important conclusions of NORTON and TAYLOR (1979) is that most of the sub–solidus hydrothermal exchange in the Skaergaard pluton took place at very high temperatures (400°–800°C); this is compatible with the general absence of hydrous alteration products in the mineral assemblages, and with the presence of clinopyroxenes and high–temperature amphiboles in the veins and fractures; the latter were deposited both before and after the intrusion of cross-cutting granophyre dikes and sills (NORTON *et al.*, 1984). These granitic magma bodies are not chilled at their margins, so

emplacement clearly took place at quite high temperatures (WAGER and DEER, 1939). Of course, outside in the country–rock basalts the average temperature of hydrothermal alteration was much lower (200°–400°C), and there was extensive development of hydrous minerals such as chlorite, epidote, and prehnite; locally, along late–stage veins, these minerals also developed in the pluton.

Water/rock ratios in the Skaergaard intrusion

The results of the calculations of NORTON and TAYLOR (1979) for the Skaergaard intrusion illustrate the relationship between material–balance water/rock ratios and the calculated amounts of water that physically flow through the system and thus might actually interact with the rocks. Figure 5 shows the integrated mass flux of fluid at various times as a function of stratigraphic position in the Skaergaard system. Over the 500,000–year lifetime of this hydrothermal system, and above the level of the unconformity, integrated amounts of 100 to 5000 kg of H$_2$O have flowed through each square centimeter cross–section of rock. This means that *along* the flow path each cm^3 of rock has been exposed to something on the order of 100,000 to 5,000,000 cm^3 of fluid (remember that the permeability is only 10^{-12} to 10^{-13} cm^2!). However, the water/rock (W/R) ratios for various parts of the system, calculated either by material balance or by choosing a sufficiently large-sized mass of rock, are much smaller. For example, the overall, integrated W/R ratio for that part of the intrusion lying above the unconformity is only 0.52 in weight units (NORTON and TAYLOR, 1979), and the instantaneous W/R ratio is probably no more than 0.01 to 0.03 (the latter is essentially completely controlled by the numerical value of the interconnected porosity of the crystalline gabbro at the appropriate pressure and temperature).

The above "discrepancy" in W/R is a simple consequence of the fact that each cm^3 of rock along a given flow path sees the same packet of water that passed through all of the previous cm^3 volumes of rock encountered along this flow path. Because each cm^3 of fluid has chemically and isotopically exchanged with all of these earlier volumes of rock, it will have very little effect on the next cm^3 that it encounters. The overall isotopic shift is produced by adding up the effects of all these individual fluid packets in cumulative fashion as they cross each cm^2 cross–section of rock, whereas in the overall material–balance calculation for the whole body, each cm^3 packet of fluid entering the intrusion is, of course, only counted once.

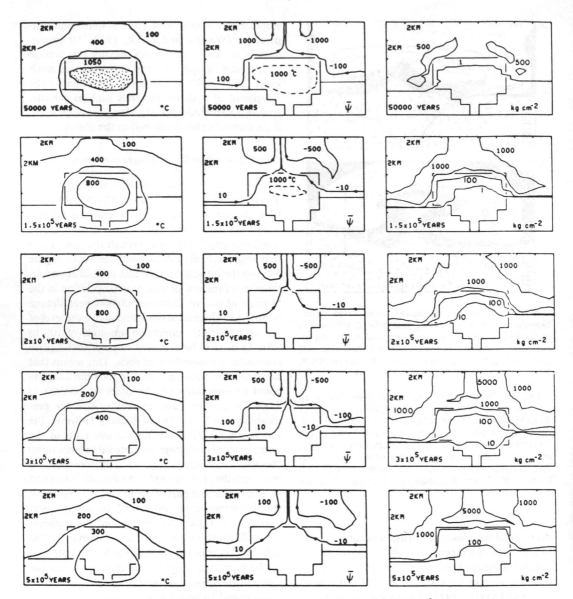

FIG. 5. Temperature (°C), streamlines (ψ), and integrated fluid flux (kg cm^{-2}) determined as a function of time by NORTON and TAYLOR (1979) in their numerical modeling study of the Skaergaard intrusion. In the streamline plots, note that at any instant, the largest amounts of H$_2$O are flowing through zones where the streamlines are closest together. Stippled pattern represents magma and dashed contour in streamline plot shows region where intrusion is assumed to be impermeable, *e.g.,* where $T > 1000$°C. After 400,000 years, this history of groundwater flow leads to the calculated δ^{18}O pattern shown in Figure 4b.

It is very important to understand that these seemingly gigantic "actual" water/rock ratios of 10^5 to 10^7 along a flow path are not only real, but also are perfectly compatible with the overall material–balance water/rock ratios of about 0.5 to 1.0 that are required both by the energy balance and by the mass balance of oxygen isotopes (or strontium isotopes, for example, if the initial and final concen-

trations and isotopic ratios of Sr are known for both the fluids and the rocks, see McCULLOCH *et al.,* 1981). When one realizes that for a hand–specimen sized volume of rock, the actual, cumulative, fluid/rock ratios in hydrothermal systems (measured over the lifetime of the system) are commonly as high as 1,000,000 to 1, even in such low-permeability systems, it may be easier to understand the thorough "soaking" experienced by these rock systems, as well as the pervasiveness and uniformity of the ^{18}O–depletions that are observed in the feldspars of such hydrothermally altered rocks.

Other layered gabbro bodies

Other layered gabbro plutons that have established meteoric–hydrothermal convective systems are Jabal at Tirf in Saudi Arabia, Stony Mountain in Colorado, the Islands of Skye and Mull in Scotland, and Ardnamurchan in Scotland (TAYLOR and FORESTER, 1971; 1979; TAYLOR, 1980, 1983; FORESTER and TAYLOR, 1976, 1977, 1980). Analogous examples involving ocean water hydrothermal systems are observed in all ophiolite complexes, including Cyprus and the Samail ophiolite, Oman (GREGORY and TAYLOR, 1981), as well as in dredged samples from the Indian Ocean (STAKES *et al.,* 1983). In all these areas, large portions of the layered gabbro complexes display marked ^{18}O/^{16}O disequilibrium between coexisting pyroxene and plagioclase (see Figures 1 and 6).

Although in these gabbro bodies there is local development of chlorite, epidote, actinolite, talc, sphene, prehnite, and other low–temperature (greenschist facies) minerals, particularly in areas of diking, heavy fracturing and veining, or multiple intrusion, the bodies in general are astonishingly free of *any* petrographic or mineralogic features indicative of hydrothermal alteration. This statement, of course, does not apply where an older gabbro body has been invaded and hydrothermally metamorphosed by a younger intrusion (*e.g.,* the heavily altered Broadford gabbro body on Skye, FORESTER and TAYLOR, 1977), but it definitely applies to *all* gabbro bodies involved in a single cycle of meteoric hydrothermal activities (*i.e.,* the convection cells set up by that particular gabbro magma chamber itself).

Petrographic evidence for hydrothermal alteration can usually be observed through a careful study of such ^{18}O–depleted gabbros, although the ^{18}O/^{16}O "reversals" between plagioclase and pyroxene always represent the most definitive relationship. These petrographic features are typically fairly subtle and include: (1) clouding or turbidity of some of the plagioclase; (2) development of minor talc–magnetite rims on olivine; (3) coarsening of exsolution lamellae around microfractures in clinopyroxene grains; (4) macroscopic veins easily seen on good, glaciated outcrops (on careful examination these veins often prove to contain pyroxene, magnetite, and high–temperature amphiboles such as hornblende); and (5) development of minor

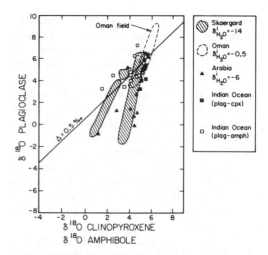

FIG. 6. Comparison of δ^{18}O data on clinopyroxene-plagioclase pairs from the Skaergaard intrusion (diagonal lined pattern) with analogous data from Indian Ocean dredge hauls (STAKES *et al.,* 1983), the Oman ophiolite (GREGORY and TAYLOR, 1981), and the Jabal at Tirf complex, Arabia. The Cuillin gabbros of Skye (data not shown) display similar relationships indicative of alteration by meteoric–hydrothermal fluids (FORESTER and TAYLOR, 1977). The $\delta^i_{H_2O}$ values (including δ^{18}O = 0 for the Indian Ocean) indicate the most probable initial δ^{18}O values of the surface waters involved in these hydrothermal systems. Seven out of the ten data points at Jabal at Tirf correspond very well with the data from Oman, the main Layered Series rocks at Skaergaard, and the Cuillin Gabbro Series. The other 3 samples from Jabal at Tirf contain unusually low–^{18}O pyroxenes and, like the Eucrite Series gabbros from Skye, and the eastern rift–zone basalts from Iceland, these probably formed from low–^{18}O magmas (see text).

amounts of actinolite, biotite, chlorite, and epidote in zones transitional to lower temperature hydrothermal activity (*e.g.,* late–stage veins). All of the above features are described in the references given above, or in NORTON *et al.* (1984), or both. In addition, FERRY (1985b) has demonstrated that secondary magnetite (relatively pure Fe_3O_4) is ubiquitous in the Skye gabbros, and also that the calculated temperatures of formation of the talc–olivine–orthopyroxene assemblages are in the range of 525°–545°C. The early veins throughout practically the entire section of layered gabbro in the Skaergaard intrusion contain hydrothermal clinopyroxenes with *minimum* solvus temperatures of 500° to 750°C (MANNING and BIRD, 1986).

It is thus an inescapable conclusion that layered gabbro bodies typically undergo meteoric–hydrothermal alteration at very high temperatures, in large part in the range 500°–900°C. This conclusion is totally at odds with the statements made by CATHLES (1983) on this matter (see above). Only at such high temperatures could we simultaneously obtain the clear $^{18}O/^{16}O$ evidence for intense hydrothermal alteration and the virtually complete absence of low–temperature hydrous minerals. The temperatures (500°–900°C) and P_{H_2O} values (200–800 bars) are clearly not within the stability fields of chlorite, epidote, serpentine, actinolite, and clay minerals during the bulk of the hydrothermal activity. In fact, much of the mineralogical alteration that *does* occur in these gabbros takes place at 450–550°C, considerably higher than the average temperature of alteration in the granitic plutons described below, but still *lower* than the temperatures at which the bulk of the ^{18}O depletion occurred in these gabbros (TAYLOR and FORESTER, 1979; FERRY, 1985b).

Why then do we mainly see only the effects of 350°–400°C H_2O, for example, at the mid–ocean ridge spreading centers and on land in places like the Salton Sea? As NORTON and KNIGHT (1977) and NORTON (1984) have emphasized, convective systems are strongly controlled by the approximate coincidence of a viscosity minimum, together with a maximum in the isobaric coefficient of thermal expansion in critical to supercritical H_2O (350° < T < 450°C; 200 < P < 800 bars). These are also the conditions where the density of H_2O undergoes the most rapid change as a function of a temperature. Furthermore, the heat capacity of the fluid under these pressure and temperature conditions is quite large, and is maximized at the critical point. Thus, in this general pressure-temperature range, the buoyancy and heat transport properties of the fluid are maximized and the drag forces minimized (NORTON and KNIGHT, 1977). In fact, it is very likely these physical properties of H_2O that control the observed upper limit of 350°–400°C observed during surface venting of modern hydrothermal systems, *not* the "fact" that all rocks hotter than 400°C are impermeable, as proposed by CATHLES (1983).

Granitic plutons

There are no known examples of *mineralogically unaltered* granitic rocks that also exhibit marked disequilibrium $\delta^{18}O$ values in the coexisting quartz and feldspar. The only low–^{18}O granites that are free of such mineralogical alteration products are those that crystallized from low–^{18}O magmas, and these rocks formed as equilibrium assemblages (*e.g.*, the Seychelles granites, see TAYLOR, 1977; also see below). The most conspicuous and quantifiable petrographic change observed in the hydrothermally altered granitic plutons is chloritization of mafic minerals, particularly the biotite, and a major in-

crease in the turbidity of the feldspars. These changes are easily seen with the petrographic microscope, which additionally shows that the conversion to chlorite first proceeds along cleavage planes and grain boundaries of the biotite (CRISS and TAYLOR, 1986). Other common secondary minerals are montmorillonite, sericite, calcite, epidote, prehnite, and zeolites. In most extremely ^{18}O-depleted rocks the alteration is so intense that it can be seen in hand specimen. In such cases the K-feldspar is also strongly clouded, and albite twinning in plagioclase may be absent.

A good correlation exists between the degree of feldspar turbidity and $\delta^{18}O$ in the hydrothermally altered Skye granites (FORESTER and TAYLOR, 1977; FERRY, 1985a). The temperature of much of the alteration in these Skye granites was about 350°–450°C, based on various cation exchange geothermometers and phase equilibria (FERRY, 1985a); locally temperatures as low as 200°C are recorded.

Somewhat higher temperature alteration can occur immediately adjacent to and within the central intrusions of meteoric–hydrothermal systems in granites (CRISS and TAYLOR, 1983). Coarse–grained epidote and secondary amphiboles can occur in this zone, and in many cases intruded country rocks develop hornfelsic textures. Changes in the structural state of feldspar commonly occur, particularly inversion of microcline to orthoclase.

Contrasting effects in gabbros and granites

As indicated above, there are very distinctive mineralogical features observed in ^{18}O–depleted layered gabbros as opposed to those typically observed in ^{18}O–depleted granitic plutons. To recapitulate some of these differences: (1) Gabbros with reversed, nonequilibrium $\Delta^{18}O$ plagioclase–pyroxene values but practically no mineralogic alteration are quite common, whereas (2) Granodiorite, quartz monzonite, and tonalite plutons with large, nonequilibrium $\Delta^{18}O$ quartz–feldspar values are almost invariably strongly altered, containing abundant chlorite, epidote, sericite, turbid feldspars etc. A significant question is, why does most of the *external* (hydrostatic) meteoric–hydrothermal fluid move through layered gabbro bodies at much higher temperatures than in the case of granitic plutons?

There are several aspects to this problem. Seven of the major differences between gabbros and granites that probably contribute to the observed isotopic contrasts are listed in Table 1. It is interesting that each of these seven different igneous rock properties seem to favor a higher temperature of meteoric–

Table 1. Contrasting Properties of Granitic and Gabbroic Plutons

	Granitic plutons	Gabbro plutons
Latent Heat of Fusion	low (40 cal/g)	high (100 cal/g)
Magma Temperature	650°–900°C	1000°–1200°C
Initial H_2O Content of Magma	2–5 weight percent (2nd boiling common)	<1 weight percent (2nd boiling localized in late–stage granophyres)
Initial % Crystallized	20–60% (reduces latent heat)	<10% (little effect on latent heat)
Fracture Network	may be very dense within the pluton because of hydraulic fracturing (e.g., porphyry Cu bodies) and caldera collapse; also very large volume change associated with α-β quartz transition	usually simple contraction cooling and jointing, although occurrence in rift–zone environments is common, suggesting a very deep–seated and pervasive extensional fracture network in the country rocks.
Presence of Magmatic H_2O Envelope at Lithostatic Pressure	very common for H_2O–rich magmas at shallow depths in the crust	non–existent, except perhaps in H_2O–rich alkalic gabbro magmas rich in biotite and hornblende
Geometry of Crystallization	homogeneous solidification of the entire body (only local separation of late–stage melt from crystals); forms a single, integrated meteoric–hydrothermal system, or overlapping systems when complicated by multiple intrusions; these systems cannot migrate into the pluton until the magmatic H_2O envelope is dissipated	strong separation of cumulate crystals from silicate melt, with crystallization upward from the floor of the magma chamber; typically forms two decoupled meteoric-hydrothermal systems, a lower-T system in the country rocks and upper border zone rocks above the late-crystallizing sheet of magma near the roof of the chamber, and a higher-T system in layered cumulates below this magma sheet

hydrothermal alteration in the gabbro as opposed to the granites.

The most obvious reason why the meteoric–hydrothermal systems of gabbros can exist at much higher temperatures than in granites is that gabbros solidify at much higher temperatures, 1000°–1050°C. Granitic materials are still liquid at these temperatures and are thus unable to sustain fractures that would allow penetration by hydrothermal fluids. In addition, the latent heats of crystallization of gabbros are higher and the initial percentage of crystals in the gabbroic magmas is typically lower at the time of intrusion. Both features indicate that there is much more energy available in a gabbro for raising external H_2O to a high temperature than there is in a granite.

Possibly of equal importance to the above is the geometry of crystallization of the magma body. The typical granitic magma probably is intruded with a higher percentage of crystals, and being relatively viscous, both these crystals and newly–formed crystals are probably distributed throughout the mass, with final crystallization taking place in the center and at the lowest levels of the body. This is decidedly not how layered gabbros crystallize. As shown schematically in Figure 5 for the Skaergaard

intrusion, these igneous bodies characteristically solidify upward from the bottom, and the last liquid to crystallize is a sheet–like layer of liquid near the roof of the body. This sheet of late liquid is crystallizing very slowly, because crystallization proceeds as the square root of time and the body is at that stage surrounded by a mass of very hot, insulating rock. This means that the crystalline gabbro *underneath* the sheet of late-stage magmatic liquid will remain *very* hot for a long time. When such material fractures, it will be penetrated by very hot aqueous fluids which flow inward underneath the magma sheet, which is itself impermeable to the fracture-controlled hydrothermal system. As indicated schematically for the Oman ophiolite magma chamber (Figure 7), this characteristically seems to produce two decoupled hydrothermal systems in layered gabbros (see NORTON and TAYLOR, 1979; NORTON et al., 1984; and GREGORY and TAYLOR, 1981): (a) A relatively low–temperature, vigorous hydrothermal system above the magma sheet, where water/rock ratios are high and T = 350°–400°C, and (b) A much higher temperature system below the impermeable magma sheet where T = 500°–900°C and water/rock ratios are much lower (0.1–0.3). Only after the final liquid sheet crystallizes

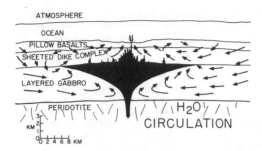

FIG. 7. Cartoon sketch, modified after GREGORY and TAYLOR (1981), showing probable seawater circulation patterns in a cross–section through a fast–spreading mid-ocean ridge. The solid black zone indicates the characteristic, funnel–shaped cross–section of a basaltic magma chamber (*e.g.,* Muskox intrusion, Great Dyke). Note that in the fractured rocks outside the magma chamber there are two essentially isolated circulation regimes, separated by a thin sheet of magma sandwiched between the layered gabbros and the overlying dike complex. The magma is essentially impermeable to the fluids in the two hydrostatic, sea–water convective systems.

can the temperature of the layered gabbro begin to fall rapidly, and at this stage, the two decoupled systems become connected and the temperature of the intrusion sharply declines as strong upward fluid flow is now possible out through the top of the intrusion.

Another major difference is the contrasting effects of magmatic H_2O in the two cases. Granitic magmas typically contain much higher concentrations of H_2O than tholeiitic gabbro magmas. This magmatic water in fact is thought to provide the force that causes abundant fracturing in prophyry copper deposits (BURNHAM, 1979). Any strong fracturing event will increase the permeability enormously, allowing correspondingly larger amounts of fluid to enter the system on a shorter time scale; the pluton would thus be cooled down to 200°–300°C fairly rapidly. These types of effects will not occur in the gabbroic systems. Another contributing aspect is the major volume change that accompanies the α-β quartz transition in granites that is absent in gabbro.

The final feature that may be important is the fact that any magma body that releases significant H_2O under lithostatic pressure can produce a magmatic H_2O envelope that will keep the external, meteoric–hydrothermal system outside the intrusion until this magmatic H_2O envelope is dissipated. This effect seems to occur in porphyry copper systems (SHEPPARD *et al.,* 1971; TAYLOR, 1974). Thus, no low-^{18}O effects would be seen in the pluton until the very late, low–temperature stages. Because of

their low magmatic H_2O contents, these effects would not typically be observed in tholeiitic gabbros. However, they might possibly be seen in volatile-rich alkali gabbros, and this may be the reason for the contrasting isotopic behavior of the Lilloise alkalic hornblende gabbro intrusion as compared to the tholeiitic Skaergaard intrusion (Figure 3). Although both intrusions were emplaced into similar country rocks in East Greenland about 55 Ma ago, strong ^{18}O depletions are not observed in the Lilloise body (SHEPPARD *et al.,* 1977).

To conclude this section, we summarize the factors that tend to keep an external (hydrostatic) marine- or meteoric–hydrothermal system from penetrating and causing $\delta^{18}O$ changes in either a gabbroic or a granitic pluton:

(1) Intrusion into relatively impermeable country rocks, for example into either (a) limestones and evaporites, which are susceptible to ductile deformation at pressure-temperature conditions where most other rocks undergo brittle fracture, or into (b) ordinary silicate rocks at sufficient depth in the crust for recrystallization (metamorphism) to occur, particularly in the absence of extensional tectonics (rifting) or strong, brittle deformation.

(2) Presence of the silicate melt itself, which also is essentially impermeable as long as the strain rates are low enough and the percent of melt high enough so that an interconnected fracture network cannot develop. This is very important in layered gabbros because the late–stage magma sheet at the roof of the intrusion provides a barrier between the very hot, lower hydrothermal system in the layered cumulates and a cooler system at 250°–400°C and higher water/rock ratios in the roof rocks. The hydrothermal fluids in both systems are externally derived and are under hydrostatic pressures.

(3) Evolution of magmatic fluids from late–crystallizing portions of a (granitic) pluton, producing a magmatic H_2O envelope under lithostatic pressure that fills all available fractures outward from the crystallization front. This keeps the low-^{18}O meteoric waters outside the pluton until the temperatures have fallen into the range of stability of sericite, chlorite, etc. Thus, two decoupled hydrothermal systems may also commonly occur around H_2O–rich granitic magma chambers, but in such cases the two types of fluids are genetically very different and are under different pressures. Also, the magmatic H_2O system does not produce the kinds of drastic $^{18}O/^{16}O$ changes in the mineral assemblages that are produced by the meteoric–hydrothermal systems.

ORIGIN OF LOW–^{18}O MAGMAS

General statement

The preceding discussion shows that it is relatively easy to demonstrate sub–solidus, meteoric–hydrothermal, ^{18}O/^{16}O exchange in ancient, deeply–eroded igneous centers, and there is definitive evidence that aqueous fluids penetrate in abundance into gabbro intrusions at very high temperatures, well into the temperature range at which granitic melts can be formed. Because of the pervasive hydrothermal effects, it is very difficult to prove the existence of low–^{18}O *magmas* in such environments. Considerable effort is required to "see through" the alteration effects and find out how much of the ^{18}O depletion was an original magmatic phenomenon. On the other hand, if one can analyze fresh, unaltered lava flows that have not been buried and therefore have not suffered any sub–solidus hydrothermal alteration, this is not a problem.

MUEHLENBACHS *et al.* (1974) and FRIEDMAN *et al.* (1974), respectively, made the important discoveries that such low–^{18}O volcanic magmas had in fact been formed in large volumes in two of the major late Cenozoic volcanic fields of the world, namely in Iceland and in the Yellowstone Plateau, Wyoming. This phenomenon is one of the truly unique aspects of the geochemistry of igneous rocks that was developed through the utilization of stable isotope techniques. The existence of such magmas, and the necessity for a special mechanism to explain them, was not even suspected before ^{18}O/^{16}O analyses of silicates became commonplace.

In the past decade, a fairly intensive search has been made for new occurrences of low–^{18}O magmas. Although a few isolated occurrences have been found, two conclusions can be made from this decade–long search: (1) Such magmas are much less abundant than was originally thought; and (2) no new giant occurrences comparable to Yellowstone or Iceland have been found among Late Cenozoic lavas *anywhere* in the world. The following discussion addresses the reasons for the rarity of these types of magmas, which is clearly also related to the problem of their origin.

Iceland

The data of MUEHLENBACHS *et al.* (1974) and CONDOMINES *et al.* (1983) can be briefly summarized as follows (Figure 8): (1) The alkali olivine basalts on Iceland have distinctly higher and more uniform δ^{18}O values than any of the other Icelandic basalts, +5.3 to +5.7; these values are similar to

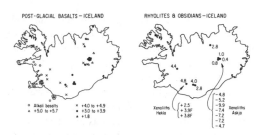

FIG. 8. Maps of Iceland, on the left showing the range of δ^{18}O values determined for post-glacial basalts, and on the right showing the range of δ^{18}O values of rhyolites, obsidians, and siliceous xenoliths (data from MUEHLENBACHS *et al.*, 1974, and MUEHLENBACHS, 1973).

those found in most basalts throughout the world. These alkali olivine basalts are found only on the periphery of the island, at the edges of any deep meteoric–hydrothermal circulation systems. (2) Relative to the alkali olivine basalts and the tholeiites, the transitional alkali basalts occupy an intermediate geographic position in the eastern rift zone (Figure 8) and also have intermediate δ^{18}O values, +3.9 to +4.9. (3) The olivine tholeiites from the western rift zone (Figure 8) also have intermediate δ^{18}O values, +4.0 to +5.7, with the highest values closest to the coast on the SW part of Reykjanes peninsula. (4) In the basalts, the lowest and most heterogeneous δ^{18}O values, +1.8 and +5.4, are all confined to the quartz tholeiites of the eastern rift zone. (5) There are some crude correlations between δ^{18}O and chemical composition; increasing SiO_2 and K_2O tend to be accompanied by decreasing δ^{18}O for each petrologic class of basalts. In particular, the tholeiite with by far the most extreme δ^{18}O value (+1.8) also has by far the highest K_2O content. (6) The rhyolites and obsidians tend to have even lower δ^{18}O values than the basalts (Figure 8), and the rhyolites with the lowest and most variable δ^{18}O values all come from the eastern rift zone. (7) There is a weak correlation between δ^{18}O and ^{3}He/^{4}He, with the intermediate and silicic volcanic glasses typically having ^{3}He/^{4}He ratios close to the atmospheric value.

The isotopic relationships described above strongly support the idea that assimilation or partial melting, or both, of hydrothermally altered country rocks in the deeper parts of the rift zone is the most likely mode of formation of the low–^{18}O magmas from Iceland. This explanation is compatible with the Pb isotope data of WELKE *et al.* (1968) and the ^{87}Sr/^{86}Sr and rare–earth data of O'NIONS and GRONVOLD (1973), and it readily explains why the most contaminated magmas are either the rhyolites, or those basalts that have probably most strongly interacted with roof rocks or rhyolite melts, namely, the K–rich and Fe–rich tholeiites. Direct exchange with meteoric waters would not be expected to produce these relationships.

It is probably significant that all of the extremely low–^{18}O basaltic and rhyolitic magmas are confined to the eastern rift zone, which has only been active during the past 3–4 m.y. (SAEMUNDSSON, 1974). The magmas in this rift zone are penetrating upward through the lower parts of volcanic and plutonic rocks that presumably were intensively hydrothermally altered in an earlier episode of magmatic activity at the time of their original formation in the western rift zone. Thus, the magmas coming up through the eastern rift zone would be interacting with country rocks that had already suffered heterogeneous ^{18}O depletions, and which, through subsidence, have been brought down into a much higher temperature regime (15–20 km depth?).

Approximately 200 km^3 of low–^{18}O tholeiite has been erupted in the eastern rift zone in the last 12,000 years (JACOBSSON, 1972). It was the scale of this process that most bothered MUEHLENBACHS et al. (1974) when they rejected the meteoric–hydrothermal explanation for the origin of these types of magmas. This mechanism for the origin of the Icelandic volcanic rocks has, however, been strongly favored over the past few years by TAYLOR (1974; 1977; 1979), and recently HATTORI and MUEHLENBACHS (1982) and CONDOMINES et al. (1983) provided strong new support for this mechanism.

The model of CONDOMINES et al. (1983), based on combined He, O, Sr, and Nd isotopic relationships, is shown in Figure 9. They propose that the primary ^{3}He/^{4}He and ^{18}O/^{16}O ratios of mantle–derived magmas were changed in a deep magma reservoir by exchange or contamination between the magma and the surrounding meteoric–hydrothermally altered basaltic crust. Such processes to not appear to have introduced much water into the magma because Icelandic volcanism, except for subglacial eruptions, is usually not explosive, and hydrous minerals in plutonic ejecta are rare. The rhyolitic magmas are assumed to have been produced at relatively shallow crustal levels, either by melting of hydrothermally altered rocks or as the deeper magmas moved upward and underwent further contamination processes. These processes caused introduction of atmospherically–derived He into the magmas.

Catastrophic isotopic changes in magmas during caldera collapse, Yellowstone volcanic field

The title of this section and much of the following discussion is taken from a very significant paper by HILDRETH et al. (1984), who followed up the original discovery by FRIEDMAN et al. (1974) with a detailed chemical and isotopic study of the Quaternary rhyolites of the Yellowstone Plateau, Wyoming.

This 17,000–km^2 volcanic field (Figure 10) consists of three large, overlapping, rhyolitic calderas, apparently a 115 km long extension of the axis of the Snake River Plain. Rhyolitic magmatism has migrated northeastward in this area at a rate of 2–4 cm/year for at least the last 10 Ma (ARMSTRONG et al., 1975; CHRISTIANSEN and MCKEE, 1978). Former rhyolitic centers have progressively subsided and been buried by the basalts that now make up the floor of the plain.

Brief, caldera-forming, ash–flow eruptions occurred at 2.0, 1.3, and 0.6 Ma (CHRISTIANSEN, 1983), with minimum volumes of 2500, 280, and 1000 km^3, respectively. The δ^{18}O values of the first-cycle Huckleberry Ridge Tuff (HRT) are among the highest in the Yellowstone area (δ^{18}O quartz = +7.1 to +7.6, giving a calculated coexisting magma δ^{18}O ≈ +6.4 to +6.9). The subsequent caldera-forming eruptions (Figure 11) have slightly lower calculated δ^{18}O magma values of +4.8 to +5.2 (Mesa Falls Tuff) and +5.3 to +6.0 (Lava Creek Tuff).

Almost all of the Yellowstone rhyolites are somewhat depleted in ^{18}O relative to the 'normal' δ^{18}O values of +7 to +10 usually observed in silicic volcanic rocks throughout the world (TAYLOR, 1968, 1974). However, the postcaldera rhyolites of the first and third caldera cycles include some *extraordinarily low–*^{18}O eruptive units (Figure 11).

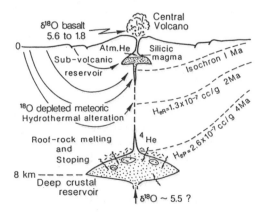

FIG. 9. Schematic section summarizing the model of CONDOMINES et al. (1983) for the evolution of the Icelandic magmas. (He$_R$ = radiogenic helium).

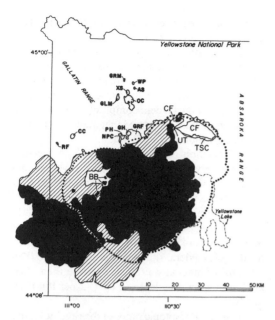

FIG. 10. Generalized map of Yellowstone National Park, showing the outline of the caldera rim (outer fault scarp) of Yellowstone caldera (dotted), and the locations of third–cycle post–caldera and extracaldera rhyolites. The extreme low–¹⁸O post-caldera rhyolites ($\delta^{18}O$ quartz < +2.0) are indicated by symbol: CF, TSC, and UT in the northeast part of the caldera, and BB in the southwest. The moderately low–¹⁸O post-caldera rhyolites (+4.0 < $\delta^{18}O$ quartz < +5.0) are shown in solid black, and the slightly ¹⁸O–depleted post-caldera rhyolites ($\delta^{18}O$ quartz ≥ +5.0) are indicated by the diagonal–lined pattern. The normal–¹⁸O extracaldera rhyolites north of the caldera are the small outcroppings indicated by various symbols. The first-cycle caldera lies about 15 km SW of the western park boundary. Abbreviations: GRM—Gardner River mixed lavas, WP—Willow Park dome, AS—Apollinaris Spring dome, XS—Crystal Spring flow, OC—Obsidian Cliff flow, GLM—Grizzly Lake mixed lavas, CC—Cougar Creek flow, RF—Riverside flow, PH—Paintpot Hill dome, GH—Gibbon Hill dome, GRF—Gibbon River flow, NPC—Nez Perce Creek flow. Modified after HILDRETH *et al.* (1984).

The combined areal extent of the two first–cycle low–¹⁸O flows ($\delta^{18}O$ = +2.9 to +3.6) is ≈66 km²; their original volume was at least 10 km³, and they were erupted 30,000 to 350,000 years after the >2500 km³ Huckleberry Ridge Tuff. These low–¹⁸O lavas are similar to most other Yellowstone rhyolites, and contain ≈76% SiO_2 and 15–20% phenocrysts of quartz, sanidine, plagioclase, clinopyroxene, fayalite, and Fe–Ti oxides.

The second cycle was unique, in that there was no major change in the $\delta^{18}O$ of the magma body after eruption of the Mesa Falls Tuff (MFT) and collapse of its 380–km² caldera. However, during the third cycle, enormous ¹⁸O depletions were observed (Figure 11) in lavas that vented in two sep-

arate areas ≈45 km apart, soon after eruption of the >1000 km³ Lava Creek Tuff (LCT), along the compound ring fracture zone of the Yellowstone caldera. Some of these units are petrographically distinct in containing plagioclase in excess of sanidine. In the NE part of the caldera, there are three major low–¹⁸O units, which together cover ≈140 km² and represent >40 km³ of magma, perhaps as much as 70 km³, all of it with $\delta^{18}O$ between about +0.6 and +1.2 (calculated from $\delta^{18}O$ of the quartz). The second area of very low ¹⁸O third–cycle rhyolites is in the southwestern ring fracture zone, where scattered exposures in a 25–km² area have calculated $\delta^{18}O$ values between −0.1 and +0.6, and a *minimum* volume of 2.5 km³. Chemically, the low–¹⁸O third–cycle magma completely overlaps the compositional range of the Lava Creek Tuff, but it also extends to less evolved compositions (HILDRETH *et al.*, 1984).

Pb and Sr isotope data by DOE *et al.* (1982) and HILDRETH *et al.* (1984) are compared in Figure 12, scaled to reveal the correlations between $\delta^{18}O$, ²⁰⁶Pb/²⁰⁴Pb, and ⁸⁷Sr/⁸⁶Sr, specifically between lower $\delta^{18}O$ and more radiogenic Pb and Sr. The Pb and Sr isotope data define a zigzag pattern that reflects abrupt caldera-forming events and longer intervals of partial recovery toward precaldera isotopic ratios very much like the $\delta^{18}O$ patterns. The low–¹⁸O early postcaldera rhyolites of the third cycle are particularly enriched in radiogenic Pb and Sr. The rhyolitic units with the most radiogenic ²⁰⁶Pb/²⁰⁴Pb ratios also have the highest ratios of ²⁰⁷Pb/²⁰⁴Pb, suggestive of collapse–related mixing of the rhyolitic magmas with pre-Cenozoic upper crustal components, most

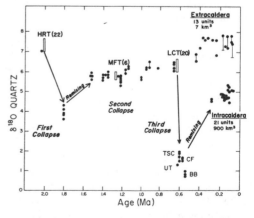

FIG. 11. Plot of $\delta^{18}O$ quartz vs. K–Ar age for the rhyolites of the Yellowstone Plateau volcanic field. HRT, MFT, and LCT are (with numbers of analyzed samples) the major ash–flow sheets referred to in the text. Modified after HILDRETH *et al.* (1984).

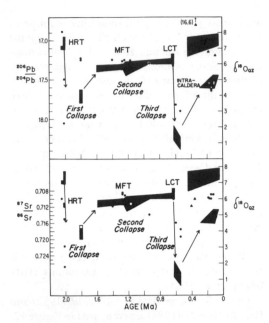

FIG. 12. Initial Sr and Pb isotopic values for rhyolites of the Yellowstone volcanic field (dots, left axis), plotted against ages of the units and superimposed on shaded fields (right axis) representing the $\delta^{18}O$ quartz data of Figure 11. Modified after HILDRETH *et al.* (1984).

of which probably contain more radiogenic Pb and Sr than the precaldera or extracaldera rhyolites (DOE *et al.*, 1982).

Origin of the low-^{18}O magmas at Yellowstone

Following is a summary of the critical features pertinent to the development of low-^{18}O magmas in the Yellowstone caldera complex, modified after HILDRETH *et al.* (1984):

(1) The narrow range of $\delta^{18}O$ in the three major ash–flow sheets (+5 to +7) contrasts sharply with the wide variation in the postcaldera rhyolitic lavas. The earliest postcaldera lavas are depleted in ^{18}O by as much as 3 to 6 per mil relative to the immediately preceding ash flow tuffs. The biggest ^{18}O effects follow immediately after the two biggest ash–flow eruptions.

(2) The ^{18}O depletions were geologically short-lived events (<300,000–500,000 years) that followed caldera subsidence in some cases by less than 50,000–100,000 years. The earliest post–collapse lavas are the most ^{18}O–depleted units, and successively younger postcaldera lavas show partial recovery of the magma toward precaldera $\delta^{18}O$ values, presumably by exchange and mixing with deeper levels of the magma reservoir. However, no intra-

caldera rhyolite is ever again as ^{18}O–rich as the ash–flow tuff that preceded it.

(3) The pattern of stepwise ^{18}O depletions is reflected in lavas erupted as far apart as 115 km, indicating that these are not isolated events. Particularly for the third cycle, the ^{18}O–depleted rhyolites cover most of the caldera complex (Figure 10), and they are part of an integrated magmatic system that evolved for more than 2.0 Ma.

(4) Depletion of ^{18}O occurred *only* in the sub-caldera reservoir; contemporaneous rhyolites that vented just outside the caldera have the highest $\delta^{18}O$ values in the volcanic field (Figure 11).

(5) Voluminous postcaldera rhyolites, all exhibiting the drastic 5 per mil ^{18}O depletion, erupted at opposite ends (45 km apart) of the newly–formed third–cycle caldera, showing that much more than 100 km^3 of magma was affected. More than 1000 km^3 of magma was depleted by 1–2 per mil (Figure 10).

(6) Sr and Pb isotopic ratios of the rhyolites jump to more radiogenic values immediately subsequent to caldera formation. The long-term pattern for Pb and Sr isotopic ratios of rhyolites erupted from the subcaldera magma reservoir is similar to the zigzag pattern displayed by $\delta^{18}O$. Depletion in ^{18}O is clearly related to caldera collapse, and this structural disruption was also important for introducing radiogenic Pb and Sr into the magma.

To explain all the above effects, HILDRETH *et al.* (1984) strongly favored a mechanism of ^{18}O depletion involving influx and solution of low–^{18}O meteoric *water* in enormous amounts directly into the rhyolitic magma. They realized that unless such waters included deep brines exceptionally rich in Pb and Sr (and concentrated brines are *not* low in ^{18}O because they invariably have undergone large ^{18}O shifts), the isotopic shifts of Pb and Sr would require additional exchange with large quantities of foundering roof rocks. Nevertheless, HILDRETH *et al.* (1984) believe that water itself is by far the predominant contaminant.

Although congratulating HILDRETH *et al.* (1984) for a truly magnificent study, this writer is in serious disagreement with their ultimate conclusion concerning the mechanism of ^{18}O depletion of the magmas. A much more likely mechanism(s) would seem to be (1) partial melting of hydrothermally altered rhyolitic country rocks, with subsequent uprise and mixing of this re–melted material back into the roof zone of the magma chamber and/or (2) foundering of such ^{18}O–depleted roof rocks into the magma chamber, accompanied by melting, assimilation, or exchange of this foreign material with the shallow magma reservoir. Based on constraints imposed by the physics and chemistry of H_2O transport through ductile rocks and silicate melts, which are generally thought to be virtually impermeable to convecting aqueous fluids (see above), these mechanisms seem much more plausible than catastrophic influx of water directly into the magma chamber. These concerns are elaborated on in several papers by TAYLOR (1974; 1977; 1979; 1983), and in fact, this writer believes that the data of HILDRETH et al. (1984) actually provide strong new support for those conclusions.

The most abundant country rocks above and along the margins of the Yellowstone magma chamber prior to each major ash–flow eruption were certainly hydrothermally altered, ^{18}O–depleted earlier–cycle rhyolites. In the Yellowstone Plateau, pre-Cenozoic rocks are very subordinate to the volcanic rocks. Such young volcanic rocks cannot represent the contaminant responsible for the Pb and Sr isotope signatures that abruptly appear in the Yellowstone rhyolites; therefore, it is an absolute certainty that if foundering of roof rocks or melting of older country rocks is necessary to explain the Sr and Pb isotope effects, then this must have been accompanied by prodigious amounts of assimilation or melting of the much more abundant rhyolitic country rocks. The only observable chemical effects of this latter process on the magma reservoir would be: (1) lower δ^{18}O values, (2) lower δD values, and possibly (3) higher H_2O concentrations. This is because these rhyolitic country rocks are chemically and isotopically almost identical to their original parent magmas, except for the subsolidus hydrothermal changes they have undergone, which to a first approximation only involve hydration and depletion in ^{18}O and D. Note also that such hydrated rhyolites will be much more easily melted than any other type of country rock.

The other important feature of the HILDRETH et al. (1984) study is the catastrophic and abrupt interval over which the isotopic changes occur. The effects are clearly related to caldera collapse. There is no better way of changing the composition of a magma on a short time–scale than catastrophic failure and massive collapse of the roof material into the magma chamber, or rapid subsidence of such rocks into the deep melting zone along the edges of the magma chamber. Not only does the low–^{18}O, hydrothermally altered rhyolite undergo catastrophic subsidence, but new, hotter magma also almost immediately moves upward into the resurgent domes and ring intrusions. Similar, short–time scale, catastrophic processes had in fact been invoked previously by TAYLOR (1974; 1977), but

up until the time of the HILDRETH et al. (1984) study there was little direct support for the idea that such extreme changes could take place on such a rapid time scale. Even though diffusive transport of enormous amounts of H_2O directly into the magma a priori seemed unlikely, there was always the possibility that given enough time it perhaps could occur. The data of HILDRETH et al. (1984) show clearly that δ^{18}O changes in the magma chamber do not occur by such a long-term process! In fact, the long-term changes are in the opposite direction, toward recovery of the original magmatic δ^{18}O value, after the intense ^{18}O depletion event that accompanied the foundering of the roof. On such a short geological time–scale, there is no physical way of separating pore waters from their rock matrix, particularly considering the slow rates of diffusion of H_2O through silicate melts or through ductilely deforming rocks. To this writer, it is obvious that water is indeed involved in the process, but this water is only the pore water that permeates the rock matrix and which is essentially carried along in any foundering or re–melting process that affects the rocks themselves.

HILDRETH et al. (1984) raised several objections against the partial melting or bulk assimilation process, among which are the fact that there is no evidence of the type of massive cooling or increase in phenocryst content that might be expected in the contaminated magmas, nor is there any obvious change in major or trace element composition. The first objection loses its validity when it is realized that mixing or exchange with enormous volumes of liquid H_2O would have an even more profound cooling effect on the magma than would melting or assimilation of hot, altered country rock. The specific heat of water is three to five times higher than that of the rock. Also, the temperatures in these types of systems are probably controlled by the large reservoir of more mafic, higher–temperature magma underneath the rhyolite sheet. The second argument is refuted by the fact that the rocks being assimilated or melted are predominantly older–cycle rhyolites with essentially identical chemical compositions to the younger magmas.

To this writer, the only observations by HILDRETH et al. (1984) that may be in disagreement with the roof-rock melting or assimilation mechanism are: (1) The material–balance argument that the effects are too large to be explained by incorporation of altered rocks, which are unlikely to have an average δ^{18}O any lower than about -5 to -8; and therefore that it is much easier to produce the observed effects with pure H_2O, which could have δ^{18}O $= -19$ (today's water) or perhaps even -25

(Pleistocene H_2O?). (2) The apparent absence of xenocrysts in the lavas.

The material–balance problem is a difficult one, no matter what mechanism one chooses, but it is really not much more of a problem for one process than for the other, as shown in Figure 13. Even at shallow levels of less than 200 m depth, hydrothermally altered rocks with $\delta^{18}O = -8$ to -9 are not uncommon at Yellowstone (HILDRETH et al., 1984). At depths of several km, these values could be expected to be as low as -12 or lower, judging by the data from the deep Krafla drill hole on Iceland (HATTORI and MUEHLENBACHS, 1982). Note that young volcanic rocks are also very porous, so the foundered roof rocks that fall into the magma chamber or which subside into a deep melting zone during such a catastrophic process are not just rock, but H_2O–saturated rock with perhaps 20–35% pore space. Therefore, the water–saturated rocks could easily have an average $\delta^{18}O = -13$ to -15, or perhaps even lower during the Pleistocene glacial period.

The coupled assimilation-fractional crystallization (AFC) process modeled in Figure 13 is based on R values (ratio of cumulates to assimilated rocks) of only 1.5 to 5, which might seem to be unusually low for such a shallow assimilation process (TAYLOR, 1980). However, these calculations are concerned with the *chemical effects* on the rhyolite magma layer, whereas the *heat effects* in these types of magma systems are dominated by the underlying, higher–temperature, more mafic magma reservoir. Also, the latent heats of fusion of the assimilated glassy rhyolites are probably very low (<30 cal/g).

The explosive shattering that would accompany the engulfment of such water–saturated rocks into the magma chamber would enhance complete as-

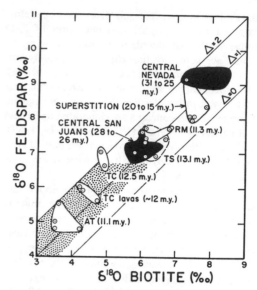

FIG. 14. Plot of $\delta^{18}O$ feldspar vs. $\delta^{18}O$ biotite, modified after LARSON and TAYLOR (1986), for various ash–flow tuff complexes in the western U.S.A. AT = Ammonia Tank, TC = Tiva Canyon, TS = Topopah Springs, and RM = Rainier Mesa; these are different ash—flow units of the southwestern Nevada caldera complex studied by LIPMAN and FRIEDMAN (1975). The stippled pattern indicates the field of Yellowstone rhyolites (HILDRETH et al., 1984). The data on the Superstition volcanic field are from STUCKLESS and O'NEIL (1973).

similation. The lack of xenocrysts may be related to this shattering event or possibly to a super-heated condition in the rhyolite melt, which is being heated from below by the underlying, more mafic magma. In any case, virtually no xenocrysts would be expected to be derived from such very fine–grained to glassy rhyolites in the first place. It is thus conceivable that there were very few xenocrysts, and those that were present were either largely dissolved, or they were so finely shattered and dispersed that they are not recognizable. In any case, this single observation is in no way sufficient to tip the scales against the re–melting/assimilation mechanism. Actually, probably the biggest objection to this mechanism is the fact that low–^{18}O magmas are *not* observed in most rhyolitic caldera complexes (LARSON and TAYLOR, 1986). This problem is discussed in more detail below.

Other low–^{18}O rhyolite magmas

The only other area in the world where low–^{18}O magmas have been *proven* to have erupted in large quantities is in the Oasis Valley-Timber Mountain caldera complex (Figure 14) in southwestern Nevada (FRIEDMAN et al., 1974; LIPMAN and FRIED-

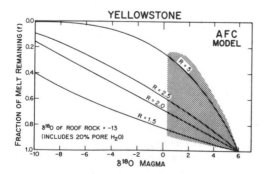

FIG. 13. The curves show AFC model calculations (TAYLOR, 1980) for the Yellowstone caldera system for different assumed values of R, the ratio of cumulates to assimilated rock. The stippled pattern shows the range of $\delta^{18}O$ of the post–caldera rhyolites.

MAN, 1975). A search for low–^{18}O magmas was specifically made by LARSON and TAYLOR (1986) in two of the largest and most complex caldera complexes in the United States, in Central Nevada and in the Central San Juan Mountains of Colorado, as well as in the Lake City caldera in the Western San Juan Mountains. No evidence of any significant magmatic ^{18}O depletion was found in any of these areas (Figure 14). In fact, in each of these areas, LARSON and TAYLOR (1986) observed a remarkably uniform set of δ^{18}O magma values that persisted from 31 to 25 Ma in Central Nevada and from 28 to 26 Ma in the San Juans. The Central Nevada magmas had δ^{18}O = +9.5 to +10.0, whereas the Central San Juan complex is much lower in ^{18}O but just as uniform (δ^{18}O = +7.0 to +7.5). These are all primary magmatic values; the two complexes differ because the source-rock materials of these magmas at depth were more than 2 per mil different from one another.

Because of the rarity of low–^{18}O magmas, the one other caldera complex in addition to Yellowstone that does show these low–^{18}O magmatic effects is of considerable interst (LIPMAN and FRIEDMAN, 1975). The major δ^{18}O changes in the southwestern Nevada complex (Figure 14) are as follows. In the first cycle, the Topopah Spring eruption (>300 km^3, 13.1 Ma) had δ^{18}O feldspar = +6.8 to +8.1, and was compositionally zoned from rhyolite (77 weight percent SiO$_2$, 1% phenocrysts) to quartz latite (69 weight percent SiO$_2$, 21% phenocrysts). This eruption was followed by the chemically similar, but much bigger Tiva Canyon eruption (>1000 km^3, 12.5 Ma), which had distinctly lower δ^{18}O feldspar = +6.4 to +7.0. Late rhyolitic lavas associated with the margins of the Tiva Canyon caldera have even lower δ^{18}O feldspar = +5.1 to +6.1.

After a hiatus of about 1,000,000 years, a second eruptive cycle began starting with the Rainier Mesa eruption (>1200 km^3, 11.3 Ma), which is zoned from rhyolite (76–77 weight percent SiO$_2$, 20–25% phenocrysts) to quartz latite (66–68 weight percent SiO$_2$, 35–50% phenocrysts), and has δ^{18}O feldspar = +6.6 to +7.7 (i.e., the magma system had essentially recovered from the Tiva Canyon ^{18}O depletion event). This eruption was, however, quickly followed only about 200,000 years later by a very ^{18}O–depleted magma, the giant Ammonia Tank eruption (>2000 km^3, 11.1 Ma), which has δ^{18}O feldspar = +4.7 to +5.5. The Ammonia Tank unit is also very phenocryst-rich, but is chemically and petrographically almost identical to the higher–^{18}O Rainier Mesa unit.

In each of the two cycles, a normal–^{18}O eruption was followed by a chemically similar, but ^{18}O–de-

pleted eruption. The low–^{18}O magma in each case exhibits some *very slight* differences from the immediately preceding normal–^{18}O eruption, namely: (1) the low–^{18}O magmas lack plagioclase; (2) they contain sphene; (3) they contain more sodic alkali feldspar; and (4) they have very low ilmenite–magnetite ratios. These differences were interpreted by LIPMAN (1971) and LIPMAN and FRIEDMAN (1975) as indicating that the low–^{18}O units crystallized at higher P_{H_2O} and higher f_{O_2} than the immediately preceding units. These features, of course, would be a natural consequence of the low–^{18}O contamination process by either hydrous roof rocks or direct influx of meteoric H$_2$O. Evidence of an open system involving older, ^{87}Sr-rich rocks is seen in the Sr isotope measurements of NOBLE and HEDGE (1969). Thus, all of the arguments made above favoring a partial melting or AFC model for the ^{18}O depletions at Yellowstone are also valid for these southwestern Nevada occurrences, and the material–balance 'problem' is not as serious because the ^{18}O depletions are much less (counterbalanced, of course, by the fact that the Ammonia Tank eruption was much larger than any of the low–^{18}O eruptions at Yellowstone).

The low–^{18}O magma problem

There are many as yet unanswered questions about the origin of low–^{18}O magmas. This problem should in fact be generalized further to include a variety of other types of magmas generated in sub–volcanic terranes. It cannot be too strongly stressed that our knowledge of the existence of this phenomenon is entirely attributable to our ability to make precise ^{18}O/^{16}O measurements, because at the present time there is really *no other way of recognizing* the existence of such magmas. Thus, exactly the same mechanism may be operating in areas of higher–^{18}O hydrothermal waters (*e.g.*, sea waters, metamorphic waters, formation waters, etc.), but then we would not be able to discern that a unique process was involved, because the δ^{18}O signature would be obscured.

Clearly, a major problem that needs to be solved is why these low–^{18}O magmas are formed in some volcanic fields but not in others. For example, why should some of the eruptive units in the southwest Nevada and Yellowstone volcanic fields show an order of magnitude greater ^{18}O depletion than any of those from the central Nevada, San Juan, and Superstition volcanic fields? This is the most striking problem raised by the work of LARSON and TAYLOR (1986). These dramatic ^{18}O/^{16}O differences cannot be attributed to variations in the δ^{18}O of the associated meteoric ground waters, which certainly were quite low, particularly in Nevada and Colorado ($<$ -10). The differences must be due to some differences in the way the magma chambers at Yellowstone and the southwest Nevada complex interacted with their environment. For some reason, those particular sub–volcanic magma chambers interacted much more strongly with hydrothermally altered roof rocks or with the meteoric ground waters themselves. Based on the new data obtained by LARSON and TAYLOR (1986), this process (whatever it is) must be much less common than heretofore believed (*e.g.,* LIPMAN and FRIEDMAN, 1975, p. 701).

LARSON and TAYLOR (1986) concluded that the [18]O depletions in caldera–related rhyolite magmas are *not* directly related to: (1) the size of the eruption; (2) the duration and intensity of magmatic activity; (3) the complexity of the eruption cycle; or (4) overlapping collapse of a number of different calderas. However, (3) and (4) may be a *necessary* condition for development of low–[18]O ash-flow tuff magmas (see below). LARSON and TAYLOR (1986) were able to discern only two factors that may be significantly different in separating the low–[18]O and normal–[18]O caldera complexes, namely emplacement into an extensional tectonic setting, and the existence of a wide range of chemical compositions in the silicic differentiates. It is also possible, however, that the *depth* of the magma chamber may also be important. Obviously, because these low–[18]O effects involve hydrothermal systems connected to the surface, such low–[18]O magmas cannot be produced in extremely deep magma chambers where the country rocks are relatively impermeable.

Emplacement into an extensional tectonic setting. This is the most important of the two factors. The two caldera complexes in the United States that exhibit the greatest $\delta^{18}O$ shifts (Yellowstone and Southwest Nevada) are both younger than 15 Ma. The period between 15 and 20 Ma corresponds to the time of transition into the brittle, fracture-dominated extensional tectonic regime of the Basin–Range province (for example, see STEWART, 1978; ZOBACK *et al.,* 1981). The Yellowstone caldera, in fact, lies on the eastern end of the currently–active Snake River Plain rift system, and the Southwest Nevada caldera complex lies in the midst of abundant Basin–Range extensional features. Such region-wide extension must produce fractures that penetrate deeply into the crust. These fractures could allow meteoric water to circulate very deeply, as they clearly have in Iceland (HATTORI and MUEHLENBACHS, 1982). Therefore, the regions around the Yellowstone and Southwest Nevada

caldera complexes conceivably could have been subjected to much greater rifting and regional extension than the central Nevada or central San Juan caldera complexes, allowing for much deeper penetration of low–[18]O meteoric ground waters into the crust. This hypothesis could be tested by specifically carrying out more comparative ^{18}O/^{16}O studies of caldera complexes developed in rift and non–rift environments. Nevertheless, at least with the presently available data-set, it is important to realize that the only localities on Earth where there has been clear-cut development of low–[18]O magmas is in major rift–zone environments. Low–[18]O magmas are identified with much less certainty in plutonic environments, but virtually all of the examples that have been found to date are also all emplaced into rift–zone settings, as in the Scottish Hebrides (FORESTER and TAYLOR, 1977) and on the margin of the Red Sea (TAYLOR, 1979; 1983).

Chemical composition. Examination of Figure 1 in HILDRETH (1981) shows that many of the known low–[18]O rhyolites belong to his Groups I and II (high–SiO₂ rhyolites and high–SiO₂ rhyolites zoned to intermediate compositions). These extremely ^{18}O–depleted ash-flow tuffs commonly exhibit SiO₂ contents up to and above 77 weight percent, indicative of some type of very strong chemical differentiation (or fractional melting?). The normal–[18]O ash-flow tuffs from the central Nevada and central San Juan complexes are either monotonous intermediates (*e.g.,* Fish Canyon and Monotony Tuffs), or, if they do include rhyolites, the rhyolites are zoned to SiO₂ contents no higher than 74–76 weight percent (LARSON and TAYLOR, 1986).

CONCLUSIONS

We have shown that layered gabbros typically undergo hydrothermal interaction with externally-derived aqueous fluids (meteoric water, sea water, etc.) at *much* higher temperatures than in the case of granitic plutons (typically 500°–900°C vs. 200°–450°C). This is manifested in the absence or rarity of hydrous alteration minerals (amphibole, chlorite, epidote, etc.) in gabbros that show clear-cut $\delta^{18}O$ signatures of having interacted with very large quantities of external H_2O. In such rocks, the $\delta^{18}O$ effects may be the *only* obvious indications of intense hydrothermal exchange.

Although a number of characteristics contribute to the higher alteration temperatures of the gabbros, such as higher solidus temperature, greater latent heat of crystallization, higher melt/crystal ratio, and characteristically lower fracture density, the two most important factors appear to be:

(1) *The geometry of crystallization,* wherein the layered gabbro cumulates crystallize upward from the floor of their magma chamber, and thus are overlain by a sheet of magma at $T > 1000°C$ for a protracted period of time. The cumulates are therefore held at a very high temperature long after they are rigid enough to be fractured and made susceptible to penetration by external waters. This goes on for a long period because crystallization proceeds approximately proportional to the square root of the time (NORTON and TAYLOR, 1979), and thus during the late stages, the final sheet of silicate melt takes a very long time to crystallize. This leads to two essentially decoupled hydrothermal systems in layered gabbros: A relatively low–temperature system in the roof rocks operates at high water/rock ratios, and is separated by a thin sheet of impermeable magma from the very high temperature system that operates at much lower water/rock ratios in the cumulates.

(2) *The higher magmatic H_2O contents of granitic plutons,* which is important because as long as a magmatic H_2O envelope under lithostatic pressure is present in and around the pluton, filling all the fractures, the system is totally impermeable to the external, hydrostatic system. The latter can only gain entry at very late stages after the magmatic water system has dissipated.

We also conclude that low–^{18}O magmas are formed mainly by re–melting of hydrothermally altered country rocks (and their pore waters) or by large-scale assimilation of such materials. In conjunction with the new data by HILDRETH *et al.* (1984) and LARSON and TAYLOR (1986), this suggests that low–^{18}O magmas can be formed: (1) as a result of catastrophic caldera collapse; (2) on very short time–scales (<100,000 years); (3) in enormous amounts (>1000 km^3 in several localities); (4) however, these giant occurrences are found only in rift–zone tectonic settings; and (5) such magmas are only rarely formed in other types of plutonic–volcanic environments.

In retrospect, the connection between a rift–zone tectonic setting and low–^{18}O magmas may seem to be a fairly obvious one. After all, where on Earth is there a better chance of bringing into close juxtaposition the two geological materials that are essential to make very high–temperature hydrothermal systems and low–^{18}O magmas? Only in rift zones and spreading centers do we find the large-scale extensions and brittle fracturing that are necessary to allow massive amounts of magma to come upward into the crust, as well as providing the greatly increased fracture permeability that allows

surface waters to penetrate to depths of at least 10 or 15 km. Such environments represent the best way to attain the required combination of very high temperatures together with large quantities of low–^{18}O, hydrothermally altered rocks and meteoric pore waters. This is also the environment of essentially all of the layered gabbro bodies that have so far been shown to have interacted with surface waters at very high temperatures.

Acknowledgments—I am grateful for discussions and collaborations over the past few years on problems of meteoric–hydrothermal alteration and the origin of low–^{18}O magmas with Denis Norton, Robert E. Criss, Robert G. Coleman, G. C. Solomon, M. Magaritz, R. T. Gregory, and P. B. Larson. The manuscript was improved by a critical review by Robert E. Criss. Financial support for this research was provided by the National Science Foundation, Grant No. EAR 83-13106.

REFERENCES

ARMSTRONG R. L., LEEMAN W. P. and MALDE H. E. (1975) K–Ar dating, Quaternary and Neogene volcanic rocks of the Snake River Plain, Idaho. *Amer. J. Sci.* **275,** 225–251.

BRACE W. F. (1984) Permeability of crystalline rocks: New *in situ* measurements. *J. Geophys. Res.* **89,** 4327–4330.

BURNHAM C. W. (1979) Magmas and hydrotherm fluids. In *Geochemistry of Hydrothermal Ore Deposits,* 2nd ed., (ed. H. L. BARNES), Chapter 3, pp. 71–136 John Wiley and Sons.

CATHLES L. M. (1983) An analysis of the hydrothermal system responsible for massive sulfide deposition in the Hokuroku Basin of Japan. In *The Kuroko and Related Volcanogenic Massive Sulfide Deposits, Econ. Geol. Monog. 5,* pp. 439–487.

CHRISTIANSEN R. L. (1983) Yellowstone magmatic evolution: Its bearing on understanding large-volume explosive volcanism. In *Explosive Volcanism,* (ed. F. R. BOYD), pp. 84–95 National Academy of Sciences, Washington, D.C.

CHRISTIANSEN R. L. and McKEE E. H. (1978) Late Cenozoic volcanic and tectonic evolution of the Great Basin and Columbia Intermontane regions. *Geol. Soc. Amer. Mem.* **152,** 283–311.

CONDOMINES M., GRONVOLD K., HOOKER P. J., MUEHLENBACHS K., O'NIONS R. K., OSKARSSON N. and OXBURGH E. R. (1983) Helium, oxygen, strontium and neodymium isotopic relationships in Icelandic volcanics. *Earth Planet. Sci. Lett.* **66,** 125–136.

CRISS R. E., GREGORY R. T. and TAYLOR H. P., JR. (1986) Kinetic theory of oxygen isotopic exchange between minerals and water. *Geochim. Cosmochim. Acta.* (submitted).

CRISS R. E. and TAYLOR H. P., JR. (1983) An ^{18}O/^{16}O and D/H study of Tertiary hydrothermal systems in the southern half of the Idaho batholith. *Bull. Geol. Soc. Amer.* **94,** 640–663.

CRISS R. E. and TAYLOR H. P., JR. (1986) Meteoric-hydrothermal systems. In *Reviews in Mineralogy 16,* (eds. J. W. VALLEY, H. P. TAYLOR, JR. and J. R. O'NEIL), Chap. 11, pp. 491–560 Mineralogical Society of America.

DOE B. R., LEEMAN W. P., CHRISTIANSEN R. L. and HEDGE C. E. (1982) Lead and strontium isotopes and related trace elements as genetic tracers in the Upper Cenozoic rhyolite-basalt association of the Yellowstone Plateau volcanic field. *J. Geophys. Res.* **87**, 4785–4806.

FERRY J. M. (1985a) Hydrothermal alteration of Tertiary igneous rocks from the Isle of Skye, northwest Scotland: II. Granites. *Contrib. Mineral. Petrol.* **91**, 283–304.

FERRY J. M. (1985b) Hydrothermal alteration of Tertiary igneous rocks from the Isle of Skye, northwest Scotland I. Gabbros. *Contrib. Mineral. Petrol.* **91**, 264–282.

FORESTER R. W. and TAYLOR H. P., JR. (1976) ^{18}O depleted igneous rocks from the Tertiary complex of the Isle of Mull, Scotland. *Earth Planet. Sci. Lett.* **32**, 11–17.

FORESTER R. W. and TAYLOR H. P., JR. (1977) $^{18}O/^{16}O$, D/H and $^{13}C/^{12}C$ studies of the Tertiary igneous complex of Skye, Scotland. *Amer. J. Sci.* **277**, 136–177.

FORESTER R. W. and TAYLOR H. P., JR. (1980) Oxygen, hydrogen, and carbon isotope studies of the Stony Mountain complex, western San Juan Mountains, Colorado. *Econ. Geol.* **75**, 362–383.

FRIEDMAN I., LIPMAN P., OBRADOVICH J. D., GLEASON J. D. and CHRISTIANSEN R. L. (1974) Meteoric water in magmas. *Science* **184**, 1069–1072.

GARLICK G. D. and EPSTEIN S. (1967) Oxygen isotope ratios in coexisting minerals of regionally metamorphosed rocks. *Geochim. Cosmochim. Acta* **31**, 181–214.

GREGORY R. T. and TAYLOR H. P., JR. (1981) An oxygen isotope profile in a section of Cretaceous oceanic crust. Samail Ophiolite, Oman: evidence for ^{18}O-buffering of the oceans by deep (>5 km) seawater-hydrothermal circulation at mid–ocean ridges. *J. Geophys. Res.* **86**, 2737–2755.

GREGORY R. T. and TAYLOR H. P., JR. (1986) Non-equilibrium, metasomatic $^{18}O/^{16}O$ effects in upper mantle mineral assemblages. *Contrib. Mineral. Petrol.* **93**, 124–135.

GREGORY R. T., TAYLOR H. P., JR. and COLEMAN R. G. (1980) The origin of plagiogranite by partial melting of hydrothermally altered stoped blocks at the roof of a Cretaceous mid–ocean ridge magma chamber, the Samail ophiolite, Oman. *Geol. Soc. Amer. Abstr. with Progr.* **12**, 437.

HATTORI K. and MUEHLENBACHS K. (1982) Oxygen isotope ratios of the Icelandic crust. *J. Geophys. Res.* **87**, 6559–6565.

HILDRETH W. (1981) Gradients in silicic magma chambers: implications for lithospheric magmatism. *J. Geophys. Res.* **86**, 10153–10192.

HILDRETH W., CHRISTIANSEN R. L. and O'NEIL J. R. (1984) Catastrophic isotopic modification of rhyolitic magma at times of caldera subsidence, Yellowstone Plateau volcanic field. *J. Geophys. Res.* **89**, 8339–8369.

JACOBSSON S. P. (1972) Chemistry and distribution pattern of recent basaltic rocks in Iceland. *Lithos* **5**, 365–386.

LARSON P. B. and TAYLOR H. P., JR. (1986) $^{18}O/^{16}O$ ratios in ash–flow tuffs and lavas erupted from the central Nevada caldera complex and the central San Juan caldera complex, Colorado. *Contrib. Mineral. Petrol.* **92**, 146–156.

LIPMAN P. W. (1971) Iron-titanium oxide phenocrysts in compositionally zoned ash-flow sheets from southern Nevada. *Jour. Geol.* **79**, 438–456.

LIPMAN P. W. and FRIEDMAN I. (1975) Interaction of meteoric water with magma: An oxygen–isotope study of ash-flow sheets from southern Nevada. *Bull. Geol. Soc. Amer.* **86**, 695–702.

MAGARITZ M. and TAYLOR H. P., JR. (1986) Oxygen 18/ Oxygen 16 and D/H studies of plutonic granitic and metamorphic rocks across the Cordilleran batholiths of southern British Columbia. *J. Geophys. Res.* **91**, No. B2, 2193–2217.

MANNING C. E. and BIRD D. (1986) Hydrothermal clinopyroxenes from the Skaergaard intrusion. *Contrib. Mineral. Petrol.* **92**, 437–447.

MCCULLOCH M. T., GREGORY R. T., WASSERBURG G. J. and TAYLOR H. P., JR. (1981) Sm–Nd, Rb–Sr, and $^{18}O/^{16}O$ isotopic systematics in an oceanic crustal section: Evidence from the Samail ophiolite. *J. Geophys. Res.* **86**, 2721–2735.

MUEHLENBACHS K. (1973) The oxygen isotope geochemistry of siliceous volcanic rocks from Iceland. *Carnegie Inst. Wash. Yearb.* **72**, 593–597.

MUEHLENBACHS K., ANDERSON A. T. and SIGVALDASON G. E. (1974) Low-^{18}O basalts from Iceland. *Geochim. Cosmochim. Acta* **38**, 577–588.

NOBLE D. C. and HEDGE C. E. (1969) $^{87}Sr/^{86}Sr$ variations within individual ash–flow sheets. *U.S. Geol. Survey Prof. Paper 475-B*, B52–B55.

NORTON D. (1984) Theory of hydrothermal systems. *Ann. Rev. Earth Planet. Sci.* **12**, 155–177.

NORTON D. and KNIGHT J. (1977) Transport phenomena in hydrothermal systems: Cooling plutons. *Amer. J. Sci.* **277**, 937–981.

NORTON D. and TAYLOR H. P., JR. (1979) Quantitative simulation of the hydrothermal systems of crystallizing magmas on the basis of transport theory and oxygen isotope data: An analysis of the Skaergaard intrusion. *J. Petrol.* **20**, 421–486.

NORTON D., TAYLOR H. P., JR. and BIRD D. K. (1984) The geometry and high–temperature brittle deformation of the Skaergaard intrusion. *J. Geophys. Res.* **89**, 10178–10192.

O'NIONS R. K. and GRONVOLD K. (1973) Petrogenetic relationship of acid and basic rocks in Iceland: Sr isotopes and rare earth elements in late post-glacial volcanics. *Earth Planet. Sci. Lett.* **19**, 397–409.

RYE R. O., SCHULING R. D., RYE D. M. and JANSEN J. BEN. H. (1976) Carbon, hydrogen, and oxygen isotope studies of the regional metamorphic complex at Naxos, Greece. *Geochim. Cosmochim. Acta* **40**, 1031–1049.

SAEMUNDSSON K. (1974) Evolution of the axial rifting zone in northern Iceland and Tjornes fracture zone. *Bull. Geol. Soc. Amer.* **85**, 495–504.

SHEPPARD S. M. F., NIELSEN R. L. and TAYLOR H. P., JR. (1971) Hydrogen and oxygen isotope ratios in minerals from porphyry copper deposits. *Econ. Geol.* **66**, 515–542.

SHEPPARD S. M. F., BROWN P. E. and CHAMBERS A. D. (1977) The Lilloise intrusion, East Greenland: Hydrogen isotope evidence for the efflux of magmatic water into the contact metamorphic aureole. *Contrib. Mineral. Petrol.* **68**, 129–147.

STAKES D. S., TAYLOR H. P., JR. and FISHER R. L. (1983) Oxygen–isotope and geochemical characterization of hydrothermal alteration in ophiolite complexes and modern oceanic crust. In *Ophiolites and Oceanic Lithosphere*, (eds. I. G. GASS, S. J. LIPPARD and A. W. SHELTON), pp. 199–214 Blackwell Scientific Publications.

STEWART J. H. (1978) Basin–Range structure in western North America. *Geol. Soc. Amer. Mem.* **152**, 131 pp.

STUCKLESS J. S. and O'NEIL J. R. (1973) Petrogenesis of

the Superstition—Superior volcanic area as inferred from strontium and oxygen isotope studies. *Bull. Geol. Soc. Amer.* **84,** 1987–1997.

TAYLOR H. P., JR. (1968) The oxygen isotope geochemistry of igneous rocks. *Contrib. Mineral. Petrol.* **19,** 1–71.

TAYLOR H. P., JR. (1974) The application of oxygen and hydrogen isotope studies to problems of hydrothermal alteration and ore deposition. *Econ. Geol.* **69,** 843–883.

TAYLOR H. P., JR. (1977) Water/rock interactions and the origin of H_2O in granitic batholiths. *J. Geol. Soc. London* **133,** 509–558.

TAYLOR H. P., JR. (1979) Stable isotope studies of spreading centers and their bearing on the origin of granophyres and plagiogranites. *Colloques Internationataux du CNRS no. 272—Associations Mafiques Ultra-Mafiques dans les Orogenes,* pp. 149–165.

TAYLOR H. P., JR. (1980) The effects of assimilation of country rocks by magmas on $^{18}O/^{16}O$ and $^{87}Sr/^{86}Sr$ systematics in igneous rocks. *Earth Planet. Sci. Lett.* **47,** 243–254.

TAYLOR H. P., JR. (1983) Oxygen and hydrogen isotope studies of hydrothermal interactions at submarine and subaerial spreading centers. In *NATO Symposium Volume on Hydrothermal Processes at Seafloor Spreading Centers,* (eds. P. A. RONA, K. BOSTROM, L. LAUBIER, and K. SMITH, JR.), pp. 83–139 Plenum, New York.

TAYLOR H. P., JR. (1986) Igneous rocks. II. Isotope case studies of circum-pacific magmatism. In *Reviews in Mineralogy 16,* (eds. J. W. VALLEY, H. P. TAYLOR, JR. and J. R. O'NEIL), Chap. 9, pp. 273–318 Mineralogical Society of America.

TAYLOR H. P., JR. and EPSTEIN S. (1963) $^{18}O/^{16}O$ ratios in rocks and coexisting minerals of the Skaergaard intrusion. *J. Petrol.* **4,** 51–74.

TAYLOR H. P., JR. and FORESTER R. W. (1971) Low–^{18}O igneous rocks from the intrusive complexes of Skye, Mull, and Ardnamurchan, western Scotland. *J. Petrol.* **12,** 465–497.

TAYLOR H. P., JR. and TURI B. (1976) High–^{18}O igneous rocks from the Tuscan magmatic province, Italy. *Contrib. Mineral. Petrol.* **55,** 33–54.

TAYLOR H. P., JR. and FORESTER R. W. (1979) An oxygen and hydrogen isotope study of the Skaergaard intrusion and its country rocks: a description of a 55-M.Y. old fossil hydrothermal system. *J. Petrol.* **20,** 355–419.

WAGER L. R. and DEER W. A. (1939) Geological investigations in East Greenland, Pt. 3. The petrology of the Skaergaard intrusion, Kangerdlugssuag region. *Medd. om Gronland* **105,** no. 4, 1–352.

WELKE H., MOORBATH S., CUMMING G. L. and SIGURDSSON H. (1968) Lead isotope studies on igneous rocks from Iceland. *Earth Planet. Sci. Lett.* **4,** 221–231.

ZOBACK M. L., ANDERSON R. E. and THOMPSON G. A. (1981) Cenozoic evolution of the state of stress and style of tectonism of the Basin and Range province of the western United States. *Philos. Trans. Roy. Soc. London,* Ser. A, **300,** 407–434.

Part E.
Crustal Felsic Magma Properties and Processes

Magmatic Processes: Physicochemical Principles
© The Geochemical Society, Special Publication No. 1, 1987
Editor B. O. Mysen

The Macusani glasses, SE Peru: evidence of chemical fractionation in peraluminous magmas

Michel Pichavant,[1] Jacinto Valencia Herrera,[2] Suzanne Boulmier,[1] Louis Briqueu,[3]
Jean-Louis Joron,[4] Martine Juteau,[1] Luc Marin,[1] Annie Michard,[1]
Simon M. F. Sheppard,[1] Michel Treuil[4] and Michel Vernet[1]

[1] Centre de Recherches Pétrographiques et Géochimiques, B.P. 20, 54501 Vandoeuvre les Nancy, France
[2] Instituto Peruano de Energia Nuclear, Av. Canada 1470, Lima 13, Peru
[3] Universite des Sciences et Techniques du Languedoc, Laboratoire de Geochimie Isotopique,
34060 Montpellier, France
[4] Groupe des Sciences de la Terre, Laboratoire P. Sue, C.E.N. Saclay, B.P. 12, 91191, Gif/Yvette, France

Abstract—The Pliocene non-hydrated obsidian glasses from Macusani, SE Peru, are associated with felsic peraluminous ash–flow tuffs of crustal origin. Several glasses have been analysed for their major, trace (37 elements) and isotopic compositions ($^{87}Sr/^{86}Sr$, $^{18}O/^{16}O$, D/H). They have a unique composition for obsidian glasses, similar to some rare–element pegmatites. The glasses represent a rare example of the preservation of lithophile–rich peraluminous melts as quenched glasses. The major element composition is characterized by moderately high SiO_2 (~ 72 weight percent), very high Al_2O_3 (~ 16 weight percent), high alkalies and particularly elevated concentrations in F (≥ 1.3 weight percent), B_2O_3 (0.6 weight percent), Li_2O (0.7 weight percent) and P_2O_5 (0.5 weight percent). CaO, FeO_t, MgO and TiO_2 are very low. Trace elements show marked enrichments in Rb, Ta, U, As, Sn, Sb, W, Cs and also marked depletions in Sr, Ba, Pb, Mo, Th, LREE, Eu, Zr, Hf, Y. Two groups of glasses can be defined on the basis of field relations, major/trace elements and Sr and O isotopes. The Rb–Sr isochron age (4.9 Ma, initial $^{87}Sr/^{86}Sr$ about 0.7309, Chilcuno Chico samples) is in agreement with K–Ar and fission-track ages. Oxygen isotopic compositions are tightly grouped between 11.9 and 12.4‰. Hydrogen isotopic compositions (δD between -140 and $-155‰$) and their low H_2O^+ (≤ 0.4 weight percent) suggest degassing. The discovery of inclusions of glasses in the ash–flow tuffs, the nearly identical mineralogy and mineral chemistry between tuffs and glasses and the chemical and isotopic data demonstrate a genetic filiation between these two rocks.

Experimental data in the Qz–Ab–Or system with added F, B and Li suggest that the glasses are residual liquids from the fractional crystallization of the tuffs. Enrichment/depletion patterns between tuffs and glasses are for the most part consistent with crystal fractionation of the observed phenocrysts. However, the behaviour of Al and P is unusual and relates to the specific structure of strongly peraluminous, F–, B–, P–, and H_2O–rich melts. The role of these elements in the formation of complexes in the liquid and in promoting depolymerization is emphasized.

INTRODUCTION

THE PRESENT DEBATE about chemical fractionation in felsic magmas is dominated by examples taken from the peralkaline and metaluminous series (*e.g.,* HILDRETH, 1979; BACON *et al.,* 1981; CRECRAFT *et al.,* 1981; MAHOOD, 1981; MICHAEL, 1983; CHRISTIANSEN *et al.,* 1984) with very few examples from peraluminous series (MITTLEFEHLDT and MILLER, 1983; PRICE, 1983; WHALEN, 1983; MICHAEL, 1984). However, felsic peraluminous rocks, such as the two-mica granites, are known for their markedly fractionated and distinctive chemical signatures, with enrichments in B, F, Li and trace elements Sn, W, U (*e.g.,* PICHAVANT and MANNING, 1984). Because most of these magmas crystallized as plutonic rocks, mechanisms and trends of fractionation patterns in felsic peraluminous magmas are not well defined. This situation could be significantly improved by studying felsic peraluminous volcanic series because they may provide naturally

quenched eruption products (phenocrysts and glasses) that have not suffered postmagmatic alteration. Obsidian glasses are particularly interesting as they furnish relatively direct information regarding the composition of the silicate liquid in the magma chamber.

The Macusani volcanics, from SE Peru, are a rare example of a felsic peraluminous volcanic series. Here, we concentrate on the Macusani glasses (or macusanites). These are obsidian glasses spatially associated with the Pliocene ash–flow tuffs of the Macusani area. Although considered earlier as tektites (*e.g.,* LINCK, 1926 and discussion in BARNES *et al.,* 1970), recent studies have provided evidence of the volcanic origin of these glasses and of a close link with the ash–flow tuffs (BARNES *et al.,* 1970; FRENCH and MEYER, 1970; FRENCH *et al.,* 1978; KONTAK *et al.,* 1984b; NOBLE *et al.,* 1984; VALENCIA HERRERA *et al.,* 1984). The aim of this paper is to discuss the genetic relation between the ash–flow tuffs and the glasses. The following questions

are specifically addressed: 1) are the ash–flow tuffs
and the Macusani glasses comagmatic? 2) by what
magmatic fractionation process are the glasses derived? 3) what are the implications for chemical
fractionation in silicic magmas? The discussion is
based on a comprehensive set of chemical data obtained on several glass samples, and on new field
and mineralogical data.

THE MACUSANI GLASSES

The existence of the Macusani glasses (or macusanites)
has been known for over half a century (BARNES *et al.*,
1970 and references therein; FRENCH and MEYER, 1970;
FRENCH *et al.*, 1978; KONTAK *et al.*, 1984b; NOBLE *et al.*,
1984; VALENCIA HERRERA *et al.*, 1984). Because they were
considered as possible tektites up to the late fifties, they
have been principally discussed in the tektite literature
(see *Geochim. Cosmochim. Acta*, 14, n° 4, 1958, *Special
Issue on Tektites;* TAYLOR and EPSTEIN, 1962).

Macusanite occurrences are restricted to the region near
Macusani, SE Peru. LINCK's (1926) "Paucartambo glass"
probably comes from the Macusani area (see MARTIN and
DE SITTER–KOOMANS, 1955). Geologically, the area be-
longs to the Western Cordillera in the Central Andes and
forms part of the Inner Arc environment (CLARK *et al.*,
1984). This area differs from the Main Arc system because
its magmatism is of dominantly crustal origin (CLARK *et
al.*, 1984; KONTAK *et al.*, 1984a). In particular, the Western
Cordillera experienced Miocene–Pliocene volcanic activity
that produced peraluminous ash–flow tuffs deposited in
three small tectonic basins: Macusani, Crucero and Ananea
(LAUBACHER, 1978; see also BARNES *et al.*, 1970; NOBLE
et al., 1984; VALENCIA HERRERA *et al.*, 1984; PICHAVANT
et al., in prep.). BARNES *et al.* (1970) first suggested a field
relation between the glass pebbles and the tuffs from the
Macusani basin.

There are two main glass deposits, Caluyo Mayo and
Chilcuno Chico, both of which lie at the eastern edge of
the Macusani volcanic field. The first is localized within
glacial moraine deposits (see map in BARNES *et al.*, 1970),
whereas the second is situated in the tuff outcrop area.
These two deposits are about 10 km apart. In both localities, macusanites are found as *pebbles* in stream gravels
without any clear relation with the neighbouring tuffs.
However, we have discovered for the first time several
inclusions of macusanite in ash–flow tuffs from different
localities. One of these inclusions was collected in a boulder
of tuff near Chilcuno Chico (VALENCIA HERRERA, 1982,
unpublished). Several other inclusions (Figure 1) have been
found *in situ* in tuffs from the Chapi area, about 30 km
from Chilcuno Chico, in the northwestern part of the volcanic field (BRIQUEU *et al.*, 1985, unpublished). These inclusions have been thus far discovered only in tuffs that
belong to the upper cooling units. The glasses occupy small
cavities lined with a white, kaolinite–rich layer (Figure 1).

Several petrographic descriptions of the Macusani glasses
have been published (*e.g.*, MARTIN and DE SITTER–KOO-
MANS, 1955; BARNES *et al.*, 1970). The pebbles from the
stream gravels have an ellipsoidal shape and are most frequently rounded. They have a variable but limited size
and samples larger than 10 × 5 cm have not been found.
One of the striking petrographical features of the Macusani
glasses is their etched surface, which led to the confusion
with tektites (see BARNES *et al.*, 1970). The glasses are

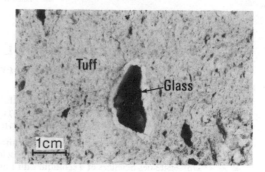

FIG. 1. Glass inclusion (sample CH0) in an ash–flow
tuff, Chapi area. Note the kaolinite–rich white layer lining
the cavity.

either translucent–green, opaque milky–green or opaque
red–brown in colour. In the opaque milky–green varieties,
opacity is due to the presence of abundant fluid inclusions
of strongly irregular shape. The red–brown varieties consist
of a dense arrangement of red spots irregularly dispersed
in a translucent green matrix. These red spots contain numerous minute opaque phases (size less than 1 μm), presumably Fe-oxides. Stratified specimens, with either red
or milky–green bands and streaks in a translucent green
matrix have been encountered. The inclusions found in
the tuffs are similar in shape to the pebbles, although of
generally smaller size ($\leqslant$2 cm). They also display a finely
etched surface and are all of the translucent–green variety.

The Macusani glasses can be defined as non-hydrated
obsidians (ROSS and SMITH, 1955). Although nearly
aphyric, the amount of crystalline phases varies both within
and between samples. All the glass pebbles studied contain
a homogeneous distribution of andalusite needles (up to
3 mm long). The andalusite is euhedral, prismatic, pleo-
chroic (colourless-pink) and optically and chemically is
similar to that present in the ash–flow tuffs (BARNES *et
al.*, 1970; KONTAK *et al.*, 1984b; NOBLE *et al.*, 1984; VAL-
ENCIA HERRERA *et al.*, 1984). Its modal abundance is low
(a few crystals per thin section). Additional phases have
been reported to occur (LINCK, 1926; FRENCH and MEYER,
1970; FRENCH *et al.*, 1978). All are discrete microphenocrysts (a few tens to hundred μm.). The andalusites, the
microphenocrysts and the fluid inclusions define a flow
banding parallel to the stratification of the stratified specimens. Although we have not attempted to determine the
complete mineralogy of the glasses, the presence of virgilite
(FRENCH *et al.*, 1978), quartz, sanidine ($Or_{68-71}Ab_{32-29}$),
plagioclase ($Ab_{83-84}An_{11}Or_{5-6}$), and biotite (Al-, Fe- and
F-rich) has been confirmed (detailed analyses in PICHAV-
ANT *et al.*, in preparation). Sillimanite, ilmenite, zircon
and monazite are mainly found as inclusions in the andalusites. The mineralogy of the glasses is nearly identical
to that of the tuffs (KONTAK *et al.*, 1984b; NOBLE *et al.*,
1984; VALENCIA HERRERA *et al.*, 1984). In particular, there
is a close similarity between the compositions of sanidines,
biotites and andalusites in the glasses and in the tuffs. The
plagioclase composition in the glasses is identical to the
composition of the latest plagioclase phenocrysts in the
tuffs. The glass inclusions in the tuffs are generally poorer
in microphenocrysts than the pebbles; no andalusite has
been found in our glass inclusions.

A consistent age of 4.3 Ma has been obtained for the
Macusani glasses, both by K–Ar on the Caluyo Mayo

samples (BARNES et al., 1970, recalculated to new constants), and by the fission–track method (FLEISCHER and PRICE, 1964). Therefore, the Macusani glasses mark one of the younger volcanic events in the area because K–Ar dates of the tuffs (BARNES et al., 1970; KONTAK et al., 1984a; NOBLE et al., 1984) indicate continuous (or semi-continuous) volcanic activity from about 10 Ma to 4 Ma. The young ages found for the glasses agree with the discovery of macusanite inclusions in tuffs from the upper units.

The most comprehensive analytical work has been carried out on two pebbles from Chilcuno Chico, JV1 of the translucent–green variety and JV2 of the opaque milky-green variety. CC2 is an additional sample (red variety) from the same deposit. JV3 is the glass inclusion in the tuff collected in the same district. Samples from the Caluyo Mayo deposit are CM1 (translucent–green), CM2 (red) and VB1 and VB2 (both of the translucent–green varieties, kindly provided by V. E. Barnes). CH0 and CH1 are two of the glass inclusions found in the Chapi area.

ANALYTICAL TECHNIQUES

Electron microprobe analyses were performed with a Camebax electron microprobe (University of Nancy I). Analytical conditions are: acceleration voltage 15 KV, sample current 6–8 nA, counting time on peak 6 sec., beam defocussed to 12–15 μm., silicate crystals as standards, and ZAF correction procedures. Because of the problem of loss of alkalies, Na_2O and K_2O were corrected slightly upwards. The correction factors were determined by analysing synthetic glasses of comparable compositions and volatile contents (PICHAVANT, 1984). Fluorine was analysed separately with longer counting time (10 sec. on peak) and higher sample current (40 nA) than for the major elements. To avoid complex ZAF corrections procedures for light elements, the raw counts for a given sample were averaged (usually from between 5 and 10 spots), then plotted against a calibration curve constructed from two standards (glass JV1 and an F-apatite) and the concentrations thus determined.

Several wet-chemical techniques, used routinely at the Centre de Recherches Pétrographiques et Géochimiques (CRPG), Nancy (CRPG Annual Report, 1984–1985), were employed for major elements, for Li (analytical error 1%), B (5%), F (2%) and for a number of trace elements, including Be (2%), Sn (3%), Zn (8%), Rb (3%), Cs (3%), Mo (25%), W (5%), As (4%), Cl (6%), and S (20%).

The instrumental neutron activation analyses were carried out at the Groupe des Sciences de la Terre, Laboratoire P. Sue, Saclay for Sc, Rb, Sb, Cs, Hf, Ta, W, Th, U, La (CHAYLA et al., 1973; JORON, 1974, accuracy 5%). V, Cr, Co, Ni, Cu, Ba (microwave plasma emission spectroscopy, GOVINDARAJU et al., 1976, accuracy 10%), Y, U, Th and the REE (ICP, GOVINDARAJU and MEVELLE, unpublished, 1983–1986, accuracy 10%) were analysed at CRPG. Ga was analysed by arc-emission spectroscopy at the Ecole Nationale de Géologie de Nancy (ENSG) (ROUILLIER, unpublished, accuracy 30%). Zn, Zr, Nb, Sn, W, Pb, were analysed by XRF at the University of Lyon (GERMANIQUE, unpublished, accuracy 5%).

Rb and Sr concentrations and Sr isotopic compositions were determined at CRPG (samples JV1, JV2, VB2) and at the University of Monpellier (samples CH0 and CH1). Details about techniques and accuracies can be found in ALIBERT et al. (1983), JUTEAU et al. (1984), and BRIQUEU and LANCELOT (1977).

Oxygen and hydrogen isotopic compositions were also determined at CRPG. Oxygen was extracted using the BrF_5 method with the conversion of the liberated oxygen to CO_2 gas. The data are reported as $\delta^{18}O$ values relative to SMOW. Reproductibility is better than 0.2‰. NBS-28 gives 9.60‰. The H_2O^+ content of the sample is derived from the manometrically measured yield of H_2 released after the removal of adsorbed water (e.g., SHEPPARD and HAR-RIS, 1985). For the hydrogen isotopic compositions, reproducibility is better than 2‰.

RESULTS

Major elements

The new whole–rock analyses for two glass pebbles (Table 1) are more complete and also more accurate than previous analyses (BARNES et al., 1970; ELLIOTT and MOSS, 1965; see also MARTIN and DE SITTER–KOOMANS, 1955 and LINCK, 1926). The older analyses, however, are relatively similar to our results. The two new whole–rock analyses are identical within analytical error. The electron microprobe analysis of JV1 (Table 2) is also in good agreement with the whole–rock data. All of the analysed glass pebbles, from either Chilcuno Chico or Caluyo Mayo yielded essentially identical electron microprobe results within error, except for the opaque red–brown varieties (Table 2). In the latter glasses, the translucent–green zones are compositionally similar to the other pebbles (Table 2, columns 2 and 4). However, the red spots show an important, though variable increase of FeO_t relative to the other oxides (Table 2, column 5), interpreted to result from variable contamination of the glass because of the heterogeneous distribution of the opaque phases. Apart from these red patches, all of the glass pebbles define a chemically homogeneous population, both at the macroscopic and microscopic scale.

The composition of the analysed glass inclusions (Table 2) contrasts with that of the pebbles. JV3 is only slightly different (lower SiO_2, FeO_t, higher Al_2O_3, Na_2O, F, Table 2, column 6), but the compositions of the two glass inclusions from Chapi (Table 2, column 7 and 8) are markedly distinct from all the other analysed glasses (including pebbles and JV3), principally by their higher (K_2O + Na_2O), lower molar $Al_2O_3/(CaO + Na_2O + K_2O)$, or A/CNK, and their variable (and generally higher) F contents. In addition, there are marked differences between the composition of CH0 and CH1 (reversed Na_2O/K_2O ratio, different F contents).

Taken as a whole, the major element composition of the Macusani glasses are characterized by moderate to low SiO_2, high to very high Al_2O_3, high alkalies, F, Li, B, P, and low FeO_t, MgO, TiO_2, CaO,

Table 1. Whole–rock analyses (weight percent)

	JV1[1] pebble	JV2[1] pebble	3[2] pebble	4[3] pebble	MH3[4] tuff
SiO_2	72.26	72.32	72.8	71.6	73.43
Al_2O_3	15.83	15.85	16.3	16.7	14.35
Fe_2O_3	0.04	0.05	0.29	—	—
FeO	0.57	0.56	0.30	0.6[5]	1.13[5]
MnO	0.06	0.06	0.04	0.05	0.04
MgO	0.02	0.02	0.00	tr.	0.20
CaO	0.22	0.21	0.16	0.4	0.50
Na_2O	4.14	4.13	4.1	4.7	3.37
K_2O	3.66	3.64	3.7	3.6	4.54
TiO_2	0.04	0.04	0.02	0.04	0.13
P_2O_5	0.53	0.53	0.55	0.4	0.36
B_2O_3[6]	0.62	0.60	—	0.4	0.031
Li_2O[7]	0.74	0.74	0.8	—	0.17
F[8]	1.33	1.30	—	1.4	0.41
CO_2	0.09	0.06	0.10	—	
H_2O^+	⎰ 0.46	⎰ 0.40	0.70	0.2	⎱ 1.51
H_2O^-	⎱	⎱	0.00	—	⎰
Total	100.61	100.51	99.06	100.89	100.17
O ≡ F	0.56	0.55	—	0.59	0.17
Total	100.05	99.96	—	100.30	100.00

[1] New analyses; average of two duplicates–Glasses from Chilcuno Chico.
[2] From BARNES *et al.* (1970)–Glass from Caluyo Mayo.
[3] From ELLIOTT and MOSS (1965)—Glass of unspecified provenance.
[4] From PICHAVANT *et al.* (in preparation).
[5] Total Fe as FeO.
[6] For VB1, B_2O_3 = 0.63 weight percent; for VB2, B_2O_3 = 0.62 weight percent.
[7] For VB1, Li_2O = 0.73 weight percent; for VB2, Li_2O = 0.75 weight percent.
[8] For VB1 and VB2, F = 1.30 weight percent.

Table 2. Electron microprobe analyses of the Macusani glasses (weight percent)

	JV1[1] pebble	CC2[2] pebble	CM1[3] pebble	CM2[4] pebble	CM2[5] pebble	JV3[6] inclusion	CH0[7] inclusion	CH1[8] inclusion
SiO_2	72.26(0.51)	72.69(0.25)	72.29(0.46)	72.31(0.20)	71.95	70.54(0.36)	70.38(0.20)	69.69(0.23)
Al_2O_3	15.79(0.22)	16.00(0.16)	15.92(0.05)	15.75(0.05)	15.62	16.31(0.04)	16.26(0.06)	16.30(0.13)
FeO	0.54(0.05)	0.56(0.04)	0.59(0.05)	0.47(0.02)	2.75	0.38(0.03)	0.30(0.05)	0.38(0.06)
MnO	0.03(0.03)	0.03(0.03)	0.08(0.08)	0.04(0.03)	0.12	n.d.	n.d.	0.09(0.05)
MgO	0.02(0.03)	0.01(0.00)	0.02(0.01)	0.01(0.01)	0.03	0.02(0.01)	0.01(0.01)	0.00(0.01)
CaO	0.19(0.03)	0.10(0.10)	0.19(0.03)	0.15(0.01)	0.26	0.17(0.02)	0.14(0.02)	0.14(0.02)
Na_2O	4.29(0.14)	4.30(0.09)	4.32(0.04)	4.27(0.14)	3.93	4.42(0.05)	5.15(0.05)	4.63(0.08)
K_2O	3.83(0.06)	3.77(0.01)	3.74(0.09)	3.88(0.01)	3.56	3.70(0.03)	4.41(0.06)	5.36(0.09)
TiO_2	0.07(0.06)	0.00(0.00)	0.02(0.01)	0.04(0.01)	0.00	0.01(0.02)	0.02(0.02)	0.00(0.01)
F[10]	1.33[9]	1.27(0.03)	1.35(0.02)	1.33(0.02)	1.27(0.03)	1.92(0.04)	1.88(0.04)	1.68(0.06)
Total	98.35	98.73	98.52	98.25	99.49	97.47	98.55	98.27
O ≡ F	0.56	0.53	0.57	0.55	0.53	0.81	0.79	0.71
Total	97.79	98.20	97.95	97.70	98.96	96.66	97.76	97.56

[1] Average of 5 spots. Translucent green glass from Chilcuno Chico.
[2] Average of 2 spots. Opaque red–brown glass from Chilcuno Chico, preserved green zones.
[3] Average of 2 spots. Translucent–green glass from Caluyo Mayo.
[4] Average of 2 spots. Opaque red–brown glass from Caluyo Mayo, preserved green zones.
[5] Red zone. Same glass as[4].
[6] Average of 4 spots. Glass inclusion in ash–flow tuff, Chilcuno Chico area.
[7] Average of 4 spots. Glass inclusion in ash–flow tuff, Chapi area.
[8] Average of 5 spots. Glass inclusion in ash–flow tuff, Chapi area.
[9] From Table 1.
[10] Analyzed separately (see text). Average of between 5 and 10 spots per sample.

Fe₂O₃/FeO. Also note the low FeO_t/MnO ratio and the Na₂O/K₂O > 1 (CH1 being excepted). Such a composition resembles that of a felsic, strongly per-aluminous, volatile–rich granitic magma, similar to some leucogranites (*e.g.*, LE FORT, 1981; PICHAV-ANT and MANNING, 1984; BENARD *et al.*, 1985) and granitic pegmatites (STEWART, 1978; CERNY, 1982). The felsic character of the macusanites is best evidenced by their CIPW composition (more than 90 weight percent normative Qz + Ab + Or). Normative corundum (Co) reaches 5 weight percent with only a very small part accounted for by the presence of andalusite phenocrysts.

A unique feature is the concentration of F, B, Li and P. The glass pebbles, from either Chilcuno Chico or Caluyo Mayo, are homogeneous in terms of F, B and Li. The F contents reach 1.3 weight percent in the pebbles (Tables 1 and 2) and increases up to 1.9 weight percent (Table 2) in the glass in-clusions, a range similar to those found for some ongonites (KOVALENKO and KOVALENKO, 1976), topaz rhyolites (CHRISTIANSEN *et al.*, 1983; 1984), topaz granites (PICHAVANT and MANNING, 1984) and rare–element pegmatites (*e.g.*, CERNY, 1982). The Li₂O contents (0.73–0.75 weight percent, Table 1) are only available for 5 glass pebbles; they are in a range comparable to the rare–element pegmatites (STEWART, 1978; CERNY, 1982; CERNY *et al.*, 1985). The Macusani glasses (data for 5 pebbles) have the highest B₂O₃ content ever reported for a natural glass (0.60–0.63 weight percent, Table 1). Elevated, though lower, B₂O₃ concentrations are found in tourmaline granites, aplites and pegmatites (*e.g.*, NEMEC, 1978; PICHAVANT and MANNING, 1984; CERNY, 1982). The very high P₂O₅ (0.53 weight percent, Table 1) is comparable to that found in topaz granites (PICHAVANT and MANNING, 1984), and rare–element pegmatites (CERNY, 1982; LONDON, 1987) and exceeds the amount of CaO (about 0.2 weight percent). On the other hand, the H₂O content (H₂O⁺: 0.23 weight percent, FRIED-MAN, 1958; 0.16 to 0.40 weight percent, Table 5; H₂O_t: 0.40–0.46 weight percent, Table 1) is low, in agreement with that expected for a non–hydrated obsidian (ROSS and SMITH, 1955). LONDON (1987 and written communication) reports higher H₂O contents (1.1 weight percent) for a translucent–green glass analysed by ion microprobe.

As pointed out by previous workers (BARNES *et al.*, 1970; NOBLE *et al.*, 1984; VALENCIA HERRERA *et al.*, 1984), the major element composition of the adjacent ash–flow tuffs (Table 1) compares in many aspects with that of the macusanites: moderate SiO₂, very high Al₂O₃, high alkali, Li, B, F and P, low FeO_t, MgO, TiO₂, and CaO. However, significant

differences appear (inversion of the Na₂O/K₂O ratio from the tuffs to the glasses). The ash–flow tuffs (data from PICHAVANT *et al.*, in preparation) and the glasses (Tables 1 and 2) plot in separate fields on a Qz–Ab–Or diagram (Figure 2).

Trace elements

The new analyses (Tables 3, 4 and 5) present a nearly complete trace element data base. Where available, earlier data (BARNES *et al.*, 1970) are also reported for comparison. Except for a few elements, the agreement between the two sets of data is rather poor. In the present study, several elements (Zn, Rb, Sn, Cs, W, Th, U) have been duplicated or triplicated with good agreement among the various techniques.

The analysed glass pebbles yield strikingly similar values (Tables 3, 4 and 5). A further check of the homogeneity of these glass pebbles is provided by the Rb data; the high values all lie within a narrow bracket (Table 5). In contrast, Rb analyses of two glass inclusions from Chapi yielded distinctly higher values (Table 5).

Compared to normal silicic rocks (*e.g.*, low–Ca granites, TUREKIAN and WEDEPOHL, 1961), the macusanites are enriched in As, Rb, Sn, Sb, Cs, Ta, W, Zn, Nb, U, Be, Ga, and are in the same range

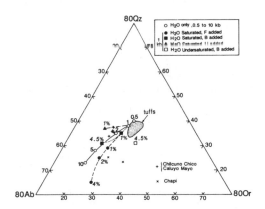

FIG. 2. CIPW normative composition of the Macusani glasses (+ : samples from Chilcuno Chico and Caluyo Mayo; × : samples from Chapi). The CIPW normative compositions of the tuffs (stippled field) are shown for comparison (data from PICHAVANT *et al.*, in preparation). For the glasses, both whole–rock (Table 1) and electron microprobe data (Table 2) have been plotted. Minima and eutectic compositions in the Qz–Ab–Or system taken from LUTH (1976), MANNING (1981), MARTIN (1983), PI-CHAVANT (1984), PICHAVANT and RAMBOZ (1985a,b). The experimental points for H₂O only (LUTH, 1976) are labelled with pressure from 0.5 to 10 kbar. All the other experimental points are labelled with concentrations (weight percent) of either F, Li₂O, or B₂O₃ in the melt.

Table 3. Trace element composition (ppm)

	JV1 pebble	JV2 pebble	3[7] pebble	MH3[8] tuff	1[9] tuff
Be	41.1[1]	41.3[1]	—	—	37
S	60[1]	70[1]	—	80	—
Cl	427[1]	446[1]	—	49	—
Sc	2.2[2]	2.4[2]	<3	3.0	2.76
V	<10[3]	14[3]	—	13	4.5
Cr	<10[3]	<10[3]	10	<10	15–1.7
Co	0.71[2]	0.71[2]	—	1.5	1.13
Ni	<10[3]	<10[3]	—	<10	—
Cu	<10[3]	<10[3]	4	<10	3.7
Zn	92[4]–97[1]	96[4]–95[1]	80	69	111–83
Ga	42.4[5]	43.0[5]	45	—	40
As	314[1]	325[1]	130	—	73
Rb	see Table 5	see Table 5	890	549	488
Sr	see Table 5	see Table 5	—	88	115
Y	5.16[6]	4.89[6]	—	8.89	—
Zr	39[4]	34[4]	15	72	23
Nb	44[4]	47[4]	—	20	—
Mo	0.4[1]	0.4[1]	<5	0.2	2–<1
Ag	—	—	0.17	—	—
Sn	155[4]–194[1]	203[4]–202[1]	—	46	45–60
Sb	3.6[2]	3.5[2]	—	0.66	2.0
Cs	566[1]–516[2]	571[1]–524[2]	400	91.8	118
Ba	<10[3]	<10[3]	<50	246	340–270
Hf	<2[2]	<2[2]	—	2.7	2.99
Ta	26.9[2]	26.7[2]	—	5.1	6.28
W	62[4]—73[1]–59[2]	69[4]–73[1]–60[2]	—	9.3	<4
Tl	—	—	18	—	2.0
Pb	7[4]	10[4]	14	30	52–35
Bi	—	—	1	—	<5
Th	2.3[6]–0.064[2]	1.7[6]–0.11[2]	—	9.1–8.6	14.6–11
U	18.8[6]–23.1[2]	20.3[6]—31.7[2]	—	6–11.3	10.2

[1] Wet chemical techniques.
[2] Instrumental neutron activation.
[3] Microwave plasma emission spectroscopy.
[4] X-ray fluorescence.
[5] Arc emission spectrometry.
[6] Inductively coupled plasma emission spectroscopy.
[7] From BARNES *et al.* (1970).
[8] From PICHAVANT *et al.* (in preparation).
[9] From NOBLE *et al.* (1984).

Table 4. REE composition (ppm)

	JV1 pebble	JV2 pebble	MH3[3] tuff	1[4] tuff
La	1.92[1]–1.3[2]	1.21[1]–1.7[2]	14.45	19.8
Ce[1]	4.52	3.06	29.57	—
Sm[1]	0.86	0.51	2.95	3.47
Eu[1]	0.02	0.03	0.41	0.62
Gd[1]	0.72	0.63	2.35	—
Dy[1]	0.79	0.85	1.66	—
Er[1]	0.35	0.40	0.87	—
Yb[1]	0.39	0.40	0.70	0.81
Lu[1]	0.06	0.06	0.11	—

[1] Inductively coupled plasma emission spectrometry.
[2] Instrumental neutron activation.
[3] From PICHAVANT *et al.* (in preparation).
[4] From NOBLE *et al.* (1984).

or depleted in Cl, Y, V, Cr, Ni, Cu, Mo, S, Sr, Zr, Ba, Hf, Pb, Th. The data of BARNES *et al.* (1970 and Table 3) also suggest enrichments in Ag, Tl and Bi. COHEN (1960) found a concentration of 5 ppm Ge for a Macusani glass which also implies a Ge enrichment compared to normal silicic rocks.

The Rb, Ta (26–27 ppm, same value given by NOBLE *et al.*, 1984), U (18–32 ppm, 18 ppm according to FRIEDMAN, 1958), Ga (42–43 ppm) and Be (41–42 ppm) values (Table 3) are in the range found for some of the most differentiated silicic rocks on Earth such as ongonites (KOVALENKO and KOVALENKO, 1976), topaz rhyolites (CHRISTIANSEN *et al.*, 1983, 1984) and rare-element pegmatites (CERNY *et al.*, 1985). The Nb (44–47 ppm) and Zn (92–97 ppm) both lie in the lower range of values

Table 5. Rb, Sr concentrations and isotopic compositions

	JV1 pebble	JV2 pebble	VB1 pebble	VB2 pebble	CM2 pebble	CHO inclusion	CH1 inclusion	Qtz[1]
Rb (ppm)	1151[2]–1177[3]–1166[4]	1153[2]–1168[3]–1171[4]	1166[4]	1179[2]–1179[4]	—	1424[2]	1377[2]	—
Sr (ppm)	1.62[2]	1.30[2]	—	1.41[2]	—	4.13[2]	3.97[2]	—
$^{87}Sr/^{86}Sr$	0.87356 ± 6	0.91060 ± 9.5	—	088260 ± 6	—	0.76875 ± 5	0.76928 ± 24	—
$\delta^{18}O$ (‰)	12.2 ± 0.2	12.1	12.4 ± 0.1	—	11.9	12.1 ± 0.1	—	12.7 ± 0.1
H_2O^+ (weight percent)	0.22	0.16	0.40	—	—	—	—	—
δD (‰)	−155 ± 6	−143	−140 ± 3	—	—	—	—	—

[1] Quartz separated from the tuff hosting the glass inclusion CHO.
[2] Isotopic dilution.
[3] Instrumental neutron activation.
[4] Wet chemical techniques.

for ongonites or topaz rhyolites but remain lower than in peralkaline rhyolites (e.g., NICHOLLS and CARMICHAEL, 1969; BARBERI et al., 1975; MAHOOD, 1981). The As (310–330 ppm), Sn (150–200 ppm range), Sb (3–4 ppm), Cs (510–570 ppm, 500–540 ppm, NOBLE et al., 1984), and W (60–90 ppm range) all display values well above those found in topaz rhyolites and ongonites; some of these elements are in the range found in rare–element pegmatites (CERNY, 1982; CERNY et al., 1985).

Unlike the above elements, Sr (1–2 ppm for the pebbles, about 4 ppm for the Chapi inclusions), Ba (<10 ppm), Pb (7–10 ppm), Sc (2–3 ppm), Y (4–6 ppm) as well as Co, V, Cr, Ni and Cu (Table 3) are all depleted to levels similar to those found in ongonites, topaz rhyolites and peralkaline rhyolites. The S content (60–70 ppm) is very low, and the Cl content (420–450 ppm) is not as high as expected from the F contents; it is in the range found for ongonites (KOVALENKO and KOVALENKO, 1976) and calc–alkaline rhyolites (HILDRETH, 1979) but remains lower than in topaz rhyolites (CHRISTIANSEN et al., 1984). The Cl is one order of magnitude lower than in peralkaline rhyolites (e.g., CARMICHAEL, 1962). The Mo (0.4 ppm), Th (1–3 ppm), Zr (30–40 ppm) and Hf (<2 ppm) all are depleted in the macusanites compared to the ongonites, topaz rhyolites, calc–alkaline rhyolites (Bishop Tuff) and peralkaline rhyolites. In particular, the behaviour of Mo contrasts with that of W. F/Cl (30–40), Rb/Sr (1000), and W/Mo (150–225) are markedly elevated while Nb/Ta (1.7), Th/U (0.03–0.1) and Zr/Hf (17–20) are notably low.

The chondrite–normalized REE patterns (Table 4) are shown on figure 3 where they are compared with an ash–flow tuff (MH3, Table 4), a topaz rhyolite (CHRISTIANSEN et al., 1984), and an aplite from the Manaslu leucogranite massif (VIDAL et al., 1982). The two analysed glass pebbles have very low total REE (below 10 × chondrites) and their normalized patterns are similar. They are flat (La/Lu = 2–4) and have a marked Eu anomaly (Eu/Eu* = 0.08–0.16). Such a pattern is similar to the Manaslu aplite but contrasts with both the topaz rhyolites (higher total REE, lower Eu/Eu*) and the Macusani tuffs (higher total REE, Eu/Eu* and La/Lu).

Patterns of chemical fractionation between the tuffs and the glasses have been calculated by dividing the concentration in JV1 by the corresponding concentration in MH3 or 1 (Tables 1, 3 and 4). This is a somewhat arbitrary method as the tuffs probably are not exactly representative of their parental magma. Nevertheless, major element ratios calculated from two glasses (JV3/JV1) conform with

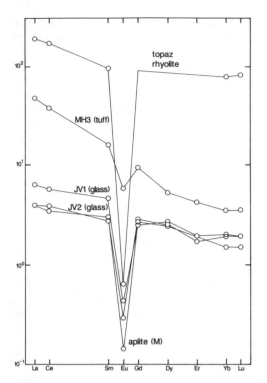

FIG. 3. REE composition of two Macusani glasses (Table 4), compared to a topaz rhyolite (CHRISTIANSEN *et al.*, 1984) and a Manaslu aplite (VIDAL *et al.*, 1982). Concentrations are normalized to 0.83 Leedey chondrite.

those calculated from JV1/MH3. Such patterns (Figure 4) show both marked similarities and differences with those defined for calc–alkaline rhyolites (Bishop Tuff, HILDRETH, 1979) and topaz rhyolites (CHRISTIANSEN *et al.*, 1984). Elements that conform to the variations observed in calc–alkaline and topaz rhyolites include Li, B, F, Cl, Be, Na, Mn, Rb, Nb, Sn, Sb, Cs, Ta, W, Mo, U (enriched), and Mg, K, Ca, Ti, Fe, Co, Sr, Zr, Ba, LREE, Eu and Hf (depleted). On the other hand, Al and P (enriched), Y, Th and the HREE (depleted) are reversed compared to both Bishop Tuff and topaz rhyolites. The Sc and Pb are depleted in Macusani glasses (also Sc in topaz rhyolites) whereas they are enriched in the Bishop Tuff. The Zn appears slightly enriched in the Macusani glasses as opposed to the Bishop Tuff. Thus, fractionation patterns are, for most elements, identical in the peraluminous Macusani volcanics and in the calc–alkaline and topaz rhyolites. However, there is poor agreement (Figure 4) in the relative enrichments (see for example Be, Cl, Nb, Sn, Ta, W, U), whereas there is good agreement in the relative depletions; in the three example considered, Sr, Ba and Eu are the most depleted

trace elements. As noted previously for the Bishop Tuff (HILDRETH, 1979) and for the topaz rhyolites (CHRISTIANSEN *et al.*, 1984), the peraluminous Macusani volcanics show large enrichments or depletions in the trace elements but only very minor changes in the major elements.

Isotopic compositions

Rb–Sr data have been obtained for both Chilcuno Chico and Caluyo Mayo pebbles and for the two Chapi inclusions (Table 5). The data can be divided into two groups. An isochron age calculated from the two pebbles from Chilcuno Chico yields 4.9 Ma (error interval 4.2–6.1 Ma, calculated with an experimental uncertainty of 2% on the $^{87}Rb/^{86}Sr$), in agreement with the 4.3 ± 1.5 Ma K–Ar age (Caluyo Mayo samples, BARNES *et al.*, 1970) and with the 4.3 ± 0.4 Ma fission track age (FLEISCHER and PRICE, 1964). There is a large error on the initial $^{87}Sr/^{86}Sr$ ratio (0.73097) due to the markedly elevated $^{87}Rb/^{86}Sr$ values. The third glass pebble (Caluyo Mayo) lies slightly outside the isochron. By assuming an initial $^{87}Sr/^{86}Sr$ ratio identical to the other pebbles, its calculated age is 4.4 Ma, in excellent agreement with the K–Ar and fission track ages. One possible hypothesis is that the glass pebbles are cogenetic, the Caluyo Mayo samples having a slightly different cooling history than the others. On the other hand, the two Chapi inclusions (nearly identical within error) plot off the isochron for the pebbles, even if account is taken of the experimental errors. Therefore, the Chapi inclusions cannot have both the same age and initial $^{87}Sr/^{86}Sr$ ratio as the pebbles. Given the field relations and the major and trace element data, it is likely that the Chapi inclusions and the Chilcuno Chico/Caluyo Mayo samples have different initial $^{87}Sr/^{86}Sr$ ratios and are not cogenetic. Preliminary Rb–Sr data on the tuffs (KONTAK *et al.*, 1984b; NOBLE *et al.*, 1984; PICHAVANT *et al.*, in preparation) indicate that they have high initial $^{87}Sr/^{86}Sr$ (0.723). Further Sr isotopic data are clearly required, for both tuffs and glasses, to constrain more tightly the genetic relation between tuffs and glasses and between the different glasses.

The oxygen isotopic composition of the glasses is listed on Table 5. The $\delta^{18}O$ values of four pebbles range from 12.4 to 12.1, mean 12.2, and the single glass inclusion at 11.9 is thus slightly less enriched in ^{18}O. These results are in agreement with an earlier analyses of a pebble ($\delta^{18}O = 12.0$, TAYLOR and EPSTEIN, 1962). Because these non-hydrated obsidians have not suffered any postmagmatic alteration, the $\delta^{18}O$ values are likely to represent the oxygen iso-

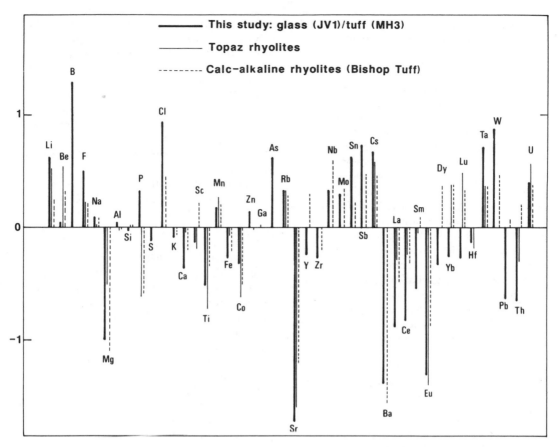

FIG. 4. Enrichment/depletion patterns for the peraluminous Macusani volcanics, calculated as log (concentration in glass JV1/concentration in tuff MH3), (data in Tables 1, 3, 4 and 5) and compared with topaz rhyolites (CHRISTIANSEN et al., 1983, 1984) and calc-alkaline rhyolites (Bishop Tuff, HILDRETH, 1979).

topic composition of the parental magma. Additionally, the fractionation of oxygen isotopes between the glass inclusion and quartz from the host tuff (Table 5) is that expected if the glass parental melt and the quartz were once at equilibrium (TAYLOR, 1968; TAYLOR and TURI, 1976). Oxygen isotope analyses, reported by NOBLE et al. (1984), of minerals from other ash–flow tuffs in the region are similarly high in ^{18}O.

The oxygen isotope data strongly support a genetic filiation between the tuffs and glasses. However, the small difference between $\delta^{18}O$ of the pebbles and inclusion are consistent with the other chemical and isotope data that the pebbles and inclusions are not cogenetic.

The Macusani glasses belong to TAYLOR's (1968) high ^{18}O HH–group of igneous rocks and such values are characteristic of peraluminous magmas (SHEPPARD, 1986). Although there are relatively few published studies on high ^{18}O volcanic rocks, com-

parable or higher oxygen isotopic compositions are known in rhyolites/rhyodacites from the Tuscan volcanic province (TAYLOR and TURI, 1976), in the Banda Arc (MAGARITZ et al., 1978) and in the Neogene volcanics from SE Spain (MUNKSGAARD, 1984).

Hydrogen isotope compositions are listed in Table 5. The δD values are very low and cover a range between -140 and $-155‰$. FRIEDMAN's (1958) value of 0.0127 mole percent D converts to a δD value of about $-175‰$ with a large uncertainty. Macusanites have not experienced any hydration by meteoric water as in perlites. Consequently, the δD numbers apply to magmatic water held in solution within the glass. Similarly low δD values in obsidians have been reported by TAYLOR et al. (1983) and were attributed to the degassing of the magma prior to and during the eruption. This interpretation is consistent with the very low H_2O^+ contents found for the Macusani glasses and mineralogical evidence that the H_2O^+ content of the parental magmas were higher (see below).

DISCUSSION

The discovery of inclusions of macusanites in ash–flow tuffs and the nearly identical mineralogy and mineral chemistry between coexisting tuffs and glasses, based on the comprehensive set of chemical and isotopic data presented here, lead us to conclude that 1) the tuffs and their glass inclusion are cogenetic, and 2) all the glass pebbles and inclusions, although not cogenetic, have shared a common magmatic evolution with the tuffs. Similar but less rigorously constrained conclusions were reached by BARNES *et al.* (1970) who suggested that the Macusani glass pebbles and ash–flow tuffs originated during the same volcanic episode, on the basis of similarities in age, chemical composition and mineralogy (*e.g.*, presence of andalusite in both rocks) and by NOBLE *et al.* (1984) who also related the glass pebbles to the tuffs that they studied. We now concentrate on fractionation processes that relate the glasses and the tuffs.

Application of experimental data in the Qz–Ab–Or system

Phase relations in the haplogranite system Qz–Ab–Or strongly depend on bulk chemical composition. The Macusani glasses contain elevated concentrations of components such as F, Li, B, Al, P that may change the phase relations in the Qz–Ab–Or system. The presence of F, Li, B can be taken into account as the individual effect of these components are known experimentally (MANNING, 1981; MARTIN, 1983; PICHAVANT, 1984; PICHAVANT and RAMBOZ, 1985a,b). Concentrations of Al_2O_3 in large excess of that required to form feldspars shift the minimum liquidus compositions towards the quartz apex (BURNHAM and NEKVASIL, 1986). In addition, rocks with high normative corundum can give misleading plots in the Qz–Ab–Or system (too low in normative quartz). On the other hand, the role of P can not be evaluated due to the lack of experimental data. Nevertheless, the low FeO_t, MgO, TiO_2 and CaO contents of the macusanites make the use of the Qz–Ab–Or system particularly appropriate.

An unknown when trying to apply the experimental data is the H_2O content of the macusanite parental liquid. The hydrogen isotope data imply that degassing occurred. This is consistent with the presence of fluid inclusions which indicate that vesiculation of the liquid took place. The red spots and streaks (oriented parallel to the planes of fluid inclusions) are zones of localized oxidation that may

possibly be related to the exsolution of vapour phase (CANDELA, 1986). The rim of whitish material around the glass inclusions (Figure 1) suggests that degassing continued after the deposition. Therefore, the low H_2O contents (Tables 1 and 5) are considered to be unrepresentative of the H_2O contents during magmatic evolution. The H_2O contents were quite high as indicated by the presence of muscovite phenocrysts in the ash–flow tuffs (KONTAK *et al.*, 1984b; NOBLE *et al.*, 1984; VALENCIA HERRERA *et al.*, 1984).

Another problem is that the individual effect of H_2O on phase relations in the Qz–Ab–Or system is not well known. Phase relations in this system have been determined for various pressures mostly under H_2O–saturated conditions (TUTTLE and BOWEN, 1958; LUTH *et al.*, 1964; LUTH, 1976). Thus, the effect of H_2O on phase relations cannot be separated from that of pressure. There is only a limited number of experiments carried out in that system under H_2O–undersaturated conditions (STEINER *et al.*, 1975; PICHAVANT and RAMBOZ, 1985a,b); they consistently show that an isobaric reduction of the melt H_2O content shifts the minimum liquidus compositions at approximately constant Qz contents towards the Qz–Or side of the diagram. In contrast, calculations indicate a displacement of these minima towards the feldspar join (BURNHAM and NEKVASIL, 1986; NEKVASIL and BURNHAM, 1987).

In spite of these restrictions, several conclusions arise from the Qz–Ab–Or diagram (Figure 2). The tuffs cluster near and below the low–pressure H_2O–saturated minima. In contrast, the Chilcuno Chico and Caluyo Mayo glasses are shifted towards the Ab corner, the more marked shift being for the more F–rich glass (JV3, Table 2, Figure 2). This displacement reflects the progressive increase of F, Li and B in the liquid, and the combined effects of these components on phase relations in the Qz–Ab–Or system. Starting from a magma compositionally similar to the tuffs, fractional crystallization of quartz and feldspar phases is able to produce liquid compositions similar to the macusanites in Figure 2. This fractionation is associated with a reduction of the liquidus temperatures, in accord with the nearly aphyric character of the glasses and in contrast with the crystal-rich nature of the tuffs (40–50 volume percent, PICHAVANT *et al.*, in preparation). Mineralogical evidence (PICHAVANT *et al.*, in preparation) also shows that the glasses quenched a lower-temperature phenocryst assemblage than in the tuffs. The Chapi inclusions (Figure 2) are different. They are significantly less peraluminous (3 weight percent normative corundum) than the other glasses. They have variable F contents and their Li and B concentrations are not known. CH0 plots close to the 2 weight percent added F minimum point; its position is compatible with fractional crystallization from a magma with the composition of the tuffs, as suggested previously. However, CH1 plots distinctly away from CH0 toward the Qz–Or side. One plausible explanation to this peculiar composition is that the H_2O contents of CH0 and CH1, like the F contents, were significantly different during the magmatic evolution. An isobaric reduction of the melt H_2O content between CH0 and CH1 would account for their relative position in Figure 2.

Origin of the Macusani glasses

The details of the mechanisms that led to the segregation and collection of the residual melts are not known, mainly because of limited field information on the relations between the tuffs and the glasses. The field relations, major elements, trace elements, Sr and O isotopes show that the investigated macusanites can be divided into two groups (the Chilcuno Chico and Caluyo Mayo samples on one hand and the Chapi inclusions on the other hand). It is likely that each group was derived from a different magma (the ash–flow tuffs are not cogenetic, even if generated in a closely similar way, see NOBLE *et al.*, 1984). Within each group, there is some indication of chemical heterogeneity [compare the F content of JV3 with the other glass pebbles (Tables 1, 2); also CH0 and CH1]. However, given the limited field information, it is not possible to relate these chemical variations with the relative position in the magma chamber. During crystallization in the magma chamber, low viscosities, as promoted by the presence of F in the residual liquids (DINGWELL *et al.*, 1985; DINGWELL, 1987) would greatly enhance their segregation and collection. These residual liquids were not simply quenched at the top of the chamber but were complexly degassed, prior to and during eruption. Pieces of the residual liquid were erupted as clasts and were air-quenched, as indicated by their shape and etched surface. The quenched glasses were included by compaction within the air-fall tephra. Later, essentially mechanical (alluvial and glacial) alteration eventually separated the inclusions from their host tuff, yielding the pebbles.

Implications for chemical fractionation in volatile–rich peraluminous magmas

We have not attempted to model quantitatively the chemical fractionation observed between the tuffs and the glasses. This is, firstly, because of un-

certainties regarding the composition of the parental magma of the tuffs. Secondly, the partition coefficients are lacking for phases such as the aluminium silicates, muscovite, cordierite, and tourmaline that are major mineralogical components in the tuffs. Thirdly, partition coefficients for rhyolites are strongly dependent on the structure and composition of the liquid (MAHOOD and HILDRETH, 1983). Crystal fractionation of quartz and feldspar phases in a liquid becoming progressively enriched in F, Li and B, as proposed above, accounts for the increase in Na_2O, Al_2O_3, Na_2O/K_2O and for the depletion in SiO_2, K_2O and CaO with differentiation. Trace element data are also consistent with fractional crystallization; Sr, Ba, Pb, Eu are depleted, whereas Rb and Cs are enriched. The very low Sr and Ba contents in the glasses show that they cannot be produced directly by partial melting unless the source was nearly devoid of feldspars. Simultaneous fractionation of other major phases such as biotite also accounts for the depletions in Fe, Mg, Ti. Continuous fractionation of zircon agrees with the depletion in Zr and with the decrease of the Zr/Hf ratio. The observed depletions in Th, LREE are attributed to the removal/fractionation of restite monazites (MITTLEFEHLDT and MILLER, 1983; MONTEL, 1986). The change in the REE patterns between the tuffs and the glasses has been modelled by monazite removal/fractionation (MONTEL, 1986). Thus, there is little doubt as to the importance of crystal fractionation in these peraluminous volcanics. CHRISTIANSEN *et al.* (1984) similarly concluded that the chemical variations observed in topaz rhyolites (which have similar fractionation patterns, Figure 4) are for the most part consistent with crystal fractionation.

The behaviour of several components, however, cannot be explained by such a model. Both P and Al are enriched during differentiation (Figure 4) and their behaviour in the Macusani volcanics contrasts with calc–alkaline and topaz rhyolites. The very high P_2O_5 (0.53 weight percent) of the glasses is incompatible with apatite fractionation. Apatite is an abundant phenocryst in the tuffs (KONTAK *et al.*, 1984b; NOBLE *et al.*, 1984; VALENCIA HERRERA *et al.*, 1984). It has not been identified in the glasses. Atomic Ca/P ratios change from 1.5 to 3 in the tuffs (1.67 in apatite) to 0.4 in the glasses. In addition, many peraluminous felsic magmas have P_2O_5 higher than 0.2 weight percent (LE FORT, 1981; BENARD *et al.*, 1985; ADAM and GAGNY, 1986). Fractionation of apatite should buffer the melt P_2O_5 content to the apatite solubility value for the prevailing intensive and compositional variables. Calculation of the melt P_2O_5 concentration from apatite solubility data of HARRISON and

WATSON (1984) yields $P_2O_5 = 30$ ppm for a temperature of 650°C.

In the same way, values of normative corundum in the range found in the Macusani glasses (from 3 to more than 5 weight percent Co) have been attained in experimental melts only at high temperatures (800–850°C), pressures (5 kbar), H_2O contents (from 4 to 10 weight percent) and in the presence of excess sillimanite (CLEMENS and WALL, 1981). Experimental liquids saturated with aluminium silicate phases at lower temperatures and pressures have much lower normative corundum (about 2 weight percent, BURNHAM and NEKVASIL, 1986). The available experimental data therefore indicate that the normative corundum of the melt should increase mainly with temperature and possibly melt H_2O content. Thus, for a melt saturated with aluminium silicates (such as the Macusani tuffs and glasses), the normative corundum is not expected to increase significantly with differentiation (decrease in temperature, increase in melt H_2O content). Nevertheless, we record a clear increase of the normative corundum from the tuffs (average 3 weight percent Co) to the glasses (from 3 to more than 5 weight percent Co). Moreover, the more fractionated glasses (JV3) have the highest normative corundum yet they contain less andalusite. This increase of the peraluminous character of the melt during differentiation contrasts with the "restite model" (WHITE and CHAPPELL, 1977), as proposed for more mafic peraluminous series ("S–type"), which predicts less peraluminous magmas with increase in differentiation.

Another problem is the boron concentration. B_2O_3 contents as high as 0.60 weight percent in the glasses are also incompatible with the presence of tourmaline, a conspicuous early phenocryst in the tuffs (KONTAK *et al.*, 1984b; VALENCIA HERRERA *et al.*, 1984). Continuous fractionation of tourmaline should buffer the melt B_2O_3 content to very low values, as shown experimentally by BENARD *et al.* (1985).

The above difficulties clearly relate to the specific structure of peraluminous, F–, Li–, B–, P– and H_2O–rich melts. One characteristic feature of the Macusani melt is its strongly peraluminous composition. There is increasing evidence that Al behaves like a network–modifying cation in peraluminous melts (MYSEN *et al.*, 1985). H_2O solubility in melts has been found to increase in peraluminous compared to metaluminous compositions (DINGWELL *et al.*, 1984). In contradiction to some previous assumptions (*e.g.*, LUTH, 1976), liquidus temperatures are significantly lowered in peraluminous compared to metaluminous compositions

(BURNHAM and NEKVASIL, 1986). Another characteristic feature is the presence of elevated concentrations of components such as F, B, Li, P and H_2O. The chemical interaction between these components and the aluminosilicate network leads to a major reorganisation of the melt structure. One important structural modification is the formation of interstitial structural units that involve cations (Al and alkalies) previously forming part of the aluminosilicate network. The existence of structural units involving F, B and the Ab–forming components of the melt is indicated by phase equilibrium studies in the H_2O–saturated Qz–Ab–Or system in the presence of either F or B (MANNING, 1981; MANNING *et al.*, 1984; PICHAVANT, 1984; MANNING and PICHAVANT, 1986; BURNHAM and NEKVASIL, 1986). In the F–bearing system, the observed changes in phase relations are interpreted to indicate the formation of interstitial aluminofluoride complexes (MANNING, 1981; MANNING *et al.*, 1984). Spectroscopic studies in the system Na_2O–Al_2O_3–SiO_2–F at 1 atm (MYSEN and VIRGO, 1985) suggest association of F with Na or Al or both. In the same way, B may coordinate with alkalies or Al, or both (MANNING *et al.*, 1984). By analogy with F and B, the observed changes in phase relations with an isobaric increase of the melt H_2O content (STEINER *et al.*, 1975; PICHAVANT and RAMBOZ, 1985a,b) suggest association of H_2O with the Ab–forming components. Complexes of Al with P have been identified spectroscopically in three-dimensional aluminosilicate melts ($AlPO_4$, MYSEN *et al.*, 1981). When complexes with Al are formed, the alkalies previously required for charge-balancing Al in tetrahedral coordination become network–modifiers. When complexes with alkalies are formed, an equivalent fraction of Al can no longer exist in tetrahedral coordination and becomes network–modifier. This results in both cases in the expulsion of Al from tetrahedral coordination and in the increased abundance of network–modifiers that will promote depolymerization of the melt. Depolymerization would be enhanced with the presence of components such as B which increase the H_2O solubility in the melt (PICHAVANT, 1981).

The preceeding discussion emphasizes the role of Al, alkalies, and of F, B, P and H_2O in the formation of complexes and in promoting depolymerization of the melt. This explains the observed increase in the peraluminous character of the melt during fractionation and also the behaviour of phosphorus in strongly peraluminous magmas. Lowering of the liquidus temperatures would allow these magmas to remain crystal-poor, even at low temperatures (600–650°C). The solubility of "refractory" minerals (*e.g.*, aluminium silicates, tourmaline, apatite) in the melt would be enhanced. The existence of Al complexes with P is consistent with the commonly observed crystallization of amblygonite (LiAlPO$_4$ (OH, F)) rather than apatite in some leucogranites and rare–element pegmatites (CUNEY et al., 1986; LONDON, 1987).

Acknowledgements—Our thanks go first to R. F. Martin, who mentioned the existence of the Macusani glass in 1982. D. Dingwell, D. London, J. M. Montel, H. P. Taylor, Jr. and B. Mysen provided constructive reviews and editorial comments. Discussions with G. Arroyo, C. Wayne Burnham, P. Cerny, B. Charoy, E. Christiansen, C. Esteyries, E. Grew, D. Kontak, J. Lancelot, S. Ludington, D. Manning, J. M. Montel, D. Strong have been very useful. G. Carlier and M. Valencia assisted during the field work in Peru. V. E. Barnes provided samples and H. P. Taylor, Jr. drew our attention to his work on tektites. K. Govindaraju and the Service des Analyses (CRPG) provided analytical data. Part of this paper has been presented at the International Conference and Field Study on the Physicochemical Principles of Magmatic Processes held in Hawaii in honor of H. S. Yoder, Jr. We thank B. Mysen and H. S. Yoder, Jr. for the invitation. This work was partly supported by CNRS, ATP "Transfert 1984".

REFERENCES

ADAM D. and GAGNY C. (1986) L'expression minéralogique du phosphore dans les leucogranites. Apport à la métallogénie de l'étain–tungstène. Cas de la mine de Ribeira (Tras os Montes, Portugal). *Bull. Mineral.* **109**, 441–460.

ALIBERT C., MICHARD A. and ALBAREDE F. (1983) The transition from alkali basalts to kimberlites: isotope and trace element evidence from melilitites. *Contrib Mineral. Petrol.* **82**, 176–186.

BACON C. R., MC DONALD R., SMITH R. L. and BAEDECKER, P. A. (1981) Pleistocene high-silica rhyolites of the Coso volcanic field, Inyo Country, California. *J. Geophys. Res.* **86**, 10223–10241.

BARBERI F., FERRARA G., SANTACROCE R., TREUIL M. and VARET J. (1975) A transitional basalt–pantellerite sequence of fractional crystallization, the Boina centre (Afar Rift, Ethiopia). *J. Petrol.* **16**, 22–56.

BARNES V. E., EDWARDS G., MC LAUGHLIN W. A., FRIEDMAN I. and JOENSUU O. (1970) Macusanite occurrence, age, and composition, Macusani, Peru. *Bull. Geol. Soc. Amer.* **81**, 1539–1546.

BENARD F., MOUTOU P. and PICHAVANT M. (1985) Phase relations of tourmaline leucogranites and the significance of tourmaline in silicic magmas. *J. Geol.* **93**, 271–291.

BRIQUEU L. and LANCELOT J. R. (1977) Nouvelles données analytiques et essai d'interprétation des compositions du strontium des laves calco-alcalines plio-quaternaires du Pérou. *Bull. Soc. Geol. Fr.* **19**, 1223–1232.

BURNHAM C. W. and NEKVASIL H. (1986) Equilibrium properties of granite pegmatite magmas. *Amer. Mineral.* **71**, 239–263.

CANDELA P. A. (1986) The evolution of aqueous vapor from silicate melts: effect on oxygen fugacity. *Geochim. Cosmochim. Acta* **50**, 1205–1211.

CARMICHAEL I. S. E. (1962) Pantelleritic liquids and their phenocrysts. *Mineral. Mag.* **33**, 86–113.

CERNY P. (1982) Anatomy and classification of granitic pegmatites. In *Short Course in Granitic Pegmatites in Science and Industry,* (ed. P. CERNY), vol. 8, pp. 1–32 Mineralogical Association of Canada.

CERNY P., MEINTZER R. E. and ANDERSON A. J. (1985) Extreme fractionation in rare–element granitic pegmatites: selected examples of data and mechanisms. *Can. Mineral.* **23**, 382–421.

CHAYLA B., JAFFREZIC H. and JORON J. L. (1973) Analyse par activation dans les neutrons épithermiques. Applications à la détermination d'éléments en trace dans les roches. *Acad. Sci., Paris, C.R.* **277**, 273–275.

CHRISTIANSEN E. H., BURT D. M., SHERIDAN M. F. and WILSON R. T. (1983) The petrogenesis of topaz rhyolites from the Western United States. *Contrib. Mineral. Petrol.* **83**, 16–30.

CHRISTIANSEN E. H., BIKUN J. V., SHERIDAN M. F. and BURT D. M. (1984) Geochemical evolution of topaz rhyolites from the Thomas Range and Spor Mountain, Utah. *Amer. Mineral.* **69**, 223–236.

CLARK A. H., KONTAK D. J. and FARRAR E. (1984) A comparative study of the metallogenetic and geochronological relationships in the northern part of the Central Andean tin belt, SE Peru and NW Bolivia. *Proc. of Quadrennial I.A.G.O.D. Symposium 6th,* 269–279.

CLEMENS J. D. and WALL V. J. (1981) Origin and crystallization of some peraluminous (S–Type) granitic magmas. *Can. Mineral.* **19**, 111–131.

COHEN A. J. (1960) Germanium content of tektites and other natural glasses. Implications concerning the origin of tektites. *Proc. Int. Geol. Cong. 21st,* Part 1, 30–39.

CRECRAFT H. R., NASH W. P. and EVANS S. H. JR. (1981) Late cenozoic volcanism at Twin Peaks, Utah: geology and petrology. *J. Geophys. Res.* **86**, 10303–10320.

CUNEY M., AUTRAN A., BURNOL L., BROUAND M., DUDOIGNON P., FEYBESSE J. L., GAGNY, C., JACQUOT T., KOSAKEVITCH A., MARTIN P., MEUNIER A., MONIER G. and TEGYEY M. (1986) Résultats préliminaires apportés par le sondage GPF sur la coupole de granite albitique à topaze-lépidolite de Beauvoir (Massif Central, France). *Acad. Sci., Paris, C.R.* **303**, 569–574.

DINGWELL, D. B. (1987) Melt viscosities in the system NaAlSi$_3$O$_8$–H$_2$O–F$_2$O$_{-1}$. In *Magmatic Processes: Physicochemical Principles,* (ed. B. O. MYSEN), The Geochemical Society Spec. Publ. 1, pp. 423–432.

DINGWELL D. B., HARRIS D. M. and SCARFE C. M. (1984) The solubility of H$_2$O in melts in the system SiO$_2$–Al$_2$O$_3$–Na$_2$O–K$_2$O at 1 to 2 kbar. *J. Geol.* **92**, 387–395.

DINGWELL D. B., SCARFE C. M. and CRONIN D. J. (1985) The effect of fluorine on viscosities in the system Na$_2$O–Al$_2$O$_3$–SiO$_2$: implications for phonolites, trachytes and rhyolites. *Amer. Mineral.* **70**, 80–87.

ELLIOTT C. J. and MOSS A. A. (1965) Natural glass from Macusani, Peru. *Mineral. Mag.* **35**, 423–424.

FLEISCHER R. L. and PRICE P. B. (1964) Fission track evidence for the simultaneous origin of tektites and other natural glasses. *Geochim. Cosmochim. Acta* **28**, 755–760.

FRENCH B. M. and MEYER H. O. A. (1970) Andalusite and "β quartz ss" in Macusani glass, Peru. *Carnegie Inst. Wash. Yearb.* **68**, 339–342.

FRENCH B. M., JEZEK P. A. and APPLEMAN D. E. (1978) Virgilite: a new lithium aluminium silicate mineral from the Macusani glass, Peru. *Amer. Mineral.* **63**, 461–465.

FRIEDMAN I. (1958) The water, deuterium gas and uranium content of tektites. *Geochim. Cosmochim. Acta* **14**, 316–322.

GOVINDARAJU K., MEVELLE G. and CHOUARD C. (1976) Automated optical emission spectrochemical bulk analysis of silicate rocks with microwave plasma excitation. *Anal. Chem.* **48**, 1325–1331.

HARRISON T. M. and WATSON E. B. (1984) The behaviour of apatite during crustal anatexis: equilibrium and kinetic considerations. *Geochim. Cosmochim. Acta* **48**, 1467–1477.

HILDRETH W. (1979) The Bishop tuff: evidence for the origin of compositional zonation in silicic magma chambers. *Geol. Soc. Am. Spec. Paper 180,* 43–75.

JORON J. L. (1974) Contribution à l'analyse des éléments en traces dans les roches et les minéraux par activation neutronique. Applications à la la caractérisation d'objets archéologiques. Thesis, Univ. of Paris.

JUTEAU M., MICHARD A., ZIMMERMANN J. L. and ALBAREDE F. (1984) Isotopic heterogeneities in the granitic intrusion of Monte Capanne (Elba Island, Italy) and dating concepts. *J. Petrol.* **25**, 532–545.

KONTAK D. J., CLARK A. H. and FARRAR E. (1984a) The magmatic evolution of the Cordillera Oriental of SE Peru: crustal versus mantle components. In *Andean Magmatism: Chemical and Isotopic Constraints,* (eds. R. S. HARMON and B. A. BARREIRO), pp. 203–219. Shiva.

KONTAK D. J., PICHAVANT M. and CLARK A. H. (1984b) Petrology of the Pliocene peraluminous volcanics from Macusani, SE Peru. *EOS* **65**, 299.

KOVALENKO V. I. and KOVALENKO N. I. (1976) *Ongonites (Topaz-Bearing Quartz Keratophyres) Subvolcanic Analogues of Rare-Metal Li-F Granites.* 124 pp. Nauka, Moscow.

LAUBACHER G. (1978) Estudio geologico de la region norte du Lago Titicaca. *Inst. Geologia Miner. Bol.* **5**, 1–120.

LE FORT P. (1981) Manaslu leucogranite: a collision signature of the Himalaya. A model for its genesis and emplacement. *J. Geophys. Res.* **86**, 10545–10568.

LINCK G. (1926) Ein neuer kristallführender Tektit von Paucartambo in Peru. *Chem. Erde* **2**, 157–174.

LONDON D. (1987) Internal evolution of rare-element pegmatites: effects of F, B, Li and P. *Geochim. Cosmochim. Acta,* (In press).

LUTH W. (1976) Granitic rocks. In *The Evolution of the Crystalline Rocks,* (eds. D. K. BAILEY and R. MC DONALD), pp. 335–417. Academic Press.

LUTH W. C., JAHNS R. H. and TUTTLE O. F. (1964). The granite system at pressures of 4 to 10 kilobars. *J. Geophys. Res.* **69**, 759–773.

MAGARITZ M., WHITFORD D. J. and JAMES D. E. (1978) Oxygen isotopes and the origin of high $^{87}Sr/^{86}Sr$ andesites. *Earth Planet. Sci. Lett.* **40**, 220–230.

MAHOOD G. A. (1981) Chemical evolution of a Pleistocene rhyolitic center: Sierra la Primavera, Jalisco, Mexico. *Contrib. Mineral. Petrol.* **77**, 129–149.

MAHOOD G. A. and HILDRETH W. (1983) Large partition coefficients for trace elements in high-silica rhyolites. *Geochim. Cosmochim. Acta* **47**, 11–30.

MANNING D. A. C. (1981) The effect of fluorine on liquidus phase relationships in the system Qz–Ab–Or with excess water at 1 kbar. *Contrib. Mineral. Petrol.* **76**, 206–215.

MANNING D. A. C. and PICHAVANT M. (1986) Volatiles and their bearing on the behaviour of metals in granitic systems. *Can. Inst. Min. Metall., Sp. Vol.,* (In press).

MANNING D. A. C., MARTIN J. S., PICHAVANT M. and HENDERSON C. M. B. (1984) The effect of B and Li on melt structures in the granite system: different mechanisms? In *Progress in Experimental Petrology,* vol. 6, pp. 36–41. The Natural Environment Research Council.

MARTIN J. S. (1983) An experimental study of the effects of lithium on the granite system. *Proc. Ussher Soc.* **5**, 417–420.

MARTIN R. and DE SITTER–KOOMANS C. M. (1955) Pseudotectites from Colombia and Peru. *Leidsche Geol. Meded.* **20**, 151–164.

MICHAEL P. J. (1983) Chemical differentiation of the Bishop tuff and other high-silica magmas through crystallization processes. *Geology* **11**, 31–34.

MICHAEL P. J. (1984) Chemical differentiation of the Cordillera Paine granite (southern Chile) by in situ fractional crystallization. *Contrib. Mineral. Petrol.* **87**, 179–195.

MITTLEFEHLDT D. W. and MILLER C. F. (1983) Geochemistry of the Sweetwater Wash Pluton, California: Implications for "anomalous" trace element behaviour during differentiation of felsic magmas. *Geochim. Cosmochim. Acta* **47**, 109–124.

MONTEL J. M. (1986) Experimental determination of the solubility of Ce-monazite in SiO_2–Al_2O_3–K_2O–Na_2 melts at 800°C 2 kbar under H_2O–saturated conditions. *Geology* **14**, 659–662.

MUNKSGAARD N. C. (1984) High $\delta^{18}O$ and possible pre-eruptional Rb–Sr isochrons in cordierite-bearing neogene volcanics from SE Spain. *Contrib. Mineral. Petrol.* **87**, 351–358.

MYSEN B. O. and VIRGO D. (1985) Structure and properties of fluorine–bearing aluminosilicate melts: the system Na_2O–Al_2O_3–SiO_2–F at 1 atm. *Contrib. Mineral. Petrol.* **91**, 205–220.

MYSEN B. O., RYERSON F. J. and VIRGO D. (1981) The structural role of phosphorus in silicate melts. *Amer. Mineral.* **66**, 106–117.

MYSEN B. O., VIRGO, D. and SEIFERT F. A. (1985) Relationships between properties and structure of aluminosilicate melts. *Amer. Mineral.* **70**, 88–105.

NEKVASIL H. and BURNHAM C. W. (1987) The calculated individual effects of pressure and water content on phase equilibria in the granite system. In *Magmatic Processes: Physicochemical Principles* (ed. B. O. MYSEN), The Geochemical Society Spec. Publ. 1, pp. 433–446.

NEMEC D. (1978) Genesis of aplite in the Ricany massif, Central Bohemia. *Neues Jahrb. Mineral. Abhand.* **132**, 322–339.

NICHOLLS J. and CARMICHAEL I. S. E. (1969) Peralkaline acid liquids: a petrological study. *Contrib. Mineral. Petrol.* **20**, 268–294.

NOBLE D. C., VOGEL T. A., PETERSON P. S., LANDIS G. P., GRANT N. K., JEZEK P. and MC KEE E. H. (1984) Rare–element enriched, S–type ash–flow tuffs containing phenocrysts of muscovite, andalusite and sillimanite, southeastern Peru. *Geology* **12**, 35–39.

PICHAVANT M. (1981) An experimental study of the effect of boron on a water–saturated haplogranite at 1 kbar pressure. Geological applications. *Contrib. Mineral. Petrol.* **76**, 430–439.

PICHAVANT M. (1984) The effect of boron on liquidus phase relationships in the system Qz–Ab–Or–H_2O at 1 kbar. *EOS* **65**, 298.

PICHAVANT M. (1986) The effects of boron and water on liquidus phase relations in the haplogranite system at 1 kbar (In preparation).

PICHAVANT M. and MANNING D. A. C. (1984) Petrogenesis of tourmaline-granites and topaz granites; the contribution of experimental data. *Phys. Earth Planet. Int.* **35**, 31–50.

PICHAVANT M. and RAMBOZ C. (1985a) Liquidus phase relationships in the system Qz–Ab–Or–B_2O_3–H_2O at 1 kbar under H_2O–undersaturated conditions and the effect of H_2O on phase relations in the haplogranite system. *Terra Cognita* **5**, 230.

PICHAVANT M. and RAMBOZ C. (1985b) Première détermination expérimentale des relations de phases dans le système haplogranitique en conditions de sous-saturation en H_2O. *Acad. Sci., Paris, C.R.* **301**, 607–610.

PICHAVANT M., KONTAK D. J., VALENCIA HERRERA J. and CLARK A. H. The Macusani volcanics, SE Peru. I mineralogy and mineral chemistry. II Evidence for the genesis of volatile-and lithophile element–rich magmas (In preparation).

PRICE R. C. (1983) Geochemistry of a peraluminous granitoid suite from north-eastern Victoria, south-eastern Australia. *Geochim. Cosmochim. Acta* **47**, 31–42.

ROSS C. S. and SMITH R. L. (1955) Water and other volatiles in volcanic glasses. *Amer. Mineral.* **40**, 1071–1089.

SHEPPARD S. M. F. and HARRIS C. (1985) Hydrogen and oxygen isotope geochemistry of Ascension Island lavas and granites: variation with crystal fractionation and interaction with sea water. *Contrib. Mineral. Petrol.* **91**, 74–81.

SHEPPARD S. M. F. (1986) Igneous rocks: III. Isotopic case studies of magmatism in Africa, Eurasia and oceanic islands. In *Reviews in Mineralogy 16, Stable Isotopes in High Temperature Geological Processes*, (eds. H. P. TAYLOR, J. R. O'NEIL and J. W. VALLEY), Chap. 10, pp. 319–371 Mineralogical Society of America.

STEINER J. C., JAHNS R. H. and LUTH W. C. (1975) Crystallization of alkali feldspar and quartz in the haplogranite system. $NaAlSi_3O_8$–$KAlSi_3O_8$–SiO_2–H_2O at 4 kb. *Bull. Geol. Soc. Amer.* **86**, 83–98.

STEWART D. B. (1978) Petrogenesis of lithium-rich pegmatites. *Amer. Mineral.* **63**, 970–980.

TAYLOR B. E., EICHELBERGER J. C. and WESTRICH H. R. (1983) Hydrogen isotopic evidence of rhyolitic magma degassing during shallow intrusion and eruption. *Nature* **306**, 541–545.

TAYLOR H. P. JR. (1968) The oxygen isotope geochemistry of igneous rocks. *Contrib. Mineral. Petrol.* **19**, 1–71.

TAYLOR H. P. JR. and EPSTEIN S. (1962) Oxygen isotope studies on the origin of tektites. *J. Geophys. Res.* **67**, 4485–4490.

TAYLOR H. P. JR. and TURI B. (1976) High $\delta^{18}O$ igneous rocks from the Tuscan magmatic province, Italy. *Contrib. Mineral. Petrol.* **55**, 33–54.

TUREKIAN K. K. and WEDEPOHL K. H. (1961) Distribution of the elements in some major units of the Earth's crust. *Bull. Geol. Soc. Amer.* **72**, 175–192.

TUTTLE O. F. and BOWEN N. L. (1958) Origin of granite in the light of experimental studies in the system $NaAlSi_3O_8$–$KAlSi_3O_8$–SiO_2–H_2O. *Geol. Soc. Amer. Mem. 74*, 153 p.

VALENCIA HERRERA J., PICHAVANT M. and ESTEYRIES C. (1984) Le volcanisme ignimbritique peralumineux plio–quaternaire de la région de Macusani, Pérou. *Acad. Sci., Paris, C.R.* **298**, 77–82.

VIDAL P., COCHERIE A. and LE FORT P. (1982). Geochemical investigation of the origin of the Manaslu leucogranite (Himalaya, Nepal). *Geochim. Cosmochim. Acta* **46**, 2279–2292.

WHALEN J. B. (1983) The Ackley City Batholith, southeastern Newfoundland: evidence for crystal versus liquid–state fractionation. *Geochim. Cosmochim. Acta* **47**, 1443–1457.

WHITE A. J. R. and CHAPPELL B. W. (1977) Ultrametamorphism and granitoid genesis. *Tectonophys.* **43**, 7–22.

Magmatic Processes: Physicochemical Principles
© The Geochemical Society, Special Publication No. 1, 1987
Editor B. O. Mysen

Magmatic silicate melts: Relations between bulk composition, structure and properties

BJORN O. MYSEN

Geophysical Laboratory, Carnegie Institution of Washington, Washington, D.C. 20008, U.S.A.

Abstract—Experimental data for silicate melt structure in iron-, phosphorous- and titanium-bearing binary metal oxide-silica and ternary metal oxide-alumina-silica systems have been employed to develop equations that describe the abundance of structural units as a function of bulk composition. These equations are used to describe the structures of natural magmatic liquids.

The structure of liquids for a collection of bulk chemical analyses of up to 2609 samples of Cenozoic volcanic rocks from rockfile RKNFSYS have been calculated. The degree of polymerization (nonbridging oxygens per tetrahedrally-coordinated cations, NBO/T) of natural magma is greater the more felsic the liquid with the NBO/T of rhyolite (from 367 analyses) = 0.031 ± 0.052, dacite (338 analyses) = 0.113 ± 0.040, andesite (2068 analyses) = 0.252 ± 0.123, tholeiite (1010 analyses) = 0.707 ± 0.250, alkali basalt (279 analyses) = 0.681 ± 0.264, basanite (206 analyses) = 0.808 ± 0.267 and nephelinite (116 analyses) = 0.909 ± 0.334. The principal structural unit in these liquids is that of a three-dimensional network (TO_2). Its abundance is positively correlated with decreasing NBO/T of the magma and ranges from an average near 50% for basaltic liquids to more than 95% for rhyolite. In all but rhyolitic and dacitic melts, the nonbridging oxygens are found predominantly in units with an average NBO/T = 2 (TO_3 units). In rhyolite and dacite, nonbridging oxygens occur mainly in structural units with NBO/T = 1 (T_2O_5 units). Units with NBO/T = 4 (TO_4 units) generally are present in all magmatic liquids and constitute between 10 and 20% of the structure in basaltic liquids.

Molar volume of natural magmatic liquids can be calculated from available experimental data. Molar volume of natural magmatic liquids is positively correlated with decreasing NBO/T and linearly correlated with the abundance of TO_2 units in the melts. Similar relationships exist for viscous properties. Both viscosity and activation energy of viscous flow, calculated with a selection of 2609 bulk chemical analyses, decrease with increasing NBO/T of the melt. As for molar volumes, the abundance of TO_2 units governs both the values of viscosity and activation energies of viscous flow. The values of these proportions appear insensitive to the mol fractions of TO_4 and T_2O_5 units in the magmatic liquids.

INTRODUCTION

THE STRUCTURE of magmatic liquids as a function of pressure, temperature and bulk composition provides a basis for characterization and prediction of the physical and chemical properties needed to describe magmatic processes. Largely as a result of extensive spectroscopic (Raman, infrared, Mossbauer, XRDF, EXAFS and XANES and NMR) studies of melt structure in binary, ternary and quarternary systems, the principal structural features likely to be encountered in the bulk compositional range of natural igneous rocks have been established. Direct determination of the structure of melts of natural igneous rock compositions generally has not been attempted (note, however, exceptions by SCARFE, 1977; HOCHELLA and BROWN, 1985). This lack of data results from limitations in resolution of spectra from melts and glasses of complex chemical systems such as natural rock compositions. This conclusion differs from that for simple binary metal oxide-silica systems where it is possible to establish the relative stability and abundance of specific structural units as a function of the type and concentration of network-modifying cations (*e.g.*, BRAWER and WHITE, 1975, 1977; VERWEIJ, 1979a,b; FURUKAWA *et al.*, 1981; MYSEN *et al.*, 1982). With more than one kind of network-modifying cations in a melt, such determinations are not possible because the structural tools are not sensitive to distinction between nonbridging oxygen bonded to specific network-modifying cations.

By addition of alumina to binary metal oxide systems, Al^{3+} generally occurs in tetrahedral coordination. In aluminosilicate melts with nonbridging oxygen Al^{3+} is distributed between the coexisting structural units (MYSEN *et al.*, 1985a; DOMINE and PIRIOU, 1986). In ternary metal oxide-alumina-silica systems, distribution coefficients for Al^{3+} between structural units may be estimated by monitoring frequency shifts of Raman bands resulting from Al^{3+} substitution for Si^{4+} (SEIFERT *et al.*, 1982; MCMILLAN *et al.*, 1982; MYSEN *et al.*, 1981a, 1985a) (Figure 1). The aluminum distribution between the structural units is, however, a function of the type of metal cation (MYSEN *et al.*, 1981a),

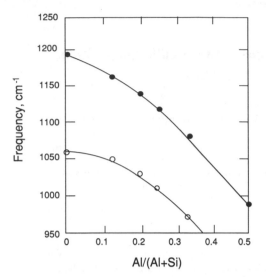

FIG. 1. Frequency shifts of $(Si,Al)-O^0$ stretch bands as a function of bulk melt $Al/(Al + Si)$ for compositions on the join $SiO_2-NaAlO_2$ (data from SEIFERT *et al.*, 1982).

but the structural tools are insufficiently sensitive to distinguish between Al^{3+} distribution governed by, for example, sodium and calcium if these two cations occur together in a melt. Both of these components play a major role in natural magmatic liquids.

Another important variable in igneous rocks is $Fe^{3+}/\Sigma Fe$, which from simple-system calibration is positively correlated with increasing oxygen fugacity, increasing total iron content, increasing $Al/(Al + Si)$ and decreasing degree of polymerization. These relationships have been determined quantitatively in simple system melts by studying each effect separately (*e.g.*, LAUER, 1977; DICKENSON and HESS, 1981, 1986; MYSEN *et al.*, 1980a, 1984, 1985b; see also Figure 2). In natural magmatic systems, these individual effects are less clear (*e.g.*, THORNBER *et al.*, 1980; SACK *et al.*, 1980; KILINC *et al.*, 1983) because bulk compositional effects cannot be separated and experimentally observed variations in $Fe^{3+}/\Sigma Fe$ cannot be uniquely ascribed to a particular change in intensive or extensive variables.

Despite the complications summarized above, one of the goals in studying the structure of silicate melts in compositionally simple systems is to apply these results quantitatively to natural magmatic liquids. In the present report, available data that quantitatively relate bulk composition, temperature and oxygen fugacity to structure and properties of silicate melts in binary, ternary and quarternary systems will be described with the aid of equations

derived from the simple system structure data. These equations will then be used to calculate the structure of magmatic liquids from published major element compositions.

DATA BASE

Bulk chemical analyses

This report serves two purposes. First, a numerical description of the structural data base will be developed. Second, this data base will be used to describe the structure of natural magmatic liquid. This model will then be applied to a selection of rock analyses from a file of 16129 bulk chemical analyses of cenozoic extrusive rocks (RKNFSYS) compiled by CHAYES (1975a,b, 1985). In these calculations the structural features and other properties will be calculated under the assumption that the whole rock analyses, as compiled in RKNFSYS, represent the bulk chemical composition of the igneous rocks in a completely molten state. It is recognized that this assumption does not take into account possible variations in $Fe^{3+}/\Sigma Fe$ during and subsequent to crystallization.

The rockfile RKNFSYS provides the opportunity to extract bulk compositional information based on rock names used in the original sources, on geographic distributions, age or by means of chemical discriminants. In the present report, calculations of structure of magmatic liquids will be done on large groups of analyses of commonly used names of extrusive igneous rocks. No attempt will be made to refine or redefine the names used in the original sources of the RKNFSYS, and the file will be employed with only one provision. It is considered unlikely that an unaltered extrusive igneous rock will contain more than 2 weight percent H_2O, and analyses with more than this amount of water have not been used. The range in concentration of each oxide together with the average value and standard error of these averages are shown in Table 1 together with the number of analyses in each group. This collection of analyses is by no means exhaustive, but it is hoped to be representative.

Structural information

The main features of silicate melt structure may be divided into four categories. These are (1) network-modifying cations (alkali metals, alkaline earths and ferrous iron), (2) aluminum and the relationship between tetrahedrally-coordinated Al^{3+} and cations required for electrical charge-balance of Al^{3+} in tetrahedral coordination, (3) ferric iron as a network-former (tetrahedrally-coordinated

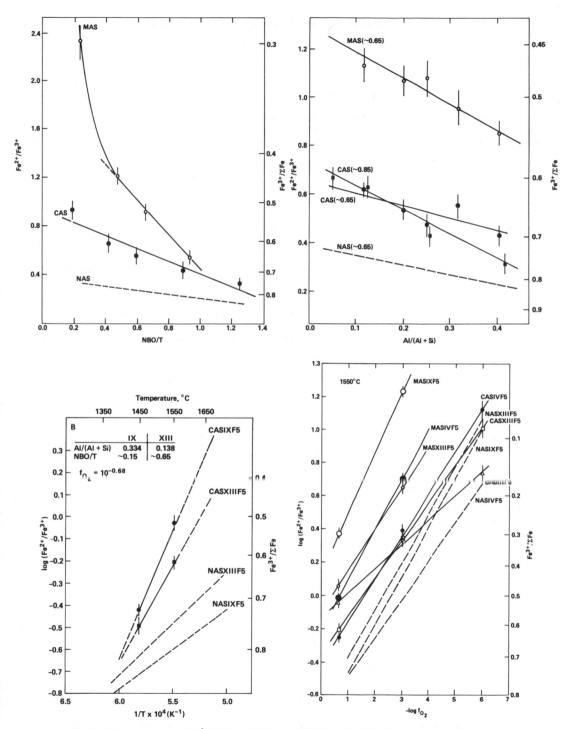

FIG. 2. Relations between $Fe^{3+}/\Sigma Fe$ and bulk melt $Al/(Al + Si)$, NBO/T (A), temperature (B) and oxygen fugacity (C). Abbreviations: NAS; system $Na_2O-Al_2O_3-SiO_2$, CAS; system $CaO-Al_2O_3-SiO_2$, MAS; system $MgO-Al_2O_3-SiO_2$. Roman numerals: IV, $Al/(Al + Si) = 0.334$ and NBO/T ~ 0.65; IX, $Al/(Al + Si) = 0.334$ and NBO/T ~ 0.15; XIII, $Al/(Al + Si) = 0.138$ and NBO/T ~ 0.65. For all compositions, 5 weight percent iron oxide was added as Fe_2O_3 (data from MYSEN et al., 1985b).

B. O. Mysen

Table 1. Bulk chemical information, common extrusive rocks

Name	Rhyolite			Dacite			Andesite		
No. of analyses	367			338			2068		
	Min.	Max.	Average	Min.	Max.	Average	Min.	Max.	Average
SiO_2	58.66	80.93	71.81 ± 3.44	50.79	79.20	65.23 ± 4.39	45.33	73.32	57.71 ± 4.19
Al_2O_3	7.97	19.10	13.65 ± 1.69	11.05	19.79	15.79 ± 1.50	10.95	23.71	17.25 ± 1.56
Fe_2O_3	0.00	7.17	1.89 ± 1.33	0.24	7.49	2.47 ± 1.24	0.00	12.16	3.09 ± 1.35
FeO	0.00	6.82	1.04 ± 1.05	0.00	8.16	2.56 ± 1.49	0.00	12.16	4.38 ± 1.77
MnO	0.00	0.31	0.06 ± 0.07	0.00	1.80	0.11 ± 0.17	0.00	1.91	0.14 ± 0.11
MgO	0.00	2.85	0.41 ± 0.43	0.08	5.97	1.75 ± 1.09	0.00	10.52	3.42 ± 1.40
CaO	0.00	6.31	1.33 ± 1.04	0.70	11.08	4.62 ± 1.80	0.11	12.37	7.09 ± 1.65
Na_2O	0.25	7.90	4.05 ± 1.18	1.83	7.56	3.76 ± 0.79	0.30	7.33	3.36 ± 0.78
K_2O	0.41	8.92	4.41 ± 1.19	0.28	8.06	2.08 ± 0.99	0.11	12.68	1.57 ± 0.82
TiO_2	0.00	1.20	0.30 ± 0.22	0.00	2.82	0.62 ± 0.32	0.00	5.10	0.88 ± 0.51
P_2O_5	0.00	0.94	0.07 ± 0.10	0.00	1.10	0.17 ± 0.12	0.00	2.32	0.23 ± 0.21

Name	Tholeiite & Olivine Tholeiite			Alkali Basalt			Basanite & Basanitoid		
No. of analyses	1010			279			206		
	Min.	Max.	Average	Min.	Max.	Average	Min.	Max.	Average
SiO_2	40.10	63.28	48.62 ± 2.94	35.69	58.20	46.33 ± 3.16	39.25	52.70	44.33 ± 2.55
Al_2O_3	3.62	24.05	15.10 ± 2.30	8.60	22.80	14.90 ± 2.03	8.31	19.53	14.40 ± 2.03
Fe_2O_3	0.04	11.91	3.46 ± 1.70	0.66	16.79	4.69 ± 2.24	0.91	10.50	4.44 ± 1.69
FeO	0.40	13.63	7.77 ± 1.79	0.39	10.94	7.12 ± 2.36	1.46	12.16	7.34 ± 2.13
MnO	0.00	1.23	0.17 ± 0.06	0.00	0.60	0.18 ± 0.07	0.00	1.48	0.17 ± 0.11
MgO	1.12	25.48	7.84 ± 2.79	2.04	20.17	8.09 ± 2.64	2.93	17.53	8.76 ± 2.79
CaO	4.36	15.12	10.17 ± 1.36	3.67	15.63	9.88 ± 1.73	5.89	15.50	10.58 ± 1.56
Na_2O	0.95	5.99	2.76 ± 0.69	1.18	7.73	3.20 ± 0.83	1.02	8.60	3.60 ± 1.03
K_2O	0.01	3.51	0.84 ± 0.55	0.16	3.09	1.28 ± 0.62	0.44	5.83	2.01 ± 1.08
TiO_2	0.00	6.50	2.00 ± 0.96	0.46	6.16	2.58 ± 0.96	0.42	5.86	2.59 ± 1.05
P_2O_5	0.00	1.71	0.35 ± 0.25	0.00	2.39	0.52 ± 0.33	0.00	1.99	0.62 ± 0.31

Name	Nephelinite		
No. of analyses	116		
	Min.	Max.	Average
SiO_2	35.68	44.75	39.95 ± 2.24
Al_2O_3	7.85	22.13	13.50 ± 2.78
Fe_2O_3	0.20	11.65	5.57 ± 2.15
FeO	0.54	11.05	6.76 ± 2.17
MnO	0.00	1.44	0.28 ± 0.22
MgO	1.03	19.45	7.86 ± 4.22
CaO	7.39	19.34	12.94 ± 2.21
Na_2O	1.26	14.24	4.54 ± 1.66
K_2O	0.54	9.00	3.32 ± 1.96
TiO_2	0.19	4.74	2.76 ± 0.75
P_2O_5	0.00	2.82	1.06 ± 0.51

Fe^{3+}) and network-modifier together with charge-balance considerations of tetrahedrally-coordinated Fe^{3+}, and (4) other cations (principally titanium and phosphorous). Experimental data from which all these aspects generally may be evaluated are available from a range of simple systems. As will be shown below, when such data do not exist, inter-polations and extrapolations may be conducted with some confidence.

(1) *Network-modifying cations.* From the frequencies, peak-heights and polarization behavior of appropriate Si-O stretch bands in Raman spectra of alkali metal-silica and alkaline earth-silica systems as a function of metal/silicon and type of metal

cation, BRAWER and WHITE (1975, 1977), VIRGO *et al.,* (1980), FURUKAWA *et al.,* (1981), MYSEN *et al.,* (1980b), MCMILLAN (1984), MATSON *et al.,* (1983) and VERWEIJ (1979a,b) concluded that the melt structures can be described in terms of a relatively small number of structural units. MYSEN *et al.,* (1982) extended the interpretations of these data by converting the Raman intensity data in the systems Na_2O-SiO_2, BaO-SiO_2, CaO-SiO_2 and $Ca_{0.5}Mg_{0.5}O$-SiO_2 to relative abundance of structural units in the melts. Those data are recast (Figure 3) as abundance of individual units as a function of Z/r^2 of the metal cation at fixed values of NBO/T

(NBO/T; nonbridging oxygen per tetrahedrally coordinated cations). The error bars in Figure 3 reflect the fitting errors for the Raman bands employed to calculate the relative abundance of the structural units. By progression of error calculations, the error in the abundance values typically is 10–15%. It is evident from those data that at a specific NBO/T value of the melt (corresponding to a specific metal/silicon), the abundance of TO_2 structural units (TO_2; three-dimensionally interconnected network) is positively correlated with Z/r^2 over the entire compositional range where TO_2 units could be detected. Because the overall polymerization of the

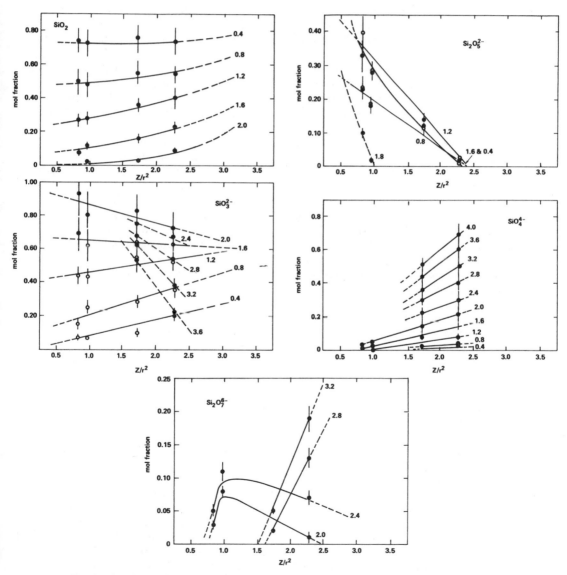

FIG. 3. Abundance of anionic structural units in binary metal oxide-silica systems as a function of Z/r^2 of the metal cation for bulk melt NBO/Si-values as indicated in the figure.

melts is not affected by the type of network-modifying metal cation, the abundance of depolymerized structural units in the melt (such units have NBO/T > 0) also varies with Z/r^2 (Figure 3). In the bulk melt NBO/T-range between 0 and about 2, the $X_{T_2O_5}$ decreases and the X_{TO_3} increases with increasing Z/r^2. For melt less polymerized than that of NBO/T ~ 2, T_2O_5 units generally are not detected, whereas X_{TO_3} and X_{TO_4} increase systematically with increasing Z/r^2 of the network-modifying cation.

By extrapolation of the curves in Figure 3 to Z/r^2 = 0.47 (potassium), 2.71 (ferrous iron) and 3.13 (magnesium), the abundance of anionic units in the systems K_2O-SiO_2, FeO-SiO_2 and MgO-SiO_2 is obtained. From this data base, the X_{TO_2}, $X_{T_2O_5}$, X_{TO_3}, $X_{T_2O_7}$, and X_{TO_4} can be expressed numerically (Table 2) as a function of bulk melt NBO/T for each major network-modifying cation in natural magmatic liquids (Figure 4). From these results (Figure 4), it is seen that the T_2O_5 units (NBO/T = 1) display decreasing abundance with increasing Z/r^2 so that for MgO-SiO_2, T_2O_5 units appear unstable and the proportion is quite small in the FeO-SiO_2 system. In those systems, TO_3 (NBO/T = 2) and TO_4 (NBO/T = 4) units are the principal depolymerized structural units. The maxima in T_2O_5 and TO_3 abundances occur near, but not necessarily at, the bulk melt NBO/T (or M/Si) corresponding to that degree of polymerization, a feature also observed by BRAWER and WHITE (1975, 1977) and FURUKAWA et al., (1981).

(2) *Aluminum.* Provided that large electropositive cations such as alkali metals or alkaline earths are available for electrical charge-balance, Al^{3+} is in tetrahedral coordination in silicate melts, at least at 1 bar pressure (*e.g.,* TAYLOR and BROWN, 1979a,b; NAVROTSKY et al., 1982, 1983; MCMILLAN et al., 1982; MYSEN et al., 1980b, 1982; SEIFERT et al., 1982; DOMINE and PIRIOU, 1986). Thus, melts on silica-aluminate joins have a three-dimensional network structure.

Substitution of Al^{3+} for Si^{4+} in tetrahedral coordination in crystalline materials results in a systematic decrease in T-O-T angle (BROWN et al., 1969; GIBBS et al., 1981) thus resulting in a lowering of the density of bonding electrons and, therefore, T-O bond strength. This decrease in bonding electron density is consistent with the 20% decrease in force constants for (Si,Al)-O stretching in the Al/(Al + Si)-range between 0 and 0.5 (SEIFERT et al., 1982; see also Figure 5) for melts on the join $NaAlO_2$-SiO_2.

Quantitative spectral data do not exist for melts on the join $KAlO_2$-SiO_2. Heat of mixing on joins such as $NaAlSi_2O_6$-$KAlSi_2O_6$ and $NaAlSi_4O_{10}$-

$KAlSi_4O_{10}$ (FRASER et al., 1983; FRASER and BOTTINGA, 1985) show only very subtle variations from 0 over most of the compositional range. It is, therefore, assumed that the structures of melt on the join $KAlO_2$-SiO_2 resemble those of melts on the join $NaAlO_2$-SiO_2. As a result, in the present discussion, K^+ and Na^+ as charge-balancing cations for Al^{3+} are treated similarly.

There is, however, a pronounced difference between the structural interpretation of vibrational spectra of melts on the join $NaAlO_2$-SiO_2 and those on the joins $CaAl_2O_4$-SiO_2 and $MgAl_2O_4$-SiO_2, where the spectra indicate significant (Si,Al)-ordering (SEIFERT et al., 1982; MCMILLAN et al., 1982; see also NAVROTSKY et al., 1982; RICHET and BOTTINGA, 1985, 1986; for detailed discussion on relationships between thermodynamic and structural data).

A detailed and quantitative model was suggested by SEIFERT et al., (1982), who from Raman spectroscopic data found that these latter melts could be described in terms of mixing of a small number of three-dimensionally interconnected structural units. Their proportions, but not their Al/(Al + Si), vary as systematic functions of the bulk melt Al/(Al + Si) (Figure 6). There are small, and perhaps insignificant, differences between the structure of melts on the joins $CaAl_2O_4$-SiO_2 and $MgAl_2O_4$-SiO_2. Coefficients from least-squares fitted curves through these data are provided in Table 3.

With compositions of melts on $MAlO_2$-$M''Al_2O_4$-SiO_2 joins (*M;* monovalent cation, *M''*; divalent cation), it is considered, therefore, that the melts can be described in terms of $(Al,Si)_3O_8^-$, $Al_2Si_2O_8^{2-}$ and AlO_2^- units. For a specific Al/(Al + Si), the abundance of these three-dimensionally-interconnected units does vary, therefore, with *M/M''*. An example is shown in Figure 7 for melts on the join $NaAlSi_3O_8$-$CaAl_2Si_2O_8$. For all but the melts near the anorthite composition, these can be described in terms of two three-dimensional network structural units. One is of the type $Al_2Si_2O_8^{2-}$ with, therefore, constant Al/(Al + Si) across the entire composition join. The other unit, $(Al,Si)_3O_8^-$, exhibits a slowly decreasing Al/(Al + Si) as the proportion of anorthite component in the melt increases.

It has been observed (*e.g.,* BROWN et al., 1969) that in structures of crystalline aluminosilicates with a range of T-O-T angles, Al^{3+} exhibits a preference for the sites associated with the smallest T-O-T angle. In silicate melts with bulk melt NBO/T > 0, structural units with different T-O-T angles coexist (FURUKAWA et al., 1981; VIRGO et al., 1980). In general, the greater the NBO/T of an individual unit in a melt, the larger the T-O-T (FURUKAWA

Table 2. Regression coefficients,* mol fraction of structural units

			TO_2		
Network-modifier	a	b	c	d	e
K	1.01 ± 0.03	-0.83 ± 0.06	0.16 ± 0.03		
Na	1.01 ± 0.02	-0.76 ± 0.05	0.12 ± 0.03		
Ca	1.00 ± 0.02	-0.76 ± 0.02	0.13 ± 0.01		
Fe^{2+}	0.99 ± 0.02	-0.60 ± 0.05	0.05 ± 0.02		
Mg	0.98 ± 0.03	-0.47 ± 0.07	0.05 ± 0.02		

			T_2O_5		
Network-modifier	a	b	c	d	e
K	-0.07 ± 0.03	1.18 ± 0.09	-0.60 ± 0.05		
Na	-0.04 ± 0.01	0.91 ± 0.04	-0.39 ± 0.06	-0.17 ± 0.03	0.08 ± 0.01
Ca	0.06 ± 0.02	0.17 ± 0.14	-0.09 ± 0.02		
Fe^{2+}		Not stable			
Mg			Not stable		

			TO_3		
Network-modifier	a	b	c	d	e
K	-0.26 ± 0.06	0.58 ± 0.04			
Na	-0.22 ± 0.07	0.57 ± 0.05			
Ca	-0.03 ± 0.01	0.17 ± 0.03	0.32 ± 0.04	-0.10 ± 0.02	-0.001 ± 0.001
Fe^{2+}	-0.08 ± 0.03	0.87 ± 0.09	-0.29 ± 0.06	0.01 ± 0.01	
Mg	-0.05 ± 0.01	0.86 ± 0.06	-0.26 ± 0.02		

			T_2O_7		
Network-modifier	a	b	c	d	e
K			No data available		
Na			No data available		
Ca	-0.120 ± 0.001	0.500 ± 0.001			
Fe^{2+}	-0.55 ± 0.09	0.27 ± 0.04			
Mg	-0.7 ± 0.1	0.35 ± 0.07			

			TO_4		
Network-modifier	a	b	c	d	e
K			No data available		
Na	0.2 ± 0.1	-0.3 ± 0.1	0.09 ± 0.04		
Ca	0.023 ± 0.001	-0.032 ± 0.003	0.049 ± 0.002	-0.001 ± 0.001	
Fe^{2+}	-0.03 ± 0.01	0.09 ± 0.02	0.035 ± 0.004		
Mg	-0.04 ± 0.02	0.11 ± 0.03	0.039 ± 0.006		

* Expression: $X = a + b(NBO/T) + c(NBO/T)^2 + d(NBO/T)^3 + e(NBO/T)^4$.

et al., 1981). This logic leads to the suggestion that aluminum should exhibit a preference for the most polymerized structural unit in the melts, a suggestion that is supported by observation (MYSEN *et al.*, 1981a, 1985a; DOMINE and PIRIOU, 1986). For alu-minosilicate melts with Al_2O_3 concentrations in the range observed in natural magmatic liquids (see Table 1), the ratio $[Al/(Al + Si)]^{TO_2}/[Al/(Al + Si)]^{T_2O_5}$ is near 2 and that of $[Al/(Al + Si)]^{T_2O_5}/[Al/(Al + Si)]^{TO_3}$ near 1.5 (MYSEN *et al.*,

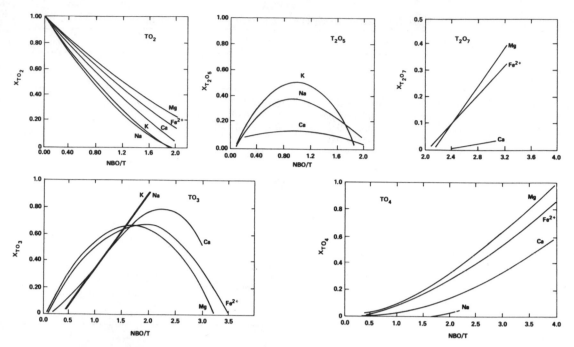

FIG. 4. Calculated abundance curves of anionic structural units in the systems K_2O-SiO_2, Na_2O-SiO_2, CaO-SiO_2, FeO-SiO_2 and MgO-SiO_2.

1981a). These values are not completely independent of bulk composition (changes in charge-balancing cations and degree of melt polymerization) and probably temperature and pressure (MYSEN et al., 1985a). Available experimental data do not, however, permit a detailed assessment of these variables. For the present discussion the values of 2.0 and 1.5 will be used.

In natural magmatic liquids, cations such as K^+, Na^+, Ca^{2+}, Mg^{2+} and Fe^{2+} are major components. All or portions of these may occur as charge-balancing cations for Al^{3+} in tetrahedral coordination. Thermodynamic data from glasses and melts on silica-aluminate joins (NAVROTSKY et al., 1980,

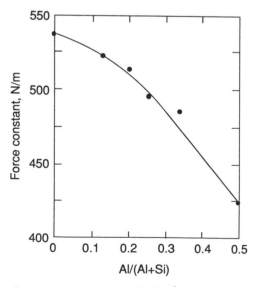

FIG. 5. Force constants for (Si,Al)-O^0 stretch vibrations as a function of bulk melt Al/(Al + Si) along the join SiO_2-$NaAlO_2$ (data from SEIFERT et al., 1982).

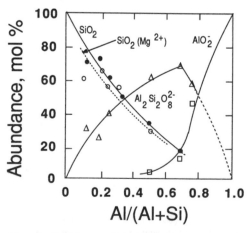

FIG. 6. Abundance of three-dimensionally-interconnected structural units in the systems SiO_2-$CaAl_2O_4$ and SiO_2-$MgAl_2O_4$ as a function of Al/(Al + Si) (data from SEIFERT et al., 1982).

Table 3. Regression coefficients,* aluminate structural units

Unit type	a	b	c	d	e
$(Al, Si)_3O_8^-$	0.93 ± 0.04	-1.5 ± 0.2	0.6 ± 0.2		
$Al_2Si_2O_8^{2-}$	0.012 ± 0.005	2.2 ± 0.8	-3.8 ± 0.4	5.6 ± 0.6	-3.9 ± 0.3
AlO_2^{2-}	0.4 ± 0.2	-2.1 ± 0.7	2.7 ± 0.5		

* Expression: $X = a + b[Al/(Al + Si)] + c[Al/(Al + Si)]^2 + d[Al/(Al + Si)]^3 + e[Al/(Al + Si)]^4$.

1982, 1983; RAY and NAVROTSKY, 1984) can be used to establish a hierarchy of relative stabilities of aluminum-bearing silicate units in the melts. Heat of solution data show that the relative stabilities are positively correlated with Z/r^2 of the charge-balancing cation (Figure 8), a suggestion first made by BOTTINGA and WEILL (1972). Thus, the hierarchy of stabilities are $K^+ > Na^+ > Ca^{2+} > Mg^{2+}$. Because the Z/r^2 of Fe^{2+} is between that of Ca^{2+} and Mg^{2+}, it is likely that Fe^{2+}-charge-balanced aluminate complexes are less stable than those of Ca^{2+}, but more so than those of Mg^{2+}.

(3) *Ferric iron.* Iron oxides in magmatic liquids is of interest in part because redox relations may be employed to deduce pressure-temperature-oxygen fugacity relations (*e.g.*, KENNEDY, 1948; FUDALI, 1965; MO *et al.*, 1982; SACK *et al.*, 1980; KILINC *et al.*, 1983; VIRGO and MYSEN, 1985; MYSEN, 1986). The principal relationships may be illustrated with the expression (MO *et al.*, 1982):

$$\ln f_{O_2}^P = \ln f_{O_2}^{1\ bar} + \frac{(2\overline{V}_{Fe_2O_3} - 4\overline{V}_{FeO})(P - 1)}{RT}. \quad (1)$$

The temperature-dependence of the redox ratio can be inferred from the RT-term and a possible pressure dependence from the volume term, where experimental data (MO *et al.*, 1982; BOTTINGA *et al.*, 1983) show that at least at 1 bar pressure, the partial molar volume of ferric oxide exceeds that of ferrous oxide in silicate melts. Thus, at constant oxygen fugacity, composition and temperature, the ferric/ferrous of a melt should decrease with increasing pressure. This suggestion has been experimentally verified (MYSEN and VIRGO, 1978, 1985).

Ferric and ferrous oxide in magmatic liquids are additionally important because a change in redox ratio affects the structure of the melt, and, therefore, melt properties that are governed by its structure (*e.g.*, MYSEN and VIRGO, 1980; MYSEN *et al.*, 1984, 1985c; FOX *et al.*, 1982; DICKENSON and HESS, 1981, 1986). Structural data for iron-bearing melts have been reported mostly for binary metal oxide-silica and ternary metal oxide-alumina-silica melts. Evidence from optical and luminescence spectroscopy (FOX *et al.*, 1982), optical spectroscopy (NOLET *et al.*, 1979) Raman and Mossbauer spectroscopy (*e.g.*, FOX *et al.*, 1982; MYSEN *et al.*, 1980b,

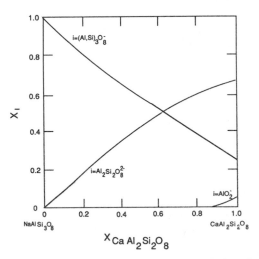

FIG. 7. Calculated abundance of structural units in melts on the join $NaAlSi_3O_8$-$CaAl_2Si_2O_8$.

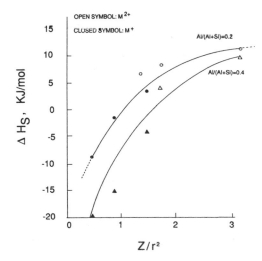

FIG. 8. Heat of solution of quenched melts along aluminate-silica join as a function of Z/r^2 of the charge-balancing cation (heat of solution data from NAVROTSKY *et al.*, 1980, 1982, 1983; RAY and NAVROTSKY, 1984).

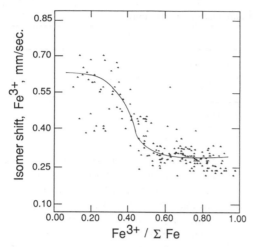

FIG. 9. Isomer shifts of ferric iron relative to that of iron metal ($IS_{Fe^{3+}}$, mm/sec) as a function of $Fe^{3+}/\Sigma Fe$ for quenched melts in the systems $CaO-SiO_2$, $(Ca_{0.5}Mg_{0.5})O-SiO_2$, Na_2O-SiO_2, $CaO-Al_2O_3-SiO_2$ and $MgO-Al_2O_3-SiO_2$ with total iron contents in the range 5 to 15 weight percent (as Fe_2O_3) (data from MYSEN *et al.*, 1984, 1985b,c; MYSEN and VIRGO, 1985).

1984, 1985b; SEIFERT *et al.*, 1979; VIRGO and MYSEN, 1985) and electron spin resonance spectroscopy (CALAS and PETIAU, 1983) indicate that whereas ferrous iron is a network-modifier in the temperature, oxygen fugacity and compositional ranges of igneous processes, ferric iron may play a dual role. From Mossbauer spectroscopy, the isomer shift of Fe^{3+} ($IS_{Fe^{3+}}$) is a sensitive indicator of the oxygen polyhedron around Fe^{3+}. For values (relative to Fe metal) greater than about 0.5 mm/sec, Fe^{3+} is a network-modifier, whereas with $IS_{Fe^{3+}} < 0.3$ mm/sec, Fe^{3+} is a network-former (VIRGO and MYSEN, 1985). In the range between 0.3 and 0.5 mm/sec, both network-forming and network-modifying Fe^{3+} coexist (see VIRGO and MYSEN, 1985, for a detailed review of the evidence). On the basis

of several hundred Mossbauer analyses of quenched melts at 1 bar and at high pressure, it appears, therefore, that the $Fe^{3+}/\Sigma Fe$ of the melts is closely related to the structural position of Fe^{3+} (Figure 9). From the compilation of chemical analyses of cenozoic volcanic rocks (CHAYES, 1975a,b, 1985), one may derive information for the structural role of Fe^{3+} in their molten state provided that the published bulk chemical $Fe^{3+}/\Sigma Fe$ values represent the redox ratio of iron in these rocks prior to crystallization (Table 4). Most likely ferric iron occurs as both a network-former and as network-modifier in natural magmatic liquids. It is also evident from those data that the proportion of tetrahedrally-coordinated ferric iron becomes increasingly important the more felsic the magmatic liquid (more polymerized the liquid) although one may suggest that increasing alkalinity also results in more tetrahedrally-coordinated ferric iron. These generalizations have been quantified in binary and ternary systems used in laboratory studies, where a clear relationship between $Al/(Al + Si)$ (Figure 2) and type metal cations have been observed (Figure 10). Increasing $Al/(Al + Si)$ and decreasing Z/r^2 of the metal cations result in enhanced $Fe^{3+}/\Sigma Fe$.

Qualitatively, the behavior of tetrahedrally-coordinated ferric iron resembles that of Al^{3+} in that it requires electrical charge-balance with alkali metals, alkaline earths or ferrous iron. VIRGO and MYSEN (1985) in a summary of spectroscopic data relevant to iron oxides in silicate melts concluded, however, that in contrast to Al^{3+}, regardless of the type of charge-balancing cation, random substitution of Fe^{3+} for Si^{4+} in melt structural units does not appear to take place. Rather, ferrisilicate or ferrite structural units with constant $Fe^{3+}/(Fe^{3+} + Si)$ are stabilized, and the proportions of such units may be functions of bulk melt $Fe^{3+}/(Fe^{3+} + Si)$. Additional structural information similar to that of aluminum is not yet available. In the present report,

Table 4. Percent of rock analyses within $Fe^{3+}/\Sigma Fe$ brackets indicated

Name	Rhyolite	Dacite	Andesite	Tholeiite & Olivine Tholeiite	Alkali Basalt	Basanite & Basanitoid	Nephelinite
No. analyses	367	338	2068	1010	279	206	116
$Fe^{3+}/\Sigma Fe$	0.63 ± 0.25	0.48 ± 0.20	0.40 ± 0.07	0.29 ± 0.13	0.38 ± 0.19	0.36 ± 0.14	0.43 ± 0.16
<0.3*	13.9	23.3	28.9	58.3	63.4	62.6	23.3
$0.3-0.5$†	17.2	43.1	49.7	34.7	40.1	46.2	43.1
>0.5‡	68.9	33.6	21.5	7.0	23.3	16.5	33.6

* With $Fe^{3+}/\Sigma Fe < 0.3$, all Fe^{3+} in melt is network-modifier.
† With $Fe^{3+}/\Sigma Fe = 0.3-0.5$, Fe^{3+} in melt is partially network-modifier and partially network-former.
‡ With $Fe^{3+}/\Sigma Fe > 0.5$, all Fe^{3+} in melt is network-former.

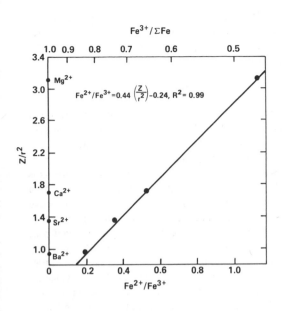

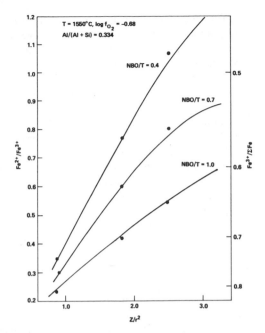

FIG. 10. $Fe^{3+}/\Sigma Fe$ as a function of Z/r^2 of metal cation in metal oxide-silica and metal oxide-alumina-silica systems as indicated on figure (data from MYSEN *et al.*, 1984, 1985b).

tetrahedrally coordinated ferric iron will be considered as a separate unit with no additional breakdown. It is suggested, however, that a hierarchy in relative stability exists. In contrast with the silica-alumina systems (Figure 8), not enough data are available for relative stabilities of ferric iron complexes. In view of the many similarities in structural behavior of tetrahedrally-coordinated ferric iron and aluminum, it is suggested that a hierarchy similar to that of aluminum charge-balance exists for tetrahedrally-coordinated ferric iron (*i.e.*, K > Na > Ca > Fe^{2+} > Mg).

(4) *Other cations.* Among the major element oxides in igneous rocks, titanium and phosphorous generally are the least abundant (Table 1). Nevertheless, these two oxides have attracted considerable attention because even in their natural concentration ranges melt properties are profoundly affected (*e.g.*, KUSHIRO, 1975; VISSER and VAN GROOS, 1979; WATSON, 1976; RYERSON and HESS, 1980; DICKINSON and HESS, 1985). Both Ti^{4+} and P^{5+} enhances the activity coefficients of SiO_2 in silicate melts. (KUSHIRO, 1975; RYERSON, 1985) in contrast to components such as alkali metals or alkaline earths. Thus, it has been frequently suggested (*e.g.*, KUSHIRO, 1975; WATSON, 1976; RYERSON and HESS, 1980; VISSER and VAN GROOS, 1979) that both cations act as network-formers in silicate melts.

Vibrational spectroscopic data of phosphorous-bearing melts support this conclusion (*e.g.*, GALEENER and MIKKELSEN, 1979; NELSON and EXHAROS, 1979; MYSEN *et al.*, 1981b; NELSON and TALLANT, 1984). Those data also indicate that in depolymerized melts ($NBO/T > 0$), phosphorous occurs in phosphate complexes either in the form of PO_3^- (NELSON and EXHAROS, 1979; MYSEN *et al.*, 1981b) or as PO_4^{3-} complexes (NELSON and TALLANT, 1984). These complexes are electrically neutralized with alkali metals or alkaline earths. Although it has not been established which cation or cations are involved in this process, free energy of formation data for crystalline analogues indicate that calcium phosphate complexing is the most likely.

The vibrational data for titanium-bearing silicate melts are less conclusive than those for phosphorous-bearing systems. TOBIN and BAAK (1968) investigated the system SiO_2-TiO_2 and suggested that titanium is in four-fold coordination and CHANDRASEKHAR *et al.*, (1979) implied that the structure of vitreous TiO_2 resembles that of vitreous SiO_2 (three-dimensionally-interconnected network). X-ray absorption data on SiO_2-TiO_2 glasses (*e.g.*, SANDSTROM *et al.*, 1980) indicate that both tetrahedral and octahedral Ti^{4+} may occur, and that tetrahedral coordination becomes more dominant with increasing Ti-content. FURUKAWA and WHITE

(1979), in their study of glass structure in the system $Li_2Si_2O_5$-TiO_2 suggested that at least some of the Ti^{4+} might not occur in tetrahedral coordination. MYSEN et al., (1980c) observed that the frequency of Raman bands from Si-O⁻ stretching decreasing somewhat as a function of increasing bulk melt Ti/(Ti + Si). They also observed that Ca- and Ca,Mg-metasilicate melts became more polymerized as TiO_2 was added. These observations suggest that Ti^{4+} may be tetrahedrally-coordinated in the melts. It was also observed, however, that vibrational spectra of crystals and quenched melts of Na_2TiO_3 composition exhibited significant similarities. Similar comparisons were conducted in the system K_2O-TiO_2 by DICKINSON and HESS (1985). The latter authors suggested that in the silica-free titanate systems, titanium occurs principally in a highly distorted octahedron. The data by MYSEN et al., (1980c) for Na_2O-TiO_2 could be interpreted similarly. From these, seemingly somewhat conflicting, data it would appear that most likely titanium occurs in more than one coordination state in silicate melts. The exact compositional control on the proportions of the individual coordination polyhedra as well as the type of coordination polyhedra remains open to further investigation.

STRUCTURE OF MAGMATIC LIQUIDS

The structural data detailed above can be combined to describe the structure of magmatic liquids on the basis of their bulk chemical composition. The first step in this procedure is to establish the proportions of tetrahedrally-coordinated cations and the types and proportions of cations required for electrical charge-balance. Phosphorous most likely occurs as orthophosphate (PO_4^{3-}) and are considered in association with Ca^{2+}. Thus, for each P^{5+} cation, 1.5 Ca^{2+} is required. This proportion of Ca^{2+} is, therefore, subtracted from the total amount present. The proportion of ferric iron in tetrahedral coordination is calculated on the basis of $Fe^{3+}/\Sigma Fe$ as indicated above (see also Figure 9). Both Al^{3+} and Fe^{3+} require charge-balance in tetrahedral coordination and the hierarchy of relative stabilities of the Al^{3+}- and Fe^{3+}-complexes is identical. The ferric iron is assigned first. Thus, for potassium, for example, if there is sufficient amount available, a proportion equivalent to the proportion of Fe^{3+} is assigned and this proportion is subtracted from the total potassium concentration before assignment of the remainder (and possibly other cations) as charge-balancing cations for Al^{3+}. If there is insufficient potassium in the melt, the sodium is also a charge-balancing cation for tetrahedrally-coordinated Fe^{3+}. In that case, no potassium can charge-balance Al^{3+}. The complete procedure is carried out in the order K^+, Na^+, Ca^{2+}, Mg^{2+} first with assignment to Fe^{3+} and then to Al^{3+}. The bulk melt NBO/T of a natural magmatic liquid now can be calculated from the expression:

$$NBO/T = 1/T \sum_{i=1}^{i} M_i^{n+}, \qquad (2)$$

where M_i^{n+} is the proportion of network-modifying cation i with electrical charge $n+$ and T is the proportion of tetrahedrally coordinated cations.

Distribution of NBO/T of melts of rhyolite, andesite, tholeiite and basanite melt compositions is shown in Figure 11. It is evident from those data that most common extrusive igneous rocks have NBO/T-values between 0 and 1 and that the more felsic the igneous rock, the more polymerized it is. There are distinct maxima in the NBO/T distributions for each rock type, but the range of values spread quite widely in particular for the basanite composition as also reflected in the standard errors of the average NBO/T calculated for seven major igneous rock types (Table 5).

As a result of the charge-balancing requirements for Al^{3+} and Fe^{3+} in tetrahedral coordination, metal cations that exist as network-modifiers in simple metal oxide-silica systems, may not always be network-modifiers in natural magmatic liquids (Table 5). For example, the alkali metal contents of most igneous extrusive rocks generally are so low that only in a fraction of the rhyolite and nephelinite compositions will Na or K occur as network-mod-

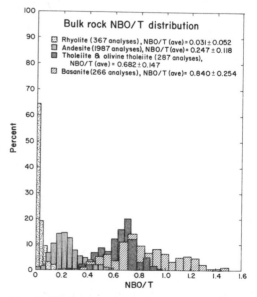

FIG. 11. Distribution of bulk melt NBO/T of magmatic liquids as indicated on figure. The rock analyses were extracted from rockfile RKNFSYS (CHAYES, 1975a,b; 1985; see also discussion in text).

Table 5. Percentage of analyses where individual metal cations are network-modifiers

Name	Rhyolite	Dacite	Andesite	Tholeiite & Olivine Tholeiite	Alkali Basalt	Basanite & Basanitoid	Nephelinite
No. analyses	367	338	2068	1010	279	206	116
Na + K	5.5	0.0	0.0	0.01	0.0	0.0	4.3
Ca	22.9	31.1	61.4	98.3	92.8	100.0	100.0
Fe^{2+}	50.4	79.3	94.9	100.0	100.0	100.0	100.0
Mg	99.2	00.4	99.9	100.0	100.0	100.0	100.0
Fe^{3+} *	100.0	100.0	100.0	100.0	100.0	100.0	100.0

* Where $Fe^{3+}/\Sigma Fe$ is sufficiently low so that some or all Fe^{3+} is network-modifier.

ifiers. In all other compositions, and in most rhyolites and nephelinites, $[(Al^{3+}(IV) + Fe^{3+}(IV)] > [Na + K]$ and all alkali metals are charge-balancing cations. In fact, for a given rock type, the more electronegative the metal cation, the larger is the proportion of melts where the metal cation occurs as a network-modifier. For a given metal cation, the more polymerized the melt (the smaller the bulk melt NBO/T), the smaller the fraction of analyses where this cation occurs as a network-modifier (Table 5).

Fractions of nonbridging oxygens in magmatic liquids associated with individual network-modifying cations are shown in Figure 12. It can be seen from those data that ferrous iron and magnesium are the most important network-modifying cations in melts of common extrusive igneous rocks. In all but the most felsic rocks (rhyolite), magnesium is the most important followed by ferrous iron. In rhyolite composition melts, the relative importance of ferrous iron and magnesium is reversed. The data in Figure 12 also illustrate the fact that Na + K is relatively unimportant as network-modifier in natural magmatic liquids.

The abundance of anionic structural units in natural magmatic liquids can be estimated with the aid of the expressions in Table 2. In order to employ those equations for complex natural melts, the mol fraction of a structural unit associated with a particular network-modifying cation must be multiplied by the atomic proportion of the network-modifying cation in question. For example, if $X_{TO_3}^{Mg}$ from Table 2 is 0.5 and the atomic fraction of the network-modifying cation is 0.25, the fraction of TO_3 associated with Mg^{2+} in the melt is 0.125. The overall abundance of each structural unit is then given by the summations:

$$X_{T_2O_5} = \sum_{i=1}^{i} T_2O_{5i}, \qquad (3)$$

$$X_{TO_3} = \sum_{i=1}^{i} TO_{3i}, \qquad (4)$$

$$X_{T_2O_7} = \sum_{i=1}^{i} T_2O_{7i}, \qquad (5)$$

$$X_{TO_4} = \sum_{i=1}^{i} TO_{4i} \qquad (6)$$

and

$$X_{TO_2} = 1.0 - (X_{T_2O_5} + X_{TO_3} + X_{T_2O_7} + X_{TO_4}), \qquad (7)$$

where T_2O_{5i} etc. is the proportion of specified structural unit associated with network-modifying cat-

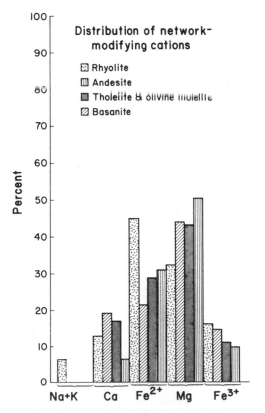

FIG. 12. Distribution of network-modifying cations in melts major extrusive rock types. (CHAYES, 1975a,b; 1985; see also discussion in text).

ion, i, and $X_{T_2O_5}$ etc. represent the total abundance of this structural unit in the melt.

The distribution of structural units in a subset of rhyolite, andesite, tholeiite and basanite (2295 analyses in total) as a function of bulk melt NBO/T is shown in Figure 13. The distribution of $X_{T_2O_7}$ is not shown because only 20 of the 2295 analyses show any amount of $X_{T_2O_7}$, and even in those cases this amount was vanishingly small. One may conclude, therefore, that pyrosilicate structural units are uncommon in melts of common extrusive igneous rocks. The distribution of T_2O_5 exhibits a distinct maximum at bulk melt NBO/T near 0.5. This NBO/T-value corresponds to quartz tholeiite and basaltic andesite. The results (Figure 13) illus-

trate that TO_2 and TO_3 are the most common structural units in most natural magmatic liquids with the possible exception of compositions in the NBO/T-range between 0.3 and 0.4 (typically corresponding to andesite), where T_2O_5 units may be more important and for compositions with bulk NBO/T < 0.2 (corresponding to rhyolite and the most felsic andesite compositions), where TO_3 units are not present in the melt and the nonbridging oxygens occur only in TO_4 and T_2O_5 units. This structure distribution results from the fact that alkalies and Ca^{2+} are associated with T_2O_5 units in these rock types (Figure 14), whereas magnesium, and to a lesser degree, ferrous iron, exhibits a pronounced preference for TO_4 units.

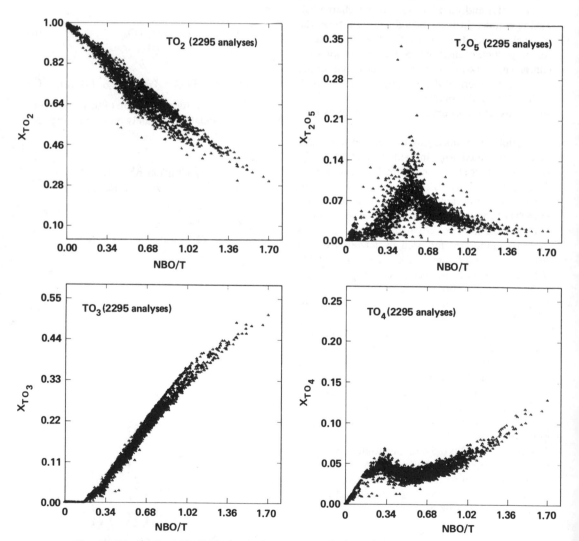

FIG. 13. Distribution of individual anionic structural units in major rock types as a function of bulk melt NBO/T. Rock analyses were extracted from rockfile RKNFSYS (CHAYES, 1975a,b; 1985; see also discussion in text).

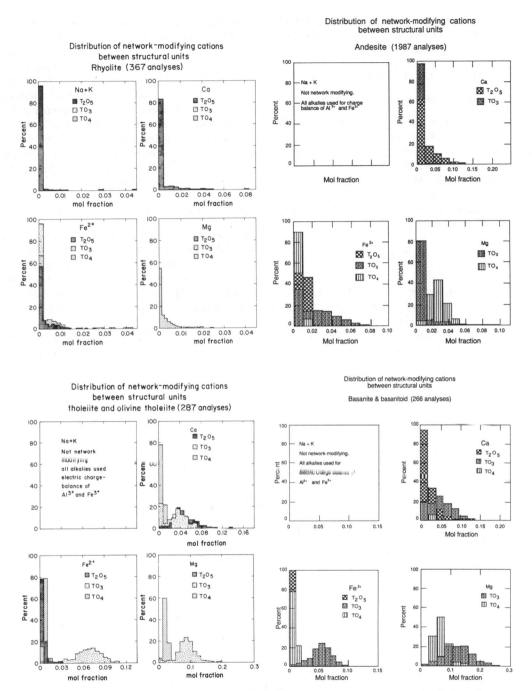

FIG. 14. Network-modifying cation distribution among individual anionic structural units in melts of major extrusive rock types. (A) Rhyolite, (B) Andesite, (C) Tholeiite and olivine tholeiite and (D) Basanite and basanitoid (CHAYES, 1975a,b; 1985; see also discussion in text).

The abundance of three-dimensional network units was obtained by mass balance [equation (8)]. With the exception of a few basanite analyses, these TO_2 units can be described as a mixture of alkali charge-balanced $AlSi_3O_8^-$ and predominantly alkaline earth balanced $Al_2Si_2O_8^{2-}$. For andesitic and more felsic rocks, a significant proportion of the Al^{3+} is also charge-balanced with ferrous iron as the summary in Table 5 indicates that for these rocks only ~20–50% of these compositions have insufficient Ca^{2+} for aluminum electrical charge-balance.

PETROLOGICAL APPLICATIONS

Structure and properties of magmatic liquids

Recent experimental data indicate that partial molar volumes of the major element oxides in magmatic liquids are independent of bulk chemical composition in the compositional range between rhyolite and basalt (MO *et al.*, 1982; BOTTINGA *et al.*, 1983). From the partial molar volume data, the molar volume of a magmatic liquid is calculated from the expression:

$$V = \sum_{i=1}^{i} X_i \overline{V_i}, \qquad (9)$$

where $\overline{V_i}$ and X_i are the partial molar volume and mol fraction of oxide component i. The distribution of molar volumes calculated from 2607 rock analyses from RKNFSYS is shown in Figure 15. This histogram exhibits a slightly skewed distribution

where the enrichment near 28 cm^3/mol reflects the rhyolite analyses (average: 28.3 ± 0.4 cm^3/mol). The maximum near 26 cm^3/mol represents a mixture of the remaining rock types (average values are: basanite, 23.8 ± 0.9 cm^3/mol; tholeiite, 24.0 ± 0.5 cm^3/mol; andesite, 26.3 ± 0.7 cm^3/mol) although there is a general positive correlation between molar volume and degree of polymerization of the melt (Figure 16). A very simple positive linear correlation exists, however, between the proportion of three-dimensional network units in the melt (Figure 16) with the least-squares fitted straight line:

$$V = 16.33 \pm 0.05 + 11.72 \pm 0.07 X_{TO_2}. \qquad (10)$$

As shown above (Figure 13), the proportions of TO_3 and TO_4 units vary inversely with TO_2. Thus, as expected, the molar volumes of natural magmatic liquids decrease systematically with increasing abundance of TO_3 and TO_4 units in the melts. No apparent correlation exists between molar volume and the abundance of T_2O_5 units in the magmatic liquids. Comparable linear relations exist between molar volume of X_{TO_2} units in binary metal oxide-

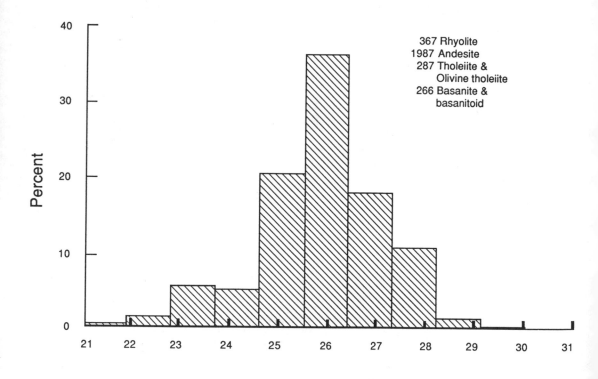

FIG. 15. Distribution of calculated molar volume (cm^3/mol) for melts of major types of extrusive rocks used in Figure 19. Molar volumes were calculated from the partial molar volume data of BOTTINGA *et al.* (1983). (CHAYES, 1975a,b; 1985; see also discussion in text).

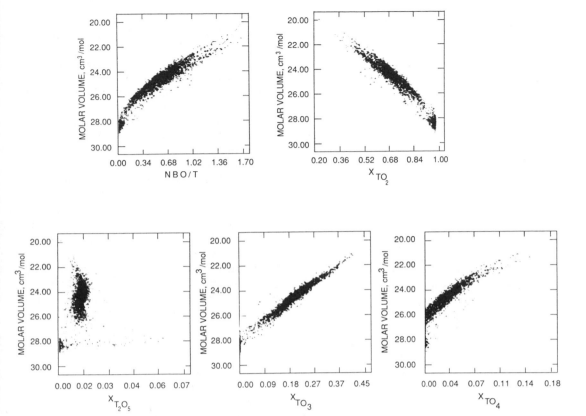

Fig. 16. Relations between bulk melt NBO/T, proportions of structural units and molar volume of melts of major extrusive rock types. Molar volumes were calculated from the partial molar volume data of BOTTINGA *et al.* (1983). (CHAYES, 1975a,b; 1985; see also discussion in text).

silica melts (BOCKRIS and KOJONEN, 1960; ROBINSON, 1969; MYSEN, 1986). It appears, therefore, that the proportion of three-dimensional network-units is the principal structural control of the molar volume of natural magmatic liquids and that the molar volumes may be estimated provided that the X_{TO_2} value is known.

One-bar viscous properties of magmatic liquids may be approximated with the model for calculation published by BOTTINGA and WEILL (1972). The distribution of viscosity η and activation energies of viscous flow (E_η) calculated for the same data base as the molar volumes are shown in Figure 17. The activation energies were calculated with the assumption that in the superliquidus temperature region, the E_η is independent of temperature. The apparent bimodal distribution of values is at least partly due to the fact that as for molar volumes, viscous properties of rhyolite melt tend to cluster in a group separated distinctly from other rock types, with significantly greater viscosities and activation energies of viscous flow. At 1300°C, for

example, the average viscosities ($\log_{10} \eta$; poise) for the groups of rocks are, 4.8 ± 0.3, 3.7 ± 0.5, 2.2 ± 0.2 and 1.7 ± 0.3 for rhyolite, andesite, tholeiite and basanite melts, respectively. The pronounced maximum near $\log \eta = 3.5$ (poise) is principally controlled by the large number of andesite analyses in the rockfile. Similar relationships exist for the activation energy of viscous flow (Figure 17).

As has been observed in simple binary and ternary systems (see summary of available data by MYSEN *et al.*, 1982; RICHET, 1984) there is a positive correlation between viscosity and NBO/T and activation energy of viscous flow and NBO/T of natural magmatic liquids, although there is a significant scatter in the data (Figure 18). The preexponential factor;

$$\ln \eta_0 = \ln \eta - E_\eta/RT, \quad (11)$$

where η is viscosity, $\ln \eta_0$ the preexponential factor, E_η activation energy of viscous flow, R the gas constant and T the absolute temperature, does not correlate well with any structural factor calculated from

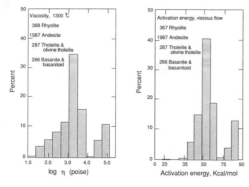

FIG. 17. Distribution of viscosity ($\log_{10} \eta$, poise at 1300°C) and activation energy of viscous flow (E_η, kcal/mol) of the melts of major extrusive rock types used in Figures 18 and 19. Viscosities and activation energies were calculated with the method of BOTTINGA and WEILL (1972) assuming Ahrrenian behavior of the melts in the superliquidus temperature region. Bulk compositions of rocktypes were extracted from rockfile RKNFSYS (CHAYES, 1975a,b; 1985).

the rock analyses. It is also noted that as has been shown for simple binary and ternary melt compositions (see BOCKRIS and REDDY, 1970; BOTTINGA and WEILL, 1972, for summary for available data), both the viscosity and the activation energies are principally functions of the proportions of three-dimensional network units in the magmatic liquids (Figure 19). As would be expected, both $\log \eta$ and E_η are negatively correlated with the proportions of TO_3 and TO_4 units in the liquids.

Redox equilibria and melt structure

It has been suggested (*e.g.,* FUDALI, 1965; SACK *et al.,* 1980) that the redox ratio of iron in natural magmatic liquids may be used to calculate the oxygen fugacity of equilibration if the temperature is known. To this end SACK *et al.,* (1980) and KILINC *et al.,* (1983) used stepwise, multiple linear regression of ferric/ferrous to incorporate temperature, oxygen fugacity, and the abundance of various oxide components. These investigators employed an expression of the form

$\ln (Fe_2O_3/FeO)$

$$= a \ln f_{O_2} + b/T + c + \sum_{i=1}^{i} d_i X_i, \quad (12)$$

where a, b, c, and d_i are regression coefficients, T is absolute temperature, $\ln f_{O_2}$ is the natural logarithm of the oxygen fugacity, and X_i are concentrations of oxide components. By fitting 57 analyses

from experimentally equilibrated liquids to this expression, SACK *et al.,* (1980) found positive correlation of $\ln (Fe_2O_3/FeO)$ with Na_2O, K_2O, and CaO, whereas MgO, Al_2O_3, and FeO were negatively correlated. In a subsequent refinement of this treatment, KILINC *et al.,* (1983) concluded that the ferric/ferrous depended only on CaO, Na_2O, K_2O, FeO (all positively correlated), and Al_2O_3 (which remained negatively correlated; Table 6). KILINC *et al.,* (1983) concluded that Fe_2O_3/FeO was independent of MgO content of the liquid. Magnesium oxide was not identified as a variable in the experimental results reported by THORNBER *et al.,* (1980).

There were significant bulk compositional differences between the samples used by THORNBER *et al.,* (1980), SACK *et al.,* (1980), and KILINC *et al.,* (1983). Whereas THORNBER *et al.,* (1980) employed mostly basaltic liquids, with selective addition of specific oxides, SACK *et al.,* (1980) and KILINC *et al.,* (1983) reported laboratory-calibrated $Fe^{3+}/\Sigma Fe$ with a wide range of bulk compositions from mafic to felsic.

The discrepancies between these data sets most probably arise from the fact that in neither treatment of the whole-rock analyses were the structural roles of the cations and the structural positions of ferric and ferrous iron considered. The limitation of this approach was evident in the disagreement between the functional relationships of the various oxide components depending on temperature, oxygen fugacity, and bulk composition itself (SACK *et al.,* 1980; THORNBER *et al.,* 1980; KILINC *et al.,* 1983). The standard errors of the regression coefficients (Table 6) were also quite large, suggesting that regression of $\ln (Fe_2O_3/FeO)$ against the metal oxides does not result in the best possible relationship between redox state and melt composition.

With the structural information discussed above, the approach suggested by SACK *et al.,* (1980) may be refined and stepwise linear regression may be applied to a rock composition after it has been recast to the relevant structural components. The structural components are those found to govern $Fe^{3+}/\Sigma Fe$ in binary and ternary systems. It is suggested that the expression of the form

$\ln (Fe^{2+}/Fe^{3+}) = a + 10^4 b/T + c \ln f_{O_2}$

$$+ d[Al/(Al + Si)] + e[Fe^{3+}/(Fe^{3+} + Si^{4+})]$$

$$+ \sum_{j=1}^{j} f_j (NBO/T)_j \quad (13)$$

can be used to describe the relationship between Fe^{2+}/Fe^{3+}, temperature, oxygen fugacity and melt

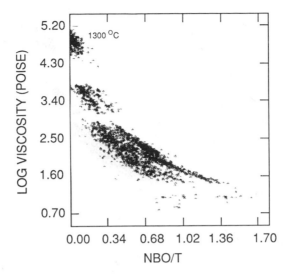

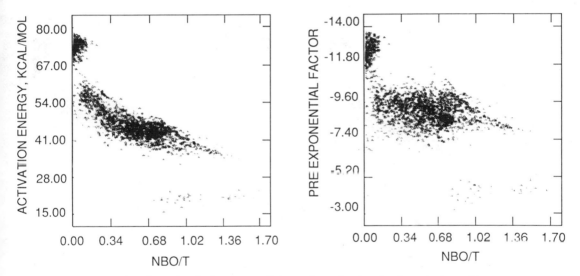

FIG. 18. Viscosity and activation energy of viscous flow of melts of major extrusive rocktypes as a function of NBO/T of the melts. Viscosities and activation energies were calculated with the method of BOTTINGA and WEILL (1972) assuming Ahrrenian behavior of the melts in the superliquidus temperature region. Bulk compositions of rocktypes were extracted from rockfile RKNFSYS (CHAYES, 1975a,b; 1985).

structure. The f_j and $(NBO/T)_j$ are the regression coefficients and NBO/T values for the individual network-modifying oxides, respectively. The coefficients a, b, c, and d together with f_j are obtained with stepwise linear regression. The expression in equation (13) takes into account each of the variables identified as independent variables affecting $Fe^{3+}/\Sigma Fe$. This treatment differs from that resulting in equation (12), where no decision was made in regard to which structural variables influence $Fe^{3+}/\Sigma Fe$.

Linear regression has been carried out with 267 experimental calibrations of $Fe^{3+}/\Sigma Fe$ in simple melt systems (only binary metal oxide and ternary metal oxide-alumina-silica systems). The resulting coefficients are shown in Table 7 and are compared with those obtained by similar regression of available data for laboratory-calibrated, natural rock

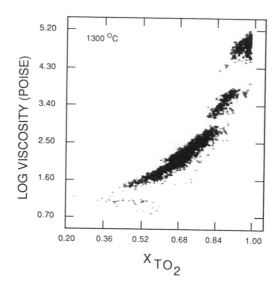

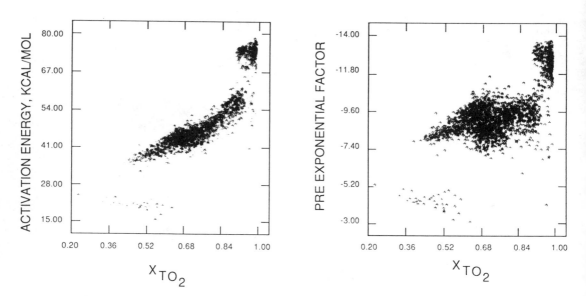

FIG. 19. Viscosity and activation energy of viscous flow of melts of major extrusive rock types as a function of abundance of structural units (A, X_{TO_2}; B, X_{TO_3}) in the melts. Viscosities and activation energies were calculated with the method of BOTTINGA and WEILL (1972) assuming Ahrrenian behavior of the melts in the superliquidus temperature region. Bulk compositions of rocktypes were extracted from rockfile RKNFSYS (CHAYES, 1975a,b; 1985).

compositions as well as compositions in simple systems (KENNEDY, 1948; FUDALI, 1965; SACK et al., 1980; KILINC et al., 1983; THORNBER et al., 1980) with a total of 460 analyses. Finally, the redox ratios of both groups of compositions were regressed against melt structural parameters according to equation (13) (a total of 460 analyses). In each of these sets of coefficients, the standard errors are sig-

nificantly smaller than those found from oxide components (SACK et al., 1980; KILINC et al., 1983). Thus, regression of ln (Fe^{2+}/Fe^{3+}) against independently established melt structural factors yields a more reliable formulation than one based on empirical relationships between redox ratio and oxide contents of the melts.

It is evident from this exercise that among the

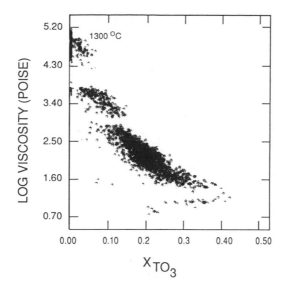

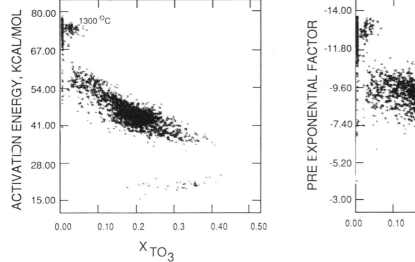

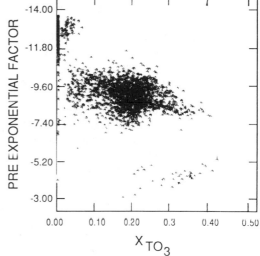

FIG. 19. (Continued)

network-modifying cations the ln (Fe^{2+}/Fe^{3+}) is negatively correlated with the proportion of non-bridging oxygen associated with Ca^{2+}, Na^+, and Fe^{2+} (Table 7). A negative but less reliable correlation also exists in some of the data summarized in Table 6 (KILINC *et al.*, 1983). All three analyses show that Fe^{2+}/Fe^{3+} *increases* with increasing NBO/T associated with Mg^{2+}. It is not clear why Mg^{2+} is an exception among the network-modifying cations in this respect. There is also a rapid decrease in ln (Fe^{2+}/Fe^{3+}) with increasing $Fe^{3+}/(Fe^{3+} + Si^{4+})$. In

the case of the simple-system calibration, as well as that based on all analyses, the ln (Fe^{2+}/Fe^{3+}) will decrease with increasing Al/(Al + Si) (with all Al^{3+} charge-balanced in tetrahedral coordination), but the coefficient obtained by regression with only the natural rock compositions has the opposite sign. Although there is no obvious explanation for this apparent difference, it should be remembered that the range in Al/(Al + Si) in the natural rock compositions is relatively small (0.15–0.25), whereas for the simple-system calibration, the range covered is

Table 6. Regression coefficients with standard errors ($\pm 1\sigma$) for Equation (12)

Coefficient	SACK et al. (1980)		KILINC et al. (1983)	
	Value	Standard error	Value	Standard error
a	0.218	0.007	0.219	0.004
b	13,184.7	959.0	12,670.0	900.0
c	−4.50	3.04	−7.54	0.55
$d_{Al_2O_3}$	−2.15	2.88	−2.24	1.03
d_{FeO} *	−4.50	3.69	1.55	1.03
d_{MgO}	−5.44	3.04	—	—
d_{CaO}	0.07	3.08	2.96	0.53
d_{Na_2O}	3.54	3.97	8.42	1.41
d_{K_2O}	4.19	4.12	9.59	1.45

* Total iron oxide as FeO.

between 0 and 0.43. More reliance is placed, therefore, on this latter analysis.

The coefficients in Table 7 may be inserted in equation (13), and this equation may be used, for example, as an oxygen-fugacity barometer. The calculated f_{O_2} values for the samples in the data set are compared with the measured values in Figure 20. From this comparison it is evident that as an oxygen-fugacity barometer of natural igneous rocks based only on the simple-system calibration, 40% of the calculated values are within ±0.5 log unit and 67% are within ±1.0 log unit of f_{O_2}. Between 85 and 90% of the calculated values are within ±1.5 log units of the measured value (Figure 20a). When the whole data set of simple-system and natural melt compositions is employed (see Table 7), the deviation from measured values is smaller (Figure 20b), and 54% of the analyses are within ±0.5 log unit

of oxygen fugacity and 85%, within ±1.0 log unit. About 95% fall within ±1.5 log units. It is suggested that this model relating redox ratios of iron to temperature, oxygen fugacity, and melt structure is an adequate description and that equation (13), with the coefficients in Table 7, can be used with confidence to calculate oxygen-fugacity conditions of natural magmatic liquids at 1 atmosphere. Although some of the principles governing the pressure dependence of ferric/ferrous have been established (MO et al., 1982; MYSEN and VIRGO, 1983), the data base is at present insufficient to extend this treatment to high pressure.

SUMMARY

Available experimental data from model system melt structure studies have been used to describe

Table 7. Regression coefficients for Equation (13)

	Simple system		Natural rocks		All analyses	
	Coefficient	Standard error	Coefficient	Standard error	Coefficient	Standard error
a (const.)	10.814	1.134	4.384	0.524	15.437	0.786
b $(1/T)$	−1.989	0.203	−0.9077	0.0915	−2.848	0.138
c $(\ln f_{O_2})$	−0.3210	0.0117	−0.1420	0.0081	−0.3484	0.0120
d $[Al^{3+}/(Al^{3+} + Si)]$	−1.535	0.467	1.621	0.418	−1.309	0.469
e $[Fe^{3+}/(Fe^{3+} + Si)]$	−4.067	0.985	−9.875	0.952	−2.121	1.055
$(NBO/T)^{Mg}$	0.494	0.134	0.8607	0.2093	0.6662	0.0966
f_i $(NBO/T)^{Ca}$	−0.5228	0.1095	−0.6560	−0.1617	−0.5255	0.1084
$(NBO/T)^{Na}$	−1.584	0.238	−1.194	0.5112	−1.125	0.1790
$(NBO/T)^{Fe^{2+}}$	−1.951	0.507	−2.310	0.422	−3.215	0.538

Number of analyses in regression: simple systems, 267; natural rocks, 193; all analyses, 460.
Experimental data for regression from KENNEDY (1948), FUDALI (1965), SACK et al. (1980), THORNBER et al. (1980), KILINC et al. (1983), SEIFERT et al. (1979), VIRGO et al. (1981), MYSEN and VIRGO (1983), VIRGO and MYSEN (1985), MYSEN et al. (1980b, 1984, 1985b,c).

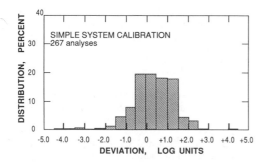

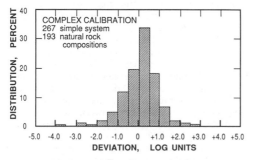

FIG. 20. Distribution of calculated oxygen fugacity from equation (13) and coefficients from Table 7 based on calibration with 267 experimental data points in simple model systems (a) and 460 experimental data points including both simple model systems and natural melt compositions (b).

the major structural features of natural magmatic liquids. These results can be used to characterize relationships between the structure of the magmatic liquid and molar volume as well as viscous behavior. The redox relations of iron can also be quantitatively described.

Acknowledgements—This study could not have been carried out without the patient guidance by F. Chayes in order to extract the relevant bulk chemical analyses from rockfile RKNFSYS. His help is greatly appreciated. Critical reviews by R. W. Luth, I. Kushiro and D. Virgo are appreciated.

REFERENCES

BOCKRIS J. O'M. and KOJONEN F. (1960) The compressibility of certain molten alkali silicates and borates. *J. Amer. Chem. Soc.* **82**, 4493–4497.

BOCKRIS J. O'M. and REDDY A. K. N. (1970) *Modern Electrochemistry,* Vol. 1, 622 pp. Plenum Press.

BOTTINGA Y. and WEILL D. F. (1972) The viscosity of magmatic silicate liquids: A model for calculation. *Amer. J. Sci.* **272**, 438–475.

BOTTINGA Y., RICHET P. and WEILL D. F. (1983) Calculation of the density and thermal expansion coefficient of silicate liquids. *Bull. Mineral.* **106**, 129–138.

BRAWER S. A. and WHITE W. B. (1975) Raman spectroscopic investigation of the structure of silicate glasses.

I. The binary silicate glasses. *J. Chem. Phys.* **63**, 2421–2432.

BRAWER S. A. and WHITE W. B. (1977) Raman spectroscopic investigation of the structure of silicate glasses. II. The soda-alkaline earth-alumina ternary and quarternary glasses. *J. Non-Cryst. Solids* **23**, 261–278.

BROWN G. E., GIBBS G. V. and RIBBE P. H. (1969) The nature and variation in length of the Si-O and Al-O bonds in framework silicates. *Am. Mineral.* **54**, 1044–1061.

CALAS G. and PETIAU J. (1983) Structure of oxide glasses: Spectroscopic studies of local order and crystallochemistry: Geochemical implications. *Bull. Mineral.* **106**, 33–55.

CHANDRASEKHAR H. R., CHANDRASEKHAR M. and MANGNHANI M. H. (1979) Phonons in titanium-doped vitreous silica. *Solid State Commun.* **31**, 329–333.

CHAYES F. (1975a) A world data base for igneous petrology. *Carnegie Inst. Wash. Yearb.* **74**, 549–550.

CHAYES F. (1978b) Average composition of the commoner cenozoic volcanic rocks. *Carnegie Inst. Wash. Yearb.* **74**, 547–549.

CHAYES F. (1985) Version NTRM2 of system RKNFSYS. Unpubl. document, Geophys. Lab., Carnegie Inst. Wash., Washington DC.

DICKENSON M. P. and HESS P. C. (1981) Redox equilibria and the structural role of iron in aluminosilicate melts. *Contrib. Mineral. Petrol.* **78**, 352–358.

DICKENSON M. P. and HESS P. C. (1986) The structural role and homogenous redox equilibria of iron in peraluminous, metaluminous and peralkaline silicate melts. *Contrib. Mineral. Petrol.* **92**, 207–217.

DICKINSON J. E. and HESS P. C. (1985) Rutile solubility and titanium coordination in silicate melts. *Geochim. Cosmochim. Acta* **49**, 2289–2296.

DOMINE F. and PIRIOU B. (1986) Raman spectroscopic study of the SiO_2-Al_2O_3-K_2O vitreous system: Distribution of silicon and second neighbors. *Amer. Mineral.* **71**, 38–50.

FOX K. E., FURUKAWA T. and WHITE W. B. (1982) Luminescence of Fe^{3+} in metaphosphate glass: evidence for four- and six-coordinated sites. *J. Amer. Ceram. Soc.* **64**, C42–C43.

FRASER D. G., RAMMENSEE W. and JONES R. H. (1983) The mixing properties of melts in the system $NaAlSi_2O_6$-$KAlSi_2O_6$ as determined by Knudsen cell mass spectrometry. *Bull. Mineral.* **106**, 111–117.

FRASER D. G. and BOTTINGA Y. (1985) The mixing properties of melts and glasses in the system $NaAlSi_3O_8$-$KAlSi_3O_8$: Comparison of experimental data obtained by Knudsen cell mass spectrometry and solution calorimetry. *Geochim. Cosmochim. Acta* **49**, 1377–1381.

FUDALI F. (1965) Oxygen fugacity of basaltic and andesitic magmas. *Geochim. Cosmochim. Acta* **29**, 1063–1075.

FURUKAWA T. and WHITE W. B. (1979) Structure and crystallization of glasses in the $Li_2Si_2O_5$-TiO_2 system determined by Raman spectroscopy. *Phys. Chem. Glasses* **20**, 69–80.

FURUKAWA T., FOX K. E. and WHITE W. B. (1981) Raman spectroscopic investigation of the structure of silicate glasses. III. Raman intensities and structural units in sodium silicate glasses. *J. Chem. Phys.* **15**, 3226–3237.

GALEENER F. L. and MIKKELSEN J. C. (1979) The Raman spectra and structure of pure vitreous P_2O_5. *Solid State Commun.* **20**, 505–510.

GIBBS G. V., MEAGHER E. P., NEWTON M. D. and SWANSON D. K. (1981) A comparison of experimental and theoretical bond length and angle variations for minerals and inorganic solids, and molecules. In *Structure and Bonding in Crystals* (editors M. O'KEEFE and A. NAVROTSKY), Vol. 2, Chap. 9, Academic Press.

HOCHELLA M. F. and BROWN G. E. (1985) Structure and viscosity of rhyolite composition melt. *Geochim. Cosmochim. Acta* **49**, 2631–2640.

KENNEDY G. C. (1948) Equilibrium between volatiles and iron oxides in rocks. *Amer. J. Sci.* **246**, 529–549.

KILINC A., CARMICHAEL I. S. E., RIVERS M. L. and SACK R. O. (1983) The ferric-ferrous ratio of natural silicate liquids equilibrated in air. *Contrib. Mineral. Petrol.* **83**, 136–141.

KUSHIRO I. (1975) On the nature of silicate melt and its significance in magma genesis: Regularities in the shift of liquidus boundaries involving olivine pyroxene, and silica materials. *Amer. J. Sci.* **275**, 411–431.

LAUER H. V. (1977) Effect of glass composition on major element redox equilibria: Fe^{2+}-Fe^{3+}. *Phys. Chem. Glasses* **18**, 49–52.

MATSON D. W., SHARMA S. K. and PHILPOTTS J. A. (1983) The structure of high-silica alkali-silicate glasses—A Raman spectroscopic investigation. *J. Non-Cryst. Solids* **58**, 323–352.

MCMILLAN P. (1984) Structural studies of silicate glasses and melts—applications and limitations of Raman spectroscopy. *Amer. Mineral.* **69**, 622–644.

MCMILLAN P., PIRIOU B. and NAVROTSKY A. (1982) A Raman spectroscopic study of glasses along the joins silica-calcium aluminate, silica-sodium aluminate, and silica-potassium aluminate. *Geochim. Cosmochim. Acta* **46**, 2021–2037.

MO X. X., CARMICHAEL I. S. E., RIVERS M. and STEBBINS J. (1982) The partial molar volume of Fe_2O_3 in multicomponent silicate liquids and the pressure dependence of oxygen fugacity in magmas. *Mineral. Mag.* **45**, 237–245.

MYSEN B. O. (1986) Structure and petrologically important properties of silicate melts relevant to natural magmatic liquids. In *Short Course in Silicate Melts,* (editor C. M. SCARFE), Chap. 7, pp. 180–209, Mineral. Soc. Canada.

MYSEN B. O. and VIRGO D. (1978) Influence of pressure, temperature, and bulk composition on melt structures in the system $NaAlSi_2O_6$-$NaFe^{3+}Si_2O_6$. *Amer. J. Sci.* **278**, 1307–1322.

MYSEN B. O. and VIRGO D. (1980) Trace element partitioning and melt structure: An experimental study at 1 atm. pressure. *Geochim. Cosmochim. Acta* **44**, 1917–1930.

MYSEN B. O. and VIRGO D. (1983) Redox equilibria, structure and melt properties in the system Na_2O-Al_2O_3-SiO_2-Fe-O. *Carnegie Inst. Wash. Yearb.* **82**, 313–317.

MYSEN B. O. and VIRGO D. (1985) Iron-bearing silicate melts: relations between pressure and redox equilibria. *Phys. Chem. Mineral.* **12**, 191–200.

MYSEN B. O., SEIFERT F. A. and VIRGO D. (1980a) Structure and redox equilibria of iron-bearing silicate melts. *Amer. Mineral.* **65**, 867–884.

MYSEN B. O., VIRGO D. and SCARFE C. M. (1980b) Relations between the anionic structure and viscosity of silicate melts—a Raman spectroscopic study. *Amer. Mineral.* **65**, 690–710.

MYSEN B. O., RYERSON F. J. and VIRGO D. (1980c) The influence of TiO_2 on structure and derivative properties of silicate melts. *Amer. Mineral.* **65**, 1150–1165.

MYSEN B. O., VIRGO D. and KUSHIRO I. (1981a) The structural role of aluminum in silicate melts—A Raman spectroscopic study at 1 atmosphere. *Amer. Mineral.* **66**, 678–701.

MYSEN B. O., RYERSON F. and VIRGO D. (1981b) The structural role of phosphorous in silicate melts. *Amer. Mineral.* **66**, 106–117.

MYSEN B. O., VIRGO D. and SEIFERT F. A. (1982) The structure of silicate melts: Implications for chemical and physical properties of natural magma. *Rev. Geophys.* **20**, 353–383.

MYSEN B. O., VIRGO D. and SEIFERT F. A. (1984) Redox equilibria of iron in alkaline earth silicate melts: Relationships between melt structure, oxygen fugacity, temperature and properties of iron-bearing silicate liquids. *Amer. Mineral.* **69**, 834–848.

MYSEN B. O., VIRGO D. and SEIFERT F. A. (1985a) Relationships between properties and structure of aluminosilicate melts. *Amer. Mineral.* **70**, 834–847.

MYSEN B. O., VIRGO D., NEUMANN E.-R. and SEIFERT F. A. (1985b) Redox equilibria and the structural states of ferric and ferrous iron in melts in the system CaO-MgO-Al_2O_3-SiO_2: Relations between redox equilibria, melt structure and liquidus phase equilibria. *Amer. Mineral.* **70**, 317–322.

MYSEN B. O., VIRGO D., SCARFE C. M. and CRONIN D. J. (1985c) Viscosity and structure of iron- and aluminum-bearing calcium silicate melts. *Amer. Mineral.* **70**, 487–498.

NAVROTSKY A., HON R., WEILL D. F. and HENRY D. J. (1980) Thermochemistry of glasses and liquids in the systems $CaMgSi_2O_6$-$CaAl_2SiO_6$-$NaAlSi_3O_8$, SiO_2-$CaAl_2Si_2O_8$-$NaAlSi_3O_8$, and SiO_2-Al_2O_3-CaO-Na_2O. *Geochim. Cosmochim. Acta* **44**, 1409–1433.

NAVROTSKY A., PERAUDEAU P., MCMILLAN P. and COUTOURES J. P. (1982) A thermochemical study of glasses and crystals along the joins silica-calcium aluminate and silica-sodium aluminate. *Geochim. Cosmochim. Acta* **46**, 2039–2049.

NAVROTSKY A., ZIMMERMANN H. D. and HERWIG R. L. (1983) Thermochemical study of glasses in the system $CaMgSi_2O_6$-$CaAl_2SiO_6$. *Geochim. Cosmochim. Acta* **47**, 1535–1539.

NELSON B. N. and EXHAROS G. J. (1979) Vibrational spectroscopy of cation-site interactions in phosphate glasses. *J. Chem. Phys.* **71**, 2739–2747.

NELSON C. and TALLANT D. R. (1984) Raman studies of sodium silicate glasses with low phosphate contents. *Phys. Chem. Glasses* **25**, 31–39.

NOLET D. A., BURNS R. G., FLAMM S. L. and BESANCON J. R. (1979) Spectra of Fe-Ti silicate glasses: Implications to remote-sensing of planetary surfaces. *Proc. Lunar Planet. Sci. Conf. 10th,* 1775–1786.

RAY B. N. and NAVROTSKY A. (1984) Thermochemistry of charge-coupled substitutions in silicate glasses: The system $M^{n+}_{1/n}$ AlO_2-SiO_2 (M = Li, Na, K, Rb, Cs, Mg, Ca, Sr, Ba, Pb). *J. Amer. Ceram. Soc.* **89**, 606–610.

RICHET P. (1984) Viscosity and configurational entropy of silicate melts. *Geochim. Cosmochim. Acta* **48**, 471–485.

RICHET P. and BOTTINGA Y. (1985) Heat capacity of aluminum-free liquid silicates. *Geochim. Cosmochim. Acta* **49**, 471–486.

RICHET P. and BOTTINGA Y. (1986) Thermochemical properties of silicate glasses and liquids: A review. *Rev. Geophys.* **24**, 1–26.

ROBINSON H. A. (1969) Physical properties of alkali silicate glasses: I. Additive relations in alkali binary glasses. *J. Amer. Ceram. Soc.* **53**, 392–399.

RYERSON F. J. (1985) Oxide solution mechanisms in silicate melts: Systematic variations in the activity coeffient of SiO_2. *Geochim. Cosmochim. Acta* **49**, 637–651.

RYERSON F. J. and HESS P. C. (1980) The role of P_2O_5 in silicate melts. *Geochim. Cosmochim. Acta* 44, 611–625.

SACK R. O., CARMICHAEL I. S. E., RIVERS M. and GHIORSO M. S. (1980) Ferric-ferrous equilibria in natural silicate liquids at 1 bar. *Contrib. Mineral. Petrol.* **75**, 369–377.

SANDSTROM D. R., LYTLE F. W., WEI P., GREEGOR R. B., WONG J. and SCHULTZ P. (1980) Coordination of Ti in TiO_2-SiO_2 glasses by X-ray absorption spectroscopy. *J. Non-Cryst. Solids* **41**, 201–207.

SCARFE C. M. (1977) Structure of two silicate rock melts charted by infrared absorption spectroscopy. *Chem. Geol.* **15**, 77–80.

SEIFERT F. A., VIRGO D. and MYSEN B. O. (1979) Melt structures and redox equilibria in the system Na_2O-FeO-Fe_2O_3-$Al_2O_3SiO_2$. *Carnegie Inst. Wash. Yearb.* **78**, 511–519.

SEIFERT F. A., MYSEN B. O. and VIRGO D. (1982) Three-dimensional network melt structure in the systems SiO_2-$NaAlO_2$, SiO_2-$CaAl_2O_4$ and SiO_2-$MgAl_2O_4$. *Amer. Mineral.* **67**, 696–711.

TAYLOR M. and BROWN G. E. (1979a) Structure of mineral glasses. II. The SiO_2-$NaAlSiO_4$ join. *Geochim. Cosmochim. Acta* **43**, 1467–1475.

TAYLOR M. and BROWN G. E. (1979b) Structure of mineral glasses. I. The feldspar glasses $NaAlSi_3O_8$, $KAlSi_3O_8$, $CaAl_2Si_2O_8$. *Geochim. Cosmochim. Acta* **43**, 61–77.

THORNBER C. R., ROEDER P. L. and FOSTER J. R. (1980) The effect of composition on the ferric-ferrous ratio in basaltic liquids at atmospheric pressure. *Geochim. Cosmochim. Acta* **44**, 525–533.

TOBIN M. C. and BAAK T. (1968) Raman spectra of some low-expansion glasses. *J. Opt. Soc. Amer.* **58**, 1459–1460.

VERWEIJ H. (1979a) Raman study of the structure of alkali germanosilicate glasses. I. Sodium and potassium metagermanosilicate glasses. *J. Non-Cryst. Solids* **33**, 41–53.

VERWEIJ H. (1979b) Raman study of the structure of alkali germanosilicate glasses. II. Lithium, sodium and potassium digermanosilicates glasses. *J. Non-Cryst. Solids* **33**, 55–69.

VIRGO D. and MYSEN B. O. (1985) The structural state of iron in oxidized vs. reduced glasses at 1 atm.: a ^{57}Fe Mossbauer study. *Phys. Chem. Mineral.* **12**, 65–76.

VIRGO D., MYSEN B. O. and KUSHIRO I. (1980) Anionic constitution of silicate melts quenched at 1 atm from Raman spectroscopy: Implications for the structure of igneous melts. *Science* **208**, 1371–1373.

VIRGO D., MYSEN B. O. and SEIFERT F. A. (1981) Relationship between the oxidation state of iron and structure of silicate melts. *Carnegie Inst. Wash. Yearb.* **80**, 308–311.

VISSER W. and VAN GROOS A. F. KOSTER (1979) Effects of P_2O_5 and TiO_2 on the liquid-liquid equilibria in the system K_2O-FeO-Al_2O_3-SiO_2. *Amer. J. Sci.* **279**, 970–988.

WATSON E. B. (1976) Two-liquid partition coefficients: Experimental data and geochemical implications. *Contrib. Mineral. Petrol.* **56**, 119–134.

Magmatic Processes: Physicochemical Principles
© The Geochemical Society, Special Publication No. 1, 1987
Editor B. O. Mysen

Determination of the mixing properties of granitic and other aluminosilicate melts by Knudsen Cell Mass Spectrometry

DONALD G. FRASER

Department of Earth Sciences, University of Oxford, Parks Road, Oxford OX1 3PR, U.K.

and

WERNER RAMMENSEE

Mineralogisch-Petrologisches Institut, Universität Gottingen, Goldschmidstrasse 1, 3400 Göttingen, F.R.G.

Abstract—The entropies, enthalpies and free energies of mixing of melts in the system NaAlSiO₄–
KAlSiO₄–SiO₂ have been determined by Knudsen Cell Mass Spectrometry (KCMS). The enthalpies
of mixing have an accuracy and precision $(1-\sigma)$ of around 1.0 and 0.5 KJ/mol, respectively (FRASER
and BOTTINGA, 1985), and are thus comparable with the best solution calorimetric data available
for glasses. The mixing properties of the melts are related to the topology of the underlying liquidus
surface so that marked changes in the entropy, enthalpy and free energy of mixing of the melts
accompany changes in the nature of the crystalline phase in equilibrium with the melt at lower
temperatures. These data and the results of X–ray diffraction, NMR and Raman spectroscopic studies
of glasses of similar composition indicate that clustering may occur in the melts to produce regions
which differ in structure and composition on the approximately 20 Å scale. The equilibrium liquids
are thus inhomogeneous on the micro–scale and may contain structures more closely related to the
crystalline phase on the liquidus than to a mineral phase of the same composition as the liquid.

INTRODUCTION

ACCURATE MEASUREMENTS of the activities of silicate components in magmas and their interpretation in terms of molecular structure are essential for the construction of predictive models for a wide range of igneous phenomena. Several independent thermodynamic studies of the mixing properties of melts and glasses in the system NaAlSiO₄–KAlSiO₄–SiO₂ (Ne–Ks–Qz) have recently been made using a range of experimental techniques. These include vapour pressure measurements made on the melts themselves at high temperature by KCMS (RAMMENSEE and FRASER, 1982; FRASER et al., 1983, 1985; ROGEZ et al., 1983), HF and borate melt solution calorimetry on quenched and annealed glasses (ROGEZ et al., 1983; HENRY et al., 1982; HERVIG and NAVROTSKY, 1984; HOVIS, 1984) and drop calorimetry on melts and glasses (STEBBINS et al., 1983; RICHET and BOTTINGA, 1984). In addition, measurements of the structures of glasses and melts in this and allied systems have been made by X–ray diffraction (TAYLOR and BROWN, 1979a, 1979b), Raman spectroscopy (MYSEN et al., 1981; MCMILLAN et al., 1982) and NMR spectroscopy (MURDOCH et al., 1985; STEBBINS et al., 1985; FRASER and CLAYDEN, 1986) so that considerable structural information is also available.

Estimates of the activities of silicate components in melts have been obtained until recently mainly from depression of freezing-point calculations (e.g., WOOD and FRASER, 1976; BOTTINGA and RICHET, 1978). These suffer from large inaccuracies resulting from the combination of uncertainties in the positions of liquidus surfaces and in the heats of fusion used in the calculations. In the present series of measurements we set out to determine the activities of silicate components in melts directly by measuring their equilibrium vapour pressures by KCMS.

The purpose of the present paper is to review the application of KCMS to the study of the thermodynamic properties of silicate melts, to report the entropies of mixing of NaAlSi₃O₈–KAlSi₃O₈ melts and to comment on possible relationships between these measurements and the atomic and molecular structures of the melts.

METHOD

The measurement of activities in solution by KCMS is a standard technique in modern physical chemistry. It is based on the fundamental relationship between activity and equilibrium vapour pressure:

$$a_i = f_i / f_i^0 = p_i / p_i^0 \qquad (1)$$

where a_i is the activity of component i, and f_i and f_i^0, p_i and p_i^0 are the fugacities and partial pressures of the component over a solution and standard state respectively.

For the very low partial pressures of components in equilibrium with most silicate melts the fugacity is equal to the partial pressure. Thus, by measuring the equilibrium vapour pressures of silicate melts, the activities of the silicate components may be determined directly using equation (1).

The volatilization of Na and, to a lesser extent, K, from silicate melts and glasses is well known to experimental petrologists (e.g., KRACEK, 1930; O'HARA et al., 1970) and to analysts of glasses using the electron microprobe. The volatility and the absolute Na/K ratio of the vapour depend on the temperature, oxygen activity and chemical properties of the melt or glass in question. Since silicate melts thus vaporize incongruently, great care was taken in all our experiments to identify all possible species present and to consider the effects of these on the results. In the case of the alkali aluminosilicate melts considered in the present paper, we have shown (RAMMENSEE and FRASER, 1982) that at temperatures up to about 1600°C the vapour species consist almost entirely of monatomic Na and K atoms (Figure 1). Measurements made with low ionization energies in the ionization chamber failed to show any evidence of Na– or K–oxide species and so the possibility of fragmentation of oxide species in the ionization chamber can be excluded. At the highest temperatures studied, very small amounts of SiO are measurable.

The equilibrium vapour pressures of alkali aluminosilicate melts are the sums of the partial pressures of all the species present. Thus the vapour pressures of, for example, $NaAlSi_3O_8$ and $KAlSi_3O_8$ melts are given by:

$$P_{NaAlSi_3O_8} = p_{Na} + p_{NaO} + p_O + p_{SiO} + \cdots \quad (2)$$

$$P_{KAlSi_3O_8} = p_K + p_{KO} + p_O + p_{SiO} + \cdots \quad (3)$$

Because the vapours above melts in this system are composed almost entirely of monatomic Na and K atoms and the oxygen fugacity is buffered at low values by the Mo Knudsen cells, then the equilibrium vapour pressures of pure $NaAlSi_3O_8$ and $KAlSi_3O_8$ melts are to a first approximation simply the partial pressures of Na and K:

$$P_{NaAlSi_3O_8} = p_{Na} \quad (4)$$

$$P_{KAlSi_3O_8} = p_K. \quad (5)$$

The activities of the $NaAlSi_3O_8$ and $KAlSi_3O_8$ components in mixed melts can therefore be obtained by combining equations (4) or (5) with (1)

$$a_{NaAlSi_3O_8} = p_{Na}/p_{Na}^0 \quad (6)$$

$$a_{KAlSi_3O_8} = p_K/p_K^0 \quad (7)$$

where p_{Na} and p_K are the partial pressures of Na and K above the mixed melts and p_{Na}^0 and p_K^0 are the partial pressures of Na and K above pure molten $NaAlSi_3O_8$ and $KAlSi_3O_8$ respectively.

It should be noted that although these melts vaporize incongruently to give vapours consisting almost entirely of alkali metal atoms, the activities obtained by taking the ratio of the partial pressure of an alkali metal atom above a mixed melt to that of the same alkali metal atom species in equilibrium with an aluminosilicate melt as standard state, are the activities of the aluminosilicate components and not, for example, Na metal or Na_2O. Activities relative to the latter standard states could be obtained if, for example, the partial pressures were compared with those above pure liquid Na metal or Na_2O respectively.

In the Knudsen effusion technique (KNUDSEN, 1909), equilibrium partial pressures are determined by measuring the flux of vapour species through a minute aperture in an inert container (the Knudsen cell) into an evacuated

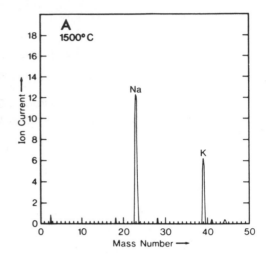

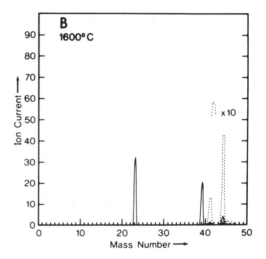

FIG. 1. Mass spectra for the 50:50 (mol) composition in the system $NaAlSi_3O_8$–$KAlSi_3O_8$ in arbitrary units of ion current. The spectra are dominated by ^{23}Na and ^{37}K. Minor amounts of ^{41}K can also be seen, and at higher temperatures small peaks due to ^{28}SiO, ^{29}SiO and ^{30}SiO are visible.

space. So long as the area of the effusion aperture is small relative to the surface area of the lid, which is bombarded inside the cell by the vapour molecules thus setting up the intrinsic vapour pressure, the rate of effusion is related to the equilibrium vapour pressure inside the cell by the Herz-Knudsen equation:

$$dN_i/dt = (p_i A K)/(2\pi R M_i T)^{1/2} \quad (8)$$

where N_i is the number of moles of component i evaporated in time t, p_i is the partial pressure, A the orifice area, K the Clausing emission efficiency factor, R is the gas constant, M_i the molecular weight, and T the temperature (PAULE and MARGRAVE, 1976).

In our experiments, the fluxes of vapour components in the resulting molecular beam were determined using a quadrupole mass spectrometer and the partial pressures obtained from the measured ion currents using the relationship (CHATILLON et al., 1975):

$$p_i = (I_i^+ T)/S_i \qquad (9)$$

where I_i^+ is the ion current of species i and S_i is the sensitivity of the spectrometer to species i.

EXPERIMENTAL

The apparatus used for making these measurements consists of a rotatable tantalum block containing four Mo Knudsen cells. The block can be heated by a Ta rod furnace to 1650°C. The molecular beam of Na and K atoms effusing from each cell passes through a rotating beam-chopper to enable automatic subtraction of background by distinguishing between atoms which effuse directly from the cells and those originating above the chopper from elsewhere in the apparatus. The chopped beam then passes into the ionization chamber of a quadrupole mass spectrometer. The apparatus has been described in detail in previous papers (RAMMENSEE and FRASER, 1981; FRASER and RAMMENSEE, 1982; FRASER et al., 1985) and a diagram is shown in Figure 2. Gel starting materials used in our earlier experiments were replaced by pre-melted mixed glasses to improve equilibration times and temperatures were measured using Pt6%Rh/Pt30%Rh thermocouples calibrated by observing changes in the vapour pressure of pure Ag at the melting point (FRASER et al., 1985).

Considerable care was taken in all experiments to ensure that equilibrium was achieved. Vapour pressure measurements were always made in both increasing and decreasing temperature sequences so that the equilibrium value was approached from both sides. In addition, because the Knudsen effusion technique depends on the loss of small amounts of material from the system, it is important to show that no systematic changes in sample composition occurred during the experiments. For this reason, the temperatures at which measurements were made were non–sequential as is shown in Figure 3, so that any change in sample composition would be immediately visible as a systematic error. Some samples quenched at the end of an experiment were also sectioned longitudinally and analyzed for Na and K using the electron microprobe. No significant surface depletion effects could be observed. The consistency of measurements made throughout any experiment and the homogeneity of glass compositions analyzed by the electron microprobe imply that no significant changes in sample composition occurred during these experiments. This conclusion is in accord with calculations of the very small weight loss which occurs during an experiment and the high diffusivities of Na and K in these melts (JAMBON, 1982) which allow any surface loss by evaporation to be immediately restored. Even for an average vapour pressure of 10^{-5} torr and the most unfavourable case in which the Clausing emission factor is 1, the total loss of material during a typical six to ten hour experiment is of the order of only 3 μg. A detailed description of the experimental technique is given in previous papers (RAMMENSEE and FRASER, 1982; FRASER et al., 1985).

MEASUREMENTS IN THE SYSTEM NaAlSiO₄–KAlSiO₄–SiO₂

We have investigated the mixing properties of four binary and pseudobinary joins in the system

$NaAlSiO_4–KAlSiO_4–SiO_2$ having fixed Si/Al ratios, and a series of compositions on the join $Na_{0.7}K_{0.3}AlSiO_4–SiO_2$ (RAMMENSEE and FRASER, 1982, 1986; FRASER et al., 1983, 1985) as shown in Figure 4. In addition, the heats of mixing of quenched and annealed glasses in this system have been measured by HF and high temperature oxide melt solution calorimetry (ROGEZ et al., 1983; HOVIS, 1984; HENRY et al., 1982; HERVIG and NAVROTSKY, 1984) and these have been reviewed recently by FRASER and BOTTINGA (1985).

In the Knudsen effusion experiments, the Na/K ratios of vapours in equilibrium with mixed Na, K aluminosilicate melts were measured as functions of bulk composition. Typical results for the system $NaAlSi_5O_{12}–KAlSi_5O_{12}$ which is near the granite minimum at 1 atmosphere pressure in the system $NaAlSiO_4–KAlSiO_4–SiO_2$ (SCHAIRER 1957), are shown in Figure 3. Note that results were obtained both down– and up–temperature as shown by the run numbers. The activities of the aluminosilicate components can, in principle, be obtained by comparing the partial pressures of Na and K above the melts with those measured above pure liquid $NaAlSi_5O_{12}$ and $KAlSi_5O_{12}$ respectively. However such "direct" activity measurements are experimentally very difficult because small differences in the temperature and size and shape of the Knudsen aperture between sample and reference, quickly lead to significant errors when the ratios are taken. These problems are very greatly reduced if the variation of the Na/K ratio with bulk composition is determined instead of the individual ion currents of Na and K. The activities and activity coefficients of the end member components may then be extracted from the rates of change with composition by applying the Gibbs-Duhem equation:

$$\ln a_{NaAlSi_3O_8} = -\int_{X_K=0}^{X_K=X_K} X_K d\ln(I_K/I_{Na}) \qquad (10)$$

$$\ln \gamma_{NaAlSi_3O_8} = -\int_{X_K=0}^{X_K=X_K} X_K d\ln(I_K/I_{Na} \cdot X_{Na}/X_K). \qquad (11)$$

In addition, because the temperature dependence of the ion current ratios is related to the heat of mixing, enthalpies of mixing may be obtained from the composition dependence of the slopes of data of the sort shown in Figure 3:

$$\Delta h_1 = -R \int_{X_2=0}^{X_2=X} X_2 d(\partial \ln(I_2/I_1)/1/T) \qquad (12)$$

where Δh_1 is the partial molar enthalpy of mixing of component 1.

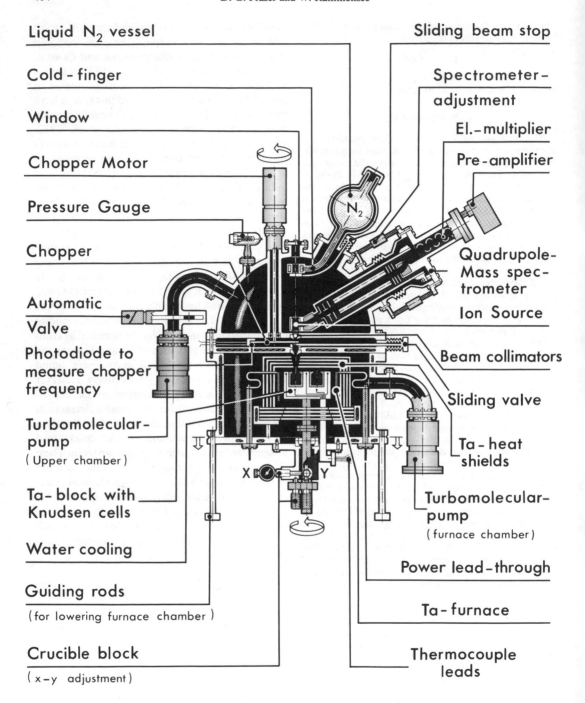

Liquid N₂ vessel

Cold - finger

Window

Chopper Motor

Pressure Gauge

Chopper

Automatic
Valve

Photodiode to
measure chopper
frequency

Turbomolecular-
pump
(Upper chamber)

Ta- block with
Knudsen cells

Water cooling

Guiding rods
(for lowering furnace chamber)

Crucible block
(x–y adjustment)

Sliding beam stop

Spectrometer-
adjustment

El.-multiplier

Pre-amplifier

Quadrupole-
Mass spec-
trometer

Ion Source

Beam collimators

Sliding valve

Ta - heat
shields

Turbomolecular-
pump
(furnace chamber)

Power lead-through

Ta- furnace

Thermocouple
leads

Knudsen Cell – Mass Spectrometer

FIG. 2. Diagram of the Knudsen Cell Mass Spectrometer apparatus.

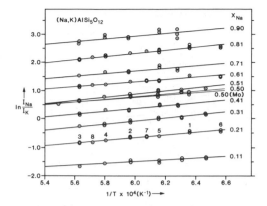

FIG. 3. Variations in ln (I_{Na}/I_K) with $1/T$. Compositions are indicated at the right hand side. Results were obtained both up– and down–temperature as a check on possible systematic errors occurring during each run and the numbers indicate the order in which the measurements were made. The data marked (Mo) were obtained in an experiment in which a Mo plate was inserted inside the Knudsen cell so as to block the direct line of flight of vapour species from the melt through the Knudsen orifice.

RESULTS

Heats of mixing

The partial molar enthalpies of mixing of melts on the joins $NaAlSi_2O_6$–$KAlSi_2O_6$ (FRASER et al., 1983), $NaAlSi_3O_8$–$KAlSi_3O_8$ (RAMMENSEE and FRASER, 1982) $NaAlSi_4O_{10}$–$KAlSi_4O_{10}$ and $NaAlSi_5O_{12}$–$KAlSi_5O_{12}$ (FRASER et al., 1985) have been determined from the temperature dependence of the Na/K ion current ratios using Equation (12) and have been reported previously. In addition, the Knudsen cell data of ROGEZ et al. (1983) for the system $NaAlSi_3O_8$–$KAlSi_3O_8$ have been analyzed to obtain an independent set of enthalpy measurements (FRASER and BOTTINGA, 1985). These data are summarized in Figure 4. The enthalpies of mixing along the three most silicic joins are all close to zero. The $NaAlSi_4O_{10}$–$KAlSi_4O_{10}$ and $NaAlSi_5O_{12}$–$KAlSi_5O_{12}$ joins have slight negative heats of mixing with symmetrical minima at the $X_{Na} = 0.5$ compositions. Heats of mixing on the $NaAlSi_3O_8$–$KAlSi_3O_8$ join vary with Na/K ratio. They are slightly negative in Na–rich compositions, but show significant positive enthalpies of mixing at the potassium rich side of the join, with a maximum at $X_K = 0.8$. The two independent Knudsen cell studies are in excellent agreement (FRASER and BOTTINGA, 1985) even though substantially different types of apparatus were used with Mo Knudsen cells used in one set of measurements and Pt cells in the other. Comparison of the two sets of data indicates precisions ($\pm 1\sigma$) of the order of 500 J and an ac-

curacy of better than 1.0 KJ/mol. The precision of our more recent experiments is a factor of two better (FRASER et al., 1985).

The heats of mixing obtained from these Knudsen cell studies may be compared with the heats of mixing obtained from calorimetric studies of glasses in the system $NaAlSi_3O_8$–$KAlSi_3O_8$ (HERVIG and NAVROTSKY, 1984; HOVIS, 1984). However this can only be done if corrections are applied to take account of the differences between the properties of glasses and liquids. In particular, enthalpies of mixing measured by solution calorimetry on glasses, must be corrected for thermal effects associated with the glass transition. This can only be done at present for one composition in the system $NaAlSi_3O_8$–$KAlSi_3O_8$ for which measurements of the fictive temperature are available. The relevant calculations indicate (FRASER and BOTTINGA, 1985) that for the $X_{Na} = 0.5$ composition in the system $NaAlSi_3O_8$–$KAlSi_3O_8$, around 2.0 KJ/mol must be added to the enthalpy of mixing reported for the particular glasses used for the calorimetric measurements. The absolute value of this correction varies with the thermal history of the glasses and with bulk composition. When this correction is made, the calorimetric data on glasses are consistent with the Knudsen cell measurements made on the equilibrium liquids at high temperature. Accordingly, care should be taken in using data obtained from glasses below their fictive temperatures to apply to the properties of melts unless measurements of the fictive temperatures of the glasses are available for all compositions and the associated thermal corrections are made.

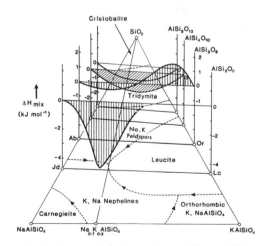

FIG. 4. Enthalpies of mixing for four different joins in the system $NaAlSiO_4$–$KAlSiO_4$–SiO_2. The values on the $NaAlSi_2O_6$–$KAlSi_2O_6$ join refer to the partial molar enthalpy of mixing of the $NaAlSi_2O_6$ component.

Free energies and entropies of mixing

The free energies of mixing of binary alumino-silicate melts may be obtained from the vapour pressure data using Equations (10) or (11) and excess entropies of mixing may then be calculated by combining these values with the heats of mixing derived from Equation (12). The free energies of mixing of melts in the system NaAlSiO$_4$–KAlSiO$_4$–SiO$_2$ have been previously reported (RAMMENSEE and FRASER, 1982; FRASER *et al.*, 1985) and behave in a closely similar way to the enthalpies of mixing shown in Figure 4 and discussed above. Melts on the two most siliceous joins examined mix essentially ideally and their excess free energies of mixing vary symmetrically with composition. The maximum values at the $X_{Na} = 0.5$ composition are 70 ± 160 J mol^{-1} on the NaAlSi$_4$O$_{10}$–KAlSi$_4$O$_{10}$ join and −240 ± 160 J/mol for NaAlSi$_5$O$_{12}$–KAlSi$_5$O$_{12}$ join. The excess free energies of mixing of NaAlSi$_3$O$_8$–KAlSi$_3$O$_8$ melts are more interesting. Significant positive deviations from ideality occur throughout with a maximum at the K–rich side of the join (RAMMENSEE and FRASER, 1982, ROGEZ *et al.*, 1983). This tendency to unmixing may be related to the structural mismatch that causes unmixing and perthite formation in the solid state. However, X–ray determinations of the radial distribution functions RDFs of KAlSi$_3$O$_8$ glass (TAYLOR and BROWN, 1979a) suggest a closer similarity to the structures of leucite or tridymite than that of sanidine, and the positive deviation from ideality in the melts is more probably related to the incongruent melting of sanidine at low pressures and the resulting large stability field of leucite and this will be discussed below.

Excess entropies of mixing of melts on the four joins can be calculated by combining the results obtained using Equations (11) and (12) and are shown in Figure 5. The excess entropies of mixing are similar in form to the heats of mixing shown in Figure 4. The NaAlSi$_4$O$_{10}$–KAlSi$_4$O$_{10}$ and NaAlSi$_5$O$_{12}$–KAlSi$_5$O$_{12}$ joins have small symmetrical excess entropies of mixing (FRASER *et al.*, 1985). Melts on the join NaAlSi$_3$O$_8$–KAlSi$_3$O$_8$ show significant positive excess entropies of mixing (ΔS^{xm}) at the potassic side of the join. Similar behaviour was observed for the high–temperature alkali feldspars (THOMPSON and HOVIS, 1979). In addition, measurements of the densities of mixed NaAlSi$_3$O$_8$–KAlSi$_3$O$_8$ glasses show a pronounced excess volume of mixing (HAYWARD, 1977) with a maximum near $X_K = 0.8$ and the viscosities of KAlSi$_3$O$_8$ melts have different Arrhenian activation

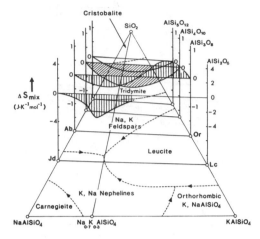

FIG. 5. Entropies of mixing for four different joins in the system NaAlSiO$_4$–KAlSiO$_4$–SiO$_2$. The values on the NaAlSi$_2$O$_6$–KAlSi$_2$O$_6$ join refer to the partial molar entropy of mixing of the NaAlSi$_2$O$_6$ component. Note the different scale for the latter compositions.

energies below and above 1550°C (URBAIN *et al.*, 1982).

The positive non–ideal entropies of mixing shown in Figure 5 cannot be interpreted on the basis of non–random clustering of Na and K ions on sites within a constant aluminosilicate network because such an effect could only produce negative values of ΔS^{xm}. Any negative contribution to ΔS^{xm} must therefore be exceeded by the effect of a positive excess in the heat capacity or, if medium range ordering effects lead to the formation of more than one different sub–network structure in the melts, additional configurational entropy terms may result. In the case of the high temperature crystalline alkali feldspars, it was suggested (THOMPSON and HOVIS, 1979) that the asymmetry in the positive values of ΔS^{xm} is consistent with the argument that the smaller Na ion would have greater vibrational freedom on a large–ion site than vice versa. The values of ΔS^{xm} reported for the NaAlSi$_3$O$_8$–KAlSi$_3$O$_8$ join in Figure 5 have a maximum positive value of 0.64 J K^{-1} mol^{-1} at the $X_K = 0.8$ composition and are thus smaller than the maximum of around 3.3 J K^{-1} mol^{-1} reported by THOMPSON and HOVIS (1979) for the feldspars taking ΔH^{xm} to be asymmetrical. This difference is unlikely to be caused by the difference in temperature between the low temperature experiments used by THOMPSON and HOVIS (1979) and the high temperatures at which the melt data were obtained because ΔC_p^{xm} is probably positive as noted above. However, mixing in the melts is complicated by the additional degrees of freedom which allow variations in the aluminosilicate network structure. The RDFs reported by TAYLOR and BROWN (1979a,b) indicate that there is a close correspondence between the local structures of NaAlSi$_3$O$_8$ glass and crystalline high albite out to 4 Å. This includes the Si–O(1), Si–Si(1) and Si–O(2) vectors. The peak at 4.0–4.5 Å characteristic of the second nearest neighbour Si–Si(2) distance across the four–membered rings of the feldspar structure is still present, but is weaker in the glass RDF whereas a new peak appears in the glass structure at around 5.1 Å. There are no interatomic distances in the high albite structure which have this value. This distance correlates instead with the Si–Si(2) distance across the six–membered rings of a stuffed tridymite structure. The correspondence between the RDFs of KAlSi$_3$O$_8$ glass and sanidine is less good. There is a clear minimum in the KAlSi$_3$O$_8$ glass radial distribution function at 4.5 Å where there is a maximum for sanidine, whereas the reverse is true of a maximum in the glass RDF at around 5.1 Å. The minimum agrees well with a minimum at 4.5 Å in the leucite structure which contains both four and six–membered rings of (Si,Al)O$_4$ tetrahedra. However, at larger distances the correspondence is weaker. Neither the leucite nor a stuffed tridymite structure describes all the features of the glass X–ray data. The glass and hence presumably also melts, may therefore be inhomogeneous on the scale of 20 Å containing regions of leucite–like and stuffed tridymite structures. Other ring configurations such as five–membered rings may also be present. This view which we originally proposed to explain the positive heats of mixing observed in NaAlSi$_3$O$_8$–KAlSi$_3$O$_8$ melts (RAMMENSEE and FRASER, 1982) is further supported by a recent Raman spectral study of KAlSi$_3$O$_8$ glasses (MATSON et al., 1986) which indicated regions of leucite–like and tridymite–like structure in KAlSi$_3$O$_8$ glasses. Similar ideas involving the presence of 4– and 6–membered rings have been proposed by MAMMONE et al. (1981) and SEIFERT et al. (1982) who have suggested that the proportion of 4–membered rings increases with increasing Al/(Al + Si) ratio in the melt.

The Raman spectra of NaAlSi$_3$O$_8$ and KAlSi$_3$O$_8$ glasses show small differences with the weak band at 1100 cm^{-1} being more prominent in the KAlSi$_3$O$_8$ composition (MCMILLAN et al., 1982). It has been suggested by the latter authors that these differences indicate increased clustering of aluminosilicate groups in the Na-bearing system than in KAlSi$_3$O$_8$ glasses and similar conclusions were reached by FLOOD and KNAPP (1968) based on

cryoscopic arguments. Using the $Q^n(x$Al)–notation familiar from NMR spectroscopy in which n represents the number of bridging oxygen atoms around the Si, and x the number of next–nearest neighbour Al atoms, this can be represented by:

$$Q^4 + Q^4(2\text{Al}) = 2Q^4(1\text{Al}). \tag{13}$$

Replacement of K by Na moves the equilibrium distribution to the left forming Al–rich and Si–rich clusters in the melt. This is similar to the mechanism proposed recently to explain deviations in the ^{29}Si NMR spectra of NaAlSi$_3$O$_8$–CaMgSi$_2$O$_6$ glasses (FRASER and CLAYDEN, 1986).

The excess entropies of mixing of NaAlSi$_2$O$_6$ and KAlSi$_2$O$_6$ melts stand in marked contrast to those on the more SiO$_2$–rich joins. NaAlSi$_2$O$_6$–KAlSi$_2$O$_6$ melts show a pronounced negative partial molar entropy of mixing of the NaAlSi$_2$O$_6$ component with a maximum deviation at the Na–rich side near the Na:K = 3:1 composition. The most likely interpretation of this negative excess entropy of mixing is that Na–K clustering takes place in these liquids (FRASER et al., 1983). The coincidence of the maximum negative value with the 3:1 composition suggests that the melts may contain regions similar in structure to that of Na$_3$K–nepheline which is stable on the liquidus on this join because of the incongruent melting of jadeite.

DISCUSSION

The enthalpies and entropies of Na–K mixing shown in Figures 4 and 5 carry important implications for the structures and general mixing properties of aluminosilicate melts and hence the theoretical modelling of crystal–liquid equilibrium in these systems.

On the basis of the mixing properties of the melts, the system NaAlSiO$_4$–KAlSiO$_4$–SiO$_2$ may be considered in three parts:

(1) Compositions near the join NaAlSi$_2$O$_6$–KAlSi$_2$O$_6$ where strongly negative partial molar entropies and enthalpies of mixing are observed near the Na:K = 3:1 composition.

(2) Compositions near the feldspar join on which a significant positive deviation from ideality is observed at the potassic side.

(3) More SiO$_2$–rich liquids near the 1–atmosphere granite minimum which mix nearly ideally and for which no abrupt changes in mixing properties are observed.

If these groups are compared with the ternary phase diagram (SCHAIRER, 1957), it can be seen that the mixing properties observed in the equilibrium liquids at high temperatures above the liquidus are broadly correlated with changes in the nature of the crystalline phase at lower temperatures. Somewhat similar suggestions have been made previously by us (RAMMENSEE and FRASER, 1982; FRASER et al., 1983) and by Burnham (BURNHAM, 1981; BURNHAM and NEKVASIL, 1986, 1987). Both jadeite and sanidine melt incongruently at low pressure so that the joins $NaAlSi_2O_6$–$KAlSi_2O_6$ and $NaAlSi_3O_8$–$KAlSi_3O_8$ are pseudobinaries and are each cut by a large field representing the crystalline phase of lowest free energy which is in equilibrium with these liquids. The $NaAlSi_2O_6$–$KAlSi_2O_6$ join is cut by the primary phase volume of Na,K–nephelines. X–ray diffraction studies of crystalline Na,K–nepheline indicate a structure containing small and large cation sites in the ratio 3:1 so that the equilibrium composition is close to $Na_3K(AlSiO_4)_4$ (BUERGER et al., 1954). The pronounced negative heats and entropies of mixing observed on the $NaAlSi_2O_6$–$KAlSi_2O_6$ join near the Na_3K composition indicate that ordering also occurs in the liquids so that the structures of Na–rich liquids on the $NaAlSi_2O_6$–$KAlSi_2O_6$ join probably contain regions of nepheline–like structure rather than that of jadeite or leucite.

The join $NaAlSi_3O_8$–$KAlSi_3O_8$ shows different but analogous behaviour. Here the sharp change in mixing properties and the positive enthalpy and entropy of mixing correlate qualitatively, but not exactly, with the composition at which the join is cut by the primary phase volume of leucite, the break in mixing properties indicated by the Knudsen cell studies of vapour pressures occurring near $X_K = 0.8$. The positive enthalpy, entropy and free energy of mixing at the potassic side of the join are supported by the positive excess volumes of mixing of glasses in the system (HAYWARD, 1977) and by the analogous behaviour of the high temperature feldspars which show positive excess entropies of mixing (ΔS^{xm}) at the potassic side of the join (THOMPSON and HOVIS, 1979). Radial distribution function studies of $KAlSi_3O_8$ glasses show a clear minimum at 4.5 Å. This is consistent with a leucite–like structure for $KAlSi_3O_8$ glass and not with a sanidine–like structure although features at higher radial distances suggest a stuffed tridymite structure (TAYLOR and BROWN, 1979a,b). We, therefore, previously suggested (RAMMENSEE and FRASER, 1982) that $KAlSi_3O_8$ melts are inhomogeneous and contain medium-range domains of leucite-like structure composed of 4– and 6–membered rings on the approximately 20 Å scale linked to regions of tridymite-like structure consisting of cross-linked

sheets of 6–membered rings. Other sub–network structures are also possible, but the whole linked three–dimensional network must be homogeneous on the optical scale. Recent support for this model has been provided by the Raman spectral study of MATSON et al. (1986) which indicates T–O–T modes which correspond with regions of leucite-like and tridymite–like structure. The effect of adding potassium across the join seems to be to stabilize 4–membered rings and a similar effect has been noted in vitreous silica in which 4–rings are stabilized by broken Si–O defect centres (SHARMA et al., 1981).

Finally the two silicic joins which lie almost entirely within the single–phase mixed feldspar field in which no change of crystallizing phase occurs on the liquidus show no analogous breaks in mixing properties.

CONCLUSIONS

The overriding conclusion of these studies is that the structures of aluminosilicate melts may bear little relationship to those of minerals of the same composition. This is seen most clearly in the case of the minerals which melt incongruently in this system, jadeite and sanidine. Melts on the $NaAlSi_2O_6$–$KAlSi_2O_6$ join are probably composed of inter–linked regions of nepheline–like and tridymite–like structure. Melts on the $NaAlSi_3O_8$–$KAlSi_3O_8$ join are unlike the feldspars, but contain regions of tridymite–like and, at the potassic side, leucite–like structure. The implication of these models is that aluminosilicate melts are inhomogeneous on the medium–range scale and may contain two or more topologically distinct domains linked to form an optically homogeneous three-dimensional network. It is likely that the crystal chemical factors which determine the structures of minimum free energy in the liquid operate similarly to determine the crystalline phase of minimum free energy on the liquidus. The consequence of these models is that the sharp changes in mixing properties observed in the melts at high temperature above the liquidus may often occur at similar compositions to those at which a change occurs in the stable phase on the liquidus.

The development of models which relate the thermodynamic properties of silicate melts to their structures requires, in particular, accurate measurements of the entropies and free energies of mixing. KCMS allows these quantities to be determined directly together with the enthalpies of mixing. The latter can be shown to have accuracy and precision comparable to or slightly better than the best measurements of the enthalpies of mixing of glasses by solution calorimetry currently available. In addition, because measurements are made at high temperature on the melts themselves, KCMS data are not affected by problems related to the glass transition or quenching.

The data now available for the system NaAlSiO$_4$–KAlSiO$_4$–SiO$_2$ suggest that medium range clustering occurs in the melts. These melts are probably inhomogeneous on the approximately 20 Å scale and may contain local structures which, in part, resemble the crystalline phase which is in equilibrium with a melt of given composition rather than that necessarily of a crystalline phase of the same composition. The distinction between the latter two is particularly important in systems containing phases which melt incongruently.

Acknowledgements—We thank Bjorn Mysen and Bob Luth for their helpful reviews and the N.E.R.C. (D.G.F.) and D.F.G. (W.R.) for financial support.

REFERENCES

BOTTINGA Y. and RICHET P. (1978) Thermodynamics of liquid silicates, a preliminary report. *Earth Planet. Sci. Lett.* **40**, 382–400.

BUERGER M. J., KLEIN G. E. and DONNAY G. (1954) Determination of the crystal structure of nepheline. *Amer. Mineral.* **39**, 805–818.

BURNHAM W. C. (1981) The nature of multicomponent aluminosilicate melts. *Phys. Chem. Earth* **13**, 197–229.

BURNHAM W. C. and NEKVASIL H. (1986) Equilibrium properties of granite pegmatite magmas. *Amer. Mineral.* **71**, 239–263.

CHATILLON C., PATTORET A. and DROUWART J. (1975) Etudes thermodynamiques des phases condensées par spectrometrie de masse á haute temperature: Analyse de la méthode et revue des résultats. *High Temp.–High Press.* **7**, 119–148.

FLOOD H. and KNAPP W. J. (1968) Structural characteristics of liquid mixtures of feldspar and silica. *J. Amer. Ceram. Soc.* **51**, 259–263.

FRASER D. G. and RAMMENSEE W. (1982) Activity measurements by Knudsen Cell Mass Spectrometry—the system Fe–Co–Ni and implications for condensation processes in the solar nebula. *Geochim. Cosmochim. Acta* **46**, 549–556.

FRASER D. G. and BOTTINGA Y. (1985) The mixing properties of melts and glasses in the system NaAlSi$_3$O$_8$–KAlSi$_3$O$_8$: Comparison of experimental data observed by Knudsen Cell Mass Spectrometry and solution calorimetry. *Geochim. Cosmochim. Acta* **49**, 1377–1381.

FRASER D. G. and CLAYDEN N. J. (1986) A high resolution ^{29}Si nuclear magnetic resonance study of ordering in silicate glasses on the join CaMgSi$_2$O$_6$–NaAlSi$_3$O$_8$. *Chem. Geol.* (In press).

FRASER D. G., RAMMENSEE W. and JONES R. (1983) The mixing properties of melts in the system NaAlSi$_2$O$_6$–KAlSi$_2$O$_6$ determined by Knudsen Cell Mass Spectrometry. *Bull. Mineral.* **106**, 111–117.

FRASER D. G., RAMMENSEE W. and HARDWICK A. (1985) Determination of the mixing properties of molten sili-

cates by Knudsen Cell Mass Spectrometry—II. The systems (Na–K)AlSi$_4$O$_{10}$ and (NA–K)AlSi$_5$O$_{12}$. *Geochim. Cosmochim. Acta* **49**, 349–359.

HAYWARD P. J. (1977) The mixed alkali effect in aluminosilicate glasses. Part 2. The effect of non–bridging oxygen content. *Phys. Chem. Glasses* **18**, 1–6.

HENRY D. J., NAVROTSKY A. and ZIMMERMAN H. D. (1982) Themodynamics of plagioclase-melt equilibria in the system albite–anorthite–diopside. *Geochim. Cosmochim. Acta* **46**, 381–391.

HERVIG R. L. and NAVROTSKY A. (1984) Thermochemical study of glass in the system NaAlSi$_3$O$_8$–KAlSi$_3$O$_8$–Si$_4$O$_8$ and the join Na$_{1.6}$Al$_{1.6}$Si$_{2.4}$O$_8$–K$_{1.6}$Al$_{1.6}$Si$_{2.4}$O$_8$. *Geochim. Cosmochim. Acta* **48**, 513–522.

HOVIS G. L. (1984) A hydrofluoric acid solution calorimetric investigation of glasses in the system NaAlSi$_3$O$_8$–KALSi$_3$O$_8$ and NaAlSi$_3$O$_8$–Si$_4$O$_8$. *Geochim. Cosmochim. Acta* **48**, 523–525.

JAMBON A. (1982) Tracer diffusion in granitic melts. Experimental results for Na, K, Rb, Cs, Ca, Sr, Ba, Ce, Eu to 1300°C and a model of calculation. *J. Geophys. Res.* **87**, 10797–10810.

KNUDSEN M. (1909) Die Molekularströmung der Gase durch Offnungen unde die Effusion. *Ann. Phys.* **28**, 999–1016.

KRACEK F. C. (1930) The system sodium oxide–silica. *J. Phys. Chem.* **34**, 1583–1598.

McMILLAN P., PIRIOU B. and NAVROTSKY A. (1982) A Raman spectroscopic study of glasses along the joins silica-calcium aluminate, silica–sodium aluminate, and silica potassium aluminate. *Geochim. Cosmochim. Acta* **46**, 2021–2037.

MAMMONE J. F., SHARMA S. K. and NICOL M. F. (1981) Ring structures in silica glass—a Raman spectroscopic investigation. *EOS* **62**, 425.

MATSON D. W., SHARMA S. K. and PHILPOTTS J. A. (1986) Raman spectra of some tectosilicates along the ortho-clase-anorthite and nepheline-anorthite joins. *Amer. Mineral.* **71**, 694–704.

MURDOCH J. B., STEBBINS J. F. and CARMICHAEL I. S. E. (1985) High resolution ^{29}Si NMR study of silicate and aluminosilicate glasses: The effects of network-modifying cations. *Amer. Mineral.* **70**, 332–343.

MYSEN B. O., VIRGO D. and KUSHIRO I. (1981) The structural role of aluminum in silicate melts—A Raman spectroscopic study at 1 atmosphere. *Amer. Mineral.* **66**, 678–701.

NEKVASIL H. and BURNHAM C. W. (1987) The calculated individual effects of pressure and water content on phase equilibria in the granite system. In *Magmatic Processes: Physicochemical Principles* (ed. B. O. MYSEN), pp. 433–446. The Geochemical Society Spec. Publ. No. 1.

O'HARA M. J., BIGGAR G. M., RICHARDSON S. W., FORD C. E. and JAMIESON B. G. (1970) The nature of seas, mascons and the lunar interior in the light of experimental studies. In *Proc. Apollo II Lunar Sci. Conf.*, 695–710.

PAULE R. C. and MARGRAVE J. L. (1976) Free evaporation and effusion techniques. In *The Characterization of High Temperature Vapors*, (ed. J. L. MARGRAVE), pp. 130–151, Wiley.

RAMMENSEE W. and FRASER D. G. (1981) Activities in solid and liquid Fe–Ni and Fe–Co alloys determined by Knudsen Cell Mass Spectrometry. *Ber. Bunsenges. Phys. Chem.* **85**, 588–592.

RAMMENSEE W. and FRASER D. G. (1982) Determination

of activities in silicate melts by Knudsen Cell Mass Spectrometry—I. The system $NaAlSi_3O_8–KAlSi_3O_8$. *Geochim. Cosmochim. Acta* **46**, 2269–2278.

RAMMENSEE W. and FRASER D. G. The effects of changing Si/Al ratio on the mixing of melts in the system $NaAlSiO_4–KAlSiO_4–SiO_2$. *Chem. Geol.* (In press).

RICHET P. and BOTTINGA Y. (1984) Glass transitions and thermodynamic properties of amorphous SiO_2, $NaAlSi_nO_{2n+2}$ and $KAlSi_3O_8$. *Geochim. Cosmochim. Acta* **48**, 453–470.

ROGEZ J. R., CHASTEL C., BERGMAN C., BROUSSE C., CASTANET R. and MATHIEU J–C. (1983) Etude thermodynamique du systéme albite–orthose par calorimetrie de dissolution et effusion de Knudsen couplée a un spectrométre de masse. *Bull. Mineral.* **106**, 119–128.

SCHAIRER F. J. (1957) Melting relations of the common rock–forming silicates. *J. Amer. Ceram. Soc.* **40**, 215–235.

SEIFERT F., MYSEN B. O. and VIRGO D. (1982) Three-dimensional network structure of quenched melts (glass) in the systems $SiO_2–NaAlO_2$, $SiO_2–CaAl_2O_4$ and $SiO_2–MgAl_2O_4$. *Amer. Mineral.* **67**, 696–717.

SHARMA S. K., MAMMONE J. F. and NICOL M. F. (1981) Raman investigation of ring configurations in vitreous silica. *Nature* **292**, 140–141.

STEBBINS J. F., CARMICHAEL I. S. E. and WEILL D. F. (1983) The high temperature liquid and glass heat contents and the heats of fusion of diopside, albite, sanidine and nepheline. *Amer. Mineral.* **68**, 717–730.

STEBBINS J. F., MURDOCH J. B., SCHNEIDER E., CARMICHAEL I. S. E. and PINES A. (1985) A high temperature high resolution NMR study of ^{23}Na, ^{27}Al and ^{27}Si in molten silicates. *Nature* **314**, 250–252.

TAYLOR M. and BROWN G. E. (1979a) Structure of mineral glasses—I. The feldspar glasses $NaAlSi_3O_8$, $KAlSi_3O_8$, $CaAl_2Si_2O_8$. *Geochim. Cosmochim. Acta* **43**, 51–75.

TAYLOR M. and BROWN G. E. (1979b) Structure of mineral glasses—II. The $SiO_2–NaAlSiO_4$ join. *Geochim. Cosmochim. Acta* **43**, 1467–1473.

THOMPSON J. B. and HOVIS G. L. (1979) Entropy of mixing in sanidine. *Amer. Mineral.* **64**, 57–65.

URBAIN G., BOTTINGA Y. and RICHET P. (1982) Viscosity of liquid silica, silicates and aluminosilicates. *Geochim. Cosmochim. Acta* **46**, 1061–1072.

WOOD B. J. and FRASER D. G. (1976) *Elementary Thermodynamics for Geologists*. 303 pp. Oxford Univ. Press.

Magmatic Processes: Physicochemical Principles
© The Geochemical Society, Special Publication No. 1, 1987
Editor B. O. Mysen

Calorimetric studies of melts, crystals, and glasses, especially in hydrous systems

ALEXANDRA NAVROTSKY

Dept. of Geological and Geophysical Sciences, Princeton University, Princeton, N.J. 08544, U.S.A.

Abstract—Hydrous minerals, melts, and glasses may be studied by a variety of calorimetric techniques. These include measurements of heat capacities and enthalpies by adiabatic, drop, transposed-temperature-drop and differential scanning calorimetry, and the measurement of enthalpies of reaction by solution, drop-solution, and decrepitation calorimetry. Some calorimetric experiments on hydrous melts at high P and T are possible. The system $NaAlSi_3O_8$-H_2O has been studied. It shows small negative heats of mixing which may be consistent with STOLPER's model of an H_2O dissociation equilibrium.

INTRODUCTION

THE PROGRESS of experimental petrology has been limited at all stages by available technology. The problems one can attack can be exemplified by the tale of the drunkard seeking a lost key under a streetlamp on a moonless night. "Is that where you lost it?" "No, but it is the only place I have light enough to find it. And I might find some other treasure." So it is with experimental thermochemical studies of magmas. Are melts too difficult to study by structural and thermodynamic methods? Then study glasses. Are hydrous multicomponent melts at high pressure even more difficult? Then keep them dry and iron-free at atmospheric pressure. Yet much interesting science, which can be applied with care to nature's crucible, has come from such studies of simplified systems. The purpose of this paper is to review calorimetric studies of silicate systems, especially those containing H_2O, to show how such thermochemical data can be applied to petrology, to point out thermochemical studies that are now feasible, and to present some recent calorimetric results for the system albite-water.

THE USES OF CALORIMETRIC DATA IN HYDROUS SILICATE SYSTEMS

In phase equilibria involving silicate melts, the most obvious influence of H_2O is to lower liquidus temperatures very substantially. In addition, crystallization sequences can be altered and hydrous crystalline phases (*e.g.*, amphiboles and micas) can form, either on the liquidus or in subsequent reactions. These phase relations reflect the free energy relations among the phases and specifically, the lowering of the activity of aluminosilicate components in the melt by the addition of H_2O. Indeed, thermodynamic models of melt-water interactions (e.g., BURNHAM, 1981; BURNHAM and NEKVASIL, 1986; SILVER and STOLPER, 1985) often focus on these activities and seek simple phenomenological expressions for them. In general, each activity term can be separated into energetic (enthalpic) and entropic components. For a given process,

$$\mu - \mu^0 = \Delta\mu = RT\ln a = \Delta\bar{h} - T\Delta\bar{s} \quad (1)$$

(where μ = chemical potential, a = activity, $\bar{h}$ = partial molar enthalpy, and $\bar{s}$ = partial molar entropy). The range of temperatures over which phase equilibria give estimates of activity is frequently too small and the uncertainty too large to permit an accurate separation of enthalpy and entropy terms from the temperature dependence of μ. Thus a major purpose of calorimetric studies is to determine the integral and partial molar enthalpies directly for specific reactions. One can measure the heat of formation, of reaction, of phase transition, or of dehydration of a crystalline silicate. One can obtain enthalpies of mixing in solid solutions, in glasses and melts, and in glass-water or melt-water systems (see below). One can determine enthalpies of vitrification (crystal → glass) and of fusion (crystal → liquid). One can study order-disorder processes. All the above are necessary for a complete thermodynamic description of phase equilibria.

Calorimetric studies also provide heat capacity (C_p) and heat content ($H_T^0-H_{298}^0$, $H_T^0-H_0^0$) data. In addition to providing information on the temperature dependence of heats of reaction, such data yield values for the absolute entropy (S_T^0) of a phase if the substance in question is a perfectly ordered crystal at low temperature. Heat capacities and heats of fusion are also needed to construct the heat budget of an ascending magma and its interactions with the surrounding rocks.

Furthermore, the magnitude of thermochemical parameters can give insight into structure and

411

bonding. Mixing models of solid and liquid solutions are based on assumptions about the configurational entropy which, although it does not imply speciation directly, puts constraints on the numbers and types of entities being mixed. Thus if one can obtain entropies of mixing from a combination of activity and enthalpy of mixing data, one can compare those to values predicted by various statistical models and, in favorable cases, distinguish among or rule out some possible models (WEILL et al., 1980; HENRY et al., 1982). Systematic trends in thermochemical parameters for heats of formation (NAVROTSKY, 1982, 1985) and vitrification and heats of fusion (HERVIG et al., 1985; STEBBINS et al., 1984) also lend insight into structural features and permit estimates of values for other real or hypothetical phases.

From the point of view of silicate-water interactions, thermochemical data in several areas are needed. First, one would like to know the enthalpies, entropies, and free energies of all important hydrated mineral phases and their solid solutions. Second, one needs data on heats of fusion of important anhydrous silicates and heats of mixing in melts of rock-forming compositions. Third, one requires knowledge of the energetics of melt-water interactions and of the enthalpies of solution of H_2O at high P and T in aluminosilicate melts. Fourth, one wishes information on thermodynamics of aqueous fluids at high P and T. The next section describes the current possibilities and limitations for calorimetric study in these related areas.

TYPES OF CALORIMETRIC STUDIES

Heat capacities

Heat capacities at 4–300 K are measured by adiabatic calorimetry (WESTRUM et al., 1968; ROBIE and HEMINGWAY, 1972). Such studies usually require 1–30 g of sample. Heat capacities and standard entropies of a number of hydrous silicates including amphiboles and micas have been measured (see for example PERKINS et al., 1979, 1980; HEMINGWAY et al., 1986; ROBIE et al., 1976; KRUPKA et al., 1979). Heat capacities of several anhydrous aluminosilicate glasses have also been reported (ROBIE et al., 1978; KRUPKA et al., 1979). In glassy materials, $\int_0^T (C_p/T)dT$ gives only the vibrational contribution to the entropy; a term arising from configurational disorder, and depending on cooling history, also contributes to the total entropy. Low temperature heat capacities could be measured on a quenched hydrated glass (to the best of my knowledge this has not been done) and would pro-

vide information on the excess vibrational entropy of mixing. This excess vibrational entropy would be related to both the speciation of H_2O (STOLPER, 1982) and the extent to which the aluminosilicate framework is perturbed by the dissolution of water. However, a relatively large amount (several grams) of homogeneous hydrous glass with no fluid inclusions of H_2O would be needed and the excess heat capacities might be quite small.

At temperatures from about 100°C to near 800°C, commercial differential scanning calorimeters can measure heat capacities on samples of 10–200 mg. Heat capacities of a number of minerals and of aluminosilicate glasses have been measured (STEBBINS et al., 1984; KRUPKA et al., 1979). The upper end of this range samples the glass transition region for anhydrous glasses and the beginning of decomposition (dehydration) for hydrated minerals and glasses. Because DSC is usually done at fairly rapid scanning rates (1–10°C per minute), kinetic factors may dominate the decompositions seen. In combination with thermogravimetric analysis (TGA), such DSC measurements provide useful data on the heat capacities, decomposition rates, and decomposition kinetics of hydrous crystals and glasses. Although no commercial DSC units currently operate above 827°C, this limitation is more a matter of market demand (mostly for polymer-related work) and the softening point of the silver block commonly used than of any other major technical obstacle. A DSC for work with small samples to at least 1000°C should be possible. Differential thermal analysis (DTA) provides transition temperatures and qualitative indication of heat effects up to ~2000°C using commercially available instruments.

Above 800°C, heat capacities are generally derived from measurements of heat contents ($H_T^0-H_{298}^0$) either by conventional drop calorimetry (dropping a sample from a furnace into a calorimeter at room temperature) (STEBBINS et al., 1982, 1983, 1984) or by transposed-temperature-drop calorimetry (dropping a sample from room temperature into a hot calorimeter). The upper temperature limit of the former is ~1700°C with normal Pt-wound furnaces (STEBBINS et al., 1982, 1984) and ~2300°C with specially designed furnaces (STOUT and PIWINSKII, 1982). The method requires 5–50 g of sample. One must be concerned about the energetic state of glassy samples quenched from different temperatures (STEBBINS et al., 1983, 1984; RICHET et al., 1982; RICHET and BOTTINGA, 1984). A commercial transposed-temperature-drop calorimeter operates to 1500°C. It requires samples of <100 mg and, because the final state of the sample is melt at high temperature, the potential variability in the glassy state can be avoided.

To study the high temperature heat capacities and enthalpies of hydrous systems one must apply pressure to maintain H_2O as a dense fluid or a dissolved phase. KASPER *et al.* (1979) measured the heat content of aqueous NaCl-H_2O solutions by dropping sealed Pt capsules of solution from room temperature into a calorimeter operating at 184, 263, or 712°C. The entire system inside the calorimeter and extending from the hot zone within the thermal sensing elements (at the bottom) to the cold pressure head containing the sample to be dropped (at the top) was held at 0.5 to 1.0 kbar argon pressure. Although the experiments were successful, the geometrical arrangement with hot zone at the bottom resulted in very substantial convection and limited the sensitivity of the method.

CLEMENS and NAVROTSKY (1986) used a variant of conventional drop calorimetry to study the heat contents of hydrous albite glass and melt. By placing an internally heated pressure vessel above a calorimeter operating near room temperature (see Figures 1 and 2), and allowing a long cold-finger of the

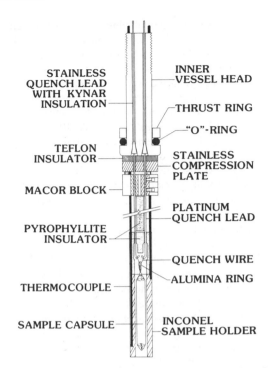

FIG. 2. Schematic section through the inner portion of the pressure head showing the sample assembly for drop calorimetry of hydrous melts (CLEMENS and NAVROTSKY, 1986).

vessel to extend down into the calorimeter, they could drop samples from temperatures of up to 1033°C to room temperature within a system maintained at pressures to 1.5 kbar. Convection was decreased by having the hot zone on top and better results were obtained. The upper temperature was limited by the location and material of seals in the internally heated vessel. It should be possible with some modification to achieve somewhat higher temperatures. At temperatures below the liquidus at each H_2O content, albite crystallized rapidly, limiting the utility of the measurements to the stable liquid range. The results for Ab-H_2O are discussed below.

Heats of fusion

Since most rock forming silicates melt above ~1000°C, their heats of fusion require the use of drop or transposed-temperature-drop methods. Drop calorimetry works most conveniently when the melt crystallizes rapidly, releasing the latent heat in the calorimeter. The heat of fusion of fayalite has been measured in this manner (STEBBINS and CARMICHAEL, 1984). When the melt forms a glass

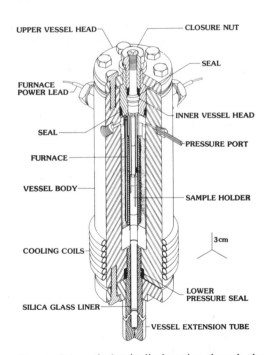

FIG. 1. Schematic longitudinal section through the pressure vessel used for high pressure drop experiments (CLEMENS and NAVROTSKY, 1986). The vessel extension tubes reaches into the calorimeter roughly 90 cm below the furnace.

on rapid cooling, the enthalpy difference between glass and crystal must be measured by solution calorimetry (see below) and a thermochemical cycle involving this heat of vitrification and the heat capacities of glass, liquid, and crystal must be used to calculate the heat of fusion. This has been done for a number of rock-forming silicates (STEBBINS et al., 1982, 1983, 1984; RICHET et al., 1982; RICHET and BOTTINGA, 1984). A general conclusion is that the heat of fusion can depend quite strongly on temperature because the heat capacity of the liquid can be substantially greater than that of the crystal. This effect can be quite important in thermochemical calculations of phase equilibria.

Transposed temperature drop calorimetry can measure heats of fusion to ~1500°C; results for diopside are shown in Figure 3 (ZIEGLER and NAVROTSKY, 1986). They confirm the previous measurements using a thermochemical cycle involving the glassy state (STEBBINS et al., 1983; RICHET and BOTTINGA, 1984) and prove the utility of the commercial instrument. Direct measurement of the heats of fusion of other silicates, of multicomponent systems and of model rock compositions and/or rocks should be feasible. The calorimeter can operate under controlled low oxygen fugacity, making the control of oxidation state feasible. However, the small size of the calorimeter makes it unsuitable for work under high pressure, so heat contents and heats of fusion of hydrous systems cannot be studied in it. Rapid dehydration reactions (decrepitation)

can be studied readily; one can pick a temperature at which decomposition occurs rapidly but not violently.

Enthalpies of formation and of mixing

To relate the energetics of one phase assemblage to that of another (oxide mixture versus compound, end-member versus solid solution) when one cannot be transformed rapidly to the other in the calorimeter, a thermochemical cycle must be used. This cycle usually involves the dissolution of reactants and products in a solvent where the final dissolved state is the same for both. The primary measurement is then of a heat of solution. Two classes of solution calorimetry have commonly been in use for silicates: dissolution in aqueous acid (usually HF) near room temperature (ROBIE and HEMINGWAY, 1972) and dissolution in a molten oxide (usually $2PbO \cdot B_2O_3$) at 700–800°C (NAVROTSKY, 1977, 1979). Each has some advantages and disadvantages. Hydrous minerals have generally been studied using aqueous HF for the following reasons. They often dissolve readily in that solvent, they frequently decompose below 700°C and the final state of the water which would contact lead borate melt at 700°C is unknown. However, HF calorimetry requires several grams of sample, alumina-rich samples do not always dissolve well, and much data on analogous anhydrous compositions have been obtained by oxide melt calorimetry. Thus more direct interrelation of the two methods and exploration of the applicability of oxide melt calorimetry to hydrous systems appears highly desirable. We have begun exploration of such possibilities and present several preliminary findings.

For phases which contain OH^-, such as micas and amphiboles, the fluoride analogues can be studied by oxide melt calorimetry and the systematics of cation-substitution energetics can be applied to the hydroxy materials. This was exploited by WESTRICH et al. (1981) for fluorpargasite, fluorapatite and fluorphlogopite and more recently by GRAHAM and NAVROTSKY (1986) for the fluor-tremolite-fluoredenite solid solution series, for which heat of solution data are shown in Figure 4.

Not all phases containing water decompose rapidly at 700°C. Well-crystallized pargasite and phlogopite persist unchanged for 24 hours in the calorimeter and appear to give reproducible heats of solution in $2PbO \cdot B_2O_3$ (N. ROSS, J. D. CLEMENS and A. NAVROTSKY, unpublished). There is some evidence that when ~30 mg of a hydrous phase containing 2–5 weight percent H_2O is dissolved in 30 g of $2PbO \cdot B_2O_3$ at 700°C, the H_2O stays in

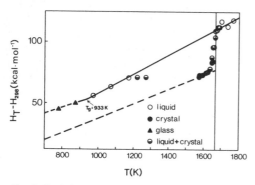

FIG. 3. Enthalpy vs. temperature for $CaMgSi_2O_6$ crystal, glass, and liquid crystal. The calorimetric data for glass have been corrected to the crystalline reference state by adding the enthalpy of vitrification. The line through the crystal data is that given by STEBBINS et al. (1983). The line through the liquid data is the best fit line for the data shown by open circles. The different symbols represent the state of the sample at the upper temperature, T. Open circles—liquid, closed circles—crystal, half-filled circles—crystal + liquid, triangles—glass. The figure shows that diopside melts incongruently. From ZIEGLER and NAVROTSKY (1986).

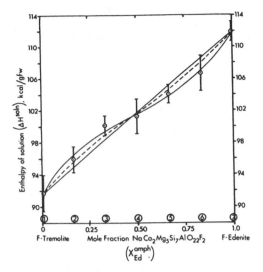

FIG. 4. Enthalpies of solution of synthetic fluortremolite-fluoredenite amphiboles. Brackets indicate standard deviations. Solid curve is least squares fitted third-order polynomial through all the data, consistent with subregular solution model. Dashed curve is for subregular model with W_{tr}^H and W_{ed}^H fixed at lower magnitude limits permitted by uncertainties at the 95% confidence interval. Solid straight line indicates ideal mixing (GRAHAM and NAVROTSKY, 1986).

solution, corresponding to an H_2O solubility on the order of 50 ppm or greater, but this needs further study. Preliminary data for the heat of formation of phlogopite using its heat of solution in lead borate (J. D. CLEMENS and A. NAVROTSKY, unpublished) look reasonable. These same studies also suggest the feasibility of studying the energetics of stacking disorder in layer silicates by solution calorimetry. An exploratory study of the heats of solution along the F-pargasite-OH-pargasite join (N. ROSS and A. NAVROTSKY, unpublished) also is promising. Problems and limitations seem to arise more from sample quality (phase purity, structural state and homogeneity) than from calorimetric difficulties.

Thermochemical cycles can be devised to deal with phases which decompose if left at 700°C in the calorimeter to equilibrate prior to dissolution. The first of these can be called "drop-solution" calorimetry. In it, the sample is dropped from room temperature into molten lead borate in the calorimeter, whereupon it dissolves. The enthalpy measured is its heat content plus its heat of solution. Subtraction of the heat content gives the heat of solution. A similar approach has been taken to calorimetry of high pressure phases which do not persist metastably at 700°C and atmospheric pressure (AKAOGI and NAVROTSKY, 1985).

"Drop-solution" calorimetry on MgO and $Mg(OH)_2$ has been used to estimate the heat of solution of H_2O in lead borate melt at 700°C and 1 atm (S. CIRCONE and A. NAVROTSKY, unpublished). An exothermic enthalpy of solution of -5.7 ± 0.7 kcal/mole has been estimated. This value can be used in other thermochemical cycles to calculate heats of formation of hydrous minerals from "drop-solution" calorimetry and to correlate data obtained by high temperature oxide melt calorimetry and by HF solution calorimetry. Though still preliminary, this work opens the way to many other studies of hydrous phases.

Another approach to high temperature calorimetry of hydrous phases not persisting near 700°C is direct measurement of heat of dehydration by transposed-temperature-drop calorimetry. This method works well when samples dehydrate rapidly and we have called it decrepitation calorimetry. The enthalpy of dehydration is measured as the difference between first and second drop experiments on an unsealed Pt capsule containing powdered material. Weight change is monitored to check for complete H_2O loss and the dehydrated product is characterized by X-ray and optical means and by a normal heat of solution measurement in $2PbO \cdot B_2O_3$ at 700°C. This heat of solution permits a comparison of the energetics of the fine-grained and often amorphous or poorly crystalline dehydration product with that of a crystalline anhydrous phase assemblage or a previously melted glass. From these data, a thermochemical cycle calculates the enthalpy of dehydration to well-characterized products. The results of such studies for glasses in the system albite-water are discussed below. The method should be applicable to a variety of synthetic and also possibly natural hydrous materials. Changes in oxidation state of iron during dehydration present a complication and experiments are best confined to samples with low iron contents.

Oxide melt solution calorimetry using $2PbO \cdot B_2O_3$ near 700°C has been used extensively in anhydrous silicate systems to study heats of formation (NAVROTSKY and COONS, 1976), phase transition (AKAOGI and NAVROTSKY, 1985), solid solution formation (CARPENTER et al., 1985) and vitrification and mixing in the glassy state (NAVROTSKY et al., 1980). It is now an essentially routine procedure for phases containing SiO_2, Al_2O_3, Fe_2O_3, GeO_2, alkali and alkaline earth oxides, CoO, NiO, ZnO and can be applied, with care under a non-oxidizing atmosphere to compounds containing Mn^{2+}, Cr^{3+}, and to some Fe^{2+}-silicates (AKAOGI and NAVROTSKY, 1985; NAVROTSKY et al., 1979; WOOD and KLEPPA, 1984). Combined with the methods discussed above, hydrous crystals and glasses containing these same oxides can be studied. More rigorous oxygen fugacity control for calorimetry of iron-bearing phases can be achieved in an alkali borate solvent near 750°C (CHATILLON-COLINET et al., 1983). This melt is also useful for calorimetry of titanates and zirconates (E. MURO-MACHI-TAKAYAMA and A. NAVROTSKY, in preparation) which form insoluble precipitates on reaction with lead borate. The calorimetric behavior of samples containing small amounts of Ti^{4+} and/or Fe^{2+} has not been explored in detail.

The growing data base of mixing properties of aluminosilicate glasses has been extrapolated to provide enthalpy of mixing data for the melts. The validity of such extrapolations from glass to melt has been discussed extensively (NAVROTSKY et al., 1982; STEBBINS et al., 1982, 1983, 1984; BOETTCHER et al., 1982; RICHET et al., 1982; RICHET and BOTTINGA, 1984). The verdict is not in yet but the evidence seems to suggest that when ΔC_p terms are known, such comparisons can be made with reasonable accuracy. Direct measurement of heats of mixing in silicate melts at 1000–1400°C would be very desirable. Conventional oxide melt solution calorimetry is not readily extended into that range because of problems of calorimeter stability, volatility of solvent, corrosion of ceramic and SiO_2-glass parts, and other technical difficulties. Transposed temperature drop calorimetry, using closed capsules of mixtures of phases which melt in the calorimeter, is potentially capable of getting heats of fusion and heats of mixing in melts. We are currently investigating the applicability of this method to melts in the system albite-anorthite-diopside at 1100–1500°C.

The simultaneous application of high pressure to contain H_2O and temperatures of 800–1500°C, necessary for direct calorimetric study of mixing in hydrous silicate melts, is very difficult. Other than the heat content measurements by drop calorimetry described above, there is little hope, in my opinion, for immediate progress in this area. Qualitative methods, such as DTA at high pressure (ROSEN-HAUER, 1976) offer some possibilities, but they are more useful in mapping out phase relations, including those in hydrous systems, than in attempts to obtain quantitative thermochemical data.

Aqueous fluids at high P and T

Despite the importance of the aqueous phase in geothermal systems, hydrothermal and magmatic ore deposits, late stage magmatism and pegmatites, and explosive volcanism, thermochemical data for aqueous fluids, especially away from the vapor saturation curve, are largely lacking. Calorimetric data are difficult to obtain above $\sim$100°C. KASPER et al. (1981), as mentioned above, obtained some heat content data on $NaCl$-H_2O at 0.5 and 1 kbar up to 700°C. H. FUKUYAMA, J. R. HOLLOWAY and A. NAVROTSKY (unpublished) designed and operated a mixing cell for use to $\sim$350°C at pressures of up to 1 kbar. Some heat of dilution data in the system $NaCl$-H_2O were obtained but the cell, which contained inconel bellows assemblies to equalize internal pressure to that of the external argon medium, proved prone to mechanical failure when used with salt solutions. Both the heat content and heat of dilution measurements were time- and labor-intensive; each calorimetric experiment had to be assembled, pressurized, equilibrated at high temperature, run, disassembled, and cleaned. Thus even under optimum conditions, productivity in such experiments was limited to one run per day. After these experiences, it appears that future calorimetric studies at high temperature should concentrate on methods which maintain a continuous high P, T regime while measuring enthalpy changes associated with aqueous solutions passing through the system, that is, flow calorimetry. Such calorimeters have been in use for a number of years for both heat capacity measurements and heats of mixing at the low end of the needed temperature range, namely up to about 200°C and at pressures either along the vapor saturation curve or up to a few hundred bars (MESSIKOMER and WOOD, 1975; PERRON et al., 1975; PICKER et al., 1971; BUSEY et al., 1984; SIMMONS et al., 1985). It should be technologically feasible to extend the range to $\sim$1 to 2 kbar and 500–600°C.

THERMOCHEMISTRY OF GLASSES AND MELTS IN THE SYSTEM ALBITE-WATER AND A MODEL GRANITE

To illustrate the application of the methods described above to a specific petrologic system, this section summarizes recently completed work on the system albite-water and a model hydrous granite (CLEMENS and NAVROTSKY, 1986). The purpose of this study was to characterize the energetics of dissolution of H_2O into framework silicate melts. Samples were prepared at appropriate high P and T conditions (900–1200°C, 1–5 kbar) to give liquids with H_2O-contents between 1.5 and 9 weight percent (mole fraction H_2O in $NaAlSi_3O_8$-H_2O in the range 0.05 to 0.59). The P-T conditions were chosen, for a given H_2O-content, to be at temperatures above the liquidus and at water contents below H_2O-saturation, so that quenched melts would not trap H_2O bubbles. Samples were quenched to glasses and H_2O-contents were determined by TGA. The quenched glasses were generally clear and showed no indication of exsolved water either prior to or after drop calorimetry at high pressure. Three types of calorimetric measurements were performed: "decrepitation" calorimetry mainly at 795°C (some experiments also at 700°C), heat of solution measurements in $2PbO \cdot B_2O_3$ at 700°C of decrepitated glasses, and drop calorimetry at 1 kbar of one anhydrous and one hydrous samples with the furnace at 983°C and 1033°C, the calorimeter at 85°C. Similar decrepitation experiments were run for a model hydrous granite.

A thermochemical cycle, shown in Figure 5, can be used to calculate the heat of mixing of Ab-H_2O at 25°C from the decrepitation and solution calorimetry. The results, for the reaction

$$X H_2O \, (\text{liquid}, 25°C) + (1 - X) \, Ab \, (\text{glass}, 25°C)$$
$$= \text{hydrous glass} \, (25°C) \quad (2)$$

are shown in Figure 6. A negative heat of mixing, with a minimum near $X = 0.25$, is seen. The results may also be represented as the heat of solution of one mole of H_2O in an amount of albite glass to form hydrous glass of the appropriate composition. This quantity is shown in Figure 7. The heat of solution, per mole H_2O, is also exothermic, but becomes less so with increasing H_2O content.

Analogous, though less complete, data are shown for a melt near the granite eutectic, where the "granite" component is $Ab_{0.35}Or_{0.26}1Q_{0.39}$ (see Figures 6 and 7). The heats of mixing and of solution of H_2O are much closer to zero and, though they show slight positive values at $X = 0.09$ and negative values at $X = 0.32$ and 0.55, these are near the resolution of the calorimetric data.

What do these data mean in terms of proposed thermodynamic model for H_2O solubility and speciation in glasses and melts? The model proposed by BURNHAM (1975, 1981) suggests that H_2O mixes ideally with the aluminosilicate component written on an 8-oxygen basis. Strictly speaking, ideality implies zero heat of mixing. The calorimetric data show small but significant non-zero heats of mixing in the glasses. These may be somewhat smaller (see below) in the melts. The Ab-H_2O and granite-H_2O systems behave somewhat differently from each other. Thus although the deviations from BURNHAM's model are small, they do suggest that in terms of energetics, the dissolution of H_2O is somewhat more complex than the model implies. Indeed, since a miscibility gap exists in Ab-H_2O at high values of X (saturated melt coexisting with aqueous fluid)

energetic complexity must exist somewhere in the system.

BURNHAM's model is based on the assumption that all H_2O in the melt is present as OH^-. STOLPER (1982) suggests an equilibrium among H_2O and OH^- species according to the reaction:

$$H_2O \, (\text{in melt}) + O^= \, (\text{in melt}) = 2OH^- \, (\text{in melt}). \quad (3)$$

The equilibrium constant for this reaction is suggested to have virtually no temperature dependence and only a weak dependence on composition (STOLPER, 1982; SILVER and STOLPER, 1985) with a value of ~ 0.17 for the system albite-water (SILVER and STOLPER, 1985).

The mixing of H_2O with albite melt may be thought of as consisting of two steps. The first is the dissolution as molecular water

$$H_2O \, (\text{fluid}) = H_2O \, (\text{in melt}). \quad (4)$$

The second is the dissociation equilibrium [Eqn. (3)] above. These reactions may be rewritten as follows, with X = mole fraction total H_2O dissolved and Y = fraction of the dissolved H_2O which dissociates (Y = degree of dissociation).

$$X H_2O \, (\text{fluid}) = X H_2O \, (\text{in melt}) \quad (\Delta H_A) \quad (5)$$

$$(YX) H_2O \, (\text{in melt}) + (YX) O^= \, (\text{in melt})$$
$$= (YX) OH^- \, (\text{in melt}) \quad (\Delta H_B). \quad (6)$$

Assume that each of these steps has an enthalpy associated with it which depends linearly on the number of moles of H_2O undergoing the process. Then

$$\Delta H = X\Delta H_A + XY\Delta H_B \approx \Delta H^{mix} \quad (7)$$

where ΔH_A is that of reaction (5) in the text and ΔH_B is that of reaction (6).

This is approximately equal to the heat of mixing if the energetics of any perturbation of the aluminosilicate framework is also linearly dependent in energy on the amount of H_2O initially dissolved (X) and is therefore incorporated in Eqn. (5). This formulation is reasonable for relatively low water contents but obviously cannot be extrapolated to $X \rightarrow 1$. Since immiscibility intervenes at high H_2O content no formulation which ignores that feature can be valid at high water contents anyhow, as STOLPER (1982) has recognized.

One may then calculate Y, the degree of dissociation, as a function of X, the mole fraction of H_2O in $NaAlSi_3O_8$-H_2O, using the formulation of STOLPER (1982) and SILVER and STOLPER (1985). The numbers of moles of species, n, per mole of $X H_2O$-$(1 - X) NaAlSi_3O_8$ are

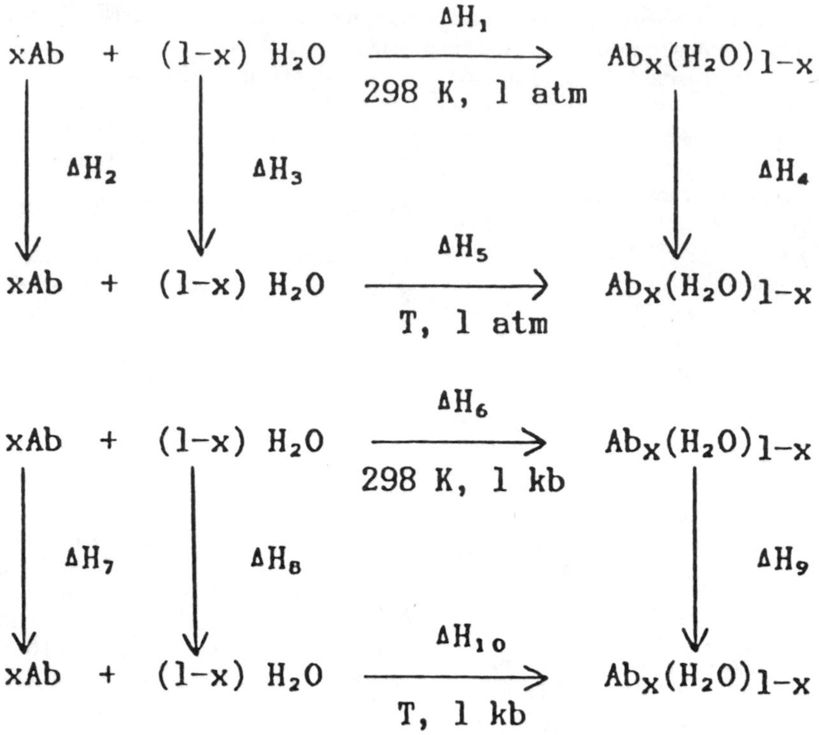

$$xAb \; + \; (1-x) \; H_2O \; \xrightarrow[\text{298 K, 1 atm}]{\Delta H_1} \; Ab_x(H_2O)_{1-x}$$

$$\Big\downarrow \Delta H_2 \qquad \Big\downarrow \Delta H_3 \qquad\qquad\qquad \Big\downarrow \Delta H_4$$

$$xAb \; + \; (1-x) \; H_2O \; \xrightarrow[\text{T, 1 atm}]{\Delta H_5} \; Ab_x(H_2O)_{1-x}$$

$$xAb \; + \; (1-x) \; H_2O \; \xrightarrow[\text{298 K, 1 kb}]{\Delta H_6} \; Ab_x(H_2O)_{1-x}$$

$$\Big\downarrow \Delta H_7 \qquad \Big\downarrow \Delta H_8 \qquad\qquad\qquad \Big\downarrow \Delta H_9$$

$$xAb \; + \; (1-x) \; H_2O \; \xrightarrow[\text{T, 1 kb}]{\Delta H_{10}} \; Ab_x(H_2O)_{1-x}$$

Measure: $\Delta H_4 - \Delta H_3$ by decrepitation
calorimetry and $\Delta H_{solution}$ of anhydrous glass
ΔH_9 by high P drop calorimetry

Know: $\Delta H_2 = H_T^\circ - H_{298}^\circ$ of Ab glass
ΔH_3, ΔH_8 from properties of H_2O

Assume: $\Delta H_2 = \Delta H_7$, $\Delta H_1 = \Delta H_6$

Then

Calculate ΔH_1 = heat of mixing of Ab (glass) and H_2O (liquid)
at 298 K and 1 atm

$$\Delta H_1 = \Delta H_2 + \Delta H_3 + \Delta H_5 - \Delta H_4$$

Calculate ΔH_{10} = heat of mixing of Ab (glass or melt) and H_2O (fluid)
at T and 1 kb (or other operating P)

$$\Delta H_{10} = -\Delta H_7 - \Delta H_8 + \Delta H_6 + \Delta H_9$$

FIG. 5. Thermochemical cycles used to calculate enthalpies of mixing in Ab-H_2O glasses and liquids.

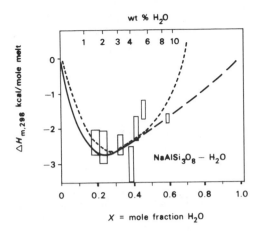

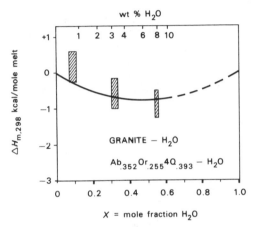

$$\Delta H/X = \Delta H_A + Y\Delta H_B. \qquad (10)$$

A least squares fit of $\Delta H^{mix}/X$ versus Y gives, $\Delta H_A = 10.71 \pm 2.54$ kcal, and $\Delta H_B = -33.8 \pm 4.74$ kcal, with a correlation coefficient of 0.946. Using the same approach with $K = 0.085$ and 0.34 (half and twice STOLPER's values) gives different parameters with a slightly worse fit. The calculated heat of mixing is shown as the dotted curve in Figure 6; the values of $\Delta H^{mix}/X$ calculated from the fit are shown by the solid curve in Figure 7. Note that although $\Delta H^{mix}/X$ is assumed to be a linear function of Y (Eqn. 10), $\Delta H^{mix}/X$ varies nonlinearly with X because X and Y are related by the nonlinear function given by the equilibrium constant K (Eqn. 9).

The calorimetric data for mixing in Ab-H_2O glasses thus appear to be consistent with STOLPER's proposed speciation reactions. The minimum in the heat of mixing appears to arise from the balance of two factors: an endothermic heat of dissolution of molecular H_2O and an exothermic heat of dissociation, with the degree of dissociation decreasing with increasing H_2O content. For $X > \sim 0.7$, Eqns. (9) and (10) suggest positive heats of mixing (see Figure 6). It may be coincidental that the compo-

FIG. 6. Enthalpies of mixing at 25°C for the systems albite-water and "granite"-water. Rectangles are data points, showing uncertainties, from CLEMENS and NAVROTSKY (1986). Solid and long-dashed curves are visual estimates of smooth curves which approach $\Delta H^{mix} = 0$ at $X = 0$ and $X = 1$. Short-dashed curve is calculated using STOLPER's (1982) model and $K = 0.17$ (SILVER and STOLPER, 1985). See text.

$$nOH^- = 2XY$$

$$nH_2O = X(1 - Y) \qquad (8)$$

$$nO^= = 8(1 - X) - XY = 8 - 8X - XY$$

and the equilibrium constant, K, for the dissociation (STOLPER's K_2) is:

$$K = \frac{(2XY)^2}{X(1 - Y)(8 - 8X - XY)}$$

$$= \frac{4XY^2}{(1 - Y)(8 - 8X - XY)} \approx 0.17. \qquad (9)$$

For any value of X one can solve for Y, then Eqn. (7) can be rewritten, per mole of H_2O dissolved as

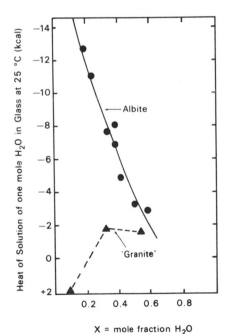

FIG. 7. Enthalpy of solution of one mole of water in glass at 25°C for system albite-water and "granite"-water. Data are from CLEMENS and NAVROTSKY (1986). Circles are data points for $NaAlSi_3O_8$-H_2O, triangles for $Ab_{0.352}Or_{0.255}4Q_{0.393}$-$H_2O$. Solid curve is calculated using STOLPER's (1982) model, see text. Dashed line simply connects "granite" data points.

sition where ΔH^{mix} becomes positive corresponds to the two-phase (saturated melt plus vapor) region, but the trend toward positive enthalpies is interesting.

However, this general interpretation suffers one serious problem. For reaction (3), STOLPER proposes an equilibrium constant of 0.17 to 0.2; we suggest an enthalpy on the order of -34 kcal. Since $\Delta G^0 = -RT \ln K$, and $K < 1$, ΔG^0 is positive regardless of what temperature (that of melt, quenched glass, or some intermediate temperature at which the speciation is frozen in) one associates with STOLPER's values of K. A small positive ΔG^0 and a large negative ΔH^0 imply a large negative ΔS^0 (on the order of -30 to -40 cal K^{-1} mol^{-1}, depending on the value of T chosen for the dissociation equilibrium). Though such large negative ΔS^0 and negative ΔH^0 values are seen for some acid dissociation reactions in aqueous solutions which produce charged species (DASENT, 1981), it is not clear that the reaction of H_2O in silicate melts bears any real analogy. Furthermore, the large magnitudes of ΔH^0 and ΔS^0 are very difficult to reconcile with a temperature-independent dissociation constant. Indeed, the values of ΔH and ΔS would suggest that the equilibrium shifts toward molecular water with increasing temperature. Further study is needed.

STOLPER's model is quite simple and neglects both positive interactions leading to immiscibility and energetic differences among differently bonded oxygen species in the melt. Though the present calorimetric data appear to be described adequately, except for the problem discussed above, one might expect that higher resolution in future calorimetric studies could require more complex models. At present the overall quality of models and data are well matched in that both are fairly crude first approximations.

The fact that negative heats of mixing are seen argues for chemical interaction between glass and H_2O. Were the water present only as species having the characteristics of pure H_2O (e.g., as microbubbles) then a zero heat of mixing (that of a mechanical mixture) would be expected. Thus, regardless of details of speciation, our data suggest that H_2O is interacting energetically on a microscopic (atomistic) scale as a truly dissolved entity.

The drop calorimetric experiments from high P and T allow estimates of the heat of mixing at high P and T for the composition studied by using the thermochemical cycle in Figure 5. The results for $x = 0.388$ are shown in Figure 8. The heat of mixing becomes less exothermic with increasing temperature. However, this is at least partly a result of increasing the molar volume of H_2O with increasing T (at 1000°C and 1 kbar, the volume of H_2O is

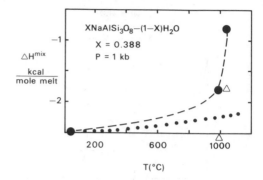

FIG. 8. Enthalpy of mixing of Ab-H_2O at $X_{H_2O} = 0.388$ at different temperatures. Solid circles are points calculated using thermochemical cycle in Figure 5. Triangles are calculated from these by subtracting the term $P(V_{H_2O(P,T)} - V_{H_2O(1\,atm,\,25°C)})$ with $P = 1$ kbar. These then refer to mixing of water at a constant volume. Uncertainties are about ± 0.4 kcal at 25°C and ± 0.8 to ± 1.0 kcal at high temperature.

roughly 5 times the volume of liquid water at 25°C). When the PV work done to compress the H_2O back to its original volume is subtracted, the resulting heat of mixing (open circles in Figure 8) shows a temperature dependence which is not outside experimental error. This suggests that the interactions within the melt or glass phase may not be very temperature dependent.

CONCLUSIONS

Calorimetric study of hydrous crystals and glasses is possible utilizing thermochemical cycles involving solution calorimetry, drop and transposed-temperature-drop calorimetry, and direct study of water release. Direct calorimetric data on silicate melts may be obtainable by transposed temperature drop calorimetry. Calorimetric study of hydrous melts at high P and T is feasible for fairly limited heat content studies. Calorimetry of aqueous fluids at 200–600°C can be developed. Many of these methods are new and exploratory. The next few years should provide many new advances.

Acknowledgements—This work has been supported by the U.S. Dept. of Energy (Grant DE-FG02-85ER-13437) and the National Science Foundation (Grant EAR-8513916).

REFERENCES

AKAOGI M. and NAVROTSKY A. (1985) Calorimetric study of high pressure polymorphs of $MnSiO_3$. *Phys. Chem. Min.* **12**, 317–323.
BOETTCHER A. L., BURNHAM C. W., WINDOM K. E. and BOHLEN S. R. (1982) Liquids, glasses, and the melting of silicates at high pressure. *J. Geol.* **90**, 127–138.
BURNHAM C. W. (1975) Water and magmas: a mixing model. *Geochim. Cosmochim. Acta* **39**, 1077–1084.

BURNHAM C. W. (1981) The nature of multicomponent aluminosilicate melts. In *Chemistry and Geochemistry of Solutions at High Temperatures and Pressures* (eds. D. T. RICKARD and F. E. WICKMAN), *Physics and Chemistry of the Earth,* Vol. 13–14, 197–229, Pergamon Press.

BURNHAM C. W. and NEKVASIL H. (1986) Equilibrium properties of granite pegmatite magmas. *Amer. Mineral.* **71**, 239–263.

BUSEY R. H., HOLMES H. F. and MESMER R. F. (1984) The enthalpy of dilution of aqueous sodium chloride to 373 K using a new heat flow and liquid flow microcalorimeter. Excess thermodynamic properties and their pressure derivatives. *J. Chem. Thermo.* **16**, 343–372.

CARPENTER M. A., MCCONNELL J. D. C. and NAVROTSKY A. (1985) Enthalpies of Al-Si ordering in the plagioclase feldspar solid solution. *Geochim. Cosmochim. Acta* **49**, 947–966.

CHATILLON-COLINET C., KLEPPA O. J., NEWTON R. C. and PERKINS D. (1983) Enthalpy of formation of $Fe_3Al_2Si_3O_{12}$ (almandine) by high temperature alkali borate solution calorimetry. *Geochim. Cosmochim. Acta* **47**, 439–444.

CLEMENS J. D. and NAVROTSKY A. (1986). Mixing properties of $NaAlSi_3O_8$ melt-H_2O: New calorimetric data and some geological applications. *J. Geol.* (In press).

DASENT W. E. (1981) *Inorganic Energetics.* 167–171. Cambridge Univ. Press.

GRAHAM C. M. and NAVROTSKY A. (1986) Thermochemistry of the tremolite-edenite amphiboles using fluorine analogues, and applications to amphibole-plagioclase-quartz equilibria. *Contrib. Mineral. Petrol.* **93**, 18–32.

HEMINGWAY B. S., BARTON M. D., ROBIE R. A. and HASELTON H. T. (1986) Heat capacities and thermodynamic functions for beryl, phenakite, euclase, bertrandite and chrysoberyl. *Amer. Mineral.* **71**, 557–568.

HENRY D. J., NAVROTSKY A. and ZIMMERMANN H. D. (1982) Thermodynamics of plagioclase-melt equilibria in the system albite-anorthite-diopside. *Geochim. Cosmochim. Acta* **46**, 381–391.

HERVIG R. L., SCOTT D. and NAVROTSKY A. (1985) Thermochemistry of glasses along joins of pyroxene stoichiometry in the system $Ca_2Si_2O_6$-$Mg_2Si_2O_6$-Al_4O_6. *Geochim. Cosmochim. Acta* **49**, 1497–1501.

KASPER R. B., HOLLOWAY J. R. and NAVROTSKY A. (1979) Direct calorimetric measurement of enthalpies of aqueous sodium chloride solutions at high temperatures and pressures. *J. Chem. Thermo.* **11**, 13–24.

KRUPKA K. M., ROBIE R. A. and HEMINGWAY B. S. (1979) High-temperature heat capacities of corundum, periclase, anorthite, $CaAl_2Si_2O_8$ glass, muscovite, pyrophillite, $KAlSi_3O_8$ glass, grossular, and $NaAlSi_3O_8$ glass. *Amer. Mineral.* **64**, 86–101.

MESSIKOMER E. E. and WOOD R. H. (1975) The enthalpy of dilution of aqueous sodium chloride at 298.15 to 373.15 K, measured with a flow calorimeter. *Jour. Chem. Thermo.* **7**, 119–130.

NAVROTSKY A. (1977) Recent progress and new directions in high temperature calorimetry. *Phys. Chem. Min.* **2**, 89–104.

NAVROTSKY A. (1979) Calorimetry: its application to petrology. *Ann. Rev. Earth Planet. Sci.* **7**, 93–115.

NAVROTSKY A. (1982) Trends and systematics in mineral thermodynamics. *Berichte Bunsengesellschaft Phys. Chem.* **86**, 994–1001.

NAVROTSKY A. (1985) Crystal chemical constraints on thermochemistry of minerals. In *Microscopic to Mac-*

roscopic (eds. KIEFFER S. W. and A. NAVROTSKY), Chap. 7, pp. 225–276 Min. Soc. Amer.

NAVROTSKY A. and COONS W. E. (1976) Thermochemistry of some pyroxenes and related compounds. *Geochim. Cosmochim. Acta* **33**, 1281–1288.

NAVROTSKY A., CAPOBIANCO C. and STEBBINS J. (1982) Some thermodynamic and experimental constraints on the melting of albite at atmospheric and high pressure. *J. Geol.* **90**, 679–698.

NAVROTSKY A., HON R., WEIL D. F. and HENRY D. J. (1980) Thermochemistry of glasses and liquids in the systems $CaMgSi_2O_6$-$CaAl_2Si_2O_8$-$NaAlSi_3O_8$, SiO_2-$CaAl_2Si_2O_8$-$NaAlSi_3O_8$ and SiO_2-Al_2O_2-CaO-Na_2O. *Geochim. Cosmochim. Acta* **44**, 1409–1423.

NAVROTSKY A., PINTCHOVSKI F. S. and AKIMOTO S. (1979) Calorimetric study of the stability of high pressure phases in the systems CoO-SiO_2 and "FeO"-SiO_2 and calculation of phase diagrams in MO-SiO_2 systems. *Phys. Earth Planet. Int.* **19**, 275–292.

PERKINS D., ESSENE E. J. and WESTRUM E. T. (1979) New thermodynamic data for diaspore and their application to the system Al_2O_3-SiO_2-H_2O. *Amer. Mineral.* **64**, 1080–1090.

PERKINS D., WESTRUM E. F. and ESSENE E. J. (1980) The thermodynamic properties and phase relations of some minerals in the system CaO-Al_2O_3-SiO_2-H_2O. *Geochim. Cosmochim. Acta* **44**, 61–84.

PERRON G., FORTIER J. and DESNOYERS J. E. (1975) The apparent molar heat capacities and volumes of aqueous NaCl from 0.01 to 3 mol kg^{-1} in the temperature range 274.65 to 318.15 K. *J. Chem. Thermo.* **7**, 1177–1184.

PICKER P. K., LUDUC P., PHILIP P. and DESNOYERS J. E. (1971) Heat capacity of solutions by flow microcalorimetry. *J. Chem. Thermo.* **3**, 631–642.

RICHET P. and BOTTINGA Y. (1984) Anorthite, andesine, wollastonite, diopside, cordierite, and pyrope: thermodynamics of melting, glass transitions, and properties of the amorphous phases. *Earth Planet. Sci. Lett.* **67**, 415–422.

RICHET P., BOTTINGA Y., DANIELOU L., PETITET J. P. and TEQUI C. (1982) Thermodynamic properties of quartz, cristobalite and amorphous SiO_2: drop calorimetry measurements between 1000 and 1800 K and a review from 0 to 2000 K. *Geochim. Cosmochim. Acta* **46**, 2639–2658.

ROBIE R. A. and HEMINGWAY B. S. (1972) Calorimeters for heat of solution and low-temperature heat capacity measurements. *U.S. Geol. Survey Prof. Paper 755,* U.S. Govt. Printing Office, Washington, DC.

ROBIE R. A., HEMINGWAY B. S. and WILSON W. H. (1976) The heat capacities of calorimetry conference copper and of muscovite, pyrophillite, and illite between 15 and 375 K and their standard entropies at 298.15 K. *J. Res. Nat. Bur. Stand.* **4**, 631–644.

ROBIE R. A., HEMINGWAY B. S. and WILSON W. H. (1978) Low-temperature heat capacities and entropies of feldspar glasses and of anorthite. *Amer. Mineral.* **63**, 109–123.

ROSENHAUER M. (1976) Effect of pressure on the melting enthalpy of diopside under dry and H_2O-saturated conditions. *Carnegie Inst. Wash. Yearb.* **75**, 648–650.

SILVER L. and STOLPER E. (1985) A thermodynamic model for hydrous silicate melts. *J. Geol.* **93**, 161–178.

SIMMONS J. M., BUSEY R. H. and MESMER R. E. (1985) Enthalpies of dilution of aqueous calcium chloride at low molalities at high temperatures. *J. Phys. Chem.* **89**, 557–560.

STEBBINS J. F. and CARMICHAEL I. S. E. (1984) The heat of fusion of fayalite. *Amer. Mineral.* **69,** 292–297.

STEBBINS J. F., CARMICHAEL I. S. E. and MORET L. K. (1984) Heat capacities and entropies of silicate liquids and glasses. *Contrib. Mineral. Petrol.* **86,** 137–148.

STEBBINS J. F., WEILL D. F. and CARMICHAEL I. S. E. (1983) The high temperature liquid and glass heat contents and the heats of fusion of diopside, albite, sanadine, and nepheline. *Amer. Mineral.* **68,** 717–730.

STEBBINS J. F., WEILL D. F. and CARMICHAEL I. S. E. (1983) The high temperature liquid and glass heat contents and the heats of fusion of diopside, albite, sanidine, and nepheline. *Amer. Mineral.* **68,** 717–730.

STOLPER E. (1982) The speciation of water in silicate melts. *Geochim. Cosmochim. Acta* **46,** 2609–2620.

STOUT N. D. and PIWINSKII A. J. (1982) Enthalpy of silicate melts from 1520 to 2600 K under ambient pressure. *High Temp. Sci.* **15,** 275–292.

WESTRUM E. F. JR., FURUKAWA G. T. and McCULLOUGH J. P. (1968) Adiabatic low temperature calorimetry. In *Experimental Thermodynamics, Vol. I, Calorimetry of Non-Reacting Systems,* (eds. J. P. McCULLOUGH and D. W. SCOTT), pp. 133–214, Plenum, New York.

WEILL D. F., HON R. and NAVROTSKY A. (1980) The igneous system $CaMgSi_2O_6$-$CaAl_2Si_2O_6$-$NaAlSi_3O_8$: Variations on a classic theme by Bowen. In *Physics of Magmatic Processes,* (ed. R. B. HARGRAVES), pp. 49–92, Princeton Univ. Press.

WESTRICH H. R., HOLLOWAY J. R. and NAVROTSKY A. (1981) Some thermodynamic properties of fluorapatite, fluorpargasite, and fluorphlogopite. *Amer. J. Sci.* **281,** 1091–1195.

WOOD B. J. and KLEPPA O. J. (1984) Chromium-aluminum mixing in garnet: a thermochemical study. *Geochim. Cosmochim. Acta* **48,** 1373–1375.

ZIEGLER D. and NAVROTSKY A. (1986) Direct measurement of the enthalpy of fusion of diopside. *Geochim. Cosmochim. Acta.* **50,** 2461–2466.

Magmatic Processes: Physicochemical Principles
© The Geochemical Society, Special Publication No. 1, 1987
Editor B. O. Mysen

Melt viscosities in the system $NaAlSi_3O_8-H_2O-F_2O_{-1}$

DONALD B. DINGWELL*

Geophysical Laboratory, Carnegie Institution of Washington,
2801 Upton St. N.W., Washington, D.C. 20008, U.S.A.

Abstract—The viscosities of eight melt compositions in the system albite–H_2O–F_2O_{-1} have been determined at high pressure and temperature. Measurements were performed using the falling sphere method at 1000–1600°C and 2.5 to 22.5 kbar.

Fluorine and water strongly decrease the viscosity of albite melt at all pressures and temperatures investigated. The viscosity of a fluorine-bearing albitic melt decreases from 1 bar to 22.5 kbar, whereas there is a viscosity maximum near 7.5 kbar for a water–bearing albite melt. At constant pressure the effects of fluorine and water, compared on an equimolar basis of added F_2O_{-1} and H_2O (or F and OH), are broadly similar and both F_2O_{-1} and H_2O are less effective than Na_2O in reducing the viscosity of albite melt. Log viscosity exhibits a positive curvature as a function of mole fraction of added fluorine or water.

The effect of HF on albite melt viscosity is greater than that predicted from a linear interpolation of the effects of water and fluorine. This negative deviation from additivity may result in extremely low viscosities for granitic magmas enriched in both fluorine and water.

INTRODUCTION

THE MOST IMPORTANT melt structural parameter determining silicate melt viscosity is the degree of polymerization of the aluminosilicate tetrahedra which form the basic building blocks of silicate melts. Model granitic melts of the dry haplogranite system (SiO_2–$NaAlSi_3O_8$–$KAlSi_3O_8$) are fully polymerized liquids with all oxygens shared between two tetrahedral (T) cations (Si or Al) to form a three-dimensional network structure. This fully polymerized structure results in extremely high viscosities for melts of albite, orthoclase and SiO_2 and mixtures of these haplogranite components (HÖF-MAIER and URBAIN, 1968; RIEBLING, 1966; URBAIN et al., 1982). Previous work on the viscosity of silicate melts has confirmed a decrease in melt viscosity accompanying the depolymerization of silicate melts due to the incorporation of water into the melt structure (BURNHAM, 1963; FRIEDMAN et al., 1963; SHAW, 1963; BURNHAM, 1975; STOLPER, 1982).

Preliminary investigations of the effects of fluorine on melt structure have also illustrated a positive correlation between viscosity and polymerization in fluorine–bearing silicate melts (DINGWELL et al., 1985; MYSEN and VIRGO, 1985). A direct comparison of the effects of fluorine and water on the viscosity of a simple silicate melt would be very useful for placing constraints on the relative importance of fluorine and water in determining the viscosities

of granitic melts and on the relationship between structure and viscosity in silicate melts. Until now, such a comparison has not been possible because the limited viscosity data available for water- and fluorine–bearing melts have been obtained under differing conditions of temperature, pressure, bulk composition, and fluorine or water concentration. In order to provide a comparative data base for the effects of water and fluorine on the viscosity of a fully polymerized melt, DINGWELL and MYSEN (1985) initiated a study of the temperature–, pressure–, and composition–dependence of viscosity in the binary systems albite–H_2O and albite–F_2O_{-1}. The present study extends the investigation of melt viscosities in the system albite–H_2O–F_2O_{-1} with direct comparisons of the temperature–, pressure– and concentration-dependence of the effects of fluorine and water on albite melt viscosity. The additive effect of fluorine and water on albite melt viscosity has also been investigated.

Stoichiometry of fluorine and water substitutions

The additions of fluorine and water to albite melt are discussed below in terms of the anionic mol fraction, $X/(X + O)$ where $X = F$, OH. This parameter was calculated from the bulk stoichiometry of the melt composition assuming complete dissociation of water to form OH units. This computation is not intended to reflect the proportion of dissociation of water that occurs in albite melt under the conditions of these experiments but rather to provide a basis for comparison of water with fluorine in terms of the variable $X/(X + O)$. Thus the quantitative comparison of water and fluorine is based

---

* Now at: Earth and Planetary Sciences, Erindale Campus, University of Toronto, Mississauga, Ontario, L5L 1C6, Canada.

on a stoichiometry derived from the known added concentrations of fluorine and water.

The incorporation of fluorine into the albite glasses involves the exchange of two moles of fluorine for one mole of oxygen, denoted by the exchange operator F_2O_{-1} (molar). The substitutions are described by the following equations:

$$NaAlSi_3O_8 + yF_2O_{-1} = NaAlSi_3O_{8-y}F_{2y} \quad (1a)$$

$$NaAlSi_3O_8 + yH_2O = NaAlSi_3O_{8-y}OH_{2y}. \quad (1b)$$

The fluorine and water contents of this study correspond to $X/(X + O) = 0.1, 0.2$ and 0.3. The melts containing these quantities of F or OH are denoted AB1F, AB2F and AB3F; and AB1H, AB2H and AB3H, respectively. The one melt investigated that contains both water and fluorine has $X/(X + O) = 0.2$ and is denoted ABHF (Figure 1).

EXPERIMENTAL

The starting materials for viscometry experiments were water–bearing and fluorine–bearing albitic composition glasses. The water–bearing glasses were synthesized by high–pressure fusion of water + previously prepared albite glass as described by DINGWELL and MYSEN (1985). Small, known amounts of water were added to 200–250 mg of powdered albite glass, sealed in 5 mm diameter platinum capsules and fused in a solid–media high–pressure apparatus at 1400°C and 15 kbar for 1 hour. This method was also used for the preparation of the single hydrous, fluorine–bearing albite glass. The resulting, water–undersaturated, albite glasses were bubble–free and were used in the high–pressure viscosity experiments. The synthesis and analysis of the fluorine–bearing albite glass containing 5.8 weight percent fluorine has been previously described by DINGWELL et al. (1985). Attempts to synthesize albite

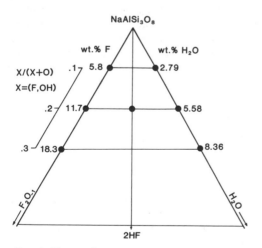

NaAlSi$_3$O$_8$

wt.% F   wt.% H$_2$O

$X/(X+O)$ .1   5.8   2.79
$X=(F,OH)$

.2   11.7   5.58

.3   18.3   8.36

2HF

FIG. 1. The starting compositions used in this study, plotted in the system NaAlSi$_3$O$_8$–H$_2$O–F$_2$O$_{-1}$ (molar). The weight percent additions of fluorine and water listed are equivalent in terms of the molar variable $X/(X + O)$ as defined in the text.

glasses with higher fluorine contents at 1 bar were unsuccessful due to significant volatilization of Si and Na with F during fusion. Therefore, the fluorine–bearing albite glasses containing 11.7 and 18.3 weight percent (nominal) fluorine were synthesized at high pressure. Powder mixes of Al_2O_3, SiO_2, NaF, and AlF_3 were mixed under alcohol in an agate mortar for 2 hours and dried at 110°C. The samples were fired at 400°C for 1 minute immediately prior to sealing in platinum capsules. Then, the 200–250 mg batches of mix were sealed in platinum capsules and fused at 1400°C and 15 kbar for 1 hour. The quenched melts were used for the AB2F and AB3F viscometry runs.

The anhydrous fluorine–free and fluorine–bearing albite glasses were analysed by electron microprobe with wavelength dispersive analysis. Microprobe analyses of the anhydrous fluorine–free and fluorine–bearing glasses are presented in Table 1 along with the operating conditions and standard compositions.

Platinum capsules for viscosity experiments were tightly packed with powdered glass and spheres of platinum were placed near the top of the charge. A small amount of the fine (<1 μm) platinum powder was placed at this level to mark the starting location of the spheres and a thin layer of powdered glass was placed above this level to prevent the spheres from sticking to the capsule.

The viscometry experiments at 2.5 kbar were conducted in an internally–heated pressure vessel using Argon as the pressure medium. The capsules were pressurized to 2.2 kbar and then heated to run temperature at a rate of 35–40°C/min for the final 400°C. The pressure climbed during heating to final run pressure. The viscometry experiments at 7.5, 15 and 22.5 kbar were conducted in a solid–media high–pressure apparatus using the technique described by DINGWELL and MYSEN (1985). Sealed platinum capsules were placed within ¾″–diameter talc–pyrex furnace assemblies with tapered graphite heaters to reduce temperature gradients along the length of the capsule to <10°C (KUSHIRO, 1976). The temperatures were measured with a Pt-Pt$_{90}$Rh$_{10}$ thermocouple with no correction for the effect of pressure on e.m.f. The experimental charges were raised to final pressure plus 15–20% (except for 7.5 kbar experiments which were overpressured to 15 kbar during initial heating) and then heated to 400°C below the final temperature and held there for 1 minute while the pressure was trimmed to the final value. The pressure correction for this procedure is -3%, as calibrated with the quartz-coesite transition. The final temperature was approached at 400°C/min using an automatic controller.

The experiments ranged in duration from 8 to 200 minutes and were quenched by turning off the power to the furnace. The quench rate for the solid–media apparatus is greater than 250°C/sec whereas that for the internally–heated argon vessel is 350°C/min for the first minute.

After an experiment, the capsule was sawn lengthwise on opposite sides to remove part of the platinum capsule walls. The resulting plate of glass was taped to the X–ray film holder of a flat-plate camera, and the film was exposed to Mo white radiation (35 kV, 12 mA) for 10 seconds (HAZEN and SHARPE, 1983). The sphere positions were measured from the developed film with a microscope.

The results of time series experiments by DINGWELL and MYSEN (1985) indicated that the time versus distance relationship for this type of experiment is a straight line passing through the origin. In this study, a time series of two experiments has been performed for each viscosity determination and the relationship reported above has been confirmed.

Table 1. Analyses of starting glasses.

Element	Albite	AB1F	AB2F	AB3F	Stoichiometric Ab
Na	8.79	8.31	9.19	8.03	8.74
Al	10.36	9.15	9.98	9.39	10.29
Si	31.95	31.54	28.90	28.55	32.12
O	48.76	44.5	39.78	35.71	48.84
F	—	5.8	12.4	18.9	—
Total	99.86	99.30	100.25	100.58	100.00

Analysis of Albite is from DINGWELL and MYSEN (1985); analysis of AB1F is from DINGWELL *et al.* (1985). Analyses of AB2F and AB3F by wavelength dispersive analysis using a JEOL JSM-35 instrument and Krisel control system. Operating conditions were 15 kV accelerating voltage, 150 nA beam current on carbon, and 200 sec maximum count times using a 10 × 10 micron raster and moving the sample under the beam continuously. Standards were a synthetic diopside-jadeite glass for Na, Al and Si and AB1F for F. Oxygen by stoichiometry. Microprobe uncertainties quoted at 3σ are: ±3% relative for Na, Al and Si and ±7% relative for F.

The experiments yield sphere position data relative to the platinum powder datum. Settling velocity data (accuracy = ±5%) derived from the sphere positions versus time were related to the viscosity of the silicate melt by Stokes law for the settling of a sphere in a less dense media:

$$\eta = 2r^2 g \Delta\rho / 9v \qquad (2)$$

where η is the viscosity, v is settling velocity, $\Delta\rho$ is density contrast, g is acceleration due to gravity and r is the sphere radius (all in cgs units). The density contrast, $\Delta\rho$, between platinum metal and silicate melt in this system is so large ($\Delta\rho = 19.1 \pm 0.1$ g/cm^3 at 1 bar and 20°C) (DINGWELL and MYSEN, 1985) that corrections to this term due to the different compressibilities and thermal expansion coefficients of metal and melt yield insignificant changes (<1%) in the value of $\Delta\rho$ in equation (2). Therefore, $\Delta\rho$ was assigned a value of 19.1 for the calculation of viscosities with Equation (2). The final measured quantity in Equation (2) is the sphere radius, r, which was measured with a microscope to an accuracy of ±0.0005 cm, corresponding to a maximum of ±5% uncertainty for the range of sphere sizes used in this study (0.0100 to 0.0200 cm radii).

The Faxen correction for extra drag exerted on the settling sphere by capsule walls (SHAW, 1963; KUSHIRO, 1976) was applied to the results of Equation (2) with the term:

$$1 - 2.104(r/R) + 2.09(r/R)^3 - 0.95(r/R)^5 \qquad (3)$$

where r is the sphere radius and R is the capsule radius. In practice, this correction term results in a 15% reduction in calculated viscosities for the sphere and capsule dimensions of this study. Table 2 lists the measured sphere radii and calculated settling velocities and viscosities obtained in this study along with the preliminary data of DINGWELL and MYSEN (1985) in this system.

RESULTS

The temperature dependence of viscosity has been expressed as a log–linear, Arrhenius function of reciprocal absolute temperature using an equation of the form;

$$\log_{10} \eta = A + B/2.303RT, \qquad (4)$$

where R is the gas constant, $\log_{10} \eta$ is the logarithm of viscosity at temperature, T (in K) and A and B

are termed the pre-exponential or frequency factor and the activation energy of viscous flow, respectively. There is ample experimental evidence that the viscosity-temperature relationships of most silicate melts are non–linear over larger temperature ranges (>500°C), exhibiting, instead, a positive curvature ($\partial^2(\log \eta)/\partial(1/T)^2 > 0$) versus reciprocal absolute temperature (RICHET, 1984). The data in this study, however, are not extensive enough to permit the calculaton of non–linear functions for the temperature dependence of viscosity and the linear approximations to the temperature-viscosity relationships provide extrapolations to lower temperature which must be viewed as lower limits for the low temperature viscosities of these melts.

The pressure dependence of melt viscosity has been studied for two melt compositions, AB1H and AB1F. The data for AB1F have been obtained at 1200 and 1400°C and pressures of 7.5, 15 and 22.5 kbar. These data are presented in Figure 2 along with the data of KUSHIRO (1978) for albite melt at high pressure and the 1 bar data of DINGWELL *et al.* (1985) for AB1F.

The viscosity of AB1F melt decreases with increasing pressure from 1 bar to 22.5 kbar. The viscosity decrease is not linear but describes instead a positive curvature with respect to pressure (Figure 2). Comparison of the 1400°C curves of Figure 2 for albite and AB1F reveals that the effect of F$_2$O$_{-1}$ on the viscosity of albite melt is largest in the pressure range of 6–8 kbar. The Arrhenian activation energy of viscous flow for AB1F is constant within the uncertainties of the data from 1 atm to 22.5 kbar (Figure 2).

The pressure dependence of viscosity of AB1H is complex compared with that of AB1F (Figures 2,3). The 1200°C data exhibit an increasing viscosity from 2.5 to 7.5 kbar and a decreasing viscosity

Table 2. Viscosity data for the system Albite–H_2O–F_2O_{-1}

Composition	Temperature (°C)	Pressure (kbar)	Radius ($\times 10^{-2}$ cm)	Speed ($\times 10^{-4}$ cm/sec)	$\log_{10}\eta$ (poise)
Ab	1600	7.5	1.0	0.789	3.65
AB1F	1200	7.5	—	—	3.70
	1400	7.5	—	—	3.11
	1200	15.0	—	—	3.46
	1400	15.0	—	—	2.91
	1200	22.5	—	—	3.20
	1400	22.5	—	—	2.63
AB2F	1200	7.5	1.00	1.75	3.30
	1400	7.5	1.00	9.85	2.55
AB3F	1200	7.5	1.33	8.43	2.85
AB1H	1100	2.5	1.67	0.253	4.55
	1200	2.5	1.67	1.27	3.85
	1200	7.5	1.33	0.474	4.10
	1400	7.5	—	—	2.81
	1200	15.0	1.55	1.57	3.70
	1400	15.0	—	—	2.57
	1400	22.5	—	—	2.60
AB2H	1000	7.5	1.33	1.06	3.75
	1200	7.5	1.00	6.18	2.76
AB3H	1000	7.5	1.00	1.25	3.45
	1200	7.5	1.00	15.7	2.35
ABHF	1000	7.5	1.33	2.38	3.40
	1200	7.5	1.00	8.00	2.64

Uncertainties are as follows: T, $\pm 10°C$; P, ± 0.5 kbar; radii, $\pm .0005$ cm; velocity, $\pm 5\%$; viscosity, $\pm 15\%$.

from 7.5 to 15 kbar, resulting, therefore, in a viscosity maximum for AB1H at 1200°C and a pressure of 7.5 kbar. The 1400°C data indicate a decreasing viscosity with increasing pressure from 7.5 to 15 kbar and a constant viscosity with further pressure increase to 22.5 kbar. The activation energy for AB1H from 2.5 to 15 kbar is constant.

The composition dependence of viscosity along the binary joins albite–H_2O and albite–F_2O_{-1} has been investigated at 7.5 kbar pressure in the range of 1000 to 1600°C (Figure 4). The viscosity of albite melt at 1600°C and 7.5 kbar was measured to determine the temperature dependence of albite melt at 7.5 kbar in combination with interpolated data from KUSHIRO (1978) at 1400°C. The data for AB1H, AB2H and AB3H illustrate that the isothermal decrease of melt viscosity with addition of H_2O to albite melt is a strongly nonlinear, positively curved function (see below). The activation energy of viscous flow decreases from 89.6 kcal/mol for albite melt to 72.7 kcal/mol for AB1H, whereas the values for AB2H and AB3H are constant within error at 42.5 kcal/mol and 47.2 kcal/mol, respectively. Similarly, the isothermal dependence of vis-

cosity on F_2O_{-1} content for melts on the join albite–F_2O_{-1} exhibits a very strong positive curvature.

The viscosity of ABHF has been determined at 1000 and 1200°C and 7.5 kbar (Figure 5). The viscosity of ABHF is lower than that of AB2H. The viscosity of ABHF is lower than that of either AB2F or AB2H at 1000°C and lower than the viscosity of AB2F but equal to the viscosity of AB2H at 1200°C.

DISCUSSION

As noted above the equivalence of the stoichiometry of F– and OH–bearing melts of this study, in terms of the molar variable $X/(X + O)$, permits a comparison of the effects of F_2O_{-1} and H_2O on the viscosity of albite on an equimolar basis. Qualitatively, several aspects of the viscosities of albite melts with equimolar additions of F_2O_{-1} and H_2O are similar. The activation energy and viscosity of albite melt decrease strongly with the addition of F_2O_{-1} or H_2O and secondly, and the rates of decrease of activation energy and viscosity diminish as a function of added F_2O_{-1} or H_2O. On closer inspection, however, it is clear that the viscosities

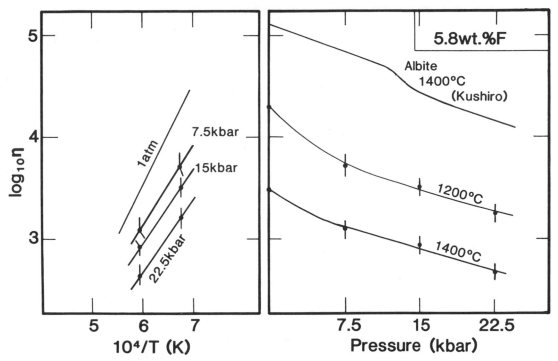

FIG. 2. The viscosity of AB1F melt at 1200–1400°C and 1 bar to 22.5 kbar: (a) viscosity data for AB1F melt at 1 bar, 7.5, 15 and 22.5 kbar plotted versus reciprocal absolute temperature; (b) the pressure dependence of AB1F melt viscosity at 1200 and 1400°C. (1 bar AB1F data from DINGWELL *et al.*, 1985; Ab viscosity data at 1400°C from KUSHIRO, 1978).

of AB1F and AB1H, AB2F and AB2H, and AB3F and AB3H are not equivalent. The relationship between viscosity and composition for the albite–H_2O and albite–F_2O_{-1} systems is demonstrated in Figure 6 where log η is plotted vs. $X/(X + O)$ for experiments at 1200°C. Also included in Figure 6 are data for the system albite–Na_2O that were interpolated from the systems $Na_2Si_2O_5$–$Na_4Al_2O_5$ (DINGWELL, 1986) and $Na_2Si_4O_9$–$Na_6Al_4O_9$ (DINGWELL *et al.*, in prep.) which intersect the join albite–Na_2O at 0.67 and 0.50 mol fraction Na_2O, respectively. There is evidence from studies of peralkaline melts in the system Na_2O–Al_2O_3–SiO_2 that the effect of adding excess Na_2O to albite melt results in depolymerization due to coordination of nonbridging oxygens (NBOs) by Na atoms. Such a solution mechanism involves no change in the number of tetrahedral (T) cations (*i.e.*, there is no evidence for non-tetrahedral coordination of Al in these melts). Therefore $X/(X + O)$ is linearly proportional to NBO/T or bulk polymerization for this join. Figure 6 illustrates therefore that in the system albite–Na_2O, log η is not a linear function of bulk polymerization. This observation is consistent with the more general observation that log η varies greatly

in binary systems of constant bulk polymerization (RIEBLING, 1966; DINGWELL, 1986).

A previous discussion of the compositional dependence of log viscosity in the system granite–water by STOLPER (1982) suggested that the curvature of log viscosity as a function of added water was due to the increasing proportion of water dissolved in the form of undissociated, molecular water as total water content increased. As discussed by DINGWELL and MYSEN (1985) the coincidence of dissociation and log viscosity curves illustrated in Figure 14 of STOLPER (1982) implies that the functional form of log viscosity corrected to the anionic proportion of OH units (and therefore NBO/T by the solution mechanism of Stolper (1982)) would be linear whereas the data of Figure 6 indicate that the system albite–Na_2O exhibits a slightly curved relationship between log viscosity and $X/(X + O)$.

The system albite–F_2O_{-1} also exhibits a strongly curved log η–$X/(X + O)$ relationship. It is unlikely that this viscosity behavior can be explained in a manner analogous to that of STOLPER (1982) for the granite–H_2O system simply because there is no evidence for molecular fluorine in these melts (MYSEN and VIRGO, 1985). What alternative explana-

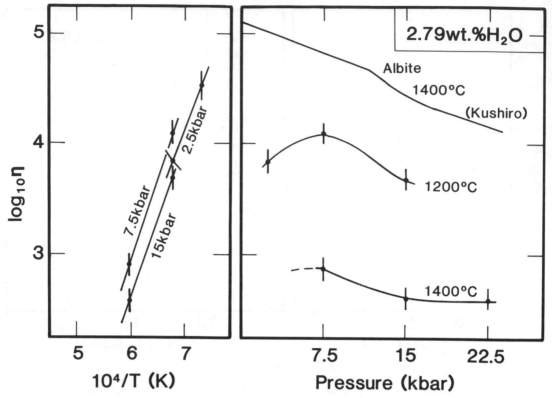

FIG. 3. The viscosity of AB1H melt at 1100–1400°C and 2.5 to 22.5 kbar: (a) viscosity data for AB1H melt at 2.5, 7.5 and 15 kbar plotted versus reciprocal absolute temperature; (b) the pressure dependence of AB1H melt viscosity at 1200 and 1400°C. (albite viscosity data at 1400°C from KUSHIRO, 1978).

tions exist to explain the extreme curvature in log viscosity versus $X/(X + O)$? Two possibilities are (1) the expulsion of tetrahedral cations during fluorine solution such that $X/(X + O)$ is no longer linearly related to bulk polymerization expressed as NBO/T and (2) variation in the types of cations coordinating fluorine as a function of fluorine concentration such that the rate of depolymerization per mole of fluorine is not constant. Both of these concepts have been put forth by MYSEN and VIRGO (1985) in their interpretation of Raman spectra in the system albite–F_2O_{-1}.

There is evidence to suggest that the configurational entropy theory of viscous flow (ADAM and GIBBS, 1965) provides a self–consistent and independently verifiable framework for interpreting the temperature and composition dependence of the viscosity of silicate melts (RICHET, 1984; HUMMEL and ARNDT, 1985; BREARLEY et al., 1986; RICHET et al., 1986; TAUBER and ARNDT, 1986). This theory states that the viscosity of a silicate melt is related to the temperature and configurational entropy of the melt by the following relationship:

$$\log_{10} \eta = Ae + Be/TS_{conf}, \qquad (5)$$

where Ae and Be are constants for a given melt composition and S_{conf} (configurational entropy) varies with temperature in a manner that may be computed where sufficient thermodynamic data exist (specifically, C_p data for liquid, crystal and glass; the configurational entropy of the crystal and the fusion temperature and entropy of fusion of the crystal). Thus, the composition–dependence of viscosity of a silicate melt at temperature, T is determined by three independent parameters, Ae, Be and S_{conf}. The slope parameter, Be, is interpreted as a potential energy barrier to cooperative rearrangements of melt units that must occur during viscous flow. If this energy barrier is an expression of the average bond strength of silicate melts then the value of Be should be positively correlated with melt polymerization. It follows from Equation (5) that it is the temperature dependence of S_{conf} which produces the non–linear or non–Arrhenian behavior of melt viscosity as a function of reciprocal absolute temperature.

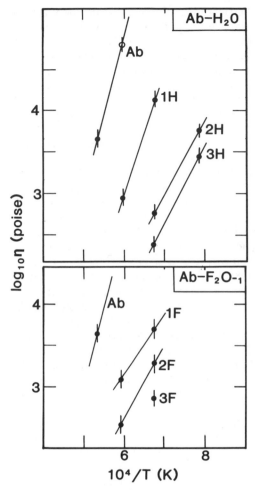

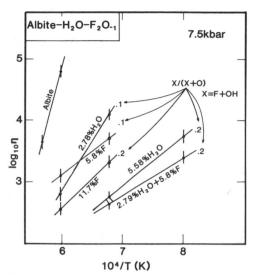

FIG. 5. Viscosities of melts in the system NaAlSi₃O₈–H₂O–F₂O₋₁ with $X/(X + O) = 0.1$ and 0.2 plotted versus reciprocal absolute temperature.

The negative deviation of the viscosity of ABHF from the weighted mean of the viscosities of AB2F and AB2H (the straight lines in Figure 7) increases with decreasing temperature such that the exchange

FIG. 4. The composition dependence of melt viscosity in the systems NaAlSi₃O₈–H₂O and NaAlSi₃O₈–F₂O₋₁ at 7.5 kbar. (albite data at 1400°C from KUSHIRO, 1978).

The composition–dependence of viscosity along some binary silicate joins that exhibit diminishing negative deviations of viscosity from linear interpolation are explained for systems of constant polymerization such as albite-anorthite and K₂Si₃O₇–Na₂Si₃O₇ by the increase in configurational entropy in these systems due to the entropy of mixing term (HUMMEL and ARNDT, 1985; RICHET, 1984).

Figure 7 illustrates the viscosity temperature relationships for the binary join AB2F–AB2H. This join represents the exchange of F₂O₋₁ for H₂O (or F for OH) under conditions of constant $X/(X + O)$. Clearly, the presence of up to 0.1 anionic mol fraction of F or OH does not diminish the viscosity-reducing effect of a further addition of either OH or F, respectively, when compared with the viscosity of AB2H or AB2F.

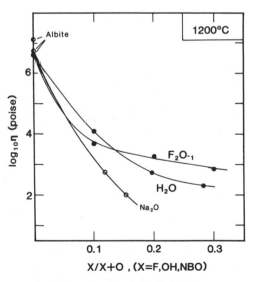

FIG. 6. The relative effects of three depolymerizing components, F₂O₋₁, H₂O and Na₂O on the viscosity of albite melt compared on the equimolar basis of $X/(X + O)$ where X = F, OH, NBO (P = 7.5 kbar). (albite–Na₂O data (1 bar) from DINGWELL, 1986 and DINGWELL et al., in prep.; albite data (1 bar) from RIEBLING, 1966; SCARFE and CRONIN, 1986; URBAIN et al., 1982). NBO—nonbridging oxygen. In the system NaAlSi₃O₈–Na₂O, one nonbridging oxygen is formed per Na added to the system.

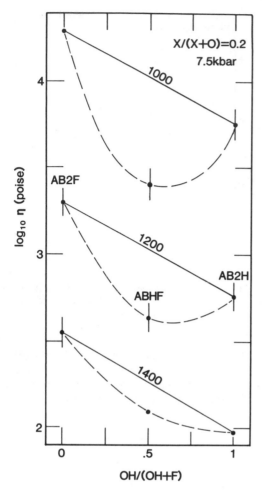

FIG. 7. The variation of melt viscosity with composition along the join AB2F–AB2H. The lower activation energy of ABHF in the temperature range investigated results in a negative deviation of viscosity from additivity, which increases with diminishing temperature. (Points without error bars are extrapolated.)

of F for OH results in a decrease in melt viscosity at temperatures below 1400°C.

As noted above, a negative deviation of viscosity from additivity that increases with decreasing temperature is the behavior predicted from the configurational entropy concept of viscous flow (ADAM and GIBBS, 1965; RICHET, 1984). According to this concept, the mixing of AB2F and AB2H melts to form ABHF could result in an additional contribution to the total configurational entropy of the mixture due to a configurational entropy of mixing term arising from F–OH mixing on melt structural sites.

The general behavior exhibited in Figure 7 will result in extremely large negative deviations from additivity and correspondingly low viscosities for F

+ OH–bearing albite melts at the lower temperatures relevant to the petrogenesis of F– and OH– rich granitic magmas. The data of Figure 7 are the first clear indication the fluorine and water in combination will be even more effective in reducing melt viscosities than previous data on fluorine–bearing and water–bearing melt viscosities have indicated (BURNHAM, 1963; SHAW, 1963; DINGWELL et al., 1985; DINGWELL and MYSEN, 1985).

The data available on the pressure dependence of silicate melt viscosity includes examples of several types of behavior. Both $\delta(\log \eta)/\delta P$ and $\delta^2(\log \eta)/\delta P^2$ may be positive, negative or equal to zero depending on melt composition. Only two studies provide data on the pressure dependence of melt viscosity as a systematic function of melt composition; one by KUSHIRO (1981) in a system of constant polymerization and one by BREARLEY et al. (1986) in a system of varying polymerization. KU-SHIRO (1981) investigated the pressure dependence of the viscosities of melts on the join $SiO_2–CaAl_2O_4$ observing a linear pressure dependence of melt viscosity that changed from a negative value $(\partial(\log \eta)/\partial P < 0)$ at mol fraction $CaAl_2O_4 = 0.15$ and 0.2, through pressure invariance $(\partial(\log \eta)/\partial P = 0)$ at $CaAl_2O_4 = 0.33$ to a positive pressure dependence of viscosity $(\partial(\log \eta)/\partial P > 0)$ at mol fraction $CaAl_2O_4 = 0.4$.

BREARLEY et al. (1986) have investigated the pressure dependence of viscosity of melts on the join albite-diopside. The transition from negative to positive pressure dependence of viscosity occurs at 10–15 kbar in the intermediate composition $Di_{75}Ab_{25}$ which exhibits a viscosity minimum in this pressure range. Thus, the transition from negative to positive pressure–dependence of melt viscosity may occur in a different manner along this join of variable polymerization than that which is observed for joins of constant polymerization. Probably the non–linearity of the pressure–dependence of viscosity of many relatively depolymerized silicate melts is a response to variations of the proportions of various polyanionic units in these melts (SCARFE et al., 1979).

Although it is premature to reach conclusions on the structural significance of the pressure–dependence of the viscosities of silicate melts; the contrasting pressure dependence of the viscosities of AB1F and AB1H indicates that some structural parameter is significantly different for these two melts. One possible explanation of the positive pressure dependence of viscosity of AB1H at low pressure is that the speciation of dissolved water in this melt may be pressure–dependent. If the speciation of water in albite melt is quenchable (STOLPER et al., 1983; MCMILLAN et al., 1983) then the pressure dependence of water speciation in this undersaturated melt may require further investigation.

SUMMARY

Investigation of the pressure–, temperature– and composition–dependence of melt viscosity in the system albite–H_2O–F_2O_{-1} has provided an experimental framework for further evaluation of structural models of melt viscosity and has provided a model system for the estimation of viscosities of fluorine– and water–rich granitic magmas.

The viscosity–composition relationships observed in the system albite–Na_2O, albite–H_2O and albite–F_2O_{-1} are qualitatively similar. If the diminishing effectiveness of H_2O in reducing melt viscosity is attributed to the increasing proportion of H_2O dissolved as the molecular species then alternative, more and less effective mechanisms must be provided to explain the viscosity reductions due to F_2O_{-1} and Na_2O, respectively. If current models for the structure of peralkaline melts in the system Na_2O–Al_2O_3–SiO_2 are correct, then the variation of log viscosity with melt polymerization shows a positive curvature. Thus attempts to rationalize melt structure to obtain a linear relationship between log viscosity and polymerization may lead to erroneous conclusions.

The additive effect of fluorine and water on the viscosity of silicate melts is greater than that predicted from a linear interpolation from the albite–H_2O and albite–F_2O_{-1} systems. This negative deviation from additivity results in the estimation of extremely low viscosities for granitic magmas enriched in both water and fluorine.

Acknowledgements—I am grateful to Greg Muncill for operating the internally–heated pressure vessel during the 2.5 kbar runs. M. Dickenson, M. Pichavant, and B. Mysen provided useful reviews. P. Denning assisted with manuscript preparation.

REFERENCES

ADAM G. and GIBBS J. H. (1965) On the temperature dependence of cooperative relaxation properties in glass forming liquids. *J. Phys. Chem.* **43**, 139–146.

BOTTINGA Y. and WEILL D. F. (1972) The viscosity of magmatic liquids: a model for calculation. *Amer. J.Sci.* **272**, 438–475.

BREARLEY M., DICKINSON J. and SCARFE C. M. (1986) Pressure dependence of melt viscosity on the join albite–diopside: implications for melt structure. *Geochim. Cosmochim. Acta* (In press).

BURNHAM C. WAYNE (1963) Viscosity of a water–rich pegmatite. (abstr.) *Geol. Soc. Am. Special Pap.* **76**, 26.

BURNHAM C. WAYNE (1975) Water and magmas: a mixing model. *Geochim. Cosmochim. Acta* **39**, 1077–1084.

DINGWELL D. B. (1986) Viscosity–temperature relationships in the system $Na_2Si_2O_5$–$Na_4Al_2O_5$. *Geochim. Cosmochim. Acta* **50**, 1261–1265.

DINGWELL D. B. and MYSEN B. O. (1985) Effects of fluorine and water on the viscosity of albite melt at high pressure: a preliminary investigation. *Earth Planet. Sci. Lett.* **74**, 266–274.

DINGWELL D. B., SCARFE C. M. and CRONIN D. J. (1985) The effect of fluorine on viscosities in the system Na_2O–Al_2O_3–SiO_2: implications for phonolites, trachytes and rhyolites. *Amer. Mineral.* **70**, 80–87.

FRIEDMAN I., LONG W. and SMITH R. L. (1963) Viscosity and water content of rhyolite glass. *J. Geophys. Res.* **68**, 6523–6535.

HAZEN R. M. and SHARPE M. R. (1983) Radiographic determination of the positions of platinum spheres in density–viscosity studies of silicate melts. *Carnegie Inst. Wash. Yearb.* **82**, 428–430.

HOFMAIER G. and URBAIN G. (1968) The viscosity of pure silica. *Sci. Ceram.* **4**, 25–32.

HUMMEL W. and ARNDT J. (1985) Variation of viscosity with temperature and composition in the plagioclase system. *Contrib. Mineral. Petrol.* **90**, 83–92.

KUSHIRO I. (1976) Changes in viscosity and structure of melt of $NaAlSi_2O_6$ composition at high pressures. *J. Geophys. Res.* **81**, 6347–6350.

KUSHIRO I. (1978) Viscosity and structural changes of albite ($NaAlSi_3O_8$) melt at high pressures. *Earth Planet. Sci. Lett.* **41**, 87–90.

KUSHIRO I. (1981) Change in viscosity with pressure of melts in the system SiO_2–$CaAl_2O_4$. *Carnegie Inst. Wash. Yearb.* **80**, 339–341.

McMILLAN P. F., JAKOBSSON S., HOLLOWAY J. R. and SILVER L. A. (1983) A note on the Raman spectra of water-bearing albite glasses. *Geochim. Cosmochim. Acta* **47**, 1937–1943.

MYSEN B. O. and VIRGO D. (1985) Structure and properties of fluorine–bearing aluminosilicate melts: the system Na_2O–Al_2O_3–SiO_2–F at 1 atm. *Contrib. Mineral. Petrol.* **91**, 205–220.

RICHET P. (1984) Viscosity and configurational entropy of silicate melts. *Geochim. Cosmochim. Acta* **48**, 471–483.

RICHET P., ROBIE R. A. and HEMINGWAY B. S. (1986) Low-temperature heat capacity of glassy diopside ($CaMgSi_2O_6$): A calorimetric test of the configurational-entropy theory applied to the viscosity of liquid silicates. *Geochim. Cosmochim. Acta* **50**, 1521–1533.

RIEBLING E. F. (1966) Structure of sodium aluminosilicate melts containing at least 50 mole% SiO_2 at 1500°C. *J. Phys. Chem.* **44**, 2857–2865.

SCARFE C. M. and CRONIN D. J. (1986) Viscosity–temperature relationships of melts at 1-atm in the system albite-diopside. *Amer. Mineral.* **71**, 767–771.

SCARFE C. M., MYSEN B. O. and VIRGO D. (1979) Changes in viscosity and density of melts of sodium disilicate, sodium metasilicate and diopside composition with pressure. *Carnegie Inst. Wash. Yearb.* **78**, 547–551.

SHAW H. R. (1963) Obsidian-H_2O viscosities at 1000 and 2000 bars in the temperature range 700° to 900°C. *J. Geophys. Res.* **68**, 6337–6343.

STOLPER E. (1982) Water in silicate glasses: an infrared spectroscopic study. *Contrib. Mineral. Petrol.* **81**, 1–17.

STOLPER E., SILVER L. A. and AINES R. (1983) The effects of quenching rate and temperature on the speciation of water in silicate glasses. (abstr.) *EOS* **64**, 339.

TAUBER P. and ARNDT J. (1986) Viscosity-temperature relationship of liquid diopside. *Phys. Chem. Earth Plan. Inter.* **43**, 97–103.

URBAIN G., BOTTINGA Y. and RICHET P. (1982) Viscosity of liquid silica, silicates and aluminosilicates. *Geochim. Cosmochim. Acta* **46**, 1061–1072.

Magmatic Processes: Physicochemical Principles
© The Geochemical Society, Special Publication No. 1, 1987
Editor B. O. Mysen

The calculated individual effects of pressure and water content on phase equilibria in the granite system

HANNA NEKVASIL

Chemistry Department, Arizona State University, Tempe, AZ 85287 U.S.A.

and

C. WAYNE BURNHAM

Department of Geosciences, The Pennsylvania State University, University Park, PA 16802 U.S.A.

Abstract—The revised quasi–crystalline model has been used to calculate phase relations in the granite (An–Ab–Or–Qz–(H_2O)) system and bounding binary and ternary subsystems for pressures and water contents of relevance to felsic magma crystallization. The predicted phase relations have been compared with available H_2O–saturated experimental phase equilibrium data on the subsystems Ab–Or–H_2O, An–Ab–H_2O and An–Or–H_2O and indicate good agreement. Comparisons of experimental data against calculated phase relations for the bounding subsystems Ab–Or–Qz–H_2O, Ab–An–Qz–H_2O, Ab–Or–An–H_2O and An–Or–Qz–H_2O also indicate favorable agreement, particularly for the An–poor regions.

Use of the model in a predictive capacity to calculate H_2O–undersaturated phase relations for the granite system and relevant subsystems indicates that the effect of decreasing pressure or increasing H_2O content is the differential contraction of the Qz + L field and concommitant shift of the feldspar/quartz cotectics towards more Qz–rich melt compositions. Although H_2O–saturated phase relations also indicate a shift of the feldspar/quartz cotectics toward the Qz apex with decreasing pressure, the magnitude of this shift is less than for H_2O–undersaturated conditions. The shift in cotectic position under H_2O–saturated conditions is attributable to the combination of the opposing effects of changing both pressure and H_2O content.

INTRODUCTION

DEDUCTION of the crystallization and melting histories of igneous rocks purely from studies of natural rocks is complex because natural rocks are end products of numerous possible magmatic processes. In order to be able to evaluate the effects of natural magmatic processes on the crystallization behavior of magmas, the individual effects of pressure, volatile content and melt composition on phase equilibria must be understood. These variables can be isolated through experimentation under controlled laboratory conditions. Information obtained through experimental investigations on the effects of these variables on phase equilibria can be combined with data on natural rocks (*i.e.*, compositional, modal, and textural) to obtain more specific information regarding possible magmatic histories.

Experimental petrologic research has followed two main directions. Natural rock compositions have been melted under pressure and temperature conditions, and volatile contents which approximate those believed to exist within the earth's crust and mantle (*e.g.*, PIWINSKII and WYLLIE, 1968; 1970; EGGLER, 1972; CLEMENS and WALL, 1981). Such experimental investigations can effectively isolate variables such as pressure and volatile content; however, they do not effectively isolate compositional variables. It is, thus, difficult to generalize quantitatively the results of such experimental investigations and apply them to rocks much different in composition.

An alternate experimental approach was exemplified by the pioneering efforts of BOWEN (1913) and TUTTLE and BOWEN (1958) and more recently by workers such as BOETTCHER *et al.* (1984). This approach focuses on compositionally simple systems in order to evaluate systematically the effects of varying compositional variables in addition to the intensive variables pressure and temperature. Attempts to use information on the melting behavior of simple systems to explain the behavior of more complex systems, have led to the use of thermodynamic formalisms (*e.g.*, a Margules formalism). These formalisms have been applied to the interpretation of phase equilibrium data in order to model melt component mixing behavior (*e.g.*, BOTTINGA and RICHET, 1978; BERMAN and BROWN, 1984). Alternatively, direct calorimetric measurements of heats of mixing in simple systems have been conducted (*e.g.*, NAVROTSKY *et al.*, 1980; HERVIG and NAVROTSKY, 1984) which obviate the need for use of phase equilibrium data in the evaluation of melt component mixing behavior. Attempts have been made to use calorimetric results to calculate liquidus surfaces for systems in which insufficient calorimetric data are available (*e.g.*, WEILL *et al.*, 1980; NAVROTSKY *et al.*, 1980) with varying degrees of success. However, none of the above approaches have been able to predict phase equilibria in systems more complex than ternary using data from simple systems. GHIORSO *et al.* (1983) avoided the problems involved in extending the results of such approaches to complex natural compositions by using a Margules formalism to obtain mixing parameters from experimental data on natural multicomponent mafic compositions in a massive regression approach (see also GHIORSO and KELEMAN, 1987). Such an approach has intrinsic problems in terms of predictive capabilities for compositions lying outside the range used in the regression. Although the final goal of all modelling attempts on silicate melts is the accurate prediction of phase relations for natural compositions, insight into the chemical principles influencing melt behavior (*e.g.*, mixing behavior of melt components) can most readily be obtained through study of the behavior of compositionally simple systems as well as of the variations in phase relations induced by systematically increasing the number of compositional variables. Such an approach was adopted by BURNHAM (1981) in his development of the "quasi–crystalline" model.

The quasi–crystalline model of silicate melts has been extensively revised and refined in order to permit calculation of phase equilibria in the system Ab–Or–An–Qz–H_2O (granite system) for a variety of pressures and H_2O contents of relevance to felsic magma genesis (NEKVASIL, 1986). The modifications and refinements of the model as well as the sources of data used in the development of the revised model have been outlined in BURNHAM and NEKVASIL (1986). Calculated phase relations based on the revised model differ to the greatest extent from those based on the model of BURNHAM (1981) in the plagioclase-bearing subsystems of the granite system and differ only slightly in the haplogranite system. Through application of the revised model (BURNHAM and NEKVASIL, 1986), it was implied that use of experimental data for the fusion of the pure aluminosilicates as well as for the aluminosilicate-quartz (A + Qz) solidi in combination with the speciation model, permits calculation of phase equilibria throughout the granite system. In the following discussion, the model will be used to demonstrate the calculated individual effects of pressure and H_2O content on phase equilibria in the granite system. However, as a preliminary step to the utilization of the predictive capabilities of the model,

such capabilities will be assessed by comparing calculated phase relations with those experimentally determined for subsystems of the granite system.

THE REVISED QUASI–CRYSTALLINE MODEL

The quasi–crystalline model (BURNHAM, 1981; BURNHAM and NEKVASIL, 1986) treats igneous rock melts as multicomponent solutions of chemically discrete and thermodynamically distinct neutral complexes or species that are correlated with thermodynamic components. It is considered in the model that these species mimic the stoichiometry of the solid phases that crystallize from the melts. Additionally, further species may be produced by homogeneous speciation reactions (*i.e.*, dissociation of a component or interaction between two or more components). The melt species (components) produced by speciation reactions have different stoichiometries and different thermodynamic properties relative to the components involved in the homogeneous reactions.

The revised quasi–crystalline model (BURNHAM and NEKVASIL, 1986; NEKVASIL, 1986) retains the basic premises of the model as outlined by BURNHAM (1981). The refinements presented by the revised model are mainly constrained to the quantified aspects of the model. Refined internally-consistent expressions for the standard state free energies of fusion for the components *ab* (albite), *an* (anorthite), *or* (sanidine) and *qz* (Si_4O_8) were obtained. This was done by using phase equilibrium data for the systems Ab–An (BOWEN, 1913) at 1 atm, Or–H_2O (LAMBERT *et al.*, 1969), Qz–H_2O (KENNEDY *et al.*, 1962) and Qz (JACKSON, 1976), calorimetric heats of fusion, and available data on molar volumes, thermal expansions and compressibilities for the components. (See BURNHAM and NEKVASIL, 1986 for details on the data used and their sources.) The resulting expressions for the standard state free energies of fusion of the four components (listed in BURNHAM and NEKVASIL, 1986) were used to calculate the activities of the melt components *ab*, *an* and *or* along their experimentally-determined anhydrous solidi. The albite–H_2O model (BURNHAM, 1975) was used in addition, to evaluate the activities of these components along their H_2O–saturated solidi. In the selection of experimental solidi for this purpose, preference was given to the recent data of workers such as BOETTCHER *et al.* (1982, 1984).

As was predicted by BURNHAM (1981) and noted by BOETTCHER *et al.* (1982, 1984), with increasing pressure, the activities of the components *ab* and *an*, calculated using the revised model, decrease from 1.0 along their anhydrous and H_2O–saturated solidi. For the *or* component, however, the calculated activity along its dry and H_2O–saturated solidi decreases from 1.0 with *decreasing* pressure, reflecting the direction of increased proximity to the incongruent melting region (for details on the experimental data considered, see BURNHAM and NEKVASIL, 1986). The extent of lowering of the activities (relative to the experimental mol fractions) along the solidi has been interpreted as indicative of the extent of dissociation of the components in the melt through homogeneous reactions (BURNHAM, 1981; BOETTCHER *et al.,* 1984), and has been quantified as a function of pressure. In order to interpret the melt component behavior in the aluminosilicate-quartz binaries, the calculated lowering of the melt component activities (relative to their experimental mol fractions) at the eutectics was assumed to be attributable solely to interactions between the two relevant components. These interactions were considered describable by homogeneous stoichiometric reactions (NEKVASIL, 1986; BURNHAM and NEKVASIL, 1986). By using experimental pressure-temperature data on the aluminosilicate (A) + Qz solidi, the activity of each silicate component was calculated. A stoichiometric relation describing the lowering of the activities of each component from its experimental mole fraction entirely by interaction of the two components was formulated for each A–Qz binary. The stoichiometric relations permit calculation of the eutectic composition and its variation with pressure for each Qz–bearing binary system. Each stoichiometric relation then yielded the extent of interaction of *qz* (silica melt component) with the aluminosilicate component *a*. The extent of interaction of *a* with *qz* in each of the anhydrous and H_2O–saturated Qz–bearing "binaries" was quantified as a function of pressure using the calculated eutectic compositions. A detailed discussion is presented in NEKVASIL (1986) for each binary and the resulting equations have been summarized in BURNHAM and NEKVASIL (1986).

It was assumed that the lowering of the activity of *qz* and *a* in the region $X_a \leq X_a(E)$ (where $X_a(E)$ is the binary eutectic composition) is solely a result of interaction (and is not a combined result of dissociation of *a* as well as interaction of *a* with *qz*). This assumption was adopted due to the remarkable agreement of the predicted variation in eutectic composition with pressure with available compositional data in the A + Qz systems and the observation that the activity of *qz* along its pure solidus does not vary from 1.0 with pressure. The amount of lowering of the activities of the components from

their experimental mole fractions in this region is obtained by weighting the value at the eutectics (from which the compositional dependence has been removed) by the product of the mol fractions of the two interacting components (in accordance with the mass action principle). In order to calculate H_2O–undersaturated phase relations it was also assumed that the extent of interaction is linearly proportional to the H_2O content from the anhydrous to the H_2O–saturated "binaries." These assumptions permit calculation of H_2O–saturated and H_2O–undersaturated relations within the granite system. The region $X_a > X_a(E)$ is of interest primarily in the calculation of complete binary A_1–A_2 (aluminosilicate–aluminosilicate) liquidus relations. Inasmuch as interaction was assumed to contribute solely to the lowering of the activity of the *a* component at the eutectics, yet at the pure aluminosilicate sidelines the lowering was attributed to dissociation, within this region, both dissociation and interaction are likely to be taking place. Due to the complexities involved in trying to assess the amounts of each type of speciation, the total deviation of the activity from the mol fractions has been quantified in this region.

A mixing model was adopted for the two ternary solid solutions, plagioclase and alkali feldspar, for use in the calculation of phase equilibria in the system Ab–Or–An–Qz(–H_2O). An excess free energy formalism was used which yields a Margules expansion of the ternary excess free energy. This formalism takes into account the resulting interdependence of the asymmetric binary interaction parameters in a ternary solid solution (based on the discussion of ANDERSON and LINDSLEY, 1981). For the binary Ab–Or interaction parameters, the asymmetric regular solution model of THOMPSON and HOVIS (1979) was used. For binary plagioclase, the results of the polynomial fit of BLENCOE (personal communication, 1981) to G^{ex} (excess free energy of mixing of crystalline An and Ab components) based on the ion exchange data of SEIL and BLENCOE (1979 and personal communication) were refit to yield asymmetric regular solution parameters which are linear with temperature. These interaction parameters in turn yield activity coefficients for binary plagioclase which remain greater than 1.0 and fall within the range given by CARPENTER and FERRY (1984) when extrapolated to the temperatures of the 1 bar melting loop. Use of these parameters obviates the need for use of Al-avoidance entropies (HENRY *et al.,* 1982). In order to obtain interaction parameters for the mixing behavior of the crystalline components An and Or consistent with the newly derived regular solution interaction parameters for binary plagioclase, the method of GHIORSO (1984) was adopted (albeit in a modified form). Expressions for the activities for the ternary asymmetric regular solution were combined with the ternary solid solution data of SECK (1971a,b) in order to obtain, through regression, interaction parameters for the An–Or binary solid solution. The interaction parameters of all three binary regular solutions were then adopted for use in the calculation of ternary feldspar/melt equilibria using modified versions of the activity-composition expressions of GHIORSO (1984). A summary of the asymmetric regular solution interaction parameters used in the calculations of phase relations in the granite system is presented in Table 1. The complete activity–composition relations for ternary feldspar solid solutions are not presented here but are discussed in NEKVASIL (1986).

The refinements summarized above have resulted in the formulation of new activity–composition relations for the melt components, the selection and development of solid solution mixing models for the binary and ternary feldspars and the derivation of pressure and temperature–dependent expressions for the free energy of fusion of the components constrained by available calorimetric data.

COMPARISON OF CALCULATED AND EXPERIMENTAL PHASE EQUILIBRIA

Complete aluminosilicate–aluminosilicate (A_1–A_2) binary liquidus relations have been computed using the equation of equilibrium (obtained by equating chemical potentials of a in solid and melt) shown below.

Table 1. Interaction parameters W_{A1-A2}, $W_{A2-A1} = a + bT$ for the feldspar components according to the asymmetric binary regular solution formulation

$$G^{ex}_{A1,A2} = W_{A1,A2}X_{A1}X^2_{A2} + W_{A2,A1}X^2_{A1}X_{A2}*$$

Interacting component			Coefficients	
A1	A2	Parameter W(cal)	a	b
Ab	Or	A1, A2	7404	−5.120
		A2, A1	4078	—
Ab	An	A1, A2	3377	−1.476
		A2, A1	2683	−1.882
An	Or	A1, A2	5980	2.581
		A2, A1	17931	−5.494

* For Ab–Or interaction:

$$G^{ex}_{A1,A2} = G^{ex}_{A1,A2} + 0.086PX_{Ab}X_{Or}.†$$

† See text for sources.

$$\text{Ln} \frac{a_i^{hm(am)}}{a_i^s} = -\frac{\Delta G^0_{mi}(1,T) + \int_1^P \Delta V_{mi}dP}{RT}, \quad (1)$$

where $a_i^{hm(am)}$ is the activity of component i (ab, or, an, or qz) in the hydrous or anhydrous melt evaluated from the quantified activity-composition relations for the pure components (i.e. functions of pressure and H_2O content), and a_i^s is the activity of component i in crystalline solution, evaluated using the ternary asymmetric solution model briefly summarized above. $\Delta G^0_{mi}(1,T)$ and ΔV_{mi} are the standard state free energy of fusion at 1 bar and T and the volume of fusion of the component i, respectively. The calculation of ternary feldspar/melt equilibria involves the linking of three of the above expressions (one for each feldspar component) in the following manner

$$\frac{a_{ab}^{hm(am)}}{\gamma_{ab}^{Pl(Af)}} \exp \left[\frac{\Delta G^0_{mab}(1,T) + \int_1^P \Delta V_{mab}}{RT} \right]$$
$$+ \frac{a_{an}^{hm(am)}}{\gamma_{an}^{Pl(Af)}} \exp \left[\frac{\Delta G^0_{man}(1,T) + \int_1^P \Delta V_{man} \cdot dP}{RT} \right]$$
$$+ \frac{a_{or}^{hm(am)}}{\gamma_{or}^{Pl(Af)}} \exp \left[\frac{\Delta G^0_{mor}(1,T) + \int_1^P \Delta V_{mor} \cdot dP}{RT} \right] = 1,$$
$$(2)$$

where $\gamma_i^{Pl(Af)}$ refers to the activity coefficient of component i in plagioclase or alkali feldspar. This expression includes the constraint that the sum of the mol fractions of the crystalline components Ab, An, and Or must be unity. The cryoscopic equations $((\Delta G^0_{mi}(1,T) + \int_1^P \Delta V_{mi} dP)/RT)$ for the components ab, or, an, and qz are a function of temperature and pressure and are listed in BURNHAM and NEKVASIL (1986) as are the activity-composition relations for the melt and crystalline components. The calculation of complete binary A_1–A_2 liquidus relations thus involves solving the relevant equations for the saturation temperature.

Figures 1a–c show the calculated binary liquidus relations for plagioclase at 1 bar and 2 and 5 kbar under H_2O-saturated conditions. The 1 bar melting loop was included for reference. It, however, provides no test of the model in that it was used in the derivation of internally-consistent expressions for the standard state free energies of fusion of ab and an (NEKVASIL, 1986). For the 2 kbar melting loop, the calculated melting relations can be compared

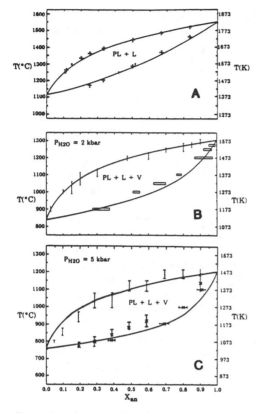

FIG. 1. Calculated plagioclase melting loop at (A) 1 bar, (B) P(H₂O) = 2 kbar and (C) P(H₂O) = 5 kbar. Experimental data, plotted for comparison, are from (A) BOWEN (1913); (B) ERIKSON (1979) (open rectangles indicate the compositional range obtained experimentally for each solidus datum) and (C) YODER et al. (1957) (vertical bars) and JOHANNES (1978) (horizontal bars). Error bars designated with a cross refer to solidus data.

against the experimental data of ERIKSON (1979). The calculated liquidus agrees well with the experimental data. The solidus also agrees in the regions close to the endmembers but indicates some discrepancies at intermediate compositions. Within this region, however, lie the greatest differences between equilibrium crystal and melt compositions and therefore, the intermediate compositions mark the compositional region in which the probability is greatest that the co-existing phases did not equilibrate in the run durations of the experiments. The greatest discrepancy occurs at 1000°C, a temperature at which ERIKSON (1979) noted direct evidence for disequilibrium in that he obtained inconsistent crystal compositions. Figure 1c shows the calculated melting loop at 5 kbar pressure and under H₂O-saturated conditions and, for comparison, the data of YODER et al. (1957) supplemented by solidus

data of JOHANNES (1978). The liquidus data and sparse solidus data of YODER et al. (1957) indicate a narrower melting loop than that determined at P(H₂O) = 2 kbar by ERIKSON (1979), which would imply the unlikely condition of a positive temperature dependence of the solid solution interaction parameters for plagioclase. The solidus data of JOHANNES (1978), on the other hand, does indeed indicate a wider melting loop. Once again, the major discrepancies appear to occur in the region of intermediate plagioclase compositions. JOHANNES (1978) noted the difficulty in attaining equilibrium at temperatures below 900°C, therefore, the calculated melting loop may more realistically represent the true melting relations of plagioclase at this pressure.

Figures 2a–c show the calculated melting loops for the subsystem Ab–Or(–H₂O) at 1 bar, and 2 and 5 kbar pressure for H₂O-saturated conditions. As was the case for the calculation of the plagioclase melting relations, only the melting behavior of the pure components and the cryoscopic equations for

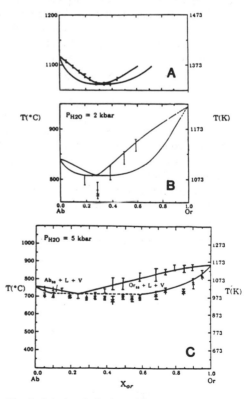

FIG. 2. Calculated Ab–Or melting relations at (A) 1 bar; (B) P(H₂O) = 2 kbar and (C) P(H₂O) = 5 kbar. Experimental data, plotted for comparison, are from (A) SCHAIRER (1950); (B) BOWEN and TUTTLE (1950) and (C) YODER et al. (1957). Solidus data are indicated by crosses.

ab and *or* were used in combination with a mixing model for binary alkali feldspar (*i.e.,* the asymmetric regular solution model of THOMPSON and HOVIS, 1979) in the calculations. The calculated 1 bar azeotropic relations of the system Ab–Or are shown in Figure 2a and the experimental data of SCHAIRER (1950) have been plotted for comparison. As evidenced by this figure, the agreement is very good. Figures 2b and 2c show the calculated melting loop and eutectic relations at 2 and 5 kbar (H₂O–saturated), respectively. Plotted for comparison in Figures 2b,c are the 2 kbar data of BOWEN and TUTTLE (1950) and the 5 kbar data of YODER *et al.* (1957), respectively. The calculated liquidus relations at each pressure agree with the experimental data. The calculated solidus temperature at both 2 and 5 kbar, however, is approximately 20 degrees higher than that experimentally determined.

The calculated phase relations for the H₂O–saturated system Or–An at 5 kbar are shown in Figure 3 along with the experimental data of YODER *et al.* (1957). Once again, the agreement is good. The slight discrepancy at high An contents is attributable to use of an H₂O–saturated melting curve based on the data of YODER (1965) and ERIKSON (1979) in the development of the revised model. [This curve lies at lower temperatures than the data of YODER *et al.* (1957).]. The calculated solidus once again appears to lie at a temperature about 20 degrees higher than that suggested by the experimental data.

Liquidus relations for the A–Qz(–H₂O) bounding binaries have been calculated. These calculations use the cryoscopic equation for each component and the activity-composition relations obtained from available pressure-temperature data on the A + Qz solidi (see BURNHAM and NEKVASIL, 1986 for details). Few experimental data are available for the liquidus limbs in these binary subsystems. Therefore, evaluation of the predicted liquidus limbs can not be readily undertaken. Consistency can be assessed, however, by comparison of the calculated and experimentally based phase relations of the ternary subsystems.

Figure 4 shows the phase relations in the haplogranite system at P(H₂O) = 2 kbar as determined by TUTTLE and BOWEN (1958). In general, the calculated liquidus relations (Figure 4b) agree with the experimental data in the bounding binaries as well as within the ternary. It is interesting to note that the 800°C isotherm appears to be drawn inconsistently, relative to the other isotherms in Figure 4a, a direct result of attempts by TUTTLE and BOWEN (1958) to incorporate their binary Ab–Or data. The calculated 800°C isotherm on the other hand, is consistent with the other calculated isotherms and the higher calculated binary minimum temperature (see above discussion of binary relations). The discrepancies between the positions of the calculated and experimentally based isotherms on the very steep quartz liquidus surface could be corrected by a small compositional change of 1–2 mol percent Qz. Such a correction of TUTTLE and BOWEN's (1958) compositions in this region is not unreasonable because they used an open-capsule technique in many of these experiments and may have preferentially lost silica to the fluid (BURNHAM, personal communication, 1983). The calculated 2 kbar minimum composition of Qz₃₄Ab₄₂Or₂₅ agrees well with that of TUTTLE and BOWEN (1958) (Qz₃₅Ab₄₀Or₂₅). TUTTLE and BOWEN (1958) also determined the compositions of feldspars coexisting with quartz and melt for two melt compositions (along the cotectic). Figure 4a shows a comparison of the data versus the calculated melt composition with which Or₆ and Or₅₆ (determined by 690°C by TUTTLE and BOWEN (1958) and calculated at 712°C) and quartz are in equilibrium. The compositional agreement is within 2 mol percent. The calculated minimum temperature of 709°C, however, is higher than the <700°C approximated from TUTTLE and BOWEN (1958).

Figures 4c,d show the experimentally based phase relations for the haplogranite system as determined by LUTH *et al.* (1964) at 5 kbar under H₂O–saturated conditions and the calculated phase relations for comparison. As determined experimentally, the model predicts that above about 3.5 kbar, the feldspar solvus will be intersected by the liquidus surface and an additional cotectic will appear separating two feldspar + L fields. At P(H₂O)'s only slightly above that of the initial intersection of the ternary solvus, the binary solvus will not necessarily have been intersected because the binary Ab–Or mini-

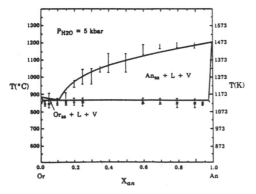

FIG. 3. Calculated An–Or melting relations at P(H₂O) = 5 kbar. Experimental data, plotted for comparison are from YODER *et al.* (1957). Solidus data are indicated by crosses.

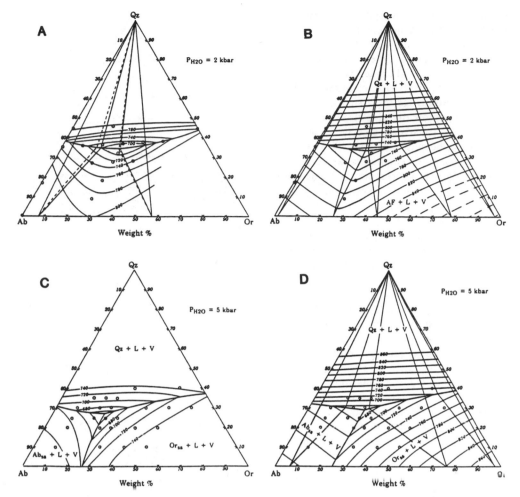

FIG. 4. Experimental data and calculated phase relations in the haplogranite system at $P(H_2O)$ = 2 and 5 kbar. (A) Isotherms and experimental data (open circles) at $P(H_2O)$ = 2 kbars from TUTTLE and BOWEN (1958). The solid three–phase triangles are based on the data of TUTTLE and BOWEN (1958); the dashed three–phase triangles are calculated and shown for comparison. (Note that the 700°C isotherm on the quartz liquidus surface as drawn by TUTTLE and BOWEN (1958) is thermodynamically implausible in curvature.) (B) Calculated phase relations at $P(H_2O)$ = 2 kbar. Experimental data of TUTTLE and BOWEN (1958) (open circles) are plotted to facilitate comparison with (A). (C) Isotherms and experimental data of LUTH et al. (1964) (open circles) at $P(H_2O)$ = 5 kbar. (D) Calculated phase relations at $P(H_2O)$ = 5 kbar. Experimental data of LUTH et al. (1964) have been plotted to facilitate comparison with (C).

mum lies at higher temperatures than the ternary minimum. Therefore, at such pressures the feldspar cotectic will not extend fully to the Ab–Or sideline from the ternary eutectic. At $P(H_2O)$ = 5 kbar, however, both the calculated and experimentally based phase relations indicate that the solvus has also been intersected in the binary. There is good agreement (within the stated experimental error of ±10°C) between most of the experimental points and the calculated relations. The calculated ternary eutectic composition of $Qz_{29}Ab_{46}Or_{25}$ compares well with the value of $Qz_{27}Ab_{50}Or_{23}$ of LUTH et al. (1964)

as does the calculated eutectic temperature of 662°C with the experimentally determined ternary eutectic temperature of 650° ± 10°C. It is interesting to note that the 700°C isotherm as drawn by LUTH et al. (1964) (700°C) is inconsistent in curvature with respect to that of their other isotherms. This may have been induced by forcing agreement of their ternary data with their earlier determined binary Ab–Or eutectic temperature. If this isotherm were redrawn to be consistent with the others the binary eutectic temperature would be closer to the 725°C calculated. Calculated three-phase triangles are shown in Figure 4d and indicate the changes in alkali feldspar composition along the cotectic.

Data available for the other bounding ternary subsystems of the granite system are sparse. Figure 5 shows the cotectic and isotherms as determined by YODER (1968) at 5 kbar pressure under H_2O–saturated conditions for the system Ab–An–Qz–H_2O. For comparison, the calculated cotectic is also indicated. The calculated cotectic differs only slightly compositionally from that of YODER (1968) indicating a maximum compositional difference of 6 weight percent. The temperatures calculated along the cotectic agree with the experimentally based isotherms. However, there are significant differences in the calculated cotectic temperature for a given X_{ab}/X_{an} ratio of the melt. JOHANNES (1978) noted that the experiments of YODER (1968) were unreversed and he was unable to reproduce the cotectic temperatures obtained by YODER (1968).

Phase relations in the system Ab–Or–An–H_2O at $P(H_2O) = 5$ kbar were determined by YODER et al. (1957); their results are shown in Figure 6a. Figure 6b shows the calculated liquidus relations under the same conditions of pressure and H_2O content. At high An contents agreement between the positions of the calculated isotherms and those drawn of YODER et al. (1957) is good. The greatest divergence occurs in the region of the cotectic. According to YODER et al. (1957), the liquidus surface has a fairly constant slope until very close to the sideline. This is in large part a result of their acceptance of

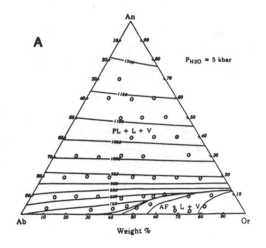

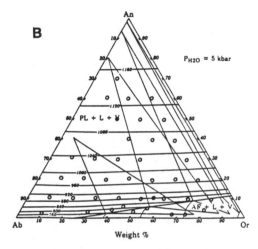

FIG. 6. Calculated and experimentally based phase relations and isotherms in the system Ab–Or–An–H_2O at $P(H_2O) = 5$ kbar. Experimental data (open circles) are from YODER et al. (1957) in (A); calculated relations are shown in (B). The symbols PL and AF refer to plagioclase and alkali feldspar respectively.

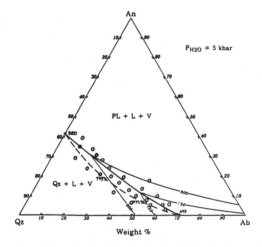

FIG. 5. Experimental data (open circles), isotherms and the plagioclase/quartz cotectic in the system Ab–An–Qz–H_2O at $P(H_2O) = 5$ kbar as determined by YODER (1968). The calculated cotectic is shown by the dashed curve. Several calculated cotectic temperatures have been shown for ease of comparison with the experimentally-based isotherms of YODER (1968).

their binary melting loop (see Figure 4c). The calculated liquidus relations on the other hand, indicate more curvature in the sloping surface. The isotherms in the alkali feldspar field agree in the binary (Ab–Or) with those calculated, however, the strong temperature depression and marked change in curvature of the eutectic portrayed by YODER et al. (1957) within the ternary (Figure 6a) were not detected during calculation. However, this is a compositionally complex region with high Ab contents and low temperatures where the possibility of disequilibrium in the experiments is great.

The position of the ternary eutectic in the system An–Or–Qz–H_2O for H_2O–saturated conditions has been determined by WINKLER and LINDEMANN

(1972) and WINKLER and GHOSE (1973). The eutectic temperatures that they obtained have been verified by JOHANNES (1984) as lying at 738°C and 700°C at 2 and 5 kbar, respectively. The calculated eutectic temperatures are 752°C and 720°C, respectively. The calculated and experimentally determined eutectic compositions, however, differ markedly in An content from the 14 weight percent An determined by Winkler at 5 kbar to the 5 weight percent calculated. This great discrepancy may arise from imprecise quantification of the interaction behavior of qz and an which is based on sparse experimental data (although the differences may also be due to experimental problems as noted by JOHANNES, 1978).

From the above examples, it can be concluded that the model is able to predict the general topology of the liquidus surfaces of the subsystems of the granite system. Additionally, it is reasonable to conclude that although the calculated phase diagrams may not be correct in every quantitative detail, the model can certainly be used in a predictive capacity to obtain information regarding general quantitative trends.

Quantification of the activity–composition relationships was also undertaken for the anhydrous subsystems using the same approach as taken in the H_2O–saturated systems (NEKVASIL, 1986; BURNHAM and NEKVASIL, 1986). Evaluation of the model's predictive capabilities in the anhydrous granite sytem, however, is more problematic due to the paucity of experimental data arising from the difficulty of overcoming kinetic problems in the very viscous melts that characterize the anhydrous region of this system. However, the phase assemblage data of WHITNEY (1972) can be used for comparison of calculated anhydrous phase relations for compositions within the granite system (NEKVASIL-CORAOR and BURNHAM, 1983; 1984). If the activity–composition relations obtained from the anhydrous and H_2O–saturated unary solidi and A + Qz solidi are assumed linear with H_2O content from these two limiting conditions, very good agreement can be obtained between WHITNEY's (1972) experimental data in the H_2O–undersaturated region and calculated saturation curves. On the basis of this agreement (NEKVASIL, 1986; NEKVASIL-CORAOR and BURNHAM, 1983), it is concluded that the model can provide at least a first order quantitative assessment of the phase relations in the H_2O–undersaturated regions. This capability is of great importance in that it permits the calculation of the individual effects of pressure and H_2O content on phase equilibria in the granite system. It is through the isolation of the effects of these variables that more insight can be gained into the crystallization and melting histories of silicic magmas.

THE EFFECTS OF PRESSURE AND H_2O CONTENT

The effects of pressure and H_2O content on the positions of the cotectic surfaces within the granite tetrahedron have direct bearing on the crystallization path of a silicic magma. The pioneering efforts of TUTTLE and BOWEN (1958) and LUTH et al. (1964) on the determination of the haplogranite liquidus relations at several pressures (under H_2O–saturated conditions) clearly elucidated the effects of increasing $P(H_2O)$ on the haplogranite cotectic, that is, on the intersection of the quartz/alkali feldspar cotectic surface with the Ab–Or–Qz ternary subsystem. Inasmuch as the solubility of H_2O in silicic melts increases with pressure, two variables were being changed simultaneously in these experiments and the resulting data represented the effect of both variables. Natural felsic melts are generally H_2O–undersaturated until the late stages of crystallization; therefore, the results of these experiments do not simulate common natural magmatic conditions and it is very important that the individual effects of each variable be evaluated.

The following discussion will focus on the predicted effects of pressure and X_w^m (the mol fraction of water in the melt) on the cotectic surfaces within the granite system. As can be seen in Figure 7, the granite system contains three cotectic surfaces. The quartz/plagioclase cotectic surface separates the cri

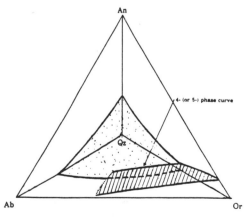

FIG. 7. General phase relations in the granite tetrahedron. The stippled surface denotes the plagioclase + quartz + L cotectic surface whereas the ruled region denotes the alkali feldspar + plagioclase cotectic surface. The intersection of the three cotectic surfaces defines the 4- (or 5-) phase curve. At high $P(H_2O)$, the alkali feldspar/plagioclase cotectic surface intersects the haplogranite base.

tensive Pl (plagioclase) + L field from the less extensive Qz + L field. The plagioclase/alkali feldspar
cotectic surface is much flatter and separates the Pl
+ L field from the AF (alkali feldspar) + L field.
Only at high P(H₂O) does this surface intersect the
haplogranite subsystem. The intersection of the two
cotectic surfaces marks the emanation curve of a
third minor cotectic surface separating the Qz + L
field from the AF + L field. The emanation or intersection curve of these three surfaces, is the 4-
phase curve (or 5-phase curve if fluid is present).
This curve is of great importance in that it indicates
the compositions of melts formed in equilibrium
with quartz and two feldspars in the source region.

Figure 8 shows the calculated cotectics in the
granite system for 2, 3 and 5 kbar at $X_w^m = 0.20$
(1.9 weight percent H₂O) as a demonstration of the
effects of pressure. It is apparent that the effect of
increasing pressure is to expand preferentially the
Qz + L field at the expense of the Pl(AF) + L fields.
The feldspar cotectics, on the other hand, show only
very slight differential pressure effects. The pressure
sensitivity of Qz + L field can be readily attributed
to the larger ΔV of fusion for Si_4O_8 relative to that
for the aluminosilicates (NEKVASIL, 1986).

The calculated effects of increasing X_w^m isobarically are shown in Figure 9. Increasing H₂O content
results in a contraction of the Qz + L field and thus
induces an effect opposite to that resulting from
increasing pressure at constant H₂O content. Once
again this differential effect is much stronger for the

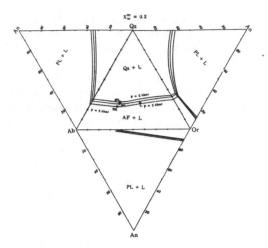

FIG. 9. Calculated cotectic relations within the granite
tetrahedron at 2 kbars pressure for $X_w^m = 0.0$, 0.20 (1.9
weight percent H₂O) and 0.40 (3.7 weight percent H₂O),
demonstrating the effect of variable H₂O content on the
cotectic surfaces. The tetrahedron has been unfolded as
described in Figure 8.

Qz + L field than for the Pl(AF) + L fields. The
smaller ΔH of fusion of Si_4O_8 relative to that of the
feldspars indicates that an equal the change in the
activity of the *qz* and *a* melt components such as
upon the addition of H₂O will result in a greater
change in the quartz/melt equilibrium temperature
than in the feldspar/melt equilibrium temperature.

The individual effects of pressure (Figure 8) and
H₂O content (Figure 9) can be compared with Figure 10 which shows the effects of changing both
pressure and H₂O content (*i.e.*, changing P(H₂O)

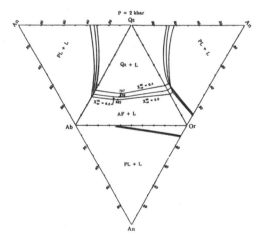

FIG. 8. Calculated cotectic relations in the granite tetrahedron at $X_w^m = 0.20$ (1.9 weight percent) and 2, 3 and
5 kbar pressure, demonstrating the effects of pressure variation on the cotectic surfaces. The granite tetrahedron has
been unfolded to facilitate visualization of the intersections
of all cotectic surfaces with the bounding ternary subsystems (*i.e.*, faces of the tetrahedron).

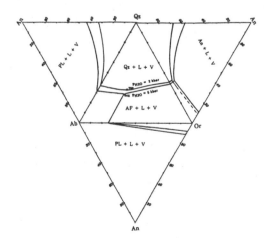

FIG. 10. Calculated cotectic relations within the granite
tetrahedron at P(H₂O) = 2 and 5 kbar (H₂O-saturated),
demonstrating the combined effects of varying pressure
and H₂O content. The tetrahedron has been unfolded as
described in Figure 8.

under H_2O-saturated conditions). In this case, the effects of increasing H_2O content (due to increased H_2O solubility with pressure) are offset by the effects of increased pressure. The net effect is a slight expansion of the $Qz + L$ field with increased $P(H_2O)$. This illustrates the importance of understanding the individual effects of pressure and H_2O content on phase equilibria in the granite system and the inadequacy of sole use of H_2O-saturated phase relations for the interpretation of magmatic history.

DISCUSSION

The results of phase equilibria calculations in the granite system using the revised quasi-crystalline model indicate that decreasing pressure and increasing H_2O content result in a contraction of the $Qz + L$ field at the expense of the $Pl(AF) + L$ fields. On a compositional basis, the feldspar fields adjacent to each other are only slightly affected by changes in these two variables. The effects of these variables on the liquidus temperatures, however, are considerable.

A given Qz-poor, Or-rich hydrous (but H_2O-undersaturated) magma in which one feldspar is forming will begin crystallizing a second feldspar (at the two feldspar cotectic surface) when the remaining melt has attained a specific composition. For such a magma, the composition at which the second feldspar begins to appear is approximately constant, that is, it is relatively independent of initial H_2O content of the magma. Differences in H_2O content, however, will strongly affect the crystallization history after this surface has been reached. With higher H_2O contents, the $Qz + L$ field contracts, shifting the 4-phase curve (*i.e.*, the intersection curve of the cotectic surfaces within the tetrahedron) toward the Qz apex and away from the An and Or apices. Therefore, higher H_2O contents for a given bulk composition will required that more plagioclase and alkali feldspar components be removed from the melt in order to enrich it in silica before the 4-phase curve is intersected. This, in turn, implies that the percentage of these crystals will be higher (and the fraction of silicate melt remaining lower) at the onset of crystallization of quartz than would be expected for lower bulk H_2O contents. It is important to note that the 4-phase curve continuously shifts during crystallization in response to the increase in H_2O content in the melt upon the crystallization of anhydrous phases. Because of this shift, crystallization paths for these melts follow an infinite succession of tetrahedra and the system must be considered truly quinary.

For tonalitic compositions (*i.e.*, poor in Or), the differences in crystallization path of a melt at different H_2O contents will be more pronounced at an earlier stage of crystallization, because of the sensitivity of the position of the quartz/plagioclase cotectic surface to the H_2O content of the melt. Melts with higher H_2O contents must crystallize more plagioclase to enrich the remaining melt in silica before quartz will appear (in reflection of the shift of the cotectic surface toward the Qz apex). Therefore, the H_2O content strongly affects the composition of the melt at the intersection of the melt composition path with the cotectic surface and the onset of crystallization of quartz. As was mentioned above, the position of the 4-phase curve also shifts towards the Qz apex with increasing H_2O content. The contraction of the plagioclase/quartz cotectic surface resulting from higher H_2O contents implies that the 4-phase curve will be intersected 'earlier' at higher H_2O contents, that is, alkali feldspar will appear after less quartz has crystallized than would be the case for the same bulk composition but lower initial H_2O content.

The strong, but opposing, effect of pressure indicates that crystallization at high pressures will result in a strongly increased likelihood of quartz appearing as the liquidus phase for compositions low in An. It also indicates that decompression, such as during ascent of the magma, will preferentially affect earlier crystallized quartz relative to any early crystallized feldspar. Such quartz will be resorbed to various extents in attempt to reestablish an equilibrium assemblage with a more silica-rich melt.

It is apparent from the strong differential effect of pressure and H_2O content on the $Qz + L$ field that for bulk compositions lying close to the cotectic surfaces, the pressure and H_2O content of the system will determine the identity of the liquidus phase. Therefore, for such magmatic compositions, if the liquidus phase can be identified texturally, the model can be used to place constraints on the depth of initial emplacement as well as on the H_2O content of the magma. (For an example of such application, see LONG *et al.*, 1986.) It is hoped that such calculations will prove invaluable to the interpretation of the crystallization and melting histories of natural felsic rocks.

Acknowledgements—This work was greatly facilitated by the very generous research grant support provided by the National Science Foundation (grants EAR-7812957 and EAR-8212492 awarded to C. W. Burnham). Additional support was provided by Corning Glassworks Foundation in the form of a fellowship awarded to the first author which is gratefully acknowledged. We thank J. R. Holloway and an anonymous reviewer for the improvements that have resulted from their efforts.

REFERENCES

ANDERSON D. J. and LINDSLEY D. H. (1981) A valid Margules formulation for an asymmetric ternary solid solution: revision of the olivine ilmenite geothermometer, with applications. *Geochim. Cosmochim. Acta* **45**, 847–853.

BERMAN R. G. and BROWN T. H. (1984) A thermodynamic model for multicomponent melts with application to the system CaO–Al$_2$O$_3$–SiO$_2$. *Geochim. Cosmochim. Acta* **48**, 661–678.

BOETTCHER A. L., BURNHAM C. W., WINDOM K. E. and BOHLEN S. R. (1982) Liquids, glasses, and the melting of silicates to high pressures. *J. Geol.* **90**, 127–138.

BOETTCHER A. L., GUO Q., BOHLEN S. R. and HANSEN B. (1984) Melting in feldspar-bearing systems to high pressures and the structures of aluminosilicate liquids. *Geology,* **12**, 202–204.

BOTTINGA Y. and RICHET P. (1978) Thermodynamics of liquid silicates. *Earth Planet. Sci. Lett.* **40**, 382–400.

BOWEN N. L. (1913) The melting phenomena of the plagioclase feldspars. *Amer. J. Sci.* (Fourth Series) **35**, 577–599.

BOWEN N. L. and TUTTLE O. F. (1950) The system NaAlSi$_3$O$_8$–KAlSiO$_4$–SiO$_2$. *J. Geol.* **58**, 489–511.

BURNHAM C. W. (1975) Thermodynamics of melting in experimental silicate-volatile systems. *Geochim. Cosmochim. Acta* **39**, 1077–1084.

BURNHAM C. W. (1981) Nature of multicomponent aluminosilicate melts. *Phys. Chem. Earth* **13 & 14**, 191–227.

BURNHAM C. W. and NEKVASIL H. (1986) Equilibrium properties of granite pegmatite magmas. *Amer. Mineral.,* Jahns Memorial Volume, 239–263.

CARPENTER M. A. and FERRY J. M. (1984) Constraints on the thermodynamic mixing properties of plagioclase feldspars. *Contrib. Mineral. Petrol.* **87**, 138–148.

CLEMENS J. D. and WALL V. J. (1981) Origin and crystallization of some peraluminous (S-type) granitic magmas. *Can. Mineral.* **19**, 111–131.

EGGLER D. H. (1972) Water-saturated and undersaturated melting relations in a Paricutin andesite and an estimate of water content in the natural magma. *Contrib. Mineral. Petrol.* **34**, 261–271.

ERIKSON R. L. (1979) An experimental and theoretical investigation of plagioclase melting relations. M.S. Thesis, The Pennsylvania State University.

GHIORSO M. S. (1984) Activity-composition relations in ternary feldspars. *Contrib. Mineral. Petrol.* **87**, 282–296.

GHIORSO M. S. and KELEMEN P. B. (1987) Evaluating reaction stoichiometry in magmatic systems evolving under generalized thermodynamic constraints: Examples comparing isothermal and isenthalpic assimilation. In *Magmatic Processes: Physicochemical Principles,* (ed. B. O. MYSEN), The Geochemical Society Spec. Publ. 1, pp. 319–335.

GHIORSO M. S., CARMICHAEL I. S. E., RIVERS M. L. and SACK, R. O. (1983) The Gibbs free energy of mixing of natural magmatic liquids, an expanded regular solution approximation for the calculation of magmatic intensive variables. *Contrib. Mineral. Petrol.* **84**, 107–145.

HENRY D. J., NAVROTSKY A. and ZIMMERMAN H. D. (1982) Thermodynamics of plagioclase-melt equilibria in the system albite-anorthite-diopside. *Geochim. Cosmochim. Acta* **46**, 381–391.

HERVIG R. L. and NAVROTSKY A. (1984) Thermochemical study of glasses in the system NaAlSi$_3$O$_8$–KAlSi$_3$O$_8$–Si$_4$O$_8$ and the join Na$_{1.6}$Al$_{1.6}$Si$_{2.4}$O$_8$–K$_{1.6}$Al$_{1.6}$Si$_{2.4}$O$_8$. *Geochim. Cosmochim. Acta* **48**, 513–522.

JACKSON I. (1976) Melting of the silicate isotypes SiO$_2$, FeF$_2$, and GeO$_2$ at elevated pressures. *Phys. Earth Planet. Inter.* **13**, 218–223.

JOHANNES W. (1978) The melting of plagioclase in the system Ab–An–H$_2$O and Qz–Ab–An–H$_2$O at P(H$_2$O) = 5 kbar, an equilibrium problem. *Contrib. Mineral. Petrol.* **66**, 295–303.

JOHANNES W. (1984) Beginning of melting in the granite system Qz–Or–Ab–An–H$_2$O. *Contrib. Mineral. Petrol.* **86**, 264–273.

KENNEDY G. C., WASSERBURG G. J., HEARD H. C. and NEWTON R. C. (1962) The upper three–phase curve in the system SiO$_2$–H$_2$O. *Amer. J. Sci.* **260**, 501–521.

LAMBERT I. B., ROBERTSON J. K. and WYLLIE P. J. (1969) Melting relations in the system KAlSi$_3$O$_8$–SiO$_2$–H$_2$O to 18.5 kbars. *Amer. J. Sci.* **267**, 609–626.

LONG L. E., SIAL A. N., NEKVASIL H. and BORBA G. S. (1986) Origin of granite at Cabo de Santo Agostinho, Northeast Brazil. *Contrib. Mineral. Petrol.* **92**, 341–350.

LUTH W. C., JAHNS R. and TUTTLE F. (1964) The granite system at pressures of 4 to 10 kilobars. *J. Geophys. Res.* **69**, 759–773.

NAVROTSKY A., HON R., WEILL D. F. and HENRY D. J. (1980) Thermochemistry of glasses and liquids in the systems CaMgSi$_2$O$_6$–CaAl$_2$Si$_2$O$_8$–NaAlSi$_3$O8, SiO$_2$–CaAl$_2$Si$_2$O$_8$–NaAlSi$_3$O$_8$ and SiO$_2$–Al$_2$O$_3$–CaO–Na$_2$O. *Geochim. Cosmochim. Acta* **44**, 1409–1423.

NEKVASIL H. (1986) A theoretical thermodynamic investigation of the system Ab–Or–An–Qz(–H$_2$O) and implications for melt speciation. Ph.D. Diss., The Pennsylvania State University.

NEKVASIL-CORAOR H. and BURNHAM C. W. (1983) Thermodynamic modelling of crystallization paths of felsic melts (abstr.). *Geol. Soc. Amer. Abstr. Prog.* **15**, 651.

NEKVASIL-CORAOR H. and BURNHAM C. W. (1984) Thermodynamic modelling of crystallization paths of felsic melts: The haplogranite and haplogranodiorite systems (abstr.). *Geol. Soc. Amer. Abstr. Prog.* **16**, 609.

PIWINSKII A. J. and WYLLIE P. J. (1968) Experimental studies of igneous rock series: a zoned pluton in the Wallowa batholith, Oregon. *J. Geol.* **76**, 205–234.

PIWINSKII A. J. and WYLLIE P. J. (1970) Experimental studies of igneous rock series: 'Felsic Body Suite' from the Needle Point Pluton, Wallowa Batholith, Oregon. *J. Geol.* **78**, 52–76.

SCHAIRER J. F. (1950) The alkali feldspar join in the system NaAlSiO$_4$–KAlSiO$_4$–SiO$_2$. *J. Geol.* **58**, 512–517.

SECK H. A. (1971a) Koexistierende Alkalifeldspate und Plagioklase im System NaAlSi$_3$O$_8$–CaAl$_2$Si$_2$O$_8$–H$_2$O bei Temperaturen von 650°C–900°C. *Neues Jahrb. Mineral. (Abhand.)* **115-3**, 315–345.

SECK, A. (1971b) Der Einfluss des Druckes auf die Zussamensetzung koexistierende Alkalifeldspate und Plagioklase im System NaAlSi$_3$O$_8$–KAlSi$_3$O$_8$–CaAl$_2$Si$_2$O$_8$. *Contrib. Mineral. Petrol.* **31**, 67–86.

SEIL M. K. and BLENCOE J. G. (1979) Activity–composition relations NaAlSi$_3$O$_8$–CaAl$_2$Si$_2$O$_8$ feldspars at 2 kbars, 600–800°C. *Geol. Soc. Amer. Abstr. Prog.* **11**, 513.

THOMPSON J. B. and HOVIS G. L. (1979) Entropy of mixing in sanidine. *Amer. Mineral.* **64**, 57–65.

TUTTLE O. F. and BOWEN N. L. (1958) Origin of granite in light of experimental studies in the system NaAlSi$_3$O$_8$–KAlSi$_3$O$_8$–SiO$_2$–H$_2$O. *Geol. Soc. Amer. Mem. 74,* 1–154.

WEILL D. F., HON R. and NAVROTSKY A. (1980) The igneous system CaMgSi$_2$O$_6$–CaAl$_2$Si$_2$O$_8$–NaAlSi$_3$O$_8$: Variations on a classic theme. In *Physics of Magmatic Processes,* (ed. R. B. HARGRAVES), Chap. 2, pp. 49–92 Princeton University Press.

WHITNEY J. A. (1972) History of granodioritic and related magma systems: an experimental study. Ph.D. Diss., Stanford University.

WINKLER H. G. F. and LINDEMANN W. (1972) The system Qz–Or–An–H$_2$O within the granite system Qz–Or–An–H$_2$O. Application to granitic magma formation. *Neues Jahrb. Mineral. (Monatsh.)* **1972,** 49–61.

WINKLER H. G. F. and GHOSE N. C. (1973) Further data on the eutectics in the system Qz–Or–An–H$_2$O. *Neues Jahrb. Mineral. (Monatsh.)* **1973,** 481–484.

YODER H. S. (1954) The system diopside-anorthite-H$_2$O. *Carnegie Inst. Wash. Yearb.* **53,** 106–107.

YODER H. S. JR. (1965) Diopside-anorthite-water at five and ten kilobars and its bearing on explosive volcanism. *Carnegie Inst. Wash. Yearb.* **64,** 82–94.

YODER H. S. JR. (1968) Albite-anorthite-quartz-water at 5 kb. *Carnegie Inst. Wash. Yearb.* **66,** 477–480.

YODER H. S. JR., STEWART D. B. and SMITH J. R. (1957) Ternary feldspars. *Carnegie Inst. Wash. Yearb.* **56,** 206–214.

Magmatic Processes: Physicochemical Principles
© The Geochemical Society, Special Publication No. 1, 1987
Editor B. O. Mysen

Evolution of granitic magmas during ascent: A phase equilibrium model

MARTHA L. SYKES

Department of Geology, Arizona State University, Tempe, AZ 85287, U.S.A.*

and

JOHN R. HOLLOWAY

Departments of Chemistry and Geology, Arizona State University, Tempe, AZ 85287, U.S.A.

Abstract—Evolution of granitic magma during ascent through the crust is modelled by using phase relationships and thermodynamic properties of the system albite–H_2O, by assuming fluid–absent conditions. Pressure–temperature ascent trajectories are calculated for particular crystallization rates and H_2O contents using an expression for the albite liquidus temperature as a function of pressure and H_2O activity. Increased H_2O content causes a decrease in initial temperature of trajectories, and a decrease in dP/dT for a crystallization rate (CR) of 10% per kbar and an increase in dP/dT for all melting rates (MR). The H_2O activity in the magma continuously increases during ascent for MR < 4% per kbar and all CR's. Ascent trajectories for all CR's and MR's < about 2% have positive dP/dT. Adiabatic ascent trajectories have positive but steep dP/dT and require fusion of pre–existing (restite) crystals when present. These results suggest that restite–bearing magmas commonly undergo continued melting (resorption) during ascent, thus avoiding superheating even for nearly adiabatic ascent trajectories.

Thermodynamic and physical properties of the magma system are calculated along ascent trajectories. Volume increases during ascent for CR ≤ 5 to 6% per kbar and, at high H_2O contents, is more strongly a function of crystallization rate than H_2O content; change in volume may exceed 10% overall. The enthalpy change is exothermic during ascent provided MR < 7% per kbar; internal energy change during ascent is exothermic for MR ≤ 3% per kbar. Magma (melt + crystal) density is nearly constant along ascent trajectories, whereas melt density decreases during ascent. For intermediate to high H_2O contents, magma and melt viscosities are about constant along ascent trajectories for CR > 0% per kbar. At low total H_2O contents and CR = 10% per kbar, magma viscosity increases by up to four orders of magnitude; melt viscosity remains nearly constant. Energy available from an ascending granitic magma body as calculated from enthalpy and internal energy changes can reach two to three times the values calculated by using heat capacity data; this necessitates reconsideration of certain magma transport processes.

INTRODUCTION

THE "GRANITE CONTROVERSY" of the first part of this century (READ, 1957; WALTON, 1955) has been largely resolved: most researchers agree that granitic[1] rocks crystallize from parental melts formed by partial fusion of source rocks in the upper mantle or crust. Since CHAPPELL and WHITE (1974) championed the concept that granitic magmas inherit the characteristics of their source regions, great progress has been made in understanding chemical and mineralogical variations among suites of granitic rocks. However, the mechanisms for transport of granitic magmas between source and emplacement level, and the processes that occur during such transport, remain largely unstudied and unconstrained. For example, restite segregation has been

proposed as a geochemically viable process occurring between source and emplacement level, but its physical feasibility has not been demonstrated.

To address such problems, thermodynamics and phase equilibria of the system albite–H_2O are used to model the evolution of granitic magma during transport from source towards the surface. For given crystallization or melting rates, calculated pressure–temperature trajectories for a constant mass of magma allow evaluation of variations in magma properties (*e.g.*, volume, density, enthalpy, entropy, internal energy, viscosity) during ascent. Implications for processes important during transport of granitic magmas are presented in this paper.

Few theoretical studies of granitic magma evolution during ascent exist. Isentropic adiabats valid for multiphase, multicomponent systems calculated for albite–H_2O by RUMBLE (1976), show that adiabatic decompression along the univariant curve defined by the reaction

$$\text{albite crystal} + \text{vapor} = \text{albite melt}$$

---

* Now at Texas Wesleyan College, Department of Physics/Geology, Fort Worth, Texas 76105-9989.

[1] The terms "granitic" and "granite" here signify any calc–alkaline intrusive rock of composition from tonalite to granite, *sensu stricto*.

results in crystallization, whereas adiabatic decompression through the crystal + melt field results in melting. WALDBAUM (1971) calculated isenthalpic temperature rise during decompressive ascent of magmas by using Joule–Thompson coefficients, but neglected the change in gravity potential of the vertically moving masses, which invalidates his conclusions (RAMBERG, 1971).

Fluid dynamic studies of magma transport have been largely restricted to mantle conditions. Reviews by SPERA (1980) and TURCOTTE (1982) serve to emphasize the lack of quantitative descriptions of crustal magma transport. Diapiric transport of andesitic magma from the upper mantle through the crust has been evaluated using heat transfer models (MARSH, 1978; MARSH and KANTHA, 1978; MARSH, 1982), and fluid dynamic models. Elastic crack propagation, proposed as a major mechanism for basaltic magma transport (*e.g.*, SHAW, 1980; see also TURCOTTE, 1987), has not been directly applied to granitic rocks. Stoping can be a major transport mechanism for magmas only near the earth's surface (MARSH, 1982).

Models for granitic magma transport must explain differences between deep and shallow emplacement of granitic magma, and any inherent differences in the physical, chemical, or mineralogical properties of such magma systems.

Symbols used in thermodynamic and phase equilibrium calculations are as given in Table 1, unless otherwise defined in the text.

THE MODEL

Evolution of granitic magma during ascent is modelled using phase relationships for the system albite–H_2O (BURNHAM and DAVIS, 1971, 1974; BURNHAM, 1979). Model granitic magma consists of albite melt, dissolved H_2O, and variable amounts of albite crystals. Pressure–temperature ascent trajectories are calculated for a specified crystallization or melting rate (denoted CR and MR, respectively; given as weight percent crystals formed or melted per kbar of ascent) and volatile content by using a relationship for albite liquidus temperature as a function of P and H_2O activity (a_w).

Specification of a particular crystallization or melting rate defines the amount of heat and energy transfer from the system, once an ascent mechanism and body size and shape are defined. Certain of the ascent trajectories may therefore be considered unlikely due to excessively slow velocities required for heat transfer during crystallization, or because they require input of energy for some MR > 0%/kbar paths. Variations of enthalpy and internal energy

Table 1. Symbols and abbreviations used.

a	—activity of component
ab	—albite
C_p	—heat capacity, J/g–K
CR	—crystallization rate, in weight percent per kbar
E	—internal energy, J/g
G	—free energy, J/g
g	—constant of gravitational acceleration, 980 cm/s^2
H	—enthalpy, J/g
K	—thermal diffusivity, cm/s^2
k	—Henry's law constant
m	—mass of system, g
MR	—melting rate, in weight percent per kbar
n	—mole
P	—pressure, kbar
q	—heat energy
r	—radius, cm
R	—gas constant, 8.314 J/mol–K
S	—entropy, J/g–K
T	—temperature, K
V	—volume, cm^3
v	—velocity, cm/s
W	—weight fraction
w	—work energy
X	—mole fraction
z	—vertical coordinate
Δ	—difference between two quantities
μ	—viscosity, poise
ρ	—density, g/cm^3
ϕ	—volume fraction
β	—adiabatic gradient, K/km
α	—coefficient of thermal expansion, K^{-1}
Pe	—Peclet number, vr/K

Superscripts and subscripts	
w	—H_2O
xl	—crystal
o	—initial condition or standard state
wr	—wall rock
s	—solid
m	—melt
fl	—fluid
fus	—fusion
gl	—glass

for the magma system during ascent, given in a later section, reflect these heat transfer requirements. If a model for heat transfer is assumed, the enthalpy or internal energy data can be incorporated, allowing limitations to be made on ascent velocity and body size (or shape) for a particular trajectory (SYKES, 1986).

Constant values of CR and MR were chosen for ease of calculation. Ascent trajectories for nonlinear rates can be interpolated between those calculated.

Assumptions

In the present model for granitic magma ascent, the system albite–H_2O is used as an analogue for hydrous granitic magma. Solubility relationships of H_2O in albite melt adequately represent those for most granitic magmas (quartz– or hypersthene–normative, calc–alkaline composition), provided that melt compositions are expressed in an eight–oxygen albite–equivalent form (BURNHAM, 1979). Volume and thermodynamic properties of hydrous albite melt are also applicable to granitic magmas (BURNHAM and DAVIS, 1974); values for these properties are most accurate in the range 1 to 10 kbar.

Liquidus curves at constant X_w^m for the system albite–H_2O have similar slopes, dT/dP, to those for actual granitic magmas. For albite–H_2O, dT/dP decreases with both increasing pressure and X_w^m. This relationship also holds for crystalline phases of fixed composition in granitic magmas; the magnitude of the decrease in dT/dP depends on $S_{gl} - S_{xl}$, such that the smaller the difference, ΔS, the greater the decrease in dT/dP (BURNHAM and DAVIS, 1974).

Absolute temperatures of liquidus and solidus curves are greater for albite–H_2O than for actual granitic magmas. However, this temperature discrepancy has little effect on H_2O solubilities. BURNHAM (1979) notes that equimolal solubilities for the Harding pegmatite are duplicated by albite–H_2O within experimental error ($X_w^m \pm 0.02$) despite a 450°C difference in temperature. It will be shown in the following section that the higher temperatures for albite–H_2O have similarly small effect on ascent trajectories.

The major limitations of using albite–H_2O as a granite analogue are that the role of changing composition during differentiation, and the effects of hydrous phase crystallization (or melting), cannot be rigorously evaluated. The former limitation can be addressed by incorporation of BURNHAM's (1981) activity-composition relationships for multicomponent aluminosilicate melts into the liquidus temperature calculations. Such an approach is not used in the present model because details of phase equilibria effects, although of importance in considering evolution of particular granitic magmas, tend to obscure the effects of H_2O and simple crystal–melt equilibria on ascent trajectories and physical properties of the magmas (see also NEKVASIL and BURNHAM, 1987). The effect of hydrous phase crystallization on X_w^m and ascent trajectories depends on the rate of crystallization during ascent, as will be shown after discussion of ascent trajectory calculation.

In these model calculations, H_2O is assumed to be the dominant volatile component. Abundant evidence for the presence of H_2O exists, including the presence of hydrous phases such as micas and amphiboles, and experimentally determined phase relationships for particular granitic rocks. Evidence for the presence of volatile species relatively insoluble in silicic melts, such as CO_2, and the existence of a separate fluid phase in the early stages of granitic magmatism, is generally lacking.

Calculated values for physical and thermodynamic properties represent spatial averages for a particular batch of magma; i.e., a homogeneous distribution of crystals and volatiles is assumed. Any vertical extent of the magma body is ignored: no external pressure gradients are assumed to exist over the magma body. Ascending magma is considered to be closed with respect to addition or loss of aluminosilicate material. The major processes assumed to be affecting the magma are crystallization, melting, and dissolution of H_2O.

A major assumption of the model is that of chemical equilibrium in the ascending batch of magma. This assumption requires that processes occurring during magma ascent be reversible. For closed magma systems this assumption is good: irreversible physical processes such as viscous dissipation are insignificant compared to reversible energy sources and sinks, such as crystallization, melting and volatile exsolution. Irreversible heat and momentum transfer processes are excluded from the present model for granitic magma ascent and evolution; their effects will be considered in later papers.

Calculation of albite liquidus temperature

Albite liquidus temperature is calculated as a function of pressure and X_w^m with the relationship:

$$\ln (a_{ab}^m/a_{ab}^s) = -\frac{\Delta G_{ab}^{mo} + (P-1)\Delta V_{ab}^m}{RT}, \quad (1)$$

where a_{ab}^m and a_{ab}^s are the activities of anhydrous albite in the melt (m) and solid (s) phases, respectively, relative to $a_{ab}^m = 1$ in melt of pure albite, and $a_{ab}^s = 1$ in pure albite solid (BURNHAM, 1979). P is pressure in bar, T is in K, and ΔV_{ab}^m is the volume of melting of pure albite. The quantity $\Delta G_{ab}^{mo} + (P-1)\Delta V_{ab}^m$ can be expressed as a function of pressure and temperature. This allows an explicit equation for albite liquidus temperature for $X_w^m \leq 0.5$:

$$T = \frac{b}{R\{0.58 - \ln [(1 - X_w^m)^2]\} - c}, \quad (2)$$

and for $X_w^m > 0.5$:

$$T = \frac{(b/R) - 2667[\ln (1 - X_w^m) + X_w^m] - 515}{0.707 - (c/R) - 6.52[\ln (1 - X_w^m) + X_w^m]}, \quad (3)$$

where $b = 229.2P + 1.3075 \times 10^4$; $c = -2.0597 \times 10^{-3}P^2 - 6.6918 \times 10^{-3}P - 8.3172$; R is in cal/mol-K, T in K, and P in kbar. Parameters b and c derive from fitting the quantity $\Delta G + P\Delta V$ as a function of pressure and temperature. The quantity 0.58 in Equation (2) is an empirical correction factor required for differences in tabulated free energy data used in BURNHAM's (1979) and the current analysis of Equation (1). Activity relationships for H_2O in albite melt are taken from BURNHAM (1979; his equations 16–3 and 16–4), with $a_w^m = a_w^{fl} = 1$ at H_2O saturation. For $X_w^m \leq 0.5$:

$$a_w = k(X_w^m)^2, \tag{4}$$

and for $X_w^m > 0.5$:

$$a_w = 0.25k \exp(6.52 - 2667/T)(X_w^m - 0.5), \tag{5}$$

where T is in K, and k is a Henry's law activity constant analogue, and a function of pressure, and to a lesser extent temperature. Activity relationships for the albite component in hydrous melt are given by BURNHAM (1979; his Equations 16–13 and 16–14). Calculated liquidus temperatures agree closely with those of BURNHAM (1979) and BOETTCHER et al. (1982).

Calculation of ascent paths

The general method of calculating ascent trajectories using the model system albite–H_2O discussed by HOLLOWAY (1976) has been modified to include the calculation of the liquidus temperature of albite given in Equations (2) and (3) and additional model parameters. The sequence of calculation follows:

(1) Specify initial pressure, H_2O content (weight percent), and initial and final crystal content (this determines CR or MR).

(2) Calculate initial $X_w^m = n_w/(n_w + n_{ab}^m)$.

(3) Albite liquidus temperature is calculated as a function of pressure and X_w^m.

(4) Pressure is decreased by a small amount, P, and the calculation repeated. The ΔP must be small enough so that $dP/dT \simeq \Delta P/\Delta T$. Convergence of dP/dT and $\Delta P/\Delta T$ is almost reached at $\Delta P = 0.005$ kbar, but the differences in calculated values are less than the uncertainty in thermodynamic values, thus a ΔP of 0.01 kbar is used. The calculation is stopped when the H_2O–saturated solidus of the system has been reached: the system completely crystallizes and evolves a fluid phase.

Typical ascent trajectories for different crystallization rates are shown in Figure 1. The shape and position of the ascent path is independent of initial crystal content for equivalent CR's. In standard

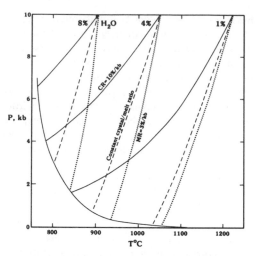

FIG. 1. Typical ascent trajectories calculated using the present model, for 1, 4, and 8 weight percent H_2O in the system. Solid lines represent a crystallization rate of 10%/kbar (CR = 10%/kbar) dashed lines represent a constant crystal/melt ratio and dotted lines represent a melting rate of 3%/kbar (MR = 3%/kbar). The light solid line that terminates ascent trajectories is the H_2O–saturated solidus for albite.

phase diagrams for albite–H_2O, the field to the right of the CR = 0%/kbar trajectory (liquidus isopleth) would consist only of melt at high pressures, and become melt + fluid at some lower pressure. Ascent trajectories for MR > 0%/kbar apparently cross the melt + fluid region. However, for MR > 0%/kbar, crystals are defined as being initially present in the magma, so the field is actually crystal + melt, the amount of melt being limited by the H_2O available. As pressure decreases during ascent and melting of these initially present crystals proceeds, X_w^m decreases, so that the melt + fluid (or melt + crystal + fluid) field is never reached.

All trajectories shown have positive dP/dT, including those for fusion of pre–existing crystals during ascent. Trajectory slope is positive for MR's < 3 to 4%/kbar for 1 to 8 weight percent H_2O in the melt.

The H_2O content affects both shape and position of ascent trajectories. Increased total H_2O content causes the initial temperature to decrease by 30 to 60°C per weight percent H_2O added to the system in the range 1 to 8 weight percent H_2O for a given initial pressure. Increased total H_2O content causes a decrease in dP/dT for CR > 0%/kbar and an increase in dP/dT for MR $\geq$ 0%/kbar, as expected from the albite–H_2O model. CR = 0%/kbar curves correspond to liquidus isopleths.

The most important characteristic for comparing ascent trajectories of model and actual granitic systems is trajectory shape. Ascent trajectories for higher CR's allow greater discrimination among possible volatile contents due to their more eccentric shapes. A decrease in initial pressure has no effect on trajectory shape: the CR = 0%/kbar trajectory remains the same, and the CR = 10%/kbar trajectory is lowered.

Trajectory shape depends largely on X_w^m, which in turn is a function of temperature. Because liquidus temperatures for albite–H_2O are greater by up to several hundred degrees than liquidus temperatures for actual granitic compositions, the effect of temperature on X_w^m and trajectory shape was tested. In the present model, the effect of temperature on X_w^m is governed by the Henry's law constant, k, used in Equations (4) and (5). Temperature dependence of k is greatest at low pressure. Test calculations were made by calculating k at 200°C lower than trajectory temperature with 1 and 8 weight percent H_2O. The X_w^m was increased by 1 and 5% at 10 kbar and 1 kbar, respectively. Ascent trajectories shift to temperatures lower by less than 3°C.

Crystallization of hydrous phases is not part of the albite–H_2O model system. The effect of hydrous phase crystallization on X_w^m can, however, be estimated by computing the amount of H_2O that would be removed from the melt by crystallizing a particular amount of the hydrous phase. For example, crystallization of 200 g of hornblende or biotite would remove about 4 or 8 g of H_2O from the system. The effect on ascent trajectory shape depends

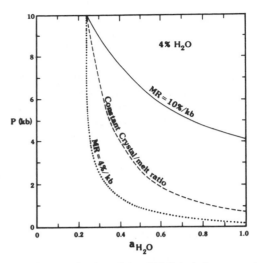

FIG. 3. Variation in activity of H_2O (a_w) along ascent trajectories for 4 weight percent H_2O in the system at various crystallization rates. Curves designated as in Figure 1. Note that the intersection of a curve with $a_{H_2O} = 1$ signifies that the H_2O–saturated solidus has been reached, and no liquid can exist at lower pressure.

on the crystallization rate for the hydrous phase and H_2O content of the system. For 1 weight percent H_2O in the system, crystallization of 20 weight percent hydrous phase at a single pressure would reduce X_w^m by about 75%, causing a shift in ascent trajectory to a higher temperature. For 8 weight percent H_2O in the magma, crystallization of 20% magma at a single pressure along the CR = 0%/kbar trajectory causes a 6% *increase* in X_w^m: the decrease in moles of melt due to crystallization has a greater effect on X_w^m than loss of H_2O to the hydrous phase. Crystallization of a hydrous phase at CR = 5%/kbar over the entire ascent distance results in ascent trajectories higher by only 1 to 2°C than those crystallizing anhydrous phases.

The change in proportions of phases along ascent trajectories reflected by X_w^m (Figure 2) also affects a_w (Figure 3). The H_2O activity in the magma continuously increases during ascent for all CR's, and for MR < 4%/kbar.

VARIATION IN MAGMA PROPERTIES DURING ASCENT

Using the pressure–temperature–composition data generated by the ascent trajectory calculations, any thermodynamic or physical property of magma that is a function of P, T, X_w or a_w can be evaluated. Partial molar thermodynamic properties for H_2O and ab components of the melt and ab crystal are calculated from equations given in BURNHAM and DAVIS (1974), by using relationships of the type:

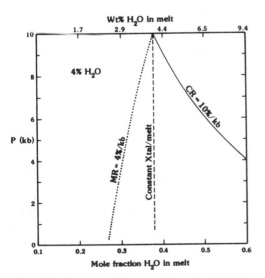

FIG. 2. Variation along ascent trajectories in X_w^m (and weight percent H_2O in melt) for 4 weight percent H_2O in the system at various crystallization rates. Curves designated as in Figure 1.

$$F_{\text{magma}} = n_{\text{m}} F_{\text{m}} + n_{\text{xl}} F_{\text{xl}}$$

$$= (n_{\text{ab}}^{\text{m}} + n_{\text{w}}^{\text{m}})(X_{\text{ab}}^{\text{m}} F_{\text{ab}}^{\text{m}} + X_{\text{w}}^{\text{m}} F_{\text{w}}^{\text{m}}) + n_{\text{xl}} F_{\text{xl}},$$

where F is any extensive thermodynamic property. Properties involving the component H_2O are calculated with a standard state based on the triple point of H_2O, necessitated by use of thermodynamic relationships from BURNHAM and DAVIS (1974).

Volume

Variation in volume, given as percent change from initial volume ($\Delta V/V_0$) is shown in Figure 4 as a function of volatile content for various crystallization rates. Important points to note for volume variation include:

(1) For a given volatile content, $\Delta V/V_0$ varies more strongly as a function of CR than pressure. Increase in total V_{w}^{m} for the system with increasing X_{w}^{m} compensates for volume decrease during crystallization. In general there will be some CR that results in a zero volume change for a given initial H_2O content. This zero volume change occurs at about CR = 6%/kbar for 1 weight percent H_2O and about CR = 5%/kbar for 8 weight percent H_2O.

(2) Volume increases do not exceed 10% from initial (10 kbar) values to final total pressures as low as 1 kbar even for H_2O–rich systems.

Enthalpy

Change in magma enthalpy for the system during ascent (ΔH), calculated as the difference in H between successive pressure–temperature–composition points, is shown in Figure 5 as a function of crystallization rate for 4 weight percent H_2O in the

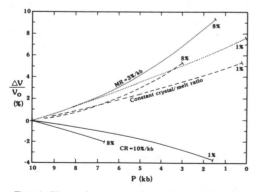

FIG. 4. Change in system volume from initial value at 10 kbar along ascent trajectories as a function of H_2O content and crystallization rate. Curves designated as in Figure 1. The figures in percent indicate system H_2O content. The intersection with the solidus is indicated by a bar perpendicular to the curves.

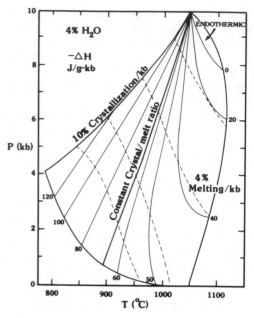

FIG. 5. Change in magma enthalpy along ascent trajectories for 4 weight percent H_2O in the system as a function of crystallization rate. Heavy solid lines are ascent trajectories as labelled. Light solid lines are contours of ΔH. Dashed lines are adiabats as discussed in the text.

system. ΔH is dependent on the ascent path, and approximately constant along any particular trajectory, ranging from about -120 to -70 J/g–kb, from 10 to 0%/kbar CR. The ΔH is usually exothermic and becomes endothermic only at MR $> 7\%$/kbar.

Variations in ΔH as a function of volatile content and crystallization rate are shown in Figure 6. The ΔH becomes more exothermic with increased volatile content at CR = 10%/kbar; the reverse is true for CR = 0%/kbar. ΔH varies from -50 to -90 J/g–kb for CR = 10%/kbar and from about -20 to -35 J/g–kb for CR = 0%/kbar in the range 1 to 8 weight percent H_2O.

Typical isenthalpic adiabatic paths defined by

$$dH = -mgdz, \qquad (6)$$

where m is mass, g is acceleration due to gravity, and z is a vertical coordinate, are shown by the dashed lines in Figure 5. For a particular depth interval $mgdz$ is constant; thus an infinite number of parallel adiabats exist. For all H_2O contents, isenthalpic adiabats require melting of crystals during ascent when crystals are initially present.

Entropy

Variation in S along ascent trajectories for 4 weight percent H_2O in the system is shown in Figure

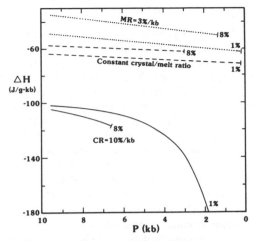

FIG. 6. Change in magma enthalpy along ascent trajectories as a function of crystallization rate and total H_2O content. Curves designated as in Figure 1. The figures in percent indicate system H_2O content. The intersection with the solidus is indicated by a bar perpendicular to the curve.

7. The S increases by about 0.3 J/g–K over the 10 kbar of ascent. The ΔS along ascent trajectories (not shown) is very small, usually less than -0.1 J/g–K–kbar. The constant entropy contours of Figure 7 define isentropic adiabats ($dS/dP = 0$) for that system. These differ from isenthalpic adiabats by

the amount $mgdz$ given in Equation (6). As for isenthalpic adiabats, melting of crystals, when present, must occur during isentropic adiabatic ascent.

Heat capacity

Heat capacities at constant pressure (C_p) for melt and magma are nearly constant. For all tested H_2O contents (0 to 8 weight percent) and crystallization rates (0 to 10%), the C_p varies only from about 1.38 to 1.51 J/g–K for the melt, and from about 1.26 to 1.46 J/g–K for magma. CR = 0%/kbar trajectories exhibit less variation in C_p than do paths with higher crystallization rates.

Internal energy

The change in internal energy, ΔE, along ascent trajectories is calculated by:

$$\Delta E = \Delta H - \Delta(PV). \qquad (7)$$

Variation in ΔE during ascent is shown as a function of crystallization rate for 4 weight percent H_2O in the system (Figure 8); again ΔE is path dependent. Trends in ΔE are similar to those for ΔH, but ΔE is less exothermic due to the contribution of $\Delta(PV)$. The effect of volatile and crystal contents on ΔE are shown in Figure 9. ΔE is exothermic

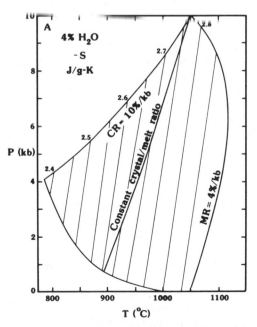

FIG. 7. Variation of entropy along ascent trajectories for 4 weight percent H_2O in the system. Heavy solid lines are ascent trajectories as labelled. Light solid lines are contours of entropy, which are also adiabats for the system.

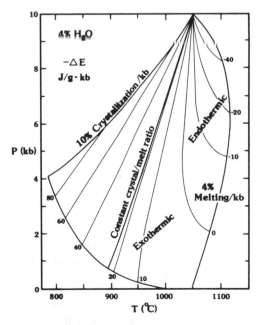

FIG. 8. Change in magma internal energy along ascent trajectories for 4 weight percent H_2O in the system as a function of crystallization rate. Heavy solid lines are ascent trajectories as labelled. Light solid lines are contours of constant ΔE.

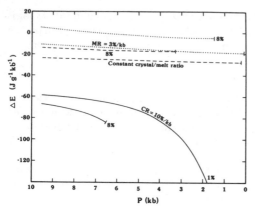

FIG. 9. Change in magma internal energy along ascent trajectories as a function of total magma H_2O content and crystallization rate. Curves designated as in Figure 1. The figures in percent indicate system H_2O content. The intersection with the solidus is indicated by a bar perpendicular to the curve.

for all CR's and for MR $\leq$ 3%/kbar for 4 weight percent H_2O in the system. As for ΔH, ΔE becomes increasingly exothermic with increasing H_2O contents at CR = 10%/kbar; the reverse is true for CR = 0%/kbar.

Density

Density of individual phases is calculated as the inverse of specific volume; magma density is calculated as:

$$\frac{1}{\rho_{magma}} = \frac{W_m}{\rho_m} + \frac{W_{xl}}{\rho_{xl}} + \frac{W_{fl}}{\rho_{fl}}, \qquad (8)$$

where W is weight fraction of the phase. The absolute values of density may be lower than the densities of actual granitic compositions because mafic phases are excluded from the model. Magnitude and direction of variation of magma and melt density with temperature, pressure and H_2O content calculated in the present model should accurately reflect the competing effects of decompression, temperature decreases, crystallization, and melting for actual granitic magmas.

Density variation is expressed as ρ/ρ_0, and is shown as a function of crystallization rate and volatile content in Figure 10 for melt and magma density. The ρ_{magma} is nearly constant along CR's, whereas ρ_m decreases during ascent. Near–constant ρ_{magma} is due to the compensating effects of crystallization and increasing X_w^m. An increase in total H_2O content of 1 weight percent results in a decrease in density of about 1% relative (*i.e.*, about 0.02 g/cm³ absolute).

Viscosity

Melt viscosity can be calculated as a function of T and X_w^m by using SHAW's (1972) empirical method, provided that an actual granitic magma composition is used. The melt composition used (Table 2) is the assumed initial (high pressure) composition of the Strathbogie granitoids (CLEMENS, 1981). The anhydrous melt composition is held constant; only the proportion of H_2O varies, thus "crystals" have the same composition as the melt. The resulting variations in melt viscosity, therefore, reflect only the effects of varying X_w^m. Pressure is assumed to have no effect on viscosity (but see also DINGWELL, 1987).

Albite liquidus temperatures are higher than those experimentally determined for granitic rocks, which would result in low melt viscosity values. Ascent trajectory temperatures are therefore arbitrarily reduced by 150°C for viscosity calculations.

The viscosity of magma (melt + crystal) is calculated by using the relationship:

$$\frac{\mu_{magma}}{\mu_m} = \left(1 - \frac{\pi}{4\alpha^2}\right) + \left(\frac{\pi}{4} - \frac{\pi}{6\alpha}\right)\left(\frac{1}{\alpha^2 - 1}\right)$$

$$\times \left(1 + \frac{2}{\sqrt{\alpha^2 - 1}} \tan^{-1} \sqrt{\frac{\alpha + 1}{\alpha - 1}}\right), \quad (9)$$

(ACKERMANN and SHEN, 1979) where $\alpha = (\phi_\infty/\phi)^{1/3}$, and ϕ is volume fraction of crystals in the magma. ϕ_∞ is taken as 0.675, a value intermediate to the range of critical ϕ values found by VAN DER MOLEN and PATERSON (1979) to separate granular-flow from suspension-like behavior in partially melted granitic rocks. The maximum allowed μ_{magma} is 10^{15} poise, based on viscosity data for rhyolite summarized in SPERA *et al.* (1982). Calculated magma viscosity approximates non–Newtonian behavior in that the magma attains an effective yield strength (μ_{magma} goes to infinity) once the volume fraction of crystals in the magma reaches ϕ_∞ (0.675 in our calculation).

Viscosity as a function of CR and volatile content is shown in Figure 11 for melt and magma. For a given pressure, both μ_{magma} and μ_m increase with decreasing CR, equivalent to increasing temperature. The effect of increasing X_w^m is greater than increasing the amount of crystals on viscosity for these systems.

During ascent, μ_m increases slightly then decreases for CR = 10%/kbar; the approximation that μ_m is constant during ascent is good. For CR < 0%/kbar, μ_m increases, by up to two orders of magnitude. Magma viscosity always increases during ascent, initially by a factor of about 0.1 per kbar.

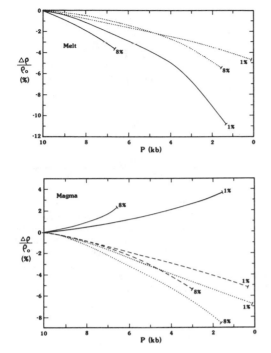

FIG. 10. Variation in density along ascent trajectories as a function of crystallization rate and H_2O content normalized to initial (10 kbar) density value. (a) Melt density. (b) Magma (crystal + melt) density. Curves designated as in Figure 1. The figures in percent indicate system H_2O content. The intersection with the solidus is indicated by a bar perpendicular to the curve.

Magma viscosity may increase by greater than three orders of magnitude over initial values for CR = 10%/kbar trajectories.

The presence of initial crystals results in an increase of the overall viscosity of the magma, and buffers changes in magma viscosity brought about by crystallization and changing X_w^m.

MODEL VALIDITY: THE STRATHBOGIE BATHOLITH AND VIOLET TOWN VOLCANICS

The major obstacle in the test of the present model for ascent trajectories and variation in magma properties is the paucity of experimental and physical data for actual granitic rocks: source region pressure and temperature, initial melt and magma composition, and amount of crystallization or fusion during ascent are adequately constrained for very few granitic rocks. Two of the best case histories found so far by the authors are for the Strathbogie batholith and Violet Town Volcanics of southeastern Australia (CLEMENS, 1981; CLEMENS and WALL, 1981, 1984). A combination of experimental, geochemical, and petrographic tech-

niques were used by CLEMENS (1981), together with field relationships, to provide tight constraints for initial (source) and final (emplacement) conditions for these "granitic" rocks.

The Strathbogie batholith

This late Devonian composite batholith discordantly intrudes folded Siluro–Devonian sedimentary rocks and cogenetic volcanic rocks. The batholith is subdivided into four mappable units based on grain size and composition. The Strathbogie granitoids are peraluminous S–types, generally containing cordierite, garnet and biotite in addition to plagioclase, quartz and K–feldspar. The most mafic variant of the Strathbogie batholith, a porphyritic microgranite (CLEMENS' sample number 889) contains 50 volume percent coarse phenocrysts; this composition is thought to represent that of initially emplaced magma (CLEMENS, 1981). Initial, ascent, and emplacement conditions deduced for the Strathbogie batholith by CLEMENS and WALL (1981) are given in Table 3. Based on H_2O content and sequence of crystallization, ascent trajectories for the Strathbogie magma are constrained to lie on the low temperature side of the quartz

Table 2. Compositions for Violet Town Volcanics and Strathbogie Granite (CLEMENS and WALL, 1984; CLEMENS, 1981)

Sample:	9443	9414	9402	1	889
SiO_2	72.10	67.65	69.19	69.06	70.25
TiO_2	0.38	0.64	0.49	0.62	0.58
Al_2O_3	13.66	15.28	15.25	15.31	14.56
FeO	2.87	3.94	3.13	4.01	4.14
MnO	0.02	0.06	0.06	0.05	0.06
MgO	0.97	1.46	1.32	1.59	1.24
CaO	1.63	3.12	2.84	2.76	1.89
Na_2O	2.71	2.60	2.72	2.75	2.66
K_2O	4.05	3.58	3.99	3.76	4.46
P_2O_5	0.14	0.17	0.16	0.18	0.18
	98.53	98.50	99.15	100.09	100.02
Qz	34.16	27.58	28.17		28.47
Co	2.19	1.86	1.66		2.41
Or	23.94	21.16	23.58		28.19
Ab	22.93	22.00	23.01		22.17
An	7.17	14.37	13.09		7.99
Hy	7.10	9.93	8.34		9.32
Il	0.72	1.22	0.93		1.08
Ap	0.33	0.40	0.38		0.45
	98.54	98.52	99.16		100.08

9443	rhyolite ignimbrite, Violet Town Volcanics
9414	rhyodacite ignimbrite, Violet Town Volcanics
9402	"schlier" from rhyodacitic ignimbrite, Violet Town Volcanics
1	calculated VTV parent magma composition
889	Strathbogie granite, parent composition

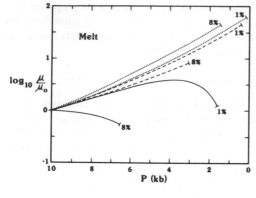

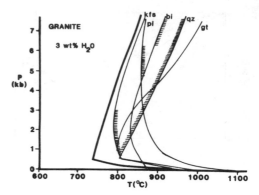

FIG. 12. Simplified experimentally determined phase relationships for the Strathbogie granite with 3 weight percent H_2O in the melt (after CLEMENS, 1981). Hatchures to the right and left of a stability curve indicate that the phase was observed to be present at higher and lower temperature, respectively (CLEMENS, 1984, personal communication). The ascent trajectory must lie within the region bounded by the hatchured curves. See text for discussion.

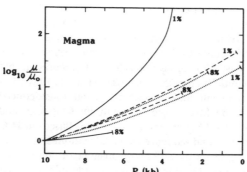

FIG. 11. Variation in viscosity along ascent trajectories as a function of crystallization rate and H_2O content normalized to initial (10 kbar) viscosity value. (a) Melt viscosity. (b) Magma viscosity. Curves designated as in Figure 1. The figures in percent indicate system H_2O content. The intersection with the solidus is indicated by a bar perpendicular to the curve.

The Violet Town Volcanics

The Violet Town Volcanics (VTV) consist of approximately 100 km^3 of peraluminous intracaldera ignimbrites of late Devonian age (CLEMENS and WALL, 1984). Compositions range from rhyolite to rhyodacite (Table 2) with phenocrysts of quartz + plagioclase + biotite + garnet + cordierite ± hypersthene. Crystal contents range from 0 to 65 modal percent, with biotite forming approximately 20 weight percent, plagioclase 20 to 60 weight percent, and quartz 20 to 40 weight percent of the totals. Crystallization sequence, inferred from textural evidence, was early garnet + quartz, plagioclase and orthopyroxene, followed by biotite, cordierite, and late-stage K-feldspar. CLEMENS and WALL (1984)

stability curve and to the high temperature side of the biotite, plagioclase, and K-feldspar stability curves, as shown in Figure 12 (CLEMENS, 1984, personal communication).

Table 3. Model conditions for ascent

Strathbogie Granite	Source	Emplacement
Pressure	5 to 7 kbar	0.5 to 1.0 kbar
T (°C)	>850	>800
Weight percent H_2O in Melt	2.8 to 4.0	2.8 to 4.0
Crystal Content	0 to 15%	0 to 50%
Experimental T		
Range at 7 kbar: 3 Weight percent H_2O:	905 to 950°C	800 to 817°C
4 Weight percent H_2O:	895 to 905°C	745 to 775°C
Violet Town Volcanics	Source	Emplacement
Pressure	3.6 to 6.0 kbar	0.5 to 1.0 kbar
T (°C)	avg. 850	750 to 850
Crystal Content	5%	5%
Experimental T		
Range at 7 kbar: 3 Weight percent H_2O:	875 to 900°C	780 to 825°C
4 Weight percent H_2O:	825 to 860°C	750 to 780°C

infer early magmatic temperatures of greater than 800 to 830°C and pressures of 3.6 to 6 kbar, with 2.5–4.5 weight percent H_2O. The total restite content is estimated at less than 2 volume percent of easily distinguished garnet, biotite, cordierite, sillimanite, and other phases. Inferred source and emplacement conditions are summarized in Table 3. These results suggest ascent in a largely liquid state and subsequent crystallization in a high level magma chamber. CLEMENS and WALL (1984) infer fluid–absent melting reactions in a source region dominated by weakly to mildly peraluminous quartzo–feldspathic metasedimentary material.

The observed chemical variation in the VTV deposits is interpreted by CLEMENS and WALL (1984) to be a result of about 35% fractional crystallization of parent magma, producing a zoned magma chamber.

Comparison with model ascent trajectories

For comparison of observed and model trajectories, the weight percent H_2O in the melt was first converted to albite–equivalent amount of H_2O, as outlined by BURNHAM (1979), resulting in 3.0 to 4.3 weight percent H_2O in the Strathbogie magma and 3.0 to 4.2 weight percent H_2O in Violet Town Volcanics magma. For purposes of ascent trajectory calculation, the H_2O content of the Strathbogie and VTV magmas are identical; ascent distance and crystal content differ.

Using the conditions given in Table 3, ascent trajectories were calculated for the Strathbogie and Violet Town magmas. Only model ascent trajectories that fit within the maximum and minimum pressure–temperature gradients obtained from Figure 12 were permitted. The 4 weight percent H_2O test case fits well for $P_0 = 7$ kbar for CR = 1.2 to 3.5%/kbar, and at $P_0 = 5$ kbar for CR = 0.3 to 3.5%/kbar. With the minimum H_2O content of 3.0 weight percent, the model trajectories fit from CR = 0 to 3 or 3.5%/kbar at 5 and 7 kbar P_0. Trajectories for many other H_2O contents and CR's were calculated, but did not fit within the observed minimum and maximum pressure–temperature gradients. The assumed H_2O contents and CR's are held to be correct.

Results for the VTV are similar to those of the Strathbogie, with the exception that crystallization rate is firmly constrained at zero percent. Calculated ascent paths with CR = 0%/kbar fit the experimentally determined ascent regions only for melt H_2O contents of less than about 4 weight percent. The model trajectories fit the experimentally constrained ascent region remarkably well.

Application of model calculations to VTV and Strathbogie magma bodies

The model ascent trajectory which best fits CLEMENS' (1981) results is that for 3 weight percent H_2O, with CR = 0, 5%/kbar initial crystals, initial $P = 5$ to 7 kbar, and initial T about 850°C. For the Strathbogie magmas, crystallization rate is not constrained, so a variety of ascent trajectories can be used to model the system.

By using physical properties calculated for magma along the above ascent trajectories, the possibility of crystal settling and restite segregation in the VTV magma can be examined. Calculations for crystal settling in a convecting magma chamber were performed after MARSH and MAXEY (1985):

$$S = 0.86 \Delta \rho a^2 \left(\frac{g}{\rho \alpha \Delta T L K \mu} \right)^{0.5}, \qquad (10)$$

where S is the ratio of crystal settling velocity to convection velocity, K is 10^{-2} cm^2/s, L is magma chamber length scale, $\Delta \rho = \rho_{magma} - \rho_{xl}$ is magma density, $\alpha = 5 \times 10^{-5}$ deg^{-1}, $g = 980$ cm/s^2, a = crystal radius, $L = 1$ km, μ = magma viscosity, and $\Delta T = (T_{magma} - T_{wr})/2$. Restitic garnet crystals in the VTV have a density of about 3.5 g/cm^3 and are about 1 cm in diameter (CLEMENS and WALL, 1984). The T_{wr} is calculated as a function of depth by using;

$$T_{wr} = T_0[1 - (z/L)^n], \qquad (11)$$

with $n = 2$ and 4 (MARSH, 1982), with $T_0 = 850$°C (temperature at initial pressure) and remaining trajectory temperatures lowered accordingly. For $S < 1$, no crystal settling occurs in the magma chamber. By using the above conditions, no crystal settling would occur in a convecting magma chamber for VTV or Strathbogie magmas.

The presence of restite inclusions in the VTV of up to 10 cm diameter also suggests that crystal settling was not an important process during ascent of the magma. By comparing terminal settling velocities for such inclusions with ascent velocities, implies that ascent velocities must have been greater than about 10^{-5} to 10^{-4} cm/s for crystal–magma density contrasts of 0.3 to 0.4 g/cm^3. Diapirism was thus probably not the major ascent mechanism for these magmas, because current models for diapiric ascent infer ascent velocities on the order of 10^{-7} to 10^{-5} cm/s (MARSH, 1982).

The relatively high viscosities, particularly at low pressures, imply that the VTV and Strathbogie magmas would be efficient at forcible emplacement and roof lifting. However, Rayleigh numbers calculated for both vertical pipe and horizontal layer configurations, with $T > 10^3$ degrees, are greater than 10^6, suggesting that convection would occur for the VTV and Strathbogie magmas both during ascent and after emplacement.

DISCUSSION

Ascent trajectories calculated by using the present model are unique because the ascent trajectory is constrained by the phase equilibria of the magma, rather than the ascent mechanism or physical properties of the magma or wall rocks. The physical properties of the magma (or the magnitudes of their variations) are known along the ascent trajectory as a function of pressure, temperature, and H_2O content of the magma. Specification of crystallization or melting rate does, however, determine the magnitude of heat transfer from the system. To determine whether the rates of heat transfer implied by particular crystallization or melting rates are probable, requires specification of a heat transfer model for a particular ascent mechanism and body geometry. Enthalpy and internal energy data from the present model has been used to constrain ascent velocity and magma body size or shape (SYKES, 1986).

Ascent trajectories

Few other examples exist of pressure–temperature ascent trajectories for magmas of any type. The most common assumption used for pressure–temperature trajectories is that of adiabatic ascent of magma through the crust. Such trajectories are usually calculated using either an expression for the adiabatic gradient,

$$\beta = -g\alpha T/C_p, \qquad (12)$$

or adiabatic temperature

$$T = T_0 \exp[\alpha g(z - z_0)/C_p], \qquad (13)$$

where α is the coefficient of thermal expansion. Such expressions are strictly applicable only to single phase, single component systems but are often used for magma, assuming average magma property values (Figure 13, curve A). RUMBLE (1976) developed an expression valid for multicomponent, multiphase adiabatic gradients. For the two component system ab–H_2O, with the two phases melt + crystal, RUMBLE's (1976) adiabats are identical to those determined using the present model (Figure 13, curve B), assuming adiabatic ascent.

Comparison of single phase and multiphase adiabats shows that the gradients in the latter are about 5 times greater than in the former. For example, for 4 weight percent H_2O, the two gradients are 2.5

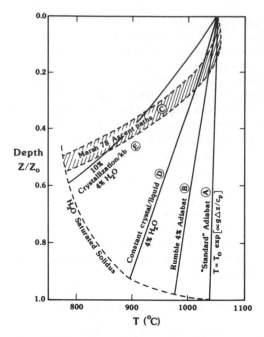

FIG. 13. Comparison of ascent trajectories calculated by using the present model and other models for magma pressure–temperature paths. Dashed line is the H_2O–saturated solidus. A: Standard one-phase adiabat calculated from Equation (13) of text. B: Multiphase adiabat from RUMBLE (1976) for 4 weight percent H_2O in the system. C: General shape of trajectories calculated after MARSH (1978) for diapiric ascent. D: Ascent trajectory, this paper, constant crystal/melt ratio (CR = 0%/kbar). E: Ascent trajectory, this paper, CR = 10%/kbar.

and 0.5 deg/km, respectively. Furthermore, the adiabat calculated in the present model requires fusion of pre–existing crystals (if present) to occur during ascent, at a rate of about 4 to 4.5 weight percent per kbar. This adiabat is not equivalent to an ascent trajectory for MR = 4 to 4.5%/kbar because the melting rate along the adiabat is not linear.

By comparing the model adiabats to model ascent trajectories, temperature gradients for the latter are much greater, on the order of 5 to 14 deg/km for CR = 10%/kbar. The larger value is for low magmatic H_2O contents. Use of adiabatic ascent trajectories in fluid dynamic models of ascent may result in inaccurate temperatures and poor estimates of properties such as viscosity and density, if crystals are initially present in the system. The assumption of adiabatic ascent may be good only for deep-seated diapiric upwelling of partially melting material with high crystal (restite) content, or for melt at superliquidus temperatures.

Non–adiabatic ascent trajectories have been calculated for diapiric ascent of granitic magma using the heat transfer model of MARSH (1978), which uses the assumption that internal energy change of the ascending magma equals heat transferred away from the body. Calculations were made using a total ascent distance of 30 to 40 km, and $K = 0.01$ cm/s^2, with ascent velocities of 10^{-8} to 10^{-4} cm/s and bodies of radius 10^4 to 10^6 cm. Resulting trajectories are shown in Figure 13. The absolute position of trajectories calculated by Marsh's method varies considerably depending on body radius and ascent velocity, but the curvature is nearly constant. It can be seen that the curvature resulting from MARSH's (1978) model is quite distinct from that of the trajectories calculated in the present model (curves D and E). Thus, care must be exercized in using ascent trajectories such as those of MARSH (1978) because they may imply inaccurate assumptions about the physical characteristics of the system. The case Marsh wished to illustrate, CR = 0%/kbar, has a trajectory intermediate to that of the multiphase adiabat and his trajectories. MARSH's trajectories are also sensitive to choice of geotherm and assumptions of initial wall rock and magma temperatures.

Energetics of magma ascent

Most heat transfer models for magma ascent or evolution assume that energy production in an evolving magma body is proportional to $C_p \Delta T$, where ΔT is the temperature difference between the magma and its surroundings, with C_p constant. Energy production within the magma body during ascent is more accurately represented by ΔH or ΔE for the system, as calculated in previous sections. The ΔE includes an amount due to $\Delta(PV)$ work, most often assumed to be zero due to constant system volume. To determine the error resulting from use of $C_p \Delta T$ for magma energy production, effective heat capacities are calculated as $C_p(H) = \Delta H / \Delta T$ and $C(E) = \Delta E / \Delta T$, where ΔH and ΔE are the change in enthalpy and internal energy of the magma system from one (P, T) point to another along the ascent trajectory. The $C_p(H)$ and $C(E)$ are up to three or two times greater than C_p, for CR = 0 to 10%/kbar and 1 to 8 weight percent H_2O. The amount of energy available from a magma body may thus be two to three times greater than that calculated by using $C_p dT$. Results of previous studies of ascent mechanisms where the $C_p dT$ approximation for magma energy is used may, therefore, be in error.

Constancy of magma properties

Fluid dynamic and heat transfer models of evolving or ascending magma bodies often assume either constant or temperature–dependent magma density, volume, or viscosity. The validity of such assumptions can be evaluated by using data from the present model.

Density is most often assumed to be constant or a function of temperature, such that;

$$\rho = \rho_0 - \rho_0 \alpha (T - T_0), \qquad (14)$$

where T_0 and ρ_0 are initial temperature and density, and α is the coefficient of thermal expansion. Numerous theoretical and experimental studies of magma chamber dynamics have demonstrated that compositional variation and crystallization effects on density can dominate the dynamics of a magma chamber (*e.g.*, HUPPERT and SPARKS, 1984). Furthermore, density variations of less than 0.02 g/cm^3 are required for the compositional effects to be important.

In equations estimating velocity of magma ascent or size of magma bodies, the density term is usually to the first power, and variation in magma density is far less than variation in other parameters, so that the assumption of constant ρ_{magma} is not a bad one. From our model calculations, ρ_{magma} is actually constant at CR = 5 to 6%/kbar for 1 to 8 weight percent H_2O, and for other CR's varies by less than 0.015 g/cm^3 per kbar. However, ρ_{melt} can vary from -0.03 to -0.087 g/cm^3 per kbar depending on the H_2O content of the magma and cannot be considered constant.

Equation (14) above yields the temperature dependence of density. By using $\alpha = 5 \times 10^{-5}$ deg^{-1}, model ascent trajectory temperatures, T, and initial conditions at 10 kbar, calculations yield increasing ρ_{magma} during ascent, whereas when compositional effects are included, ρ_{magma} decreases during ascent for CR < 5 to 6%/kbar. Even at CR = 10%/kbar, Equation (9) yields a value for ρ_{magma} only half that calculated by the present model. These differences are due to the effects of H_2O, pressure, and temperature on density; the former completely dominates temperature and pressure effects.

The effect of crystals on density is also important: $\rho_{\text{magma}} = \rho_{\text{melt}}$ only for CR = 0%/kbar with no initial crystals. The difference between ρ_{magma} and ρ_{melt} is greatest for systems with large positive crystallization or melting rates. Crystal segregation processes (or other buoyancy–driven processes) will be most effective at low pressures for CR > 0%/kbar and at high pressures for CR < 0%/kbar.

The volume of magma bodies is most often assumed to be constant in fluid dynamic models of magma ascent and evolution. Volume variations calculated for model granite systems show that volume is not constant for ascending magma bodies.

The greatest variation in V_{magma} derives not from ΔV_{xl} but from ΔV_w^m. By assuming constant V_w^m would result in a 15 to 30% underestimate in total ΔV of the system over an ascent trajectory. By assuming ΔV is due only to crystallization (using 1 bar data), the values for V_{magma} 2 to 4 times too small.

The volume increases predicted by the present model are generally too great to be accommodated by magma bodies or wall rocks. Magma internal pressures can exceed external pressure by 200 bar (an average value for wall rock yield strength) in only 7 to 10 km of ascent for CR = 0%/kbar. Fracturing of wall rocks during ascent in a brittle crust must ensue from such volume increases.

Viscosity of magma and melt plays an important role in internal evolution of magmas during ascent. For examining processes occurring during magma evolution, e.g., crystal settling or convection, variations in viscosity will be important. Three common assumptions used to calculate magma viscosity in previous studies have been (1) μ is constant, (2) μ is a function of temperature only, and (3) use of empirical viscosity relationships. Viscosities calculated by using these assumptions are compared to those calculated using the present model for varying H_2O contents and crystallization rates.

The assumption of constant magma viscosity is good for high crystallization rates and high H_2O contents, because the decrease in viscosity due to increased H_2O is offset by the increase in viscosity due to crystallization. However, at low H_2O contents, a constant value for magma viscosity could be up to four orders of magnitude too low. Melt viscosity can be assumed approximately constant for CR = 10%/kbar and high H_2O contents. The assumption of constant μ_{melt} is worst for CR $\leq$ 0%/kbar and low H_2O contents, yielding a value for viscosity at most 50 times too small.

Viscosity can be assumed to vary as a function of temperature alone, with no contribution by crystals or H_2O content. For low crystallization rates, this is a good assumption, resulting in values for viscosity half those of actual model viscosities. For higher crystallization rates, the assumption that viscosity varies only with temperature results in values for viscosity that differ by up to two orders of magnitude from model viscosities.

SPERA et al. (1982) use an empirical relationship for viscosity

$$\mu/\mu_\infty = \exp[a(T_\infty - T)], \qquad (15)$$

where a is a function of H_2O content, crystal content, and composition. The rheological parameter, a, usually varies between 0.02 and 0.10 K^{-1} (SPERA

et al., 1982). The T is magma temperature, T_∞ is liquidus temperature, and μ_∞ is apparent melt viscosity at T. Results from this formulation can be directly compared with model results, because model T's at P_o are liquidus temperatures. For CR = 10%/kbar, μ_m cannot be duplicated with $a > 0$. For CR = 10%/kbar, μ_{magma} can be approximated by $a \simeq 0.02$ for low to intermediate H_2O contents, and by $a \simeq 0$ for high H_2O contents. For ascent with CR = 0%/kbar and no initial crystals, μ_{magma} = μ_m and can be approximated by $a \simeq 0.02$ for all H_2O contents, by using the temperature from model ascent trajectories.

CONCLUSIONS

The model presented in this paper for calculating ascent trajectories for granitic magmas is consistent with observed petrologic variations, and may be used for calculating $P-T-X$ ascent paths for calc-alkaline, hypersthene- or quartz-normative magmas provided crystallization or melting rate is specified. In addition to ascent trajectories, variation of magma properties, such as volume, energy loss, density, and viscosity, during ascent can be calculated. This information can then be used to test the validity of assumptions made about magma properties, or used directly in model calculations for ascent mechanisms or magma evolution.

Ascent trajectories for MR < about 2%/kbar have positive dP/dT. Adiabatic ascent paths have steeper dP/dT (but are still positive) and require fusion of pre-existing crystals (restite or previously crystallized magma or wall rocks) when present. The usual calculated adiabatic gradient [Equation (12)] is up to 5 times too small for ascending, crystallizing bodies of intermediate H_2O content, because effects of H_2O are ignored.

Water has a large effect on magma properties, especially when crystallization or fusion occur. By neglecting H_2O in granitic magmas, large errors in magma properties can result and physically unrealistic results may be obtained. Melt density and viscosity are important quantities in calculations of crystal separation, crystal settling, or convection; these properties are not constant and cannot be approximated by using standard approaches. Magma density and viscosity can often be correctly assumed constant, or modelled after an empirical relationship such as given in Equation (9). Volume increases in ascending granitic magma bodies for CR < 5 to 6%/kbar, even at low H_2O contents, require expansion of the magma body and concurrent shear deformation of surrounding ductile wall rocks. In the brittle regime, volume increases require the generation of excess internal magma pressure which will rapidly exceed the strength of wall rocks. Consequent fracturing of wall rocks can lead to partial to complete quench of the magma, or initiation of magma ascent by fracture propagation. Only for CR > 5 to 6%/kbar is the assumption of constant volume consistent with these calculations.

The energetics of ascending granitic magma depends greatly on the contribution of the enthalpy of the hydrous melt component as pressure and degree of crystallization vary, and on the $P-V$ work associated with magma expansion. The energy available from an ascending granitic magma body can reach two to three times that calculated by using heat capacity data. Careful evaluation of ascending magma energetics is necessary for modelling ascent mechanism, ascent distance, and determining the importance of mechanisms such as stoping or wall rock contamination.

Acknowledgments—Financial support for this work was provided by NSF grants EAR8108748-01 and EAR-8407742. This manuscript greatly benefited from comments by Robert W. Luth and an anonymous reviewer. Discussions with John D. Clemens and Alan J. R. White about the ascent model also significantly improved the work.

REFERENCES

ACKERMANN N. L. and SHEN H. T. (1979) Rheological characteristics of solid–liquid mixtures. *Amer. Inst. Chem. Eng. J.* **25**, 327–332.

BOETTCHER A. L., BURNHAM C. W., WINDOM K. E. and BOHLEN S. R. (1982) Liquids, glasses and the melting of silicates to high pressures. *J. Geol.* **90**, 127–138.

BURNHAM C. W. (1979) The importance of volatile constituents. In *The Evolution of the Igneous Rocks,* (ed. H. S. YODER), pp. 439–482 Princeton Univ. Press.

BURNHAM C. W. (1981) The nature of multicomponent aluminosilicate melts. *Phys. Chem. Earth* **14–15**, 197–229.

BURNHAM C. W. and DAVIS N. F. (1971) The role of H₂O in silicate melts. I. $P-V-T$ relations in the system NaAlSi₃O₈–H₂O to 10 kilobars and 1000°C. *Amer. J. Sci.* **270**, 54–79.

BURNHAM C. W. and DAVIS N. F. (1974) The role of H₂O in silicate melts: II. Thermodynamic and phase relations in the system NaAlSi₃O₈–H₂O to 10 kilobars, 700° to 1100°C. *Amer. J. Sci.* **274**, 902–940.

CHAPPELL B. W. and WHITE A. J. R. (1974) Two contrasting granite types. *Pac. Geol.* **8**, 173–174.

CLEMENS J. D. (1981) The origin and evolution of some peraluminous acid magmas (experimental, geochemical and petrological investigations). Ph.D. Dissertation, Monash University, Australia.

CLEMENS J. D. and WALL V. J. (1981) Origin and crystallization of some peraluminous (S-type) granitic magmas. *Can. Mineral.* **19**, 111–131.

CLEMENS J. D. and WALL V. J. (1984) Origin and evolution of a peraluminous silicic ignimbrite suite: The Violet Town Volcanics. *Contrib. Mineral. Petrol.* **88**, 354–371.

DINGWELL D. B. (1987) Melt viscosities in the system NaAlSi₃O₈–H₂O–F₂O₋₁. In *Magmatic Processes: Physicochemical Principles,* (ed. B. O. MYSEN), The Geochemical Society Spec. Publ. No. 1, Chap. 00, pp. 000–000.

HOLLOWAY J. R. (1976) Fluids in the evolution of granitic magmas: Consequences of finite CO₂ solubility. *Bull. Geol. Soc. Amer.* **87**, 1513–1518.

HUPPERT H. E. and SPARKS R. S. J. (1984) Double-diffusive convection due to crystallization in magmas. *Ann. Rev. Earth Planet. Sci.* **12**, 11–37.

MARSH B. D. (1978) On the cooling of ascending andesitic magma. *Phil. Trans. Roy. Soc. London* **A288**, 611–625.

MARSH B. D. (1982) On the mechanics of igneous diapirism, stoping and zone melting. *Amer. J. Sci.* **282**, 808–855.

MARSH B. D. and KANTHA L. H. (1978) On the heat and mass transfer from an ascending magma. *Earth Planet. Sci. Let.* **39**, 435–443.

MARSH B. D. and MAXEY M. R. (1985) On the distribution and separation of crystals in convecting magma. *J. Volcanol. Geotherm. Res.* **24**, 95–150.

NEKVASIL H. and BURNHAM C. W. (1987) The calculated individual effects of pressure and water content of phase equilibria in the granite system. In *Magmatic Processes: Physicochemical Principles,* (ed. B. O. MYSEN), The Geochemical Society Spec. Publ. No. 1, pp. 433–445.

RAMBERG H. (1971) Temperature changes associated with adiabatic decompression in geological processes. *Nature* **234**, 539–540.

READ H. H. (1957) *The Granite Controversy.* T. Murby.

RUMBLE D. (1976) The adiabatic gradient and adiabatic compressibility. *Carnegie Inst. Wash. Yearb.* **75**, 651–655.

SHAW H. R. (1972) Viscosities of magmatic silicate liquids: an empirical method of prediction. *Amer. J. Sci.* **272**, 870–893.

SHAW H. R. (1980) The fracture mechanisms of magma transport from the mantle to the surface. In *Physics of Magmatic Processes,* (ed. R. B. HARGRAVES), pp. 201–264 Princeton Univ. Press.

SPERA F. J. (1980) Aspects of magma transport. In *Physics of Magmatic Processes,* (ed. R. B. HARGRAVES), pp. 265–323 Princeton Univ. Press.

SPERA F. J., YUEN D. A. and KIRSCHVINK S. J. (1982) Thermal boundary layer convection in silicic magma chambers: Effects of temperature dependent rheology and implications for thermogravitational chemical fractionation. *J. Geophys. Res.* **87**, 8755–8767.

SYKES M. L. (1986) Ascent of granitic magma: Constraints from thermodynamics and phase equilibria. Ph.D. Dissertation, Arizona State University.

TURCOTTE D. L. (1982) Magma migration. *Ann. Rev. Earth Planet. Sci.* **10**, 397–408.

TURCOTTE D. L. (1987) Physics of magma segregation processes. In *Magmatic Processes: Physicochemical Principles,* (ed. B. O. MYSEN), The Geochemical Society Spec. Publ. No. 1, Chap. 00, pp. 000–000.

VAN DER MOLEN I. and PATERSON M. S. (1979) Experimental deformation of partially melted granite. *Contrib. Mineral. Petrol.* **70**, 229–318.

WALDBAUM D. R. (1971) Temperature changes associated with adiabatic decompression in geological processes. *Nature* **232**, 545–547.

WALTON M. (1955) The emplacement of granite. *Amer. J. Sci.* **253**, 1–18.

Subject Index

465

472

473

474

475

491

494

of fusion, 442
    partial, 26, 29, 30, 32, 33,
        47, 48, 209, 217, 219
    olivine-hypersthene-plagioclase-
        diopside, 148
    phase, 151
    specific, 25, 32, 38, 40, 454
Volumetric, 73
Vorb, 244, 245, 252, 254, 256

W

Wall, 425
Washington, 279, 281
Water, 7, 8, 15, 62, 64, 65, 92, 93,
    94, 103, 108, 109, 111,
    122, 123, 124, 126, 127,
    128, 133, 134, 135, 139,
    142, 143, 146, 151, 156,
    158, 159, 160, 165, 167,
    168, 170, 176, 183, 189,
    191, 203, 218, 221, 238,
    240, 253, 264, 265, 274,
    277, 291, 301, 302, 303,
    337, 339, 340, 341, 343,
    344, 346, 348, 350, 351,
    352, 353, 354, 355, 370,
    404, 411, 412, 413, 414,
    415, 416, 417, 418, 419,
    420, 423, 424, 425, 426,
    427, 428, 430, 431, 433,
    434, 435, 436, 437, 438,
    439, 440, 441, 442, 443,
    447, 448, 449, 450, 451,
    452, 453, 454, 455, 456,
    457, 458, 460
    activity, 142
    concentration, 424
    critical, 344
    depth, 7
    formation, 353
    ground, 103, 337, 354
    hydrothermal, 353
    magmatic, 191, 203, 337, 346,
        355
    meteoric, 340, 346, 348, 350,
        353, 354
        low-^{18}O, 350
    metamorphic, 339, 353
    meteoric, 253
    molecular, 128, 133, 135, 427
    ocean, 343
    saturated, 370, 433, 435, 436,
        437, 438, 440, 441, 443
    saturated, 93, 352
    sedimentary formation, 339
    solubility, 370
    speciation, 431
    specific heat, 351
    supercritical, 344
    surface, 343, 355
    table, 265, 274
    transport, 351
    undersaturated, 168, 431, 433,
        435, 441, 443
    undersaturation, 160

Wave, 3, 4, 7, 8, 9, 10, 25, 26, 47,
    49, 54, 55, 165, 168,
    169, 170, 171, 175, 232,
    235, 244, 259, 260, 262,
    263, 264, 265, 267, 273,
    274, 277, 282
    body, 3, 8, 10
    elastic, 168
    shear, 5, 8, 9, 169, 232, 235, 244
    shock, 25, 26, 47, 49, 54, 55,
        171
    surface, 3, 4, 7, 8
    velocity, 5, 165, 168, 169, 170,
        175
        compressional, 5, 169, 170
Wavelength, 8, 9, 63, 125, 291,
    292, 299, 425
    dispersive, 125
    shift, 63
Wavenumber, 133, 289, 293, 294,
    297, 301, 303
Websterite, 185
Wedge, 147
Wehrlite, 171, 185
Weight, 38, 39, 448, 454
    atomic, 38, 39
    fraction, 448, 454
Welded, 92, 94
Well, 103
Western Cordillera, 360
Whiteschist, 156
Willow Park, 349
    dome, 349
Window, 404
Withdrawal, 273, 274
Wyoming, 337, 347, 348

X

Xenocryst, 13, 352
    garnet, 13
        subcalcic, 13
Xenolith, 11, 13, 14, 15, 18, 19,
    20, 21, 22, 23, 59, 65,
    73, 81, 116, 141, 147,
    148, 157, 161, 162, 212,
    213, 214, 215, 219, 220,
    243, 302, 340, 347
    basalt, 22, 340
    basaltic, 21
    eclogite, 14, 23
    lherzolite, 14, 15, 18, 21, 81
        garnet, 14
        spinel, 21
    peridotite, 13, 14, 19, 20, 21
        garnet, 14, 19, 20
    pyroxenite, 14
    siliceous, 347
    suite, 23
    transport, 59, 65
    ultramafic, 215, 220
    upper mantle, 116
X-ray, 64, 123, 129, 133, 159, 191,
    196, 401, 406, 408, 415,
    424

    diffraction, 123, 129, 401, 408
    diffractogram, 159
    fluorescence, 191, 196
    photoelectron, 133
    radial distribution, 64
XRDF, 64, 375, 406, 407
XRF, 15, 196

Y

Yakutian, 210
    kimberlite, 210
Yamoto, 213
Yellowstone, 302, 337, 347, 348,
    350, 351, 352, 353, 354
    caldera, 349, 352, 354
    geothermal area, 302
    magma chamber, 351
    plateau, 337, 347, 348, 349, 351
        volcanic field, 349
    rhyolite, 348
    volcanic field, 348, 350
Yellowstone National Park, 349
Yoneyama, 184, 185, 186, 187

Z

Zeleny, 220
Zeolite, 344
Zircon, 176, 201
Zonation, 253, 290
    chemical, 253
Zone, 3, 8, 11, 15, 23, 69, 70, 94,
    103, 104, 108, 117, 140,
    143, 145, 147, 152, 155,
    157, 160, 161, 162, 191,
    193, 224, 225, 241, 244,
    250, 252, 253, 259, 265,
    273, 274, 275, 277, 282,
    310, 340, 342, 343, 344,
    346, 349, 351, 352
    border, 94
    collision, 155, 161, 162
    cumulate, 253
    fault, 147
    fracture, 145, 277, 340, 349
    low velocity, 140
    marginal, 15
    melt, 70
    molten, 244, 250
        partial, 244, 250
    neovolcanic, 274, 275, 277
    rift, 117
    slip, 69
    subduction, 3, 8, 69, 147, 152,
        157, 160, 162, 193, 225
    transition, 143, 273, 282
    velocity, 3, 11, 277
        low, 3, 11, 277
    volcanic, 274, 277
Zoned, 103, 353
Zoning, 15, 18, 93, 202, 241, 245,
    250
    compositional, 202
    marginal, 15